Seismic Design of Reinforced Concrete Buildings

About the Author

Jack Moehle is the T.Y. and Margaret Lin Professor of Engineering in the Department of Civil and Environmental Engineering at the University of California, Berkeley. He received his Ph.D. from the University of Illinois and joined the U.C. Berkeley faculty in 1980. His research and teaching activities are mainly in structural engineering, with emphasis on reinforced concrete and earthquake engineering. He was founding director of the Pacific Earthquake Engineering Research Center, a multi-campus research center that advanced the concepts and practice of performance-based earthquake engineering. As a licensed Civil Engineer in the State of California, Dr. Moehle works regularly as a consulting engineer, offering advice and expert peer review on highway and mass transit systems, water distribution systems, existing construction, and high-rise buildings. He has played a leading role in developing professional guidance and design standards, including *Improved Seismic Design Guidelines for California Highway Bridges* (ATC 32); *Guidelines for Evaluation and Repair of Masonry and Concrete Walls* (FEMA 306); *Guidelines for Seismic Rehabilitation of Buildings* (FEMA 273 and ASCE 356); *Development of Next-Generation Performance-Based Seismic Design Procedures for New and Existing Buildings* (FEMA P-58); and *Guidelines for Performance-Based Seismic Design of Tall Buildings* (Tall Buildings Initiative, PEER). He has served on the Boards of Directors of the Structural Engineers Association of Northern California, the Earthquake Engineering Research Institute, and the American Concrete Institute. His awards include the Lindau Award, the Siess Award, and the Boase Award from the American Concrete Institute; the Huber Research Prize from the American Society of Civil Engineers; the Annual Distinguished Lecturer and Outstanding Paper Award from the Earthquake Engineering Research Institute; and Honorary Member and College of Fellows of the Structural Engineers Association of California. He is an elected member of the U.S. National Academy of Engineering. He has been a member of the ACI 318 Building Code Committee since 1989, chair of ACI 318H (Seismic Provisions) from 1995 to 2014, and is chair of the ACI 318 Building Code Committee for the 2014–2019 code cycle.

Cover image courtesy of Magnusson Klemencic Associates

Seismic Design of Reinforced Concrete Buildings

Jack Moehle

Mc Graw Hill Education

New York Chicago San Francisco
Athens London Madrid
Mexico City Milan New Delhi
Singapore Sydney Toronto

Cataloging-in-Publication Data is on file with the Library of Congress.

Seismic Design of Reinforced Concrete Buildings

1 2 3 4 5 6 7 8 9 0 DOC/DOC 1 2 0 9 8 7 6 5 4

ISBN 978-0-07-183944-0
MHID 0-07-183944-5

The pages within this book were printed on acid-free paper.

Sponsoring Editor
Bridget Thoreson

Editorial Supervisor
Donna M. Martone

Project Manager
Vipra Fauzdar, MPS Limited

Copy Editor
Mohammad Taiyab Khan, MPS Limited

Proofreader
Alekha Jena

Production Supervisor
Pamela A. Pelton

Composition
MPS Limited

Art Director, Cover
Jeff Weeks

To Melissa,
For time, encouragement, diversions.

Contents

Preface

This book emphasizes the behavior and design requirements for earthquake-resistant reinforced concrete buildings. Design of a building for earthquake effects requires a different perspective than is required for other load effects. Earthquake loads are mainly absent during the life of a building, but suddenly may be applied with an intensity that drives the structure beyond the linear range of response in multiple loading cycles. Earthquake response of a structure is dynamic, with distributed inertial forces that act in all directions simultaneously. To meet established performance objectives under earthquake loading, a building requires a structural system that is appropriately configured, proportioned, and detailed. These complicated design conditions are beyond the scope of traditional reinforced concrete or earthquake engineering textbooks. This book aims to provide the focused and in-depth treatment necessary to fully understand the design requirements for earthquake-resistant concrete buildings.

The content emphasizes the mechanics of reinforced concrete behavior and the design requirements applicable to buildings located in "highly seismic" regions. The content will also be of value to engineers interested in the seismic evaluation of existing structures, design in regions of lower seismicity, and the general design of concrete structures for routine and extreme loading conditions.

Although the mechanics of reinforced concrete is universal, the performance expectations and associated design requirements may vary by region. This book mainly follows the requirements of the 2014 edition of the American Concrete Institute's *Building Code Requirements for Structural Concrete (ACI 318-14) and Commentary*. Those requirements are augmented by additional recommendations derived from other codes, guidelines, and the general literature, as deemed appropriate by the author. Dual units [U.S. customary units and International System of Units (SI)] are used throughout.

The target audience is twofold: (1) graduate students with structural engineering emphasis and (2) practicing structural engineers. For graduate students, this book provides a logical progression of content that builds knowledge of reinforced concrete construction, including design methods, behavior of structural materials and members, and the assembly of structural members into complete buildings capable of resisting strong earthquake shaking. This content has been developed and honed through years of graduate student instruction. For the practitioner, this book can build knowledge and serve as a reference resource to help solve challenging design problems. The book draws extensively from research literature and the experience of the author working with practicing structural engineers. The presentation emphasizes practical aspects, with numerous illustrations of concepts and requirements.

Topics are organized in four main parts. The first part (Chapter 1) reviews design methods applicable to the earthquake-resistant design of reinforced concrete buildings. The second part (Chapters 2 to 4) discusses material properties of steel, concrete, and confined concrete that are important for seismic performance and design. The third part (Chapters 5 to 10) covers the behavior of structural concrete components, including tension and compression members, beams, columns, walls, beam-column connections, and slab-column and slab-wall framing subjected to axial force, moment, shear, and imposed inelastic deformations. The last part of the book (Chapters 11 to 16) addresses seismic design of moment-resisting frames, structural walls, gravity frames, diaphragms, and foundations. Taken together, these four parts provide comprehensive coverage of the mechanics and design of earthquake-resistant concrete buildings.

The book is suitable for advanced undergraduate or graduate courses in structural engineering. At the University of California, Berkeley, it serves as a resource for a first-semester graduate course on seismic design of reinforced concrete buildings, touching on selected subjects in most of the chapters, but leaving the remaining chapter content for individual study. The book could also be used in a two-semester sequence, the first semester covering design methods, materials, and structural components (Chapters 1 to 10) and the second semester covering the design of earthquake-resistant structural systems (Chapters 11 to 16).

Numerous graduate students have read early drafts of this book in graduate classes, and individual experts have reviewed individual chapters. Readers are encouraged to send further suggestions for improvements, clarifications, and corrections to my attention at moehleRCSeismic@gmail.com.

Jack Moehle
August 2014

Acknowledgments

An early text on seismic design of concrete buildings[1] begins with the line: "Considerable knowledge has been gained in the last three decades about the phenomena of ground motion, the characteristics of structures, and their behavior in earthquakes." In the intervening five decades, knowledge and methods for earthquake-resistant concrete buildings have grown at an increasing rate. The key contributions to this book are acknowledged in an extensive list of references at the end of each chapter.

I am grateful for the contributions of several individuals and organizations as noted below.

Instructors at the University of Illinois at Urbana–Champaign introduced me to reinforced concrete and inspired a lifelong study of the subject. Especially, Professors William Gamble and Mete Sozen emphasized the mechanics of reinforced concrete and instilled an appreciation of the role and the limits of mechanics in engineering practice and the building code. Professor Sozen has continued as a lifelong mentor.

The University of California, Berkeley, has extended to me the privilege of teaching courses and conducting research on the subject of this book over three decades. An extraordinary group of faculty members and graduate students provided me with challenges, ideas, solutions, and a testing ground for much of the content of this book.

Many structural and geotechnical engineers have collaborated with me on research, code and guideline development efforts, and structural/earthquake engineering design and assessment projects. These interactions have revealed engineering problems and solutions that served as the basis for many practical recommendations presented in the book.

Several individuals contributed directly to this book. Nicholas Moehle processed the data in support of the confined concrete models of Chapter 4. Santiago Pujol of Purdue University, while on leave at UC Berkeley in 2014, led developments on panel zone shear that are presented in Chapters 7 and 13.

[1]Blume, J.A., N.M. Newmark, and L.H. Corning (1961). *Design of Multistory Reinforced Concrete Buildings for Earthquake Motions*, Portland Cement Association, Evanston, IL, 318 pp.

Ian McFarlane, Michael Valley, and John Hooper of Magnusson Klemencic Associates; Jay Love and Wayne Low of Degenkolb Engineers; and Dom Campi of Rutherford & Chekene discussed and provided examples of foundation design. Steve Kramer of the University of Washington provided extensive references on geotechnical earthquake engineering and foundation design, and Marshall Lew of AMEC, Los Angeles, provided references on retaining wall design.

The National Institute of Standards and Technology, under the auspices of the U.S. National Earthquake Hazard Reduction Program and the leadership of John (Jack) Hayes, supported the development of three technical briefs that were the starting point of Chapters 12, 13, and 15. Co-authors of these technical briefs were John Hooper, Dave Fields, and Chris Lubke of Magnusson Klemencic Associates; Dominic Kelly of Simpson Gumpertz & Heger; and Tony Ghodsi and Rajnikanth Gedhada of Englekirk Structural Engineers.

Many individuals and organizations permitted the use of copyrighted images and tables that added considerably to the presentation. The American Concrete Institute was especially generous in facilitating the use of numerous images and other content.

Several individuals reviewed various chapters and example problems, including Ron Hamburger of Simpson Gumpertz & Heger; David Gustafson of the Concrete Reinforcing Steel Institute; Julio Ramirez and Santiago Pujol of Purdue University; Gustavo Parra of the University of Wisconsin, Madison; Ian McFarlane and Michael Valley of Magnusson Klemencic Associates; Dom Campi of Rutherford & Chekene; and Professors Paulo Monteiro and Yousef Bozorgnia, and Graduate Student Researchers Carlos Arteta, John N. Hardisty, and Ahmet Tanyeri of the University of California, Berkeley.

This book would not have been possible without the support of my wife, Melissa, who gave ideas on content and organization, proofread the chapters, and provided continual encouragement through many lost evenings and weekends.

CHAPTER 1

Seismic Design and Performance Verification

1.1 Earthquake Resistance in Concrete Buildings

Cast-in-place reinforced concrete construction is naturally well suited to earthquake resistance. As a monolithic construction form, it can readily provide a continuous load path to resist forces and maintain structural integrity during earthquake shaking. Concrete structures can also have high rigidity to protect nonstructural elements and contents from the deformations that are caused by earthquake shaking. To be earthquake-resistant, however, a reinforced concrete building should have an appropriate and clearly defined lateral-force-resisting system that is proportioned and detailed to resist the expected earthquake demands. Without these features, reinforced concrete buildings can be susceptible to localized and relatively brittle failures. The aim of modern design procedures is to produce a structural system having the stiffness, strength, and deformation capacity necessary to resist earthquake shaking with acceptable performance. The aim of this book is to describe the requirements of earthquake-resistant concrete buildings, starting from a fundamental materials level and ending with conceptual considerations and the detailed requirements for design and construction of complete structural systems.

This book is organized into a series of chapters that sequentially build the knowledge required for seismic design of concrete buildings. Chapter 1 introduces basic concepts of building performance and the methods commonly used to verify performance. Chapters 2 to 4 present common structural materials used in reinforced concrete buildings. Chapters 5 to 10 present elements and connections of reinforced concrete construction, including methods for modeling, design, and verification. Finally, Chapters 11 to 16 present requirements for seismic design and construction of complete structural systems. Together, these chapters provide a strong foundation for conceiving, designing, and verifying reinforced concrete buildings for seismic resistance.

This book emphasizes conventionally reinforced, cast-in-place construction. Structural systems that use precast or prestressed concrete, or that use specialized "self-centering" systems, are not emphasized. However, a thorough understanding of the subjects covered in this book will serve as an effective basis for the design of such systems.

1.2 Early Developments

Reinforced concrete was introduced around the middle of the 19th century. The earliest forms of reinforced concrete construction included many patented systems that are unfamiliar today. By the beginning of the 20th century, publication of papers, books, and codes introduced the mechanics and construction requirements of reinforced concrete to an increasing number of engineers, leading to acceleration in the use of reinforced concrete. Many engineers of this period understood the importance of detailing for reinforcement continuity, but general requirements for earthquake resistance were little understood, and construction in seismically active regions did not differ significantly from construction in apparently non-seismic regions.

Concepts of seismic design for concrete buildings, including proportioning and detailing for ductile response, were introduced by Blume et al. (1961). Borrowing from technologies developed for blast resistance and with a nascent understanding of earthquake design requirements, that book introduced flexural ductility concepts, capacity design for shear, requirements for reinforcement continuity, and the use of transverse reinforcement to confine heavily strained sections. Concurrent activities by the Structural Engineers Association of California (SEAOC, 1963 and later editions) further contributed to knowledge about seismic design requirements for buildings in California. By the mid-1960s, a wide audience of practicing structural engineers was learning about the requirements for earthquake-resistant concrete buildings (Figure 1.1). Salient requirements for such buildings included use of transverse reinforcement to make the strength in shear greater than the shear occurring at flexural strength; hoops with 135° and 180° hooks to keep hoops closed after spalling of cover concrete, and with tight spacing to confine concrete in potential yielding regions; and effectively continuous longitudinal reinforcement developed within the beam-column joints, with splices located away from yielding regions.

The recommendations of Blume et al. (1961) and the Structural Engineers Association of California (SEAOC, 1963 and later) were not immediately adopted as requirements by U.S. building codes. It took the 1971 San Fernando earthquake (NOAA, 1973), and its demonstration of the vulnerabilities of some concrete buildings, to instigate code changes. By 1976, the Uniform Building Code (UBC, 1976) introduced many of the recommendations of Blume et al. and SEAOC as building code requirements. Early developments in other countries have been reported by Park (1986), Otani (1995), and Fardis (2013).

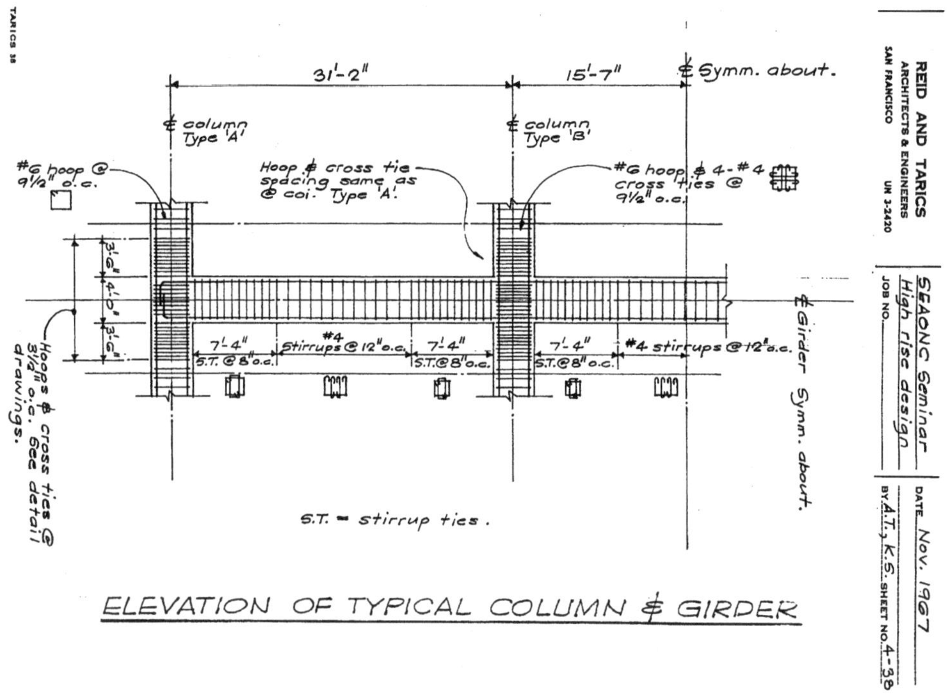

Figure 1.1 Page from 1967 Structural Engineers Association of California seminar notes (SEAOC, 1967). *Used with permission from A. Tarics and SEAONC.*

1.3 Current Practices

Experience, research, computational capabilities, and developments in conceptual thinking have led to important advances in the practice of earthquake engineering since the 1970s. Today, an engineer has available a wide variety of procedures that can be used for seismic assessment and design of buildings. These are contained in building codes, standards, guidelines, and the general literature on structural and earthquake engineering.

Earthquake-resistant reinforced concrete construction practices have by now become widely adopted. Figure 1.2 shows photographs of three recently constructed buildings designed according to current seismic design principles. Figure 1.2*a* shows a mid-rise building on the University of California, Berkeley campus. Located less than one kilometer from the active trace of the Hayward earthquake fault, the building features reinforced concrete structural walls for lateral force resistance. Figure 1.2*b* illustrates a high-rise frame-wall building under construction in the Pacific Northwest of the United States. Figure 1.2*c* illustrates a 300-m-tall core-wall building that was under construction at the time of the 2010 Chile earthquake. Each of these structural systems relies on earthquake-resistant structural systems designed using methods that will be featured in this book.

(a) Mid-rise concrete building, UC Berkeley campus. (*STUDIOS Architecture; Photograph courtesy of University of California.*)

(b) High-rise moment frame-wall building in Pacific Northwest of the United States. (*Photograph courtesy of Cary Kopczynski & Co. Cary Kopczynski.*)

(c) Gran Torre Santiago. (*Photograph courtesy of René Lagos Engineers.*)

Figure 1.2 Modern buildings designed for earthquake resistance in regions of high seismicity.

Design of any building begins with a *conceptual design*, in which the structural systems and materials are identified and configured. Once the structural system has been identified and approximately proportioned, structural analysis and design are used to confirm that the building design is capable of meeting intended performance objectives. Generally this is done following the requirements of the *building code*, using either prescriptive or performance-related provisions. In a *prescriptive design*, the structural analysis and design are implemented in strict accordance with the prescriptive requirements of the building code, with the implicit assumption that a code-conforming building will automatically meet the performance objectives. In contrast, a *performance-based design* can deviate from the prescriptive provisions and use structural analysis and design to demonstrate that the building nonetheless meets or exceeds the performance objectives of the building code.

Regardless of the design approach, competent *construction inspection* is required to ensure that the project is constructed in accordance with the design intent. The following subsections discuss aspects of building codes, conceptual design, prescriptive design, performance-based design, and construction inspection. Performance objectives are discussed in Section 1.4.

1.3.1 Building Codes

A building code is a set of minimum regulations intended to safeguard public health, safety, and general welfare of the occupants. The development, adoption, and enforcement of building codes vary widely from one country to another. In some countries, building codes are developed by governmental agencies and are enforced nationwide. In other countries, including the United States, the authority to regulate building construction is delegated to local jurisdictions. Ideally, the local jurisdictions in such countries adopt model building codes by reference, making these model codes part of the law governing construction in that jurisdiction.

In the United States, most jurisdictions adopt and use the *International Building Code* (IBC, 2012), which establishes minimum regulations for building systems using prescriptive and performance-related provisions. The IBC, in turn, adopts by reference the standard *Minimum Design Loads for Buildings and Other Structures* (ASCE 7-10, 2010), which establishes minimum requirements for design loads including those associated with earthquakes. The IBC also adopts by reference the standard *Building Code Requirements for Structural Concrete (ACI 318-11) and Commentary* (ACI 318-11, 2011),[1] which establishes minimum requirements for structural concrete design.

It should be emphasized that the purpose of a building code is to establish the minimum requirements to safeguard the public health, safety, and general welfare. These are generally established through strength, serviceability, durability, and

[1]Building codes are in a continual state of development, such that several different editions of each code or standard may be available. A jurisdiction generally adopts a specific edition of the code. In the United States, most jurisdictions adopt the IBC, but not necessarily the latest edition. At the time of this writing, the latest edition was IBC (2012), which incorporates by reference ASCE 7-10 (2010) and ACI 318-11 (2011). This book, however, references ACI 318-14 (2014).

other requirements of the building code. It is also permitted to design a building to exceed the minimum requirements of the code, including design for enhanced performance or sustainability. Various design guidelines and standards have been developed by professional organizations to present recommended practices that may exceed minimum building code requirements. Performance-based design can also be used to target performance exceeding the building code performance intent.

1.3.2 Conceptual Design

Conceptual design refers to the early design phase in which the structural systems are selected, configured, and approximately proportioned. The structural system must fit within the space and function of the building, while at the same time providing a suitable load path for anticipated loads, including gravity, wind, and earthquake loads. Selection of the structural concept is a key responsibility of the structural engineer. By selecting a good structural concept, the structural engineer usually can simplify the structural analysis, design, and review process, while providing a high degree of confidence that the performance objectives will be achieved.

Figure 1.3 illustrates typical elements of a reinforced concrete structural system. The gravity load-carrying system comprises the roof and floor system, columns and bearing walls, and the foundation. The lateral-force-resisting system comprises diaphragms, vertical elements, and the foundation. The diaphragms tie the building system together into a rigid, three-dimensional unit. Diaphragms also transmit lateral forces to the vertical elements of the lateral-force-resisting system. Chapter 15 describes diaphragm design in detail. The building codes limit the seismic-force-resisting vertical elements to moment-resisting frames (Chapter 12), structural walls (Chapter 13), or combinations of these elements. Various foundation elements (Chapter 16) are sized to transmit vertical, horizontal, and overturning forces to the supporting soils.

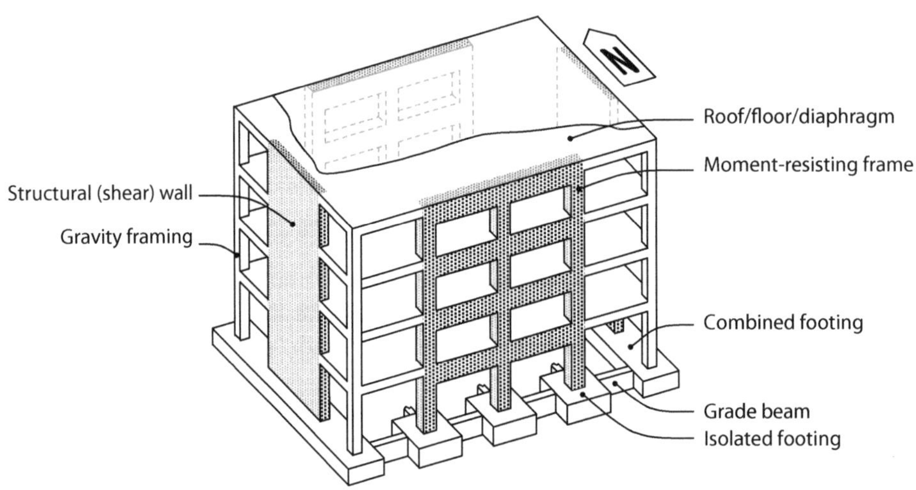

Figure 1.3 Typical elements of a reinforced concrete structural system.

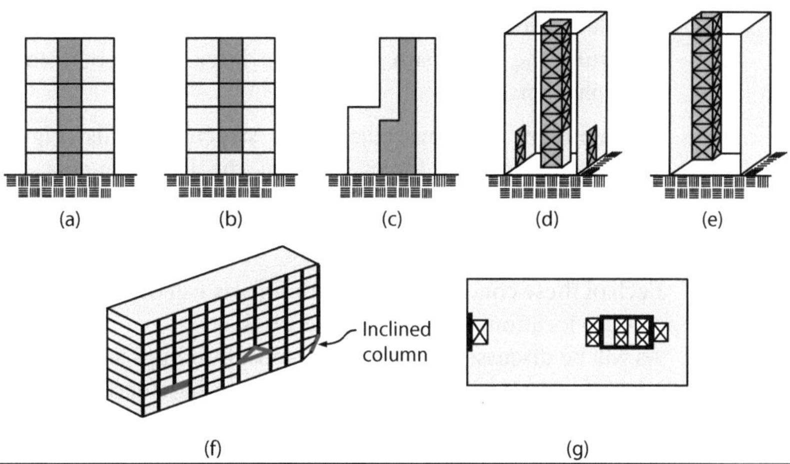

Figure 1.4 Considerations in building configurations: (*a*) vertically regular system; (*b*) discontinuous shear wall; (*c*) vertical irregularities in stiffness and mass; (*d*) horizontal offsets in lateral-force-resisting system; (*e*) torsionally unbalanced system; (*f*) isometric view of column transfers and offsets; (*g*) plan view showing limited connectivity between vertical elements and diaphragm. In (*d*) and (*e*), ⊠ is used to indicate location of a structural wall; a concrete braced frame is not a recognized seismic-force-resisting system in contemporary U.S. building codes.

The ideal structural system for an earthquake-resistant building is compact and symmetric, with stiffness and strength that are uniformly distributed over the height and across the plan, and without irregularities caused by discontinuous or offset structural elements. The structural system illustrated in Figure 1.3 has the desired attributes. Figure 1.4 illustrates a range of building configurations, some of which create design challenges that are better avoided through good conceptual design.

1. Figure 1.4*a* illustrates a regular structural system with continuous lateral resistance provided by a wall extending full height, and represents a good conceptual design.
2. A discontinuous structural wall, especially one that leaves an open first story as shown in Figure 1.4*b*, can create a weak and soft story that can be difficult to protect from excessive damage, and should be avoided.
3. Buildings with large changes in stiffness and strength over height (Figure 1.4*c*) may develop concentrations of inelastic response and damage, especially near the discontinuity.
4. Horizontal offsets in lateral systems (Figure 1.4*d*) create large force transfers across floor diaphragms, while also creating overturning problems for columns supporting discontinuous walls.
5. Eccentricity between center of resistance and center of mass (Figure 1.4*e*) results in torsion that creates design challenges. Such eccentricities should be minimized.

6. Column transfers and offsets (including inclined columns) (Figure 1.4*f*) disturb the load path and create large forces in transfer elements and diaphragms.
7. Diaphragm openings adjacent to structural walls (Figure 1.4*g*) limit the ability to transfer forces between the two elements, and are especially problematic near the base of a building where forces commonly must be transferred out of the structural walls.

Each of these conceptual design problems will be addressed in greater detail at appropriate locations throughout this book.

As will be discussed subsequently, most buildings are designed such that some inelastic response is anticipated during a design-basis earthquake. Hence, conceptual design also involves selection of a target yielding mechanism. For the structural system depicted in Figure 1.3, the intended mechanism might include flexural yielding of the walls for north-south loading and flexural yielding of the beams over the height and the columns at the base of the frames for east-west loading. The capacity design method is commonly used to proportion the structure for the intended mechanism (Section 1.5.4).

Once the structural system has been configured and approximately proportioned, preferably using a regular and symmetric layout, it can be analyzed and designed using either the prescriptive or the performance-based design approach.

1.3.3 Prescriptive Design Approach

A prescriptive design is one that adheres strictly to the prescriptive provisions of the building code, such as those contained in the IBC. Most building designs follow this approach. The provisions are prescriptive in the sense that they prescribe required analysis procedures, strengths, stiffnesses, and component and system details, with little leeway for deviating from the prescription. A typical prescriptive design includes the following steps:

- The building code specifies the intensity of the design loads for dead, live, wind, earthquake, and other effects, and spells out how the loads are to be combined for determining worst effects. In reference to Figure 1.5, dead load (D) is calculated from the weight of the building components and live load (L) is prescribed based on the building occupancy[2]. Earthquake load (E) is determined through a set of prescribed calculations set forth in the building code. Unlike other loads, E as specified in the codes is not intended to be a realistic estimation of actual earthquake loads, but instead is used to set a minimum strength, such that excessive ductility is not required under anticipated ground shaking. The code also prescribes several different load combinations, that is, ways in which to apply the specified loads, only one of which is shown in the figure. See Section 1.5.3 for load combinations.

[2]See the Notation section for definitions of all notation.

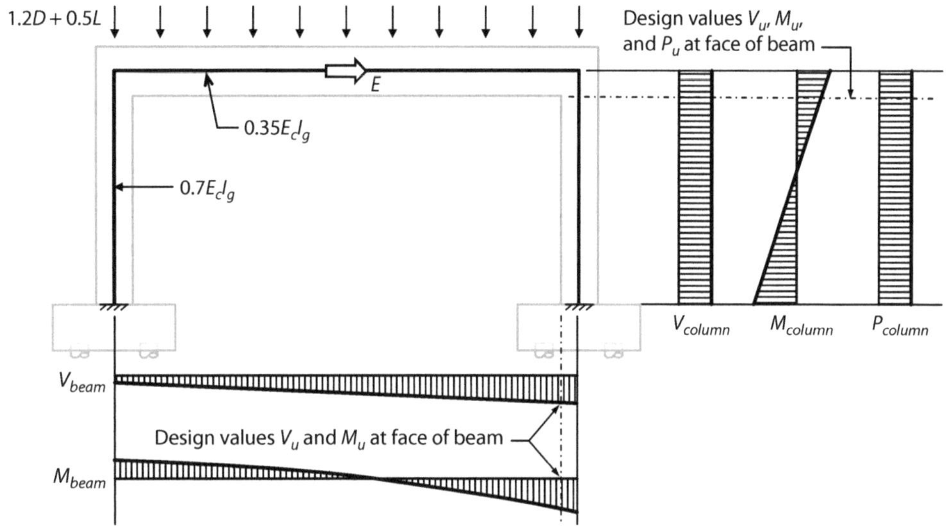

FIGURE 1.5 Determination of design shears, moments, and axial forces in a prescriptive design.

- The building code specifies how structural analysis is to be done, including the required stiffness and strength models. In most cases, the structural analysis is strictly linear, although stiffness may be reduced to approximately account for nonlinear effects, and some allowance for redistribution of internal actions may be permitted. In Figure 1.5, member stiffnesses are modified to approximate the effects of cracking and axial force, and then linear analysis is conducted to obtain the required strengths in shear (V_u), moment (M_u), and axial force (P_u).
- The building code specifies how to calculate member design strengths, and requires that these be at least equal to required strengths as determined from the structural analysis. See Section 1.5.3.
- The code also specifies requirements for member dimensions and reinforcement detailing. Figure 1.6 illustrates the types of details that will be prescribed, including requirements for continuity of reinforcement, locations and types of splices, and spacing and configuration of transverse reinforcement. Some of the reinforcement quantities and details will be dictated by calculated member actions, and others will be specified as requirements that are independent of the magnitude of the calculated member actions.
- The building code specifies displacement limits for members and for the building as a whole.

The prescriptive approach has the advantage that it uses well-established and familiar analysis and design techniques, with all of the requirements spelled out in the building code in such a way that an engineer can reliably implement them and

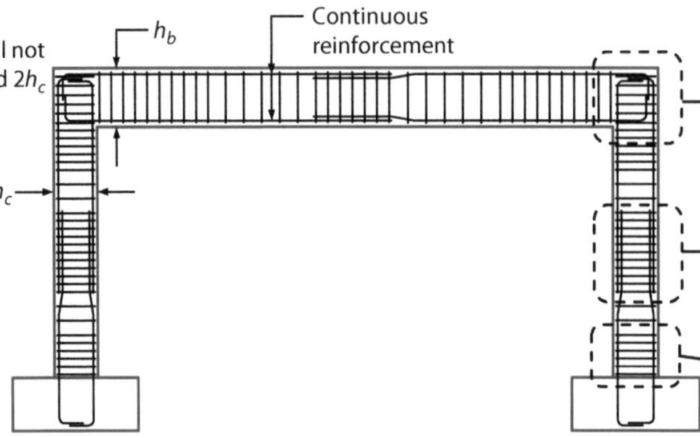

Figure 1.6 Examples of types of prescriptive dimensional limits and reinforcement details.

a building code official can check the final design for compliance with the building code requirements. A disadvantage of the prescriptive approach is that there is no calculation of expected building performance for future events.

In the United States, prescriptive requirements for proportioning and detailing of concrete buildings are contained in the seismic provisions of ACI 318 Building Code Requirements for Structural Concrete (ACI 318). Provisions relevant to other countries include Eurocode 8 (2004), NBC (2005), and NZS3101 (2006). This book describes member behavior and design in depth so that the reader can gain a full understanding of these building code requirements and the performance they are likely to deliver.

The traditional approach for proportioning and detailing of cast-in-place reinforced concrete structures intends that flexural yielding of members will be the primary source of inelastic response. That approach is emphasized in this book. Alternative approaches have introduced prestressed and precast concrete with nonlinear force-displacement mechanisms that may differ from those of cast-in-place concrete. See, for example, PRESSS (various) and PCI (2013).

1.3.4 Performance-Based Design Approach

In addition to the prescriptive design approach, most building codes also contain provisions that permit alternative designs. The IBC, for example, contains the following language:

> **104.11 Alternative materials, design and methods of construction and equipment.** The provisions of this code are not intended to prevent the installation of any material or to prohibit any design or method of construction not specifically prescribed by this code, provided that any such alternative has been approved. An alternative material, design or method of construction shall be approved where the building official finds that the proposed design is satisfactory and complies with the intent of the provisions of this code, and that the material, method or work offered is, for the purpose intended, at least the equivalent of that prescribed in this code in quality, strength, effectiveness, fire resistance, durability and safety.

Such alternative design approaches are often referred to as *performance-based approaches*, because the main basis for accepting the alternative method is a demonstration (through testing, analysis, or both) that the resulting building meets or exceeds the performance intent of the building code. Several conditions can trigger the use of performance-based design, including:

- A proposed building uses structural materials, elements, or systems that are not covered by the prescriptive building code provisions. The performance-based design may require physical tests to verify equivalent performance of the new materials. Structural analysis may also be required to demonstrate that a building using the new material will have equivalent performance.
- A proposed building height exceeds the prescriptive limits of the building code. In this case, structural analysis is used to demonstrate that the taller building can be safe and serviceable even though it exceeds the prescribed height limit.
- An owner or other responsible entity desires building performance that exceeds the minimum performance objectives of the building code. As a first step in this process, the structural engineer works with the owner or responsible entity to define the enhanced performance objective. Innovative structural materials may be introduced, or structural analysis may be used to demonstrate enhanced performance, or both.

Most performance-based designs rely on the prescriptive building code provisions, with specific exceptions to those provisions that emphasize the unique aspects of the proposed design. The performance evaluation can then focus mainly on those aspects of the design that are exceptions, greatly simplifying the process.

For buildings located in seismically active regions, performance-based seismic design generally involves a seismic hazard analysis to determine site-specific shaking levels, and usually includes the selection of representative earthquake ground motions by which to "test" the structure. A nonlinear computer model of the building structure is then subjected to these ground motions to determine the building response. Key response quantities are analyzed to establish whether the design meets the performance criteria that have been adopted for the building. TBI (2010) and LATBSDC (2011) contain detailed guidelines for performance-based seismic design of tall buildings. These same guidelines can also serve as a basis for performance-based design of other building types.

Performance-based design gives the engineer much greater flexibility in the choice of the structural system and its design method. Such designs, however, typically require additional design effort and time, advanced engineering capabilities, and a building official who is willing to accept designs not conforming strictly to the prescriptive provisions of the building code. Most building officials will not have the expertise necessary to judge the adequacy of a design falling under the alternative methods clause of the building code. Therefore, independent peer review is usually required to advise the building official as to whether a design is satisfactory.

1.3.5 Construction Inspection

Reinforced concrete seismic-force-resisting systems are complex systems whose performance depends on proper implementation of design requirements during construction. Therefore, inspection of the placement of the reinforcement and concrete should be done by a qualified inspector. The inspector should be under the supervision of the licensed design professional responsible for the structural design or a licensed design professional with demonstrated capability for supervising inspection of the construction. The inspector is required to check work for conformance to the approved design drawings and specifications; to submit interim reports identifying discrepancies and their resolution; and to submit a final report stating whether the work was in conformance with the approved plans and specifications and the applicable workmanship provisions of the building code.

1.4 Building Performance

1.4.1 Anticipated Response of Buildings to Earthquake Ground Shaking

Ground shaking is the main focus of most earthquake-resistant building designs. Ground failure associated with surface fault rupture, landslides, and soil liquefaction can also be concerns. These latter effects can be mitigated through building siting, ground improvement, or a foundation designed to support the structure despite the ground failure. Other secondary effects such as tsunami, fire, and lifeline disruption can also be considered in exceptional cases.

Effects of ground shaking can be represented through a linear response spectrum, which plots the acceleration of a linear-elastic oscillator as a function of its vibration period. Figure 1.7 plots the design earthquake response spectrum for a site on the University of California, Berkeley campus based on the provisions of ASCE 7, for the Design Earthquake (DE) shaking level. This response spectrum is a smoothed representation of the ground motion, having spectral ordinates equal

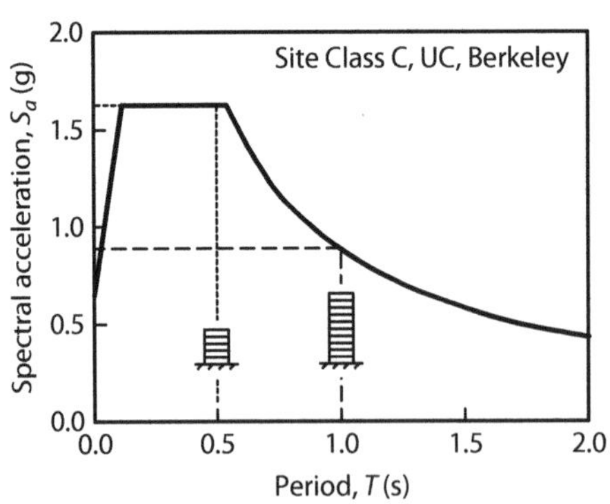

Figure 1.7 Design response spectrum for Design Earthquake (DE) for 5% damped linear response in accordance with ASCE 7.

to two-thirds of the Maximum Considered Earthquake (MCE), and serves as a basis for building design. (For additional details on the design approach, see Chapter 11.) Using the approximation that the fundamental vibration period of a building is approximately $T = 0.1N$, we would estimate the period for a 5-story building to be approximately 0.5 s, while that for a 10-story building would be approximately 1 s. Corresponding spectral accelerations are approximately $1.6g$ and $0.9g$, respectively. Thus, crude estimates for the design base shears, assuming linear-elastic response, are approximately $1.6W$ and $0.9W$, respectively, where W is the building weight.

Although it is possible to design buildings to have strengths corresponding to these design base shears, to do so would require very robust lateral-force-resisting systems. Economic and functional constraints would make such designs impractical except in unusual cases. Thus, most buildings are designed with base-shear strength lower than the strength required for linear-elastic response. A consequence is that inelastic response and corresponding damage must be anticipated for buildings subjected to DE-level ground motions. Expected building performance capability can be determined, in part, by the degree of inelastic response anticipated and by how that inelastic response is manifest in damage to the structural system.

1.4.2 Performance Concepts

Building performance can be expressed in multiple ways. In building design practice today, the most common approach is to define a series of performance objectives. A *performance objective* is a statement of the expected building performance conditioned on it having been subjected to a particular loading. For example, TBI (2010) recommends that a tall building be designed to satisfy the following two performance objectives:

1. The building shall have a small probability of life-threatening collapse given that it has been subjected to rare earthquake ground shaking defined as the Maximum Considered Earthquake (MCE) shaking level.
2. The building shall have a small probability of damage requiring repair given that it has been subjected to more frequent ground shaking defined as the Service Level Earthquake (SLE) shaking level.

According to this procedure, the building must be analyzed for two different performance objectives and it must satisfy both to be considered to have code-equivalent performance.

The concept of discrete performance objectives became firmly established in the 1990s with the introduction of the Vision 2000 Committee report on performance-based seismic design of buildings (SEAOC, 1995) and the development of performance-based assessment procedures for existing buildings (ATC 40, 1996; FEMA 273, 1997). Figure 1.8 illustrates the performance objectives suggested in SEAOC (1995), but using performance level designations of ASCE 41 (which supersedes FEMA 273). For the *Basic Objective*, which would

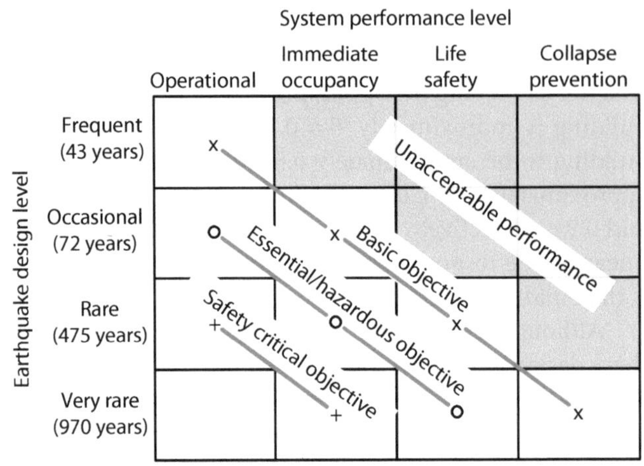

Figure 1.8 Performance objectives suggested by SEAOC (1995).

apply to the vast majority of buildings, the performance objectives would be *Operational* for *Frequent* shaking, *Immediate Occupancy* for *Occasional* shaking, *Life Safety* for *Rare* shaking, and *Collapse Prevention* for *Very Rare* shaking. Proposed return periods for these different shaking levels are shown in parentheses. For more critical structures, higher performance objectives were suggested (Figure 1.8).

An early concept was to relate performance levels to the physical condition of the building as it was subjected to increasing lateral deformation (SEAOC, 1995). Figure 1.9 illustrates three performance levels introduced in FEMA 273 (1997) and continued in ASCE 41 (2013). The performance level *Immediate Occupancy* corresponds to a state in which some damage may have occurred, but after cosmetic repairs the structure can be occupied and functional. *Collapse Prevention* is a point in the response just prior to onset of collapse. *Life Safety* is a term used to define a performance state with a "comfortable" margin below the collapse state. In ASCE 41, the margin is set at about three-quarters of the displacement corresponding to the collapse performance state, but in ASCE 7 and the building code, this margin is two-thirds. Figure 1.9 implies that performance states are a function of the deformations imposed on the structural and nonstructural systems. Performance of contents and other items that are not rigidly fixed to the structural system can instead be a function of floor acceleration or velocity.

Building performance should be defined by the performance of the building system as a whole. It can be difficult, however, to quantify performance metrics for building systems. Therefore, as a practical matter, a common practice is to define system performance based on the performance of individual structural (or nonstructural) components that compose the building system. In effect, the building performance is defined as being equal to the worst performance of any of the components of the building. This approach, which is adopted in ASCE 41, tends to be a very conservative approach.

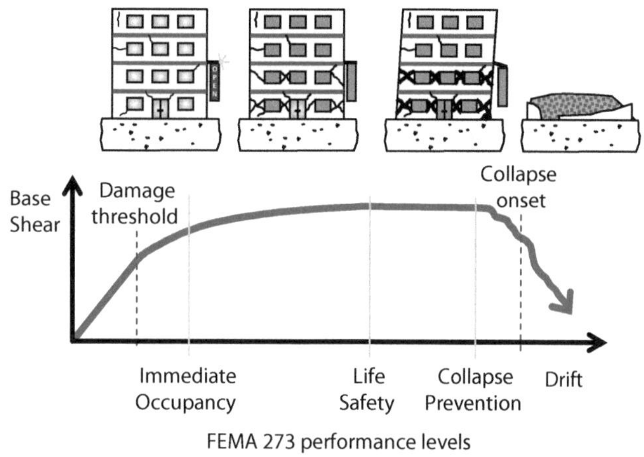

Figure 1.9 Visualization of performance levels. (*Personal communication with R. Hamburger.*)

The preceding discussion emphasizes the current approach of gauging performance using structural engineering metrics, such as displacement, story drift, floor acceleration, inelastic deformation, and component forces, all compared with values that are considered acceptable. It is also feasible to translate these engineering metrics into damage states, and from there into consequences such as casualties, repair costs, and downtime. This approach is not commonly applied today, but the capabilities exist and are occasionally applied for special buildings. The interested reader is referred to Yang et al. (2009) and FEMA P-58 (2012).

1.4.3 Use, Occupancy, and Risk Classifications

Figure 1.8 introduced the idea that building performance objectives should depend on the risk that the building poses to its occupants and the surrounding community. This idea is incorporated in current U.S. building codes. In ASCE 7, four different *Risk Categories* are defined, as summarized in Table 1.1. The vast majority of buildings correspond to Risk Category II. ASCE 7 imposes more stringent requirements for buildings with higher Risk Category, consistent with the larger population of people put at risk if the building should fail to perform. Higher design forces are also imposed through application of an importance factor, I_e, listed in Table 1.1.

1.4.4 Building Performance Expectations

The commentary to ASCE 7 quantifies the intended performance for buildings in different Risk Categories. Table 1.2 summarizes the anticipated reliability values. These values have not been validated with experience or in-depth analysis, but instead are notional values that represent the intent of the building code committee.

FEMA P-695 (2009) also suggests that the probability of collapse due to MCE ground motions should be limited to 10% for Risk Category II buildings. FEMA

Use or Occupancy of Buildings and Structures	Risk Category	Importance Factor, I_e
Buildings and other structures that represent a low risk to human life in the event of failure.	I	1.00
All buildings and other structures except those listed in Risk Categories I, III, and IV.	II	1.00
Buildings and other structures, the failure of which could pose a substantial risk to human life. Buildings and other structures, not included in Risk Category IV, with potential to cause a substantial economic impact and/or mass disruption of day-to-day civilian life in the event of failure.	III	1.25
Buildings and other structures designated as essential facilities. Buildings and other structures, including those with hazardous materials, the failure of which could pose a substantial hazard to the community.	IV	1.50

*In previous editions of ASCE 7, the term *Occupancy Category* was used instead of *Risk Category*.

Table 1.1 Risk Category* of Buildings and Other Structures (*adapted from ASCE 7*)

P-695 presents a detailed methodology for determining collapse probability of classes of buildings. Several case studies have used the methodology to benchmark performance of modern buildings, with results reported in FEMA P-695 (2009) and NIST (2010).

The collapse probabilities in Table 1.2 indicate the probabilities for an individual building given that it has been subjected to ground shaking at the MCE level. It should be noted that the term MCE, or Maximum Considered Earthquake, actually does not refer to an earthquake, but instead refers to the shaking that occurs at a site given the occurrence of an earthquake. MCE-level shaking generally indicates both a rare earthquake and unusually high ground shaking given the occurrence of that earthquake. Thus, one should not expect that all buildings in a region will be subjected to MCE-level shaking in any given earthquake. Instead, only a subset of buildings might experience MCE-level shaking, with the rest experiencing lower shaking intensity. Thus, over a region, the collapse probability for a population of new buildings is lower than the values indicated in Table 1.2.

Risk Categories	Probability of Total or Partial Collapse	Probability of Failure that Could Result in Endangerment of Individual Lives
I and II	0.10	0.25
III	0.06	0.15
IV	0.03	0.10

Table 1.2 Anticipated Reliability (Maximum Probability of Failure) for Buildings Conditioned on the Occurrence of MCE Shaking

Starting with the 2010 edition of ASCE 7, the earthquake design values for most locations in the United States have been adjusted so that structures having standard collapse fragility (thought to be a lower bound on the actual fragility of code-conforming structures) will have a collapse probability of 1% in 50 years. Near known active faults with significant slip rates and with characteristic earthquakes having magnitudes in excess of about 6.0, the design values are limited by 84th percentile spectral values associated with a characteristic earthquake on the fault. At these latter locations, the calculated probability of collapse will be somewhat higher, and in some cases much higher, than in locations not near such faults. For additional information on the derivation of the design values, see Luco et al. (2007) and NEHRP (2009).

1.5 Performance Verification

Structural engineers use a variety of different analysis and design methods to verify compliance with the prescriptive provisions of the building code, or to demonstrate performance equivalence of performance-based designs. The following paragraphs provide a brief review of some of these methods.

1.5.1 Limit State Design

A *limit state* is a condition of a structural member (or structural system) beyond which the structural member (or system) no longer satisfies a performance requirement. Two limit states commonly considered in design of buildings are the *serviceability limit state* and the *ultimate limit state*. *Limit state design* is a process by which the various limit states are identified and are designed for. Sections 1.5.2 and 1.5.3 introduce limit state design for serviceability and ultimate limit states.

1.5.2 Serviceability Limit State

One of the performance requirements for a building is that it remains functional under normal or routine load conditions. Serviceability limit states include onset of structural or nonstructural damage, visible deflections, and vibrations causing occupant discomfort. Design for a serviceability limit state can be accomplished by direct analysis for that state, by indirect analysis, or by provision of structural details, as summarized below:

- Structural analysis can be used to verify that the expected performance is within the defined performance limit. For example, a deflection limit can be established beyond which occupants discern misalignment or sense vibration in a floor, and calculations can verify that the deflections do not exceed these limits under service loads. As another example, a range of frequency and acceleration pairs can be established for wind-induced vibration beyond which building occupants experience discomfort, and these limits can be compared with values expected under service level wind loadings.

- Onset of damage to nonstructural elements can be related indirectly to the vertical or lateral distortions of a building. For example, if damageable nonstructural components are attached to a floor member, ACI 318 limits the vertical deflections after attachment of the nonstructural component to $l/480$. Similarly, common practice is to limit story drifts to $h_i/400$ and total roof drift to $h_n/500$ under service level wind loading.
- Crack width can be controlled through calculation and control of crack width indices, such as the index provided by Gergely and Lutz (1968). Alternatively, crack width can be controlled by provision of well-distributed flexural tension reinforcement, as is done in ACI 318, without direct calculation of the crack width.

For gravity loads, normal or routine loads are considered for serviceability checks. Dead load is the expected dead load without load factors. Live load is the maximum value expected over some defined period, usually 50 years, as tabulated in codes such as ASCE 7, without load factors. Procedures for establishing wind or seismic loads vary depending on the governing code. Because the consequences of exceeding wind drift or acceleration levels are not severe, it is more common to use service level wind speeds on the order of 10 or 20 years.

The serviceability limit states are generally well below the inelastic range of response, such that linear analysis is appropriate. Because even moderate-intensity earthquakes are relatively rare events, moderate excursions into the inelastic range of response may be acceptable for service-level earthquakes.

1.5.3 Ultimate Limit State (Load and Resistance Factor Design)[3]

A building should be stable under abnormal load conditions that reasonably may occur during the service life. A structure can become unstable if its materials rupture, its members buckle, or if the structural system forms a mechanism that leads to overall instability of the building frame. Such failures can lead to local collapse or overall structural collapse, both of which pose substantial risk to life safety. Such failures should have a small probability of occurring during the service life of the building. A load and resistance factor design (LRFD) method is commonly adopted to provide a structure with adequate strength so that the probability of failure is reduced to an acceptable level. MacGregor (1983) describes development of the LRFD factors and associated safety levels for reinforced concrete design.

General Approach

The LRFD method can be expressed generically through Eq. (1.1):

$$\phi S_n \geq U \tag{1.1}$$

[3]The LRFD factors and load combinations in this chapter are in accordance with the requirements of ASCE 7-10 (2010). Different factors or approaches may apply where other codes are adopted.

in which ϕS_n = design strength, ϕ = strength reduction factor, S_n = nominal strength, and U = factored load effect. In practice, Eq. (1.1) is applied more specifically to internal member forces such as shear and moment, as in

$$\phi V_n \geq V_u \tag{1.2}$$

$$\phi M_n \geq M_u \tag{1.3}$$

where V_n = nominal shear strength, M_n = nominal moment strength, V_u = shear due to factored loads, and M_u = moment due to factored loads.

Although the LRFD method refers to an ultimate limit state approaching the failure or collapse state, structural analysis for the limit state is often done using assumptions of linear-elastic behavior. The ultimate limit state for the structural system as a whole is presumed to be reached for the loading that first causes a member cross section to reach the design strength ϕS_n. Inelastic analysis would consider redistribution of internal actions, thereby permitting factored loads equal to or exceeding the factored load limit determined by linear-elastic analysis. Thus, linear-elastic analysis provides a conservative measure of the ultimate limit state performance.

Load and resistance factors for the LRFD method are established considering variability and uncertainty in different load effects and material properties, the accuracy and variability of nominal strengths, the brittleness of different failure modes, and the consequences of failure. For buildings assigned to Risk Category II of ASCE 7, the intended annual probabilities of failure for load conditions that do not include earthquake are 3×10^{-5} per year for failure that is not sudden and does not lead to widespread progression of damage, 3×10^{-6} per year for failure that is either sudden or leads to widespread progression of damage, and 7×10^{-7} per year for failure that is sudden and results in widespread progression of damage (ASCE 7).

ASCE 7 Factored Load Combinations

The factored load effect is represented by U in Eq. (1.1). In practice, the quantity U is the maximum (or minimum) load effect determined through a series of load combinations. Each load combination considers one or more load cases, with load factors adjusted to achieve approximately uniform reliability.

The main *load cases* are listed below, and refer to the load itself or to its effect on internal moments and forces:

D = dead load

E = earthquake load

F = load due to fluids with well-defined pressures and maximum heights

H = load due to lateral earth pressure, ground water pressure, or pressure of bulk materials

L = live load

L_r = roof live load

S = snow load

W = wind load

The basic *load combinations* consider different combinations of the load cases as follows:

1. $1.4D$
2. $1.2D + 1.6L + 0.5(L_r \text{ or } S \text{ or } R)$
3. $1.2D + 1.6(L_r \text{ or } S \text{ or } R) + (\alpha_L L \text{ or } 0.5W)$
4. $1.2D + 1.0W + \alpha_L L + 0.5(L_r \text{ or } S \text{ or } R)$
5. $1.2D + 1.0E + \alpha_L L + 0.2S$
6. $0.9D + 1.0W$
7. $0.9D + 1.0E$

In combinations 3, 4, and 5, the factor α_L applied to L is equal to 1.0 for garages, for areas occupied as places of public assembly, and for any occupancies in which $L > 100$ psf (4.8 kPa). Otherwise, $\alpha_L = 0.5$.

Where fluid loads F are present, they are to be included with the same load factor as dead load D in combinations 1 through 5 and 7.

Where loads H are present, they are to be included as follows:

1. Where the effect of H adds to the primary variable load effect, include H with a load factor of 1.6.
2. Where the effect of H resists the primary variable load effect, include H with a load factor of 0.9 where the load is permanent or a load factor of 0.0 for all other conditions.

In any of the load combinations, effects of one or more loads not acting or effects of loads acting in the opposite direction (where possible) are to be investigated. The most unfavorable effects from both wind and earthquake loads are to be investigated, where appropriate, but they need not be considered to act simultaneously. Additional effects of flood, atmospheric ice loads, and self-restraining loads are not covered in this book. See ASCE 7 for additional details.

For earthquake-resistant design, the engineer must consider the effects of earthquake directionality. In general, this includes effects of earthquake loads in two principal horizontal directions plus vertical earthquake shaking effects. Effects of overstrength on design loads must also be considered in some special cases. See Chapter 11 of this book for these additional details.

Figure 1.10 illustrates the application of the load combinations for a planar system considering the load cases D, L, and E. Basic load combinations 1 and 2 consider only D and combined D and L. In this illustration, both D and L are taken at their full intensities. To obtain the worst shear at beam midspan, however, L should be placed on only half of the beam span. The building code requires that this latter loading case also be considered.

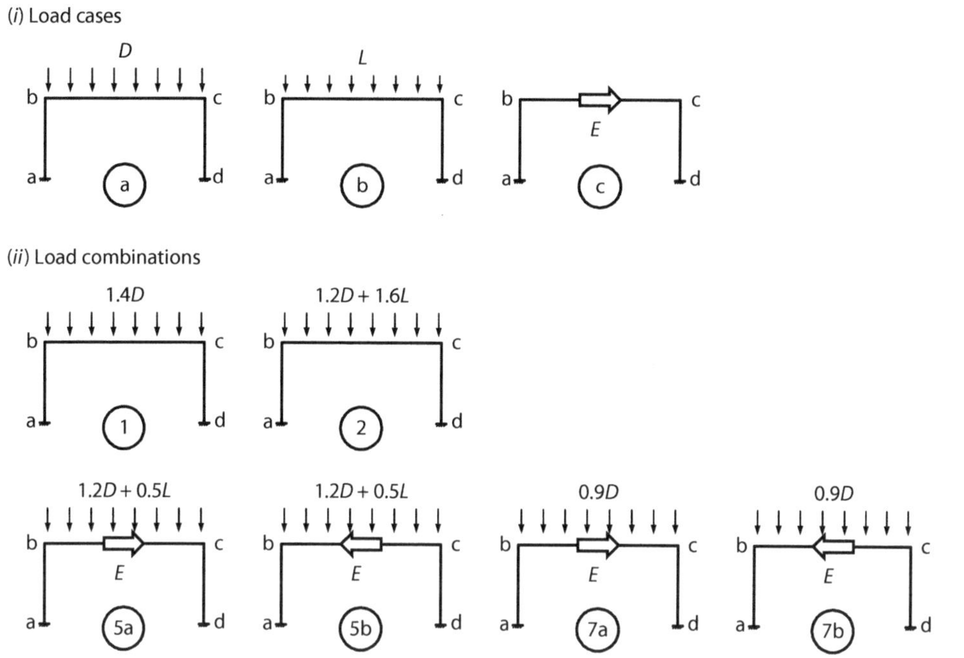

FIGURE 1.10 Load cases and load combinations in load and resistance factor design.

Diagrams 5a and 5b in Figure 1.10 illustrate ASCE 7 load combination 5; note that E must be considered both from left to right and from right to left. Illustrations 7a and 7b in Figure 1.10 illustrate load combination 7. In a typical structure, load combination 5 results in higher axial compression in columns while load combination 7 results in higher axial tension in columns. Both load combinations must be considered in design. Not shown in these diagrams is the effect of vertical earthquake loads, which must be considered in accordance with ASCE 7. See Chapter 11 for those additional details.

ACI 318 Resistance Factors

In Eq. (1.1), the term ϕS_n is referred to as the *design strength*, which is the product of strength reduction factor ϕ and nominal strength S_n. Nominal strength is determined using nominal strength equations (which are covered in later chapters of this book), and using specified material properties for concrete and reinforcement. The strength reduction factors have numerical values less than 1.0, and are provided (1) to allow for the possibility of under-strength members due to variations in material strengths and dimensions, (2) to allow for inaccuracies in the design equations, (3) to reflect the available ductility and required reliability of the member under the load effects being considered, and (4) to reflect the importance of the member in the structure.

Table 1.3 presents strength reduction factors from ACI 318.

Action or Structural Element		ϕ	Exceptions
(a)	Moment, axial force, or combined moment and axial force	0.65 to 0.9. See below	For value of ϕ near ends of pretensioned members where strands are not fully developed, see ACI 318
(b)	Shear	0.75	Additional requirements are defined for structures designed to resist earthquake effects. See below
(c)	Torsion	0.75	
(d)	Bearing	0.65	
(e)	Post-tensioned anchorage zones	0.85	
(f)	Brackets and corbels	0.75	
(g)	Struts, ties, nodal zones, and bearing areas designed in accordance with the strut-and-tie method	0.75	
(h)	Components of connections of precast members controlled by yielding of steel elements in tension	0.9	
(i)	Plain concrete elements	0.6	

Table 1.3 Strength Reduction Factors, ϕ

For slabs, beams, columns, walls, and other similar members subjected to moment, axial force, or combined moment and axial force, the strength reduction factor ϕ depends on the net tensile strain ε_t in the extreme tension reinforcement when the section develops nominal strength. For flexural members with zero axial force or with axial compression, nominal strength is reached when the extreme fiber compressive strain reaches 0.003. Values for ϕ are presented in Table 1.4.

According to ACI 318, for deformed reinforcement, ε_{ty} is taken as f_y/E_s. For Grade 60 (420)[4] deformed reinforcement, it is permitted to take ε_{ty} equal to 0.002. For all prestressed reinforcement, ε_{ty} is to be taken as 0.002. Note that the net tensile strain in the longitudinal reinforcement ε_t is the strain at the centroid of the extreme tension reinforcement of the cross section, rather than at the centroid of all of the tension reinforcement, and excludes strains due to prestress, creep, shrinkage, and temperature. Figure 1.11 illustrates the values of ϕ.

The variation of ϕ with ε_t is because sections with larger ε_t are more ductile, giving ample warning of failure with excessive deflection and cracking, whereas sections with smaller ε_t are prone to more brittle compression failure, with little

[4]The reinforcement grade refers to the minimum required yield strength. The first number gives the yield strength in ksi units and the number in parentheses gives the strength in MPa units.

Net Tensile Strain, ε_t	Classification	ϕ			
		Type of Transverse Reinforcement			
		Spiral		Other	
$\varepsilon_t \le \varepsilon_{ty}$	Compression-controlled	0.75	(a)	0.65	(b)
$\varepsilon_{ty} \le \varepsilon_t < 0.005$	Transition*	$0.75 + 0.15\dfrac{(\varepsilon_t - \varepsilon_{ty})}{(0.005 - \varepsilon_{ty})}$	(c)	$0.65 + 0.25\dfrac{(\varepsilon_t - \varepsilon_{ty})}{(0.005 - \varepsilon_{ty})}$	(d)
$\varepsilon_t \ge 0.005$	Tension-controlled	0.90	(e)	0.90	(f)

*For sections classified as transition, it is permitted to use the strength reduction factor corresponding to compression-controlled sections.

Table 1.4 Strength Reduction Factor, ϕ, for Moment, Axial Force, or Combined Moment and Axial Force

warning of impending failure. The higher ϕ factors for spiral reinforced sections are because such sections are inherently more ductile than conventionally reinforced tied columns. For the factor to apply, the spiral must satisfy the minimum spiral reinforcement requirements of the building code. See Chapter 6 for additional discussion on application of the ϕ factors for moment and axial force.

Earthquake-resistant structures in regions of high seismicity will rely on specially detailed structural walls and frames for lateral force resistance. These special systems are detailed to enable them to deform into the nonlinear range of response without excessive strength loss or failure. It is important that the different parts of the building system have a strength hierarchy that promotes yielding of the ductile frames or walls while protecting those elements intended for essentially elastic response. ACI 318 partially accomplishes this goal by setting limits on the strength reduction factors ϕ of the various elements in the load path. The specific code requirement is described next.

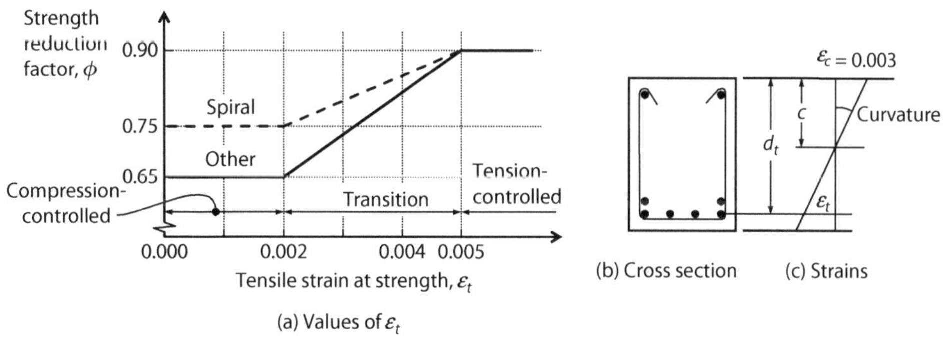

Figure 1.11 Variation of ϕ with net tensile strain in extreme tension reinforcement, ε_t, for member with Grade 60 (420) reinforcement or prestressing reinforcement. (*After ACI 318.*)

For structures designed to resist earthquake effects using (a) special moment frames, (b) special structural walls, or (c) intermediate precast structural walls in structures assigned to Seismic Design Category[5] D, E, or F, the value of ϕ for shear from Table 1.3 shall be modified in accordance with (i), (ii), and (iii):

i. For any structural member that is designed to resist *E*, ϕ for shear shall be 0.60 if the nominal shear strength of the member is less than the shear corresponding to the development of the nominal moment strength of the member. The nominal moment strength shall be determined considering the most critical factored axial loads and including *E*.

ii. For diaphragms, ϕ for shear shall not exceed the least value of ϕ for shear used for the vertical elements of the primary seismic-force-resisting system.

iii. For beam-column joints and diagonally reinforced coupling beams, ϕ for shear shall be 0.85.

Clause (i) addresses shear-controlled members, such as low-rise walls, portions of walls between openings, and diaphragms, for which nominal shear strength is less than the shear corresponding to development of nominal moment strength for the pertinent loading conditions. The intent is to provide additional strength in shear relative to the strength provided for moment resistance. Clause (i) does not apply to special moment frames. This provision is discussed more thoroughly in Chapters 13 and 15.

Clause (ii) addresses the relative strength of diaphragms and structural walls. Short structural walls were the primary vertical elements of the lateral-force-resisting system in many of the parking structures that sustained damage during the 1994 Northridge, CA earthquake. In some cases, walls remained essentially elastic, while diaphragms responded inelastically. The requirement of (ii) is intended to increase strength of the diaphragm and its connections in buildings for which the shear strength reduction factor for walls is 0.60, as those structures tend to have relatively high overstrength. Additional building code provisions may apply other factors to increase diaphragm strength. Those aspects are discussed more thoroughly in Chapter 15.

Clause (iii) notes that beam-column joints and diagonally reinforced beams have strength reduction factor ϕ for shear that differs from the factor for other members. This is because of the unique force-resisting mechanisms of these members. Additionally, the demands on beam-column joints are determined less by the load combinations than by the strengths of the members that frame into the joints. These members are discussed more fully in Chapters 7, 9, 12, and 13.

The use of strength reduction factors ϕ to control yielding mechanisms is only partially effective. Section 1.5.4 introduces the capacity design method as an alternative method for this purpose.

Example 1.1. A weightless, one-bay, one-story frame has configuration and loading shown in Figure 1.12. Concrete has $f'_c = 4000$ psi (28 MPa) and A706 Grade 60 (420) reinforcement. Dead load *D* is 3 klf (44 kN/m), live load *L* is 1.8 klf (26 kN/m), and earthquake load *E* is 45 kips

[5]Seismic Design Category is a classification assigned to a structure based on its Risk Category and the severity of the design earthquake ground motion at the site.

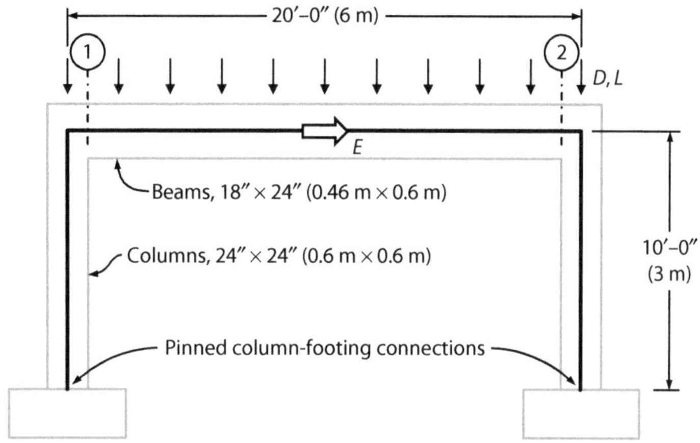

FIGURE 1.12 Frame geometry and loading.

(200 kN). Use the LRFD method to determine the required beam strengths at the faces of the beams (sections 1 and 2).

Solution

The load cases and load combinations are shown in Figure 1.10. The structure is modeled using flexural stiffness equal to $0.3EI_g$ for beams and columns (see Chapter 6 for stiffness recommendations) and analyzed for the load cases. The results of the load cases are then combined using the load combinations. Calculated moments and shears at sections 1 and 2 are tabulated on the next page.

	Moments, k-ft (kN-m)		**Shears, kips (kN)***	
	Section 1	Section 2	Section 1	Section 2
Load case				
D	−51.6 (−70.2)	−51.6 (−70.2)	27.0 (120)	−27.0 (−120)
L	−31.0 (−42.1)	−31.0 (−42.1)	16.2 (72.1)	−16.2 (−72.1)
E	203 (275)	−203 (−275)	−22.5 (−100)	22.5 (100)
Load combination				
$1.4D$	−72.2 (−98.2)	−72.2 (−98.2)	37.8 (168)	−37.8 (−168)
$1.2D + 1.6L$	−111 (−152)	−111 (−152)	58.3 (259)	−58.3 (−259)
$1.2D + 0.5L + E$	125 (170)	−280 (−381)	18.0 (80.0)	−63 (−280)
$1.2D + 0.5L - E$	−280 (−381)	125 (170)	63.0 (280)	−18 (−80)
$0.9D + E$	156 (212)	−249 (−339)	1.8 (8.0)	−46.8 (−208)
$0.9D - E$	−249 (−339)	156 (212)	46.8 (208)	−1.8 (−8.0)
Minimum	*−280 (−381)*	*−280 (−381)*	*1.8 (8.0)*	*−63 (−280)*
Maximum	*156 (212)*	*156 (212)*	*63 (280)*	*−1.8 (−8.0)*

*Note that the required shear strength is determined by capacity design rather than by the usual LRFD load combinations. See Example 1.3.

Example 1.2. Select longitudinal reinforcement to satisfy the LRFD strength requirement.

Solution

Beam axial force will be ignored because it is small compared with the axial strength. For sections that are tension-controlled, which will be the case here, the strength reduction factor is $\phi = 0.9$. Based on the results of Example 1.1, the required negative and positive nominal moment strengths are $M_n^+ = M_u^+/0.9 = 173$ k-ft (236 kN-m) and $M_n^- = M_u^-/0.9 = -311$ k-ft (−423 kN-m). The section shown in Figure 1.13 provides $M_n^+ = 205$ k-ft (279 kN-m) and $M_n^- = 392$ k-ft (532 kN-m), and therefore satisfies the strength requirement. Note that the design could be refined to result in provided strengths that are closer to required strengths. However, it will be useful to have a substantial overstrength to demonstrate capacity design in Example 1.3. See Example 6.5 for partial moment strength calculations.

1.5.4 Capacity Design

Capacity design is a design method for controlling the yielding mechanism of a structure that is expected to respond inelastically to a design loading or an overload. Capacity design was introduced in structural earthquake engineering in Blume et al. (1961). Since that time, it has become widely adopted in earthquake engineering practice.

The capacity design method involves the following steps:

1. Select a target yielding mechanism for the structural system, identifying all the member sections that are intended to yield. The selected mechanism should be one that can be detailed for ductile response.
2. Apply the design loads to the structural system, and proportion the selected yielding sections for required strength.
3. Determine the internal forces that will develop within the structure when the structure, as designed in step 2, forms the intended mechanism with each yielding section developing the expected member strength.

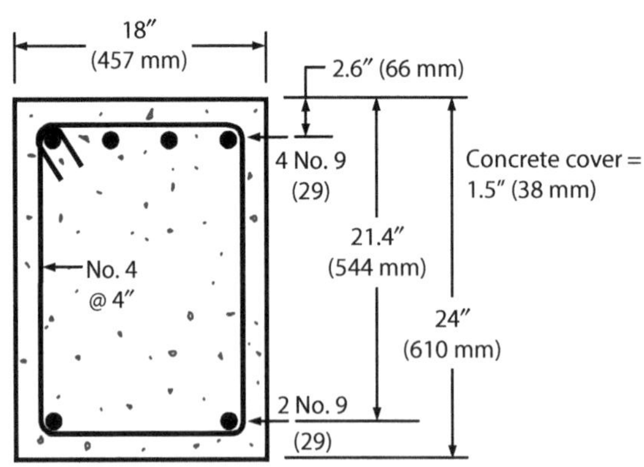

FIGURE 1.13 Beam cross section.

4. Design the yielding regions for ductile response. Design the remainder of the structure to have strength necessary to resist the internal forces determined in step 3.

Figure 1.14 illustrates the steps of the capacity design method for a one-story, one-bay frame. In step 1, a yielding mechanism is selected; for this structure, the mechanism involves yielding of columns at the fixed base and yielding of the beam at the faces of the columns. In step 2, the factored design loads are applied in combinations as required by the governing building code, and the yielding regions are designed to have design strengths at least equal to

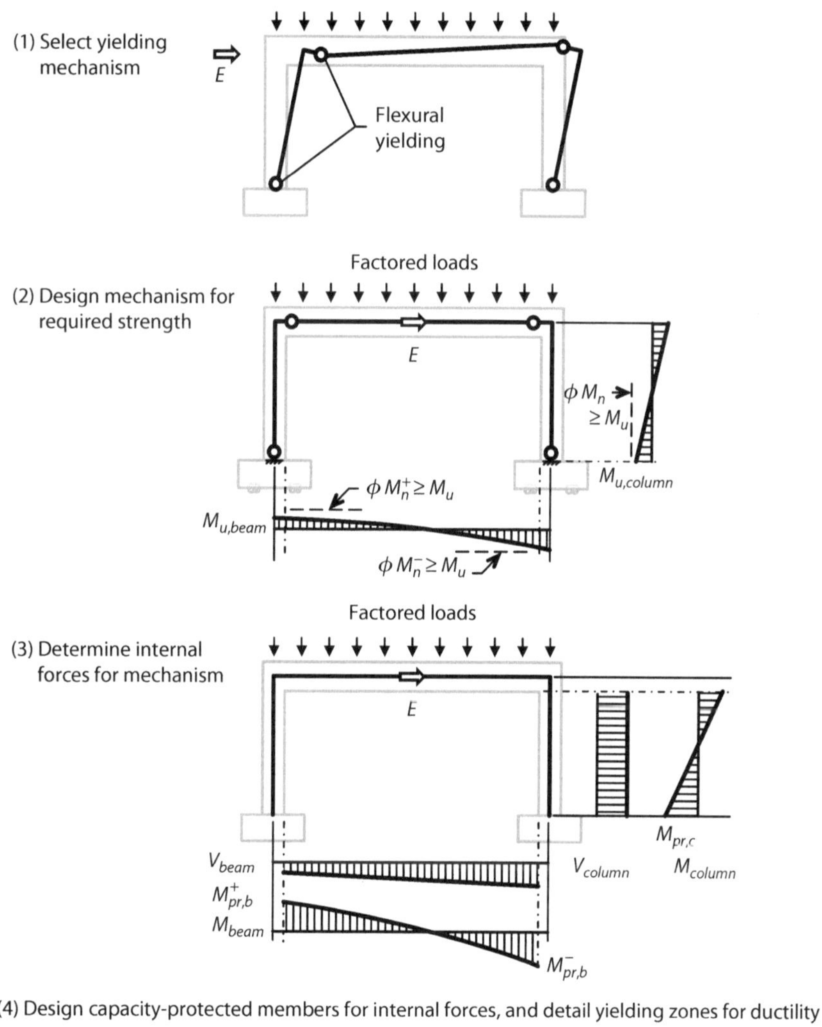

Figure 1.14 Capacity design steps.

required strengths, in this case, $\phi M_n \geq M_u$. The expected strengths of the yielding sections are then determined, considering the as-designed cross sections and the expected material behaviors. For this structure, the expected moment strengths are designated as probable moment strengths, M_{pr}. In step 3, the structure develops the yielding mechanism with M_{pr} at yielding sections, and the associated internal forces are determined. Finally, in step 4, the yielding sections are detailed for ductility and the remainder of the structural system is proportioned and detailed to have strength as required to resist the internal forces determined in step 3.

Step 3 requires determination of the internal forces associated with development of the intended yield mechanism. For this structure shown, the moments at each yielding location are known to be equal to the probable moment strengths. This renders the structure statically determinate. Thus, given the external forces, the internal forces can be established from equilibrium considerations. Consider, for example, the beam of Figure 1.14. Figure 1.15 shows a free-body diagram of that beam. Given the external loading and the probable moment strengths, the shears at the beam ends can be established as

$$V_{u,b1} = \frac{M^{+}_{pr,b1} + M^{-}_{pr,b2}}{l_n} - \frac{w_u l_n}{2}$$
$$V_{u,b2} = \frac{M^{+}_{pr,b1} + M^{-}_{pr,b2}}{l_n} + \frac{w_u l_n}{2} \tag{1.4}$$

All other internal forces can also be found by equilibrium considerations.

If the beam of Figure 1.15 is designed to satisfy Eq. (1.2), then, theoretically, the beam behavior will be controlled by flexural yielding, with a safe margin

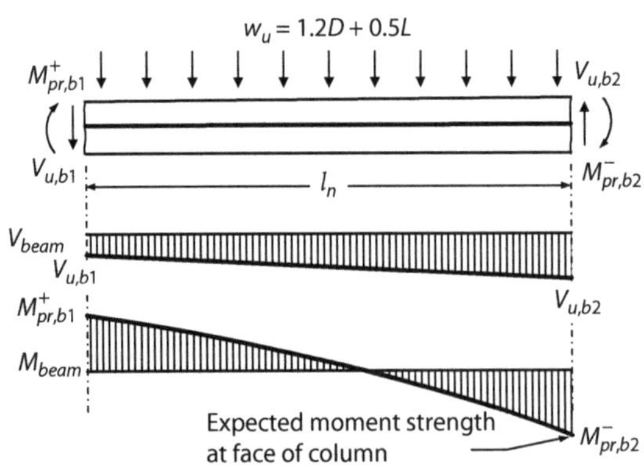

FIGURE 1.15 Determination of internal forces for capacity design.

against shear failure. As we will see in Chapters 7 and 12, however, additional steps may be required to ensure that the beam is safe from shear failure.

Example 1.3. Use capacity design to determine requirement shear strengths at sections 1 and 2 of the beam of Example 1.1.

Solution

The beam cross section is shown in Figure 1.13 of Example 1.2. Probable moment strengths M_{pr}, calculated using procedures of Chapter 6, are M^{+}_{pr} = 252 k-ft (342 kN-m) and M^{-}_{pr} = 483 k-ft (655 kN-m). Using the load combination shown in Figure 1.15, moments can be summed about the left end to obtain the value of the shear at the right end. The resulting design shear force is $V_{u,b2}$ = 81.4 kips (362 kN). The probable moments are reversed and the procedure is repeated, this time summing moments about the right end, to obtain the maximum design shear force at the left end, which is also equal to 81.4 kips (362 kN).

Recall from Example 1.1 that the shear from the conventional LRFD load combinations was 63 kips (280 kN). Thus, the design shear from the capacity design method is 1.29 times the design shear from the conventional LRFD method. The additional strength provided by capacity design is intended to ensure that the beam can develop flexural plastic hinges before reaching the shear strength. See Chapter 12 for additional discussion of this design approach for special moment frames.

1.5.5 Displacement-Based Design

Earthquake engineering performance assessment commonly involves consideration of nonlinear response, as illustrated in Figure 1.9. For a yielding system, performance may be better gauged by the imposed deformations than by the imposed forces. *Displacement-based design* is a process by which earthquake-induced displacement demands are first estimated, followed by an assessment of the local deformation demands and capacities. The concept was first expressed by Muto et al. (1960) and later presented as a general design and assessment approach (Moehle, 1992). Displacement-based concepts are now firmly embedded as design features of codes (e.g., ACI 318) and assessment standards (e.g., ASCE 41).

In displacement-based design, the first step is to analyze a structure under earthquake shaking to establish the earthquake-induced lateral displacements (e.g., δ in Figure 1.16). Structural analysis is then used to estimate the local inelastic deformation demands. Figure 1.16 illustrates a simplification in which the elastic deformations are ignored, such that lateral displacement is accommodated entirely by flexural hinge rotations. Flexural hinge rotations determined by this approach can be compared with rotation capacities at various limit states. See Chapter 6 for additional discussion.

Common practice is to use structural analysis methods that are more sophisticated than that shown in Figure 1.16. For example, in a static nonlinear analysis, instead of considering the members to have infinite elastic stiffness, the elastic flexibilities can be incorporated, resulting, in general, in a reduction in the calculated plastic hinge rotations. Alternatively, the nonlinear load-deformation characteristics of a complete structural system can be modeled using computer software, and local inelastic deformations can be evaluated through inelastic

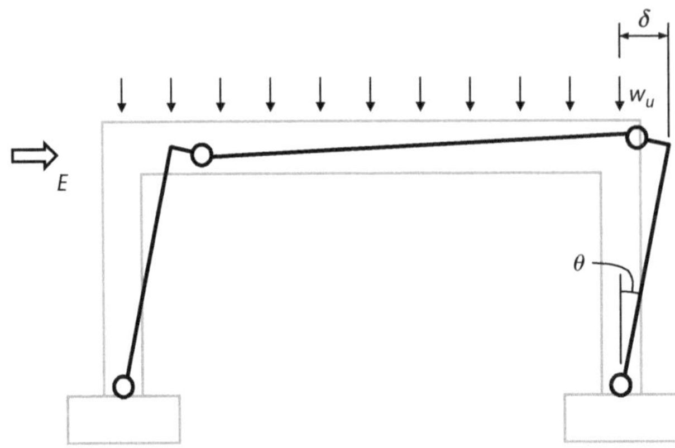

FIGURE 1.16 Displacement shape of a yielding planar frame with rigid members.

dynamic analysis under earthquake ground shaking. This approach is formalized in documents such as TBI (2010).

1.5.6 Performance Evaluation under Earthquake Ground Shaking

Nonlinear dynamic analysis can be used to gauge the performance capabilities of a building under earthquake ground shaking, or to verify that performance is at least equivalent to that obtained using the prescriptive building code provisions. In this approach, a series of ground motions is selected to represent the seismic demands associated with a selected hazard level. A computer model representing the nonlinear response characteristics of the structural framing is then assembled and used to simulate dynamic response to the selected earthquake ground motions. The calculated structural responses to the series of ground motions are then assessed to gauge the performance implications. See Figure 1.17.

The procedure outlined in the preceding paragraph forms the basis of performance-based design of tall buildings (TBI, 2010). In the usual application, two sets of earthquake ground motions are selected, one to represent an SLE level and another to represent an MCE level. Calculated responses include maximum transient and residual displacements and story drifts; inelastic rotations and strains in components that are intended to yield as part of the inelastic mechanism; and internal forces in components or in behaviors that are intended to remain within the linear-elastic range of response. Each ground motion produces its own response, requiring a statistical evaluation of the results to determine whether the structural system meets the performance criteria.

For both the SLE and the MCE shaking levels, it is necessary to establish a nonlinear computer model that will appropriately represent effective stiffness, strength, nonlinear load-deformation characteristics, and energy dissipation characteristics. The performance implications of the calculated responses must be determined. Methods for defining structural models and interpreting the calculated results can be found throughout this book.

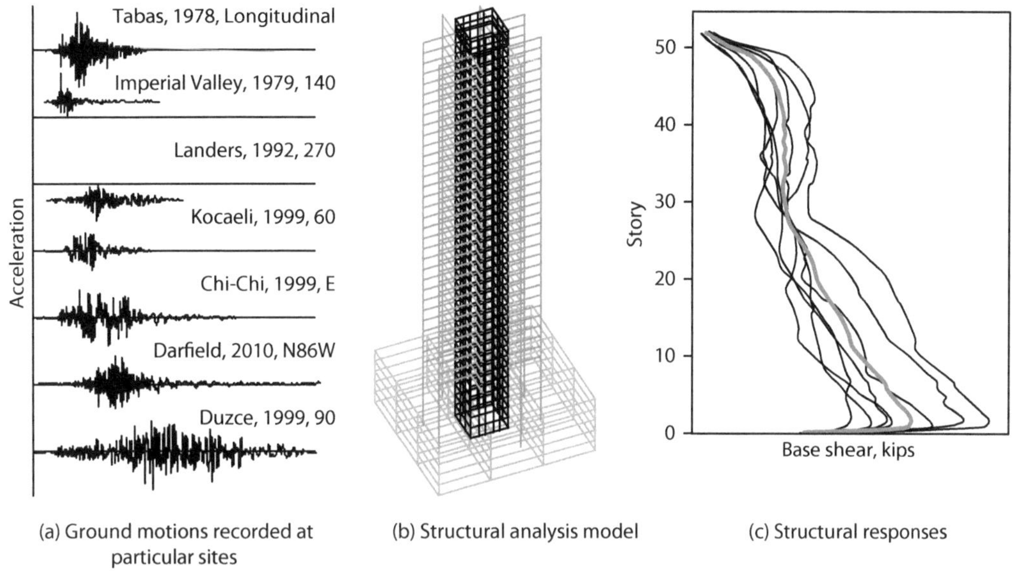

Figure 1.17 Evaluation of performance under earthquake ground shaking. (*Structural model and results courtesy of Magnusson Klemencic Associates.*)

1.6 The Purpose and Organization of This Book

The purpose of this book is to present the theory and practice for the design and evaluation of earthquake-resisting reinforced concrete buildings. The reader will gain an advanced understanding of the behavior of reinforced concrete materials, components, and systems subjected to routine and extreme loads, with emphasis on response to earthquake loading. The reader will also learn about design methods, both at a basic level as required by current building codes and at an advanced level as may be required for special problems including seismic performance assessment. The structural analyst responsible for seismic performance verification will also find data and models that will be useful for analyzing reinforced concrete structures.

The book is organized to build required knowledge in a logical sequence.

- Chapter 1 introduces basic concepts of structural design and performance verification, with emphasis on concepts that are used in earthquake engineering.
- Chapters 2 to 4 describe reinforced concrete materials. Chapters 2 and 3 consider physical characteristics and mechanical behavior of steel reinforcement and concrete, with emphasis on expected behavior under earthquake loading. Chapter 4 explains confined concrete and how it is used in earthquake-resistant buildings.
- Chapters 5 to 10 introduce structural members and connections that are used in earthquake-resisting concrete buildings. Chapter 5 explores the requirements for ductile response of axially loaded members, including

the use of transverse reinforcement to confine core concrete and restrain longitudinal reinforcement buckling. Chapter 6 describes behavior of beams, columns, and structural walls under moment or combined moment and axial force, and introduces procedures for calculating the nonlinear load-deformation response of flexural members. Chapter 7 describes the shear behavior of beams, columns, walls, and coupling beams. Chapter 8 covers requirements for development and splicing of reinforcement, including straight development lengths, lap splices and mechanical splices, and hooked and headed bar anchorages. Chapters 9 and 10 describe the behavior of beam-column, slab-column, and slab-wall connections.

- The remaining chapters (11 to 16) bring together the knowledge gained from previous chapters to describe the design and behavior of complete structural systems. Chapter 11 presents an overview of seismic design principles, including a review of earthquake shaking effects and how these are incorporated in building design. Chapters 12 to 14 cover the main vertical elements of buildings, including special moment frames, special structural walls, and gravity framing systems. Chapter 15 covers structural diaphragms and their role in completing a three-dimensional framing system capable of resisting gravity and lateral loads. Chapter 16 completes the book with a review of common foundation elements and their design for seismic resistance.

References

ACI 318-14 (2014). *Building Code Requirements for Structural Concrete (ACI 318-14) and Commentary*, American Concrete Institute, Farmington Hills, MI.

ASCE 7-10 (2010). *Minimum Design Loads for Buildings and Other Structures*, American Society of Civil Engineers, Reston, VA, 608 pp.

ASCE 41 (2013). *Seismic Evaluation and Retrofit of Existing Buildings*, American Society of Civil Engineers, Reston, VA.

ATC 40 (1996). *Seismic Evaluation and Retrofit of Buildings*, Report No. ATC-40, Applied Technology Council, Redwood City, CA.

Blume, J.A., N.M. Newmark, and L.H. Corning (1961). *Design of Multistory Reinforced Concrete Buildings for Earthquake Motions*, Portland Cement Association, Chicago, IL, 318 pp.

Eurocode 8 (2004). *Eurocode 8: Design of Structures for Earthquake Resistance, Part 1: General Rules, Seismic Actions and Rules for Buildings*, ComitéEuropéen de Normalisation, European Standard EN 1998-1:2004, Brussels, Belgium.

Fardis, M. (2013). "The European Approach to Seismic Engineering and Codification for Concrete Structures," Closing Plenary, *fib Symposium*, Tel Aviv.

FEMA 273 (1997). *NEHRP Guidelines for the Seismic Rehabilitation of Buildings, FEMA-273*, Federal Emergency Management Agency, Washington, DC.

FEMA P-58 (2012). *Seismic Performance Assessment of Buildings, Methodology and Implementation*, Vols. 1, 2, and 3, Federal Emergency Management Agency, Washington, DC.

FEMA P-695 (2009). *Quantification of Building Seismic Performance Factors*, Federal Emergency Management Agency, Washington, DC, 421 pp.

Gergely, P., and L.A. Lutz (1968). "Maximum Crack Width in Reinforced Flexural Members," *Causes, Mechanism, and Control of Cracking in Concrete, SP-20*, American Concrete Institute, pp. 87–117.

IBC (2012). *International Building Code*, International Code Council, Country Club Hills, IL.

LATBSDC (2011). *An Alternative Procedure for Seismic Analysis and Design of Tall Buildings Located in the Los Angeles Region*, Los Angeles Tall Building Structural Design Council, Los Angeles, CA.

Luco, N., B.R. Ellingwood, R.O. Hamburger, J.D. Hooper, J.K. Kimball, and C.A. Kircher (2007). "Risk-Targeted versus Current Seismic Design Maps for the Conterminous United States," *Proceedings, SEAOC 2007 Convention*, Structural Engineers Association of California, Sacramento, CA, 13 pp.

MacGregor, J.G. (1983). "Load and Resistance Factors for Concrete Design," *ACI Structural Journal*, Vol. 80, No. 4, pp. 279–287.

Moehle, J.P. (1992). "Displacement-based Design of RC Structures Subjected to Earthquakes," *Earthquake Spectra*, Vol. 8, No. 3, pp. 403–428.

Muto, K., et al. (1960). "Non-linear Response Analyzers and Application to Earthquake Resistant Design," *Proceedings, 2nd World Conference on Earthquake Engineering, Japan*, Vol. 2, pp. 649–668.

NBC (2005). *The National Building Code of Canada*, Canadian Commission on Building and Fire Codes, National Research Council, Ottawa.

NEHRP (2009). *NEHRP Recommended Provisions (and Commentary)*, National Earthquake Hazard Reduction Program.

NIST (2010). "Evaluation of the FEMA P-695 Methodology for Quantification of Building Seismic Performance Factors," NIST GCR 10-917-8, National Institute of Standards and Technology, Gaithersburg, MD, 268 pp.

NOAA (1973). *San Fernando, California, Earthquake of February 9, 1971*, U.S. Dept. of Commerce, National Oceanic and Atmospheric Administration, Washington, DC, 3 volumes.

NZS3101 (2006). *Concrete Design Standard, NZS3101:2006, Part 1 and Commentary on the Concrete Design Standard, NZS 3101:2006, Part 2*, Standards Association of New Zealand, 2006, Wellington, New Zealand.

Otani, S. (1995). "A Brief History of Japanese Seismic Design Requirements," *Concrete International*, Vol. 17, No. 12, pp. 46–53.

Park, R. (1986). "Ductile Design Approach for Reinforced Concrete Frames," *Earthquake Spectra*, Vol. 2, No. 3, pp. 565–619.

PCI (2013). *PCI Journal*, Summer issue.

PRESSS (various). Precast Seismic Structural Systems (PRESSS), http://pci.org/cms/index.cfm/PRESSSresources.

SEAOC (1963). *Recommended Lateral Force Requirements* (commonly referred to as the "Blue Book"), Seismology Committee, Structural Engineers Association of California, Sacramento, CA.

SEAOC (1967). "Design of Earthquake Resistant High-Rise Buildings," Lecture notes from SEAOC seminar, October–December 1967, figure by A. Tarics, Reid & Tarics, Structural Engineers Association of Northern California Sacramento, CA.

SEAOC (1995). *Performance-Based Seismic Engineering of Buildings* (also known as the *Vision 2000 Report*), Structural Engineers Association of California, Sacramento, CA, 538 pp.

TBI (2010). *Guidelines for Performance-Based Seismic Design of Tall Buildings*, Report No. 2010/05, Pacific Earthquake Engineering Research Center, University of California, Berkeley, CA, 84 pp.

UBC (1976). *Uniform Building Code*, International Conference of Building Officials, Whittier, CA.

Yang, T.Y., J.P. Moehle, B. Stojadinovic, and A. Der Kiureghian (2009). "Seismic Performance Evaluation of Facilities: Methodology and Implementation," *Journal of Structural Engineering*, Vol. 135, No. 10, pp. 1146–1154.

CHAPTER 2

Steel Reinforcement

2.1 Preview

Chapter 2 is the first of three chapters on the properties of reinforced concrete materials. The main emphasis of this chapter will be on aspects of steel reinforcement that are directly relevant to earthquake response of structures. Important aspects include expected material strengths and deformation capacities, and how these are affected by reversed-cyclic loading that occurs during earthquake response. Readers interested in behavior under other loads (e.g., fire or high-cycle fatigue) should refer to sources that emphasize those subjects.

2.2 Steel Reinforcement Used in Buildings

Reinforced concrete buildings came into widespread use around the start of the 20th century, as portland cement became commercially available and more individuals became familiar with its characteristics. Early examples of buildings from that era used a variety of proprietary reinforcement systems that were not standardized. As the construction form evolved, standard reinforcing bars having smooth surfaces began to be used. Cross sections were either round, square, or square sections twisted along the length. Deformed bars began to see prominent use in the United States and other countries starting in the 1950s, largely displacing smooth bars and proprietary systems. Today, nonprestressed steel reinforcement can be deformed bars, deformed bars in the form of welded mats, deformed wire, welded wire reinforcement, or plain bars and wires used for spirals, in addition to certain proprietary products.

2.2.1 Standard Steel Reinforcement

Steel reinforcement is produced in standard sizes. The standard sizes used in the United States are listed in Tables 2.1 and 2.2.

Bar Size, No. U.S. (Metric)	Nominal Diameter, in (mm)	Nominal Area, in² (mm²)
3 (10)	0.375 (9.5)	0.11 (71)
4 (13)	0.500 (12.7)	0.20 (129)
5 (16)	0.625 (15.9)	0.31 (199)
6 (19)	0.750 (19.1)	0.44 (284)
7 (22)	0.875 (22.2)	0.60 (387)
8 (25)	1.000 (25.4)	0.79 (510)
9 (29)	1.128 (28.7)	1.00 (645)
10 (32)	1.270 (32.3)	1.27 (819)
11 (36)	1.410 (35.8)	1.56 (1006)
14 (43)	1.693 (43.0)	2.25 (1452)
18 (57)	2.257 (57.3)	4.00 (2581)

Table 2.1 ASTM Standard Reinforcement Bar Sizes (*after ASTM A615*)

Eleven different sizes of deformed reinforcing bars are produced in the United States, with nominal diameters ranging from 3/8 in (10 mm) to 2.257 in (57 mm). Table 2.1 lists the bar size designations and nominal dimensions. The bars can be identified in one of two ways, using either U.S. customary in-lb units or metric units. In the U.S. customary unit system, the bar size number is the nominal diameter in eighths of an inch; actual diameter deviates from this definition for No. 9 to No. 18 bars. The same bar can also be identified by its nominal diameter in millimeter. Thus, a No. 3 bar (3/8-in diameter) in the U.S. customary system is the same as a No. 10 bar (10-mm diameter) in the metric system.

Standard wire is either plain (designated W) or deformed (designated D), with a number indicating the nominal cross-sectional area in hundredths of a square inch (Table 2.2). Thus, wire reinforcement designated D31 is deformed with nominal cross-sectional area of 0.31 in^2 (200 mm^2).

ASTM standards control the deformation patterns on deformed bars to ensure appropriate bond characteristics while avoiding sharp-cornered deformations that reduce fatigue life (Figure 2.1). In addition, standard markings identify the bar characteristics. Figure 2.2 shows bar markings used in the United States for Grades 60, 75, and 80 bars.[1] The top letter or symbol identifies the producing mill. The next marking is the bar size. The third marking symbol designates the type of steel—S for carbon steel (A615) or W for low-alloy steel (A706). Finally there is a grade marking [60, 75, or 80 for Grades 60, 75, or 80, respectively (or 4, 5, or 6 for metric grades 420, 520, or 550, respectively)] or the addition of one line (Grades 60 and 420), two lines (Grades 75 and 520), or three lines (Grades 80 and 550) that must be at least five deformation spaces long.

[1]See Section 2.2.2 for discussion of reinforcement grades and ASTM designations.

W & D Size		Nominal Diameter		Nominal Area	
Plain	Deformed	in	mm	in²	mm²
W1.4		0.134	3.40	0.014	9
W2		0.160	4.06	0.020	13
W2.5		0.178	4.52	0.025	16
W2.9		0.192	4.88	0.029	19
W3.5		0.211	5.36	0.035	23
W4	D4	0.226	5.74	0.040	26
W4.5		0.239	6.07	0.045	29
W5	D5	0.252	6.40	0.050	32
W5.5		0.265	6.73	0.055	35
W6	D6	0.276	7.01	0.060	39
W8	D8	0.319	8.10	0.080	52
W10	D10	0.357	9.07	0.100	65
W11	D11	0.374	9.50	0.110	71
W12	D12	0.391	9.93	0.120	77
W14	D14	0.422	10.7	0.140	90
W16	D16	0.451	11.5	0.160	103
W18	D18	0.479	12.2	0.180	116
W20	D20	0.505	12.8	0.200	129
W22	D22	0.529	13.4	0.220	142
W24	D24	0.533	13.5	0.240	155
W26	D26	0.575	14.6	0.260	168
W28	D28	0.597	15.2	0.280	181
W30	D30	0.618	15.7	0.300	194
W31	D31	0.628	16.0	0.310	200
W45	D45	0.757	19.2	0.450	290

*This table presents a soft conversion of the wire available in in-pound units. See ASTM A1022 for hard metric wire sizes.

Table 2.2 Standard Wire Reinforcement (*after ASTM A1022*)*

Various types of reinforcement can provide enhanced corrosion resistance in highly corrosive environments. Epoxy-coated reinforcement is one option (Figure 2.1). In the United States, green epoxy coating indicates the epoxy was applied before fabricating (cutting, bending) reinforcement. This practice ensures a uniform coating with minimal cost, but fabrication after coating can cause damage to the epoxy coating. Purple or gray epoxy coating indicates the epoxy was applied after the bars were fabricated—these bars should not be bent after

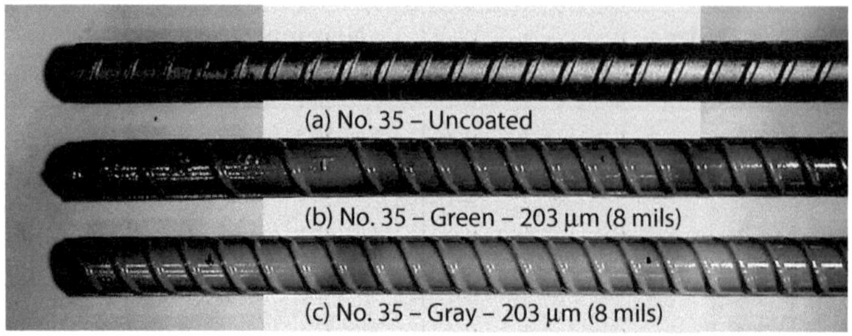

Figure 2.1 Photograph of No. 11 (No. 35 in metric) reinforcing bars: (*a*) uncoated (A615), (*b*) green epoxy coating (ASTM A775), and (*c*) gray (alternately purple) epoxy coating (ASTM A934). Bar deformation patterns other than those shown are also used.

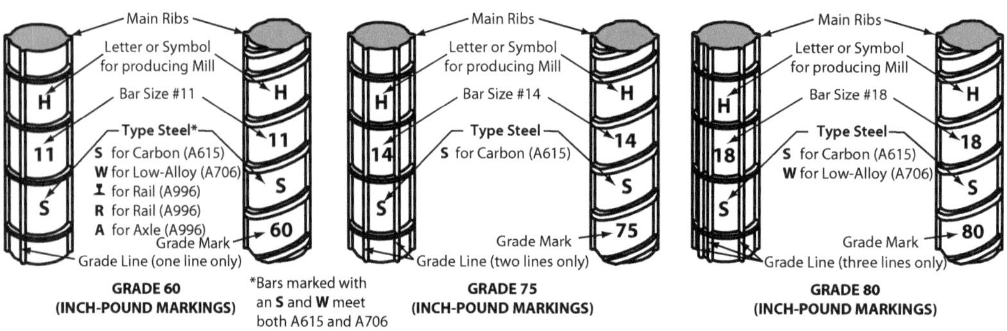

Figure 2.2 Bar markings for deformed reinforcing bars in the United States. (*Courtesy of Concrete Reinforcing Institute.*)

coating. ASTM specifications A775, A884, and A934 cover epoxy-coated reinforcement. ASTM 1055 covers zinc and epoxy dual-coated reinforcing bars. Other alternatives are stainless steel reinforcement (A955 and A1022), galvanized reinforcement (A767 and A1060), and other proprietary systems. All these corrosion resistance systems have an initial cost premium relative to regular reinforcement.

Headed bars are used to decrease development length, confine concrete, and alleviate congestion of reinforcement. Heads are attached at one or both ends of a reinforcing bar. Methods of attachment include welding, integrally hot forging a head from the reinforcing bar end, and threaded attachments. In the United States, ASTM A970 classifies headed bars as either Class A (develops the minimum specified tensile strength of the reinforcing bar) or Class B (develops the minimum specified tensile strength and the minimum specified elongation of the reinforcing bar). Additionally, Class HA heads have net bearing area not less than $4A_b$, where A_b = nominal bar area, as well as limits on obstructions due to the head attachment (Figure 2.3).

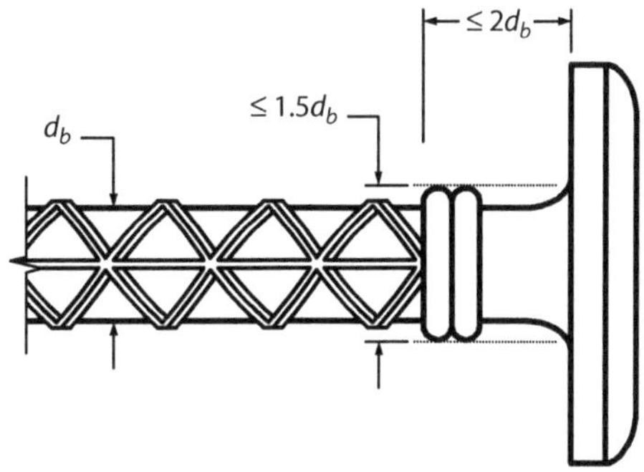

Figure 2.3 Requirements for ASTM A970, Class HA headed bars. (*After ASTM A970.*)

Steel fibers may be permitted for use in fiber-reinforced concrete. In the United States, ASTM A820 covers minimum requirements. Allowable fibers include smooth or deformed cold-drawn wire, smooth or deformed cut sheet, and mill-cut or modified cold-drawn wire steel fibers. See additional discussion of fiber-reinforced concrete in Chapter 3.

2.2.2 Reinforcement Grades and Availability

Reinforcement grade refers to the nominal yield strength of the reinforcement. In the United States, deformed reinforcing bars are available in Grades 40 (280), 50 (350), 60 (420), 75 (520), 80 (550), 100 (690), and 120 (830), where the number refers to the nominal yield strength in ksi (MPa). Grade 40 (280) bars are sometimes used for transverse reinforcement, and are available only in smaller bar sizes. Grade 100 (830) is permitted only for spiral and confinement reinforcement. In general, only Grade 60 (420) bars are widely available. See Table 2.3 for information on grades, minimum yield and tensile strengths, and elongation requirements for various types of reinforcement.

In the United States, most producers can provide 60-ft (18-m) lengths of stock without special order. Stock material in 20, 30, and 40-ft lengths is also usually available. Coils are also available for the smaller bar sizes; these are used by fabricators with automatic bending equipment for the fabrication of ties, stirrups, and spirals.

Certified mill test reports typically accompany each shipment of reinforcing bars. A mill report certifies that the reinforcing bars conform to the construction documents by indicating bar size, grade, ASTM specification (e.g., A615M), heat number, chemical composition, and results of tension and bend tests on bar samples.

For a general discussion on U.S. specifications and availability, see ACI 439. R-09 (2009).

Type of Steel	ASTM Designation	Grade Designation	Minimum Yield Strength, psi (MPa)	Minimum Tensile Strength, psi (MPa)	Minimum Elongation in 8 in (200 mm)	
					Bar No.	Elongation, %
Carbon-steel deformed bars	A615	40 (280)	40,000 (280)	60,000 (420)	3 (10) 4–6 (13-19)	11 12
		60 (420)	60,000 (420)	90,000 (620)	3–6 (10-19) 7, 8 (22, 25) 9–18 (29–57)	9 8 7
		75 (520)	75,000 (520)	100,000 (690)	3–8 (10–25)	7
		80 (550)	80,000 (550)	105,000 (725)	9–18 (29–57)	6
Low-alloy steel deformed bars	A706	60 (420)	60,000 (420) [78,000 (540) maximum]	80,000* (550)	3–6 (10–19) 7–11 (22–36) 14, 18 (43, 57)	14 12 10
		80 (550)	80,000 (550) [98,000 (675) maximum]	100,000* (690)	3–6 (10–19) 7–11 (22–36) 14, 18 (43, 57)	12 12 10
Deformed and plain stainless-steel bars	A955	60 (420) 75 (520)	60,000 (420) 75,000 (520)	90,000 (620) 100,000 (690)	3–18 (10–57)	20
Rail-steel and axle-steel deformed bars	A996	40 (280) 50 (350) 60 (420)	40,000 (280) 50,000 (350) 60,000 (420)	70,000 (500) 80,000 (550) 90,000 (620)	See ASTM A996	
Deformed and plain stainless steel wire†	A1022	Deformed wire Plain wire	75,000 (515) 70,000 (485)	85,000 (585) 80,000 (550)	None	
Deformed and plain carbon-steel wire†	A1064	Deformed wire Plain wire	75,000 (515) 70,000 (485)	85,000 (585) 80,000 (550)	None	
Deformed low-carbon, chromium, steel bars	A1035	100 (690) 120 (830)	100,000 (690) 120,000 (830)	150,000 (1030) 150,000 (1030)	3–11 (10–36) 14, 18 (43, 57)	7 6‡

*Tensile strength shall not be less than 1.25 times the actual yield strength.
†ASTM A1022 and A1064 also cover welded wire. Mechanical properties for welded wire differ from those listed here.
‡Not applicable to Grade 120 (830).

Table 2.3 ASTM Designations and Requirements

2.2.3 Permitted Reinforcement

Responsible authorities may specify the types of steel reinforcement that can be used. In the United States, ACI 318 (2014) permits the use of reinforcement in accordance with Tables 2.4 and 2.5. In these tables, the term *special seismic systems* refers to structural systems designated as part of the seismic force-resisting system in regions of highest seismicity.

Reinforcement materials satisfying the specifications in Table 2.3 can also be used in various other products covered by product-specific ASTM specifications, including deformed bar mats (ASTM A184), zinc-coated (galvanized) bars (ASTM A767), zinc-coated (galvanized) welded wire (ASTM A1060), epoxy-coated bars ASTM A775 and A934), zinc and epoxy dual-coated bars (ASTM A1055), and headed reinforcing bars (ASTM A970). As an example, galvanized bars satisfying ASTM A767 for which the zinc coating is applied to deformed bars satisfying ASTM A615 would be permitted to be used wherever ASTM A615 deformed bars are permitted by Tables 2.4 and 2.5.

Some of the reinforcement restrictions for special seismic systems are noteworthy. For example:

- Reinforcement that resists moment, axial force, and shrinkage and temperature effects in special seismic systems is required to be either A706 Grade 60 (420) or A615 Grades 40 (280) or 60 (420) satisfying additional requirements for yield strength, tensile strength, and elongation. ASTM A706 covers low-alloy steel deformed bars intended for applications where controlled tensile properties, restrictions on chemical composition to enhance weldability, or both, are required. The specification requires that the bars be marked with the letter W for type of steel. Note that A706 Grade 80 (550) reinforcement is permitted to resist flexural or axial forces in special moment frames or special structural walls only if tests and analytical studies are presented in support of its use.
- For the design of transverse reinforcement providing concrete confinement or lateral support of longitudinal bars, the maximum value of f_y permitted in design calculations is 100,000 psi (690 MPa). For the design of transverse reinforcement providing shear strength, however, the maximum value of f_{yt} permitted in design calculations is either 60,000 psi (420 MPa) or 80,000 psi (550 MPa), depending on the reinforcement type.
- Welded wire reinforcement is permitted as transverse reinforcement in special seismic systems where the welded wire wraps around the member. It is not permitted to rely on the weld to support longitudinal bars or confine concrete. This is because the specification for welds does not require that the weld be capable of developing the strength of the welded wire, with the result that brittle failures can occur.

Welding of reinforcing bars should be done only in accordance with approved procedures, and generally is not preferred. In U.S. practice, welding of reinforcing

Usage	Application	Maximum value of f_y or f_{yt} Permitted for Design Calculations, psi (MPa)	Applicable ASTM Specification			
			Deformed Bars	**Deformed Wires**	**Welded Wires**	**Welded Bar Mats**
Moment, axial force, and shrinkage and temperature	Special seismic systems	60,000 (420)	See below*	Not permitted	Not permitted	Not permitted
	Other	80,000 (550)	A615, A706, A955, A996	A1064, A1022	A1064, A1022	A184
Lateral support of longitudinal bars or concrete confinement	Special seismic systems	100,000 (690)	A615, A706, A955, A996, A1035	A1064, A1022	A1064†, A1022†	Not permitted
	Spirals	100,000 (690)	A615, A706, A955, A996, A1035	A1064, A1022	Not permitted	Not permitted
	Other	80,000 (550)	A615, A706, A955, A996	A1064, A1022	A1064, A1022	Not permitted
Shear‡	Spirals	60,000 (420)	A615, A706, A955, A996	A1064, A1022	Not permitted	Not permitted
	Shear friction	60,000 (420)	A615, A706, A955, A996	A1064, A1022	A1064, A1022	Not permitted
	Stirrups, ties, hoops	60,000 (420)	A615, A706, A955, A996	A1064, A1022	A1064, A1022	Not permitted
		80,000 (550)	Not permitted	Not permitted	A1064, A1022 (welded deformed wire)	Not permitted
Torsion	Longitudinal and transverse‡	60,000 (420)	A615, A706, A955, A996	A1064, A1022	A1064, A1022	Not permitted

*Deformed nonprestressed longitudinal reinforcement resisting earthquake-induced moment, axial force, or both in special moment frames, special structural walls, and all components of special structural walls including coupling beams and wall piers, is to be in accordance with (1) or (2):

1. ASTM A706, Grade 60 (420).
2. ASTM A615 Grades 40 (280) and 60 (420) reinforcement if the actual yield strength based on mill tests does not exceed f_y by more than 18,000 psi (124 MPa), ratio of the actual tensile strength to the actual yield strength is at least 1.25, and elongation meets requirements for A706 reinforcement.

†ASTM A1064 and A1022 are not permitted in special seismic systems where the weld is required to resist stresses in response to confinement, lateral support of longitudinal bars, shear, or other actions.

‡Nonprestressed bars and wires permitted for torsion categories also apply for special seismic systems.

Table 2.4 Permitted Nonprestressed Deformed Reinforcement (*after ACI 318*)

Usage	Application	Maximum Value of f_y or f_{yt} Permitted for Design Calculations, psi (MPa)	Applicable ASTM Specification	
			Plain Bars	Plain Wires
Lateral support of longitudinal bars or concrete confinement	Spirals in special seismic systems	100,000 (690)	A615, A706, A955, A1035	A1064, A1022
	Spirals	100,000 (690)	A615, A706, A955, A1035	A1064, A1022
Shear	Spirals	60,000 (420)	A615, A706, A955, A1035	A1064, A1022
Torsion in nonprestressed beams	Spirals	60,000(420)	A615, A706, A955, A1035	A1064, A1022

Table 2.5 Permitted Nonprestressed Plain Spiral Reinforcement (*after ACI 318*)

bars is to conform to AWS D1.4-11 (2011). Type and location of welded splices and other required welding of reinforcing bars are to be indicated in the construction documents. Except for A706 bars, bar specifications should be supplemented to require a report of material properties necessary to conform to the requirements in AWS D1.4. A706 bars have chemical content specially suited for welding. Cross-welding of bars, or welding of attachments to structural bars, can result in embrittlement and should be avoided.

2.3 Steel Reinforcement under Monotonic Loading

2.3.1 General Characteristics of the Stress–Strain Relation

The stress–strain relation of steel reinforcement is measured in a standardized tension test of a full-size bar. The stress–strain relation is in terms of engineering stress and engineering strain, where engineering stress is the force divided by the original cross-sectional area and engineering strain is the change in length divided by the original length of a gauge length of the test specimen.

Figure 2.4 shows a typical stress–strain relation for nonprestressed steel and defines several parameters and behavior ranges. The steel responds linearly at first, yields at an upper yield point, then responds at a lower yield stress along a yield plateau. In some steels there is no yield plateau. Yielding is followed by a strain-hardening region. Strain-hardening is an important characteristic to induce yielding to spread along the length of the bar and produce ductile response. At the peak of the stress–strain relation, necking occurs, causing localized reduction in the cross-sectional area and leading to failure in the necked region. A gauge length including the necked region will show continued straining as the bar is elongated to failure, whereas a gauge length outside the necked region will show unloading with reduced strain.

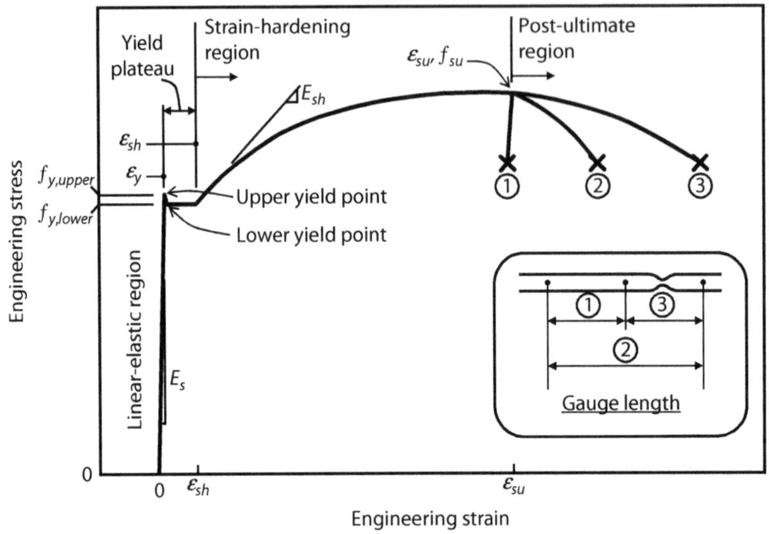

FIGURE 2.4 Monotonic stress–strain relation for mild steel in tension.

Elongation after rupture generally is measured along an 8-in (200-mm) gauge length including the fracture zone. Thus, elongation includes the plastic strain including the strain in the necked region minus the recovered elastic strain after unloading. The strain ε_{su} corresponding to the peak stress f_{su} is sometimes referred to as the *uniform elongation* or *uniform strain limit*, as this is the largest deformation in the test bar for which the tensile strains are uniform throughout the length. It marks the onset of necking in a bar. It is a useful property for design of earthquake-resistant buildings because it is the maximum strain that should be relied on in a location of yielding. This property, however, is not generally reported.

2.3.2 Tensile Properties of Steel Reinforcement

Figure 2.5 plots characteristic stress–strain relations for A615 Grades 40, 60, and 75 (280, 420, and 520), A706 Grade 60 (420), and A1035 Grade 100 (690) reinforcement. The salient properties are as follows:

- The initial modulus E_s is approximately 29,000 ksi (200,000 MPa).
- The grade number refers to the minimum yield strength in ksi (MPa). ASTM also specifies minimum tensile strengths. See Table 2.3. For A706 bars, the actual yield strength must not exceed the minimum value by more than 18 ksi (124 MPa), and the actual tensile strength must be at least 1.25 times the actual yield strength.
- Producers generally aim for an actual yield strength higher than the minimum value so that, given variations in properties, there is only a small

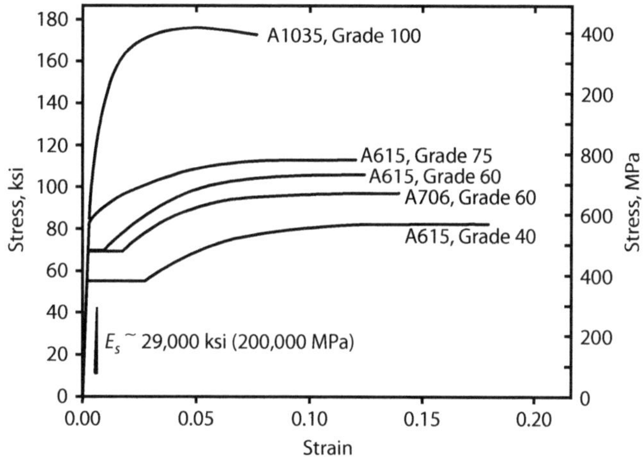

Figure 2.5 Characteristic engineering stress versus engineering strain relations for A615, A706, and A1035 deformed bars in tension. Strengths and elongations shown for A615 and A706 based on mean values from Bournonville et al. (2004). Stress–strain relation for A1035 is for MMFX-2 steel, with data from MMFX (2009) except elongation is modified based on observed values.

chance of the actual yield strength falling below the minimum value. Bournonville et al. (2004) report data on actual bar properties from mill tests (Table 2.6).

- The length of the yield plateau is not specified in the ASTM specifications and is variable. The general trend is that bars with lower strength have longer yield plateaus, and bars with higher strengths may or may not have yield plateaus. A706 Grade 60 (420) bars tend to have longer yield plateaus than A615 Grade 60 (420) bars.

	Yield Strength, $f_{y,actual}$		Tensile Strength, $f_{su,actual}$		$\frac{f_{y,actual}}{f_y}$	$\frac{f_{su,actual}}{f_{y,actual}}$	No. 4 (No. 13) Bar Elongation in 8 in (200 mm)		No. 11 (No. 36) Bar Elongation in 8 in (200 mm)	
Designation	Mean, ksi (MPa)	CoV	Mean, ksi (MPa)	CoV			Mean, %	CoV	Mean, %	CoV
A615, Grade 40	56.0 (386)	0.088	81.8 (564)	0.087	1.40	1.46	19.0	0.15	–	–
A615, Grade 60	69.6 (480)	0.072	106 (728)	0.063	1.16	1.52	13.0	0.15	12.3	0.23
A615, Grade 75	81.5 (562)	0.055	114 (786)	0.040	1.09	1.40	11.9	0.13	10.4	0.083
A706, Grade 60	69.1 (477)	0.053	95.2 (636)	0.052	1.15	1.33	15.4	0.088	14.9	0.13

Table 2.6 Mean and Coefficient of Variation (CoV) of Reinforcement Mechanical Properties (*after Bournonville et al., 2004*)

- Initial strain-hardening modulus tends to be around 1000 ksi (7000 MPa), although the value is not specified in the ASTM specifications and is variable.
- ASTM specifies minimum required percentage elongations in an 8-in (200-mm) gauge length including the fractured section. Table 2.3 summarizes required elongations, and Table 2.6 summarizes measured elongation statistics from mill tests.

The monotonic stress–strain behavior of A615 and A706 bars in the strain-hardening range can be approximated by (Mander et al., 1984)

$$f_s = f_{su} + \left(f_y - f_{su}\right)\left(\frac{\varepsilon_{su} - \varepsilon_s}{\varepsilon_{su} - \varepsilon_{sh}}\right)^P \tag{2.1}$$

in which

$$P = E_{sh}\left(\frac{\varepsilon_{su} - \varepsilon_{sh}}{f_{su} - f_y}\right) \tag{2.2}$$

2.3.3 Compressive Properties of Steel Reinforcing Bars

When reinforcing bars are loaded in tension, the cross section decreases because of the Poisson effect. In contrast, the cross section increases under compressive loading. As a result, the relations between engineering stress and strain are different in tension and compression. Additionally, bars in compression may buckle, causing further deviations in behavior. This section will address the differences in tensile and compressive behavior associated with the Poisson effect. The problem of reinforcing bar buckling is considered in detail in Chapter 5.

The Poisson effect can be evaluated by assuming that the volume of a reinforcing bar is constant as it is strained beyond the yield point. Conservation of volume for small strains requires a Poisson's ratio of $\nu = 0.5$. For this value of Poisson's ratio, the cross-sectional area varies as $\left(1-0.5\varepsilon_l\right)^2$, where the longitudinal strain ε_l is taken positive in tension. Thus, the ratio of cross-sectional areas for compressive loading and tensile loading is $\left(1+0.5|\varepsilon_l|\right)^2/\left(1-0.5|\varepsilon_l|\right)^2$. At strain $\varepsilon_l = 0.05$, this area ratio has value 1.10, suggesting 10% higher engineering stress in compression than in tension.

To demonstrate this effect, consider the stress–strain data in Figure 2.6. The engineering stress versus engineering strain relations were measured in tension and compression tests on nominally identical reinforcing bar specimens (Dodd and Restrepo-Posada, 1995). As expected, the engineering stress–strain relations diverge with increasing strain.[2] We can convert from engineering stress to true

[2] *Engineering stress* is defined as the ratio of the applied force to the initial area A_s. *True stress* is the ratio of the applied force to the instantaneous cross-sectional area $A_s\left(1-0.5\varepsilon_l\right)^2$.

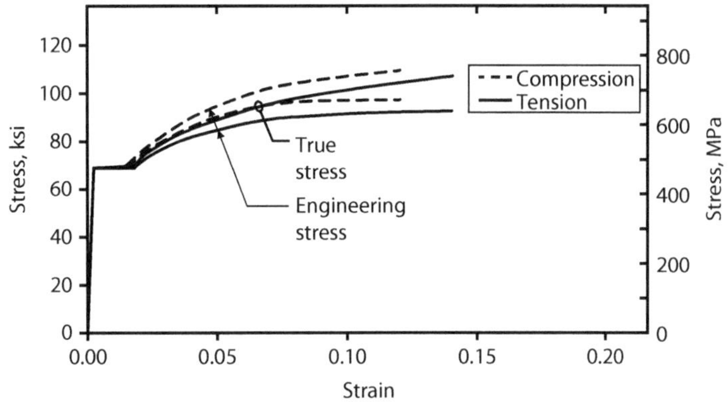

FIGURE 2.6 Stress–strain relations of reinforcing bars in tension and compression. (*After Dodd and Restrepo-Posada, 1995.*)

stress by dividing the engineering stress by the quantity $(1-0.5\varepsilon_t)^2$. The converted relations are nearly identical up to relatively large strain values (Figure 2.6). At very large strains, instability reduces the apparent resistance of the bar loaded in compression.

Although the Poisson effect is generally known, it is not taken into consideration in routine engineering calculations. Instead, it is more common to assume the tensile stress–strain relation represents behavior in both tension and compression. Effects of reinforcement buckling in compression, however, should be taken into consideration. See Chapter 5 for additional discussion.

2.3.4 Strain Rate Effect

Strain rate increases both the yield and ultimate strengths of reinforcing bars (Figure 2.7*a*). Malvar (1998) presents dynamic increase factors (ratios of apparent strength for dynamic loading to strength for near-static loading) (Figure 2.7*b*). As shown, the strain-rate effect is higher for yield strength than for ultimate strength, and higher for lower-strength reinforcement than for higher-strength reinforcement. The dynamic effect is greater for the upper yield point than for the lower yield point.

ASTM A370 describes test requirements for reinforcement. It permits strain rate as high as 0.001/s through the yield point, though it is more common for mill tests to follow the alternative procedure in which the rate is around 0.00006/s. The Concrete Reinforcing Steel Institute, Materials Properties Committee[3] recommends a rate around 0.00003/s. Based on the data in Figure 2.7*b*, at these latter rates the measured strength can be assumed to be equal to the static

[3]Personal communication from Robert Smith, ERICO, 2008.

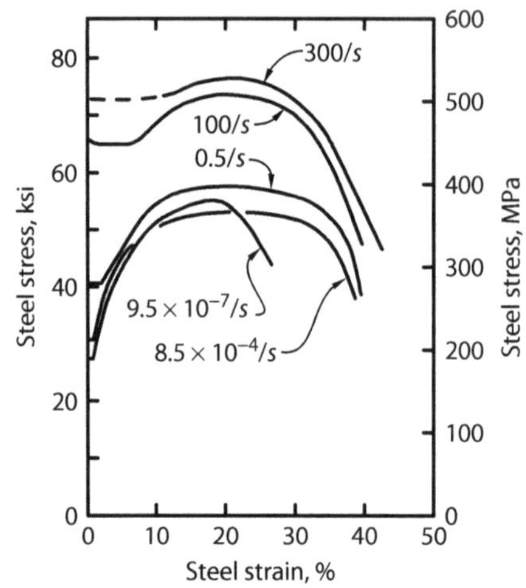

(a)

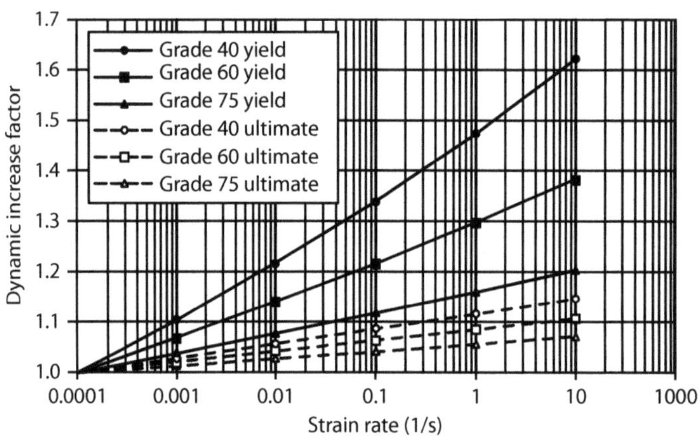

(b)

FIGURE 2.7 Strain-rate effect on tensile properties of reinforcing bars: (*a*) stress–strain relations (*After Manjoine, 1944, with permission of ASME*); (*b*) dynamic increase factors for upper yield and ultimate strengths for various reinforcement grades (*After Malvar, 1998, with permission of ASCE*).

strength appropriate for dead loads and most live loads. A slightly higher apparent strength might result from earthquake loading rates. For example, considering a structure with vibration period of 1 s and Grade 60 bars reaching maximum strain $3\varepsilon_y$. the strain rate would be approximately $3 \times 0.002/0.25$ s = 0.024/s. According to Figure 2.7*b*, the dynamic increase factors for this strain rate are 1.17 and 1.05 for yield and ultimate strengths, respectively. Though appreciable, the dynamic rate effect is not routinely considered for earthquake analysis or design.

2.4 Reinforcing Bars under Cyclic Loading

2.4.1 Stress–Strain Response

Figure 2.8 illustrates the stress–strain behavior of steel reinforcing bar under stress and strain reversals. The strain history is representative of the response of top longitudinal bars in a beam subjected to deformation reversals under earthquake loading.[4] Salient characteristics are as follows:

- The unloading slope is essentially equal to the initial Young's modulus for steel. If the steel is reloaded before significant yielding in compression, it reloads again essentially at the same slope.
- The monotonic stress–strain relation is essentially an envelope to the reversed loading history.
- Once tensile yielding has occurred, the stress–strain relation under compressive stress becomes nonlinear for stresses below the monotonic compressive yield stress. This phenomenon is known as the Bauschinger effect.
- The stress is not uniquely related to the strain. Note that compressive stress can occur for tensile strain.

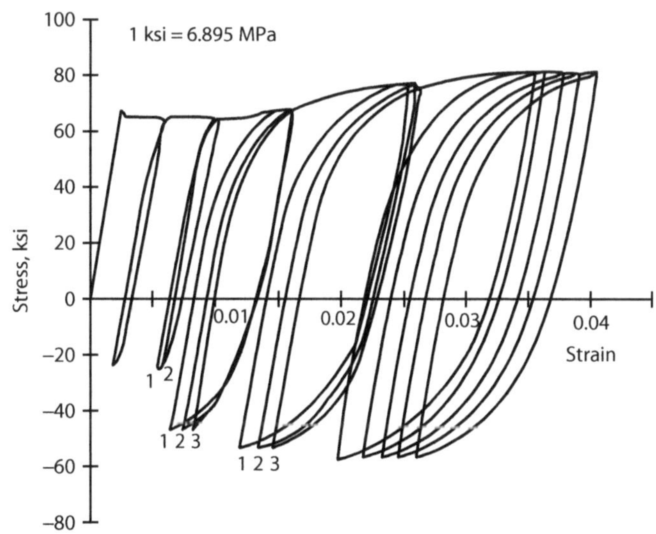

Figure 2.8 Cyclic stress–strain relation of a steel reinforcing bar. (*After Ma et al., 1976, with permission of the University of California, Berkeley.*)

[4]In many beams, the area of top reinforcement exceeds the area of bottom reinforcement at the face of the column. Under earthquake loading, the top reinforcement yields in tension for loading in one direction. Upon moment reversal, the bottom reinforcement yields in tension, but its force capacity is insufficient to yield the larger area of top reinforcement. Therefore, the top reinforcement primarily experiences tensile strain with only moderate compressive stress reversals.

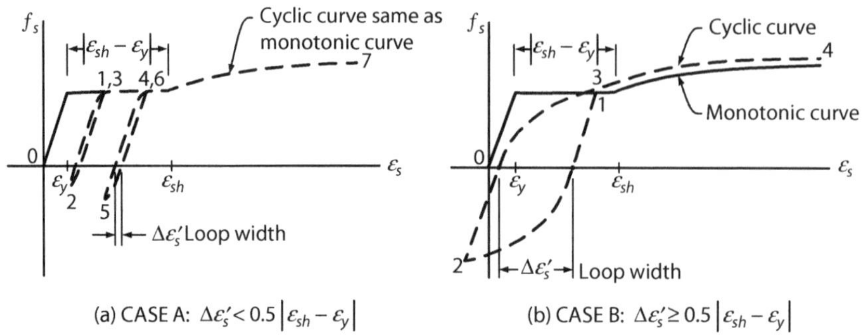

FIGURE 2.9 Effect of loop width $\Delta\varepsilon_s'$ on strain-hardening under cyclic loading. (*After Ma et al., 1976, with permission of the University of California, Berkeley.*)

Figure 2.9*a* illustrates the same basic behavior as in Figure 2.8. Figure 2.9*b* illustrates that if significant compressive yielding occurs, the monotonic stress–strain relation no longer is the envelope to the cyclic behavior.

Figure 2.10 illustrates stress–strain relations for the case of nearly equal compressive and tensile strains. For loadings with such large strain reversals, the stress envelope is outside the monotonic stress–strain relation, and stresses exceeding the monotonic yield stress can occur for strains smaller than the yield strain.

It is apparent that, for steel reinforcing bars subjected to reversed cycling loading, the stress depends on the strain path up to the current loading point. This

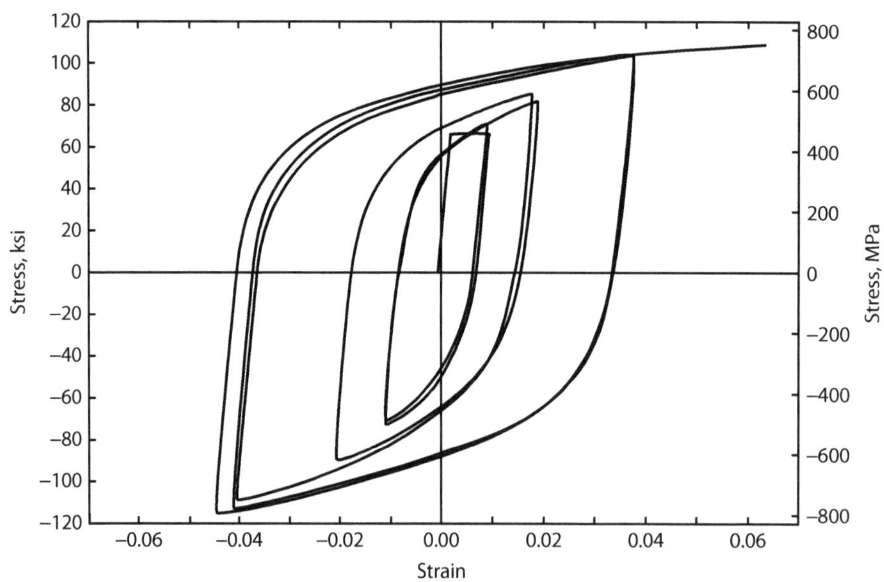

FIGURE 2.10 Stress–strain relation for full strain reversals. (*After Aktan et al., 1973, courtesy of University of Illinois.*)

history dependence is known as *hysteresis*. Analytical models for hysteretic response of steel reinforcing bars are available in the literature (e.g., Kunnath et al., 2009), but not further considered here.

Steel that has been previously strained beyond the initiation of strain-hardening, then left for an extended period, shows a moderate increase in the apparent yield strength and decrease in strain capacity.[5] *Strain-aging* depends on the chemical content of the steel, and is not observed for some steels. This behavior is not considered in routine analysis or design.

2.4.2 Low-Cycle Fatigue

Repeated and reversed cyclic loading of steel reinforcing bars results in reduced strain capacity. High-cycle fatigue, referring to many thousands to millions of cycles, is applicable to structures such as bridges where the design may be controlled by millions of applications of vehicle loadings. The interested reader is referred to ACI 215R (1997) for information on this subject. This chapter is concerned with low-cycle fatigue behavior relevant to earthquake loading, for which 100 or fewer cycles are expected.

Figure 2.11 presents stress–strain histories from low-cycle fatigue tests. The cycles show strength degradation associated with fatigue damage, followed eventually by bar fracture. The fracture process typically initiates with a small crack at the intersection of the main shaft of the bar and a rolled-on deformation, which propagates with cycling until failure. The sharpness of the rolled-on deformations affects crack initiation, so bars with different deformation patterns may have different fatigue lives. Bar buckling during cycling increases local strains because of the curvature resulting from the buckled shape, so buckling accelerates crack

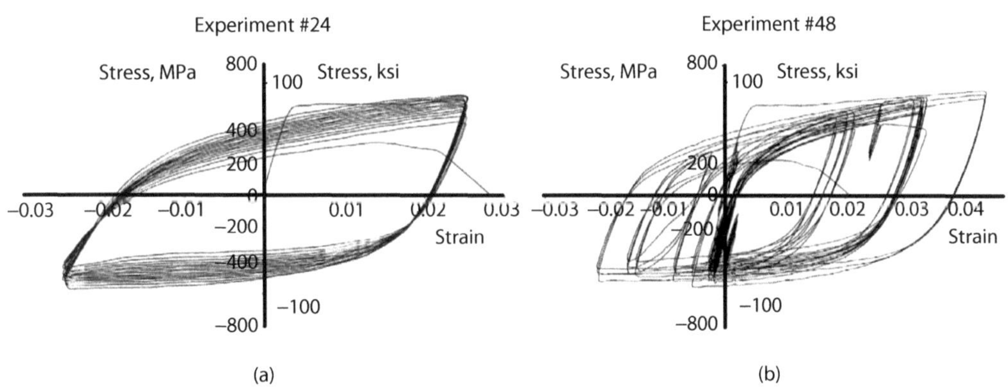

FIGURE 2.11 Low-cycle fatigue tests of A615 reinforcing bars under (*a*) constant strain amplitude and (*b*) variable strain amplitude. (*After Brown and Kunnath, 2000.*)

[5]Strain-aging was noted by Bauschinger (Martens, 1899) and Richart et al. (1929). Restrepo-Posada et al. (1994) provide more recent discussion.

initiation. The tendency for buckling of a bar in air (no cover concrete) depends on the deformation history, the boundary conditions, and the unrestrained length of the bar. In the tests reported in Figure 2.11, the bars met ASTM standards for bar deformations and the unrestrained length between fixed ends during the test was $6d_b$, which is in the range commonly used for earthquake-resistant buildings. Thus, the tests can be considered representative for bars in a member with spalled cover concrete, but they do not necessarily cover the range of possible conditions.

Figure 2.12 shows fatigue-life data for constant-amplitude cycling of the type shown in Figure 2.11*a*. Note that the trends are slightly different for the different bar sizes, indicating that fatigue life varied slightly with bar size.

We can use the Coffin–Manson relation (Manson, 1953; Coffin, 1954) to represent the data trend, as

$$\varepsilon_a = M(2N_f)^m \tag{2.3}$$

In Eq. (2.3), ε_a is the strain amplitude [defined as half the strain range, that is, $\varepsilon_a = 0.5(\varepsilon_{max} - \varepsilon_{min})$] during a cycle, N_f is the number of full cycles (therefore, $2N_f$ = number of half cycles = number of reversals) to failure, and M and m are material constants to be determined from experiments. Brown and Kunnath (2004) reported different values for M and m for different bar sizes. Here we adopt $M = 0.11$ and $m = -0.44$, which was reported as the best fit for No. 7 (25) bars.

The fatigue-life data indicate that the maximum strain capacity under cyclic loading is less than it is under monotonic loading. We can use Eq. (2.3) to calculate

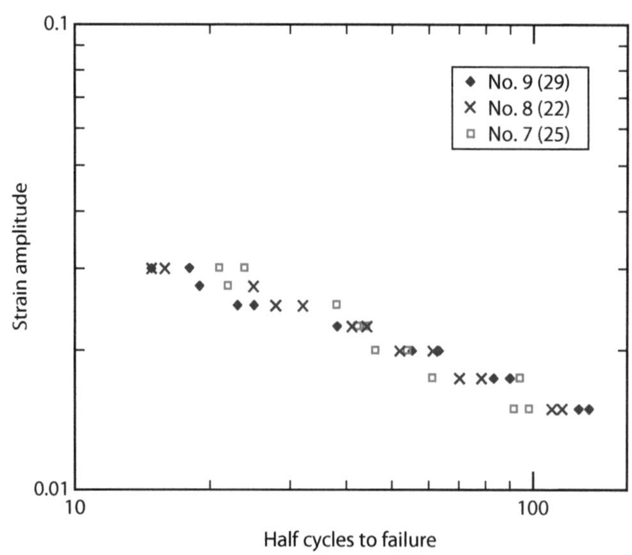

Figure 2.12 Fatigue-life data for A615 Grade 60 (420) reinforcing bars in constant-amplitude tests. Strain amplitude is equal to half the strain range [that is, $\varepsilon_a = (\varepsilon_{max} - \varepsilon_{min})/2$]. (*After Brown and Kunnath, 2000.*)

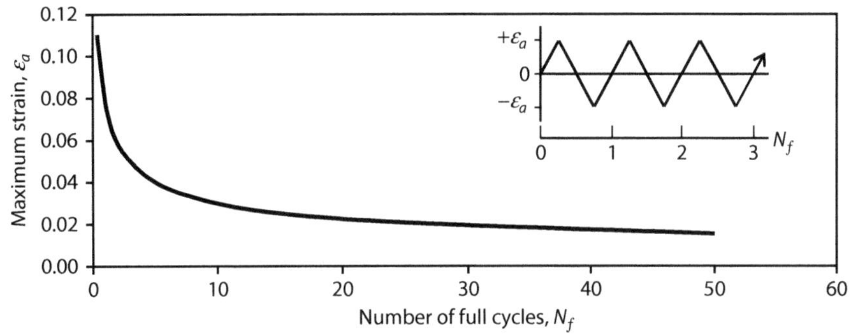

Figure 2.13 Number of full cycles to failure as function of strain amplitude.

the strain amplitude ε_a at fracture as a function of the number of full cycles N_f. Figure 2.13 plots the result.

Earthquake loading involves cyclic histories that are more complicated than the constant-amplitude example of Figure 2.11*a*. For more complicated histories, we can adopt Miner's rule (Miner, 1945), which is based on the assumption that damage accumulates linearly during cyclic loading. Accordingly, if a reinforcing bar can sustain $2N_f$ half cycles under fully reversed strain cycles to amplitude ε_a, then the fraction of total damage in half cycle i at amplitude ε_{ai} is given by

$$D_i = \frac{1}{2N_{fi}} \tag{2.4}$$

in which $2N_{fi}$ = number of half cycles to failure at strain ε_{ai} and D_i = damage accumulated during a half cycle at strain ε_{ai}. Total damage is the linear sum of damage in the individual half cycles, as

$$D = \sum D_i \tag{2.5}$$

Failure is predicted when the damage index $D = 1$. To use Eq. (2.4), we must first calculate the quantity $2N_{fi}$. We do this by solving Eq. (2.3) for $2N_{fi}$ as

$$(2N_{fi}) = \left(\frac{\varepsilon_{ai}}{M}\right)^{\frac{1}{m}} \tag{2.6}$$

For each half cycle to strain amplitude ε_{ai}, the value of $2N_{fi}$ is calculated and inserted into Eq. (2.4) to determine the damage D_i for that half cycle, then values of D_i for different half cycles are summed according to Eq. (2.5) to determine total damage. According to the model, failure occurs for the half cycle in which the damage index D reaches or exceeds 1.0.

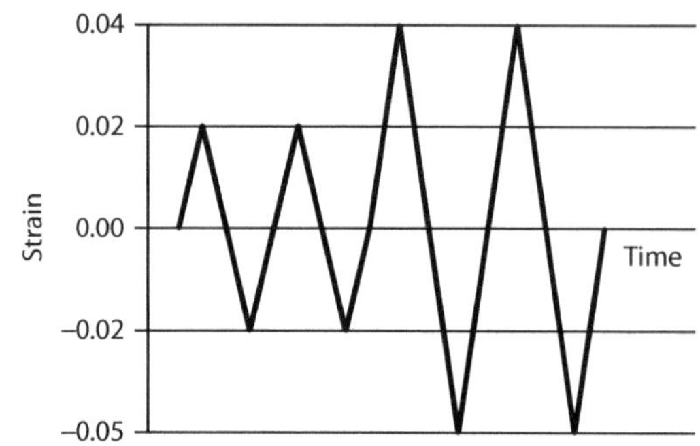

Figure 2.14 Strain history.

Example 2.1. A reinforcing bar has strain history as shown in Figure 2.14. Is the bar likely to fracture?

Solution

For the first two cycles, $\varepsilon_a = 0.02$. Therefore, from Eq. (2.6), the number of half cycles to fracture is $2N_{f0.02} = \left(\frac{0.02}{0.11}\right)^{\frac{-1}{0.44}} = 48$. Therefore, the damage at the end of four half cycles is $D_{0.02} = 4 \times \frac{1}{48} = 0.08$. For the second two cycles, $\varepsilon_a = 0.04$, and the number of half cycles to fracture at that amplitude is $2N_{f0.04} = \left(\frac{0.04}{0.11}\right)^{\frac{-1}{0.44}} = 10$. Therefore, the accumulated damage at the end of these cycles is $D = D_{0.02} + D_{0.04} = 0.08 + 4 \times \frac{1}{10} = 0.48$. Because $D < 1.0$, fracture is unlikely.

We can use Eq. (2.5) to investigate more complicated loadings. For example, a common laboratory testing protocol is to cycle a test specimen through three deformation cycles at constant amplitude Δ, followed successively by three cycles at each of 2Δ, 3Δ, 4Δ, etc. until failure. Strain amplitude in such tests generally does not scale linearly, but let us assume that it does as a simplifying assumption. Suppose we start with an initial strain amplitude $\varepsilon_a = 0.002$, with three full cycles. For each half cycle, we use Miner's rule to calculate damage accumulation. We then increment the strain amplitude to 0.004, 0.006, 0.008, etc. until failure ($D \geq 1$). We can repeat the process for different values of the strain amplitude increment ε_a. Figure 2.15 presents the results for strain amplitude increments $\varepsilon_a = 0.002$, 0.005, 0.01, and 0.02. Interestingly, the analytical model tells us that bar fracture could occur at strains ranging from approximately 0.025 to 0.045. From such results we can see that the maximum deformation capacity of a structural component subjected to cyclic loading will depend on the loading history if its response is sensitive to reinforcing bar fracture.

An important observation in fatigue problems is that fatigue life depends on two aspects of the strain history. One is the number of cycles. The other is the *strain range*, that is, the difference between the maximum and minimum strain in

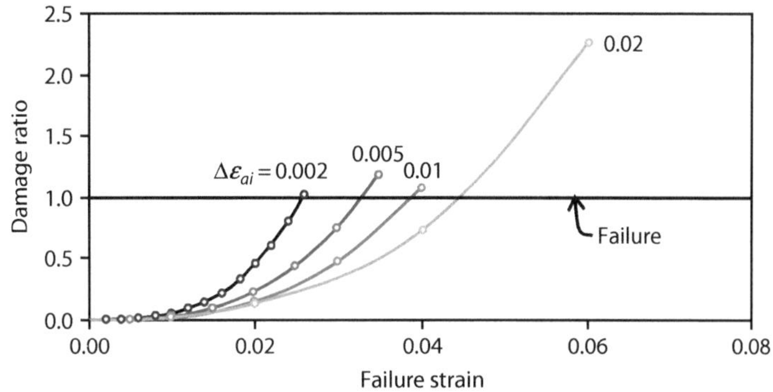

Figure 2.15 Failure strain predictions from Eq. (2.5).

a cycle. In Eq. (2.6), we defined ε_{ai} as half the strain range, that is, $\varepsilon_{ai} = 0.5(\varepsilon_{max} - \varepsilon_{min})$. Thus, according to Eq. (2.6), a bar cycled between strains 0.0 and +0.02 (ε_a = 0.01) will have a longer fatigue life than one cycled between strains −0.02 and +0.02 (ε_a = 0.02).

To extend the procedures to more complicated loadings, another algorithm will be required. Consider the reinforcing bar strain history of Figure 2.16*a* and corresponding stress–strain history of Figure 2.16*b*. As the bar is strained from **a** to **e**, we assume that it accumulates an amount of damage associated with a half cycle having a strain range along path **ae**. We also need to account for the damage resulting from the relatively smaller-amplitude cycle **bcd** that occurs along the path **ae**; we will assume that the damage is equal to the sum of damage for a half cycle having strain range **bc** and for another half cycle having strain range **cd**. Strain reversal from **e** to **f** causes damage associated with a half cycle having strain range **ef**. Finally, another small half cycle of strain range **fg** finishes the history. For each of these half cycles, we can calculate the strain amplitude ε_{ai} = half the strain range for that half cycle, use Eqs. (2.6) and (2.4) to calculate damage in that half cycle, and then sum over all half cycles using Eq. (2.5) to calculate the damage index.

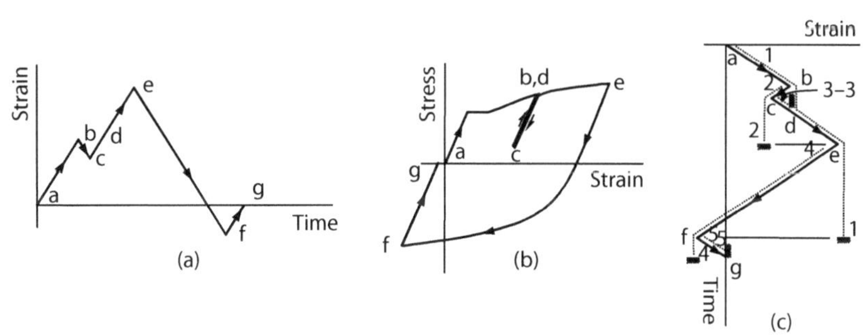

Figure 2.16 Illustration of the rainflow-counting method: (*a*) strain history; (*b*) stress–strain history; (*c*) rainflow-counting diagram.

The approach just described is known as the rainflow-counting method (ASTM E1049-85, 2005), because it is analogous to rain flowing over a series of rooftops. To use the analogy, first rotate the strain history so it appears as a series of inclined surfaces (or pagoda rooftops) and imagine rain flowing down each one (Figure 2.16*c*). Rain flow begins at each peak and ends when one of the following conditions is met:

1. The end of the strain history is reached.
2. Rain flows opposite a peak having amplitude greater than the one from which it started.
3. Rain flow is interrupted by a flow that started at an earlier peak.

For example, a raindrop starting at peak **a** (Figure 2.16*c*) would flow along rooftop **ab**, drop down to rooftop **ce**, flow along **de**, drop off **e**, and stop at point **1** because it is opposite a peak (**f**) having amplitude in the opposite direction greater than the one from which it started. Another flow starts at **b** and must stop opposite peak **e** for the same reason. A third flow starts at **c** and must stop at **d** because it is interrupted by flow coming from above, and so on. The strain amplitude of each of these paths (half cycles) is determined, and then Miner's rule is used as described above to sum the damage. Computer algorithms for automatic computation using the rainflow-counting method are available.

We can apply the rainflow-counting algorithm to any strain history. In reference to Figure 2.17, the algorithm predicts failure at strains of 0.035 for Case A and 0.06 for Case B, whereas Case C is essentially the same as monotonic loading.[6]

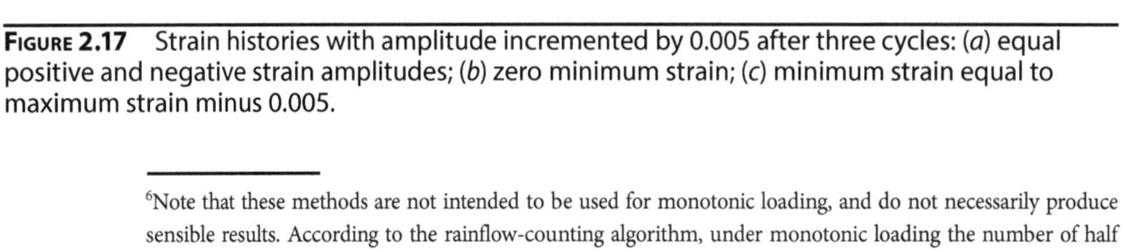

FIGURE 2.17 Strain histories with amplitude incremented by 0.005 after three cycles: (*a*) equal positive and negative strain amplitudes; (*b*) zero minimum strain; (*c*) minimum strain equal to maximum strain minus 0.005.

[6]Note that these methods are not intended to be used for monotonic loading, and do not necessarily produce sensible results. According to the rainflow-counting algorithm, under monotonic loading the number of half cycles is $2N_f = 1$ and the strain amplitude, defined as half the strain range, is $\varepsilon_a = \varepsilon_u/2$. Using $M = 0.11$ as recommended here, Eq. (2.3) gives $\varepsilon_a = 0.11$, or $\varepsilon_u = 0.22$. Usually the elongation capacity under monotonic loading is considerably less than this value.

References

ACI 215R (1997). "Considerations for Design of Concrete Structures Subjected to Fatigue Loading," Reported by ACI Committee 215 (revised 1992, reapproved 1997), *ACI Manual of Concrete Practice*, American Concrete Institute, Farmington Hills, MI, 24 pp.

ACI 318 (2014). *Building Code Requirements for Structural Concrete (ACI 318-14) and Commentary*, American Concrete Institute, Farmington Hills, MI.

ACI 439.R-09 (2009). "Report on Steel Reinforcement—Material Properties and U.S. Availability," Reported by ACI Committee 439, *ACI Manual of Concrete Practice*, American Concrete Institute, Farmington Hills, MI, 19 pp.

Aktan, A.E., B.I. Karlsson, and M.A. Sozen (1973). *Stress-Strain Relationships of Reinforcing Bars Subjected to Large Strain Reversals*, Civil Engineering Studies, Structural Research Series No. 397, University of Illinois, Urbana–Champaign, IL.

AWS D1.4-11 (2011). "Structural Welding Code—Reinforcing Steel," American Welding Society, Doral, FL. (*Note:* Append M to designation to designate the metric version.)

ASTM E1049-85. (2005, reapproved 2011). *Standard Practices for Cycle Counting in Fatigue Analysis.*

ASTM standards (append M to designation to designate the metric version).

A184 (2011). *Standard Specification for Welded Deformed Steel Bar Mats for Concrete Reinforcement.*

A370 (2013). *Standard Test Methods and Definitions for Mechanical Testing of Steel Products.*

A615 (2013). *Standard Specification for Deformed and Plain Carbon-Steel Bars for Concrete Reinforcement.*

A706 (2013). *Standard Specification for Low-Alloy Steel Deformed and Plain Bars for Concrete Reinforcement.*

A767 (2009). *Standard Specification for Zinc-Coated (Galvanized) Steel Bars for Concrete Reinforcement.*

A775 (2007). *Standard Specification for Epoxy-Coated Steel Reinforcing Bars.*

A820 (2011). *Standard Specification for Steel Fibers for Fiber-Reinforced Concrete.*

A884 (2012). *Standard Specification for Epoxy-Coated Steel Wire and Welded Wire Reinforcement.*

A934 (2013). *Standard Specification for Epoxy-Coated Prefabricated Steel Reinforcing Bars.*

A955 (2012). *Standard Specification for Deformed and Plain Stainless-Steel Bars for Concrete Reinforcement.*

A970 (2013). *Standard Specification for Headed Steel Bars for Concrete Reinforcement*, American Society for Testing and Materials.

A996 (2009). *Standard Specification for Rail-Steel and Axle-Steel Deformed Bars for Concrete Reinforcement.*

A1022 (2013). *Standard Specification for Deformed and Plain Stainless Steel Wire and Welded Wire for Concrete Reinforcement.*

A1035 (2013). *Standard Specification for Deformed and Plain, Low-Carbon, Chromium, Steel Bars for Concrete Reinforcement.*

A1055 (2010). *Standard Specification for Zinc and Epoxy Dual-Coated Steel Reinforcing Bars.*

A1060 (2011). *Standard Specification for Zinc-Coated (Galvanized) Steel Welded Wire Reinforcement, Plain and Deformed, for Concrete.*

A1064 (2013). *Standard Specification for Carbon-Steel Wire and Welded Wire Reinforcement, Plain and Deformed, for Concrete.*

Bournonville, M., J. Dahnke, and D. Darwin (2004). *Statistical Analysis of the Mechanical Properties and Weight of Reinforcing Bars*, SL Report 04-1, Structural Engineering and Materials Laboratory, The University of Kansas, Lawrence, KS, 198 pp.

Brown, J., and S.K. Kunnath (2000). *Low Cycle Fatigue Behavior of Longitudinal Reinforcement in Reinforced Concrete Bridge Columns*, Report No. MCEER-00-0007, MCEER, SUNY Buffalo, Buffalo, NY.

Brown, J., and S.K. Kunnath (2004). "Low Cycle Fatigue Failure of Reinforcing Steel Bars," *ACI Materials Journal*, Vol. 101, No. 6, pp. 457–466.

Coffin, L.F. (1954). "A Study of the Effects of Cyclic Thermal Stresses on a Ductile Material," *Transactions of the American Society of Mechanical Engineers*, Vol. 76, pp. 931–950.

Dodd, L.L., and J.I. Restrepo-Posada (1995). "Model for Predicting Cyclic Behavior of Reinforcing Steel," *Journal of Structural Engineering*, Vol. 121, No. 3, pp. 433–445.

Kunnath, S.K., Y.A. Heo, and J.F. Mohle (2009). "Nonlinear Uniaxial Material Model for Reinforcing Steel Bars," *Journal of Structural Engineering*, Vol. 135, No. 4, pp. 335–343.

Ma, S.Y., V.V. Bertero, and E.P. Popov (1976). *Experimental and Analytical Studies on the Hysteretic Behavior of Reinforced Concrete Rectangular and T-Beams*, Report No. EERC 76-2, Earthquake Engineering Research Center, University of California, Berkeley, CA.

Malvar, L.J. (1998). "Review of Static and Dynamic Properties of Steel Reinforcing Bars," *ACI Materials Journal*, Vol. 95, No. 5, pp. 609–616.

Mander, J.B., M.J.N. Priestley, and R. Park (1984). *Seismic Design of Bridge Piers*, Report 84-02, Department of Civil Engineering, University of Canterbury, Christchurch, New Zealand.

Manjoine, M.J. (1944). "Influence of Rate of Strain and Temperature on Yield Stress of Mild Steel," *Journal of Applied Mechanics*, Vol. 11, No. 4, pp. 211–218.

Manson, S.S. (1953). "Behavior of Materials under Conditions of Thermal Stress," *Heat Transfer Symposium*, University of Michigan Engineering Research Institute, Ann Arbor, MI.

Martens, A. (1899). *Handbook of Testing Materials for the Constructor*, Translation by G.C. Henning, John Wiley & Sons, New York, 622 pp.

Miner,M. A. (1945). "Cumulative Damage in Fatigue," *Journal of Applied Mechanics*, Vol. 12, pp. A159–A164.

MMFX (2009). *MMFX Microcomposite (MMFX 2) Steel Bars for Concrete Reinforcement*, MMFX Steel Corporation of America, Charlotte, NC.

Restrepo-Posada, J.I., L.L. Dodd, R. Park, and N. Cooke (1994). "Variables Affecting Cyclic Behavior of Reinforcing Steel," *Journal of Structural Engineering*, Vol. 120, No. 11, pp. 3178–3196.

Richart, F.E., A. Brandtzaeg, and R.L. Brown (1929). *The Failure of Plain and Spirally Reinforced Concrete in Compression*, Bulletin No. 190, Engineering Experiment Station, University of Illinois, Urbana, IL, 74 pp.

WRI (2006). *Structural Welded Wire Reinforcement Manual of Standard Practice*, Wire Reinforcement Institute, Hartford, CT, 8th ed., 38 pp.

CHAPTER 3
Concrete

3.1 Preview

This chapter introduces the physical and mechanical properties of concrete. We begin with a discussion of the composition and structure of concrete, as this provides a basis for understanding the mechanical properties under various loading conditions. Properties of interest include modulus of elasticity; expected strength and its dependence on proportions, curing, and test specimen parameters; and behavior under multi-axial loading including cyclic loading. Our main interest is the behavior of concrete under earthquake loadings. Readers interested in other properties of concrete such as long-term shrinkage and creep effects, or behavior under other severe environments, should refer to other sources that emphasize those subjects (e.g., Mindess et al., 2003; Mehta and Monteiro, 2014).

3.2 Composition and Structure of Concrete[1]

Concrete is a composite material consisting mainly of aggregates held together by a binding agent. Aggregate is a granular material, such as sand, gravel, or crushed stone. *Coarse aggregate* refers to particles retained on the 4.75-mm (No. 4) sieve, whereas *fine aggregate* refers to particles passing the same sieve but with restrictions to avoid particles that are too fine.

In structural concrete, the binding agent usually is cement, which is a finely pulverized material that develops binding properties as a result of hydration. The cement is defined as being *hydraulic* if the hydration products are stable in an aqueous environment. *Portland cement* is the most common hydraulic cement used in construction. Portland cement is made of hydraulic calcium silicates, calcium aluminates, and gypsum. Through addition of controlled amounts of mixing water, portland cement gains an adhesive characteristic.

In addition to aggregates, cement, and water, modern concretes usually have *admixtures*, which are added before or during mixing. Some admixtures can improve workability, thereby reducing required water and improving potential strength. Others can modify setting and hardening characteristics of the plastic concrete and can improve thermal and freeze-thaw cracking resistance.

[1]See Mehta and Monteiro (2014) for detailed discussion of concrete structure and properties.

Some special types of concrete include mortar, grout, and shotcrete. *Mortar* is a mixture of sand, cement, and water. *Grout* is a mixture of cementitious material and aggregate to which sufficient water is added to achieve a pouring consistency. *Shotcrete* is a mortar or concrete that is pneumatically transported through a hose, projected, and deposited onto a surface at high velocity.

Normalweight concrete is made of naturally occurring sand and gravel, and has unit weight of about 145 pcf (2300 kg/m^3). To approximately account for the higher unit weight of reinforcement, engineers commonly assign a unit weight of 150 pcf (2400 kg/m^3) for normalweight reinforced concrete. *Lightweight concrete* uses certain natural or pyro-processed aggregates having lower bulk density; typical unit density for structural lightweight concrete is around 120 pcf (1900 kg/m^3), but lower or higher densities are possible. *Sand lightweight concrete* refers to concrete using normalweight sand with lightweight course aggregate.

We can idealize the structure of concrete as comprising three different zones, specifically the *aggregates*, the *hydrated cement paste*, and the *interfacial transition zone* between the aggregates and hydrated cement paste. Typically, the interfacial transition zone is the weakest of the three zones, and therefore its behavior controls the nonlinear properties and strength of the concrete. Figure 3.1 shows a schematic of cracking of the composite under different levels of externally applied stress. Micro-cracks at the interfacial transition zone begin to extend at stress levels as low as 40% of the compressive strength of the concrete. At higher stresses, the cracks begin to spread through the mortar, eventually forming a continuous system of cracks leading to rupture.

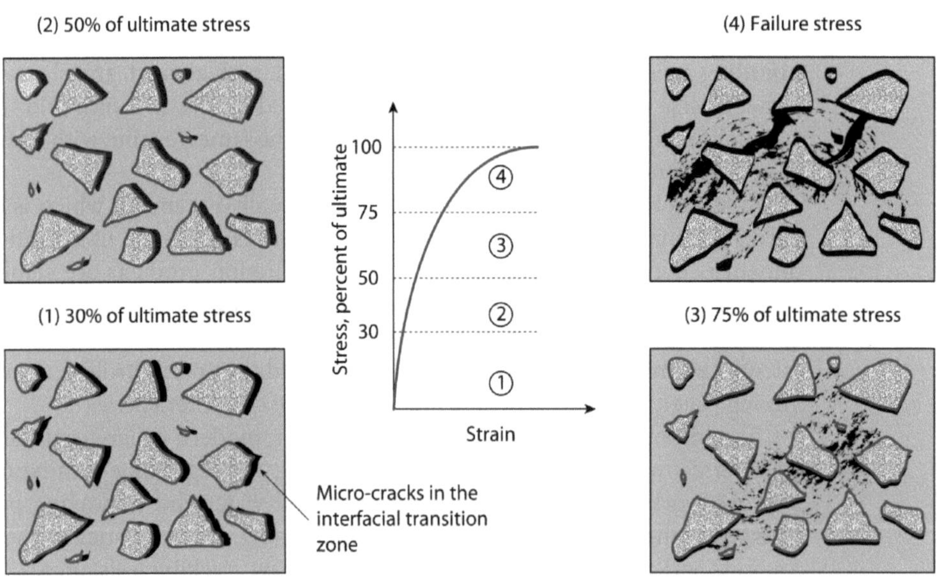

Figure 3.1 Progression of cracking and relation to stress–strain behavior. (*After Mehta and Monteiro, 2014.*)

The effect of crack spreading is apparent in the softening of the stress–strain relation under monotonic loading (Figure 3.1). At around 75% of the peak stress, the crack system becomes unstable, with spontaneous crack growth that will result in eventual failure under sustained load. For this reason, sustained loads on concrete structures must always be kept well below this critical *static fatigue* level. Micro-cracks also propagate under cyclic loading, resulting in important low-cycle and high-cycle fatigue behaviors that will be discussed later in this chapter. Migration of adsorbed water in the hydration products (not shown) along with delayed elastic response of aggregates leads to redistribution of internal stresses and may cause additional micro-crack growth, which taken together helps explain creep of concrete.

The micro-cracking behavior described above is principally applicable to normal-strength concrete. For high-strength concrete, the high strength of the mortar results in delayed cracking, such that the behavior is more nearly linear to higher stresses. Also, improvements in the interfacial transition zone make it no longer the weak link. Thus, in many high-strength concretes, the strength of the composite is limited by the strength of the aggregate rather than the mortar, and fracture through the aggregate can result in a sudden failure of the composite.

3.3 Concrete Strength

Concrete strength is the property most commonly specified by structural engineers. Concrete compressive strength is relatively easy to measure compared with other properties, and compressive strength test results can be related to other properties of interest such as permeability, weathering resistance, modulus of elasticity, and behavior under multi-axial stress states. Given these characteristics, building codes generally use specified compressive strength, f_c', both for design and for quality control tests to determine acceptance of concrete. Unless another age is indicated in the design drawings or specifications, the 28-day compressive strength determined on standard uniaxial compression test specimens is universally accepted as the general index of concrete quality and strength.

If concrete compressive strength is to be used as the general index of concrete quality, we should understand those factors that affect it. For simplicity we will consider the factors under three broad categories: (1) materials characteristics and proportions, (2) curing time and conditions, and (3) test specimen parameters. A more thorough discussion would recognize the complex interactions among many factors (Mehta and Monteiro, 2014).

3.3.1 Materials Characteristics and Proportions

Strength of solids, including concrete, is governed by the inverse relation between porosity and strength. Most aggregates have low porosity (and high strength), such that concrete strength usually is not controlled by the aggregate strength. Instead, concrete strength usually is controlled by the strength of the cement paste matrix and the interfacial transition zone. For some lightweight aggregates or for high-strength concrete, the aggregate can become the weak element in the matrix.

The discussion below addresses mainly normalweight and normal-strength concrete.

Several portland cements are available for different applications. Type I is for general use, whereas Type III is high early strength. Other types include Type II (moderate heat of hydration and moderate sulfate resistance), Type IV (low heat of hydration), Type V (high sulfate resistance), Type IS (portland blast furnace slag cement), and Type IP (portland-pozzolan cement). These other types tend to develop strength somewhat slower than Types I and III, but all eventually reach similar strengths for given mixture proportions.

The water–cement ratio, w/c, (or more properly the water–cementitious materials ratio) is one of the most important parameters affecting concrete strength (Figure 3.2). For high w/c ratios, there is more water in the mixture than required for hydration, and this increases porosity of the matrix and reduces strength. For w/c ratios below about 0.3, there is a significant increase in the strength of the interfacial transition zone leading to rapid increase in concrete strength.

Aggregate strength usually does not determine concrete strength. However, size, shape, surface texture, grading, and mineralogy can affect concrete strength. In general, larger aggregates and poorer grading lead to reduced strength.

Air entrainment can improve concrete workability, which in turn can lead to improved compaction. The increased porosity generally leads to reduced strength, but this can be compensated by positive effects of improved compaction.

Other *chemical admixtures* are classed according to function, and include (1) water-reducing admixtures that improve workability with less water, leading to reduced w/c ratio; (2) retarding admixtures that slow the setting rate of concrete; (3) accelerating admixtures to increase the rate of early strength development; (4) superplasticizers for high-slump concrete with reduced w/c ratio; and

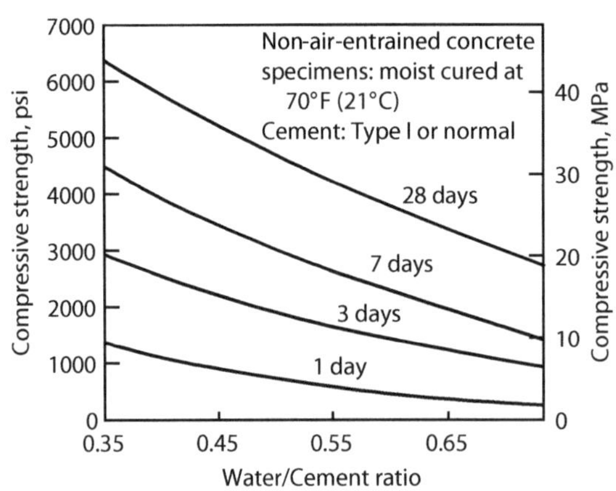

FIGURE 3.2 Compressive strength of 6 × 12 in (150 × 300 mm) cylinders as function of water–cement ratio. (*After PCA, 1988.*)

(5) corrosion-inhibiting admixtures to slow corrosion of reinforcing steel in extreme environments. In addition, air-entraining admixtures are used to increase concrete resistance against frost. *Mineral admixtures* such as fly ash, blast-furnace slag, and silica fume are used as a partial replacement for portland cement. In addition to the ecological advantages of reduced cement usage, these can lead to reduced cost, reduced permeability, and increased strength.

3.3.2 Curing Time and Conditions

Concrete curing describes the process by which concrete matures and develops its mechanical and durability properties over time as a result of continued hydration of the cement. Concrete curing is affected by the availability of water for hydration and by temperature. In the absence of deleterious effects, curing and associated concrete strength gain can continue indefinitely.

Figure 3.3 shows the influence of curing humidity on strength gain of concrete with time. If curing occurs in a relatively dry environment, it effectively ceases once the free water in the mixture is used up or evaporates from the capillaries. Because the amount of free water in a concrete mixture usually is more than sufficient for complete hydration, application of an impermeable membrane to seal in mixture water can be an effective way to maintain hydration in dry environments.

Curing temperature also affects the rate of strength gain. Concrete cast and cured in the normal temperature range [40°F to 100°F (5°C to 35°C)] shows slower strength gain at colder temperatures within the range, but eventually reaches its strength potential. For higher temperatures, the rate of hydration is increased,

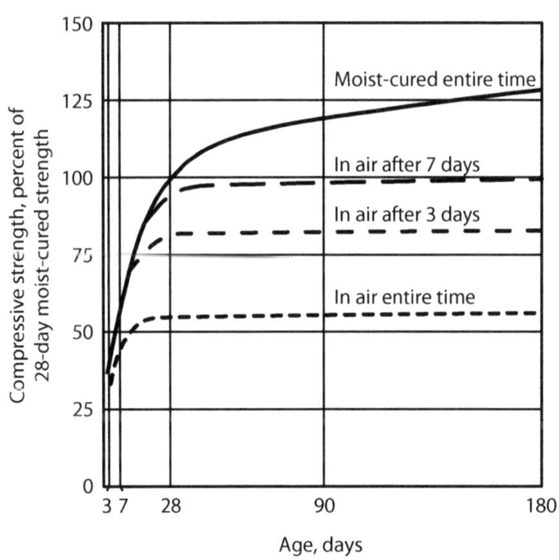

Figure 3.3 Compressive strength of 6 × 12 in (150 × 300 mm) cylinders as a function of age for a variety of curing conditions. (*After PCA, 1988.*)

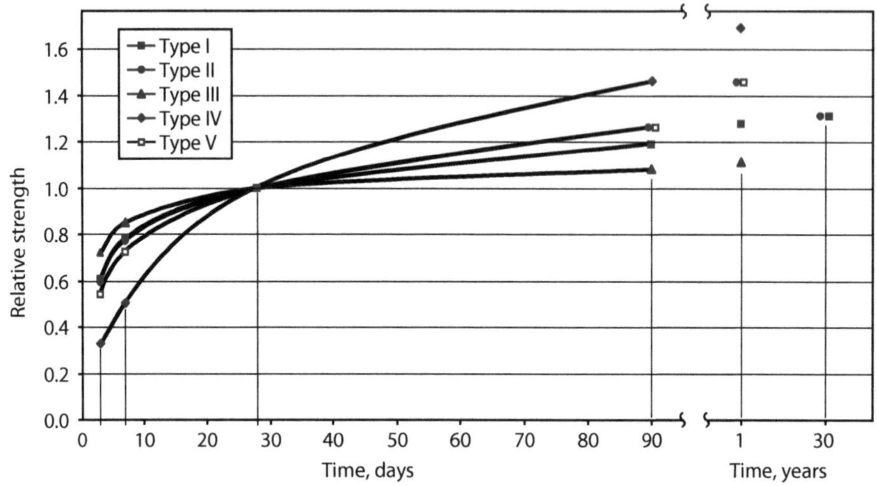

FIGURE 3.4 Concrete strength gain with time under moist curing for different types of portland cements: data up to 28 days based on mortar tests using U.S. cements from the 1990s (Mindess et al., 2003, p. 29); data for 90 days and 1 year based on concrete tests from the 1970s (Mindess et al., 2003, p. 28); data for 30 years based on concrete tests for Type I (Wood, 1991) and Type II (Monteiro and Moehle, 1995) cements.

producing a higher early strength, at the cost of some reduction in the final strength. The same applies to large structural members with high-strength concrete for which temperature rise associated with heat of hydration can be high. Cold weather concreting also can cause problems. See ACI 308R-01 (2001).

Cement fineness also affects rate of strength gain. The properties of portland cements have varied with changes in governing specifications over time, such that more recent cements gain early strength faster than older cements. Figure 3.4 presents a composite of test results from various sources including early-age strengths from more recent cements and long-term strengths for older cements. The data refer to concrete with continuous supply of moisture for curing. Long-term compressive strengths for specimens stored outdoors may develop lower strength depending on the curing environment (Wood, 1991).

3.3.3 In-Place Concrete

Acceptance testing of concrete for new structures generally is based on standardized tests of concrete that has been placed in a standard mold (usually a cylinder but in some countries a cube), compacted in a specified way, stored in a controlled curing environment, and tested under idealized boundary conditions. Concrete in actual structures has form, compaction, curing, and boundary conditions that differ from the standardized test conditions. Given the dependence of concrete strength on these variables, it should not be surprising that in-place strength differs from the strength measured in the standardized test.

Specific causes for in-place concrete strength variations include:

- *Consolidation:* In normal placement operations, good construction practice uses vibration to expel entrapped air from plastic concrete. In deep members, higher static pressure at the bottom increases consolidation and may improve strength relative to the top (ACI 214.4R-10, 2010).
- *Bleeding and segregation:* During placing and consolidation activities, segregation of concrete constituents and bleeding of water toward the top surface can result in localized differences in concrete mixtures throughout a structure. In deep members such as columns, the increased w/c ratio near the top of the member can result in reduced strength. It is common to estimate the strength of concrete at the top of a column as Cf'_c, in which $C = 0.85$. [CSA (2004) uses $C = 0.85 - \lambda_c f'_c \geq 0.67$, in which $\lambda_c = 0.00001$ (psi) or 0.0015 (MPa).][2]
- *Curing:* Curing may vary throughout the structure and generally deviates from the ideal conditions specified for control test specimens. In large sections (e.g., large columns), heat of hydration effects can cause high temperatures that permanently reduce concrete strength potential. Slabs and beams have exposed surfaces, making them sensitive to moisture loss that can result in reduced strength.
- *Micro-cracks:* Drying shrinkage, temperature change, and applied loads cause internal stresses and strains that produce micro-cracks and reduce stress-resisting capacity of in-place concrete.

These effects have been demonstrated by various types of tests on in-place concrete. *Nondestructive methods* (e.g., ultrasonic pulse velocity) determine hardened concrete properties in ways that cause no noticeable damage to concrete (ACI 228.2R-98, 1998). *In-place methods* include nondestructive methods plus other methods conducted in place that may cause observable but insignificant damage (e.g., rebound hammer) (ACI 228.1R-03, 2003). *Destructive methods* usually refer to removing cores from an existing structure for testing in a laboratory (ACI 214.4R-10, 2010). Section 3.3.4 includes additional discussion on evaluation of test results from concrete cores.

[2]This book uses both the U.S. customary units and the International System of Units (abbreviated SI). For some equations, expressions, or variables, the units are consistent, such that the same equation, expression, or variable applies to both unit systems. In others, different equations, expressions, or variables are required for the two unit systems. Where this occurs, the equation, expression, or variable is shown first in the U.S. customary units followed by the abbreviation psi (representing pounds per square inch) and second in the SI units followed by the abbreviation MPa (representing megapascal). The result of the equation, expression, or variable, however, is not necessarily in units of psi or MPa. For example, the result of Eq. (3.2) is in either psi or MPa, but the result of Eq. (6.28) is in^2 or mm^2 rather than psi or MPa.

3.3.4 Test Specimen Parameters

In the United States, the standard test specimen for determining compressive strength is a 6 × 12 in (150 × 300 mm) cylinder cast in a mold and tested in a moist condition. If any of these parameters is changed, the apparent compressive strength obtained from the test may change.

- *Moisture content:* Air-dried cylinders are on average 10% to 14% stronger than soaked cylinders, though values differ for different concretes (ACI 214.4R-10, 2010). The effect may be associated with effects of internal hydraulic pressure.
- *Length-to-diameter ratio:* During testing (Figure 3.5*a*), steel platens at the loaded ends of a compression test specimen restrain lateral expansion, confining

(a)

(b)

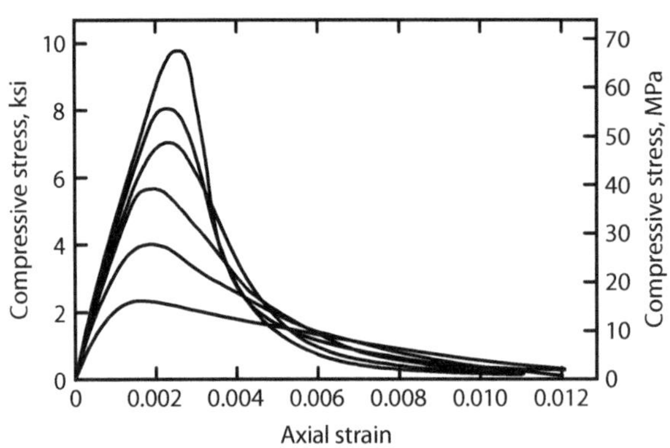

Figure 3.5 (*a*) Uniaxial compression test on 6 × 12 in (150 × 300 mm) cylinder at UC Berkeley laboratories. (*Photo courtesy of L. Stepanov.*) (*b*) Stress–strain relations of normalweight concretes under uniaxial compressive loading. (*After Wischers, 1979, as reported by ACI 363R-92, 1992.*)

the concrete and increasing the apparent strength. A length-to-diameter ratio of 2.0 is recommended so that the confinement effect near mid-length is reduced. When a shorter length-to-diameter ratio is used, as sometimes occurs when cores are taken from an existing structure, correction factors are required (ACI 214.4R-10, 2010). Many countries use a cube for concrete compressive strength tests. Cube strength is generally larger than cylinder strength, in part because of the confinement effect. The ratio of cube to cylinder strength is commonly assumed to be 1.25 for normal-strength concrete, decreasing to 1.0 for high compressive strength (around 14 ksi or 100 MPa).

- *Size:* Cylinders (or cubes) smaller than the standard sizes may be required under special circumstances such as coring between closely spaced reinforcement. For cylinders between 4 and 6 in (100 and 150 mm) there is little size effect. For smaller cylinders there is conflicting information on size effect. Minimum dimensions should not approach the size of the largest aggregate.
- *Cores taken from existing structures:* Existing structures have strength variations that have been discussed in Section 3.3.3. Cores also may have sizes and length-to-diameter ratios that differ from the standards. In addition to these aspects, drilled cores may have reduced strength because of micro-cracking that occurs during drilling. Core drilling may or may not coincide with the direction of concrete placement depending on the member being cored, so bleeding effects may manifest themselves differently. Defects such as pre-existing cracks or inclusions (including unintended reinforcing steel) also affect core test results. See ACI 214.4R-10 (2010) for detailed discussion of planning, conducting, and interpreting results of a coring program.

3.3.5 Expected Strength in Structures

Quality control requirements of building codes are written so that there is only a small chance that under-strength concrete will compromise the safety of a building. ACI 318 specifies both the frequency of concrete sampling and testing, and the strength required for acceptance. The strength requirement is as follows: (1) no individual strength test result [the average of two 6 by 12 in (150 by 300 mm) or of three 4 by 8 in (100 by 200 mm) cylinder tests] shall fall below f'_c by more than 500 psi (3.4 MPa) if $f'_c \leq 5000$ psi (34 MPa) or by more than $0.1f'_c$ if $f'_c >$ 5000 psi (34 MPa), and (2) no average of three consecutive strength tests shall be less than f'_c. To satisfy these requirements, the concrete supplier must design the concrete mixture for an average compressive strength f'_{cr} that exceeds these minimum acceptance values.

The target value of f'_{cr} will necessarily depend on the variability of the concrete strength and the accepted probability of a test not satisfying the acceptance requirement. In general, the value of f'_{cr} should be established such that nonconformance is anticipated no more than 1 in 100 times (ACI 214R-11, 2011).

Therefore, assuming a normal distribution of measured compressive strengths, the following must be satisfied:

$$f'_{cr} \geq f'_c + 2.33\sigma/\sqrt{3} \tag{3.1}$$

$$f'_{cr} \geq f'_c + 2.33\sigma - 500\text{, psi} \quad (f'_c + 2.33\sigma - 3.4\text{, MPa}) \tag{3.2}$$

where σ = standard deviation of compressive strengths. Equation (3.1) defines the required target strength so there is no more than 0.01 probability that averages of three consecutive tests will be below the specified value of f'_c. Equation (3.2) defines the target strength so there is no more than 0.01 probability of an individual test falling more than 500 psi (3.5 MPa) below the specified value of f'_c. For $f'_c >$ 5000 psi (34 MPa), substitute 0.1 f'_c for 500 psi (3.4 MPa) in Eq. (3.2).

Studies in North America have defined expected dispersions for typical concretes (ACI 214R-11, 2011). For average quality control, a coefficient of variation of 0.15 is to be expected. Excellent quality control (the upper 10% of projects studied) resulted in coefficient of variation of about 0.10. Poor quality control (the lower tenth percentile) resulted in coefficient of variation of about 0.20.[3]

Example 3.1. Concrete is specified to have f'_c = 4000 psi (28 MPa). The supplier uses Type I cement. What is the expected compressive strength after several years?

Solution

Assuming average quality control, with coefficient of variation of 0.15, Eq. (3.1) requires target compressive strength f'_{cr} = 5000 psi (34 MPa), and Eq. (3.2) requires f'_{cr} = 5400 psi (37 MPa). Assume the supplier aims to provide 5400 psi (37 MPa) concrete. Assuming the average long-term strength gain of 32% relative to the 28-day strength for Type I cement (Section 3.3.2), the expected strength at advanced age is 1.32 × 5400 psi = 7100 psi (49 MPa) or 1.8 f'_c.

The preceding example illustrates that expected compressive strength at advanced age can exceed the specified strength by a significant margin. The California Department of Transportation (Caltrans), based on limited field testing, commonly assumes compressive strength of concrete in older construction is 1.5 f'_c. ASCE 41 (2013) uses the same multiplier to obtain expected strength. Common practice for performance-based seismic design of new buildings is to estimate expected compressive strength as 1.3 f'_c (TBI, 2010). Of course, these expectations assume that the concrete is not overloaded and does not sustain durability problems.

---

[3]The values reported apply to moderate-strength concretes. For higher-strength concretes, improved quality control may result in lower coefficients of variation. This can be demonstrated on a project-by-project basis.

3.4 Behavior in Uniaxial Monotonic Loading

3.4.1 Compressive Stress–Strain Response

The stress–strain relation for concrete under monotonic uniaxial compression is readily obtained from the standard cylinder test. Figure 3.5 shows measured stress–strain relations for normalweight aggregate concrete samples of various strengths. Some noteworthy observations are as follows:

- Stiffness increases with increasing compressive strength.
- Strain at peak stress increases slightly with increasing compressive strength. Strain at peak stress is commonly assumed to be 0.002, though it can range from 0.0015 to 0.003 depending on the constituent material properties and the mixture proportions.
- The unloading branch is steeper for concrete with higher compressive strength.

Figure 3.5 shows concrete compressive strengths as high as 10 ksi (70 MPa). In some markets such as the San Francisco Bay Area, this is near the practical upper bound that can be obtained reliably using local aggregates. In other markets (e.g., Seattle and Chicago) superior aggregate qualities enable compressive strengths of around 20 ksi (140 MPa).

Section 3.2 described how micro-cracking affects the nonlinearity of the concrete stress–strain relation. For high-strength concrete, increased strength of the cement paste and interfacial transition zone results in less micro-cracking, such that stress–strain relation is more nearly linear than for lower-strength concrete. Failure of high-strength concrete may be characterized by a cleaner failure surface passing through the aggregate without well-distributed micro-cracking throughout the matrix. As a result, dilation (lateral straining) is reduced for higher-strength concrete (Figure 3.6). This behavior has important implications for confined concrete.

Figure 3.7 shows stress–strain relations for lightweight aggregate and normalweight aggregate concretes of otherwise similar mixture proportions and strengths (Bresler, 1971). Lightweight concrete tends to have lower modulus and higher strain at peak stress. Many lightweight aggregates tend to be relatively weaker than normalweight aggregates, so they can be strength-limiting and lead to relatively brittle failure at moderate strength levels.

Although the stress–strain relation for concrete is nonlinear even at low stress levels, we commonly assume it to be linear for stresses up to around $0.5f_c'$. The commonly accepted definition of concrete modulus is a chord modulus from the stress–strain point at 50 microstrain to the stress–strain point at $0.4f_c'$ (ASTM C469, 2010). ACI 318 uses Eq. (3.3) for concrete modulus.

$$E_c = 33 w_c^{1.5} \sqrt{f_c'}, \text{ psi}$$
$$\left(0.043 w_c^{1.5} \sqrt{f_c'}, \text{ MPa}\right) \tag{3.3}$$

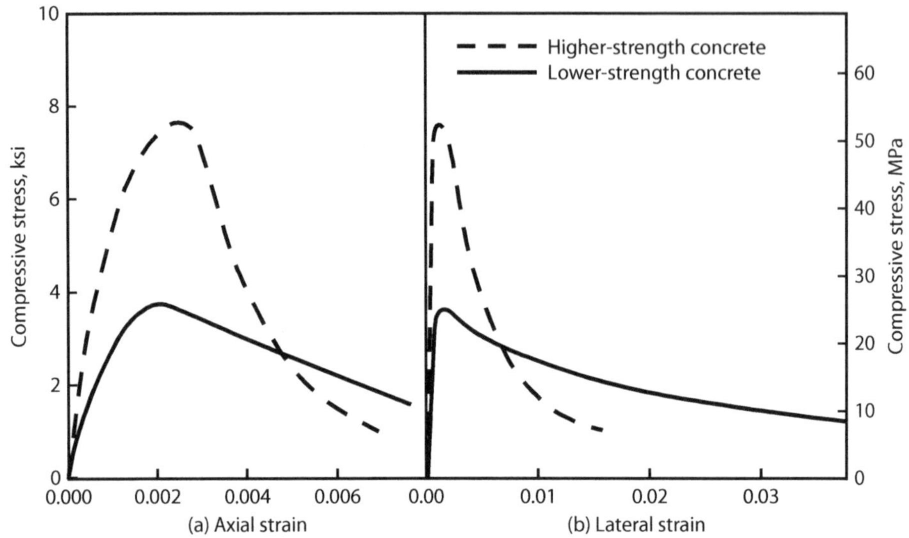

Figure 3.6 Axial stress versus axial strain and lateral strain for normalweight concrete. (*After Ahmad and Shah, 1982, with permission of ASCE.*)

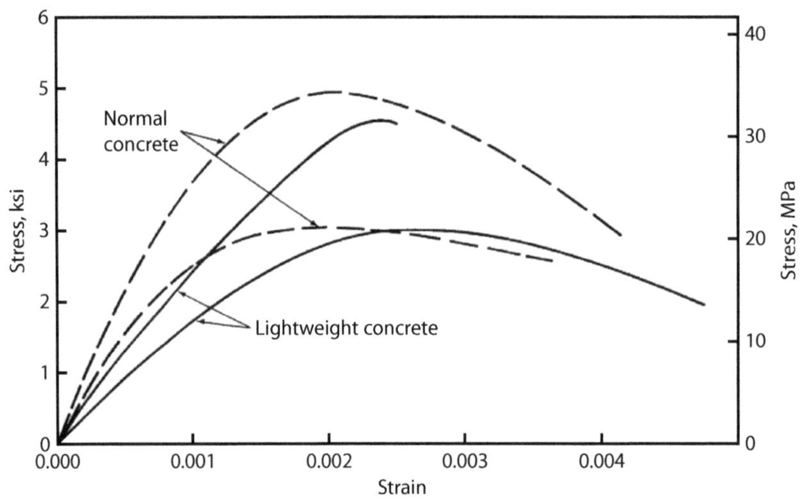

Figure 3.7 Stress–strain relations for normalweight and lightweight aggregate concretes. (*After Bresler, 1971, courtesy of American Concrete Institute.*)

in which w_c = density of concrete in lb/ft^3 (kg/m^3), which for normalweight aggregate can be simplified to

$$E_c = 57{,}000\sqrt{f_c'}, \text{ psi}$$
$$\left(4700\sqrt{f_c'}, \text{MPa}\right) \tag{3.4}$$

ACI 363R-92 (1992) reported the following equations, which provide a better fit for higher concrete strengths.

$$E_c = \left(40{,}000\sqrt{f_c'} + 1.0\times 10^6\right)\left(\frac{w_c}{145}\right)^{1.5}, \text{psi}$$

$$\left(\left(3320\sqrt{f_c'} + 6900\right)\left(\frac{w_c}{2300}\right)^{1.5}, \text{MPa}\right) \tag{3.5}$$

Figure 3.8 compares results of Eqs. (3.3) and (3.5) with test data.

3.4.2 Tensile Strength

The stress–strain relation for concrete in tension is nearly linear to the tensile strength, followed by rapid reduction in tensile resistance with increasing strain. Tensile strength of concrete, which is usually measured in the split cylinder test, is much less than the compressive strength. Based on results reported by

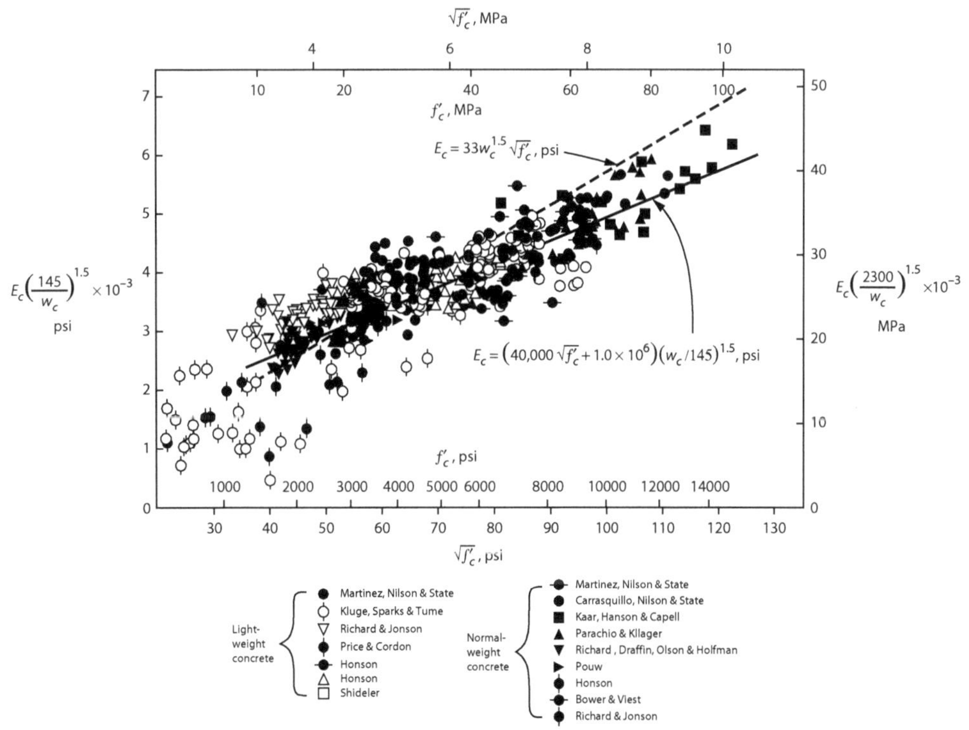

FIGURE 3.8 Modulus of elasticity versus concrete strength. (*After Martinez et al., 1982, as reported by ACI 363R-92, 1992.*)

Carrasquillo et al. (1981), ACI 363R-92 (1992) recommends Eq. (3.6) for tensile strength of concrete having compressive strength in the range 3000 to 12,000 psi (21 to 83 MPa).

$$f_{sp} = 7.4\sqrt{f'_c},\text{psi} \quad \left(0.59\sqrt{f'_c},\text{MPa}\right) \tag{3.6}$$

The apparent principal tensile strength of concrete in reinforced concrete panels subjected to in-plane normal and shear stresses is less than the strength given by Eq. (3.6). The discrepancy can be attributed to effects of multi-axial stress conditions (Section 3.6.1). Vecchio and Collins (1986) recommend tensile strength of 4 $\sqrt{f'_c}$ psi $\left(0.33\sqrt{f'_c}\,\text{MPa}\right)$ and Hsu (1993) recommends $3.75\sqrt{f'_c}$ psi $\left(0.33\sqrt{f'_c}\,\text{MPa}\right)$ for reinforced concrete sections.

3.4.3 Strain Rate Effects

Tensile and compressive strengths reported above are based on standard test specimens loaded at a relatively slow rate. Higher loading rates can increase the apparent strengths. Figure 3.9 shows the effect on compressive strengths from various sources. A simple expression that approximates the trend is

$$\text{Dynamic increase factor} = 1 + \frac{\sqrt[4]{\dot{\varepsilon}}}{2} \tag{3.7}$$

For strain rates typical of seismic loading some modest strength enhancements are indicated. Soroushian et al. (1986) showed that the rate effect depends on moisture, and Ahmad and Shah (1985) and Bing et al. (2000) found that the

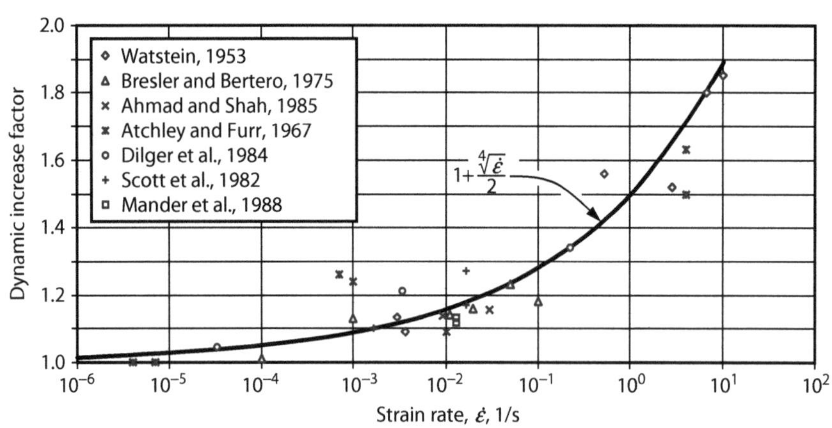

FIGURE 3.9 Strain rate effect on compressive strength of normal-strength concrete.

dynamic increase factor decreased for higher strength concrete. Ross et al. (1996) showed that the rate effect was more pronounced for concrete in tension. Given the range of variables and the relatively small effect for typical seismic loading rates, rate effects generally are ignored for routine seismic assessments and designs.

3.5 Behavior in Uniaxial Cyclic Loading

Micro-cracks in concrete grow under repeated loading, leading to continual softening of the stress–strain response under repeated loadings. For structures subjected to millions of cycles, *high-cycle fatigue* of concrete, with relatively small stress range, is an important consideration in design (ACI 215R-92, 1992). For structures subjected to large strains under earthquake loading, damage progresses more rapidly, leading to concerns for *low-cycle fatigue*. Such effects should be considered in developing nonlinear hysteretic models of concrete structures subjected to earthquake loading.

Figure 3.10 shows the uniaxial behavior of plain concrete compression members under cyclic loading. Three different loading behaviors are illustrated:

- *Cycles to the monotonic stress–strain relation* (Figure 3.10*a*): Each loading cycle causes damage that is apparent in the nonlinearity of the loading branch. Subsequent unloading occurs at a slope less than the initial modulus, and ends with residual strain at zero stress. Reloading causes additional damage, such that additional strain is required before the reloading branch reaches the monotonic stress–strain relation. After a few cycles, the accumulated strain reaches the strain capacity of the material and failure occurs. The monotonic stress–strain relation is effectively the envelope for the cyclic behavior.
- *Cycles to constant stress* (Figure 3.10*b*): The damage in each loading cycle results in a gradual accumulation of strain under cyclic loading. If the peak

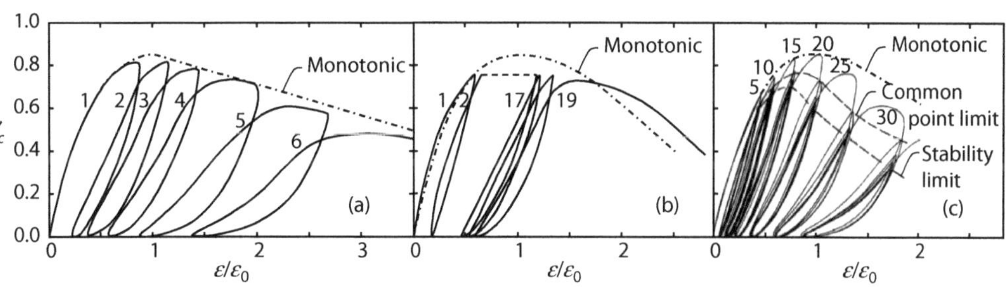

FIGURE 3.10 Uniaxial cyclic compressive behavior of normalweight concrete: (*a*) cycles to the monotonic stress–strain relation; (*b*) cycles to constant stress; (*c*) cycles to constant strain. (*After Karsan and Jirsa, 1969, with permission of ASCE.*)

stress is lower than about 75% of the monotonic compressive strength, the strain accumulation stabilizes and failure does not occur (this assumes low-cycle fatigue loading involving a limited number of cycles). If the peak stress is higher than about 75% of the monotonic compressive strength, strain accumulation leads to eventual failure.

- *Cycles to constant strain* (Figure 3.10*c*): Under this loading, the second cycle intersects the unloading branch of the first cycle at a point defined as the *common point limit*. Subsequent cycles reach progressively lower stress, but the stress eventually stabilizes at a *stability limit* (for low-cycle fatigue). If the cycles are repeated at increased strain amplitude, the pattern is repeated. Karsan and Jirsa (1969) used the monotonic envelope, the common point limit curve, and the stability limit curve to define an analytical hysteresis model for concrete under cyclic loading.

3.6 Behavior in Multi-axial Stress States

Concrete in real structures may be subjected to complex stress states resulting from restrained volume change and applied loads. For example, simple beams under transverse loading experience stresses due to combined flexure and shear, as well as stresses from bearing at the supports and from bond between longitudinal reinforcement and concrete. In columns and shear walls these actions are further complicated by the addition of axial loads. Later chapters will consider these subjects in detail. This chapter describes how concrete as a material responds to multi-axial stress and strain states.

3.6.1 Plain Concrete in Biaxial Stress State

The biaxial stress state is defined in Figure 3.11 by principal stresses f_1, f_2, and f_3, where f_1 and f_2 are positive in compression and $f_3 = 0$. Kupfer et al. (1969) report tests in which loading began in an unstressed state, and then the stresses f_1 and f_2 were increased proportionally until failure. The continuous curve in Figure 3.11 shows the strength envelope for various ratios f_1/f_2. The envelope was relatively insensitive to concrete compressive strength for the range tested [2700 to 8400 psi (19 to 58 MPa)]. Point **a** corresponds to a uniaxial compression test along axis 1. Point **b** corresponds to a uniaxial tension test along axis 1. Point **c** demonstrates how a small tensile stress in the 2 direction causes significant reduction in the compressive stress at failure in the 1 direction. Under biaxial compression with $f_1 = f_2$ (point **d**), the strength increase over the uniaxial compressive strength f_c' is about 15%, whereas the maximum strength increase is about 25% for $f_1/f_2 \approx 2$ (point **e**). Failure at points **d** and **e** is due to tensile splitting failure in the unrestrained 3 direction. To achieve greater strength, it is necessary to restrain dilation in the 3 direction. Section 3.6.3 covers behavior under triaxial loading.

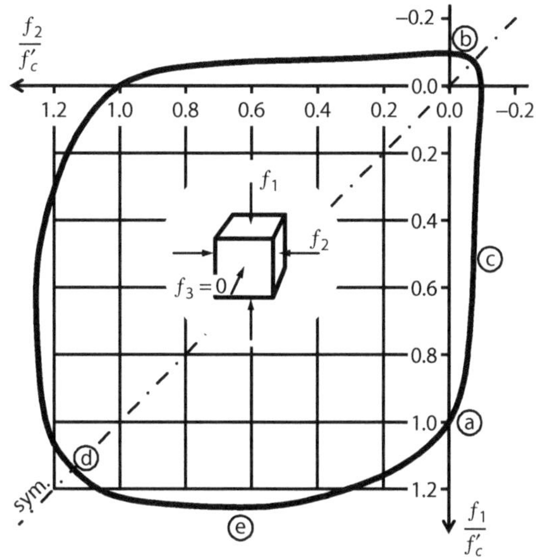

Figure 3.11 Concrete strength envelope under biaxial loading. (*After Kupfer et al., 1969, courtesy of American Concrete Institute.*)

Figure 3.12 plots relations between strain and stress ratio f_1/f'_c. The continuous curve in Figure 3.12*a* is for uniaxial compressive loading. Poisson's ratio, $\nu = \varepsilon_2/\varepsilon_1 = \varepsilon_3/\varepsilon_1$, can be scaled from the data for uniaxial loading. Typical values are $\nu = 0.15$ to 0.20. As compressive stress approaches f'_c, the rate of transverse straining increases rapidly. As shown in Figure 3.12*b*, the volumetric strain $(\Delta V/V = \varepsilon_1 + \varepsilon_2 + \varepsilon_3)$ increases beyond this point, which is a sign of more extensive micro-cracking just before failure. Similar results are shown for biaxial

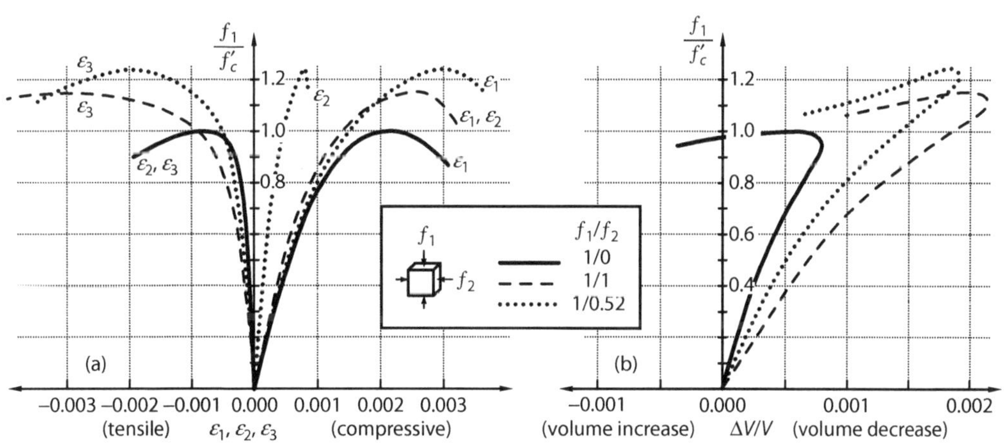

Figure 3.12 Biaxial compressive loading: (*a*) stress–strain relations; (*b*) volumetric strain. (*After Kupfer et al., 1969, courtesy of American Concrete Institute.*)

compression with $f_1/f_2 = 1/1$ and 1/0.52. The stress at which the minimum volume is achieved is sometimes referred to as the critical stress. Observed values of the critical stress usually occur in the range of $0.75f'_c$ to $1.0f'_c$, the value apparently depending on details of the load and measurement apparatus, with typical values commonly taken as $0.85f'_c$ to $0.9f'_c$.

3.6.2 Reinforced Concrete in Biaxial Loading

In a reinforced concrete section, concrete may be subjected to complex strain fields including tensile strains that cause concrete cracking. To study this behavior, reinforced concrete panels have been tested under in-plane shear and normal stresses (Vecchio and Collins, 1986; Hsu, 1993). A principal finding of these studies is that the compressive stress capacity of concrete reduces as it is subjected to increasing transverse tensile strain.

Figure 3.13 shows results from the *modified compression field theory* (Vecchio and Collins, 1986). According to this model, the concrete principal compressive strength f_1 is a function not only of the principal compressive strain ε_1 but also of

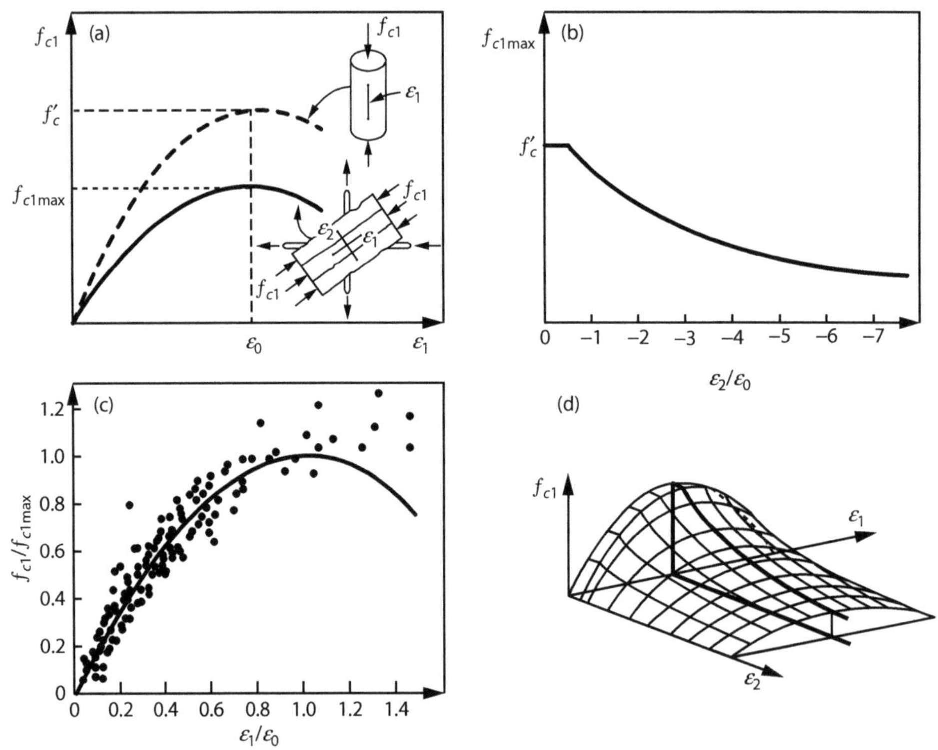

FIGURE 3.13 Biaxial response of reinforced concrete under in-plane loading: (*a*) stress–strain relationship for cracked concrete in compression; (*b*) relation between maximum compressive stress and transverse tensile strain; (*c*) correlation of test data for cracked concrete in compression; (*d*) three-dimensional representation of compressive stress–strain relationship. (*After Vechio and Collins, 1986, courtesy of American Concrete Institute.*)

the principal tensile strain ε_2 (Figure 3.13*a*). The relation between maximum principal compressive stress capacity f_{c1max} and the principal tensile strain ε_2 (defined negative for tension) is shown in Figure 3.13*b*. This results in a family of concrete compressive stress–strain relationships for different values of ε_2 as shown in Figure 3.13*d*.

In Figure 3.13*a*, the stress–strain relationship is

$$f_{c1} = \zeta \cdot f'_c \left[2\left(\frac{\varepsilon_1}{\varepsilon_0}\right) - \left(\frac{\varepsilon_1}{\varepsilon_0}\right)^2 \right] \tag{3.8}$$

in which

$$\zeta = \frac{f_{c1max}}{f'_c} = \frac{1}{0.8 - 0.34\varepsilon_2/\varepsilon_0} \tag{3.9}$$

and ε_2 is negative in tension.

The *softened truss model* (Hsu, 1993) proposes a softening coefficient in the form

$$\zeta = \frac{f_{c1max}}{f'_c} = \frac{0.9}{\sqrt{1 - 1.2\varepsilon_2/\varepsilon_0}} \tag{3.10}$$

Whereas the modified compression field theory applies the softening coefficient only to the stress, the softened truss model also applies the softening coefficient to the strain, such that the strain corresponding to peak stress f_{c1max} decreases with increasing (more negative) principal tensile strain.

The parabolic form of Eq. (3.8) is adopted from traditional stress–strain models for concrete, and is applicable for longitudinal strains up to approximately $1.75\varepsilon_0$. The softening coefficient of Eq. (3.9) or (3.10) will be useful in explaining observed shear strength of structural concrete members (see Chapter 7).

3.6.3 Plain Concrete in Triaxial Stress State

Section 3.6.1 shows that the maximum compressive strength under biaxial loading is only moderately higher than the uniaxial compressive strength, and there is little effect on strain capacity. Only a moderate effect is observed because the concrete is unconfined in the 3 direction, such that failure in that direction is not restrained. Tests show that if we can confine the concrete in all directions by applying external compressive stress, the behavior of concrete can be dramatically changed.

Figure 3.14 plots stress–strain curves obtained from concrete cylinders uniformly confined by external stress in the 2 and 3 directions. The tests were conducted by applying an external hydrostatic pressure to concrete cylinders sealed in a rubber membrane, and then loading in the longitudinal direction. Note that both the stress and the strain capacities are significantly increased by confinement.

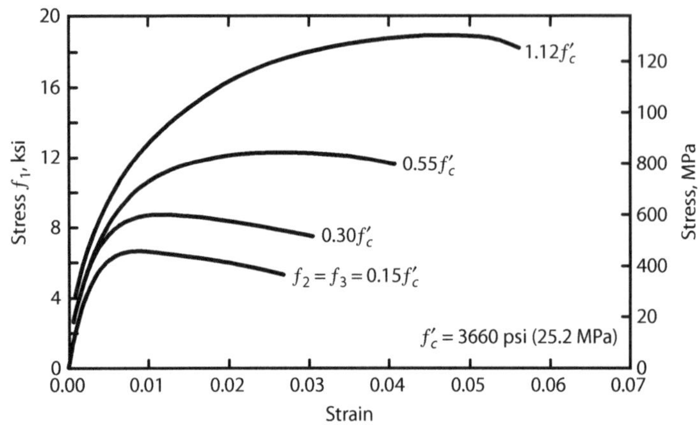

FIGURE 3.14 Stress–strain relationships for normalweight concrete confined by hydrostatic pressure and then loaded in axial compression. (*After Richart et al., 1928, courtesy of the University of Illinois at Urbana–Champaign Archives.*)

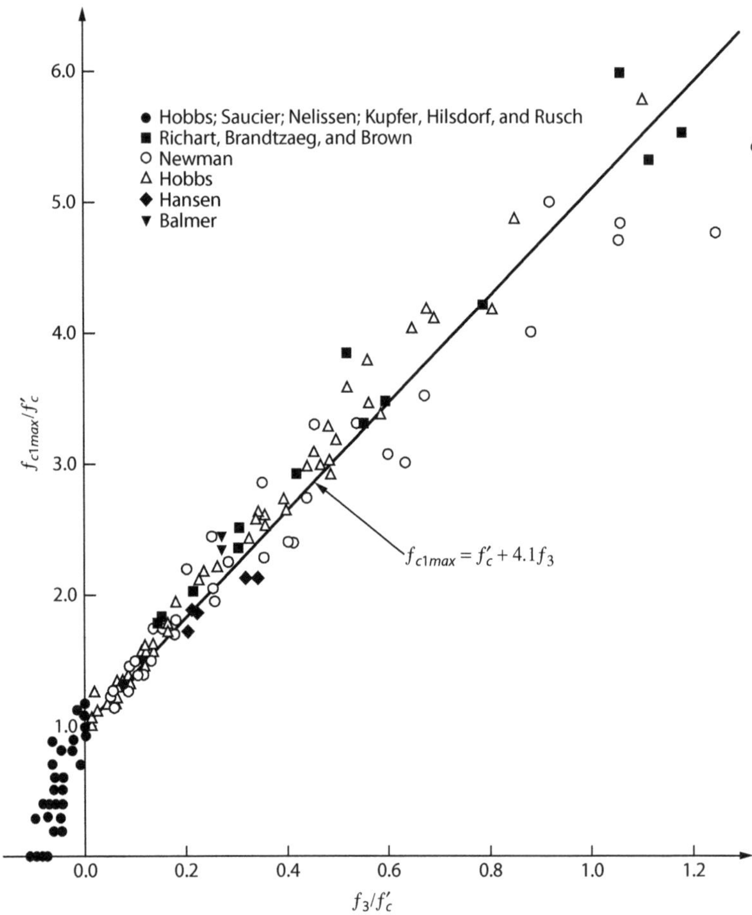

FIGURE 3.15 Comparison of measured confined concrete strength versus Eq. (3.11). (*After Hobbs et al., 1977.*)

Richart et al. (1928) proposed that the axial strength of confined concrete could be represented by

$$f_{c1max} = f'_c + kf_3 \tag{3.11}$$

in which f_3 is the smallest principal compressive stress (positive in compression). The test data show that the value of k is slightly higher for low confinement stress than for higher confinement stress. Richart et al. (1928) recommended a single value of $k = 4.1$. Subsequent comparisons with larger data sets demonstrate that Eq. (3.11) with $k = 4.1$ adequately models the data trend over a large range of confinement stresses (Figure 3.15).

3.7 Fiber-Reinforced Concrete

Plain concrete has low tensile strength and low strain capacity at fracture. In structural concrete, these weaknesses are overcome by strategic placement of reinforcing bars or prestressing steel. An alternative for improving tensile behavior of concrete is addition of fibers, creating what is known as fiber-reinforced concrete. Fibers can be of steel, glass, synthetic, or natural fiber materials. In this text we are primarily concerned with steel fibers. According to ACI 544.1R-96 (1996), these are short, discrete lengths of steel having aspect ratio (length/diameter) ranging from about 20 to 100, with any of several cross sections, and with sufficiently small length such that they can be randomly dispersed in an unhardened concrete mixture using usual mixing procedures. Steel fibers are produced by cutting and crimping wire or sheets, or by machining or melt-extraction processes (Figure 3.16).

Fiber reinforcement improves behavior of concrete by bridging across cracks. Small fibers are especially effective for micro-cracks, whereas longer fibers are more effective for cracks on a structural scale. Figure 3.17*a* shows results of tension tests on concrete reinforced with different volume ratios of steel fibers. Figure 3.17*b* shows results of beam tests for concrete using three different fibers, either all fibers of the same type in one beam or a hybrid mixture. In this example, the hybrid mixture at given volume ratio develops superior performance compared with performance for individual fibers having the same volume ratio. Fiber-reinforced concretes that develop strain-hardening behavior after cracking are commonly referred to as *high-performance fiber-reinforced concrete* (HPFRC).

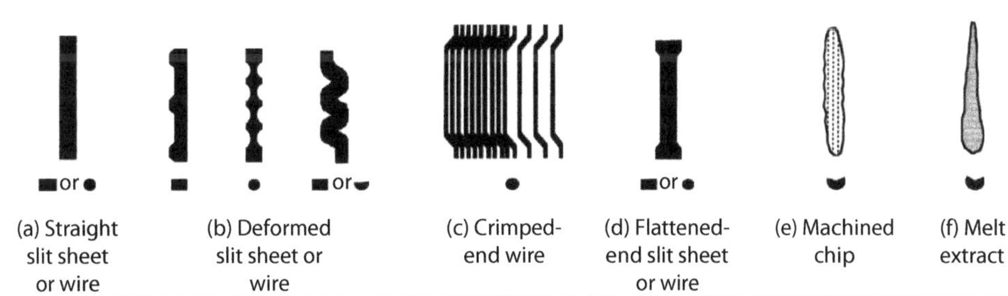

FIGURE 3.16 Various steel fiber geometries. (*After ACI 544.1R-96, 1996, courtesy of American Concrete Institute.*)

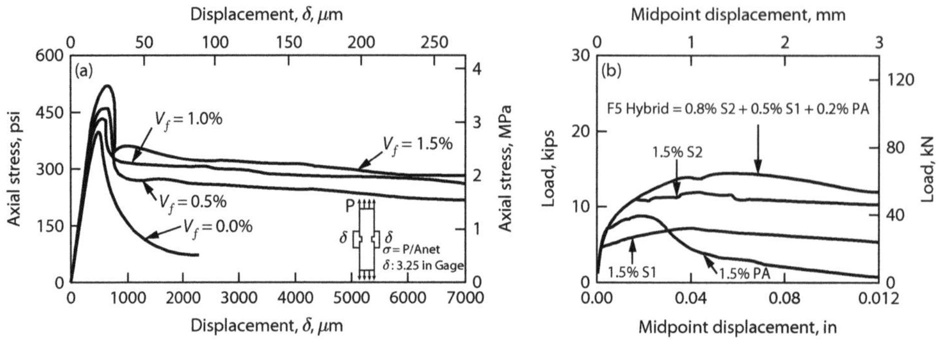

FIGURE 3.17 Load–displacement relations for fiber-reinforced concrete: (*a*) tension tests with different volume ratios of steel fibers (*after ACI 544.1R-96, 1996*); (*b*) beam load-deflection relations for hybrid fiber-reinforced concrete (*after Blunt and Ostertag, 2009, courtesy of American Concrete Institute*).

ACI 318 (2014) considers steel fiber-reinforced concrete acceptable for shear resistance if conditions (1), (2), and (3) are satisfied: (1) The weight of deformed steel fibers is at least 100 lb/yd^3 (582 N/m^3); (2) the residual strength obtained from flexural testing at a midspan deflection of 1/300 of the span length is at least 90% of the measured first-peak strength obtained from a flexural test and 90% of the strength corresponding to f_r; and (3) the residual strength obtained from flexural testing at a midspan deflection of 1/150 of the span length is at least 75% of the measured first-peak strength obtained from a flexural test and 75% of the strength corresponding to f_r. The quantity f_r is the modulus of rupture (tensile strength inferred from bending test).

3.8 Chapter Review

This chapter reviewed the composition and structure of concrete as an aid to understanding its mechanical properties. For seismic design purposes, the most important mechanical properties are stiffness, strength, and deformation capacity. Although highly variable, the modulus of elasticity of concrete can be related to the compressive strength as measured in a standard cylinder test. That compressive strength is an index for the strength potential of concrete in an actual structure, where strength varies as a function of curing conditions, time, and stress state. We should anticipate that actual concrete strength in a structure will usually exceed the specified compressive strength because of some of these effects and because concrete suppliers usually target a strength that is higher than the specified value. Under some loading conditions, however, such as under biaxial tension–compression loading, the effective compressive strength will be less than the uniaxial compressive strength. Strain capacity of concrete at the compressive strength is limited to approximately 0.002 to 0.003. That strain capacity, as well as the compressive strength, can be increased by confining the concrete along the three principal directions.

Chapter 4 will further explore the behavior of confined concrete, and later chapters will use information from Chapters 2 to 4 to derive models for behavior of structural concrete members under typical loadings.

References

ACI 214R-11 (2011). "Guide to Evaluation of Strength Test Results of Concrete," *ACI Manual of Concrete Practice*, American Concrete Institute, Farmington Hills, MI, 16 pp.

ACI 214.4R-10 (2010). "Guide for Obtaining Cores and Interpreting Compressive Strength Results," *ACI Manual of Concrete Practice*, American Concrete Institute, Farmington Hills, MI, 17 pp.

ACI 215R-92 (1992). "Considerations for Design of Concrete Structures Subjected to Fatigue Loading" (reapproved 1997), *ACI Manual of Concrete Practice*, American Concrete Institute, Farmington Hills, MI, 24 pp.

ACI 228.1R-03 (2003). "In-Place Methods to Estimate Concrete Strength," *ACI Manual of Concrete Practice*, American Concrete Institute, Farmington Hills, MI, 44 pp.

ACI 228.2R-98 (1998). "Nondestructive Test Methods for Evaluation of Concrete in Structures" (reapproved 2004), *ACI Manual of Concrete Practice*, American Concrete Institute, Farmington Hills, MI, 62 pp.

ACI 308R-01 (2001). "Guide to Curing Concrete" (reapproved 2008), *ACI Manual of Concrete Practice*, American Concrete Institute, Farmington Hills, MI, 31 pp.

ACI 318 (2014). *Building Code Requirements for Structural Concrete (ACI 318-14) and Commentary*, American Concrete Institute, Farmington Hills, MI.

ACI 363R-92 (1992). "Report on High-Strength Concrete," *ACI Manual of Concrete Practice* (reapproved 1997), American Concrete Institute, Farmington Hills, MI, 55 pp.

ACI 544.1R-96 (1996). "Report on Fiber Reinforced Concrete" reapproved 2009, *ACI Manual of Concrete Practice*, American Concrete Institute, Farmington Hills, MI, 66 pp.

Ahmad, S., and S.P. Shah (1982). "Complete Triaxial Stress-Strain Curves for Concrete," *Journal of the Structural Division, ASCE*, Vol. 108, No. ST4, pp. 728–742.

Ahmad, S., and S.P. Shah (1985). "Behavior of Hoop Confined Concrete under High Strain Rates," *ACI Journal*, Vol. 82, No. 5, pp. 634–647.

ASCE 41 (2013). *Seismic Evaluation and Retrofit of Existing Buildings*, American Society of Civil Engineers, Reston, VA.

ASTM C469 (2010). *Standard Test Method for Static Modulus of Elasticity and Poisson's Ratio of Concrete in Compression*, ASTM International, 5 pp.

Atchley, B.L., and H.L. Furr (1967). "Strength and Energy Absorption Capabilities of Plain Concrete under Dynamic and Static Loadings," *ACI Journal*, Vol. 64, No. 11, pp. 745–756.

Bing, L., R. Park, and H. Tanaka (2000). "Constitutive Behavior of High-Strength Concrete under Dynamic Loads," *ACI Structural Journal*, Vol. 97, No. 4, pp. 619–629.

Blunt, J. and C. Ostertag (2009). "Deflection Hardening and Workability of Hybrid Fiber Composites," *ACI Materials Journal*, Vol. 106, No. 3, pp. 265–272.

Bresler, B (1971). "Lightweight Aggregate Reinforced Concrete Columns," *Lightweight Concrete*, ACI Publication SP-29, American Concrete Institute, Farmington Hills, MI, pp. 81–130.

Bresler, B., and V.V. Bertero (1975). "Influence of Strain Rate and Cyclic Loading on Behavior of Unconfined and Confined Concrete in Compression," *XVII Jornadas Sudamericanas de Ingenieria Estructural, V Simposio Panamericana de Estructuras, Caracas,* 8 al 12 de Diciembre de 1975.

Carrasquillo, R.L., A.H. Nilson, and F.O. Slate (1981). "Properties of High Strength Concrete Subjected to Short-Term Loads," *ACI Journal*, Vol. 78, No. 3, pp. 171–178.

CSA (2004). *Design of Concrete Structures*, CSA A23.3-04, Canadian Standards Association, Mississauga, Canada.

Dilger, W.H., R. Koch, and R. Kowalczyk (1984). "Ductility of Plain and Confined Concrete under Different Strain Rates," *ACI Journal*, Vol. 81, No. 1, pp. 73–81.

Hobbs, D.W., C.D. Pomeroy, and J.B. Newman (1977). "Design Stresses for Concrete Structures Subject to Multi-axial Stresses," *The Structural Engineer*, The Institution of Structural Engineers, Vol. 55, No. 4, pp. 151–164.

Hsu, T.T.C. (1993). *Unified Theory of Reinforced Concrete*, CRC Press, Boca Raton, FL, 313 pp.

Karsan, I.D. and J.O. Jirsa (1969). "Behavior of Concrete under Compressive Loadings," *Journal of the Structural Division*, ASCE, Vol. 95, No. ST12, pp. 2543–2564.

Kupfer, H., H.K. Hilsdorf, and H.Rusch (1969). "Behavior of Concrete under Biaxial Stress," *ACI Journal*, Vol. 66, No. 8, pp. 656–666.

Mander, J.B., M.J.N. Priestley, and R. Park (1988). "Observed Stress-Strain Behavior of Confined Concrete," *Journal of Structural Engineering*, ASCE, Vol. 114, No. 8, pp. 1827–1849.

Martinez, S., A.H. Nilson, and F.O. Slate (1982). *Spirally-Reinforced High-Strength Concrete Columns*, Research Report No. 82-10, Department of Structural Engineering, Cornell University, Ithaca, NY.

Mehta, P.K., and P.J. Monteiro (2014). *Concrete*, 4th ed., McGraw-Hill Professional, New York, NY, 675 pp.

Mindess, S., J.F. Young, and D. Darwin (2003). *Concrete*, 2d ed., Prentice-Hall, Upper Saddle River, NJ, 644 pp.

Monteiro, P.J., and J.P. Moehle (1995). *Stiffness of Reinforced Concrete Walls Resisting In-Plane Shear—Tier 2: Aging and Durability of Concrete Used in Nuclear Power Plants*, EPRI TR-102731, T2, Electric Power Research Institute, Palo Alto, CA.

PCA (1988). *Design and Control of Concrete Mixtures*, 13th ed., Portland Cement Association, Skokie, IL.

Richart, F.E., A. Brandtzaeg, and R.L. Brown (1928). *A Study of the Failure of Concrete under Combined Compressive Stresses*, Bulletin No. 185, Engineering Experiment Station, University of Illinois, Urbana, IL, 104 pp.

Ross, C.A., D.M. Jerome, J.W. Tedesco, and M.L. Hughes (1996). "Moisture and Strain Rate Effects on Concrete Strength," *ACI Materials Journal*, Vol. 93, No. 3, pp. 293–300.

Scott, B.D., R. Park, and M.J.N. Priestley (1982). "Stress-Strain Behavior of Concrete Confined by Overlapping Hoops at Low and High Strain Rates," *ACI Journal Proceedings*, Vol. 79, No. 1, pp. 13–27.

Soroushian, P., K-B. Choi, and A. Alhamad (1986). "Dynamic Constitutive Behavior of Concrete," *ACI Journal*, Vol. 83, No. 2, pp. 251–259.

TBI (2010). *Guidelines for Performance-Based Seismic Design of Tall Buildings*, Report No. 2010/05, Pacific Earthquake Engineering Research Center, University of California, Berkeley, CA, 84 pp.

Vecchio, F.J., and M.P. Collins (1986). "The Modified Compression Field Theory for Reinforced Concrete Elements Subjected to Shear," *Journal of the American Concrete Institute*, Vol. 83, No. 2, pp. 219–231.

Watstein, D. (1953). "Effect of Straining Rate on the Compressive Strength and Elastic Properties of Concrete," *Journal of the American Concrete Institute*, Vol. 49, No. 8, pp. 729–744.

Wischers, G. (1979). "Applications and Effects of Compressive Loads on Concrete," *Betontechnische Berichte* 1978, Betone Verlag GmbH, Dusseldorf, pp. 31–56.

Wood, S.L. (1991). "Evaluation of Long-Term Properties of Concrete," *ACI Materials Journal*, Vol. 88, No. 6, pp. 630–643.

CHAPTER 4

Confined Concrete

4.1 Preview

Chapter 3 introduced the concept of confined concrete, and how confining stress can produce substantial increases in compressive strength and compressive strain capacity. In this chapter, we explore how transverse reinforcement can be strategically detailed and placed to achieve similar effects. This chapter explains the mechanism of confined concrete, introduces analytical stress–strain relations, and examines how confinement is affected by detailing and materials properties. In subsequent chapters, confined concrete will be treated as one of the fundamental materials properties that enable concrete structures to perform appropriately when subjected to earthquake ground motions.

4.2 Behavior of Confined Concrete Sections

We begin our study by reviewing the behavior of four laboratory test columns loaded to failure in axial compression. The four columns have the same gross concrete dimensions [12 × 12 × 36 in (305 × 305 × 914 mm)] and were cast from the same batch of concrete. Specimen P is a plain concrete prism. The other columns contain Grade 60 longitudinal and transverse reinforcement in various configurations (Figure 4.1). The transverse reinforcement has higher density in the end regions than in the middle, such that failure will be forced to occur in a central test region.

Figure 4.1 shows concrete stress–strain relations derived from the tests. Concrete axial force is obtained by subtracting the longitudinal reinforcement compressive force from the total axial force. Concrete stress is then defined as the concrete axial force divided by the concrete area corrected for cross section taken up by longitudinal reinforcement, with the gross concrete dimensions taken to the outside of the concrete shell before spalling and to the outside of the hoops after spalling. Strain is the average strain over the central test region.

As shown, the plain concrete specimen P reaches peak load at longitudinal strain 0.003, which corresponds to the strain at peak stress in companion test cylinders, followed by rapid loss of strength as the plain concrete shears across an inclined plane. The reinforced columns all fail in a more ductile manner. By the time the longitudinal strains reach 0.003, all instrumented transverse reinforcement

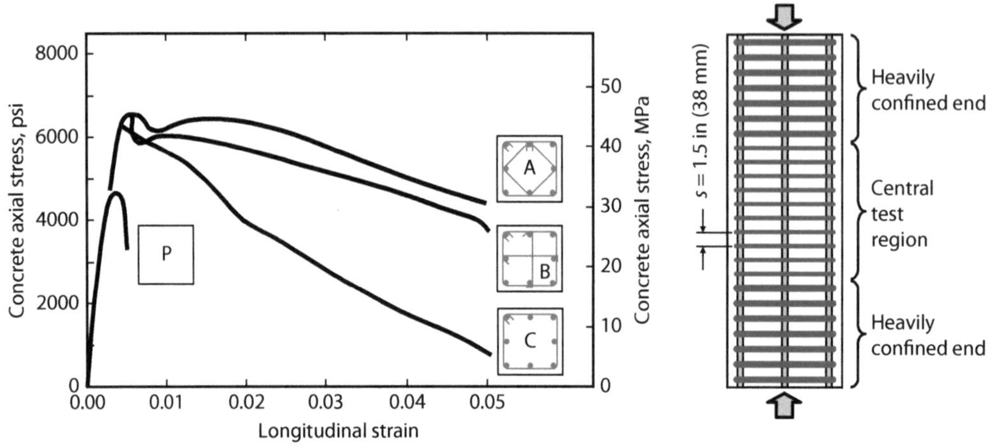

FIGURE 4.1 Stress–strain response of plain concrete (P) and three confined concrete cross sections. (*After Moehle and Cavanagh, 1985, used with permission from ASCE.*)

has yielded. Cover concrete starts to spall at longitudinal strain around 0.004, evident by vertical splitting of the cover concrete and a reduction in load resistance. Columns A and B regain some of the lost strength as strains are further increased, whereas column C continues to shed load with increasing strain. Failure occurs when the perimeter hoops fracture, accompanied by longitudinal reinforcement buckling and partial straightening of the 90° hooks on crossties of column B.

The failure sequence described above is fairly typical for confined columns loaded monotonically in compression. Closely spaced transverse reinforcement acts to confine the core, imparting enhanced longitudinal strain capacity. Columns with single perimeter hoops such as column C do not behave as well as similar columns having crossties or overlapping hoops (columns A and B). Ultimate failure typically is due to tensile failure of the hoops.

In the remainder of this chapter, we will examine the mechanisms of confined concrete, developing models for strength and stress–strain behavior. In later chapters, we will use these models to study the behavior and design requirements of structural members with confined concrete.

4.3 Mechanism of Concrete Confinement

In Chapter 3 we showed that concrete axial strength and strain capacity can be increased by applying compressive stresses in the directions transverse to the axial loading direction. This was demonstrated for plain concrete test specimens that were sealed in a rubber membrane and subjected to hydrostatic oil pressure in a testing laboratory. This form of confinement is known as *active confinement*, that is, it is achieved by actively applying an external confinement pressure. Though this approach may be effective in a laboratory setting, it is not a practical technology for application in actual buildings. An alternative approach of wrapping prestressed cylindrical columns with prestressed wire is effective (Martin, 1968), but this approach also remains impractical for most construction.

A more practical approach for application in concrete structures is to use transverse reinforcement to resist the dilation that occurs naturally when concrete is compressed. This form of confinement is known as *passive confinement* because it is activated only when the concrete is subjected to load. As shown in Figure 4.1, passive confinement by steel reinforcement can be effective in improving deformation capacity of concrete columns. We next study the mechanisms by which passive confinement occurs so that we can better understand how to use its effects in design and analysis problems.

4.3.1 Passive Confinement of Concrete

Studies of confined concrete began early in the 20th century. Richart et al. (1928) describe laboratory tests of concrete cylinders confined by closely spaced spiral reinforcement. The tests demonstrated that passive confinement is relatively ineffective in the linear range of response because the Poisson effect results in low transverse strains. As the axial stress approaches the plain concrete stress capacity, however, more extensive micro-cracking in the matrix begins to produce greater transverse strain (see Section 3.6.1), and localized plastic deformation begins to occur. This plastic material must be supported laterally if it is to continue carrying load. Initially, lateral support is provided by tensile strength of any surrounding elastic concrete and by lateral pressure from the spiral reinforcement. As support of the elastic concrete is lost through additional splitting action, the requisite support is gained in the spiral. As the action continues, the concrete increasingly behaves as a plastic mass, carrying further load in proportion to the confining stress applied by the spiral. If the spiral begins to soften due to yielding, or if the transverse strains become very large, the rate of increase in confinement stress becomes insufficient to support increasing axial load, and failure begins.

Figure 4.2 from Richart et al. (1929) shows stress–strain relations measured during one of the tests. The longitudinal stress–strain relations show the type of nonlinearity expected from standard cylinder tests. They also show that concrete in the upper portions is softer than concrete at the bottom, apparently because of differences in consolidation over the column height. The radial strains approach the yield strain (0.002) as the axial stress reaches the unconfined compressive strength f_c'. Radial strains increase rapidly as the spiral yields. Strain hardening of the spiral beyond the yield point enables the concrete to support additional axial force, but eventually it proves insufficient to support increasing force and deformation, and the concrete fails.

Richart et al. (1929) noted that spiral reinforcement did not provide uniform confining stress but instead confined the concrete through bearing at discrete locations along the column height. This led to high localized bearing stresses between spiral reinforcement and concrete, and resulted in indentations of the spiral steel into the concrete core. As a consequence, the spiral strains were moderately smaller than the concrete radial strains, though sufficient to yield. Stress concentrations can be reduced by using closely spaced, small-diameter wire/bars of moderate strength, rather than using widely spaced, large-diameter, high-strength wire/bars.

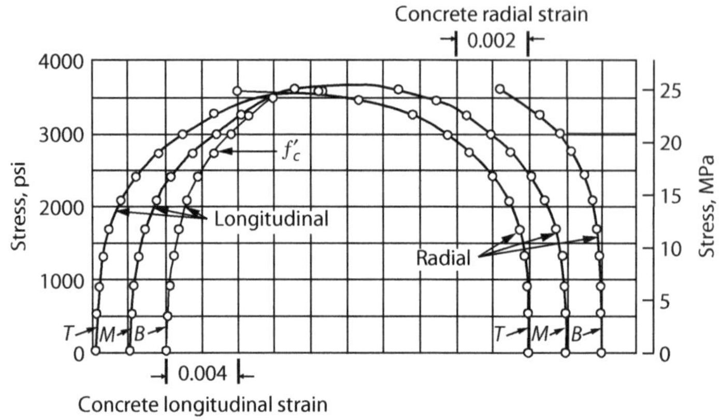

Figure 4.2 Concrete stress–strain relations for concrete cylinder confined by spiral reinforcement. T, M, and B indicate top, middle, and bottom third of cylinder height. Concrete was normalweight with compressive strength of 2725 psi (18.8 MPa). Steel yield stress was 61,000 psi (421 MPa). (*After Richart et al., 1929, courtesy of the University of Illinois at Urbana–Champaign.*)

4.3.2 Columns with Spiral and Circular Hoop Reinforcement

In the preceding section we observed that concrete could be effectively confined by using closely spaced spiral reinforcement. This section develops algebraic expressions for concrete confined by steel spiral or circular hoop reinforcement.

Confinement Stress

Figure 4.3 illustrates the column geometry. The concrete core dimension measured to the outside of the spiral or hoop reinforcement is D.[1] The core is confined by spiral or circular hoop reinforcement having cross-sectional area A_{sp} and pitch or spacing s. Spacing s is small relative to the column diameter D.[2] The cross section also contains longitudinal reinforcement of total cross-sectional area A_{st}.

[1]In the reinforced concrete literature, D is sometimes defined as the core diameter measured to the centerline or the inside of the spiral or circular hoop reinforcement. Such definitions are consistent with cover concrete spalling geometry, which commonly leaves a concrete core with diameter somewhat smaller than the outside diameter of the spiral or hoops. Here we adopt the alternative definition of D, also commonly used in the literature, because it is simpler from a design perspective and produces acceptably accurate results.

[2]The design and behavior of spiral reinforcement and circular hoop reinforcement for flexural, axial, and shear loadings are essentially identical. Therefore, to simplify the presentation in the text that follows, spiral reinforcement will be emphasized, but the discussion applies equally to circular hoop reinforcement. Distinctions between the two types of reinforcement will be necessary for reinforcement detailing and for torsional loading (the spiral pitch results in torsion directionality that does not arise for circular hoops).

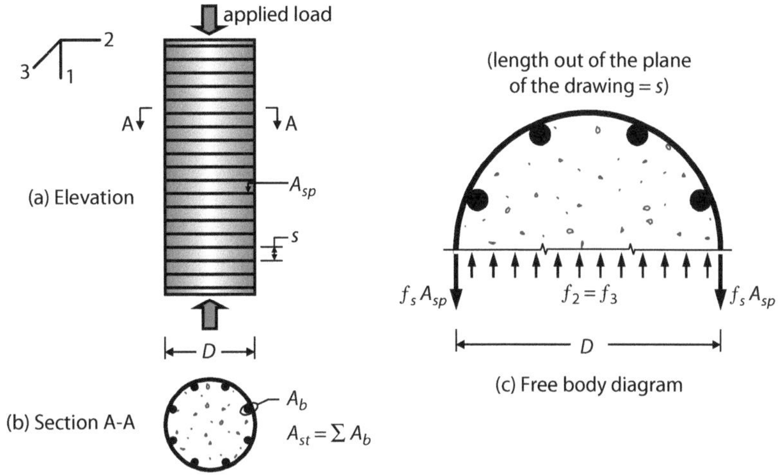

FIGURE 4.3 Concrete column confined by circular hoop or spiral reinforcement: (*a*) elevation; (*b*) section A-A; (*c*) free body diagram of a slice of the column core of thickness *s*.

Dilation under the action of axial compression results in spiral steel stress f_s and opposing confining stresses $f_2 = f_3$. Force equilibrium requires

$$f_2\, Ds = 2 f_s A_{sp} \tag{4.1}$$

The ratio of the volume of spiral reinforcement to the total volume of core confined by the spiral, ρ_s, is defined as

$$\rho_s = \frac{\pi D A_{sp}}{\pi D^2 s/4} = \frac{4A_{sp}}{Ds} \tag{4.2}$$

Solving Eq. (4.1) for f_2 and using Eq. (4.2) results in

$$f_2 = f_3 = \frac{2 f_s A_{sp}}{Ds} = \frac{\rho_s f_s}{2} \tag{4.3}$$

The confining stresses $f_2 = f_3$ defined by Eq. (4.3) will be used to determine the compressive strength of spiral or circular hoop-confined concrete. Before addressing compressive strength, however, two additional aspects require consideration. These are the effects of wide spiral pitch (or circular hoop spacing) and the limits of the confinement reinforcement stress f_s.

Spiral Spacing, Arching Action, and Confinement Effectiveness

Most building codes require that spiral reinforcement be closely spaced, such that the confining pressure of the spiral can be considered essentially uniform. For completeness, however, we also consider the case of wider spacing of spiral or circular hoop reinforcement. Although spirals and circular hoops have different configuration, and this difference results in slightly different behavior, the effects

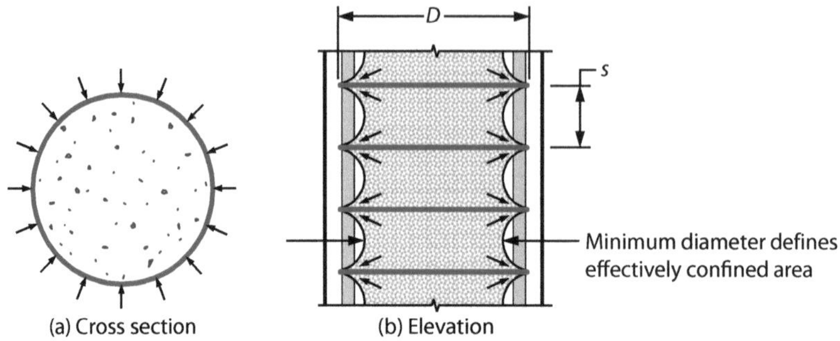

FIGURE 4.4 Confinement action in circular hoop-confined column: (*a*) cross section at level of hoop showing radial pressure from hoop; (*b*) elevation through diameter showing spread of internal stresses, arching action, and effectively confined core dimension.

are not sufficiently different to warrant a separate consideration here. The discussion here emphasizes the simpler geometry of circular hoops.

As illustrated in Figure 4.4, a circular hoop produces a concentrated ring of confinement stresses acting radially inward at the level of each hoop. Away from the hoops, the circumference of the core is a free surface that is not confined by reinforcement. Within the core, the concentrated ring pressure spreads longitudinally and radially until the internal core pressure is essentially uniform. It is this internal uniform pressure that is defined by Eq. (4.3).

The behavior described above can be viewed in an alternative way. As a column core is loaded axially, dilation of the core concrete presses outward against the confinement reinforcement. Concrete particles located near the hoop reinforcement are well supported by the hoops, whereas particles located away from the hoops do not find a solid reaction and spall away. As this action continues a series of arches develops internally, spanning vertically from one ring to another. This behavior is known as *arching action*. The wider the longitudinal spacing *s*, the deeper the arch extends into the concrete core, and the lower is the *confinement effectiveness*. We sometimes refer to the effectively confined core area as the minimum cross-sectional area of the confined core as defined by the arching action.

For small values of the ratio s/D, as is usually the case, arching action in a spiral-confined column does not reduce the confinement effectiveness significantly. As s/D increases, however, an adjustment for confinement effectiveness should be made. Various approaches have been recommended in the literature (Iyengar et al., 1970; Ahmad and Shah, 1982; Martinez et al., 1984; Mander et al., 1988a). Iyengar et al. (1970) observed that hoops were practically ineffective when the spacing was equal to the core diameter, and proposed that confinement effectiveness varies linearly from a maximum value for zero spacing to zero for hoop spacing equal to the core diameter. Based on this recommendation, we can define a confinement effectiveness factor, k_e, defined by

$$k_e = 1 - \frac{s}{D} \tag{4.4}$$

Although arching action refers strictly to a reduction in the effective core area, a more convenient and widely used alternative is to consider the full core area to be effective and instead reduce the confinement stress using factor k_e, resulting in effective confinement stress f_{2e} defined by

$$f_{2e} = k_e f_2 = k_e \frac{\rho_s f_s}{2} \tag{4.5}$$

We will use this effective confinement stress to define strength of the confined core.

Stress in the Confinement Reinforcement

Thus far we have expressed the confinement stress in terms of the stress f_s in the transverse reinforcement. The amount of transverse straining that occurs when a confined core is compressed is typically sufficient to produce yielding of the confinement reinforcement (Figure 4.2). In some cases, however, yielding of the confinement reinforcement is not observed (Sheikh and Uzumeri, 1980; Ahmad and Shah, 1982; Martinez et al., 1984; Bing et al., 2001; Paultre and Légeron, 2008). This is especially a consideration where high-strength confinement reinforcement is used.

Razvi and Saatcioglu (1999) concluded that the stress developed in the transverse reinforcement depends on the concrete strength and on the volumetric ratio and configuration of the confining reinforcement. For confinement reinforcement having f_{yt} up to 200,000 psi (1400 MPa) (the maximum value in the tests considered), they proposed

$$\begin{aligned} f_s &= E_s\left(0.0025+0.21\left(\sqrt[3]{\frac{k_e\rho_s}{f'_c}}\right)\right) \le f_{yt} \le 200{,}000,\ \text{psi} \\ f_s &= E_s\left(0.0025+0.04\left(\sqrt[3]{\frac{k_e\rho_s}{f'_c}}\right)\right) \le f_{yt} \le 1400,\ \text{MPa} \end{aligned} \tag{4.6}$$

Except where high-strength reinforcement is used, the stress f_s given by Eq. (4.6) will be equal to the yield stress f_{yt}.

Strength of the Confined Core

From Section 3.6.3, we can recall the relation between strength of confined concrete and confining pressure, that is

$$f'_{cc} = Cf'_c + 4.1 f_3 \tag{4.7}$$

Equation (4.7) introduces the term C to account for differences between in-place concrete strength and standard cylinder compressive strength f'_{cc}. For large-scale columns common in building construction, Richart and Brown (1934) recommended $C = 0.85$. This value is widely adopted in building codes, including ACI 318 (2014). CSA (2004) adjusts the value as a function of concrete compressive

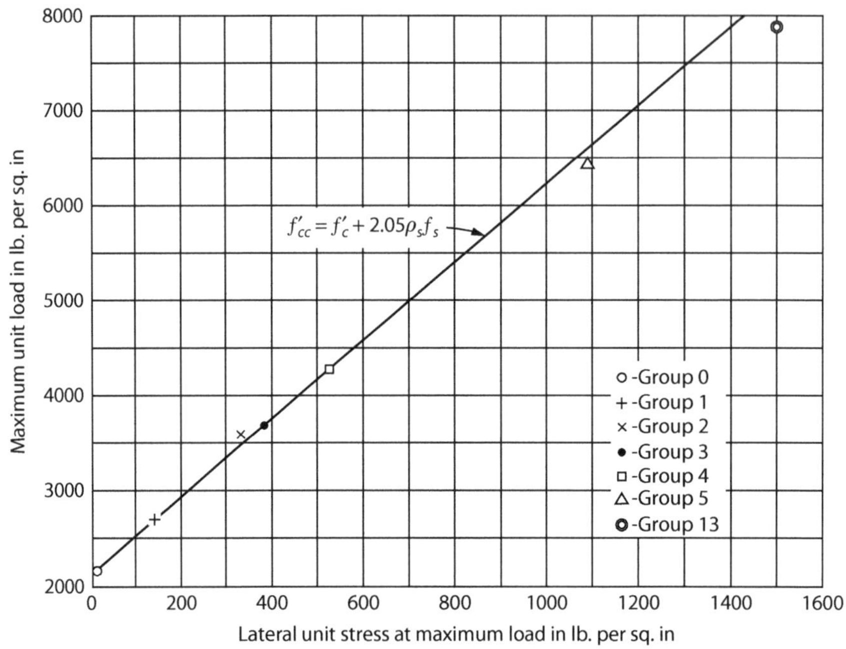

FIGURE 4.5 Comparison of test data with Eq. (4.8). Data points represent averages of multiple tests at the ultimate loading point. (*After Richart et al., 1929, courtesy of the University of Illinois at Urbana–Champaign.*)

strength, specifying $C = 0.85 - \lambda_c f'_{cc} \geq 0.67$, in which $\lambda_c = 0.00001$ (psi) or 0.0015 (MPa).

Combining Eqs. (4.5) and (4.7) results in

$$f'_{cc} = Cf'_c + 2.05k_e \rho_s f_s \tag{4.8}$$

Richart et al. demonstrated the applicability of Eq. (4.8) to spirally confined cylinders (Richart et al., 1929) and large-scale columns (Richart et al., 1934) for different stresses f_s that developed as loading progressed. Figure 4.5 compares results at the ultimate load for the smaller cylinder tests, for which the coefficient C has been taken equal to 1.0.

Several researchers have noted that the coefficient 4.1 in Eq. (4.7) underestimates the confined concrete strength for lower values of the confinement stress (Richart et al., 1928; Ahmad and Shah, 1982; Sheikh and Uzumeri, 1982; Mander et al., 1988a). Figure 4.6 presents one model that recognizes this effect (after Mander et al., 1988a). In this figure, f_{2e} and f_{3e} are the effective confining stresses in the 2 and 3 directions, with $f_{2e} \geq f_{3e}$, C is a constant to account for differences between in-place concrete strength and standard cylinder compressive strength (commonly taken as 0.85 for columns, 1.0 otherwise), and f'_{cc} is the confined concrete strength. The failure surface for biaxial loading ($f_{3e} = 0$) is the leftmost curve. The curves moving toward the right are for increasing f_{3e}, with the rightmost curve

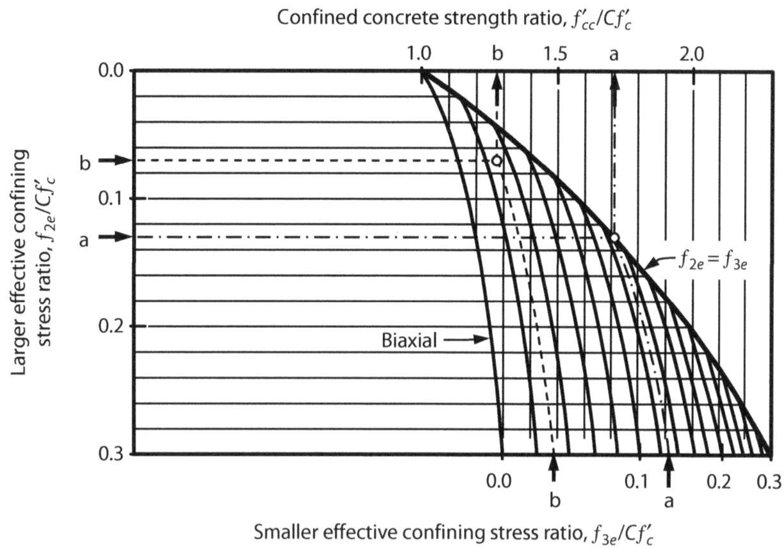

FIGURE 4.6 Confined concrete strength as function of effective confinement stresses. (*After Mander et al., 1988a.*)

corresponding to $f_{2e} = f_{3e}$. As an example, for $f_{2e}/Cf'_c = f_{3e}/Cf'_c = 0.13$, the dash-dot lines intersect at $f'_{cc}/Cf'_c = 1.72$. As another example, for $f_{2e}/Cf'_c = 0.07$ and $f_{3e}/Cf'_c = 0.03$, the dashed lines intersect at $f'_{cc}/Cf'_c = 1.27$. Use of this chart to determine confined concrete compressive strength is acceptable as an alternative to Eq. (4.8).

Example 4.1. A cylindrical column has diameter of 24 in (610 mm), 8 No. 9 (29) longitudinal bars, and No. 3 (10) spiral reinforcement at pitch of 1.75 in (44 mm). Strength of concrete measured in companion cylinders is 5000 psi (34 MPa) and reinforcement is A706 Grade 60 (420). Determine the strength of the confined core.

Solution

Assuming concrete cover of 1.5 in (38 mm) over the spiral, the core diameter is 21 in (533 mm), resulting in spiral reinforcement ratio $\rho_s = (4)(0.11\ \text{in}^2)/(21\ \text{in})(1.75\ \text{in}) = 0.012$. Confinement effectiveness factor is $k_e = (1 - 1.75\ \text{in}/21\ \text{in}) = 0.92$. Expected yield strength of A706 Grade 60 reinforcement is 1.15×60 ksi = 69 ksi (476 MPa). Using Eq. (4.8), confined concrete strength is $f'_{cc} = (0.85)(5000\ \text{psi}) + 2.05(0.92)(0.012)(69{,}000) = 5800$ psi (40 MPa).

Alternatively, we can use Figure 4.6. By this approach, from Eq. (4.5) we have $f_{2e} = (0.92)(0.012)(69{,}000\ \text{ksi})/2 - 380$ psi (2.6 MPa). Thus, $f_{2e}/Cf'_c = 380\ \text{psi}/(0.85 \times 5000\ \text{psi}) = 0.089$. From Figure 4.6, we can read $f'_{cc}/Cf_c = 1.5$. Therefore, by this method, $f'_{cc} = 6400$ psi (44 MPa).

4.3.3 Columns with Rectilinear Hoop Reinforcement

The mechanisms of confinement for columns with rectilinear hoop reinforcement are essentially the same as those for columns with spiral or circular hoop reinforcement. Rectilinear hoop reinforcement, however, does not produce uniform ring pressure on the confined core, and therefore it tends to be less

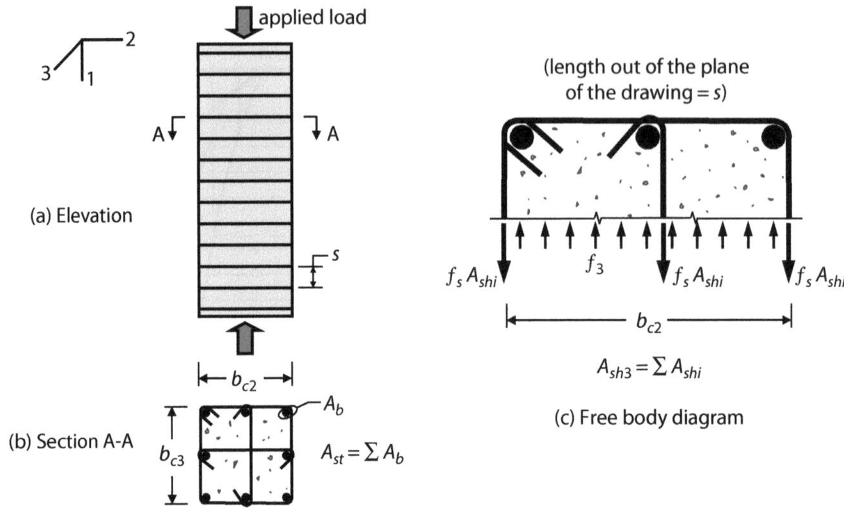

FIGURE 4.7 Concrete column confined by rectilinear hoop reinforcement: (*a*) elevation; (*b*) section A-A; (*c*) free body diagram of a slice of the column core of thickness *s*.

effective than an equivalent volume of spiral reinforcement, at least for columns with small to moderate sized diameters.[3]

Confinement Stress

Figure 4.7 illustrates the geometry for a column with square cross section. For a more general rectangular cross section, the transverse reinforcement and cross-sectional dimensions in the 2 and 3 directions will not be equal. To allow for this more general case, we define b_{c2} and b_{c3} as the cross-sectional dimensions of the core measured to the outside of the hoops in the 2 and 3 directions, and A_{sh2} and A_{sh3} as the total cross-sectional areas of confinement reinforcement parallel to the 2 and 3 directions, respectively.

Following the derivations in Section 4.3.2, force equilibrium requires

$$f_3 b_{c2} s = \sum f_s A_{shi} = f_s A_{sh3} \tag{4.9}$$

Where legs of hoops are inclined at an angle α relative to a normal to the section cut, a component of the area equal to $A_{shi} \cos \alpha$ is used rather than A_{shi}. The ratio of volume of confinement reinforcement in the 3 direction to total volume of core confined by the hoops, ρ_{s3}, is defined as

$$\rho_{s3} = \frac{b_{c3} A_{sh3}}{b_{c2} b_{c3} s} = \frac{A_{sh3}}{b_{c2} s} \tag{4.10}$$

[3]Confinement of cylindrical cores and buckling restraint for compressed longitudinal bars derives from the curvature of the circular hoop or spiral. For very large diameter columns, the hoop curvature becomes small, raising concerns about the ability of a large-diameter hoop to resist localized bursting forces from the core or from buckling longitudinal bars.

Solving Eq. (4.9) for f_3 and using Eq. (4.10), we can write

$$f_3 = \frac{f_s A_{sh3}}{b_{c2} s} = \rho_{s3} f_s \tag{4.11}$$

Similarly, confinement stress in the orthogonal direction is given by

$$f_2 = \frac{f_s A_{sh2}}{b_{c3} s} = \rho_{s2} f_s \tag{4.12}$$

Stress f_s in the confinement reinforcement can be determined using Eq. (4.6).

Arching Action and Confinement Effectiveness

Equations (4.11) and (4.12) define average confinement stresses f_2 and f_3 in the 2 and 3 directions, respectively. Unlike the case of spiral reinforcement, these stresses are not necessarily equal nor are they uniformly applied around the perimeter. Figure 4.8*a* illustrates this for a square core confined by a single perimeter hoop. As the core dilates under axial load, lateral expansion is effectively resisted by axial rigidity of the hoop reinforcement at the corners. Away from the corners, expansion must be resisted by the flexural rigidity of the hoop reinforcement. Tests have shown that, for practical hoop sizes, the flexural rigidity is insufficient to produce appreciable pressure away from the corners (Burdette and Hilsdorf, 1971). Consequently, there is high confinement pressure at the corners of the core with little confinement pressure along the sides.

To improve the confinement of rectangular cross sections, additional hoop or crosstie legs can be added (Figure 4.8*b*). Each of these added legs should engage a longitudinal bar, serving to anchor the tie and spread the confining action longitudinally.

So far, we have considered only the confining pressure acting at the perimeter of the cross section, noting the concentrated forces associated with each hoop or crosstie leg. These concentrated forces will spread laterally into the concrete core, reaching a more nearly uniform pressure distribution further within the core (Figure 4.9*a* and *b*). Likewise, the stresses must spread longitudinally between hoop sets (Figure 4.9*c*). This arching action is similar to that described for cores

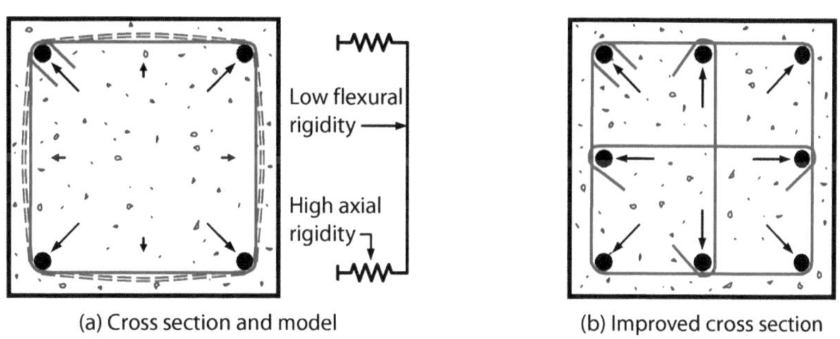

(a) Cross section and model (b) Improved cross section

Figure 4.8 Development of confinement pressure for square column with perimeter hoop.

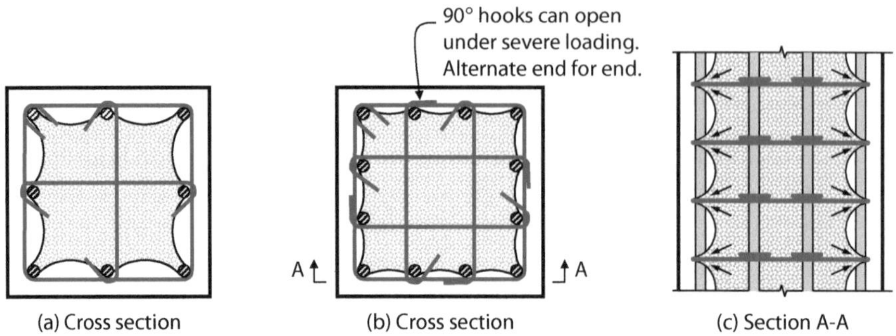

Figure 4.9 Confinement of concrete by reinforcement: (*a*) perimeter hoop with crossties; (*b*) perimeter hoop with added crossties to improve confinement effectiveness (note 90° hooks may not be fully effective under large core compressive strains); (*c*) arching action in column elevation.

confined by circular hoops (Section 4.3.2), except in the case of rectilinear confinement the arching action occurs in three dimensions.

Figure 4.10 illustrates three-dimensional arching in compression members confined by rectilinear hoop reinforcement. Where only perimeter hoops are provided at large spacing (Figure 4.10*a*), the effectively confined core is reduced considerably from the gross cross section. Additional hoop sets or crossties at reduced spacing can increase the effective cross section (Figure 4.10*b*).

Several analytical models for confined concrete include arching action and associated confinement effectiveness (Sheikh and Uzumeri, 1982; Mander et al., 1988a; Razvi and Saatcioglu, 1999; Paultre and Légeron, 2008). Here we adopt a hybrid approach of these that is simple to implement, can be physically interpreted, and applies to a wide range of materials and core geometries.

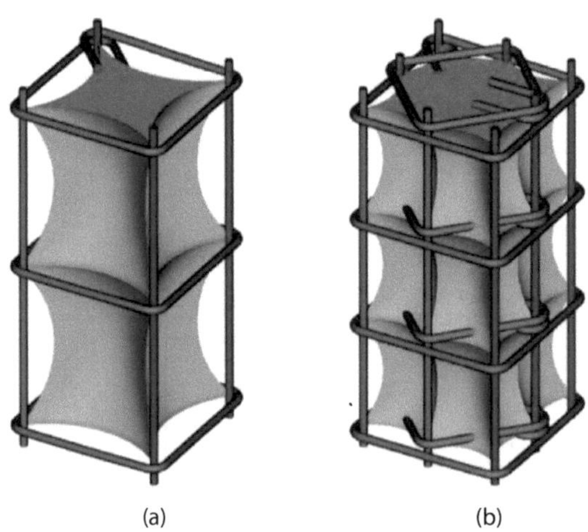

Figure 4.10 Three-dimensional arching action in rectangular prismatic column core. (*After Paultre and Légeron, 2008, used with permission from ASCE.*)

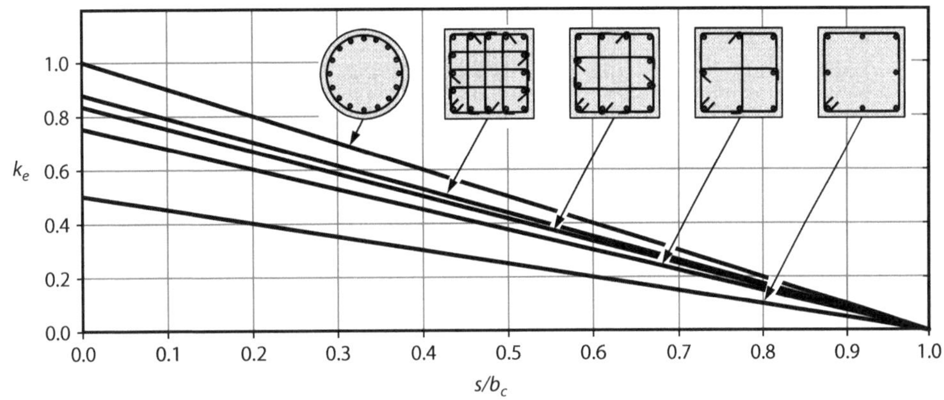

FIGURE 4.11 Confinement effectiveness for various confinement configurations and hoop spacings.

A confinement effectiveness coefficient is defined according to the configuration and longitudinal spacing of the hoops (Figure 4.11). The values of k_e vary linearly from a maximum for zero longitudinal spacing to zero for spacing equal to the core dimension. The lines were positioned to represent an average of the above-cited models in the typical range of interest for seismic designs. The line for circular hoops and spirals is from Eq. (4.4). The lines for rectilinear confinement were calculated for square cross sections only. Confinement effectiveness for other rectangular sections can be adequately estimated by averaging k_e values for the two orthogonal directions.

Confinement effectiveness as shown in Figure 4.11 can be expressed algebraically as

$$k_e = \frac{n_l - 2}{n_l}\left(1 - \frac{s}{b_c}\right) \tag{4.13}$$

in which n_l = number of longitudinal bars restrained by corners of hoops or legs of crossties around the column perimeter. For rectangular sections, for which s/b_c is different in the two orthogonal directions, the average of the s/b_c values in the two directions can be used. The expression $(n_l - 2)/n_l$ was first presented in Paultre and Légeron (2008).

Long rectangular sections, such as those that occur in structural walls, present a special case (Figure 4.12). In the long direction, the confinement stresses are nearly uniform for most of the wall length (only the end sections have arching action). Therefore, the effective confinement stress in that direction, f_{e2}, can be calculated assuming k_e = 1. In contrast, arching action in both the horizontal and vertical directions will affect confinement of the wall through its thickness. The model by Mander et al. (1988a) can be used to estimate the confinement effectiveness in the 3 direction (through the thickness). Figure 4.12 presents results for a range of variables.

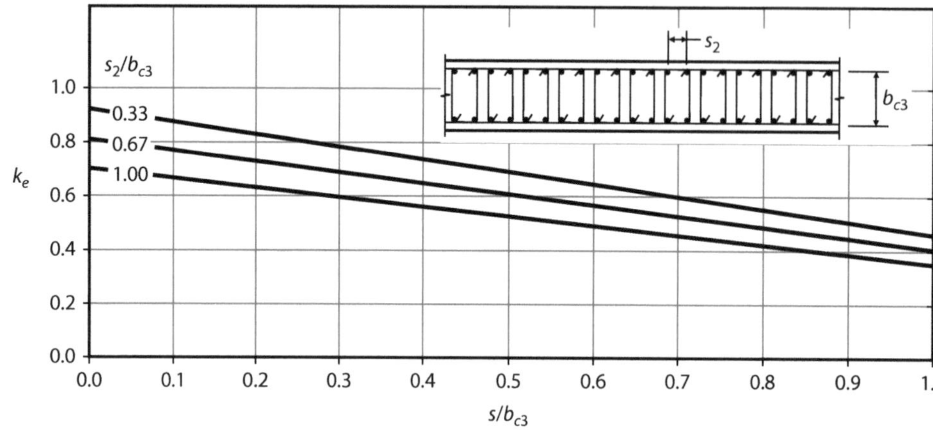

Figure 4.12 Confinement effectiveness for long rectangular sections.

Having determined the confinement effectiveness coefficient, the effective confinement stresses in the 2 and 3 directions are

$$f_{2e} = k_e f_2$$
$$f_{3e} = k_e f_3 \tag{4.14}$$

Strength of the Confined Core

Knowing f_{2e} and f_{3e} from Eq. (4.14), we can use the strength relation given by Figure 4.6 to determine the confined concrete strength f'_{cc}. The axial strength of a confined concrete core is thus $A_{ch} f'_{cc}$, in which A_{ch} is the cross-sectional area of the core measured to the outside of the hoops. The accuracy of the estimated confined concrete strength is presented in the next section following introduction of the loading rate effect.

Example 4.2. The rectangular cross section shown in Figure 4.13 has $f'_c = 5000$ psi (34 MPa) in companion cylinders and A706 Grade 60 (420) reinforcement. Calculate the strength f'_{cc} of the core concrete. Use expected material strengths. The solution is provided in tabular form.

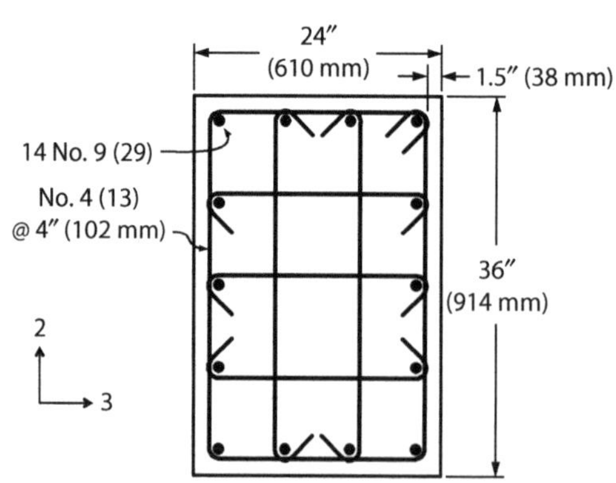

Figure 4.13 Column cross section.

Solution

Quantity	Calculation	Reference
f_3	5(0.20 in^2)(69,000 psi)/[(33 in)(4 in)] = 523 psi (3.61 MPa)	Eq. (4.11)
f_2	657 psi (4.53 MPa)	Eq. (4.12)
s/b_c average	(4 in/21 in + 4 in/33 in)/2 = 0.16	
k_e	[(14 − 2)/14] × (1 − 0.16) = 0.72	Eq. (4.13)
$f_{3e}; f_{2e}$	378 psi (2.6 MPa); 475 psi (3.3 MPa)	Eq. (4.14)
f_{3e}/Cf'_c	475 psi/[(0.85)(5000 psi)] = 0.11	
f_{2e}/Cf'_c	378 psi/[(0.85)(5000 psi)] = 0.089	
f'_{cc}/Cf'_c	1.57	Figure 4.6
f'_{cc}	(1.57)(0.85)(5000 psi) = 6700 psi (46 MPa)	

4.3.4 Loading Rate Effect

In Chapter 3 we saw how higher loading rate resulted in higher compressive strength in plain concrete. The same trend is observed for confined concrete sections (Bresler and Bertero, 1975; Scott et al., 1982; Mander et al., 1988b). The rate effect can be introduced into the previously defined models for confined concrete strength by adjusting the unconfined concrete strength f'_c by the dynamic increase factor introduced in Chapter 3, which we repeat here.

$$\text{Dynamic increase factor} = k_{dyn} = 1 + \frac{\sqrt[4]{\dot{\varepsilon}}}{2} \tag{4.15}$$

We can use the procedures of Sections 4.3.2 and 4.3.3 along with Eq. (4.15) to calculate confined concrete strength f'_{cc} for tests reported in the literature. Figure 4.14 compares measured and calculated strength ratios f'_{cc}/f'_c (confined concrete strength divided by companion cylinder strength) for a variety of confinement reinforcement configurations. Considering consolidation differences between concrete in larger columns versus standard cylinders, we normally would anticipate a strength ratio around 0.85 for plain concrete. For the data presented, measured strength ratios range from 1.15 to 2.14 times that value, indicating moderate to high levels of confinement. Dynamic increase factors were as large as 1.18. The calculated and measured strength ratios f'_{cc}/f'_c compare well for the range of confinement geometries, confinement quantities, and strain rates shown.

In Chapter 3 we also noted that the rate effect on unconfined concrete compressive strength is higher for concrete tested in "wet" condition and lower for high-strength concrete. In light of these additional variables not included in Eq. (4.15), and the relatively low effect for rates typical of seismic loadings, some writers have suggested that a reasonable approach would be to either ignore the rate effect or treat it as a constant. Dodd and Cooke (1992) suggested a factor 1.2 for normal-strength concrete, while Bing et al. (2000) suggested a factor 1.1 for

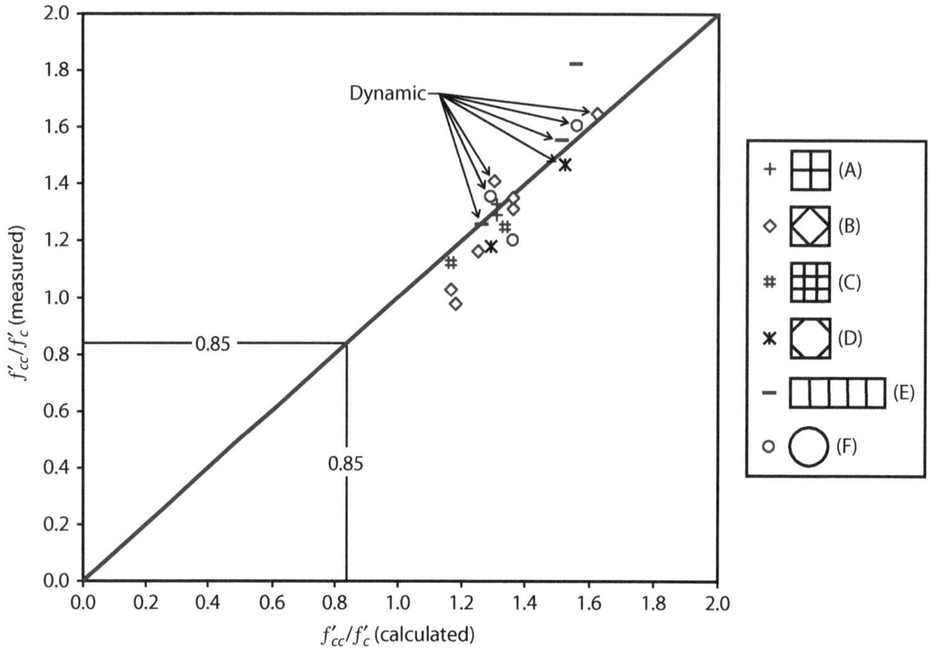

FIGURE 4.14 Measured and calculated ratios of f'_{cc}/f'_c for confined cores under static and dynamic loading rates. Dynamic tests had strain rates ranging from 0.0133/s to 0.0167/s. [*Data sources: A (Moehle and Cavanagh, 1985); B (Sheikh and Uzumeri, 1980; Scott et al., 1982; Moehle and Cavanagh, 1985); C (Sheikh and Uzumeri, 1980); D (Scott et al., 1982); E, F (Mander et al., 1988b).*]

high-strength concrete [plain concrete compressive strengths were approximately 10,000 psi (70 MPa)]. The rate effect is commonly ignored for routine seismic designs.

4.3.5 Aggregate Density Effect

Confinement is less effective for lightweight aggregate concrete than normalweight aggregate concrete (Bresler and Bertero, 1975; Ahmad and Shah, 1982; Martinez et al., 1984; Khaloo et al., 1999). Figure 4.15 shows results from one data set, for which confinement effect for lightweight concrete was about half that for normalweight concrete. In addition to aggregate density, confinement effect appears to depend on the type of lightweight aggregate (Khaloo et al., 1999) and the geometry of the confinement reinforcement (Manrique et al., 1979).

The reduction in confinement for lightweight aggregate concrete may arise from two effects. First, lightweight aggregate typically is softer than normalweight aggregate, so confinement reinforcement indents further into the softer concrete as it dilates, thereby producing less confining stress for unit dilation of the confined

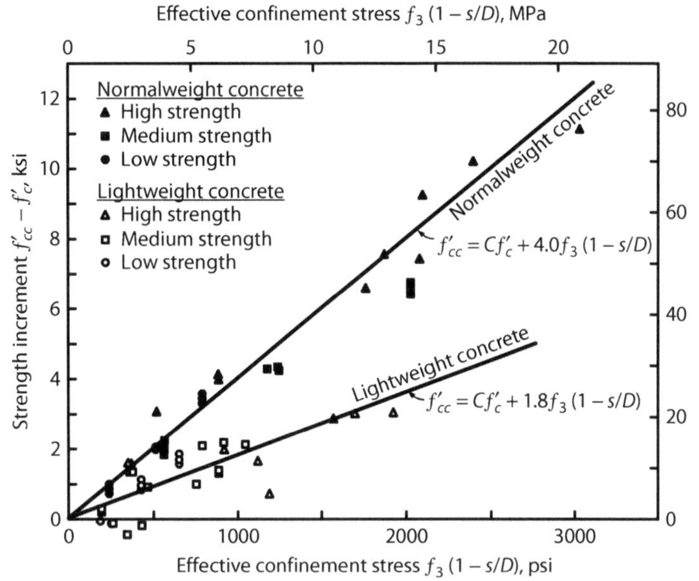

FIGURE 4.15 Confinement of normalweight and lightweight aggregate concrete. Compressive strengths of unconfined concrete ranged from 3100 to 9900 psi (21 to 68 MPa) for normalweight concrete and 3700 to 8200 psi (25 to 56 MPa) for lightweight concrete. (*After Martinez et al., 1984, courtesy of American Concrete Institute.*)

core. Second, and perhaps more important, lightweight aggregate tends to be the weak link in the concrete matrix, such that failure involves less dilation, especially for higher concrete compressive strengths.

4.3.6 Compressive Strength Effect

Passive confinement relies on dilation of the core concrete as axial stress reaches and surpasses the unconfined concrete stress capacity. It would not be surprising, then, that confinement would be less effective for high-strength concrete than for normal-strength concrete. Based on tests of high-strength confined columns, Bing et al. (2001) suggested the rate of increase in axial strength for a given amount of confinement was on the order of 70% of the rate for normal-strength concrete. Ahmad and Shah (1982) showed similar results. In contrast, Martinez et al. (1984) (Figure 4.15) and Razvi and Saatcioglu (1999) report the strength gain for normal-strength and high-strength concretes to be roughly equivalent.

Some applications of confined high-strength concrete will require large confinement stresses to achieve the intended ductile behavior, and this can lead to reinforcement congestion. High-strength transverse reinforcement can help reduce congestion in such cases. Tests show that confined concrete may reach its peak axial

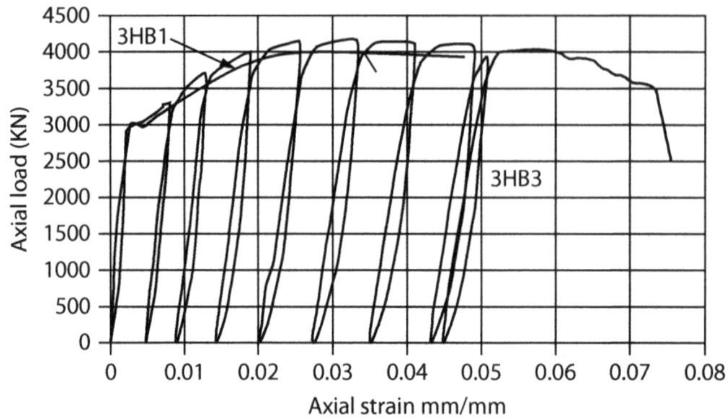

FIGURE 4.16 Stress–strain curves of specimens under cyclic and monotonic loading conditions. (*After Bing et al., 2001, courtesy of American Concrete Institute.*)

stress before the high-strength confinement reinforcement yields. Equation (4.6) is recommended to estimate the confinement steel stress corresponding to the peak stress of the confined concrete.

4.3.7 Cyclic Loading Effect

Cyclic loading tests have been reported by Bresler and Bertero (1975), Shah et al. (1983), Mander et al. (1988b), and Bing et al. (2001). With some limitations, the monotonic stress–strain curve is effectively the envelope for cyclic loading (Figure 4.16). Similar to unconfined concrete, if cycled near the monotonic stress capacity, progressive damage results in strain accumulation and eventual failure. Shah et al. (1983), Mander et al. (1988a), and Bing et al. (2001) present analytical models for cyclic loading effects.

4.3.8 Reinforcement Details

Sections 4.3.2 and 4.3.3 presented models for the passive confinement of concrete cores by spirals (or circular hoops) and rectilinear transverse reinforcement. Spirals were found to be highly effective in confining the core provided the pitch (or longitudinal spacing) was small. Rectilinear transverse reinforcement can also be effective if there are multiple hoop or crosstie legs along each face and if the hoop sets are closely spaced longitudinally.

Additionally, to be effective as confinement reinforcement, spirals, hoops, and crossties must be adequately developed in tension. Spiral reinforcement in a column should be effectively continuous along the length and extend a nominal distance into supporting members (ACI 318 requires the extension to be at least one and a half extra turns of the spiral steel at the ends). Lap splices may be permitted, but for seismic designs it is preferable to use a Type 2 mechanical splice along lengths where yielding is anticipated.

Where circular hoops are used, the ends of the hoop should be connected with a welded or mechanical splice. Alternatively, the ends should overlap by some

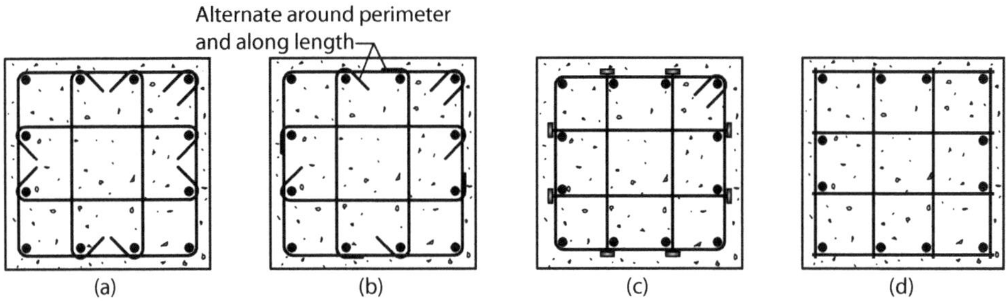

FIGURE 4.17 Confinement reinforcement detailing: (*a*) crossties with seismic hooks at both ends; (*b*) crossties with alternating seismic hooks and 90° hooks; (*c*) headed reinforcement as crossties; (*d*) welded ties (not recommended).

nominal distance [perhaps 6 in (150 mm)] and terminate with 90° hooks that engage a longitudinal column bar. In this latter case, overlaps of vertically adjacent hoops should be staggered vertically around the perimeter of the longitudinal bars. Along lengths where flexural yielding is anticipated, Type 2 mechanical splices should be used instead of lap splices.

Where rectilinear hoops and crossties are used, the perimeter hoops should be closed with standard hooks having 135° bends anchored into the confined core (Figure 4.17). Hooks with 90° bends should not be used on hoops because the tail on the hook is no longer anchored after spalling of cover concrete, allowing the hoop to open. Crossties are most effectively anchored with standard 135° or 180° hooks. To facilitate construction, some codes (e.g., ACI 318) permit the use of crossties with 135° hook on one end and 90° hook on the other end, with the ends alternated along the length and around the perimeter of the core (Figure 4.17*b*). This crosstie detail is less effective because the 90° hook is not embedded in the confined core, although it can provide adequate confinement to moderate strains (Moehle and Cavanagh, 1985; Tanaka and Park, 1987). For columns supporting axial loads higher than the balanced axial load, hooks having at least 135° bend should be used on both ends of crossties. All crossties and bends in hoops should engage longitudinal reinforcement to improve anchorage and to restrain buckling of the longitudinal bars. If a crosstie engages only the perimeter hoop, dilation of the confined core may result in a kink in the hoop at the point of engagement, thereby leading to reduced capacity (Moehle and Cavanagh, 1985).

Headed reinforcement can be used effectively as a crosstie to confine concrete (Ghali and Youakim, 2005) (Figure 4.17*c*). Although the head need not engage longitudinal reinforcement to be effective in providing confinement, it should fully engage that reinforcement for the purpose of providing support against bar buckling. Reinforcement with small heads may not be suitable for restraining large-diameter longitudinal bars. Headed reinforcement should not be used to make up perimeter hoops because the bars cannot be effectively interlocked due to interferences between the heads, and individual pieces may become disengaged under seismic loading. Some building codes (e.g., ACI 318) prohibit hoops made of interlocking bars but permit crossties made of headed bars. Welded bar mats (Figure 4.17*d*) have been used effectively to confine concrete in some tests (Saatcioglu and Grira, 1999), but fracture of welds has resulted in brittle behavior in other tests (Rood and Moehle, 2006). In welded bar mats conforming to ASTM A1022 or A1064, the welds are not required to develop the strength of the interconnected bars, with the result that weld failure could lead to brittle behavior. For this reason, ACI 318 prohibits use of welded bar mats that rely on the weld strength to confine concrete or restrain bar buckling. Thus, the detail shown in Figure 4.17*d* is not permitted.

Chapter 12 provides additional discussion and examples of transverse reinforcement detailing.

4.4 Analytical Modeling of Confined Concrete

Section 4.3 described models for the compressive strength of confined concrete cores. This section develops a complete stress–strain model for confined concrete that can be used for modeling the load-deformation behavior of structural components.

4.4.1 Strain at Peak Stress

For actively confined concrete, that is, concrete confined by externally applied hydrostatic pressure, the strain at the peak compressive stress increases with increasing confinement pressure (see Section 3.6.3). The same trend has been observed in tests of columns that are confined by transverse reinforcement (e.g., Richart et al., 1928; Martinez et al., 1984; Mander et al., 1988b; Saatcioglu and Razvi, 1992). The following expression for the strain at peak stress for confined concrete provides a reasonable approximation to the test results of Richart et al. (1928) and Balmer (1949).

$$\varepsilon_{cc} = \varepsilon_0\left(1 + 5\left(\frac{f'_{cc}}{Cf'_c} - 1\right)\right) \tag{4.16}$$

This same expression has been adopted in the stress–strain models proposed by Mander et al. (1988a) and Razvi and Saatcioglu (1999). In Eq. (4.16), the strain at peak stress in unconfined concrete, ε_0, will typically vary between 0.002 and 0.003 depending on the constituent materials. The constant C is introduced to adjust for differences in strengths of in-place concrete and standard cylinder concrete; C was not used in any of the previously cited references.

The strain at peak stress, ε_{cc}, varies inconsistently with strain rate, in some cases increasing and in others decreasing with increasing strain rate (Bresler and Bertero, 1975; Scott et al., 1982; Mander et al., 1988b). For that reason, it is considered acceptable to ignore the effect. In the stress–strain model introduced in Section 4.4.3, we will include the strain rate effect by substituting $k_{dyn}f'_c$ for f'_c and calculating f'_{cc} as described in Section 4.3.4. When these modified variables are introduced in Eq. (4.16), the effect is a slight reduction in the value of ε_{cc}. Combined with the increase in f'_{cc}, the effect on the stress–strain curve is to increase the post-peak unloading rate, consistent with laboratory tests.

4.4.2 Maximum Strain Capacity for Confined Concrete

The maximum strain capacity for confined concrete can be limited by tensile fracture of hoops or buckling of longitudinal reinforcement. In reinforced concrete members subjected to inelastic deformation reversals, maximum compressive strain commonly is limited by reinforcement buckling (and subsequent fracture when reloaded in tension). In reinforced columns under monotonic axial compression, maximum compressive strain commonly is limited by hoop fracture followed by reinforcement buckling. This section focuses on the strain at hoop fracture.

Blume et al. (1961) observed that the ratio of longitudinal to lateral strains was approximately 5:1 and therefore recommended that the maximum longitudinal strain should be $\varepsilon_{cu} = 5\varepsilon_{su} \le 0.01$, where ε_{su} = strain of transverse reinforcement at maximum stress in the concrete. The limit strain 0.01 was intentionally conservative, but believed to be adequate for most applications.

Various other models for ultimate strain capacity of confined concrete have been proposed. Most of such models have shown that ultimate strain capacity increases with increasing confining pressure, though the relations are inconsistent. For example, Scott et al. (1982) proposed that strain capacity was proportional to the confining pressure; Kaar et al. (1978) proposed that it was proportional to the square of the confining pressure; and Bing et al. (2001) proposed that it was proportional to the square root of the confining stress. In an alternative approach, Mander et al. (1988a) proposed an energy-balance model. None of the models have been shown to be universally applicable.

Figure 4.18 shows strain at hoop fracture as a function of the ratio f_{emin}/f'_c, where f_{emin} is the smaller of the effective confinement stresses f_{2e} and f_{3e}. The tests were all for normalweight aggregate, normal-strength concrete [3500 to 6000 psi (24 to 41 MPa)], and hoop reinforcement complying with A615 mechanical properties. An expression approximating the lower bound of the data is

$$\varepsilon_{cu} = 0.004 + \frac{f_{emin}}{4f'_c} \tag{4.17}$$

An alternative expression adopted from Scott et al. (1982) is

$$\varepsilon_{cu} = 0.004 + 0.075\frac{\rho_s f_{yt}}{f'_c} \tag{4.18}$$

in which ρ_s = total volume ratio of transverse reinforcement. These two expressions correlate similarly with the test data for columns with rectilinear hoops and crossties. The latter expression is more conservative than the former for spiral or circular hoop-confined columns. It is recommended that f_{yt} not exceed the value given by Eq. (4.6).

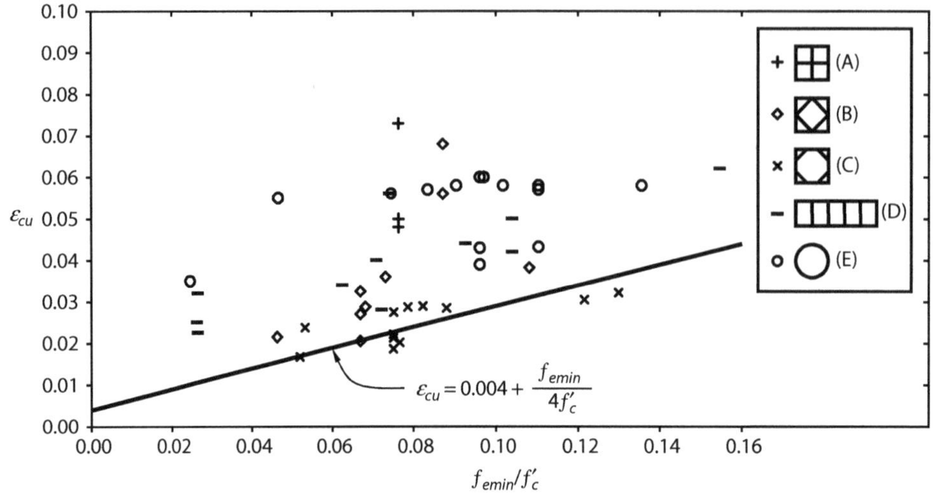

FIGURE 4.18 Strain at fracture of transverse steel for columns under monotonic concentric or moderately eccentric axial loads. The legend shows the general shape of the confinement reinforcement. [*Data sources: A (Moehle and Cavanagh, 1985); B (Scott et al., 1982; Moehle and Cavanagh, 1985); C (Scott et al., 1982); D, E (Mander et al., 1988b).*]

4.4.3 Stress–Strain Relation

Several algebraic forms have been proposed to represent the stress–strain curve for confined concrete. The models generally fall into one of three categories. Figure 4.19*a* is representative of a class of models that have an ascending branch followed by a linear descending branch with or without a residual plateau (e.g., Park et al., 1982; Razvi and Saatcioglu, 1999; Bing et al., 2001). Figure 4.19*b* is similar but with a plastic plateau at the peak stress (e.g., Sheikh and Uzumeri, 1982). Figure 4.19*c* shows an example of a single equation that gives a continuous stress–strain curve (e.g., Ahmad and Shah, 1982; Mander et al., 1988a).

Each of the models cited has been shown to produce good correlation with measured laboratory test results. Here we adopt the algebraic form proposed by

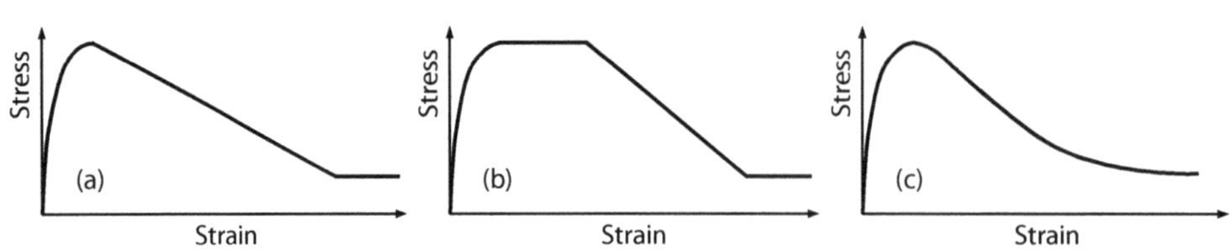

FIGURE 4.19 Different forms of confined concrete stress–strain relations.

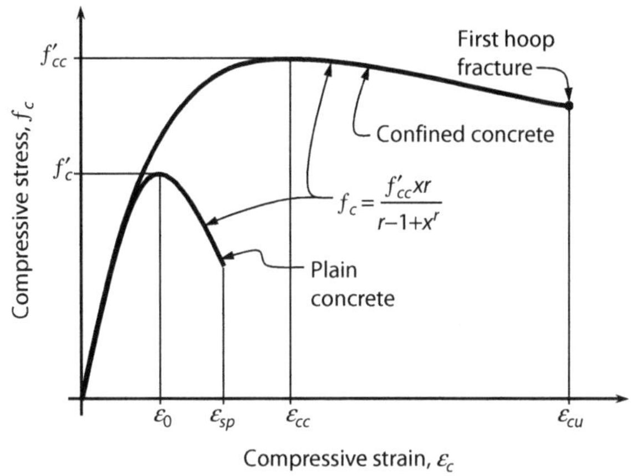

Figure 4.20 Confined concrete stress–strain relation. (*After Mander et al., 1988a.*)

Mander et al. (1988a), as shown in Figure 4.20. The longitudinal compressive stress f_c is given by

$$f_c = \frac{f'_{cc}xr}{r-1+x^r} \tag{4.19}$$

$$x = \frac{\varepsilon_c}{\varepsilon_{cc}} \tag{4.20}$$

$$r = \frac{E_c}{E_c - \dfrac{f'_{cc}}{\varepsilon_{cc}}} \tag{4.21}$$

In Eq. (4.21), the modulus of concrete, E_c, can be estimated as described in Chapter 3. The variables f'_{cc} and ε_{cc} are determined as described previously in this chapter. Note that Eq. (4.19) with appropriate input quantities can be used to describe the behavior of plain concrete as well.

Example 4.3. Construct the unconfined and confined concrete stress–strain relations for the cross section (Figure 4.21) of Example 4.3.

Solution

From Example 4.2, $k_e = 0.72$, $f_{emin} = 378$ psi (2.6 MPa), $Cf'_c = 4250$ psi (29 MPa), and $f'_{cc} = 6700$ psi (46 MPa). Taking $\varepsilon_0 = 0.002$, Eq. (4.16) gives $\varepsilon_{cc} = 0.008$ and Eq. (4.17) gives $\varepsilon_{cmax} = 0.023$. From Chapter 3, we can estimate Young's modulus as $E_c = 57{,}000\sqrt{Cf'_c}$ = 3700 ksi (16,000 MPa). Equation (4.19) is then used to plot the stress–strain relations. Note that for unconfined concrete, the material properties in Eqs. (4.19) through (4.21) are set equal to those of unconfined concrete, that is, $f'_{cc} = Cf'_c$ and $\varepsilon_{cc} = \varepsilon_0$. The results are plotted in the accompanying figure.

The proposed model has been used to compute stress–strain relations for several columns or walls tested in axial compression in the laboratory. Figure 4.22

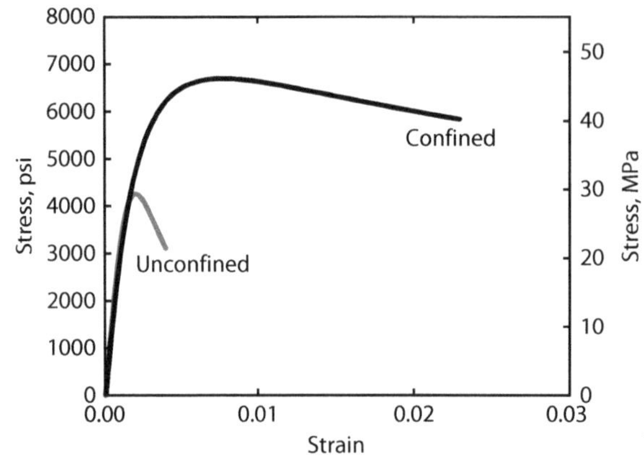

FIGURE 4.21 Stress–strain relations.

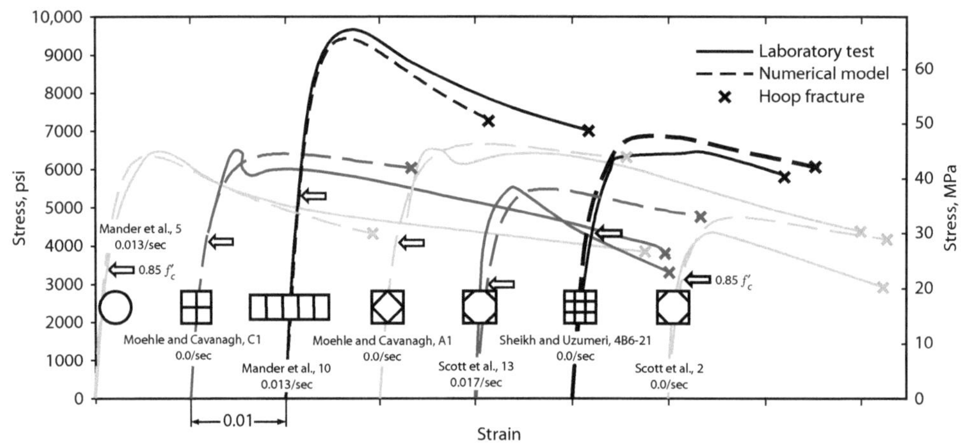

FIGURE 4.22 Measured and calculated stress–strain relations for a sample of columns tested in axial compression in different laboratories, for different confinement configurations, and at different strain rates. The arrows at 0.85 f'_c show strengths expected for unconfined concrete. **x** marks the strain (measured or calculated) at first hoop fracture. *(Data after Sheikh and Uzumeri, 1980; Scott et al., 1982; Moehle and Cavanagh, 1985; Mander et al., 1988b)*

compares calculated and measured stress–strain relations for a sample of the tests, showing generally good correlation between calculated and measured results.

References

Ahmad, S.H., and S.P. Shah (1982). "Stress-Strain Curves of Concrete Confined by Spiral Reinforcement," *ACI Journal*, Vol. 79, No. 6, pp. 484–490.

Balmer, G.G. (1949). *Shearing Strength of Concrete Under High Triaxial Stress—Computation of Mohr's Envelop as a Curve*, Structural Research Laboratory Report No. SP-23, U.S. Bureau of Reclamation, Denver, CO.

Bing, L., R. Park, and H. Tanaka (2000). "Constitutive Behavior of High-Strength Concrete under Dynamic Loads," *ACI Structural Journal*, Vol. 97, No. 4, pp. 619–629.

Bing, L., R. Park, and H. Tanaka (2001). "Stress-Strain Behavior of High-Strength Concrete Confined by Ultra-High- and Normal-Strength Transverse Reinforcements," *ACI Structural Journal*, Vol. 98, No. 3, pp. 395–406.

Blume, J.A., N.M. Newmark, and L.H. Corning (1961). *Design of Multi-Story Reinforced Concrete Buildings for Earthquake Motions*, Portland Cement Association, Chicago, IL, 318 pp.

Bresler, B., and V.V. Bertero (1975). "Influence of Strain Rate and Cyclic Loading on Behavior of Unconfined and Confined Concrete in Compression," *XVII Jornadas Sudamericanas de Ingenieria Estructural, V Simposio Panamericana de Estructuras*, Caracas, 8 al 12 de Diciembre de 1975.

Burdette, E.G., and H.K. Hilsdorf (1971). "Behavior of Laterally Reinforced Concrete Columns," *Journal of the Structural Division*, Vol. 97, No. ST2, pp. 587–602.

CSA (2004). *Design of Concrete Structures*, CSA A23.3-04, Canadian Standards Association, Mississauga, Canada.

Dodd, L.L., and N. Cooke (1992). "Dynamic Response of Circular Bridge Piers," *Proceedings*, 10th World Conference on Earthquake Engineering, A.A. Balkema, Rotterdam, pp. 3035–3039.

Ghali, A., and S.A. Youakim (2005). "Headed Studs in Concrete: State of the Art," *ACI Structural Journal*, Vol. 102, No. 5, pp. 657–667.

Iyengar, K.T., P. Desayi, and K.N. Reddy (1970). "Stress-Strain Characteristics of Concrete Confined in Steel Binders," *Concrete Research*, Vol. 22, No. 72, pp. 173–184.

Kaar, P.H., A.E. Fiorato, J.E. Carpenter, and W.G. Corley (1978). *Limiting Strains of Concrete Confined by Rectangular Hoops*, Research and Development Bulletin RD053.01D, Portland Cement Association, Skokie, IL, 12 pp.

Khaloo, A.R., K.M. El Dash, and S.H. Ahmad (1999). "Model for Lightweight Concrete Columns Confined by Either Single Hoops or Interlocking Spirals," *ACI Structural Journal*, Vol. 96, No. 6, pp. 883–890.

Mander, J.B., M.J.N. Priestley, and R. Park (1988a). "Theoretical Stress-Strain Model for Confined Concrete," *Journal of Structural Engineering*, Vol. 114, No. 8, pp. 1804–1826.

Mander, J.B., M.J.N. Priestley, and R. Park (1988b). "Observed Stress-Strain Behavior of Confined Concrete," *Journal of Structural Engineering*, Vol. 114, No. 8, pp. 1827–1849.

Manrique, M.A., V.V. Bertero, and E.P. Popov (1979). *Mechanical Behavior of Lightweight Concrete Confined by Different Types of Lateral Reinforcement*, Report No. UCB/EERC-79/05, Earthquake Engineering Research Center, University of California, Berkeley, CA, 103 pp.

Martin, C.W. (1968). "Spirally Prestressed Concrete Cylinders," *ACI Journal*, Vol. 65, No. 10, pp. 837–845.

Martinez, S., A.H. Nilson, and F.O. Slate (1984). "Spirally Reinforced High-Strength Concrete Columns," *ACI Journal*, Vol. 81, No. 5, pp. 431–441.

Moehle, J.P., and T. Cavanagh (1985). "Confinement Effectiveness of Crossties in RC," *Journal of Structural Engineering*, Vol. 111, No. 10, pp. 2105–2120.

Park, R., M.J.N. Priestley, and W.D. Gill (1982). "Ductility of Square-Confined Concrete Columns," *Journal of the Structural Division*, Vol. 108, No. ST4, pp. 929–950.

Paultre, P., and F. Légeron (2008). "Confinement Reinforcement Design for Reinforced Concrete Columns," *Journal of Structural Engineering*, Vol. 134, No. 5, pp. 738–749.

Razvi, S., and M. Saatcioglu (1999). "Confinement Model for High-Strength Concrete," *Journal of Structural Engineering*, Vol. 125, No. 3, pp. 281–289.

Richart, F.E., A. Brandtzaeg, and R.L. Brown (1928). *A Study of the Failure of Concrete under Combined Compressive Stresses*, Bulletin No. 185, Engineering Experiment Station, University of Illinois, Urbana, IL, 104 pp.

Richart, F.E., A. Brandtzaeg, and R.L. Brown (1929). *The Failure of Plain and Spirally Reinforced Concrete in Compression*, Bulletin No. 190, Engineering Experiment Station, University of Illinois, Urbana, IL, 74 pp.

Richart, F.E., and R.L. Brown (1934). *An Investigation of Reinforced Concrete Columns*, Bulletin No. 267, Engineering Experiment Station, University of Illinois, Urbana, IL, 94 pp.

Rood, M., and J.P. Moehle (2006). "Investigation of Welded Reinforcement Grids," http://nees.berkeley.edu/Projects/, 32 pp.

Saatcioglu, M., and M. Grira (1999). "Confinement of Reinforced Concrete Columns with Welded Reinforcement Grids," *ACI Structural Journal*, Vol. 96, No. 1, pp. 29–39.

Saatcioglu, M., and S.R. Razvi (1992). "Strength and Ductility of Confined Concrete," *Journal of Structural Engineering*, Vol. 118, No. 6, pp. 1590–1607.

Scott, B.D., R. Park, and M.J.N. Priestley (1982). "Stress-Strain Behavior of Concrete Confined by Overlapping Hoops at Low and High Strain Rates," *ACI Journal Proceedings*, Vol. 79, No. 1, pp. 13–27.

Shah, S.P., A. Fafitis, and R. Arnold (1983). "Cyclic Loading of Spirally Reinforced Concrete," *Journal of the Structural Division*, Vol. 109, No. 7, pp. 1695–1710.

Sheikh, S.A., and S.M. Uzumeri (1980). "Strength and Ductility of Tied Concrete Columns," *Journal of the Structural Division*, Vol. 106, No. ST5, pp. 1079–1102.

Sheikh, S.A., and S.M. Uzumeri (1982). "Analytical Model for Concrete Confinement in Tied Columns," *Journal of the Structural Division*, Vol. 108, No. ST12, pp. 2703–2722.

Tanaka, H., and R. Park (1987). "Effectiveness of Transverse Reinforcement with Alternative Anchorage Details in Reinforced Concrete Columns," *Proceedings, Pacific Conference on Earthquake Engineering*, Wairakei, New Zealand. Vol. 1, pp. 225–235.

CHAPTER 5

Axially Loaded Members

5.1 Preview

Chapter 5 is the first in a series of chapters that will investigate the behavior and design requirements of reinforced concrete members subjected to various loadings. These chapters build on knowledge of reinforced concrete materials gained in Chapters 2 through 4, and develop a comprehensive understanding of how structural members respond to static and dynamic loadings. What we learn in these chapters will enable us to develop the design requirements for complete buildings subjected to earthquake loading.

This chapter focuses on prismatic members subjected to axial compression and tension. Such members occur in columns supporting discontinuous members, in the boundaries of structural walls, and in chords and collectors of structural diaphragms. Topics of interest include stiffness, strength, deformation capacity, and stability under static or reversed cyclic loads. By the end of this chapter, the reader will understand the fundamental mechanics of these topics as well as the design requirements necessary to achieve target performance levels.

5.2 Some Observations on the Behavior of Compression Members

We begin our study of axially loaded members by reviewing the test results from two columns subjected to axial compressive loading until failure. The two columns have nominally identical circular cross sections, unrestrained lengths, and material properties (Figure 5.1). The primary difference between the two columns is the transverse reinforcement. One of the columns uses two sets of triangular ties for each hoop set to restrain six longitudinal bars, with longitudinal hoop spacing nearly equal to the column diameter (as permitted by many building codes for columns in buildings assigned to low seismic design categories and having low shear stresses). The other column uses spiral reinforcement with close spacing.

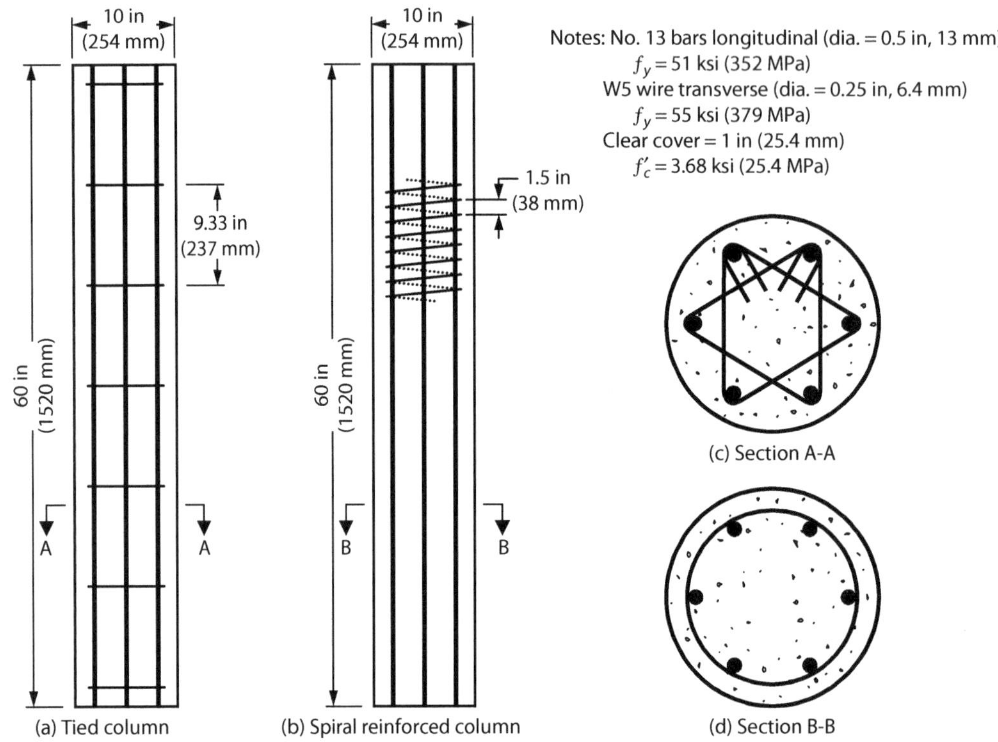

Figure 5.1 Two test columns. (*After Gamble and Thomson, 1977, courtesy of the University of Illinois at Urbana–Champaign.*)

The columns are tested in concentric axial compression at a slow loading rate, with load–displacement relation shown in Figure 5.2. Both columns develop vertical splitting cracks near the top of the column for axial force of approximately 300 kips (1300 kN). In the tied column, the splitting progresses to localized spalling of cover concrete, then buckling of the longitudinal reinforcement, with rapid reduction in axial load. In the spiral column, splitting extends farther along the column height, with subsequent cover spalling extending over three-quarters of the column height. As deformations are increased, the confined core and stable longitudinal bars continue to support axial load nearly equal to the initial splitting load. The column remains stable until the spiral reinforcement fractures, leading to longitudinal reinforcement buckling and reduction of axial resistance. Figure 5.2*b* shows the final condition of the test columns.

The distinct differences in behavior of the two columns are noteworthy. Whereas the tied column quickly exhausts its load-carrying capacity once failure begins, the spiral column remains a viable load-carrying element through relatively large displacements. Even after spiral fracture, the failure of the spiral column is comparatively gradual.

Subsequent sections in this chapter will examine the behaviors presented here, including cover spalling, instability of compressed longitudinal reinforcement,

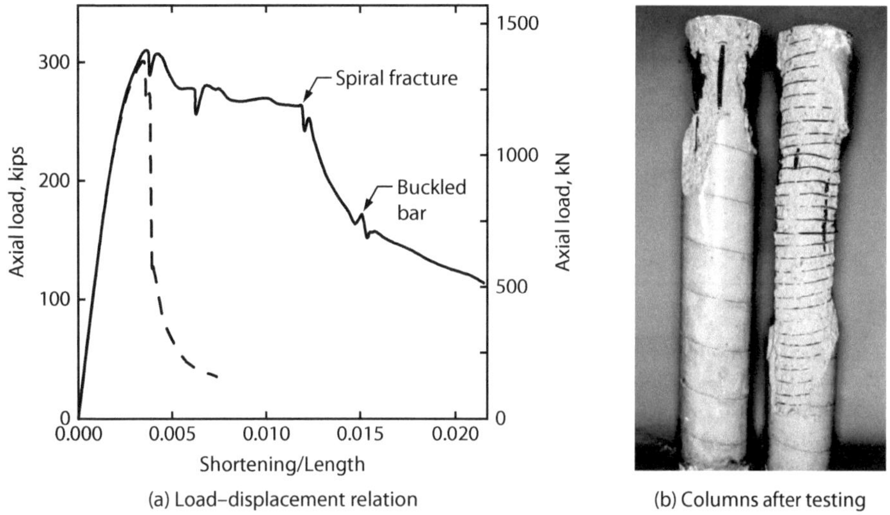

FIGURE 5.2 Behavior of test columns under concentric compression: (*a*) relation between total axial load and total column shortening; (*b*) photograph of columns after testing. (*After Gamble and Thomson, 1977, courtesy of the University of Illinois* at Urbana–Champaign.)

confined column behavior, and other aspects of axially compressed columns. The discussion will also include tension members and members subjected to tension and compression reversals.

5.3 Analysis Assumptions for Compression Members

Analysis of the load-deformation response of compression members relies on three fundamental assumptions:

1. Sections that are planar before loading remain planar after loading. The practical implication of this assumption is that the strain in concrete and steel are identical at any point in the cross section. This requires that there be no slip between steel and concrete. This assumption produces accurate results away from end regions and lap splices.
2. The stress–strain relations are known from properties measured in coupon tests on concrete and reinforcing steel. We will routinely ignore effects of creep and shrinkage, though these effects can be important, as discussed subsequently.
3. Equations of equilibrium can be used to compute the axial force and moment on the cross section given the internal stresses.

Figure 5.3 illustrates application of these assumptions to the analysis of a rectangular cross section that is symmetric about both principal axes. Under the action of concentric axial force, a uniform strain profile develops (Figure 5.3*b*).

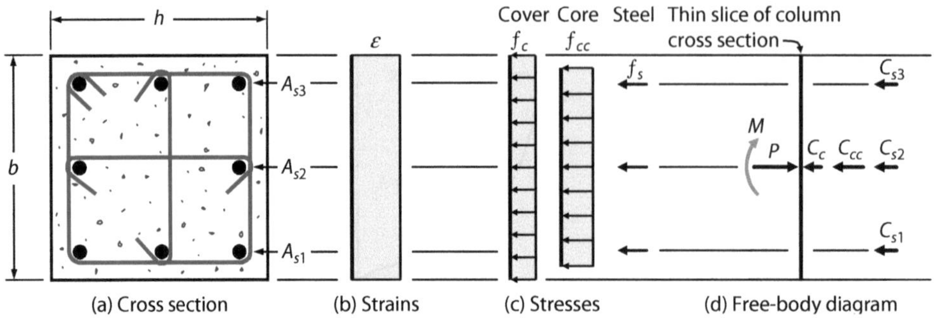

FIGURE 5.3 Concrete compression member under concentric axial compression. Moment *M* is shown in lighter shading because its value is zero for the column cross section and axial load position shown.

From the known stress–strain relations, we next establish the stress in each of the materials at every point in the cross section (Figure 5.3*c*). Integrating these stresses over the areas on which they act leads to corresponding stress resultants. Figure 5.3*d* shows a free-body diagram obtained by taking a thin slice through the column section, with the stress resultants shown on the right-hand side of the free-body diagram and the internal axial force and moment that equilibrate them on the other. Equilibrium of this free body establishes the internal forces P and M.

The approach just illustrated will be used throughout this text.

5.4 Service Load Behavior of Compression Members

5.4.1 Linear Elastic Response

The linear-elastic solution for an axially loaded column follows from the assumptions of the preceding section. We first simplify the problem by assuming confinement is inactive for low strains, consistent with the findings of Chapter 4. Thus, $f_{cc} = f_c$ in Figure 5.3. For a given strain ε within the linear range, stresses in concrete and reinforcement are $f_c = E_c\varepsilon$ and $f_s = E_s\varepsilon$, and corresponding stress resultants are $C_c = A_c f_c = (A_g - A_{st})f_c$, $C_{s1} = A_{s1}f_s$, $C_{s2} = A_{s2}f_s$, and $C_{s3} = A_{s3}f_s$. Note that we have taken C_c to represent the compression acting on the entire concrete section, with $C_{cc} = 0$ because confinement is inactive at this stage of loading. From equilibrium of the free-body diagram in Figure 5.3*d*, $P - C_c - C_{s1} - C_{s2} - C_{s3} = 0$. Combining terms, the relation between axial load and internal stresses can be written as

$$
\begin{aligned}
P &= (A_g - A_{st})f_c + A_{st}f_s \\
&= (A_g - A_{st})\varepsilon E_c + A_{st}\varepsilon E_s \\
&= \varepsilon E_c[(A_g - A_{st}) + nA_{st}] \\
&= f_c[A_g + (n-1)A_{st}]
\end{aligned}
\tag{5.1}
$$

in which $n = E_s/E_c$ is the *modular ratio*. Solving Eq. (5.1) for concrete stress f_c leads to

$$f_c = P/[A_g + (n-1)A_{st}] \tag{5.2}$$

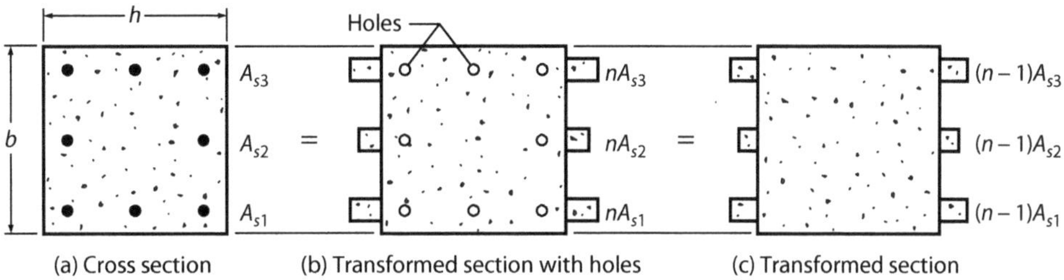

FIGURE 5.4 Transformed section for compression member.

Because steel and concrete strains are equal at any point in the cross section, we can write

$$f_s = E_s \varepsilon = \frac{E_s}{E_c} E_c \varepsilon = n f_c \tag{5.3}$$

The results of Eq. (5.2) can be obtained using the *transformed area method*, which is applicable to linear materials. Recognizing that steel has elastic modulus $E_s = nE_c$, we can achieve equivalent cross-sectional properties by replacing steel area A_{st} by concrete having area nA_{st} located at the centroid of the steel. As illustrated in Figure 5.4, when steel area A_{st} is transformed to concrete area nA_{st}, the original concrete section is left with holes having area A_{st}. To simplify the analysis, those holes are filled, leaving $(n-1)A_{st}$ for the transformed steel sections. As shown in Figure 5.4, the transformed area is $A_g + (n-1)A_{st}$; hence, the stress in concrete is given by Eq. (5.2).

5.4.2 Effects of Drying Shrinkage and Creep

When freshly cast concrete is exposed to ambient temperature and humidity conditions, it generally undergoes volume changes associated with cooling (*temperature shrinkage*) and moisture loss (*drying shrinkage*). These occur in the absence of externally applied stresses. When stresses are applied, additional *instantaneous strains* occur. Under sustained stress, concrete also experiences a gradual increase in strain known as *creep strain*. Finally, concrete that is subjected to a given level of sustained strain (e.g., due to shrinkage) experiences a gradual decrease in stress due to creep known as *stress relaxation*.

Drying shrinkage and creep strains in concrete are associated mainly with the removal of adsorbed water from the hydrated cement paste. The main difference is that the driving force in drying shrinkage is the difference in relative humidity, whereas in the case of creep the driving force is sustained applied stress. Both are affected by concrete composition, initial curing, environment, geometry, and time. Additionally, creep is affected by the intensity of applied stress and the age of concrete when stresses are applied.[1]

[1]See Mehta and Monteiro (2014) or Mindess et al. (2003) for detailed discussion of volume change characteristics.

Ultimate unrestrained shrinkage strain of concrete typically ranges from 0.0002 to 0.0008. Under normal conditions and for moderate-sized members, roughly 90% of drying shrinkage is completed by one year after casting and exposure to ambient conditions. Under long-term sustained load, creep strain, that is, the strain additional to the initial strain, can range from about 1.0 to 4.0 times the initial strain depending on materials and age at loading, with typical value around 2.0. Creep appears to continue indefinitely, but for most practical applications may be assumed to have reached its ultimate value after about five years.

We can estimate the effect of drying shrinkage on internal stresses of an otherwise unloaded reinforced concrete cross section using linear-elastic analysis methods. Figure 5.5 illustrates the procedure for a symmetric cross section. As a first step, the unrestrained shrinkage strain ε_{sh} is imposed on the section (Figure 5.5*b*). This results in compressive forces C_{s1} and C_{s2} in the longitudinal reinforcement, and a compressive force imbalance equal to $P = C_{s1} + C_{s2}$. To restore equilibrium of the unloaded cross section, superimpose an equal and opposite tension force $T = P$ (Figure 5.5*c*) to arrive at the equilibrium state (Figure 5.5*d*). Note that we are ignoring stress relaxation that would occur due to creep of the concrete under these internal stresses.

For the problem at hand, we can show that the stresses in concrete and steel are given by

$$f_c = \rho_l \varepsilon_{sh} n E_c / (1 + (n-1)\rho_l) \tag{5.4}$$

$$f_s = \varepsilon_{sh} n E_c (1 - n\rho_l / (1 + (n-1)\rho_l)) \tag{5.5}$$

For $f_c' = 6000$ psi (41 MPa), concrete modulus of elasticity and tensile strength are approximately 4000 ksi (30,000 MPa) and 400 psi (3 MPa), resulting in $n = 7$. For typical values of $\varepsilon_{sh} = 0.0004$ and $\rho_l = 0.01$, Eqs. (5.4) and (5.5) result in concrete tensile stress of 100 psi (0.7 MPa) and steel compressive stress of 10 ksi (70 MPa), which neither cracks the concrete nor yields the reinforcement. For greater values of ε_{sh} and ρ_l, the tensile stress can be sufficient to crack the concrete.

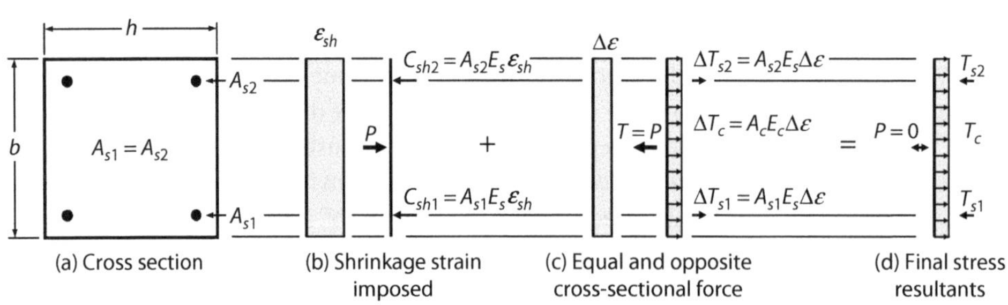

FIGURE 5.5 Internal stresses in symmetric cross section subject to longitudinal shrinkage strain.

Creep also affects internal stress distributions under service loads. In an axially loaded column, creep causes an increase in the axial compressive strain and, hence, an increase in the steel compressive stress. As the steel stress increases, the concrete stress must decrease, which, in turn, reduces the ultimate creep. This behavior is known as *restrained creep* because the longitudinal reinforcement is restraining creep in the concrete. The final steel stress can be two or more times the initial stress. The effect on steel stresses is greater for smaller steel ratios, which is one reason some building codes set a lower limit on the volume ratio of longitudinal reinforcement in compression members. For example, ACI 318 (2014) sets the minimum area of longitudinal reinforcement equal to $0.01A_g$.

Restrained creep effects can be calculated using incremental analysis methods in which the loading period is broken into small time increments, with material properties and stresses tracked for each increment. Alternatively, the effective modulus method provides a closed-form solution to approximate the effects (Dilger, 1982).

The preceding analysis suggests that creep and shrinkage can appreciably alter the internal stresses of compression members under long-term loading. This is a considerable drawback for allowable stress design methods. Strength and deformation capacity of reinforced concrete members are much less susceptible to creep and shrinkage effects. This is one of the reasons why the load and resistance factor design (LRFD) method is generally preferred for reinforced concrete.

5.5 Inelastic Behavior of Compression Members

The previous section considered elastic deformations combined with creep and shrinkage effects. Our main interest, however, is to understand the strength and inelastic deformation characteristics of structural members. This section will examine these aspects with an emphasis on monotonic behavior. Behavior under reversed cyclic loading is considered in Section 5.7.

5.5.1 Cover and Core Concrete

For strains well below the strain ε_0, cover and core concrete can be considered to behave as unconfined concrete with identical stress–strain characteristics. As the strain ε_0 is approached, and for greater strains, dilation of the concrete core activates confinement by the transverse reinforcement. Furthermore, bulging of the core pushes the cover concrete away from the core, accelerating its failure (Figure 5.6) (Bresler and Gilbert, 1961).

For concrete having $\varepsilon_0 \approx 0.002$, core and cover concretes begin to show different behaviors for strains in the range 0.0015 to 0.002 (Sheikh and Uzumeri, 1980), and cover spalling eventually removes its contribution to axial load resistance. There is no general agreement on the precise strains corresponding to these different behaviors (e.g., Blume et al., 1961; Baker and Amarakone, 1964; Park and Paulay, 1975; Sheikh and Uzumeri, 1980). Fortunately, precise definition of this behavior is not required for typical earthquake engineering applications.

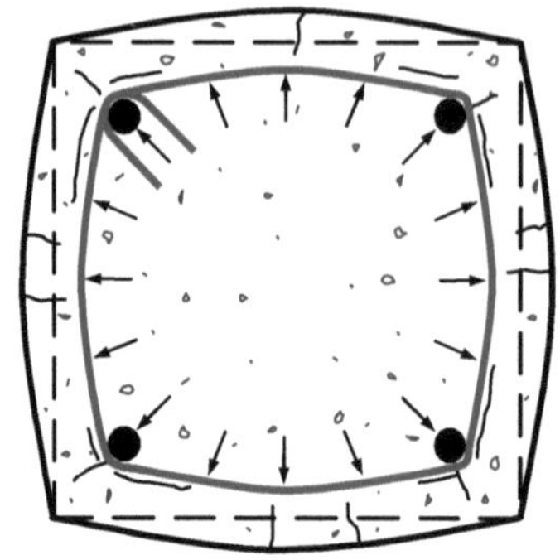

FIGURE 5.6 Bulging of core concrete results in deformations and cracking of cover concrete.

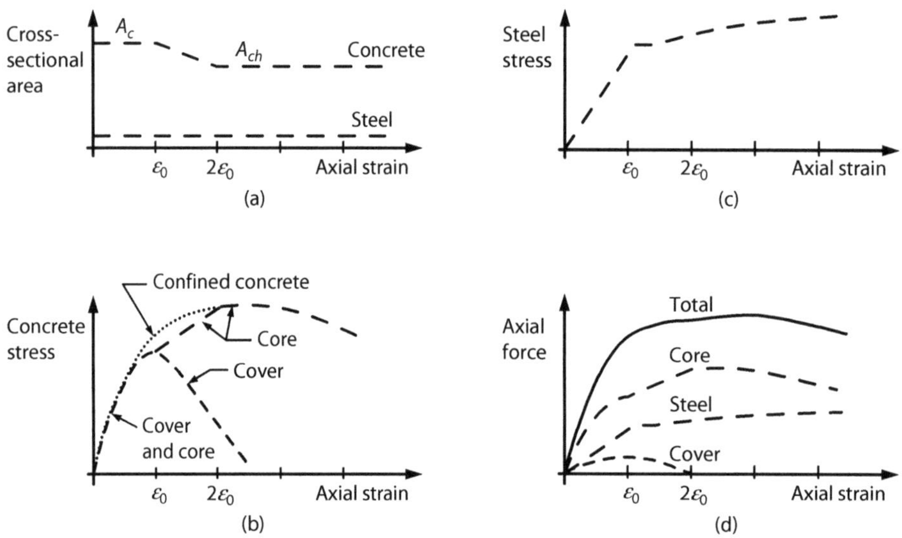

FIGURE 5.7 Assumed behavior of cover concrete, core concrete, and longitudinal reinforcement.

A reasonable approach is to assume that the cover and core concrete behaves as unconfined concrete up to ε_0, that cover concrete degrades linearly until it is completely spalled at $2\varepsilon_0$, and that core concrete gradually gains confinement and behaves as fully confined concrete for strains beyond $2\varepsilon_0$. Figure 5.7*a* and *b* illustrates this behavior.

5.5.2 Longitudinal Reinforcement

Longitudinal reinforcement behaves differently in compression and tension for two reasons. First, under compression the Poisson effect increases the cross-sectional area, whereas in tension it decreases the area, so that engineering stresses in compression exceed those measured in tension tests (see Chapter 2). This effect is ignored in routine analyses. A second effect is reinforcement instability, which can occur if longitudinal reinforcement is inadequately supported or strain ranges are relatively large. Reinforcement buckling should not be overlooked, as it can have an appreciable effect on the axial load resistance of the cross section.

Buckling of longitudinal reinforcement in reinforced concrete members is complicated by interactions with the surrounding concrete. In a compression member, dilation of the confined core exerts outward pressure on the longitudinal bars, which in combination with axial compression increases the tendency for buckling. The cover concrete initially provides some restraint against buckling, but this restraint diminishes as the cover itself becomes damaged at large strains. Transverse reinforcement thus plays a dominant role in restraining longitudinal bar buckling at large compressive strains. As will be discussed subsequently, for most practical problems, the transverse reinforcement spacing will be small enough that elastic buckling of longitudinal reinforcement is avoided. Thus, analysis of the buckling problem requires consideration of both geometric and material nonlinearities.

Bar buckling is further complicated in that different buckling modes can occur depending on the degree of restraint provided by the transverse reinforcement. The simplest mode involves buckling between two hoop sets or spiral turns (Figure 5.8*b*). In this case, the bar might be idealized as a fixed-ended column of length *s*, though the fixity at the ends is questionable. A second mode involves buckling over two or more hoop sets or spiral turns (Figure 5.8*d*). In this case, the bar might be idealized as a column of length *ns* restrained by discrete hoops. A third mode (not shown) involves the bar buckling sideways in the direction parallel to the perimeter hoop or spiral.

The buckling restraint provided by the hoop reinforcement depends on the configuration of the hoop set. For rectilinear transverse reinforcement, longitudinal bars located in the corner of a tie (e.g., bars **a** and **c** in Figure 5.8*a*) are effectively supported even if the tie has diameter as small as 15% of the longitudinal bar diameter (Bresler and Gilbert, 1961). In contrast, longitudinal bars not supported in a corner of a tie (e.g., bar **b** in Figure 5.8*a*) must rely on the flexural stiffness of the hoop for support, which is insufficient for practical hoop bar diameters. It can be argued that the unsupported bars gain indirect support from the adjacent restrained core and longitudinal bars (Pantazopoulou, 1998), but

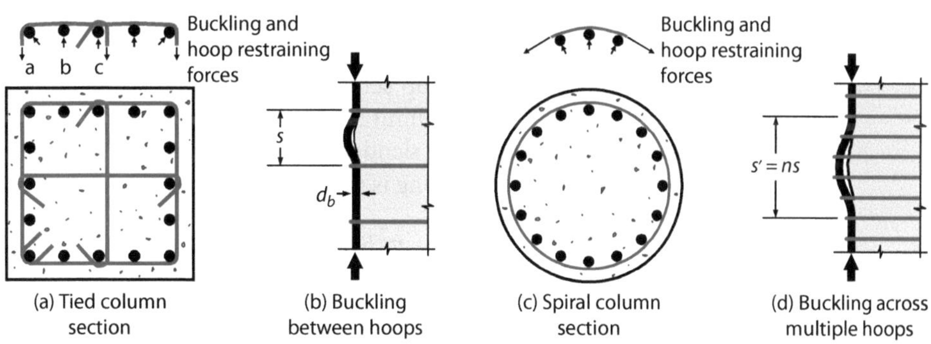

FIGURE 5.8 Restraint of longitudinal bars and idealized buckling modes.

generally the unsupported bars weaken the cross section. For circular hoops or spirals, lateral restraint is provided by the radial component of the hoop tension (Figure 5.8*c*). Whether longitudinal bars buckle between hoop sets or span multiple sets (Figure 5.8*d*) depends on the stiffness of the hoop tension action. Papia and colleagues (1988, 1989) and Brown et al. (2008) provide additional insight into the problem of buckling that spans multiple hoop sets.

Building codes include provisions for minimum size of rectilinear hoop reinforcement intended to prevent buckling over several hoop sets. ACI 318 requires that the hoop bar size be at least No. 3 (No. 10 metric) for longitudinal bars up to No. 10 (No. 32), and No. 4 (No. 13) for larger longitudinal bars. ACI 318 also requires that every corner and alternate longitudinal bar be supported by the corner of a tie with no unsupported bar more than 6 in (150 mm) clear from a supported bar. For columns of special moment frames with high concrete compressive strength or high axial load, ACI 318 requires that every longitudinal bar be supported. NZS 3101 (2006), referring to plastic hinges of beams and columns, requires that every longitudinal bar be restrained by ties, except that tie legs need not be placed closer than 8 in (200 mm) apart on centers (this permits unrestrained bars between the restrained bars). Furthermore, the diameter of ties is not to be less than 0.2 in (5 mm) and the area of a tie in the direction of potential buckling is not to be less than

$$A_{bt} = \frac{\sum A_{bl} f_y}{16 f_{yt}} \frac{s}{6 d_{bl}} \tag{5.6}$$

in which ΣA_{bl} is the sum of areas of all longitudinal bars to be restrained by the tie leg, including tributary portions of adjacent unrestrained bars. This expression was developed based on the assumption that, when the ties having spacing $s = 6d_{bl}$, the provided strength should not be less than 1/16th the strength of the restrained bars. For spiral or circular hoop confinement of column plastic hinges, NZS 3101 (2006) requires

$$\rho_s = \frac{A_{st}}{110 d'} \frac{f_y}{f_{yt}} \frac{1}{d_{bl}} \tag{5.7}$$

In Eqs. (5.6) and (5.7), f_{yt} is not to be taken greater than 116 ksi (800 MPa).

Where transverse reinforcement provides sufficient stiffness and strength to prevent buckling spanning several hoop sets, the main design parameter is the spacing of transverse reinforcement to limit bar buckling between hoops. We can use the Euler equation for buckling of slender columns to analyze this problem. Accordingly, the critical stress at buckling is given by

$$f_{cr} = \frac{\pi^2 E}{\left(\frac{kl}{r}\right)^2} \tag{5.8}$$

For longitudinal bars with circular cross section of diameter d_b, $r = d_b/4$. Recognizing that $l = s$, we can solve Eq. (5.8) for the critical hoop spacing as

$$s_{critical} = \left(\frac{\pi}{4k} \sqrt{\frac{E}{f_{cr}}} \right) d_b \tag{5.9}$$

To use Eq. (5.9), we must define the effective length factor k and the effective modulus of elasticity E. The value of k lies somewhere between 0.5 (fixed ends) and 1.0 (pinned ends). Recognizing that typical construction quality and damage after cover spalling result in less than fully fixed conditions, Bresler and Gilbert (1961) suggested a value of $k = 0.7$. For stresses up to the yield point, the elastic modulus of steel can be substituted for E. Using $k = 0.7$, $E = 29{,}000$ ksi (200,000 MPa), and yield stress $f_y = f_{cr} = 69$ ksi (480 MPa), Eq. (5.9) results in $s_{critical} = 23d_b$. This compares with the commonly recommended maximum hoop spacing of $s = 16d_b$ for non-seismic design of columns (e.g., ACI 318).

We can extend the Euler model beyond the yield point using either the tangent modulus or the double (or reduced) modulus. In the *tangent modulus approach*, the instantaneous tangent modulus E_t (Figure 5.9*a*) is substituted for E in Eq. (5.9). For example, to estimate requirements for a longitudinal bar just past the yield point, we might set E equal to the initial strain-hardening modulus, that is, $E = E_{sh} \approx 1000$ ksi (7000 MPa), along with $k = 0.7$ and $f_{cr} = 69$ ksi (480 MPa), resulting in $s_{critical} = 4d_b$. More generally, for the stress–strain relation shown in Figure 5.9*a*, which is characteristic of A706 Grade 60 (420) reinforcement, the tangent modulus predicts the $s_{critical}/d_b$ versus strain relations shown in Figure 5.9*b*. According to this model, achieving large compressive strain without buckling requires very close spacing of hoop sets.

If the reinforcement has a yield plateau, $E_t = 0$ along the plateau, and Eq. (5.9) can be interpreted as predicting buckling immediately upon reaching the yield stress. Tests, however, demonstrate that reinforcement can remain stable well past the yield plateau if the unsupported length is not large. Two factors enable this to

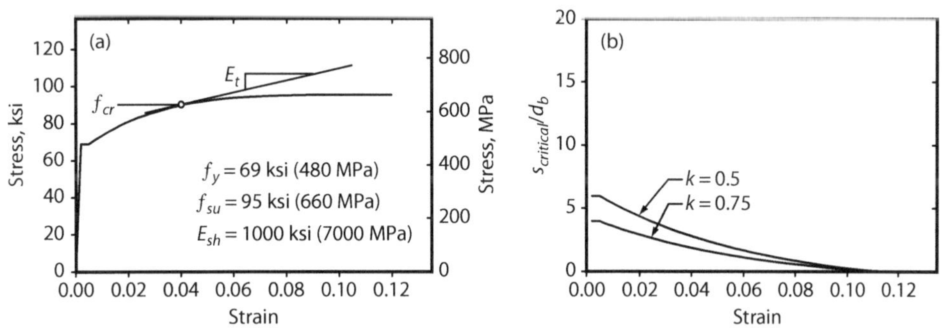

FIGURE 5.9 (*a*) Stress–strain relation and (*b*) calculated relation between $s_{critical}/d_b$ and maximum strain based on tangent modulus approach.

occur. First, the yield plateau represents the integration along a finite gauge length of slipbands occurring at discrete locations, and does not accurately represent the behavior of the entire length of bar at any strain along the plateau. Thus, the use of $E_t = 0$ in Eq. (5.9) would be inappropriate. Equally important is the effect that buckling has on curvature and internal stresses of the bar, adding stability, as explained below.

When a reinforcing bar is restrained by closely spaced hoops, buckling involves considerable curvature of the bar, leading to increased compression on one side and unloading on the other side (Figure 5.10*a*). Considering the stress–strain relation in Figure 5.10*b*, if the bar is at point A just prior to buckling, then the flexural compression side loads from A to B with initial modulus equal to the tangent modulus E_t, whereas the flexural tension side of the bar unloads from A to C with modulus equal to E_s. Thus, the bar behavior is determined by two moduli. This behavior is accounted for in the *double (or reduced) modulus approach*. Papia and Russo (1989) and Pantazopoulou (1998) describe applications of this approach to reinforced concrete construction.

Figure 5.11 compares the exact relation between tangent (E_t) and double modulus (E_r) for a circular cross section. In the range of interest, $E_r \approx 2.5E_t$. Thus, the

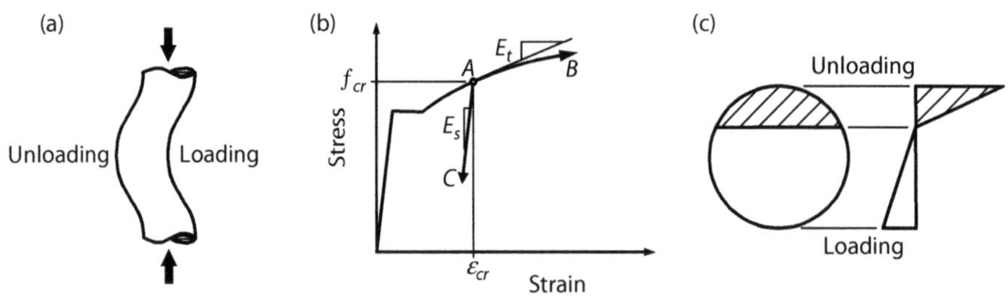

FIGURE 5.10 Double modulus buckling model: (*a*) buckled reinforcing bar; (*b*) loading and unloading stress–strain relation; (*c*) bar cross section showing change in stress from point A where buckling initiates.

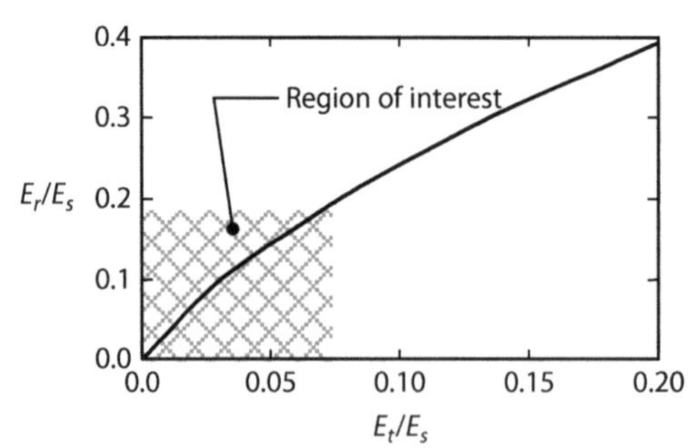

FIGURE 5.11 Relation between tangent modulus and reduced (or double) modulus. *(After Pantazopoulou, 1998, used with permission from ASCE.)*

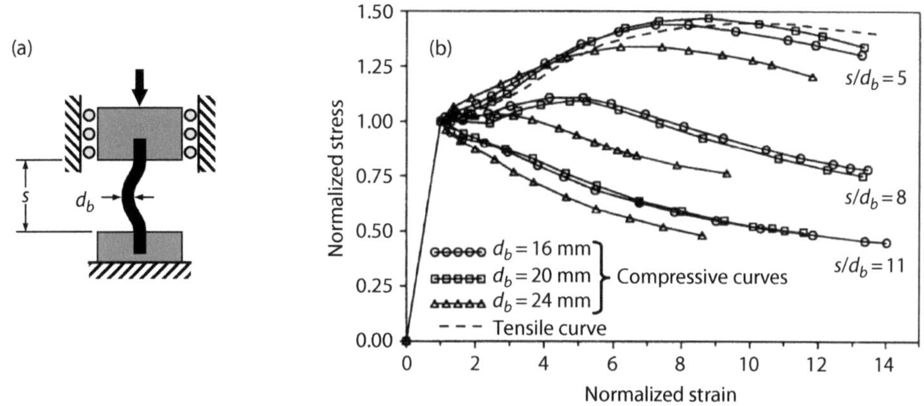

FIGURE 5.12 Stress–strain response versus slenderness ratio: (*a*) test setup idealization; (*b*) test results. (*After Monti and Nuti, 1992, used with permission from ASCE.*)

double modulus approach results in values of $s_{critical}$ approximately $\sqrt{2.5} = 1.6$ times greater than obtained by the tangent modulus approach.

Several researchers have reported tests on restrained reinforcing bars loaded in compression (e.g., Monti and Nuti, 1992; Rodriguez et al., 1999; Bae et al., 2005). Figure 5.12 shows sample results for bars with fixed ends loaded monotonically. According to the results shown, a compressed bar can achieve compressive stress–strain response essentially equivalent to the tensile stress–strain relation if the unrestrained length is $s = 5d_b$.

The onset of buckling can be defined by any of several measures. In laboratory tests, visual observations commonly are used, but such observations are inherently subjective and inaccurate. Preferred approaches include measurement of lateral displacements or measurement of strains on opposing faces of the bar. Using the latter approach, Rodriguez et al. (1999) obtained relations between compressive strain and $s_{critical}/d_b$ for reinforcing bars satisfying A706 characteristics. Figure 5.13 compares measured data with results calculated using the tangent and double modulus approaches. Both the tangent modulus approach with $k = 0.5$ and the double modulus approach with $k = 0.75$ produce results that correlate well with the observed data.

In most reported axial compression tests on confined concrete columns (see Chapter 4), hoops or crossties support all of the longitudinal bars in the cross section. Current building codes (e.g., ACI 318, NZS 3101), however, permit unsupported longitudinal bars provided that at least alternate bars are supported and no unsupported bar is more than a small distance from supported bars. Axial compression tests on such members show that the unsupported longitudinal bars can buckle over several hoop sets, deforming the perimeter hoop and enabling spalling to progress into the core (Figure 5.14). A conservative approach is to assume that unsupported bars lose all of their compressive strength following cover spalling.

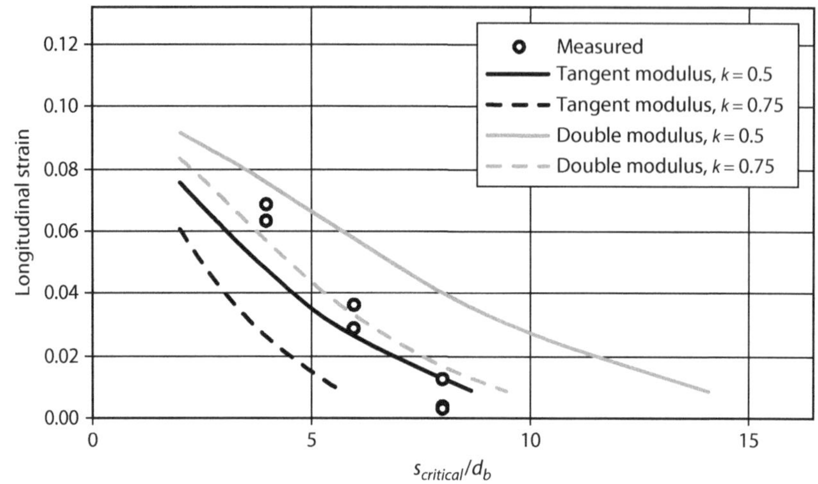

FIGURE 5.13 Longitudinal strain versus critical hoop spacing for monotonic compression. (*Data after Rodriguez et al., 1999.*)

5.5.3 Load–Displacement Response

Sections 5.5.1 and 5.5.2 described the behavior of the cover concrete, core concrete, and longitudinal reinforcement. These responses can be combined to determine the complete load-deformation response of an axially loaded, short compression member. For any given strain, the condition of the cover concrete is assessed, the stresses on the concrete and reinforcement are determined, and then the stresses are integrated across the section to determine the total force acting on the section. If we carry out this calculation for progressively increasing strains, the complete relation between axial load and axial strain is determined. The procedure is illustrated in Figure 5.7.

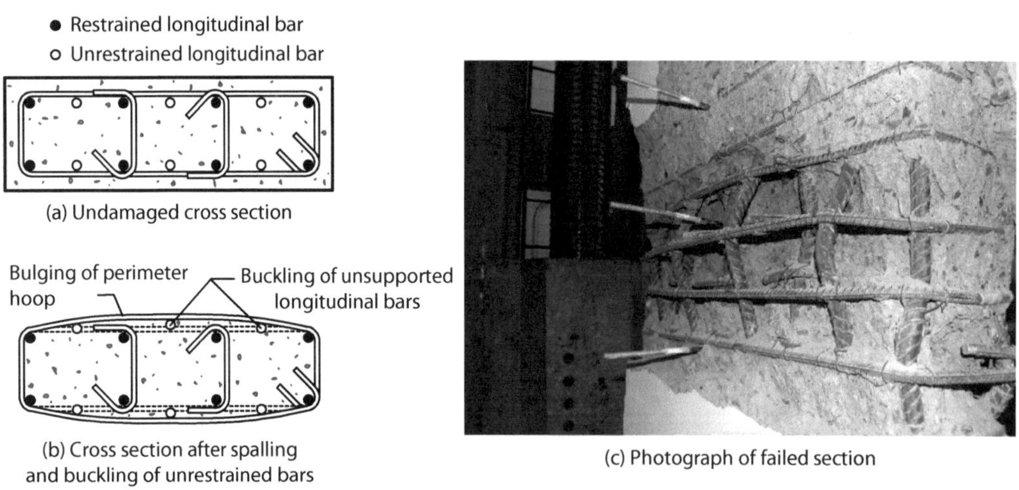

FIGURE 5.14 Failure of cross sections with unsupported longitudinal reinforcement. (*After Arteta and Moehle, 2014.*)

To determine the relation between axial load and axial shortening displacement, the axial strains must be integrated along the column length. The procedure for doing this depends on whether the relation between axial force and strain is continuously strain-hardening or whether strain-softening occurs. Consider the idealized force strain relations shown in Figure 5.15*a*. For column A, the axial compressive force increases continuously for increasing compressive strain. Thus, as the axial load increases, every cross section develops progressively increasing strain over the column height (Figure 5.15*c*). In this case, the axial shortening displacement is simply the product of the strain and the length. If the materials are relatively ductile, this leads to a relatively ductile load–displacement relation for the column (Figure 5.15*b*).

In contrast, column B initially has strain-hardening behavior to load P_0, followed by strain-softening behavior. Up to the load P_0 (point **b** in Figure 5.15*a*), the strains are uniformly distributed over the column height, so axial shortening can again be computed as the product εl_u. Beyond this point, however, the column begins to shed load as compression progresses. In a reinforced concrete column, softening could result from spalling of the cover concrete or buckling of longitudinal reinforcement. If the core at the spalled section is lightly confined, its strength will be less than that of the adjacent unspalled sections. Thus, as the column is further compressed, strains increase at the weaker spalled section, but with progressively lower axial load (e.g., loading from points **b** to **d** in Figure 5.15*a*). Simultaneously, the sections that have not spalled (most of the length of the column) will unload along **be**, leading to

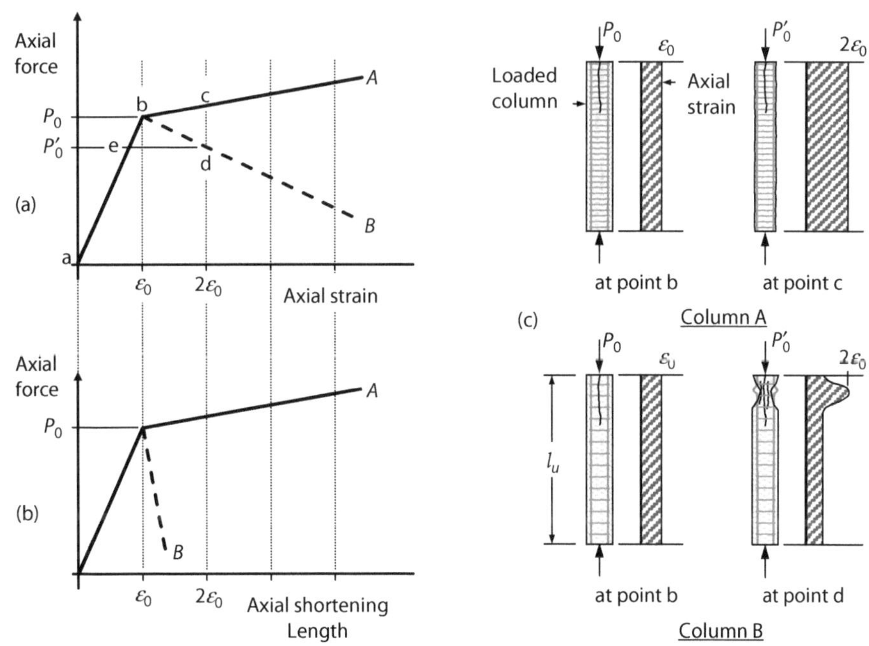

FIGURE 5.15 Axial behavior of strain-hardening (A) and strain-softening (B) compression members: (*a*) axial force versus axial strain; (*b*) axial force versus average axial shortening; (*c*) distributions of damage and strain along column length.

reduction of the axial compressive strains along the unspalled length. Because of the localized failure, the average strain along the column length is significantly reduced, leading to a steeply descending load–displacement behavior (Figure 5.15*b*).

The distinctions between columns A and B are important to recognize. Column A has a stress–strain relation that is both ductile and continuously strain-hardening. This combination produces ductile load–displacement response. In contrast, column B has a relatively ductile stress–strain relation, but strain-softening leads to strain localization, which in turn produces relatively brittle load–displacement response. In a statically determinate structure, column A would be able to support gravity load if loaded past the load P_0. In contrast, column B would experience a sudden failure if loaded with more than P_0. In a statically indeterminate structure, the reduced ductility of a strain-softening column will reduce its ability to redistribute loads to adjacent elements after overloading. In extreme cases, such columns theoretically could explode upon reaching the axial load P_0.[2]

5.5.4 Transverse Reinforcement Required for Ductility

Ductile response of an axially loaded column requires (1) ductile materials and (2) strain-hardening after spalling of the cover concrete. Ductility of the core concrete can be achieved by providing confinement reinforcement. Strain-hardening can be achieved if the strength gained through confinement of the concrete core exceeds the strength lost due to cover spalling and longitudinal reinforcement buckling, if any. The required confinement reinforcement is derived below.

The axial strength of a column at the onset of spalling can be calculated by

$$P_0 = 0.85 f_c'(A_g - A_{st}) + A_{st} f_y \tag{5.10}$$

Equation (5.10) assumes the longitudinal reinforcement is at the yield stress f_y. This is a reasonable assumption for commonly used nonprestressed reinforcement, because the strain ε_0, in addition to shrinkage and creep strains, is sufficient to yield this reinforcement.[3] The coefficient 0.85 in Eq. (5.10) is to account for the lower strength of in-place concrete relative to concrete in standard cylinder tests.[4]

Post-spalling strength is generally defined for strain just beyond the spalling strain, such that confinement is fully effective but longitudinal reinforcement strain-hardening has not occurred. Thus, the post-spalling strength is

$$P_{00} = f_{cc}'(A_{ch} - A_{st}) + \eta A_{st} f_y \tag{5.11}$$

---

[2]As the column is compressed past the strain ε_0, the rate of shortening of the spalled section may be less than the rate of elongation of the unspalled section, in which case an explosive failure theoretically occurs. In coupon tests of concrete cylinders, this same phenomenon is observed if the testing machine is too flexible. Measurement of the descending branch of a concrete cylinder requires a stiff testing machine.

[3]This assumption should be verified for high-strength longitudinal reinforcement [yield strength greater than 75 ksi (520 MPa)].

[4]The factor $C = 0.85$ is used by many codes, including ACI 318. CSA (2004) uses $C = 0.85 - \lambda_c f_c' \geq 0.67$, in which $\lambda_c = 0.00001$ (psi) or 0.0015 (MPa).

In Eq. (5.11), the term η is introduced to reduce the post-spalling strength contribution of the longitudinal reinforcement if buckling occurs. Where transverse reinforcement adequately restrains all longitudinal reinforcement, $\eta = 1.0$. Where some bars are unrestrained, as in Figure 5.14, η = area of supported bars divided by total area of bars.

Equation (5.12) expresses the requirement for ductile behavior.

$$P_{00} > P_0 \tag{5.12}$$

Combining Eqs. (5.10) through (5.12) and rearranging terms result in

$$\frac{f'_{cc}}{0.85 f'_c} > \frac{A_g - A_{st}}{A_{ch} - A_{st}} + \frac{(1-\eta)A_{st}}{A_{ch} - A_{st}} \frac{f_y}{0.85 f'_c} \approx \frac{A_g}{A_{ch}} + (1-\eta)\rho'_l \frac{f_y}{f'_c} \tag{5.13}$$

in which $\rho'_l = A_{st}/A_{ch}$. For the case that all longitudinal reinforcement is supported against buckling, Eq. (5.13) simplifies to

$$\frac{f'_{cc}}{0.85 f'_c} > \frac{A_g}{A_{ch}} \tag{5.14}$$

Spiral-Confined Columns

Repeating relations derived in Chapter 4, we can write

$$f'_{cc} = 0.85 f'_c + 4.1 f_3 = 0.85 f'_c + 2.05 \rho_s f_s \tag{5.15}$$

In Eq. (5.15), f_s is commonly the specified yield stress f_{yt}. For specified yield stress exceeding 100 ksi (700 MPa), the stress f_s in Eq. (5.15) should be limited in accordance with Eq. (4.6).

In a spiral-reinforced column, all longitudinal bars should be adequately supported by the spiral reinforcement. Thus, substituting Eq. (5.15) in Eq. (5.14) and solving for ρ_s results in

$$\rho_s > \frac{0.85}{2.05}\left(\frac{A_g}{A_{ch}} - 1\right)\frac{f'_c}{f_s} = 0.41\left(\frac{A_g}{A_{ch}} - 1\right)\frac{f'_c}{f_s} \tag{5.16}$$

To classify a column as a spiral-reinforced column, many modern building codes (e.g., ACI 318) require slightly more spiral reinforcement, as defined in the following expression:

$$\rho_s \geq 0.45\left(\frac{A_g}{A_{ch}} - 1\right)\frac{f'_c}{f_{yt}} \tag{5.17}$$

According to ACI 318, the value of f_{yt} in Eq. (5.17) is limited to 100,000 psi (689 MPa).

Recall that Eq. (5.17) was derived so that the strength after spalling of the cover concrete would not be less than the spalling load. For large-diameter columns, the term in parentheses approaches zero, for which case the required spiral ratio would decrease to zero. Clearly, this would not produce ductile column response. Some codes specify a minimum volume ratio for spiral columns subjected to earthquake loading. According to ACI 318, the minimum permitted volume ratio for spiral reinforcement in special moment frames (a seismic-force-resisting system capable of high ductility capacity) is given by

$$\rho_s \geq 0.12\frac{f_c'}{f_{yt}} \tag{5.18}$$

Rectilinear Hoop-Confined Columns

For rectilinear confinement, the hoops in each of two principal directions must provide effective confinement stress to satisfy Eq. (5.13) or (5.14). The following derivations assume that all longitudinal reinforcement is supported against buckling. Thus, Eq. (5.14) applies. From Chapter 3, recall that the confined concrete strength expressed as a function of the minimum effective confinement stress is

$$f_{cc}' = 0.85 f_c' + 4.1 f_{emin} \tag{5.19}$$

Substituting Eq. (5.19) into Eq. (5.14) and rearranging terms result in

$$f_{emin} > \frac{0.85}{4.1}\left(\frac{A_g}{A_{ch}} - 1\right) f_c' = 0.2\left(\frac{A_g}{A_{ch}} - 1\right) f_c' \tag{5.20}$$

Noting that the effective confinement stress is $f_{emin} = k_e f_{\min 2,3}$, we can write

$$f_{\min 2,3} = \min\left(\frac{A_{sh2} f_s}{b_{c3} s}, \frac{A_{sh3} f_s}{b_{c2} s}\right) > \frac{0.2}{k_e}\left(\frac{A_g}{A_{ch}} - 1\right) f_c' \tag{5.21}$$

Some building codes (e.g., ACI 318) use a form of Eq. (5.21) to define requirements for rectilinear hoop-confined columns of special moment frames. In ACI 318, the value of k_e is implicitly taken as 0.7, resulting in the following expression:

$$\min\left(\frac{A_{sh2} f_{yt}}{b_{c3} s}, \frac{A_{sh3} f_{yt}}{b_{c2} s}\right) \geq 0.3\left(\frac{A_g}{A_{ch}} - 1\right) f_c' \tag{5.22}$$

According to ACI 318, the value of f_{yt} in Eq. (5.22) is limited to 100,000 psi (689 MPa).

For large columns, the term in parentheses to the right of the inequality approaches zero, so ACI 318 introduces the following additional requirement for special moment frames:

$$\min\left(\frac{A_{sh2} f_{yt}}{b_{c3} s}, \frac{A_{sh3} f_{yt}}{b_{c2} s}\right) \geq 0.09 f_c' \tag{5.23}$$

It must be recognized that most columns in earthquake-resisting buildings do not resist pure compression, but instead resist compression combined with shear and moment. Many building codes introduce alternative requirements for confinement reinforcement based on these considerations. This subject is discussed further in Chapters 6 and 12.

Example 5.1. The rectangular cross section of Example 4.2 is used in an axially loaded column of length 100 in (2.54 m). Calculate the relation between axial compressive load and axial shortening.

Solution

Properties of unconfined and confined concrete were determined in Example 4.3. Expected longitudinal reinforcement properties are assumed to be those shown in Figure 5.9*a*. All longitudinal bars are supported by corners of hoops or crosstie legs, with s/d_b = 4 in/1.128 in = 3.5. From Figure 5.12, we conclude that the compressive behavior of the longitudinal reinforcement will be essentially the same as the monotonic behavior under tensile loading, without concern for buckling.

It will be important to check whether the post-spalling strength is at least equal to the spalling load. This can be done either directly by comparing strengths before and after spalling or by using Eq. (5.22). Here we use the latter approach. Accordingly, $\min\left(\frac{5(0.2\text{ in}^2)(69{,}000\text{ psi})}{(33\text{ in})(4\text{ in})}, \frac{4(0.2\text{ in}^2)(69{,}000\text{ psi})}{(21\text{ in})(4\text{ in})}\right) = 523\text{ psi} \geq 0.3\left(\frac{(24\text{ in})(36\text{ in})}{(21\text{ in})(33\text{ in})} - 1\right)(5\text{ ksi}) =$ 370 psi. Therefore, strain-hardening, ductile response is expected.

The axial force versus strain and shortening relations are calculated using the procedure outlined in Figure 5.7. In the table below, axial strains are selected at values of interest, leading to axial stresses, which are multiplied by associated cross-sectional areas and summed to determine the axial force. Because the system is continuously strain-hardening, axial shortening is simply the product of column strain and length. The results are plotted in Figure 5.16. Note that, as expected, the column strain-hardens after initial spalling ($\varepsilon_c = 0.002$).

ε	f_c, ksi	f_{cc}, ksi	f_s, ksi	A_{cover}, in²	A_{ch}, in²	A_s, in²	P, kips	δ, in
0.000	0	0	0	171	693	14	0	0
0.002	4.25	4.25	58	171	693	14	4480	0.2
0.004	0	6.19	69	0	693	14	5260	0.4
0.010	0	6.61	74	0	693	14	5620	1.0
$\varepsilon_{cu} = 0.023$	0	5.82	83	0	693	14	5200	2.3

Example 5.2. Qualitatively describe how Example 5.1 will change if the spacing of transverse reinforcement is increased to $s = 9$ in.

Solution

By increasing the spacing of transverse reinforcement, the provided confinement no longer satisfies Eq. (5.22). Furthermore, $s/d_b = 8$, such that, according to Figure 5.12, the longitudinal reinforcement will be susceptible to buckling and will maintain axial stress of approximately f_y as strains increase beyond the yield point. Overall, the column will have strain-softening behavior such that the

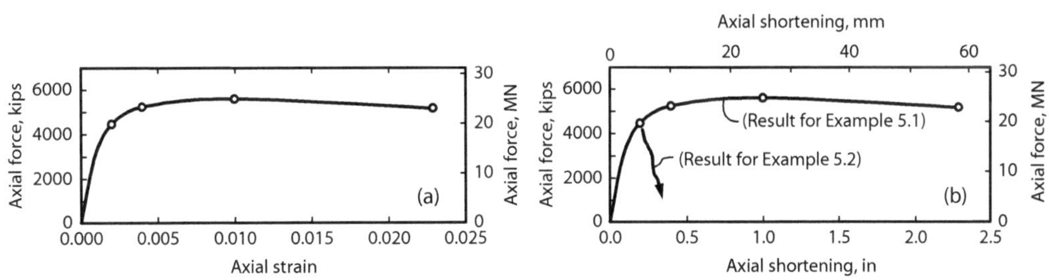

FIGURE 5.16 Calculated relations between axial force, axial strain, and axial shortening.

column will fail after the axial compressive strain reaches the strain corresponding to the peak of the unconfined concrete stress–strain relation (assumed = 0.002). This result is plotted in Figure 5.16*b*.

5.6 Tension Members

Tension members should be reinforced such that the tensile strength is determined by the reinforcement and not by tensile strength of the concrete cross section. This requirement can be expressed approximately as

$$T_n = A_{st} f_y > A_g f_t \tag{5.24}$$

in which f_t = tensile strength of concrete, which can be approximated using the split cylinder test value of $6\sqrt{f_c'}$, psi ($0.5\sqrt{f_c'}$, MPa). If Eq. (5.24) is not satisfied, a cracked section will be weaker than adjacent uncracked sections, leading to strain concentration and relatively brittle behavior if large tensile deformations are required. The inequality of Eq. (5.24) can be expressed in terms of a required tension reinforcement steel ratio as

$$\rho_l = \frac{A_{st}}{A_g} > \frac{f_t}{f_y} \tag{5.25}$$

For typical concrete and reinforcement strengths [$f_c' = 6000$ psi (40 MPa); f_y = 60,000 psi (420 MPa)], Eq. (5.25) results in $\rho_l > 0.008$. Thus, a column with minimum reinforcement ratio $\rho_l = 0.01$ as required by many building codes will satisfy Eq. (5.25) by a small margin. Additional reinforcement may be warranted for higher strength concrete.

After the onset of cracking, and assuming Eq. (5.25) is satisfied, a reinforced concrete tension member will consist of both cracked sections and uncracked sections. It is of interest to understand the development and distribution of cracking, as well as the axial stiffness, of cracked tension members.

Consider an idealized cylindrical tension member with a single, centrally located reinforcing bar (Figure 5.17). Under axial tension, cracks initially form

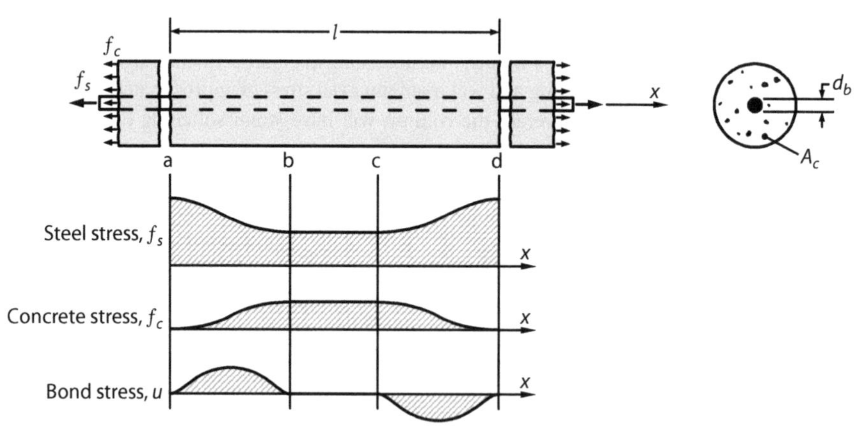

Figure 5.17 Stresses in cracked concrete tension member.

at locations that depend on the random variations of concrete strength along the length. Suppose at some loading stage there is an uncracked concrete section of length l between two cracks. At the crack locations **a** and **d**, it is reasonable to assume that steel resists all the tension. If the uncracked length l is sufficiently long, there exists a central section **bc** along which steel and concrete share the axial tension force, with equal strain, hence, with stresses f_s and f_c in proportion to the moduli of elasticity (this assumes no creep or shrinkage). The increase in concrete stress and reduction in steel stress from **a** to **b** and **d** to **c** require bond stress u acting between the steel and concrete along lengths **ab** and **cd**.

Goto (1971) tested reinforced cylinders similar to the one shown in Figure 5.17, injecting ink to record crack patterns under axial tension. The observed crack patterns suggested the deformation pattern shown in Figure 5.18, from which the existence of bond stresses can be inferred.

The relation between bond stress u, steel stress f_s, and concrete stress f_c can be derived by considering force equilibrium of a segment of a tension member having length Δx (Figure 5.19), from which we can write

$$\Delta f_s A_b = -\Delta f_c A_c = -u\pi d_b \Delta x \tag{5.26}$$

For an infinitesimally small length $\Delta x = dx$, this relation can be rearranged to express the bond stress as functions of steel and concrete stress gradients.

$$u(x) = -\frac{df_s}{dx}\frac{A_b}{\pi d_b} = -\frac{df_s}{dx}\frac{d_b}{4} \tag{5.27}$$

$$u(x) = \frac{df_c}{dx}\frac{A_c}{\pi d_b} = \frac{df_c}{dx}\frac{d_b}{4\rho_l} \tag{5.28}$$

For the central length **bc** in Figure 5.17, given the absence of bond stress and the uniform elongation of the steel and concrete, it is reasonable to conclude that

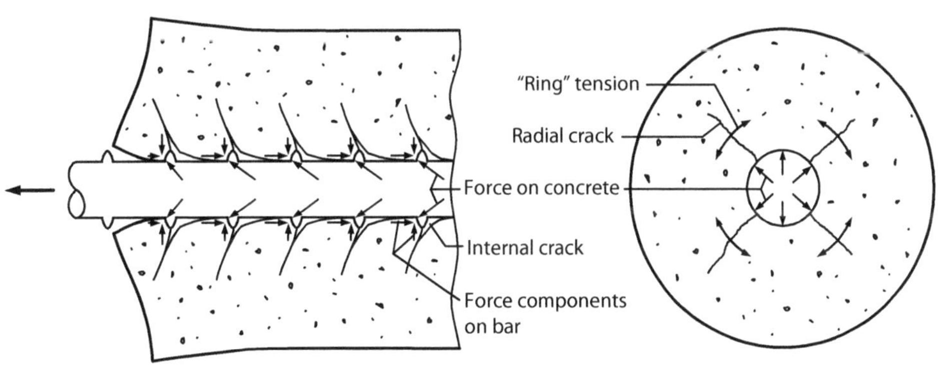

Figure 5.18 Deformation of concrete around reinforcement after formation of internal cracks. (*After Goto, 1971, courtesy of American Concrete Institute.*)

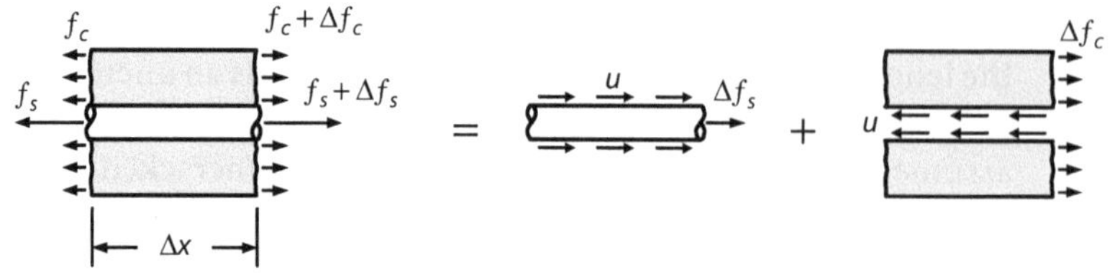

Figure 5.19 Stresses on concrete and steel in tension member.

there is no slip between concrete and reinforcement. Selecting the origin anywhere along that length, we can express the slip between reinforcement and concrete as

$$w(x)=\int_0^x \varepsilon_s\,dx-\int_0^x \varepsilon_c\,dx \tag{5.29}$$

Equation (5.29) establishes that there must be slip (relative movement) between the steel and concrete along lengths **ab** and **cd**. Furthermore, the amount of slip increases with increasing distance from the central portion **bc**.

Equation (5.27) tells us that the bond stress can be determined from the gradient of the reinforcing steel stress, which can be inferred from strain measurements in laboratory tests. It is also possible to measure the local slip between reinforcement and concrete at a crack. Bond stress–slip relations have been determined from such tests (e.g., Nilson, 1972; Eligehausen et al., 1983). Figure 5.20 shows an idealized bond stress–slip relation for monotonic loading based on a model presented by Eligehausen et al. (1998). The model depicts bond stress increasing to a peak as slip increases, followed by a descending branch, then

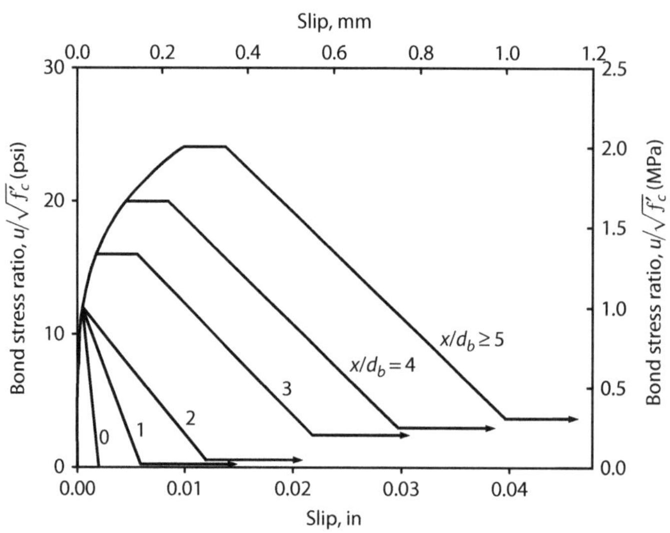

Figure 5.20 Representative monotonic relation between bond stress and slip as function of distance x from crack, bottom bar with $2d_b$ cover, light transverse reinforcement, $f_c' = 3600$ psi (25 MPa). (*After Eligehausen et al., 1998.*)

a lower plateau for larger slip. Parameters affecting the stress–strain relation include concrete material properties, bar size and deformations, proximity to cracks, transverse reinforcement, externally applied transverse compression, cyclic loading, and loading rate and duration (Eligehausen et al., 1983, Popov, 1984; Scott and Beeby, 2005).

Bond stress–slip models of the type shown in Figure 5.20 have been implemented in computer software to enable simulations of the interactions of steel and concrete in tension zones and anchorages (e.g., Eligehausen et al., 1998; Ayoub and Filippou, 1999; Lowes et al., 2004). For routine hand calculations, the bond stress–slip relation can be expressed in simplified forms, the simplest being average uniform bond stress, $\bar{u}$. Values of $\bar{u}$ ranging from $6\sqrt{f_c'}$ to $12\sqrt{f_c'}$ psi ($0.5\sqrt{f_c'}$ to $1.0\sqrt{f_c'}$ MPa) typically produce results that correlate well with observed service-level response, though values outside this range are possible for certain loadings and configurations.

Adopting the average uniform bond stress $\bar{u}$, and using Eq. (5.26), the change in concrete stress along length Δx is

$$\Delta f_c = \bar{u}\pi d_b \Delta x / A_c \tag{5.30}$$

Setting $\Delta f_c = f_{sp}$ (the splitting tensile strength of concrete) and rearranging terms, the minimum length **ab** in Figure 5.17 required to cause a new crack is

$$l_{min} = f_{sp} A_c / \bar{u}\pi d_b \tag{5.31}$$

If the length between cracks in Figure 5.17 is $l = 2l_{min}$, a new crack can form and the resulting crack spacing is l_{min}. On the other hand, if the length between cracks is slightly less than $2l_{min}$, there is insufficient bonded length to form a crack, so the crack spacing remains at slightly less than $2l_{min}$. Because the initial crack locations are essentially random, we must conclude that crack spacing will vary randomly between the values l_{min} and $2l_{min}$.

We can estimate crack width as twice the value of the slip given by Eq. (5.29) evaluated at a crack. To simplify the problem, we can ignore the concrete deformation, in which case the crack width is approximately

$$w_{cr} = 2\int_b^a \varepsilon_s \, dx \tag{5.32}$$

in which the limits of integration **a** and **b** are the points identified in Figure 5.17. We have previously determined that the spacing between cracks varies randomly between l_{min} and $2l_{min}$. Thus, the crack width according to Eq. (5.32) must also be randomly distributed between two bounds.

Concrete resists tension in the regions between cracks. Therefore, concrete stiffens a tension member even after cracks are fully developed. This so-called *tension-stiffening* effect can be beneficial in the service-load range, as it results in increased stiffness (Figure 5.21). Beyond the yield point, however, this effect can reduce the overall strain capacity of reinforced concrete (Figure 5.22). In effect,

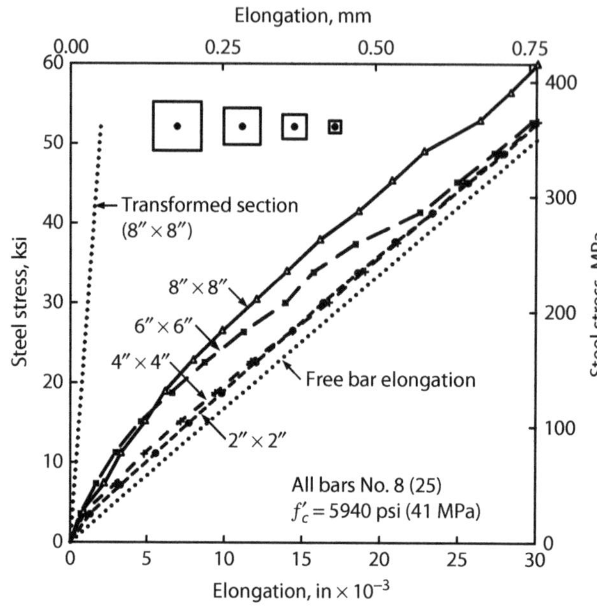

FIGURE 5.21 Effect of relative concrete and steel areas on tension-stiffening. (*After Mirza and Houde, 1979, courtesy of American Concrete Institute.*)

strains are concentrated at the cracks, such that the maximum local steel strain exceeds the average strain. These effects are especially noteworthy in lightly reinforced members and members with high concrete tensile strength.

In reinforced concrete members with transverse reinforcement, the transverse reinforcement may act as a crack initiator, in which case the crack spacing tends toward the spacing of the transverse reinforcement. This effect would modify some of the conclusions of the previous paragraphs.

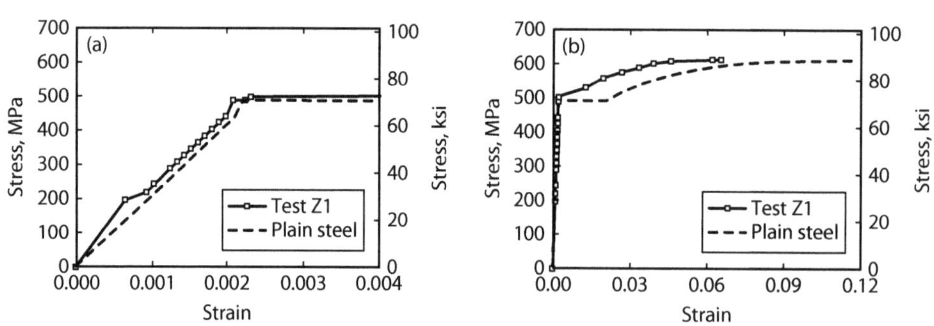

FIGURE 5.22 Steel stress as a function of strain in monotonic test on tension member. Data are for either a plain steel bar tested in air or for a reinforced concrete member having $\rho_l = 0.0098$ and $f_c' = 7.2$ ksi (50 MPa). Strain is the average along the test specimen: (*a*) strains ≤ 0.004; (*b*) total strain range. (*After Eligehausen et al., 1998, courtesy of American Concrete Institute.*)

5.7 Reversed Cyclic Loading

Yielding of a reinforced concrete tension member involves inelastic elongation of the longitudinal reinforcement, slip between reinforcement and the surrounding concrete, and cracks that widen with increasing elongation. Upon unloading, elastic strain recovery of the longitudinal reinforcement will result in partial crack closure, but cracks will remain partially open because of residual tensile strain in the yielded longitudinal reinforcement. As axial load reverses, behavior of the cracked compression member will depend on the crack width and the aspect ratio of the reinforcement and of the entire compression member. Where crack widths and aspect ratios are small, yielding of reinforcement in compression will result in stable crack closure, after which the member will behave as a composite compression member. However, if crack widths or aspect ratios are large, instability of the reinforcement or of the entire compression member may control behavior.

This section first addresses stability of longitudinal reinforcement under reversed cyclic loading. It then addresses stability of the entire reinforced concrete member.

5.7.1 Stability of Longitudinal Reinforcement

Consider a longitudinal bar that is initially yielded in tension, then unloaded, and then subjected to reversed compressive loading. Because of residual tensile strains, the bar initially will be under compressive stress but tensile strain (Figure 5.23). Furthermore, the tangent modulus will be reduced due to the Bauschinger effect. Thus, the bar may be subject to buckling instability at smaller compressive strains than would occur if the bar was loaded in monotonic compression. For cases of large tensile strains or large s/d_b, bar instability can occur while strains are still tensile (Figure 5.23*c*).

Rodriguez et al. (1999) proposed that buckling under reversed cyclic loading is controlled by the strain range ε_p from the strain at stress reversal to the maximum compressive strain (Figure 5.24). Using this concept, the monotonic strain at buckling is estimated using the methods of Section 5.5.2, and then buckling under reversed cyclic loading is estimated to occur when ε_p is equal to the monotonic buckling strain. Rodriguez et al. recommend using the double (reduced) modulus approach with effective length factor $k = 0.75$. As shown in Figure 5.25, however, similarly good correlation is obtained using the tangent modulus approach with $k = 0.5$.

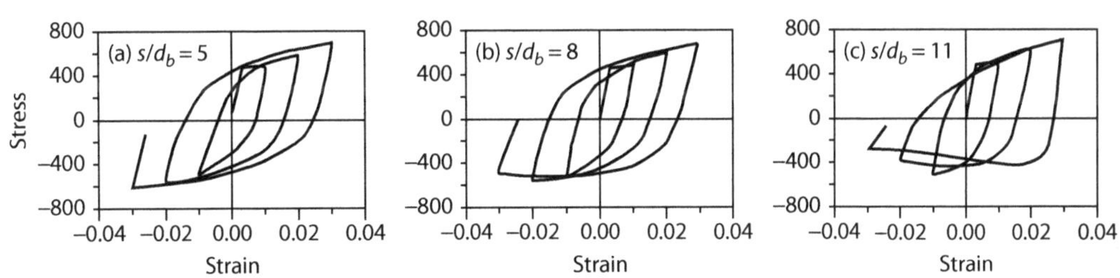

FIGURE 5.23 Reversed cyclic behavior of fixed-ended reinforcing bars: tension positive. (*After Monti and Nuti, 1992, used with permission from ASCE.*)

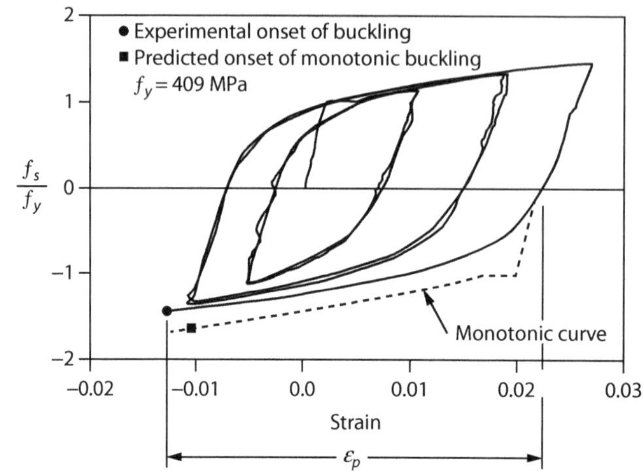

FIGURE 5.24 Cyclic stress–strain curve and shifted monotonic curve for buckling analysis. (*After Rodriguez et al., 1999, used with permission from ASCE.*)

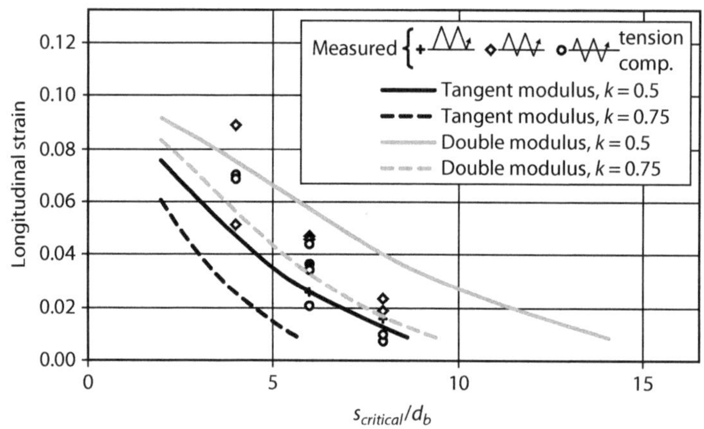

FIGURE 5.25 Comparison of measured and calculated buckling strains for reversed cyclic loading. (*Test data after Rodriguez et al., 1999.*)

The data in Figure 5.25 include tension-only cycles as well as tension–compression cycles. It is noteworthy that buckling occurs even for bars subjected to tension-only cycles. In such cases, buckling is controlled by amplitude of the tensile strains rather than amplitude of the compressive strains. Loadings similar to this occur commonly in earthquake engineering problems because tension at cracks is resisted entirely by longitudinal reinforcement whereas compression is resisted by steel and concrete following crack closure.

5.7.2 Stability of Axially Loaded Members

Behavior of axially loaded reinforced concrete members is strongly influenced by the strain history. If the loading cycles are primarily compressive, or include tensile strain excursions that do not exceed the yield strain of the longitudinal reinforcement, then behavior is primarily controlled by cyclic behavior of the concrete in compression. If, on the other hand, the loading includes tensile yielding of the longitudinal reinforcement, then stability of the member is strongly affected by the amplitude of tensile strain. It is this latter case that is the main focus of this section. Member instability was previously investigated by Paulay and Priestley (1993) and Chai and Elayer (1999). Some of the derivations here derive from those original works.

Consider a prismatic member having effective length kl, thickness b, and depth h measured perpendicular to b. The member is first subjected to axial tension T that yields the longitudinal reinforcement and reaches a peak tensile stress and strain of f_{sm} and ε_{sm} (Figure 5.26). Under load reversal, just before the member yields in compression, the longitudinal reinforcement will have unloaded by strain $\varepsilon_s = f_{sm}/E_s$ and reloaded in compression to $-\varepsilon_y$ (ignoring the Bauschinger effect), such that the residual tensile strain is approximately $\varepsilon_{res} = \varepsilon_{sm} - f_{sm}/E_s - \varepsilon_y$. We can approximate this residual strain as $\varepsilon_{res} \approx \varepsilon_{sm} - 0.005$. Invariably, in a member with two layers of reinforcement, one layer will yield before the other, producing curvature as shown in Figure 5.26*c*, and out-of-plane displacement as illustrated in Figure 5.26*b*. Alternatively, in a member with one layer of reinforcement, the member will readily rotate about the reinforcement, as shown in Figure 5.26*d*. Once the concrete interfaces on either side of a crack come into contact, the concrete compressive force may be sufficient to counteract the external moment $P\delta$, in which case the member straightens out and remains stable. If the out-of-plane displacement is sufficiently large, however, then the concrete compressive force may be unable to stabilize the member, and crushing of the compression zone will trigger

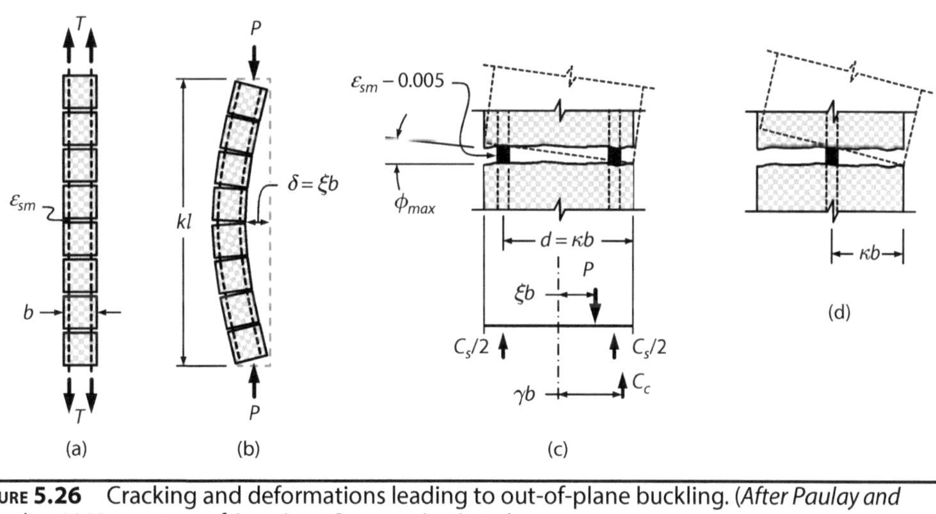

FIGURE 5.26 Cracking and deformations leading to out-of-plane buckling. (*After Paulay and Priestley, 1993, courtesy of American Concrete Institute.*)

out-of-plane failure. Thus, stability depends on the magnitude of the lateral displacement δ relative to the member thickness b, which relates to the maximum previous tensile strain ε_{sm} and the resulting curvature as illustrated in Figure 5.26*b*.

To estimate conditions for stability, we can assume that the member deforms laterally into a simple harmonic shape with maximum deflection $\delta = \xi b$ (Figure 5.26*b*). The relation between maximum deflection and maximum curvature ϕ_{max} is

$$\delta = \xi b = \phi_{max}\left(\frac{kl}{\pi}\right)^2 \tag{5.33}$$

As a first approximation, the maximum curvature from Figure 5.36*c* can be written as

$$\phi_{max} = \frac{\varepsilon_{sm} - 0.005}{d} \tag{5.34}$$

Equilibrium of forces and moments on the free-body diagram of Figure 5.26*c* results in the following two expressions:

$$\Sigma F = 0 \rightarrow P = C_s + C_c \tag{5.35}$$

$$\Sigma M = 0 \rightarrow P\xi b = C_c \gamma b \tag{5.36}$$

In Eq. (5.36), moments are taken about the centerline, such that moments of longitudinal reinforcement compressive force resultants (assumed equal) cancel. Assuming longitudinal reinforcement is stressed to f_y and assuming the concrete compressive force C_c is represented by the usual rectangular stress block with depth $\beta_1 c$ and average stress $0.85 f_c'$, we can write

$$C_s = \rho_l b h f_y \tag{5.37}$$

$$C_c = 0.85 f_c' \beta_1 c h = 0.85 f_c'(1 - 2\gamma) b h \tag{5.38}$$

Substituting Eqs. (5.35), (5.37), and (5.38) in Eq. (5.36) and manipulating the results, we obtain

$$(1 - 2\gamma)\left(\frac{\gamma}{\xi} - 1\right) = \frac{\rho_l f_y}{0.85 f_c'} = \frac{m}{0.85} \tag{5.39}$$

in which $m = \rho_l f_y / f_c'$ is the mechanical reinforcement ratio. This expression has real roots only if the following is satisfied:

$$\xi \le 0.5\left(1 + \frac{2m}{0.85} - \sqrt{\left(\frac{2m}{0.85}\right)^2 + \frac{4m}{0.85}}\right) \tag{5.40}$$

Substituting ξ from Eq. (5.40) into Eq. (5.33), solving for b/kl, and defining width b as the critical width b_{cr} result in

$$\frac{b_{cr}}{kl} = \frac{1}{\pi}\sqrt{\frac{\varepsilon_{sm} - 0.005}{\kappa\xi}} \tag{5.41}$$

The main variables appearing in Eq. (5.41) are slenderness ratio kl/b_{cr}, maximum tensile strain ε_{sm} in longitudinal reinforcement, effective depth parameter κ for longitudinal reinforcement, and ξ. Parameter κ is illustrated in Figure 5.26*c* and *d*, where it is noted that $\kappa \approx 0.8$ for thin elements with two layers of reinforcement and 0.5 for elements with single layer of reinforcement. From this, it is clear that elements with two curtains of longitudinal reinforcement are inherently more stable than elements with a single curtain. Parameter ξ relates to the mechanical reinforcement ratio [Eq. (5.40)], which is an inconvenient parameter for preliminary design. For practical construction, $0.4 \le \sqrt{\xi} \le 0.5$. Adopting values $\kappa = 0.8$ and $\sqrt{\xi} = 0.5$, Eq. (5.41) for elements with two curtains of reinforcement becomes

$$\frac{b_{cr}}{kl} = 0.7\sqrt{\varepsilon_{sm} - 0.005} \tag{5.42}$$

Chai and Elayer (1999) report results of laboratory tests on pin-ended prismatic members reinforced as wall boundary elements and subjected to tension–compression axial loading histories. Figure 5.27 compares results of Eq. (5.42) with test results for sections that buckled following tensile strain excursions to ε_{sm}. Equation (5.42) provides an acceptable approximation to the data. This expression will be used in Chapter 13 to define minimum thickness requirements of boundary elements of structural walls.

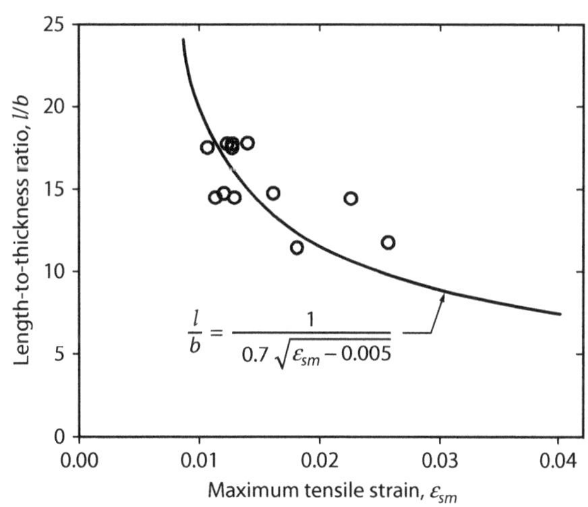

FIGURE 5.27 Buckling of prismatic sections reinforced as rectangular wall boundaries. *(Data from Chai and Elayer, 1999.)*

5.8 Chapter Review

This chapter reviewed behavior of reinforced concrete members subjected to monotonic or cyclic axial compression and axial tension. Section 5.5 presented a model for members in axial compression, suggesting that core concrete behaves as unconfined concrete until after spalling of cover concrete, and then behaves as confined concrete for strains beyond the spalling strain. Section 5.5.2 used the Euler buckling model to interpret observed instability of longitudinal reinforcement in compression members. Section 5.5.4 derived requirements for concrete confinement necessary to achieve ductile, strain-hardening response of compression members.

Section 5.6 reviewed behavior of members in tension, introducing the concept of tension-stiffening. The interaction between longitudinal reinforcement and surrounding concrete increases axial stiffness of tension members. It reduces tensile elongation capacity because reinforcement strain tends to concentrate at discrete cracks.

Section 5.7 reviewed effects of reversed cyclic loading. In reinforced concrete members for which compressive strains are limited by surrounding concrete, the tendency for buckling is mainly affected by slenderness ratio and the tensile strain amplitude.

References

ACI 318 (2014). *Building Code Requirements for Structural Concrete (ACI 318-14) and Commentary*, American Concrete Institute, Farmington Hills, MI.

Arteta, C., and J.P. Moehle (2014). Unpublished tests at the University of California, Berkeley, CA.

Ayoub, A., and F.C. Filippou (1999). "Mixed Formulation of Bond-Slip Problems under Cyclic Loads," *Journal of Structural Engineering*, Vol. 125, No. 6, pp. 661–671.

Bae, S., A.M. Mieses, and O. Bayrak (2005). "Inelastic Buckling of Reinforcing Bars," *Journal of Structural Engineering*, Vol. 131, No. 2, pp. 314–321.

Baker, A.L.L., and A.M.N. Amarakone (1964). "Inelastic Hyperstatic Frames Analysis," *Flexural Mechanics of Reinforced Concrete*, Special Publication 12, American Concrete Institute, Farmington Hills, MI, pp. 85–142.

Blume, J.A., N.M. Newmark, and L.H. Corning (1961). *Design of Multi-Story Reinforced Concrete Buildings for Earthquake Motions*, Portland Cement Association, Chicago, IL, 318 pp.

Bresler, B., and P.H. Gilbert (1961). "Tie Requirements for Reinforced Concrete Columns," *Journal of ACI*, Vol. 58, No. 5, pp. 555–570.

Brown, W.A., D.E. Lehman, and J.F. Stanton (2008). *Bar Buckling in Reinforced Concrete Bridge Columns*, Report No. PEER 2007/11, Pacific Earthquake Engineering Research Center, University of California, Berkeley, CA, 123 pp.

Chai, Y.H., and D.T. Elayer (1999). "Lateral Stability of Reinforced Concrete Columns under Axial Reversed Cyclic Tension and Compression," *ACI Structural Journal*, Vol. 96, No. 5, pp. 780–789.

CSA (2004). *Design of Concrete Structures (CSA A23.3-04)*, Canadian Standards Association, Mississauga, Canada.

Dilger, W.H. (1982). "Creep Analysis of Prestressed Concrete Structures Using Creep-Transformed Section Properties," *PCI Journal*, Vol. 27, No. 4, pp. 212–217.

Eligehausen, R., J. Ozbolt, and U. Mayer (1998). "Contribution of Concrete between Cracks at Inelastic Steel Strains and Conclusions for the Optimization of Bond," *Bond and Development of Reinforcement*, Special Publication 180, American Concrete Institute, Farmington Hills, MI, pp. 45–80.

Eligehausen, R., E.P. Popov, and V.V. Bertero (1983). *Local Bond Stress Slip Relationships of Deformed Bars under Generalized Excitations*, Report No. UCB/EERC 82/23, Earthquake Engineering Research Center, University of California, Berkeley, CA, 169 pp.

Gamble, W.L., and D. Thomson (1977). *Instructional Slides*, Produced by Civil Engineering Department and Instructional Media Division, Copyright 1977 Board of Trustees, University of Illinois, Urbana–Champaign, IL.

Goto, Y. (1971). "Cracks Formed in Concrete around Deformed Tension Bars," *ACI Journal*, Vol. 68, No. 4, pp. 244–251.

Lowes, L.N., J.P. Moehle, and S.J. Govindjee (2004). "Concrete-Steel Bond Model for Use in Finite Element Modeling of Reinforced Concrete Structures," *ACI Structural Journal*, Vol. 101, No. 4, pp. 501–511.

Mehta, P.K., and P.J. Monteiro (2014). *Concrete*, 4th ed., McGraw-Hill Professional, New York, NY, 675 pp.

Mindess, S., J.F. Young, and D. Darwin (2003). *Concrete*, 2nd ed., Prentice-Hall, Upper Saddle River, NJ, 644 pp.

Mirza, S.M., and J. Houde (1979). "Study of Bond Stress-Slip Relationships in Reinforced Concrete," *ACI Journal*, Vol. 76, No. 1, pp. 19–46.

Monti, G., and C. Nuti (1992). "Nonlinear Cyclic Behavior of Reinforcing Bars Including Buckling," *Journal of Structural Engineering*, Vol. 118, No. 12, pp. 3268–3284.

Nilson, A.H. (1972). "Internal Measurement of Bond Slip," *ACI Journal*, Vol. 69, No. 7, pp. 439–441.

NZS 3101 (2006). *Concrete Structures Standard—The Design of Concrete Structures*, Standards, New Zealand, 696 pp.

Pantazopoulou, S.J. (1998). "Detailing for Reinforcement Stability in RC Members," *Journal of Structural Engineering*, Vol. 124, No. 6, pp. 623–632.

Papia, M., G. Russo, and G. Zingone (1988). "Instability of Longitudinal Bars in RC Columns," *Journal of Structural Engineering*, Vol. 114, No. 2, pp. 445–461.

Papia, M., and G. Russo (1989). "Compressive Concrete Strain at Buckling of Longitudinal Reinforcement," *Journal of Structural Engineering*, Vol. 115, No. 2, pp. 382–397.

Park, R., and T. Paulay (1975). *Reinforced Concrete Structures*, John Wiley & Sons, Inc., New York, NY, 769 pp.

Paulay, T., and M.J.N. Priestley (1993). "Stability of Ductile Structural Walls," *ACI Structural Journal*, Vol. 90, No. 4, pp. 385–392.

Popov, E.P. (1984). "Bond and Anchorage of Reinforcing Bars under Cyclic Loading," *ACI Journal*, Vol. 81, No. 4, pp. 340–349.

Rodriguez, M.E., J.C. Botero, and J. Villa (1999). "Cyclic Stress-Strain Behavior of Reinforcing Steel Including Effect of Buckling," *Journal of Structural Engineering*, Vol. 125, No. 6, pp. 605–612.

Scott, R.H., and A.W. Beeby (2005). "Long-Term Tension-Stiffening Effects in Concrete," *ACI Structural Journal*, Vol. 102, No. 1, pp. 31–39.

Sheikh, S.A., and S.M. Uzumeri (1980). "Strength and Ductility of Tied Concrete Columns," *Journal of the Structural Division*, Vol. 106, No. ST5, pp. 1079–1102.

CHAPTER 6

Moment and Axial Force

6.1 Preview

Seismic response of a building induces internal forces and deformations of the structural framing that need to be accommodated by the components and their connections. In most cases, the framing is designed so that flexure is the predominant deformation mode, and typically this includes inelastic flexure. In this chapter we consider analysis and design for moment, including moment without significant axial load (as occurs in beams and slabs) and moment with axial load (as occurs in columns and structural walls). The presentation includes stiffness, strength, and deformation capacity under monotonic and cyclic loads. In modern reinforced concrete construction, the reinforcement to resist flexural moments can include conventional reinforcement, prestressed reinforcement, or both; in the present chapter only conventional reinforcement is considered.

6.2 Some Observations about Flexural Behavior

In a structural member, applied moments are resisted by flexural tension and compression on opposite faces of the member. If the direction of loading changes, as is common for earthquake loading, reversal of moment can lead to reversal of the tensile and compressive actions. We know from previous chapters that behavior of concrete and steel in compression and tension depends on the prior load history. Thus, behavior of a reinforced concrete flexural member also must be history-dependent. To understand some of the behaviors of interest, we begin by reviewing the performance of a reinforced concrete beam tested in a laboratory under loading that is representative of earthquake effects (Panagiotou et al., 2013).

The test specimen is a cantilever that can be considered to represent one-half of a beam span in a tall moment-resisting frame, with the left end corresponding to the connection to a column and the right end corresponding to a point of inflection in the beam (Figure 6.1). The beam is reinforced with closely spaced transverse reinforcement that provides lateral support for corner and alternate longitudinal reinforcing bars, as well as provides adequate shear strength to

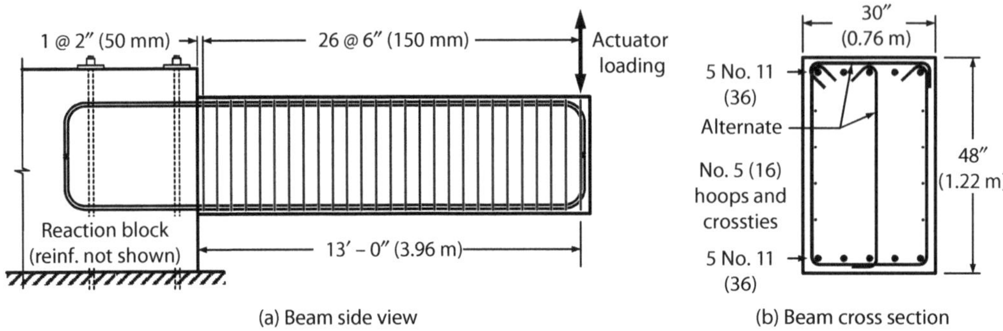

FIGURE 6.1 (*a*) Laboratory test beam. (*b*) Beam cross section. (*After Panagiotou et al., 2013.*)

preclude shear failure. Measured concrete compressive strength is 6.1 ksi (42.3 MPa) and longitudinal and transverse reinforcement yield stresses are 73 and 71 ksi (503 and 490 MPa), respectively. The beam is tested by subjecting it to multiple cycles of lateral displacements at progressively increasing amplitude.

During testing, the first signs of distress are cracks in the concrete cover, oriented mainly perpendicular to the longitudinal axis, at beam bending moment of approximately 560 k-ft (760 kN-m). The appearance of cracks is accompanied by minor reduction in stiffness (Figure 6.2). A sharp yielding point in the moment-drift ratio plot signals the onset of yielding of the longitudinal reinforcement at moment of approximately 1900 k-ft (2600 kN-m). No concrete spalling was observed until drift ratio reached approximately 0.029. Cycling at this amplitude resulted in initiation of longitudinal reinforcement buckling near the top of the beam, which can be identified by the horizontal crack at the level of the reinforcement (Figure 6.3*b*). Also of interest is a wide crack at the beam–anchorage interface, which appears to contribute significantly to total beam rotation. By drift ratio 0.039, the top reinforcement has buckled more noticeably (Figure 6.3*c*), and crushing has

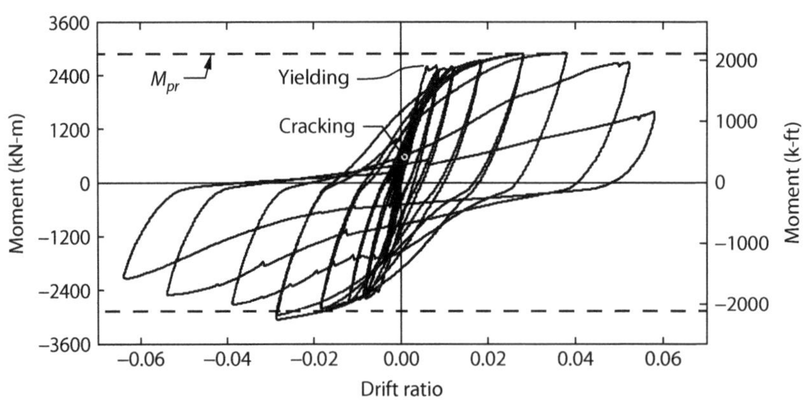

FIGURE 6.2 Measured relation between beam moment at the fixed end and drift ratio at the free end (defined as displacement at the actuator divided by horizontal distance to the fixed anchorage). Downward displacement defined as positive. (*After Panagiotou et al., 2013.*)

(a) Drift ratio = 0.010 (b) Drift ratio = 0.029 (c) Drift ratio = 0.039

FIGURE 6.3 Development of damage in test beam. (*After Panagiotou et al., 2013.*)

extended deep into the beam section. These local failures are the main source of strength degradation that occurs in subsequent loading cycles (Figure 6.2).

Figure 6.2 shows two broken lines corresponding to the *probable moment strength*, M_{pr}. This is a calculated quantity intended to approximate the moment strength for inelastic response. In this chapter, we will define procedures for calculating M_{pr}, as well as procedures for estimating other quantities of interest, such as initial stiffness, onset of spalling, and ultimate displacement capacity. These procedures will be useful for understanding and controlling the performance capabilities of earthquake-resisting building frames.

6.3 Internal and External Force Equilibrium

Structural analysis involves idealizations that must be carried through the component analysis and design in a consistent manner. As an example, consider the reinforced concrete frame and loadings shown in Figure 6.4. Following the usual convention, we idealize the foundation using simplified supports. Furthermore, the beams and columns are represented by line elements having stiffness properties based on the member cross sections (Figure 6.4*b*). We can analyze this idealized frame readily using structural analysis theory (or applicable software), producing reactions and internal actions corresponding to the member centerlines (Figure 6.4*c*). When we subsequently analyze the effect of these forces on the cross section, we must be consistent in placing them at the centerline of the member cross section (Figure 6.4*d*).

Note that Figure 6.4*d* is a free-body diagram of a thin slice of the column cross section, rotated 90° clockwise in this figure. The internal actions calculated from the frame analysis, as shown in Figure 6.4*c*, are shown on the left side of the free-body diagram in Figure 6.4*d*. The internal stresses produced by these actions are shown on the right side of the free-body diagram.

Some computer software and design aids define the axial load to be acting at a specific location of the cross section that is different from the centerline location used in the structural analysis. In this case, when the axial force is shifted by a distance *e*, the moment needs to be adjusted by the amount *Pe* so that equilibrium is maintained. Figure 6.4*e* illustrates this adjustment to the moment.

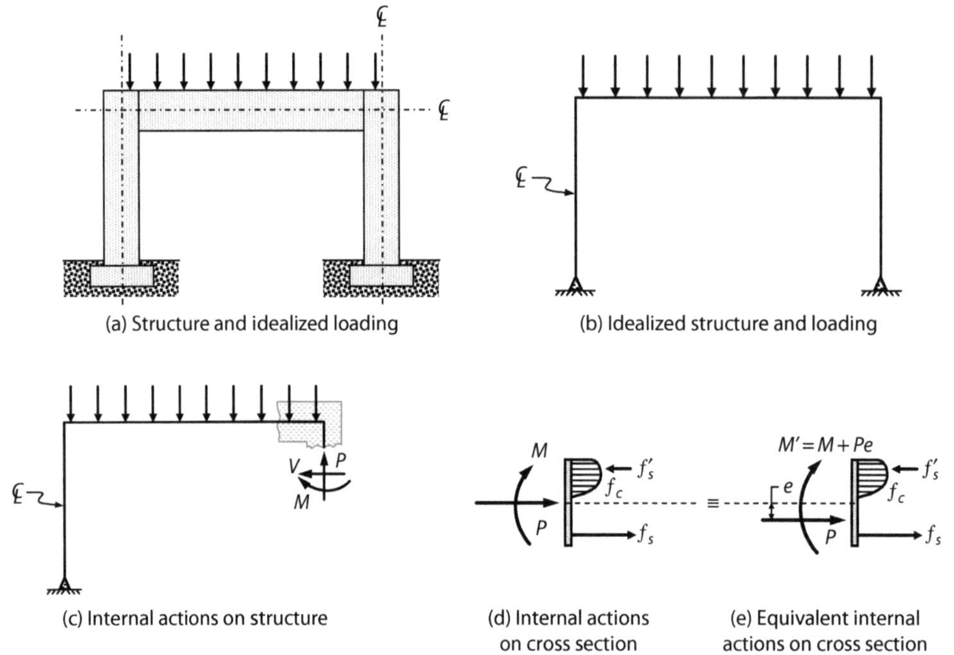

FIGURE 6.4 Structure idealization and load representation. *Note:* Shear *V* intentionally not shown in (*d*) and (*e*).

6.4 Flexural Deformations

We begin with the fundamental hypothesis of flexural theory, introduced by Bernoulli (1645–1705), which may be stated as follows: Plane sections through a flexural member, taken normal to its axis, remain plane after the member is subjected to bending. In reinforced concrete construction, this means that strains vary linearly through the member, without slip of reinforcement relative to concrete. For slender flexural members, with or without axial load, this assumption has been shown by tests to be valid in an average sense (Figure 6.5). As discussed in Chapter 5, it does not strictly hold true at crack locations. The Bernoulli assumption may also not be valid at lap splices or member ends where significant bond stresses exist, in deep or flanged members, or near ultimate loading stages after strain-softening of the flexural compression zone. We shall consider some of these conditions later in this book.

Figure 6.6 illustrates the flexural deformations of an initially straight segment of a beam-column. Curvature ϕ is the rate of angle change along the axis of the member. From the geometry shown in Figure 6.6, the relation between curvature and strain is

$$\phi = \frac{d\theta}{dx} = \frac{z\,d\theta}{z\,dx} = \frac{\varepsilon_z}{z} \tag{6.1}$$

Equation (6.1) is based solely on geometry (with the assumption that plane sections remain plane) and is valid for linear and nonlinear material

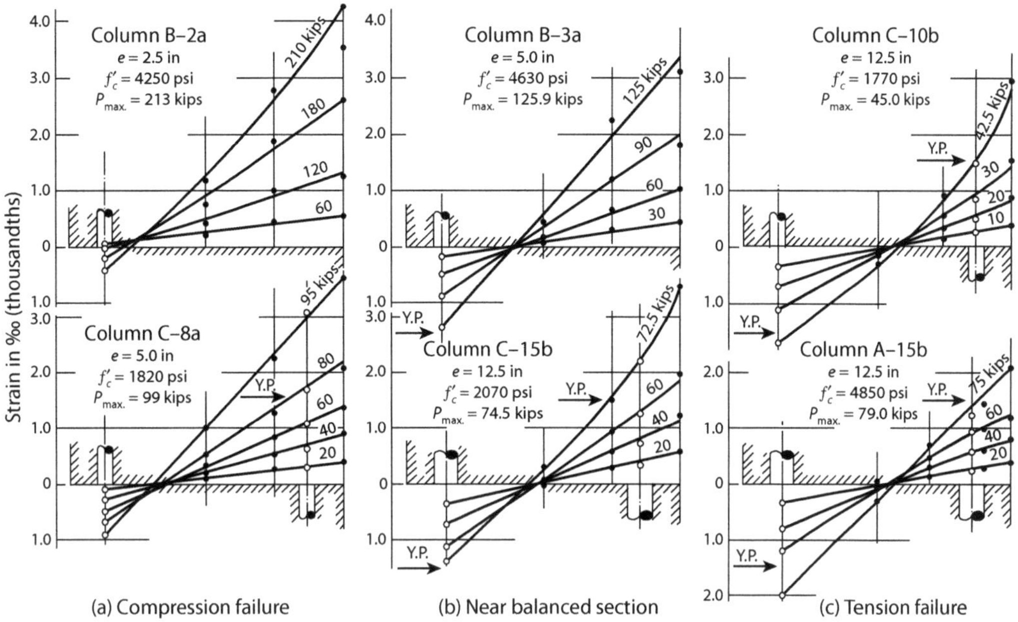

Figure 6.5 Strain distributions in tied columns. (*After Hognestad, 1951, courtesy of the University of Illinois at Urbana–Champaign Archives.*)

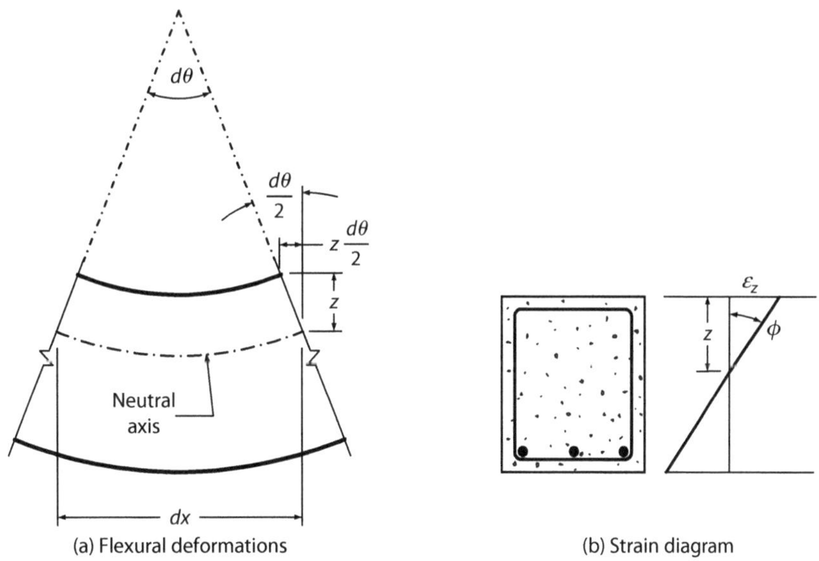

Figure 6.6 Flexural deformations and normal strains.

behavior. For the special case of linear-elastic material behavior, the following relations hold:

$$\phi = \frac{d\theta}{dx} = \frac{\varepsilon_z}{z} = \frac{\sigma_z}{E}\frac{1}{z} = \frac{Mz}{EI}\frac{1}{z} = \frac{M}{EI} \tag{6.2}$$

6.5 Flexural Behavior of Sections

6.5.1 General Observations

When a cross section is subjected to flexural moment, with or without axial load, internal stresses develop to equilibrate the external actions, with corresponding strains and curvature. Figure 6.7 depicts an idealized moment–curvature relation for a lightly reinforced concrete section. The figure also shows stress and strain states of the cross section corresponding to various stages of response.

The relation between moment and curvature is nearly linear to the point where concrete cracking occurs ($M = M_{cr}$). Although micro-cracking of the concrete matrix exists even prior to loading, cracking becomes apparent when the extreme concrete tension fiber reaches the tensile strength. When this occurs, there usually is a noticeable reduction in tangent stiffness, as shown. The reduction usually is gradual because of the tension-stiffening effect (see Chapter 5), becoming more pronounced as moment increases above the cracking moment.

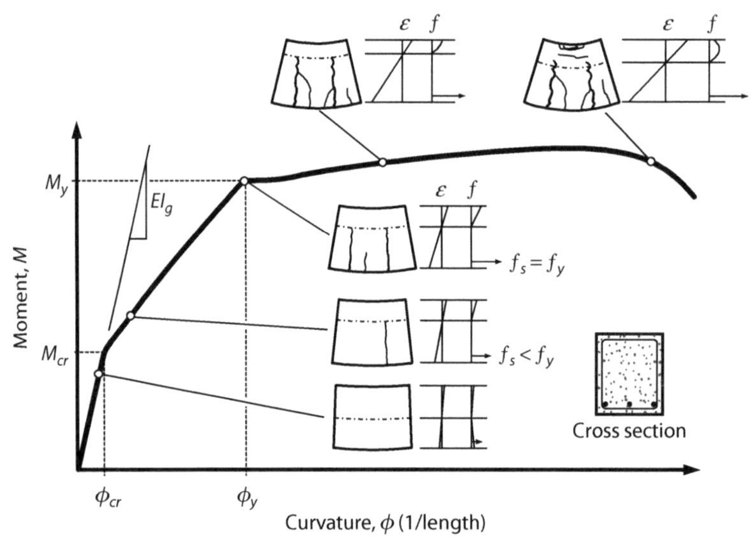

FIGURE 6.7 Moment–curvature behavior of under-reinforced beam section.[1]

[1]The amount of longitudinal reinforcement affects whether the longitudinal reinforcement yields before crushing the concrete in the compression zone. The moment–curvature relation depicted is for a lightly reinforced section for which longitudinal reinforcement yielding precedes concrete crushing. A beam with this amount of reinforcement is referred to as an *under-reinforced* beam. A heavily reinforced beam that sustains crushing failure before yielding the longitudinal tension reinforcement is said to be *over-reinforced.*

Another notable change in moment–curvature stiffness occurs upon yielding of the longitudinal tension reinforcement ($M = M_y$). The change is most noticeable in sections with longitudinal tension reinforcement concentrated in a single layer. Where the longitudinal tension reinforcement is in multiple layers, as occurs, for example, in a circular cross section, the change in stiffness due to reinforcement yielding is more gradual because only the outermost layer of reinforcement yields initially, with yielding progressing further into the section depth as curvature is progressively increased.

As curvature increases beyond the yield point, the rate by which steel stress increases is reduced because of yielding. In contrast, concrete stresses continue to increase until the stress capacity is reached. Therefore, to maintain axial force equilibrium, the depth of the compression zone must decrease with increasing curvature. This produces a slight increase in the internal moment arm and, hence, a slight increase in the moment resistance. Strain-hardening of the longitudinal reinforcement also contributes to the increase in flexural moment during this phase of response.

As curvature is further increased, concrete in the compression zone eventually reaches its in situ compressive strength. With further increases in curvature, the extreme fibers of the compression zone begin to unload. Subsequent spalling of concrete leads to reduction of the effective cross section, requiring the neutral axis to move deeper into the cross section to maintain equilibrium. For sections without confining reinforcement, this leads to irrecoverable loss in moment resistance, as shown in Figure 6.7. In a confined section, strain-hardening of the confined concrete (and of any compression longitudinal reinforcement) may be sufficient to enable the section to be deformed well beyond the stage of initial spalling of the concrete section without significant loss of moment resistance.

6.5.2 Spalling Strain

Spalling of cover concrete is a key performance point for structural concrete. For unconfined sections, spalling reduces the effective cross section and signals the onset of section failure. For confined sections, the confined core acting together with compression reinforcement may be capable of stabilizing the compression zone, in which case ductile response beyond initial spalling is possible. Regardless, spalling may require costly repairs and in some cases may trigger decisions about continued occupancy and function.

In an axial compression member, all of the fibers of the cross section reach the critical strain at the same instant in time, such that local spalling becomes apparent shortly after the section reaches the strain ε_0. In a flexural member, however, strains vary through the depth such that fibers with lower strain stabilize the cross section, in essence providing support for the extreme compression fibers. This *strain gradient effect* enables the extreme compression fibers to reach larger strain before spalling is apparent.

A similar effect occurs for flexural members with *moment gradient*. In such members, only one cross section along the length is subjected to the maximum effect, and adjacent, less strained sections may help stabilize the compression zone. A similar effect occurs in members framing into stiff adjacent supports such as walls or foundations. The adjacent support provides confinement to the

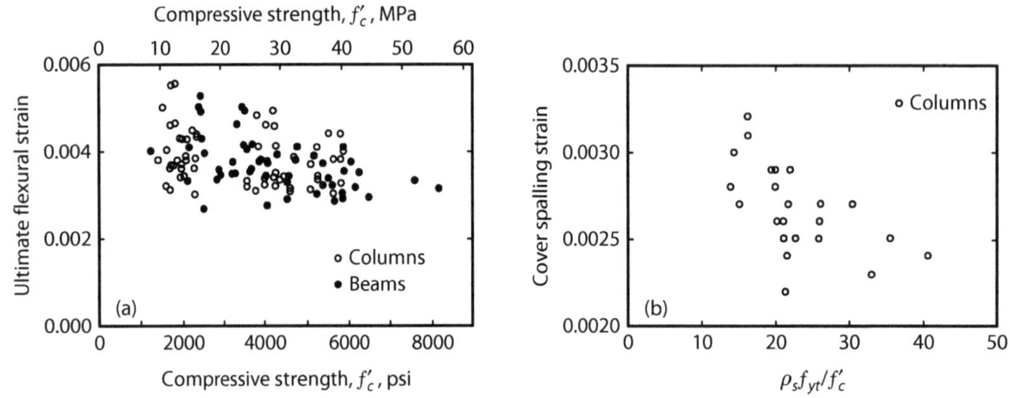

FIGURE 6.8 Spalling strains in reinforced concrete flexural members with or without axial load. [*(a) Data from Mattock et al., 1961; (b) data from Bae and Bayrak, 2003. Courtesy of American Concrete Institute.*]

member end, and in extreme cases may move the critical section away from the section of maximum moment (Lehman et al., 2004).

Laboratory tests on reinforced concrete members (Figure 6.8*a*) and plain concrete prisms indicate that the spalling strain in flexural members typically is in the range 0.003 to 0.006. Building codes commonly adopt a limiting strain of 0.003 (e.g., ACI 318, 2014; NZS 3101, 2006) or 0.0035 (e.g., CSA, 2004) as an effective lower-bound strain at failure of flexural members. Some tests on high-strength confined concrete columns (Figure 6.8*b*) have shown spalling at lower strains. Fasching and French (1998) showed that limiting strain was more sensitive to aggregate type than concrete compressive strength, with limiting values ranging from 0.002 to 0.005, and average values for all aggregates exceeding 0.003. ACI ITG-4.3R-07 (2007) recommends using 0.003 for design using high-strength concrete.

For performance-based seismic design of structures, or assessment of existing structures, a central value for the spalling strain may be sought rather than a lower bound. The value $\varepsilon_{sp} = 0.004$ is recommended in ATC 32 (1996), 0.004 to 0.005 is recommended in FEMA 306 (1999), and 0.005 is recommended in ASCE 41 (2013) and ATC 40 (1996). In this text, unless otherwise noted for specific applications, the value $\varepsilon_{sp} = 0.004$ is adopted as the expected strain at spalling in a flexural member.

6.6 Moment–Curvature Analysis

6.6.1 Analysis Assumptions and General Procedure

Analysis of the moment–curvature response of flexural members, with or without axial loads, relies on the three fundamental assumptions introduced in Chapter 5, namely:

1. Sections that are plane before loading remain plane after loading. Figure 6.5 demonstrates the validity of this assumption.

2. The stress–strain relations are known from properties measured in coupon tests on concrete and reinforcing steel. Routinely we will ignore effects of creep and shrinkage, although these effects can be important, especially for service load-deflection assessment.
3. Equations of equilibrium can be used to calculate the axial force and moment on the cross section given the internal stresses.

Another commonly applied assumption is that tension in concrete can be ignored once cracking has occurred. We will adopt this assumption for routine hand calculations. For computer calculations, however, tensile resistance of concrete is sometimes considered for any sections having tensile strain less than the cracking strain. Regardless, the effect on calculated moment strength after first cracking is negligible.

Figure 6.9 illustrates application of these assumptions to the analysis of a rectangular cross section with tension and compression longitudinal reinforcement. Under the action of concentric axial force and moment, a linear strain profile develops (Figure 6.9*b*). Stresses are obtained from strains using the stress–strain relations (Figure 6.9*c*). Integrating these stresses over the cross section leads to corresponding stress resultants. Figure 6.9*d* shows a free-body diagram obtained by taking a thin slice through the flexural member, with the stress resultants shown on the right-hand side of the free-body and the internal axial force and moment that equilibrate them on the other side. Equilibrium of this free body establishes the internal forces P and M corresponding to the assumed strain distribution. If the solution for a different axial load or moment is sought, the strain profile is adjusted and the procedure is repeated.

The analysis procedure just outlined is identical to that introduced in Chapter 5, except here the strain varies across the section. We will use this basic approach repeatedly in this chapter to analyze flexural response of cross sections.

6.6.2 Linear-Elastic Response of Uncracked Sections

Behavior before cracking can be computed according to the approach outlined in Figure 6.9 including the tensile stresses in concrete. Alternatively, we could use the transformed section analysis approach introduced in Chapter 5, transforming

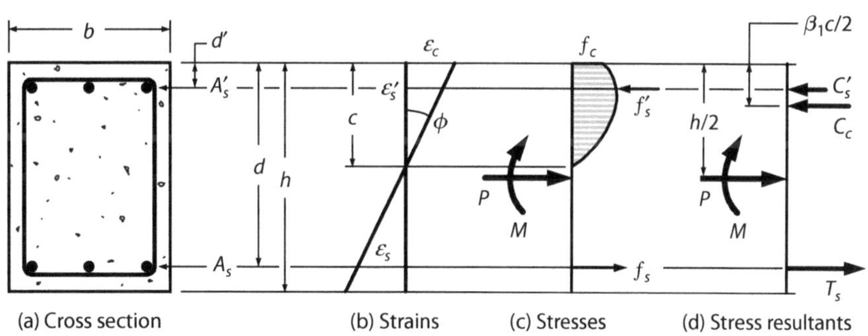

Figure 6.9 Flexural analysis of rectangular cross section.

steel areas A_s into concrete having areas nA_s, then computing the cross-sectional properties for the transformed section. This approach will suggest that increasing the steel ratio results in an increase in both stiffness and the cracking moment. As discussed in Chapter 5, however, shrinkage will produce tensile stresses in concrete, and these will increase as the steel ratio increases. Consequently, the cracking moment actually may decrease as the steel ratio is increased.

Given the complexities of including restrained shrinkage and stress relaxation in the analysis, and the uncertainties in the computed results, the expedient approach is to estimate stiffness and cracking moment using the gross section properties. Thus, for example, given a rectangular cross section of width b and total depth h, the gross moment of inertia is $I_g = bh^3/12$ and the cracking moment and curvature can be estimated as

$$M_{cr} = \frac{f_r I_g}{h/2} \tag{6.3}$$

$$\phi_{cr} = \frac{M_{cr}}{E_c I_g} \tag{6.4}$$

in which modulus of rupture can be taken as

$$\begin{aligned} f_r &= 7.5\lambda\sqrt{f'_c}, \text{psi} \\ f_r &= 0.63\lambda\sqrt{f'_c}, \text{MPa} \end{aligned} \tag{6.5}$$

6.6.3 Linear-Elastic Response of Cracked Sections

Sections with sufficient longitudinal reinforcement reach a new equilibrium state following cracking, in which the majority of flexural tension is resisted by longitudinal reinforcement with flexural compression shared by concrete and any reinforcement in the compression zone. We generally simplify the problem, without appreciable loss of accuracy, by assuming concrete resists no tension following cracking. The assumed strain and stress conditions are shown in Figure 6.10. The compression zone depth is denoted kd, where k is a coefficient depending on the axial load and section properties.

The solution for a cracked section with axial load and moment is readily solved by following the general analysis procedure outlined previously. The only additional

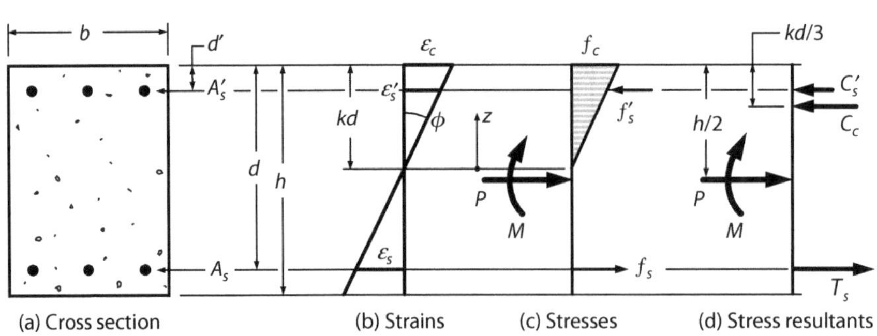

FIGURE 6.10 Stresses and strains for linear cracked sections.

requirement is to select a strain profile consistent with a cracked concrete section (tensile strain exceeding the cracking strain).

For the rectangular section shown, the sum of forces on the free-body diagram in Figure 6.10*d*, taking forces positive toward the right, is

$$\sum \vec{F} = 0 : P + T_s - C_s' - C_c = 0 \tag{6.6}$$

from which we can solve for the axial load P as

$$P = C_c + C_s' - T_s \tag{6.7}$$

Moment M is obtained by summing moments about any point. Here we select the top of the cross section, not because it is necessarily the most expedient point but instead because it serves to illustrate how all terms are included. Taking moment positive clockwise

$$\sum \widehat{M_{top}} = 0 : M - P\frac{h}{2} + C_s'd' + C_c\frac{kd}{3} - T_s d = 0 \tag{6.8}$$

from which we can solve for the moment M as

$$M = P\frac{h}{2} - C_s'd' - C_c\frac{kd}{3} + T_s d \tag{6.9}$$

The procedure outlined above produces a combination of axial force P and moment M corresponding to the selected strain distribution. If the problem at hand is for a different axial load or moment, the procedure is iterated with different strain profiles to arrive at the desired solution.

If the axial force $P = 0$, we can use the transformed section approach conveniently to develop a closed-form solution. Figure 6.11 illustrates the cracked transformed section. Note that only the concrete in the compression zone is represented, as the tension zone is assumed to be fully cracked.[2] The neutral axis

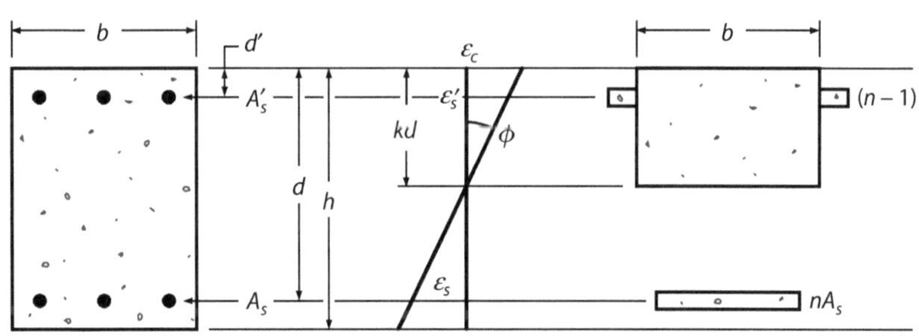

FIGURE 6.11 Cracked transformed section for the case of zero axial load.

[2]The reinforcement in the compression zone is transformed into equivalent concrete area nA_s' in this example. Under sustained loads, creep strains would increase the stress in the compression reinforcement. In the allowable stress design method of the past, it was common to transform the compression steel area to $2nA_s'$ to approximately account for this effect.

is located at the geometric centroid of the cracked section. Thus, summing moments of areas about the neutral axis we have

$$\frac{b(kd)^2}{2} + (n-1)A_s'(kd - d') = nA_s(d - kd) \tag{6.10}$$

Solving this quadratic equation for k we find

$$k = \left[(n\rho + (n-1)\rho')^2 + 2\left(n\rho + (n-1)\rho'\frac{d'}{d}\right)\right]^{1/2} - (n\rho + (n-1)\rho') \tag{6.11}$$

Equation (6.11) is usually simplified by assuming $n - 1 \approx n$, which is equivalent to ignoring the concrete displaced by the compression reinforcement, resulting in the more common form

$$k = \left[(\rho + \rho')^2 n^2 + 2\left(\rho + \rho'\frac{d'}{d}\right)n\right]^{1/2} - (\rho + \rho')n \tag{6.12}$$

The cracked section moment of inertia is then

$$I_{cr} = \frac{b(kd)^3}{3} + (n-1)A_s'(kd - d')^2 + nA_s(d - kd)^2 \tag{6.13}$$

Note that in Eq. (6.13) the moments of inertia of the transformed areas of reinforcement about their individual centroidal axes are ignored because those terms contribute little to the final result.

It is worth reiterating that the results of Eqs. (6.12) and (6.13) are valid only for rectangular sections where materials respond in the linear-elastic range, and only for the case of $P = 0$. For linear materials and $P = 0$, the neutral axis depth kd is a property of the linear-elastic cross section and is independent of the applied moment. This is not the case if $P \neq 0$.

Using the results from the transformed section (for $P = 0$), the internal stresses are

$$f_c = \frac{Mz}{I_{cr}} \tag{6.14}$$

$$f_s = n\frac{Mz}{I_{cr}} \tag{6.15}$$

in which f_c and f_s are the stresses at distance z from the neutral axis.

Curvature can be calculated using Eq. (6.1) with the strains shown in Figure 6.10, or for the case of $P = 0$ we can use the cracked section moment of inertia and write

$$\phi = \frac{M}{E_c I_{cr}} \tag{6.16}$$

Equation (6.15) can be rearranged to express moment as a function of steel stress. Substituting the yield stress f_y and setting z as the distance from the neutral axis to the centroid of the tension reinforcement, the yield moment (for $P = 0$) is

$$M_y = \frac{1}{n}\frac{f_y I_{cr}}{(d - kd)} \tag{6.17}$$

Equation (6.17) assumes the concrete and compression steel are responding in the linear-elastic range. However, the result of Eq. (6.17) may be acceptably accurate even for cases where the concrete compressive stress calculated by Eq. (6.14) is as high as f_c'. Where greater accuracy is desired, or for higher computed compressive stresses, procedures accounting for nonlinearity of the concrete stress–strain relation can be used. Later sections in this chapter present procedures for carrying out such calculations.

6.6.4 Flexural Stiffness at Service Loads

The gross section moment of inertia provides an adequate measure of flexural stiffness for members that are not cracked by service loads. Where service loads are sufficient to produce flexural cracking, some reduction in stiffness is appropriate. The fully cracked stiffness introduced in Section 6.6.3 generally will underestimate stiffness under service loads because of tension-stiffening of the flexural tension zone.[3] The degree of the underestimation depends on the longitudinal reinforcement ratio and service load intensity relative to the cracking load.

Branson (1963, 1977) proposed an effective moment of inertia to represent the flexural stiffness of cracked sections, of the form

$$I_e = \left(\frac{M_{cr}}{M_a}\right)^m I_g + \left[1 - \left(\frac{M_{cr}}{M_a}\right)^m\right] I_{cr} \le I_g \tag{6.18}$$

Equation (6.18) with $m = 4$ produces an estimate of the cross-sectional stiffness, whereas the same equation with $m = 3$ produces an estimate of the average stiffness over the entire span for deflection calculations. ACI 318 and CSA (2004) use Eq. (6.18) with $m = 3$.

Equation (6.18) was derived based on tests of beams with moderate reinforcement ratios and has been found to overestimate stiffness of lightly reinforced sections. Bischoff (2005), noting that the tension-stiffening behavior is analogous to a series of cracked and uncracked springs in series, suggested that the form of equation should be as a sum of flexibilities, rather than a sum of stiffnesses, leading to the following expression:

$$\frac{1}{I_e} = \left(\frac{M_{cr}}{M_a}\right)^m \frac{1}{I_g} + \left[1 - \left(\frac{M_{cr}}{M_a}\right)^m\right] \frac{1}{I_{cr}} \ge \frac{1}{I_g} \tag{6.19}$$

[3]See Chapter 5 for discussion of tension-stiffening in axially loaded tension members.

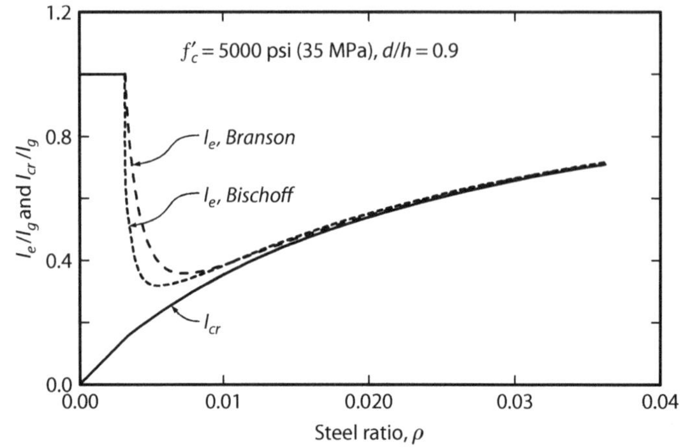

FIGURE 6.12 Relative stiffness according to Eqs. (6.18) and (6.20) for service load moment $M_a = 0.6M_n$. *(After Bischoff, 2005, used with permission from ASCE.)*

Calibrating the results of this equation with the results of Branson's original equation for beams with moderate reinforcement ratios requires $m = 2$ in Eq. (6.19). Rearranging terms, we can write

$$I_e = \frac{I_{cr}}{1 - \left(\dfrac{M_{cr}}{M_a}\right)^2 \left(1 - \dfrac{I_{cr}}{I_g}\right)} \leq I_g \tag{6.20}$$

Equation (6.20) has been found to produce better correlation for lightly reinforced sections. Figure 6.12 compares results of Eqs. (6.18) and (6.20) for service load moment $M_a = 0.6M_n$, which is in the typical range for gravity loads.

Continuous beams have multiple cross sections, each of which may have different reinforcement and different applied moment. These different lengths can be modeled as segments of different stiffness. For routine designs, however, a more practical approach is to use a weighted average moment of inertia. Deflection calculations also must take into account long-term effects associated with concrete shrinkage and creep. See ACI 318, ACI 435 (2003), and Branson (1977) for details.

The procedures outlined in the preceding paragraphs are suitable to estimate stiffness for members that are simply supported over one or more spans. Where members frame rigidly into adjacent members or supports, such as a column cantilevering from a foundation element, alternative procedures described in Section 6.11 should be used.

6.6.5 Response at Ultimate Limit States

The term *ultimate limit state* can have several interpretations. For example, cover concrete spalling can be the ultimate limit state for unconfined sections under gravity loads, or reinforcement fracture can be the ultimate limit state for a confined section

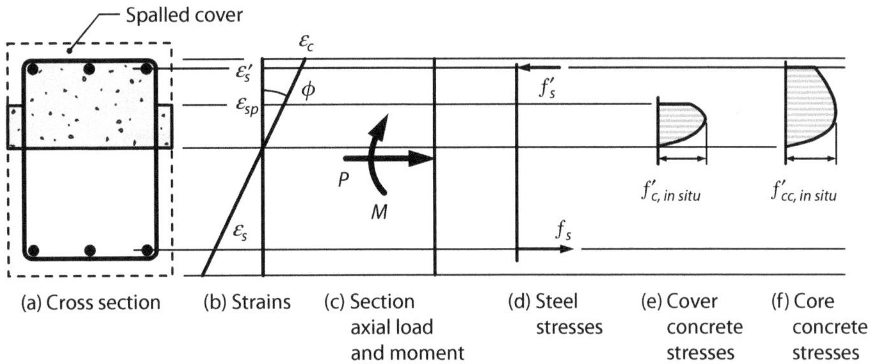

Figure 6.13 Internal actions after cover spalling.

under earthquake loading. Alternatively, in a performance-based design there can be multiple limit states corresponding to different repair actions or safety consequences such as cover spalling, longitudinal reinforcement buckling or fracture, or fracture of hoop confinement reinforcement. Flexural analysis at any of these limit states requires that we be able to model the materials well into the nonlinear range.

To model response after spalling of cover concrete requires definition of the spalling strain ε_{sp} and assumptions once that strain is exceeded. The usual approach is to assume that cover concrete behaves as unconfined concrete for strains less than ε_{sp} and that it resists no stress for larger strain. Figure 6.13 illustrates these assumptions. In this text we adopt $\varepsilon_{sp} = 0.004$ (see Section 6.5.2).

Moment–curvature analysis follows the general procedure outlined in Section 6.6.1. To facilitate the calculations, two computational procedures are introduced. The first, suitable for hand calculations, provides a shorthand for calculating the magnitude and centroidal location of compressive stress resultants (Section 6.6.6). The second procedure, more suitable for computer analysis, involves subdivision of the cross section into a grid for which axial loads and moments can be numerically computed (Section 6.6.7).

6.6.6 Compression Stress Block Parameters

To calculate internal forces on a cross section, it is useful to determine the magnitudes and centroidal locations of the internal stress resultants. This is routine for linear response (Section 6.6.3) because the concrete compressive stress block is triangular. To extend the calculations beyond the linear range, we must evaluate compressive stress blocks of a more general formulation.

Figure 6.14 illustrates the general approach for a rectangular compression zone. Strains vary linearly from zero at the bottom to maximum strain at the top. Given a mathematical relation between stress and strain (Figure 6.14*c*), we can calculate the total stress resultant as

$$C_c = \alpha_1 f'_{cc} b \beta_1 c = b \int_0^c f_c dz \tag{6.21}$$

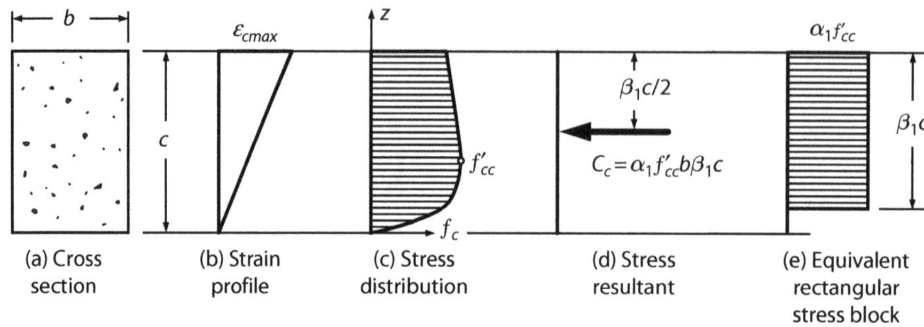

FIGURE 6.14 Compression stress block parameters.

The distance from the extreme compressive fiber to the centroid of the compressive stress is

$$\frac{\beta_1 c}{2} = \frac{b\int_0^c f_c(c-z)dz}{C_c} \tag{6.22}$$

The stress resultant and its centroidal location are defined in terms of the parameters α_1 and β_1, which relate to the geometry of an equivalent rectangular stress block, as shown in Figure 6.14*e*. The equivalent rectangular stress block has a long history of use for calculations of ultimate flexural strength of unconfined sections.[4] With a suitable mathematical model for the stress–strain relation, the results of Eqs. (6.21) and (6.22) should be equally useful for flexural moment–curvature calculations of unconfined and confined sections at limit states other than ultimate strength.

Chapter 4 introduced a mathematical relation for stress–strain behavior of unconfined and confined concrete. Substituting that relation into Eqs. (6.21) and (6.22), and carrying out the integration, we obtain the coefficients α_1 and β_1 shown in Figure 6.15.

To illustrate the application of these results, consider a rectangular compression zone of width $b = 24$ in (610 mm) and depth $c = 12$ in (305 mm), with unconfined concrete having compressive strength 6000 psi (41 MPa) and $\varepsilon_0 = 0.002$, flexed so the maximum strain at the top surface is 0.003. For this example, $K = 1.0$ and $\varepsilon_{cmax}/\varepsilon_{cc} = 0.003/0.002 = 1.5$. Hence, $\alpha_1 = 0.94$ and $\beta_1 = 0.83$. Using these results in Eqs. (6.21) and (6.22), we obtain a compressive force resultant $C_c = 0.94 \times 6$ ksi $\times$ 24 in $\times$ 0.83 $\times$ 12 in = 1350 kips (6000 kN), acting at $\beta_1 c/2 = 0.83 \times 12$ in/2 = 4.98 in (126 mm) from the extreme compression fiber.

[4]According to Hognestad (1952), the equivalent rectangular stress block was introduced by E. Svenson in 1912. Whitney (1937) demonstrated that calculations made using a rectangular stress block could produce strengths nearly identical to measured strengths. That paper served as the basis for ultimate strength building code provisions introduced in the 1956 ACI Building Code and used with modifications since that time.

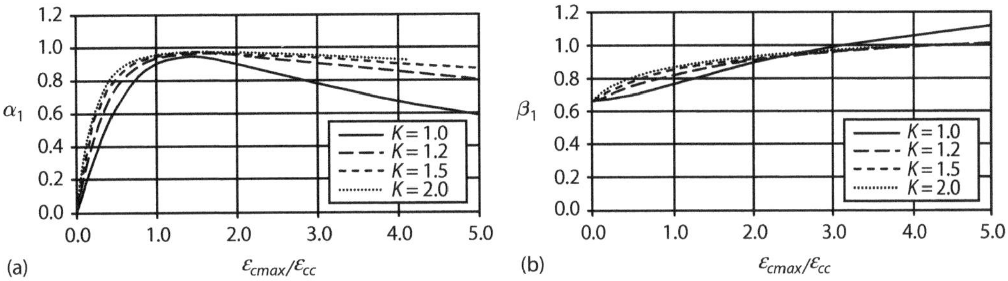

FIGURE 6.15 Compression stress block parameters α_1 and β_1. In the legend, *K* is the ratio of confined concrete compressive strength to strength of plain concrete, that is, $K = \frac{f'_{cc}}{Cf'_c}$.

Various building codes have adopted the equivalent rectangular stress block approach for calculating strength of members subjected to moment with or without axial load, but with different parameters for α_1 and β_1. For example, ACI 318 defines flexural strength at $\varepsilon_{cmax} = 0.003$, sets $\alpha_1 = 0.85$, and defines $\beta_1 = 0.85$ for $f'_c \leq 4000$ psi (27.6 MPa), 0.65 for $f'_c \geq 8000$ psi (55.2 MPa), with linear variation between these two stress values. The reduction in values of β_1 with increasing compressive strength represents the changing shape of the concrete stress–strain relation for higher strength concrete. For the example problem of the preceding paragraph, the ACI 318 stress block produces compressive force resultant $C_c = 0.85 \times 6$ ksi $\times$ 24 in $\times$ 0.75 $\times$ 12 in = 1100 kips (4900 kN), acting at $\beta_1 c/2 = 0.75 \times 12$ in/2 = 4.5 in (114 mm) from the extreme compression fiber.

Test results for high-strength concrete have led to proposals for additional modifications to the stress block parameters (Figure 6.16). The data show a steady reduction in the value of β_1 with increasing f'_c, with less clear trend for α_1. On the basis of a review of available data and proposals, ACI ITG-4.3R-07 (2007) recommends α_1 = 0.85 for $f'_c \leq$ 8000 psi (55 MPa), 0.70 for $f'_c \geq$ 18,000 psi (124 MPa), with linear variation between these two stress values. The recommended

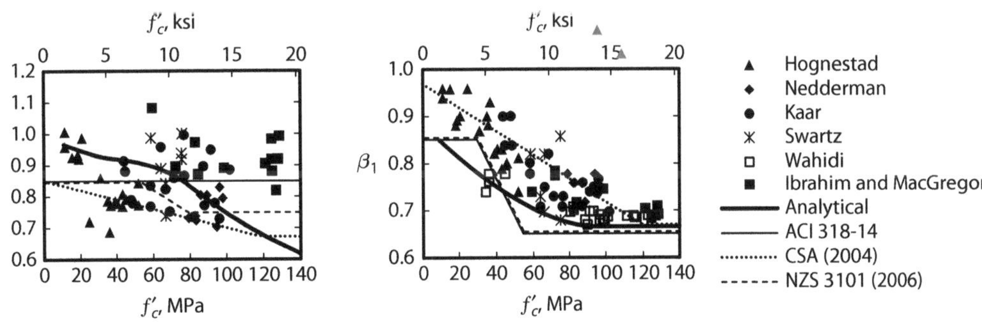

FIGURE 6.16 Stress block parameters inferred from laboratory tests and various expressions proposed for high-strength concrete. Analytical derivation based on an assumed mathematical expression for the concrete stress–strain relation. (*After Bae and Bayrak, 2003, courtesy of American Concrete Institute.*)

variation of β_1 is identical to that given in ACI 318, that is, $\beta_1 = 0.85$ for $f_c' \leq 4000$ psi (27.6 MPa), 0.65 for $f_c' \geq 8000$ psi (55.2 MPa), with linear variation between these two stress values.

Example 6.1. Calculate the moment and curvature corresponding to cracking, yielding, and spalling of cover concrete for the case of flexural tension at the bottom of the beam shown in Figure 6.17. In situ concrete compressive strength is $f_c' = 4000$ psi (28 MPa). Assume typical A706 Grade 60 (420) material properties. Clear cover over No. 4 (13) stirrups is 1.5 in (38 mm).

Solution

Cracking

The cracking moment and curvature will be based on the gross section properties. Required quantities include

$$E_c = 57{,}000\sqrt{f_c'} = 57{,}000\sqrt{4000} = 3{,}600{,}00 \text{ psi} = 3600 \text{ ksi};\ f_r = 7.5\sqrt{f_c'} = 7.5\sqrt{4000} = 474 \text{ psi};$$

$$I_g = \frac{1}{12}bh^3 = \frac{1}{12}(18'')(24'')^3 = 20{,}700 \text{ in}^4$$

From these quantities, the cracking moment and curvature are

$$M_{cr} = \frac{f_r I_g}{h/2} = \frac{(474 \text{ psi})(20{,}700 \text{ in}^4)}{24''/2} = 819 \text{ k} - \text{in}\,(92.5 \text{ kN} - \text{m}) \quad \text{and}$$

$$\phi_{cr} = \frac{M_{cr}}{E_c I_g} = \frac{819 \text{ k-in}}{(3600 \text{ ksi})(20{,}700 \text{ in}^4)} = 1.1 \times 10^{-5} \text{ in}^{-1} (4.3 \times 10^{-7} \text{ mm}^{-1})$$

Yielding

Steel areas, effective depths, steel ratios, and modular ratio are calculated as

$A_s = 4 \times 1.0 \text{ in}^2 = 4 \text{ in}^2$; $A_s' = 2 \times 1.0 \text{ in}^2 = 2 \text{ in}^2$; $d = 21.4''$; $d' = 2.6''$;
$\rho = A_s/bd = 4 \text{ in}^2/(18'' \times 21.4'') = 0.0104$; $\rho' = A_s'/bd = 2 \text{ in}^2/(18'' \times 21.4'') = 0.0052$;
$n = E_s/E_c = 29{,}000 \text{ ksi}/3600 \text{ ksi} = 8.1$

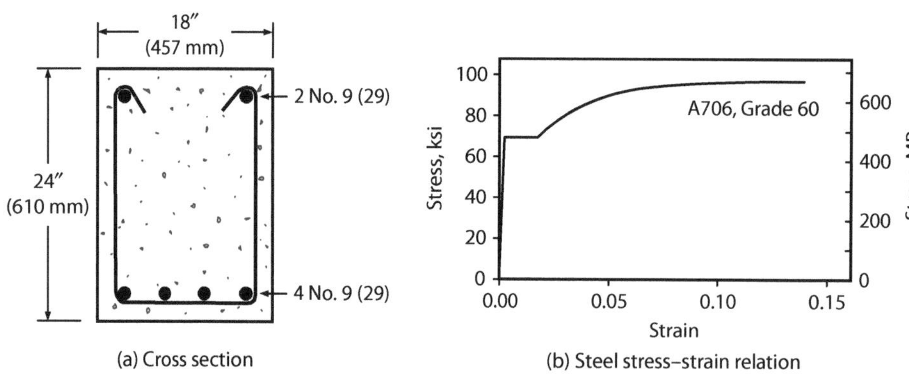

FIGURE 6.17 Beam cross section and assumed steel stress–strain relation.

The neutral axis depth is kd, where k is defined by Eq. (6.12), resulting in $k = 0.315$ and $kd = 0.315 \times 21.4'' = 6.74''$.

The moment of inertia of the cracked transformed section is obtained from Eq. (6.13) as

$$I_{cr} = \frac{b(kd)^3}{3} + (n-1)A'_s(kd-d')^2 + nA_s(d-kd)^2$$

$$= (18'')(6.74'')^3/3 + (8.1-1)(2\text{ in}^2)(6.74''-2.6'')^2 + (8.1)(4\text{ in}^2)(21.4''-6.74'')^2 = 9043\text{ in}^4$$

From Eq. (6.17) the yield moment is

$$M_y = (1/8.1)(69\text{ ksi})(9043\text{ in}^4)/(21.4''-6.74'') = 5260\text{ k-in }(594\text{ kN-m}),$$

and from Eq. (6.16) the yield curvature is

$$\phi_y = M_y/E_cI_{cr} = (5260\text{ k-in})/(3600\text{ ksi} \times 9043\text{ in}^4) = 16.1 \times 10^{-5}\text{ in}^{-1}\ (6.4 \times 10^{-6}\text{ mm}^{-1}).$$

To use linear elastic theory, both steel and concrete should be in the linear-elastic range of response. As discussed in Section 6.6.3, the results of linear theory are sufficiently accurate if the calculated concrete stresses do not exceed approximately f'_c (even though the response at this stress level is not strictly linear). Checking, the concrete stress at the yield moment is

$f_c = M_y kd/I_{cr} = 5260$ k-in $\times$ 6.74″/9043 in^4 = 3900 psi $< f'_c$. From this, we conclude that the calculated results should be reasonably accurate.

Spalling

For unconfined concrete sections, unless steel fractures at a smaller curvature, the maximum usable curvature is the curvature corresponding to concrete spalling, that is, $\varepsilon_{sp} = 0.004$. Figure 6.18 illustrates the cross section and idealized internal actions. Concrete compressive stresses are represented by the rectangular stress block introduced in Section 6.6.6.

Using linear variation of strain over the section depth (Figure 6.18b), we can write relations among the neutral axis depth and various strains as

$$\varepsilon_s = \frac{0.004}{c}(d-c)\,;\ \varepsilon'_s = \frac{0.004}{c}(c-d')$$

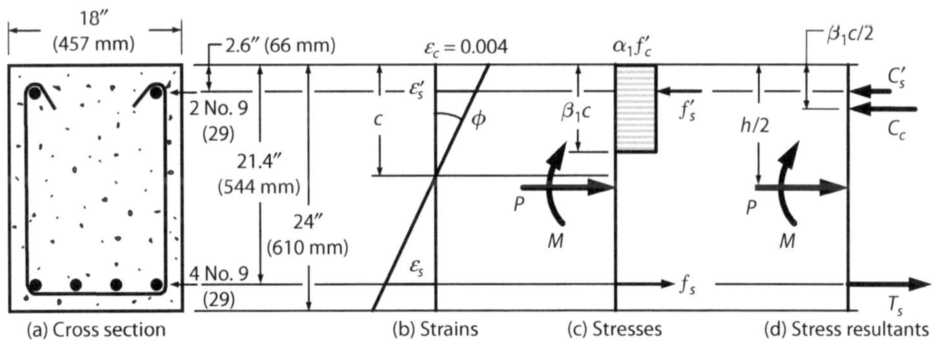

Figure 6.18 Cross section, strains, and internal actions for unconfined sections at onset of concrete spalling.

These steel strains can be used with the steel stress–strain relation (Figure 6.17*b*) to determine steel stresses f_s and f'_s. Concrete stress block parameters α_1 and β_1 are determined from Figure 6.15. Assuming the strain at peak stress for unconfined concrete is 0.002, we enter Figure 6.15 at abscissa equal to 0.004/0.002 = 2, and for $K = 1$ we read $\alpha_1 = 0.90$ and $\beta_1 = 0.89$. Thus, the stress resultants are written as

$$T_s = A_s f_s = 4f_s; \quad C'_s = A'_s(f'_s - \alpha_1 f'_c) = 2(f'_s - 0.90 \times 4 \text{ ksi}) = 2(f'_s - 3.6 \text{ ksi});$$
$$C_c = \alpha_1 f'_c b \beta_1 c = (0.90)(4 \text{ ksi})(18'')(0.89c) = 57.7c$$

In the expression for C'_s, the concrete displaced by steel is accounted for approximately by subtracting the concrete stress from the steel stress. This is a detail that does not have much impact on the final result and usually can be neglected in routine calculations.

Using the free-body diagram of Figure 6.18*d*, $P = C_c + C'_s - T_s$. For a beam, the correct strain profile results in $P = 0$. Though we could write a closed-form equation to solve this problem, it is more instructive and more generally applicable to solve this problem by iteration. Here we fix the strain at the extreme compression fiber to $\varepsilon_{sp} = 0.004$ and vary the neutral axis depth c until we find $P \approx 0$.

Iteration No.	c (in)	ε_s	ε'_s	f_s (ksi)	f'_s (ksi)	T_s (kip)	C'_s (kip)	C_c (kip)	P (kip)
1	3.0	0.0245	0.00053	76	16	304	24	173	−107
2	4.0	0.0174	0.0014	69	41	276	74	231	29
3	3.7	0.0191	0.00119	69	35	276	62	213	−1 (≈ 0)

Reviewing the results of iteration 1, the cross section has $P = -107$ kips, that is, tensile force of 107 kips. The neutral axis depth must be increased to increase the axial compression. In iteration 2, the neutral axis depth is somewhat arbitrarily increased to 4″. The result is axial compression of $P = 29$ kips. The solution clearly lies between $c = 3''$ and $c = 4''$. Linear interpolation of results suggests iteration 3, for which effectively zero axial force is found.

Moment M acting on the section can be calculated by summing moments about any convenient point. Summing moments about the centroid of the tension reinforcement results in

$M = C_c(d - \beta_1 c/2) + C'_s(d - d') = 213^k(21.4'' - 0.89 \times 3.7''/2) + 62^k(21.4'' - 2.6'') = 5370$ k-in (607 kN-m).

The corresponding curvature is

$$\phi = \varepsilon_c/c = 0.004/3.7'' = 108 \times 10^{-5} \text{ in}^{-1} \ (4.3 \times 10^{-5} \text{ mm}^{-1}).$$

It is good practice to compare the calculated results with results that can be obtained through simpler, more approximate methods. For example, the yield curvature is the yield strain divided by the depth from the neutral axis to the steel centroid. For sections with $\rho = 0.01$, the neutral axis depth is $kd \approx d/3$. Thus, the yield curvature should be approximately $\varepsilon_y/(d - kd) = \varepsilon_y/(⅔d) = (69 \text{ ksi}/29{,}000 \text{ ksi})/(⅔)(21.4'') = 17 \times 10^{-5}\text{in}^{-1}$, which is close to the calculated value of 16.1×10^{-5} in^{-1}. Similarly, yield moment should be $M_y \approx T_y(d - kd/3) = A_s f_y d(1 - 1/9) = (4 \text{ in}^2)(69 \text{ ksi})(21.4'')(8/9) = 5250$ k-in which is close to the calculated value of 5260 k-in.

Similarly, the nominal ultimate moment should be $M_n \approx A_s f_y jd$, where jd depends on the section properties. For sections with $\rho = 0.01$ and intermediate strength steel and concrete,

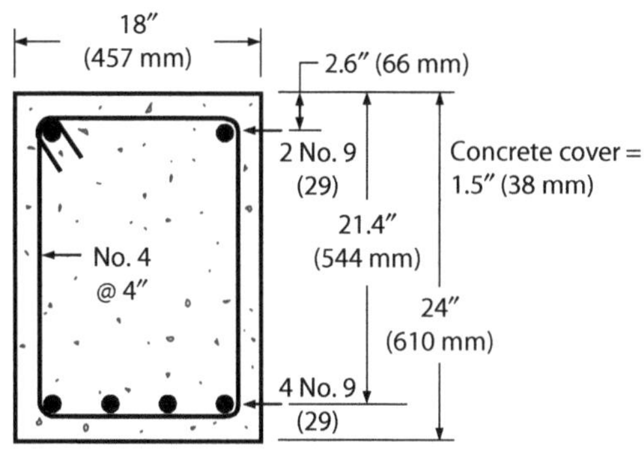

FIGURE 6.19 Cross section for Example 6.2.

$jd \approx 0.9d$. Therefore, $M_n \approx (4\ \text{in}^2)(69\ \text{ksi})(0.9)(21.4'') = 5320$ k-in which is close to the calculated value of 5370 k-in.

Example 6.2. Calculate the ultimate usable curvature and corresponding moment for the cross section shown in Figure 6.19. All material properties are the same as in Example 6.1.

Solution

The core concrete is confined by closely spaced hoops, such that confined concrete behavior should be considered for core concrete strains exceeding approximately 0.004. For practical purposes, we can consider the results of Example 6.1 to adequately represent behavior up to the spalling point. Beyond spalling, effects of confinement on the core concrete need to be considered. The ultimate moment and curvature are determined below.

Ultimate curvature could be limited by buckling of the compressive longitudinal reinforcement, fracture of the tensile longitudinal reinforcement, or failure of the core initiated by fracture of the hoops. The ratio of hoop spacing to longitudinal bar diameter is $s/d_b = 4/1.125 = 3.6$, so buckling seems unlikely until large strains. Whether tensile reinforcement fracture or compression zone failure limits the curvature depends on the longitudinal reinforcement ratio and the confinement reinforcement ratio, and is not easily determined by inspection. Therefore, we will assume one is the limiting factor, carry out the moment–curvature calculation, and then check the other. For this example, assume the ultimate curvature is limited by strain capacity of the core concrete.

On the basis of the results from Example 6.1, it seems likely that the compression zone will cover only about a quarter of the depth of the member (Figure 6.20). Therefore, the confined

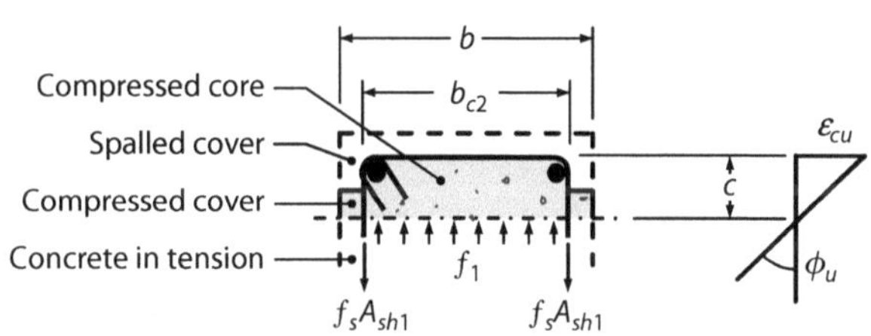

FIGURE 6.20 Geometry of the compression zone and strain profile near the ultimate curvature.

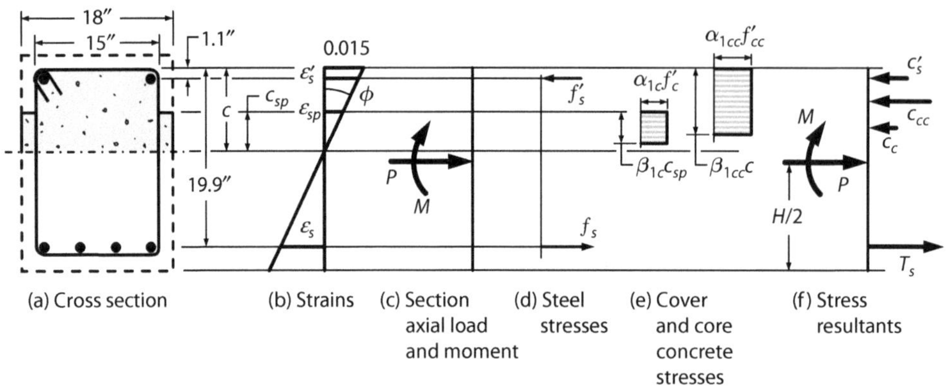

FIGURE 6.21 Cross section, strains, and internal actions near ultimate curvature.

core is subject to a steep strain gradient with only the extreme fiber at the peak strain. For such loading, behavior of the confined core is likely to be controlled by the vertical hoop legs.

The confining stress is $f_1 = 2(0.2\text{ in}^2)(69\text{ ksi})/(15'')(4'') = 0.46$ ksi. From Chapter 4, the confinement effectiveness factor for a single hoop without crossties and $s/b_{c2} = 4''/15'' = 0.27$ is $k_e = 0.37$. Thus, the effective confinement stress is $f_{1e} = k_e f_1 = 0.37 \times 0.46\text{ ksi} = 0.17$ ksi. The smaller confining stress ratio is $f_{1e}/Cf_c' = 0.17\text{ ksi}/4\text{ ksi} = 0.045$. Using the confined concrete strength chart of Figure 4.6, the confined concrete strength ratio is $f_{cc}'/Cf_c' = 1.28$. Hence, the confined concrete compressive strength is $f_{cc}' = 1.28 \times 4\text{ ksi} = 5.12$ ksi. Finally, from Chapter 4, the ultimate strain capacity is $\varepsilon_{cu} = 0.004 + 0.25 f_{emin}/f'_c = 0.004 + 0.25 \times 0.17\text{ ksi}/4\text{ ksi} = 0.015$.

Figure 6.21 illustrates the internal actions for the beam near ultimate curvature. The ultimate compressive strain has been determined to be $\varepsilon_{cu} = 0.015$. Taking the spalling strain as $\varepsilon_{sp} = 0.004$, the depth to spalling $c_{sp} = (0.004/0.015)c = 0.267c$. The relations among the strains and dimensions are

$$\varepsilon_s = \frac{0.015}{c}(19.9'' - c);\ \varepsilon_s' = \frac{0.015}{c}(c - 1.1'')$$

Steel stresses f_s and f_s' are defined by these strains and the steel stress–strain relation shown in Figure 6.17.

Concrete stress block parameters can be determined for both unconfined and confined concrete. For unconfined concrete, the peak strain is 0.004, hence, we enter Figure 6.15 at abscissa equal to 0.004/0.002 = 2, and for $K = 1$ read $\alpha_{1c} = 0.90$ and $\beta_{1c} = 0.89$ (the subscript c is used to refer to the unconfined concrete; note that the numerical values are the same as determined for Example 6.1). For confined concrete, the peak strain is 0.015 and the strain at peak of the confined concrete stress–strain relation is $\varepsilon_{cc} = \varepsilon_0\left(1 + 5\left(\frac{f_{cc}'}{Cf_c'} - 1\right)\right) = 0.002(1 + 5(1.28 - 1)) = 0.0048$. Hence, we enter Figure 6.15 at abscissa equal to 0.015/0.0048 = 3.1, and for $K = 1.28$ interpolate $\alpha_{1cc} = 0.91$ and $\beta_{1cc} = 0.96$ (the subscript cc is used to refer to the confined concrete). Thus, the stress resultants are written as

$T_s = A_s f_s = 4f_s$; $C_s' = A_s'(f_s' - \alpha_{1cc} f_{cc}') = 2(f_s' - 0.91 \times 5.12) = 2(f_s' - 4.66)$; $C_c = \alpha_{1c} f_c'(b - b_{c2})\beta_{1c} c_{sp} = 0.90(4\text{ ksi})(18'' - 15'')(0.89)(0.267c) = 2.57c$; $C_{cc} = \alpha_{1cc} f_{cc}' b_{c2} \beta_{1cc} c = 0.91(5.12\text{ ksi})(15'')(0.96)(c) = 67.1c$.

Using the free-body diagram of Figure 6.21*f*, $P = C_c + C_{cc} + C_s' - T_s$. As in Example 6.1, iterate to find the value of c that results in $P \approx 0$. We start with $c = 4$ because it is close to the ultimate neutral axis depth from Example 6.1.

Iteration No.	c (in)	ε_s	ε_s'	f_s (ksi)	f_s' (ksi)	T_s (kip)	C_s' (kip)	C_c (kip)	C_{cc} (kip)	P (kip)
1	4.0	0.0596	0.0110	90.9	69.0	364	129	10	268	44
2	3.6	0.0666	0.0106	92.0	69.0	368	129	9	242	12
3	3.5	0.0703	0.0105	92.7	69.0	371	129	9	235	2 ($\approx$ 0)

The depth of the nominally unspalled section is $c_{sp} = 0.267 \times 3.5'' = 0.9''$. Note that the steel tensile strain is $\varepsilon_s = 0.0703$, which is less than the monotonic ultimate tensile strain capacity. Thus, if the section is subjected to monotonic curvature, the curvature capacity is limited by the compressive strain capacity of the core concrete, as assumed.

Moment M acting on the section can be calculated by summing moments about any convenient point. Summing moments about the centroid of the tension reinforcement results in

$M = C_s'(d-d') + C_{cc}(d - \text{cover} - \beta_{1cc}c/2) + C_c(d - \text{cover} - c + c_{sp} - \beta_{1c}c_{sp}/2) = (129^k)(19.9'' - 1.1'') + (235^k)(19.9'' - (0.96)(3.5'')/2) + (9^k)(19.9'' - 3.5'' + 0.9'' - (0.89)(0.9'')/2) = 6860$ k-in (775 kN-m)

The corresponding curvature is

$$\phi = \varepsilon_{cu}/c = 0.015/3.5'' = 429 \times 10^{-5}\ \text{in}^{-1}\ (16.9 \times 10^{-5}\ \text{mm}^{-1})$$

In the preceding calculations note that the unspalled cover concrete does not contribute significantly to the section resistance. Thus, we could reduce the computational effort with negligible effect on accuracy by ignoring the contribution of the unconfined cover concrete. This is generally the case for well-confined sections loaded to near the ultimate curvature. Ignoring the unspalled concrete, the calculated moment and curvature are 6820 k-in (771 kN-m) and 416×10^{-5} in^{-1} (16.4×10^{-5} mm^{-1}).

A useful check of the ultimate moment strength is that it cannot exceed the product of the steel strength and the total depth to the tension reinforcement, which in this case is $M = A_s f_{su} d =$ (4 in^2)(95 ksi)(19.9″) = 7560 k-in, which indeed is greater than the calculated value of 6860 k-in. A better estimate would be $M = A_s f_{su} jd =$ (4 in^2)(95 ksi)(0.9)(19.9) = 6800 k-in. Of course, whether the steel stress reaches the ultimate value depends on the curvature achieved, which, in turn, is limited by the confinement provided.

Example 6.3. Plot the moment–curvature relation based on the results of Examples 6.1 and 6.2 and determine the curvature ductility at spalling and ultimate.

Solution

Figure 6.22 plots the moment–curvature relation. Points corresponding to cracking, yielding, spalling, and hoop fracture are shown. The moment–curvature approximation is obtained by connecting the points with straight lines. Note that immediately after cracking, the theoretical stiffness of the cross section jumps from $E_c I_g$ to $E_c I_{cr}$. We know that the average stiffness along the length does not immediately change because of the tension-stiffening effect (Section 6.6.4). In the region between cracking and yielding, we could use the results of

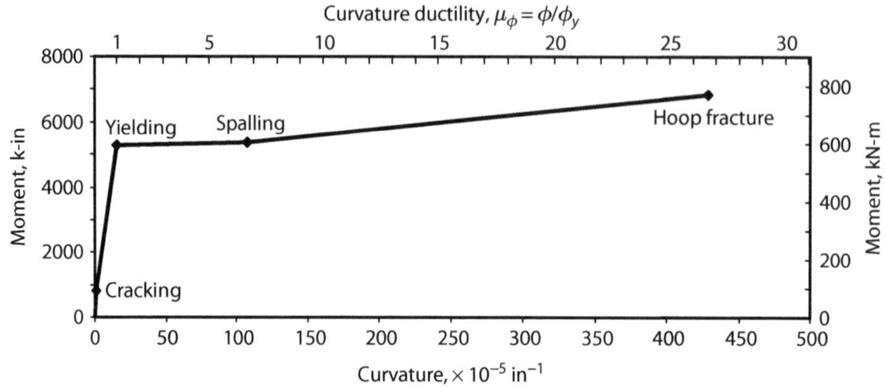

FIGURE 6.22 Moment–curvature relation for the beams of Examples 6.1 and 6.2.

Section 6.6.4 to estimate the average stiffness. In Figure 6.22, however, the cracking and yielding points are simply connected by a straight line. This straight-line approximation facilitates the calculation of displacements by direct integration, as will be done in Section 6.11.

In this example, the calculated spalling moment is only marginally larger than the calculated yielding moment. The yielding moment actually is somewhat overestimated because it was based on the assumption that the concrete behavior was linear-elastic even though maximum stress was near f'_c. A slightly better estimate of the yielding moment would be obtained by using the rectangular stress block for compressed concrete, as it correctly accounts for nonlinearity of the concrete stress–strain relation.

Curvature ductilities are $\mu_\phi = \phi/\phi_y = 108/16.1 = 6.7$ at spalling and $429/16.1 = 26.7$ at hoop fracture.

6.6.7 Automation of Moment–Curvature Calculations

Moment–curvature calculations of the type illustrated in Example 6.2 are tedious and prone to errors because of the numerous calculations involved. The calculations can be programmed to be solved by computer software. An alternative approach is preferred, however, so that generalized cross-sectional shapes, stress–strain relations, and loadings can be readily adopted.

Figure 6.23 illustrates a widely used approach in which the cross section is subdivided into fibers. In the usual implementation, the cross-sectional area and centroid of each fiber are defined and a material property is assigned. For any imposed strain profile, the strain at the centroid of each fiber can be determined. The stress is defined based on the assigned material and the centroidal strain, and a fiber stress resultant acting at the fiber centroid is then defined by the product of the stress and fiber area. Summing all the fiber stress resultants determines the corresponding axial load P, and summing moments (including the moment of the axial force P) about a predetermined point (usually the centroid of the gross section) determines moments M_x and M_y about x- and y-axes, respectively.

To use this procedure for monotonic loading, the material monotonic stress–strain relations are employed. For cyclic or reversed cyclic loading, the strain history of each fiber is stored and used to determine the appropriate stress using cyclic material models.

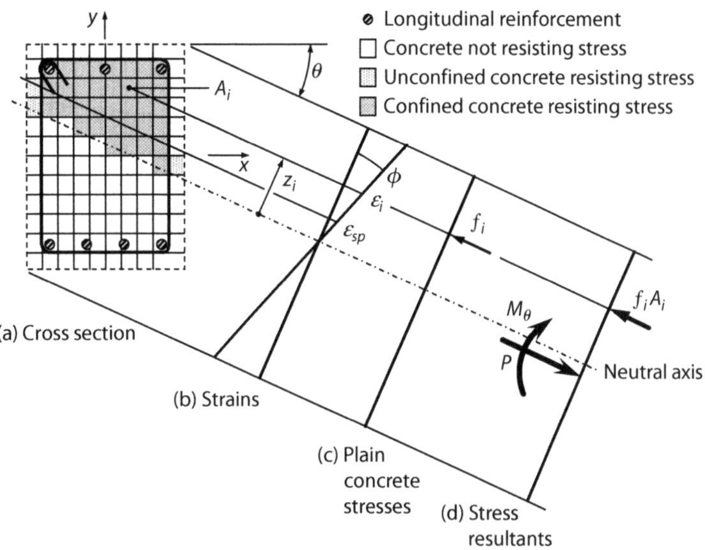

FIGURE 6.23 Fiber idealization of reinforced concrete cross section. *Note:* In general, imposed curvature about an axis inclined at an angle θ relative to the principal axes results in a moment M_θ about that axis and a moment about the orthogonal axis. To simplify the figure, only the moment M_θ is shown.

The fiber model can be implemented readily for specific cross sections using spreadsheet (or other) software. Alternatively, specialized software packages are available that enable more generalized shapes, materials, and loadings.[5]

Example 6.4. Repeat Example 6.3 using a fiber model.

Solution

For this example, the software XTRACT is selected. Figure 6.24*a* shows the discretized cross section, where different shades indicate different materials and materials states at the maximum curvature. This software uses certain default assumptions regarding the shape of the stress–strain relations and the boundary between core and cover concrete. Discretization of the cross section introduces additional approximations. Consequently, the computed moment–curvature relation does not exactly match the relation obtained by the hand calculations (Figure 6.24*b*).

6.7 Beams

The previous sections developed a general approach for flexural analysis of reinforced concrete cross sections. This section uses that procedure to explore the flexural response of beams, considering how behavior is affected by the controlling variables and introducing the concepts of nominal, design, and probable moment strengths.

[5]For example, see http://www.imbsen.com/software-XTRACT.html, http://opensees.berkeley.edu/OpenSees/manuals/ExamplesManual/HTML/3886.htm, http://www.ecf.utoronto.ca/~bentz/home.shtml.

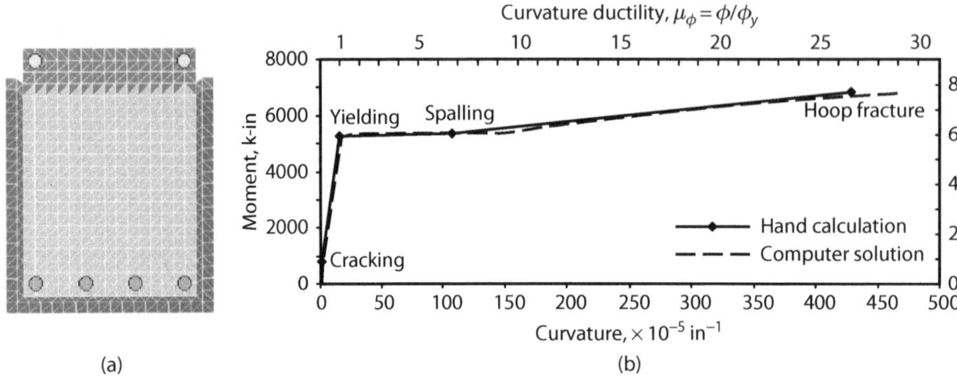

FIGURE 6.24 Fiber model analysis results: (*a*) discretized cross section showing condition of various fibers at the ultimate curvature; (*b*) moment–curvature relation.

6.7.1 Moment–Curvature Response

The moment-curvature relation of a beam depends on the section shape, the concrete and reinforcement material properties, the amount and distribution of the longitudinal reinforcement, and the amount and configuration of the transverse reinforcement. Figure 6.25 compares the calculated moment–curvature relations for singly reinforced beam sections (i.e., beams without longitudinal compression reinforcement) for different longitudinal reinforcement ratios. In reference to that figure, we can note the following:

- For lower reinforcement ratios, the longitudinal reinforcement yields before the concrete reaches the limiting strain (assumed equal to 0.004 in this analysis). Yielding is followed by a relatively long, nearly plastic plateau that

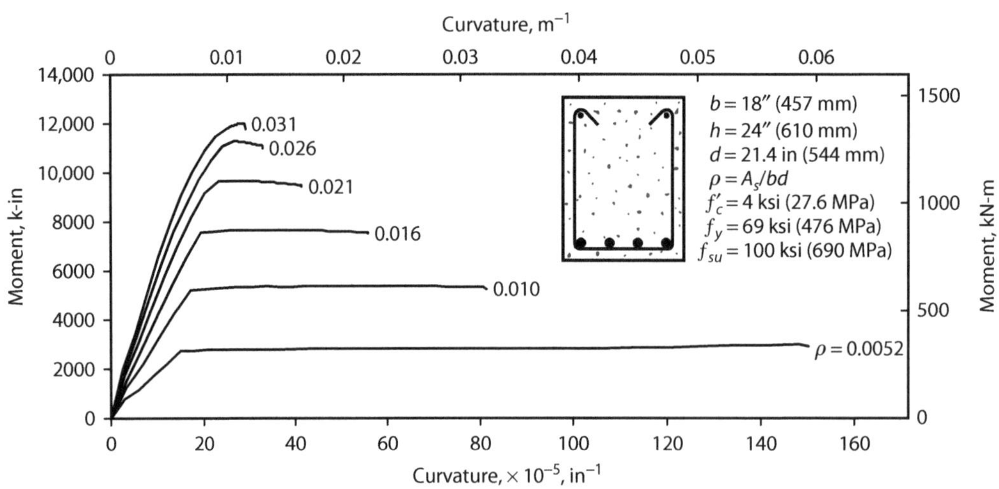

FIGURE 6.25 Moment–curvature relations for singly reinforced beam.

eventually is terminated when the concrete reaches its limiting strain. The moment strength is largely controlled by the strength of the longitudinal reinforcement. We refer to such sections as being *tension-controlled.*

- Also for lower reinforcement ratios, the moment strength is nearly (but not exactly) proportional to the area of longitudinal reinforcement. For example, if the area of longitudinal reinforcement is doubled, the moment strength is nearly (but not quite) doubled.
- As the longitudinal reinforcement area increases, axial force equilibrium of the cross section requires a deeper flexural compression zone to equilibrate the increased flexural tension force. Recall that the ultimate curvature is $\phi_u = \varepsilon_{cu}/c$. Thus, the curvature capacity decreases as the longitudinal reinforcement ratio increases.
- *Curvature ductility* is defined as the ratio of ultimate curvature to yield curvature, that is, $\mu_\phi = \phi_u/\phi_y$. Curvature ductility is relatively large for small values of the tension steel ratio ρ, and decreases as ρ increases. For some relatively large value of ρ (~0.03 in Figure 6.25), the curvature ductility reaches $\mu_\phi = 1$, that is, longitudinal tension reinforcement yielding and concrete crushing occur simultaneously. This failure condition is known as *balanced failure.* Further increases in longitudinal steel ratio beyond this point will not result in significant increase in moment strength because the section is *compression-controlled.*
- Over the range of reinforcement ratios, the yield curvature does not vary significantly. As introduced in Example 6.1, the yield curvature can be estimated by $\phi_y = \varepsilon_y/(\tfrac{2}{3}d)$. For the example beam shown in Figure 6.25, this estimate is $\phi_y = (69/29{,}000)/(\tfrac{2}{3} \times 21.4) = 17 \times 10^{-5}\ \text{in}^{-1}\ (0.0066\ \text{m}^{-1})$.

Figure 6.26 illustrates the effect of compression reinforcement on behavior of a beam. For a beam that is tension-controlled, compression reinforcement does

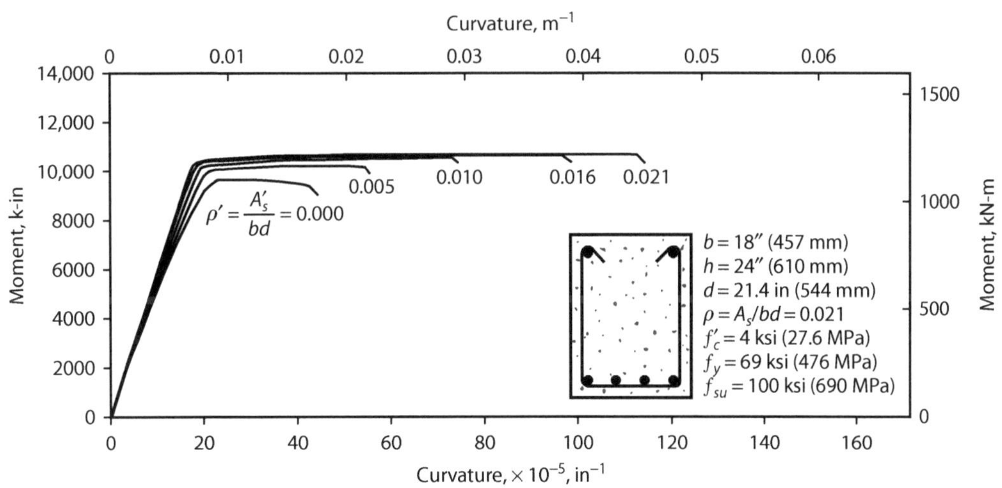

Figure 6.26 Effect of compression reinforcement on the moment–curvature relation.

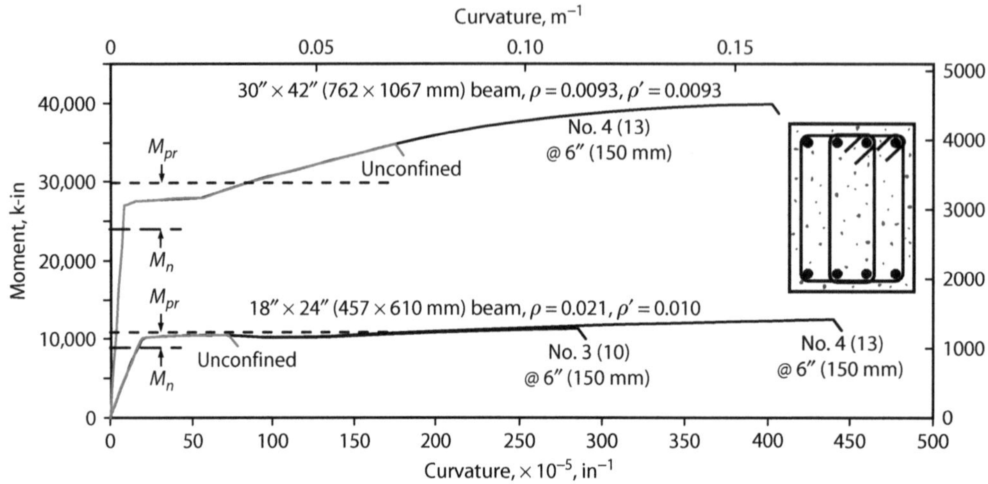

FIGURE 6.27 Effect of confinement reinforcement on the moment–curvature relation [concrete cover = 1.5″ (38 mm); $f'_c = 4$ ksi (27.6 MPa); $f_y = 69$ ksi (476 MPa); $f_{su} = 100$ ksi (690 MPa)].

not have significant effect on the flexural strength. However, compression reinforcement increases the ultimate curvature capacity. (Though not shown here, compression reinforcement also helps reduce long-term deflections associated with creep and shrinkage.)

Figure 6.27 illustrates the effect of confinement reinforcement on the theoretical moment–curvature response of beams having different gross dimensions and different ratios ρ'/ρ. For the smaller size beam, spalling of cover concrete commences at curvature around 70×10^{-5} in^{-1} (0.0028 m^{-1}). This is evident in a reduction in moment resistance for the confined section. As curvature is increased past this point, reinforcement strain-hardening along with resistance of the confined core is sufficient to recover the lost strength and produce modest strain-hardening of the cross section. Ultimate curvature is limited by the strain capacity of the confined core. The larger beam has less concrete cover relative to the overall beam depth, and the compression steel contributes more to the flexural resistance because $\rho' = \rho$. Consequently, spalling is not as apparent in the moment–curvature relation. The tension–compression couple formed by the top and bottom reinforcement results in this section having considerable ductility even without confinement reinforcement. For the confined section, ultimate curvature is limited by the tensile strain capacity of the longitudinal reinforcement. Both unconfined and confined sections show considerable strain-hardening.

Figure 6.25 shows that curvature capacity increases as tension reinforcement ratio decreases. For very small reinforcement ratios, however, tensile strains tend to concentrate at individual cracks, and this can lead to reduction in total deformation

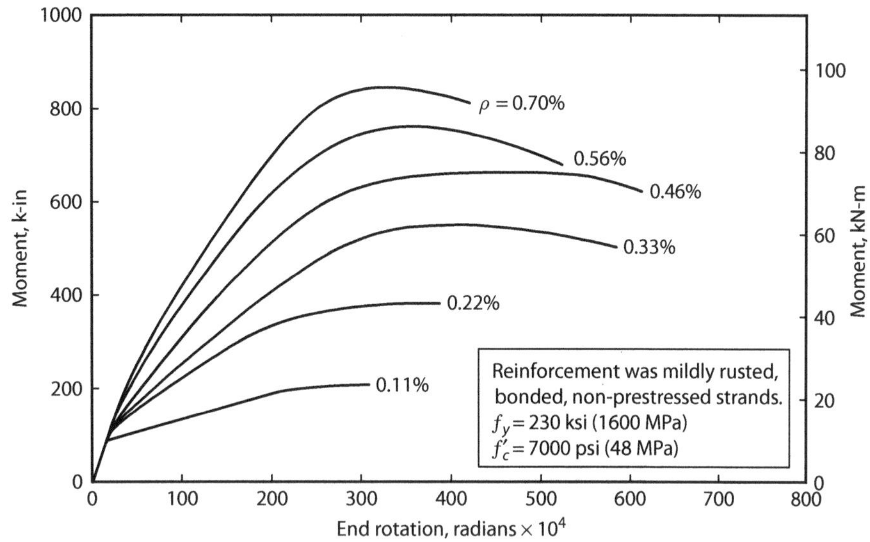

Figure 6.28 Effect of reinforcement ratio on rotation capacity. (*After Thomas and Sozen, 1965, courtesy of University of Illinois at Urbana–Champaign Archives.*)

capacity of a member. This effect was observed for tension members in Section 5.6. It also occurs for flexural members, as shown in Figure 6.28. To avoid undue loss of deformation capacity, building codes may place both lower and upper limits on the longitudinal reinforcement ratio. See, for example, Section 6.7.3.

In Chapter 5 we saw how strain-hardening of an axial compression member is important to spread inelasticity along the member length. Likewise, strain-hardening of the tension and compression zones, as well as strain-hardening of the overall moment–curvature relation, leads to spreading of inelastic curvature along the length of a flexural member, which increases rotation capacity. Figure 6.29 illustrates this for a constant-moment region of a beam. The beam with neither confinement reinforcement nor longitudinal compression reinforcement sustains a concentrated failure with limited rotation capacity, whereas beams with compression reinforcement and closely spaced hoops show distributed inelastic rotations and increased total rotation capacity.

6.7.2 Nominal, Probable, and Design Moment Strengths

Nominal moment strength, M_n, is the strength calculated for specified material properties according to prescribed flexural strength assumptions. Adopting the approach of ACI 318, the nominal moment strength of a beam is calculated using the assumptions of Figure 6.30. Maximum compressive strain of concrete is assumed to be 0.003, representing an approximate lower bound of the data in

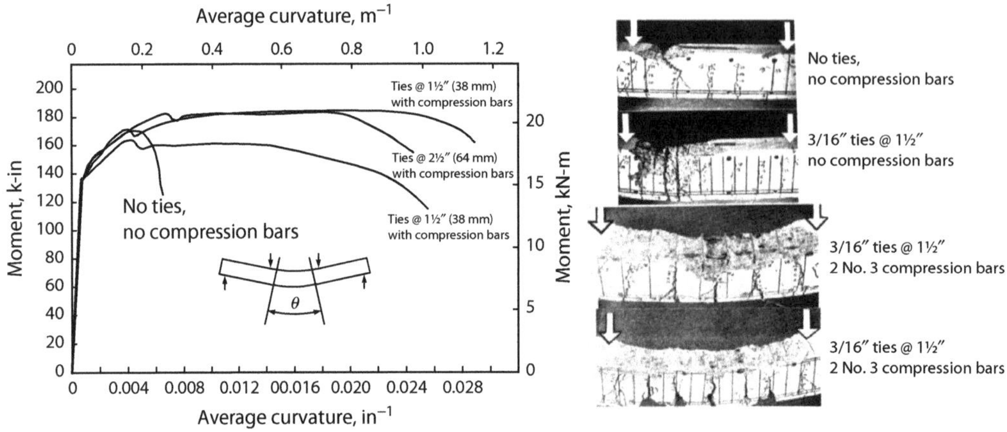

Figure 6.29 Moment-average curvature relations and spread of damage for beams with different confinement and compression reinforcement. (*After Bertero and Felippa, 1965, courtesy of American Concrete Institute.*)

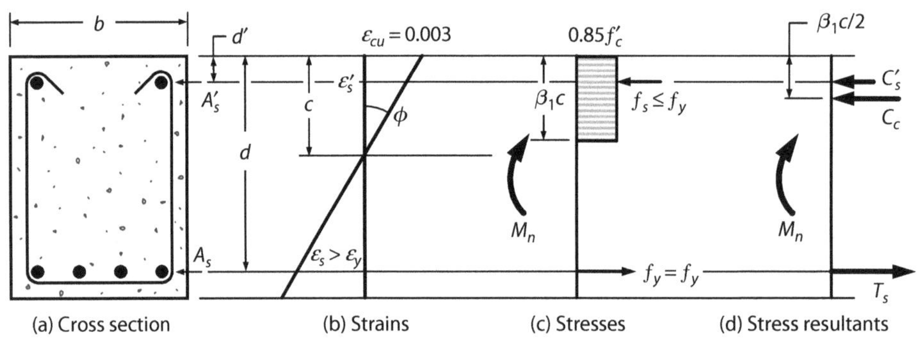

Figure 6.30 Assumptions for calculation of the ACI nominal moment strength.

Figure 6.8*a*. The value of β_1 is set equal to 0.85 for $f_c' \le 4000$ psi (28 MPa) and 0.65 for $f_c' \ge 8000$ psi (55 MPa), with linear interpolation for intermediate values of f_c'. For the case of a rectangular cross section with tension and compression reinforcement, the stress resultants are

$$T_s = A_s f_y;\ C_s' = A_s'(f_s' - 0.85 f_c');\ C_c = 0.85 f_c' b \beta_1 c \tag{6.23}$$

Force equilibrium requires $C_c + C_s' - T_s = 0$. Introducing the terms from Eq. (6.23) and solving for c

$$c = \frac{A_s f_y - A_s'(f_s' - 0.85 f_c')}{0.85 \beta_1 f_c' b} \tag{6.24}$$

Given a value for c, the strain in the compression reinforcement is

$$\varepsilon_s' = \frac{c-d'}{c}0.003 \tag{6.25}$$

To determine the neutral axis depth c, assume a value for the compressive stress f_s', solve for c from Eq. (6.24), calculate the strain ε_s' from Eq. (6.25), and iterate until the assumed stress f_s' and the resulting ε_s' are compatible.

Knowing c and f_s', the stress resultants in Eq. (6.23) can be determined and the nominal moment is then found by moment equilibrium of the actions on the free body diagram of Figure 6.30*d*. Summing moments about the centroid of the tension reinforcement

$$M_n = C_c\left(d - \frac{\beta_1 c}{2}\right) + C_s'(d-d') \tag{6.26}$$

Alternatively, we might take advantage of the observation from Figure 6.26 that moment strength is relatively insensitive to the amount of compression reinforcement. Thus, an approximation to the nominal moment is obtained by ignoring the compression reinforcement, in which case

$$M_n \approx A_s f_y d\left(1 - 0.59\frac{A_s f_y}{f_c' bd}\right) = \rho bd^2 f_y\left(1 - 0.59\rho\frac{f_y}{f_c'}\right) \tag{6.27}$$

Equation (6.27) is useful for preliminary design estimates or rapid estimates of nominal moment strength. Note that nominal moment strength always is calculated for specified material properties.

Figure 6.31 compares measured strengths with strengths calculated by these procedures for singly reinforced, unconfined beams. For this comparison, values of f_c' and f_y are based on measured properties.

Probable moment strength, M_{pr}, refers to the nominal moment strength calculated assuming a tensile stress in the longitudinal bars of at least $1.25f_y$ and $\phi = 1.0$. For practical applications using available software, M_{pr} is calculated assuming elasto-plastic steel stress–strain relation in tension and compression, with the yield stress equal to $1.25f_y$, where f_y is the specified yield stress.

The terminology *probable moment strength* should not be misunderstood to be the *expected moment strength.* Whether the ACI probable moment strength is a good estimate of the expected strength depends on the properties of the cross section and the imposed curvature. Figure 6.27 compares nominal moment strength M_n and probable moment strength M_{pr} with calculated moment–curvature relations for two beam cross sections. M_n and M_{pr} were calculated assuming $f_y = 60$ ksi (414 MPa) and $f_c' = 4$ ksi (27.6 MPa). For the smaller beam, the calculated moment–curvature relation shows that spalling of the cover concrete is compensated by strain-hardening of the longitudinal reinforcement, such that the beam cross section develops nearly elasto-plastic moment–curvature response. Consequently, M_{pr} is a reasonable estimate of the expected strength. For the larger

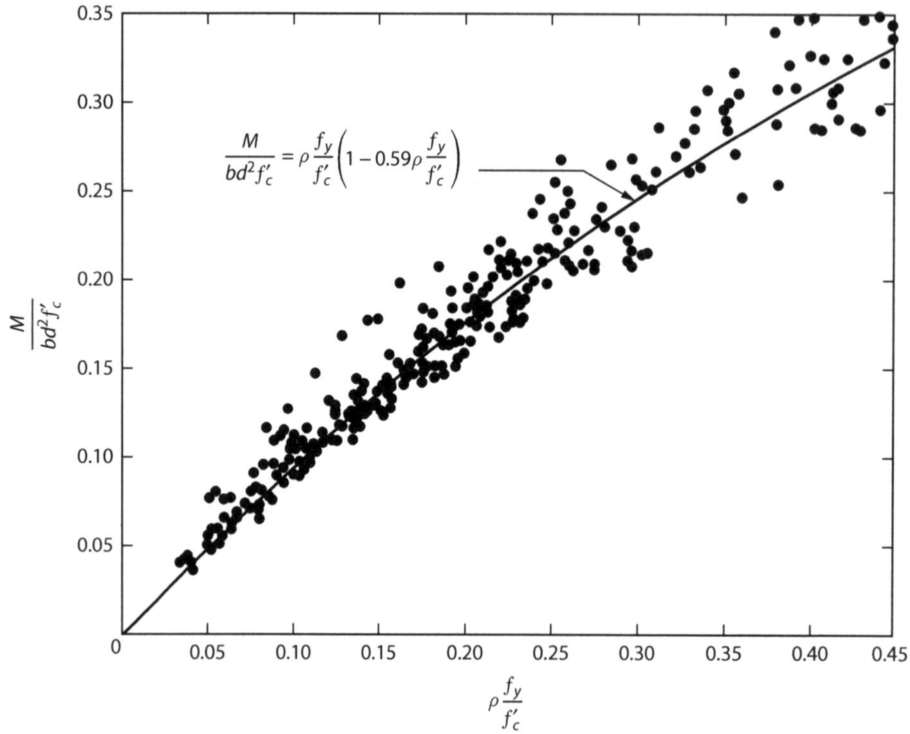

FIGURE 6.31 Tests of 364 tension-controlled beams. (*After Mattock et al., 1961, courtesy of American Concrete Institute.*)

beam, spalling does not result in much reduction in strength because the cover dimension is a small fraction of the total beam depth. Consequently, the beam develops considerable hardening, and M_{pr} underestimates the expected strength.

Design moment strength, ϕM_n, is the nominal moment strength multiplied by an appropriate strength reduction factor. According to ACI 318, the strength reduction factor ϕ depends on the net tensile strain ε_t in the extreme layer of longitudinal tension reinforcement at nominal strength, excluding strains due to effective prestress, creep, shrinkage, and temperature. Figure 6.32 illustrates the procedure for calculating ϕ for a section subjected to moment and axial force. Strain conditions are determined for the member at nominal flexural strength (Figure 6.32*c*), using assumptions of Figure 6.30. Given the strain ε_t at the extreme layer of flexural tension reinforcement, the value of ϕ is determined from Figure 6.32*a*. For most beams, $\phi = 0.90$. As described in Chapter 1, the strength design requirement is that moment determined from factored load combinations is not to exceed the design strength, that is, $M_u \le \phi M_n$.

Example 6.5. Calculate (1) M_n, (2) ϕM_n, and (3) M_{pr} for the beam of Example 6.2.

Solution

(1) For this example, the procedures of ACI 318 are strictly followed. Details of the cross section, strains, stresses, and stress resultants are as shown in Figure 6.19, except the maximum

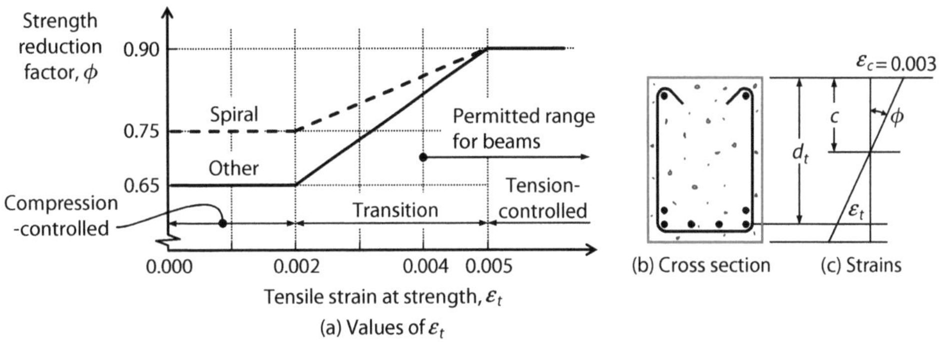

FIGURE 6.32 Strength reduction factor ϕ for moment and axial force.

compressive strain of concrete is set equal to 0.003 in accordance with ACI 318, rather than 0.004 as shown in the figure. With this minor difference, the strains in longitudinal reinforcement are

$$\varepsilon_s = \frac{0.003}{c}(d-c);\ \varepsilon_s' = \frac{0.003}{c}(c-d')$$

The stress–strain relation is assumed to follow an elasto-plastic relation, with maximum stress of 60,000 psi (413 MPa). Concrete stress block parameters are $\alpha_1 = 0.85$ and $\beta_1 = 0.85$. Thus, the stress resultants are written as

$$T_s = A_s f_s = 4 f_s;\quad C_s' = A_s'(f_s' - \alpha_1 f_c') = 2(f_s' - 0.85 \times 4\text{ ksi}) = 2(f_s' - 3.4\text{ ksi});$$
$$C_c = \alpha_1 f_c' b \beta_1 c = (0.85)(4\text{ ksi})(18'')(0.85c) = 52.0c$$

Following the procedure from Example 6.3, the neutral axis depth is iterated to find a solution near $P = 0$.

Iteration No.	c (in)	ε_s	ε_s'	f_s (ksi)	f_s' (ksi)	T_s (kip)	C_s' (kip)	C_c (kip)	P (kip)
1	3.5	0.0153	0.00077	60	22.4	240	37.9	182	−20
2	3.7	0.0144	0.00089	60	25.9	240	44.9	192	−3 (≈0)
3	3.73	0.0142	0.00091	60	26.4	240	45.9	194	0

The iteration is satisfactorily completed by iteration 2, but an additional (optional) iteration is done to obtain $P = 0$. Moment M acting on the section can be calculated by summing moments about any convenient point. Summing moments about the top of the beam $M_n = -C_c(\beta_1 c/2) - C_s'd' + T_s d = 4710$ k-in (532 kN-m). Note that the approximation of Eq. (6.27) results in $M_n = 4670$ k-in (528 kN-m), which is very close to the "exact" value.

(2) The value of the net tensile strain is $\varepsilon_t = 0.0142$, which exceeds 0.005. Therefore, according to Figure 6.32, $\phi = 0.90$, and the design moment strength is $\phi M_n = 0.9 \times 4710$ k-in $= 4240$ k-in (479 kN-m).

(3) M_{pr} is calculated using the same procedure illustrated in part (a) of this example, except the stress–strain relation of the reinforcement in tension and compression is assumed to be

elasto-plastic with peak stress of $1.25f_y = 75{,}000$ psi (517 MPa). The solution is obtained with flexural compression depth of 4.49 in (114 mm), resulting in $M_{pr} = 5800$ k-in (655 kN-m). For a tension-controlled cross section, moment strength is approximately proportional to the force in the longitudinal tension reinforcement. Thus, the approximation $M_{pr} \approx 1.25M_n = 1.25 \times 4710$ k-in = 5890 k-in (665 kN-m) is a good approximation to M_{pr}.

6.7.3 Reinforcement Limits

The longitudinal reinforcement in a beam should be sufficient to ensure that the formation of one crack will not lead to brittle failure, but not so much that it will overload the flexural compression zone before the reinforcement itself has yielded. This section addresses these limits.

Minimum Reinforcement

If the moment strength of a beam section is less than the cracking moment strength, formation of a single crack could result in brittle failure at a cross section. If the moment strength slightly exceeds the cracking moment, multiple cracks may form, but tensile strains will concentrate at cracks, reducing rotation capacity (Figure 6.28). To promote adequate rotation capacity beyond the cracking rotation, most codes require that the longitudinal reinforcement be at least equal to some limiting value. For example, ACI 318 requires

$$A_{s,min} = \frac{3\sqrt{f'_c}}{f_y} b_w d, \text{ psi}$$

$$A_{s,min} = \frac{\sqrt{f'_c}}{4 f_y} b_w d, \text{ MPa} \qquad (6.28)$$

but not less than 200 $b_w d/f_y$ psi ($1.4 b_w d/f_y$ MPa). Rectangular beams satisfying this requirement have nominal flexural strength about twice the cracking moment strength for the case of the web in tension. For an isolated T-beam or I-beam with the flange in tension, it might be more appropriate to substitute the effective width of the tension flange, or derive the minimum reinforcement requirement from first principles. For a T-beam in a monolithic floor system, cracking spreads gradually across the width of the tension flange, in which case the problem of sudden fracture is less of a concern. In this latter case, the use of b_w in Eq. (6.28) seems appropriate and is permitted by ACI 318.

Maximum Reinforcement

As shown in Figure 6.25, curvature ductility of a beam without compression reinforcement decreases as the longitudinal tension reinforcement ratio increases. Most codes impose an upper limit on the longitudinal reinforcement area so that at least some minimal curvature ductility capacity will be available. The traditional approach is to define the amount of longitudinal reinforcement corresponding to balanced failure, and permit only a fraction of that reinforcement.

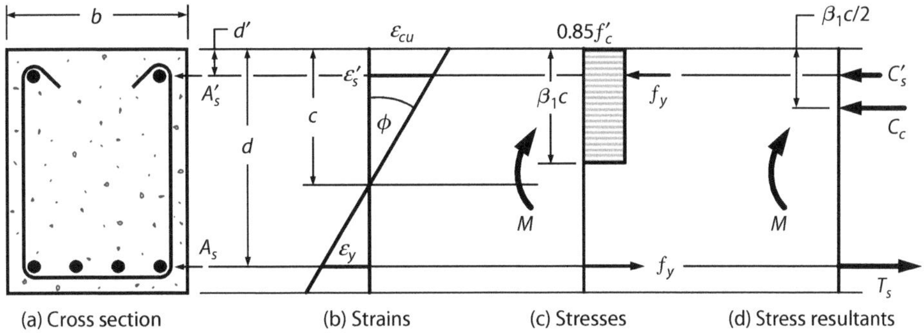

FIGURE 6.33 Balanced failure strains and internal actions.

Figure 6.33 illustrates the strains and idealized internal actions at balanced failure. Compression reinforcement, which normally is close to the extreme compression fiber, is assumed to be yielding in compression. From the geometry of strains

$$c = \frac{\varepsilon_{cu}}{\varepsilon_{cu} + \varepsilon_y} d \tag{6.29}$$

For a doubly reinforced rectangular section (i.e., one with layers of longitudinal reinforcement on opposite faces of the section), the stress resultants are

$$T_s = A_s f_y;\ C'_s = A'_s(f_y - 0.85 f'_c);\ C_c = 0.85 f'_c b \beta_1 c \tag{6.30}$$

in which the stress in the compression reinforcement is adjusted to account for the concrete it displaces. Summing forces to be equal to zero, substituting c from Eq. (6.29), and solving for A_s result in

$$A_s = \left(A'_s(f_y - 0.85 f'_c) + 0.85 \beta_1 f'_c bd \frac{\varepsilon_{cu}}{\varepsilon_{cu} + \varepsilon_y} \right) \frac{1}{f_y} \tag{6.31}$$

If we adopt the approximation $(f_y - 0.85 f'_c) \approx f_y$, Eq. (6.31) can be written in a more convenient form as

$$A_s - A'_s = \frac{0.85 \beta_1 f'_c}{f_y} \frac{\varepsilon_{cu}}{\varepsilon_{cu} + \varepsilon_y} bd \tag{6.32}$$

Equation (6.32) indicates that area of tension reinforcement corresponding to balanced failure depends on the area of compression reinforcement.

For the case of a singly reinforced rectangular beam, the balanced reinforcement area can be written as

$$A_{sb} = \frac{0.85 \beta_1 f'_c}{f_y} \frac{\varepsilon_{cu}}{\varepsilon_{cu} + \varepsilon_y} bd \tag{6.33}$$

Dividing both sides by effective area bd produces an expression for the balanced reinforcement ratio ρ_b as

$$\rho_b = \frac{0.85\beta_1 f_c'}{f_y}\frac{\varepsilon_{cu}}{\varepsilon_{cu}+\varepsilon_y} \tag{6.34}$$

In its traditional design approach, ACI 318 takes $\varepsilon_{cu} = 0.003$. For singly reinforced beams, the upper limit on the longitudinal tension reinforcement is three-quarters of the area (or steel ratio) corresponding to balanced failure. For doubly reinforced beams, the area of tension reinforcement is permitted to be the area balanced by A_s' plus three-quarters of the excess tension reinforcement $A_s - A_s'$ calculated by Eq. (6.32). For cases other than the singly reinforced beam, however, the preferred approach is to calculate the depth of compression zone for a singly reinforced beam with $0.75\rho_b$, and limit the longitudinal reinforcement so that neutral axis depth does not exceed that limiting value. For a singly reinforced rectangular beam, the neutral axis depth for $0.75\rho_b$ is 0.75 times the depth c given by Eq. (6.29).

ACI 318 no longer uses the balanced reinforcement ratio to define the upper limit of longitudinal tension reinforcement. Instead, the general requirement is that beams must have $\varepsilon_t \geq 0.004$ (Figure 6.32*a*). More restrictive limits may be imposed for beams in special moment frames (Chapter 12).

6.8 Columns

6.8.1 General Observations about Axial Force, Moment, and Curvature

As for beams, the moment–curvature relation of a column depends on the section shape, the concrete and reinforcement material properties, the amount and distribution of the longitudinal reinforcement, and the amount and configuration of the transverse reinforcement. In addition, it is strongly affected by the axial load. The effect of axial load on a symmetric section is indicated conceptually in Figure 6.34, as discussed below.

- Under zero axial load, nonlinear response is initiated by yielding of the longitudinal reinforcement, followed by relatively ductile moment–curvature response until the compressed concrete reaches its strain capacity.
- As the axial load increases, a larger compression zone is required to maintain axial force equilibrium. Therefore, curvature, defined by the ratio of the limiting concrete compressive strain to the neutral axis depth, decreases as axial load increases.
- The tensile strain in the longitudinal reinforcement decreases as the axial load increases. At the balanced axial load, the extreme longitudinal tension reinforcement yields simultaneously with reaching the limiting compressive strain of concrete. For axial loads below the balanced point, the cross section is tension-controlled and strength is limited primarily by the strength of the tension longitudinal reinforcement. For axial loads above the balanced point, the cross section is compression-controlled and strength is limited

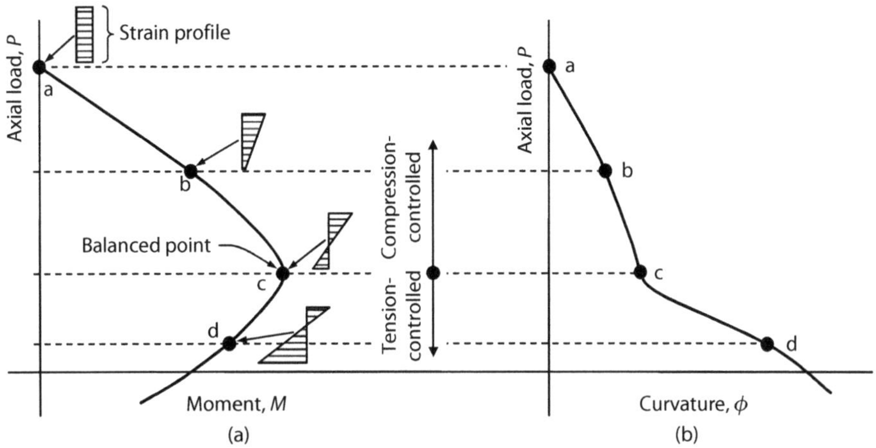

FIGURE 6.34 Internal strains and corresponding axial load, moment, and curvature in a column.

primarily by the strength of the compression zone. The balanced point identifies the axial load corresponding to maximum moment strength.

- Curvature capacity continues to decrease with increasing axial load, eventually reaching zero at the maximum axial load.

6.8.2 Construction of *P*-*M*-ϕ Relations by Hand Calculations

Calculation of *P*-*M*-ϕ relations can be accomplished using the general procedure outlined in Section 6.6.1. For example, to calculate the nominal strength of unconfined columns, the extreme compressive strain of concrete is set equal to a limiting value (0.003 according to ACI 318, or perhaps a larger value according to some guidelines and codes), a neutral axis depth is selected, and resulting stresses are integrated to determine the corresponding axial force *P* and moment *M*. For confined sections with strain-hardening longitudinal reinforcement, the calculations become more involved but the basic procedure applies.

For many practical applications, the *P*-*M* interaction diagram can be sufficiently defined by calculating only a few points. For example, to establish the nominal *P*-*M* interaction diagram according to ACI 318, the following procedure can be used:

- First calculate the axial load and moment corresponding to uniform compressive strain (point **a** in Figure 6.34). The axial load is defined by

$$P_0 = A_{st}(f_y - 0.85f'_c) + A_g(0.85f'_c) \tag{6.35}$$

  Whether there is an internal moment *M* corresponding to point **a** depends on whether the section is symmetric and whether the axial load P_0 is located at the centroid of resistance of the section. For symmetric sections with P_0 at the geometric centroid, $M = 0$.

- Next, calculate *P* and *M* corresponding to balanced failure (point **c** in Figure 6.34).

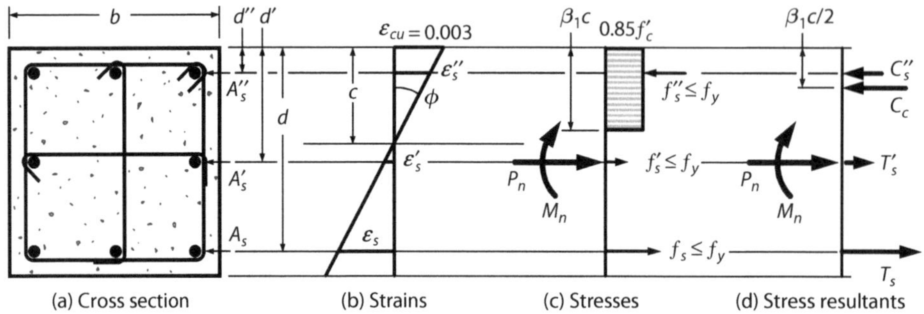

FIGURE 6.35 ACI 318 assumptions for nominal axial load and moment strength.

- Then, calculate additional points as needed. Point **d** is obtained by selecting a neutral axis depth less than the depth at balanced failure, whereas point **b** is obtained by selecting a larger neutral axis depth.

Figure 6.35 illustrates the ACI 318 assumptions for calculating nominal axial load and moment for arbitrary neutral axis depth *c*. Note that the free-body diagram in Figure 6.35*d* must be in equilibrium considering both axial force and moment. Summing axial forces results in

$$\sum \vec{F} = 0 : P_n + T_s + T'_s - C_c - C''_s = 0 \tag{6.36}$$

from which the nominal axial force P_n can be solved. Summing moments about the extreme compression fiber results in

$$\sum \widehat{M_{top}} = 0 : M_n - P_n h/2 + C''_s d'' + C_c \beta_1 c/2 - T'_s h/2 - T_s d = 0 \tag{6.37}$$

from which the nominal moment M_n can be solved.

The procedures outlined here can be used for confined concrete sections and for biaxial bending problems, but the calculations become too involved for manual work. Computer codes, some based on the fiber model idealization shown in Figure 6.23, are more suitable for such calculations.

6.8.3 Axial Force, Moment, and Curvature Response

Sample calculations were carried out on the reinforced concrete column cross section shown in Figure 6.36 using the computer program XTRACT. The column cross section was designed with specified A706 Grade 60 (420) reinforcement and concrete compressive strength $f'_c = 4000$ psi (27.6 MPa). Transverse confinement reinforcement was designed according to the ACI 318 Chapter 21 confinement requirements that were presented in Section 5.5.4. For this problem, calculations were carried out with the following assumptions:

Case 1: Concrete was modeled using the stress–strain relation of Chapter 4 assuming no confinement, with maximum compressive stress of $0.93f'_c$. The

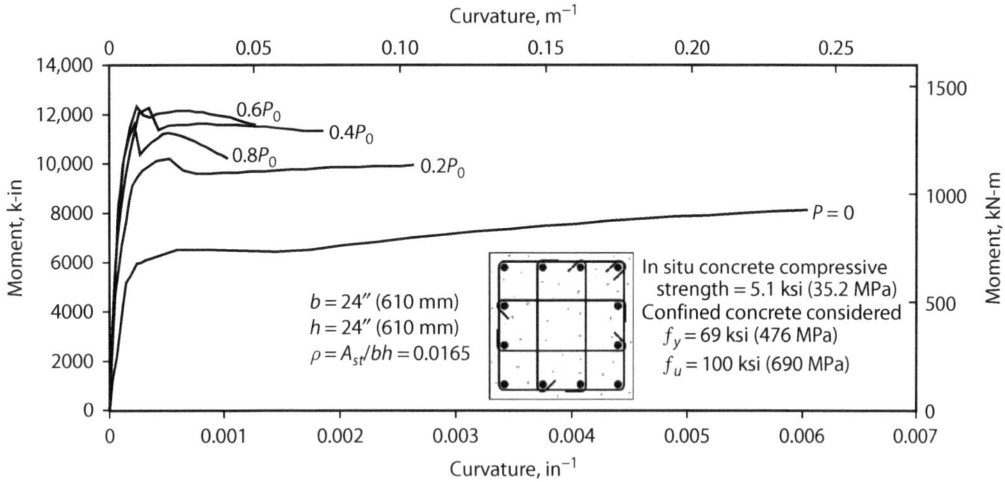

Figure 6.36 Moment–curvature relations for confined concrete column section.

coefficient 0.93 is based on the results of Hognestad et al. (1955), who found that the peak concrete compressive stress in flexure was slightly less than that measured in standard cylinder tests for compressive strengths exceeding 3000 psi (20.7 MPa). A lower coefficient of 0.85 would be more appropriate near the top of a column due to segregation and consolidation effects (Chapter 3). The elasto-plastic steel model had yield stress of 60 ksi (414 MPa).

Case 2: Concrete was modeled assuming expected strength values without confinement. For this purpose, long-term compressive strength was assumed to be $1.5f_c'$, and in situ strength was taken as 0.85 times that value. Expected steel behavior included yield stress of 69 ksi (476 MPa) and ultimate stress of 100 ksi (690 MPa).

Case 3: Materials were modeled as in case 2 except the core was modeled as confined concrete following the procedures of Chapter 4.

A reference axial load $P_0 = 2500$ k (11,000 kN) was calculated using Eq. (6.35) and nominal (specified) material strengths.

Figure 6.36 shows moment–curvature relations for case 3. As expected, curvature capacity decreases with increasing axial load.

Figure 6.37 shows relations among axial load, maximum moment strength, and ultimate curvature capacity for the three different analysis cases. (Note that the maximum moment did not necessarily occur at the time of the maximum curvature but instead may have occurred for smaller curvature.) Increasing the assumed material strengths from nominal to expected properties results in considerable strength increase, especially for higher axial loads. The effect of confinement on strength is noticeable but not very significant. In contrast, the change from nominal to expected material strength does not much affect the curvature capacity, whereas curvature capacity is considerably increased due to confinement of the core concrete.

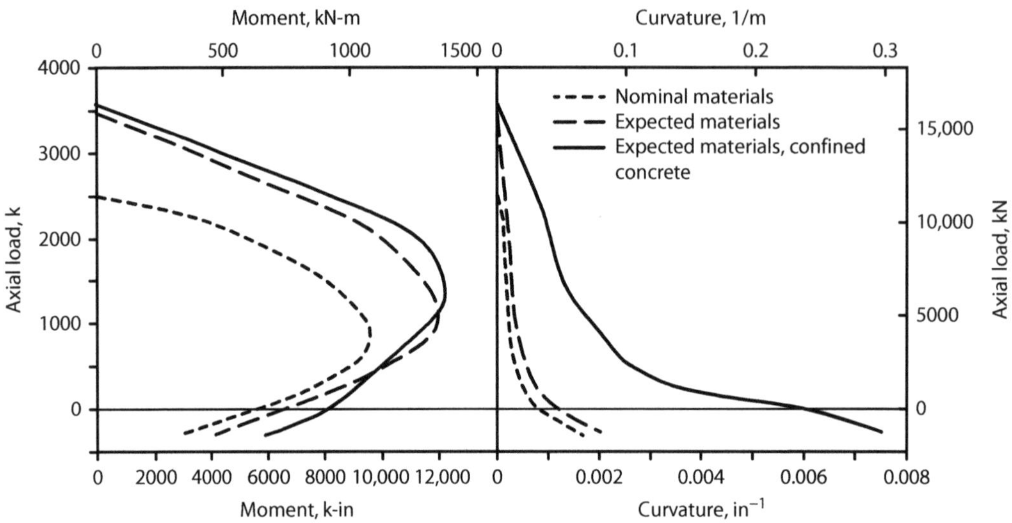

FIGURE 6.37 Relations among axial load, maximum moment strength, and ultimate curvature for various assumptions.

6.8.4 Nominal, Probable, and Design Strengths

As for beams, nominal strength is the strength calculated for specified material properties using modeling assumptions described in Section 6.7.2. The one exception is that strength under pure axial compression is calculated for maximum compressive strain ε_0 rather than ε_{cu}. For each strain profile, a corresponding pair of nominal moment and axial strength (M_n, P_n) is determined.

Similarly, probable moment strength for columns refers to the nominal moment strength calculated assuming a tensile stress in the longitudinal bars of at least $1.25f_y$ and $\phi = 1.0$. For practical applications using available software, M_{pr} is calculated assuming elasto-plastic steel stress–strain relation in tension and compression, with the yield stress equal to $1.25f_y$, where f_y is the specified yield stress. Although not commonly identified in concrete terminology, for each strain profile, the cross-sectional analysis yields a value of probable axial force in addition to M_{pr}.

To obtain design moment and axial strength values (ϕM_n, ϕP_n), the individual combinations of nominal moment and axial strength (M_n, P_n) are multiplied by a common strength reduction factor ϕ. To determine the value of ϕ, the failure strain conditions corresponding to each (M_n, P_n) pair are evaluated in accordance with Figure 6.32*c* to determine ε_t. The value of ϕ is then determined from Figure 6.32*a*. The strength design requirement is that each combination of (M_u, P_u) from the factored load combinations must fall within the design strength envelope determined by pairs of (ϕM_n, ϕP_n). ACI 318 further limits the design axial strength to

$$\begin{aligned}\phi P_{nmax} &= 0.85\phi[0.85 f'_c(A_g - A_{st}) + f_y A_{st}] \text{ for spiral-reinforced columns}\\ \phi P_{nmax} &= 0.80\phi[0.85 f'_c(A_g - A_{st}) + f_y A_{st}] \text{ for tied columns}\end{aligned} \tag{6.38}$$

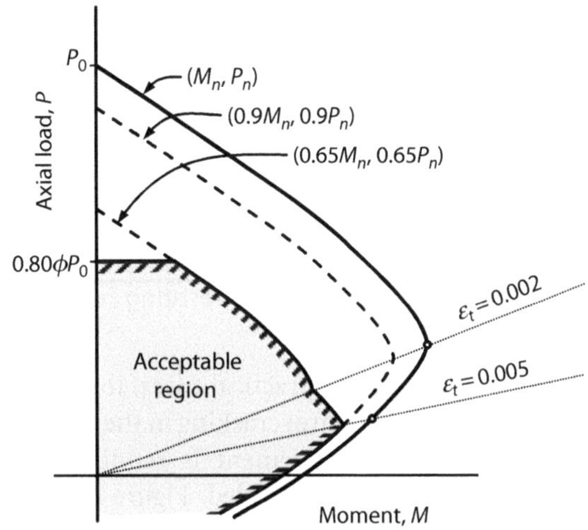

Figure 6.38 Moment–axial force interaction curves for tied column, showing both nominal strength envelope and design strength envelope.

Figure 6.38 illustrates the nominal strength envelope and the design strength envelope for a tied column.

6.8.5 Reinforcement Limits

Reinforcement is necessary to resist bending moments and axial tension, whether or not such effects are expected. Reinforcement is also required to resist effects of creep and shrinkage under sustained compressive loadings. Creep and shrinkage tend to transfer force from concrete to longitudinal reinforcement, with greater increase in reinforcement stress for smaller reinforcement ratios. Based on consideration of these effects, most codes require at least a minimum longitudinal reinforcement ratio. In ACI 318, the lower limit for compression members (columns) is $\rho_l = A_{st}/A_g \geq 0.01$.

Most codes also include an upper bound on the longitudinal reinforcement ratio of columns. In ACI 318, the limit is $\rho_l \leq 0.08$, except the limit is reduced to 0.06 in columns of special moment frames designed to resist earthquake effects. This limit is a practical maximum for reinforcement considering requirements for concrete placing. ACI 318 is unclear about how this limit applies where lap splices are used. Construction considerations would suggest that the limit of 0.08 (or 0.06) applies at sections with lap splices, including total areas of lapped bars.

6.9 Walls

6.9.1 Geometry and Reinforcement

Walls and columns are the terms commonly applied to vertical components of a seismic force-resisting system. The term *column* usually implies a limit on the cross-sectional aspect ratio. For example, for columns of special moment frames in ACI 318 the ratio of the shortest cross-sectional dimension to the perpendicular dimension must be at least 0.4. Vertical members not satisfying this requirement are

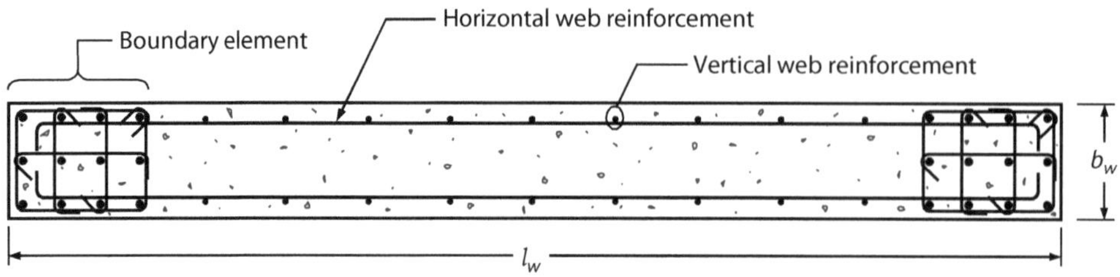

Figure 6.39 Typical configuration of earthquake-resisting concrete wall.

classified as walls. In many practical cases, the length-to-thickness ratio for walls will be 10 or greater. To control cracking in the plane of the wall, distributed vertical and horizontal web reinforcement is required. Additional boundary element reinforcement may also be provided. Figure 6.39 illustrates typical placement of reinforcement in a wall cross section in an earthquake-resisting building. Most building codes specify that the distributed reinforcement (both vertical and horizontal) must provide a minimum reinforcement ratio (typically 0.0025).

6.9.2 Axial Force, Moment, and Curvature Response

Behavior of a reinforced concrete wall under axial and in-plane lateral loading depends on aspect ratio and presence of significant openings. If the wall is sufficiently slender (aspect ratio exceeding approximately 2) and there are no significant openings, behavior can be assessed adequately using conventional flexural theory that assumes plane sections remain plane (Orakcal and Wallace, 2006). Calculation of P-M-ϕ relations can be accomplished using the general procedure outlined in Section 6.6.1.

Figure 6.40 plots examples of calculated moment–curvature relations for rectangular walls for the case of $P = 0$. In these walls the distributed reinforcement ratio was 0.0025. Boundary reinforcement ratio (area of longitudinal reinforcement in the boundary divided by the boundary area) is indicated for each plot. In each case, curvature capacity was limited by compressive strain capacity of the unconfined concrete.

A noteworthy observation is the sudden drop in moment resistance after onset of cracking, especially for walls with low boundary element reinforcement ratio. If the wall boundary reinforcement ratio is too small, it is plausible that a single crack could occur at a critical section with all subsequent deformation occurring at that crack. The concentrated curvature could result in relatively small wall displacement capacity. The unit cracking strength of the boundary element can be estimated to be equal to the modulus of rupture, f_r, and the unit strength after cracking is approximately ρf_y, where ρ refers to the local reinforcing ratio of the longitudinal reinforcement in the wall boundary. Thus, for distributed cracking to develop, the boundary element longitudinal reinforcement ratio should be at least $\rho = f_r/f_y$, or about 0.01 for typical concrete and steel strengths. This limit is the same as the lower limit for reinforced concrete columns, and should be considered as a lower limit for wall boundary element design as well. Bonelli et al. (1999) report tests of small-scale

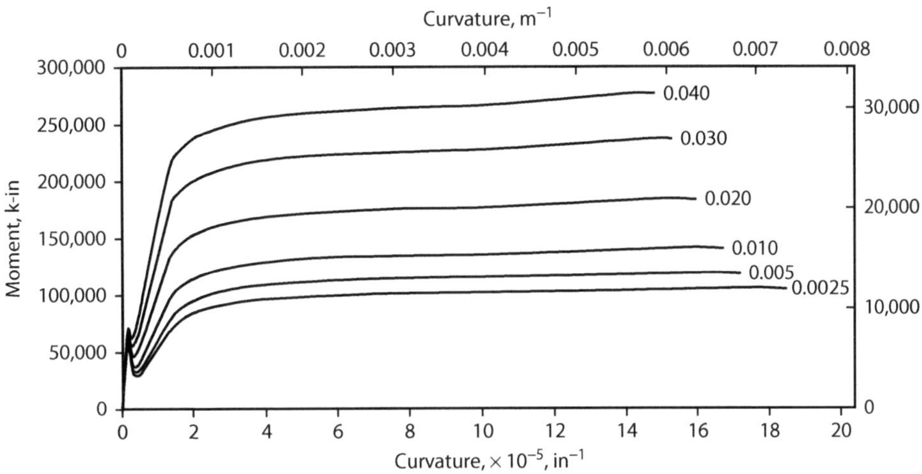

Figure 6.40 Calculated moment–curvature relations for rectangular wall with 0.0025 distributed steel plus boundary reinforcement ratios indicated adjacent to the curves. Wall length = 20 ft (6.1 m), thickness = 12 in (0.3 m), boundary element length = 2 ft (0.61 m), $f_c' = 4$ ksi (27.6 MPa), ASTM Grade 60 reinforcement ($f_y = 69$ ksi, 476 MPa), axial load $P = 0$, maximum usable compressive strain = 0.004.

walls demonstrating satisfactory behavior for walls having $\rho \geq 0.01$. A wall with half that reinforcement fractured after developing a single main flexural crack.

Wood (1989) reports wall test data for which drift capacity was limited by fracture of longitudinal reinforcement. Those data and more recent data are reassessed in Figure 6.41. Only results for large-scale cyclic tests with reinforcement

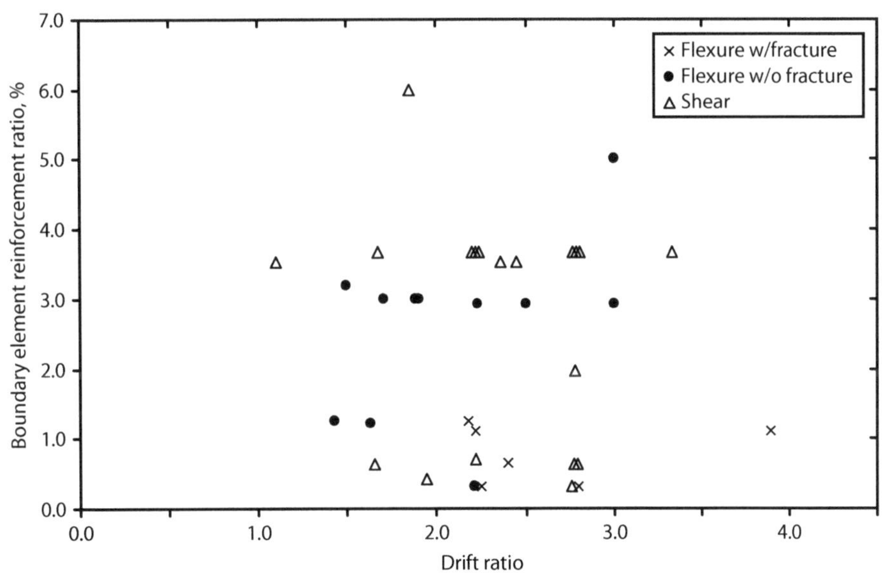

Figure 6.41 Occurrence of longitudinal reinforcement fracture for cyclic wall tests.

having elongation and strain-hardening roughly equivalent to ACI 318 requirements are included. Reinforcement fractures are reported only for boundary element reinforcement ratio near 0.01 or lower, and for drift ratios 0.02 and higher. This suggests that reinforcement ratio less than 0.01 should be avoided if expected drift ratios are roughly 0.02 or higher.

The data in Figures 6.40 and 6.41 are for walls reinforced with boundary elements in addition to 0.0025 distributed reinforcement. The boundary reinforcement improves flexural behavior by distributing flexural cracking on the tension side while assisting the concrete to resist compression on the compression side. Flexural strength, on the other hand, is little affected by placement of the vertical reinforcement, provided that it is symmetrically placed. Figure 6.42 compares calculated moment–curvature relations of two walls having equal areas of longitudinal reinforcement, one with all of the reinforcement distributed uniformly along the wall length and the other with all of the reinforcement positioned in the two boundary elements. For the wall with concentrated boundary elements, the boundary element longitudinal reinforcement resists compression, thereby reducing the depth of the neutral axis and increasing the curvature capacity relative to the wall with only distributed reinforcement.

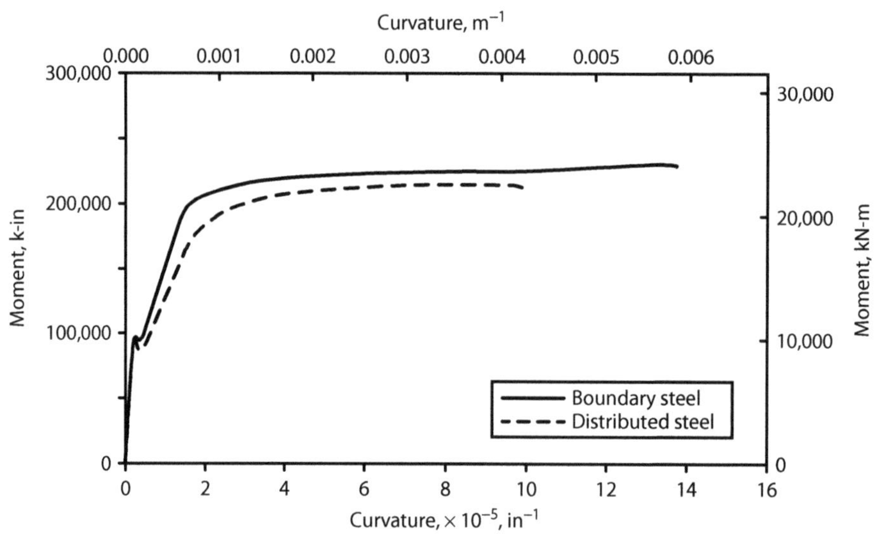

FIGURE 6.42 Calculated moment–curvature relations of walls with equal areas of distributed and concentrated reinforcement. Wall length = 20 ft (6.1 m), thickness = 12 in (0.3 m), $f_c' = 4$ ksi (27.6 MPa), ASTM Grade 60 reinforcement ($f_y = 69$ ksi, 476 MPa), axial load $P = 0.05A_g f_c'$, maximum usable compressive strain = 0.004. Walls have equal total areas of longitudinal reinforcement. One has two curtains of No. 6 (19) at 10.5 in (267 mm) and the other has two curtains of No. 4 (13) at 12.8 in (325 mm) with eight No. 9 (29) in each boundary.

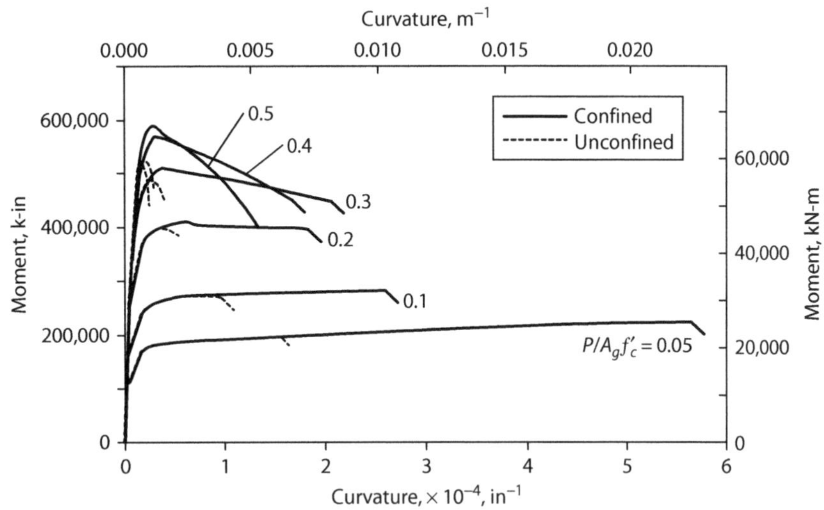

Figure 6.43 Calculated moment–curvature relations of unconfined and confined walls with different axial load ratios. Wall length = 20 ft (6.1 m), thickness = 12 in (0.3 m), $f_c' = 4$ ksi (27.6 MPa), ASTM Grade 60 reinforcement [$f_y = 69$ ksi, $f_{su} = 100$ ksi (483 MPa, 690 MPa)], maximum usable compressive strain for unconfined concrete = 0.004. Walls have boundary element reinforcement ratio 0.01 distributed along a length $0.1l_w$. Distributed reinforcement ratio is 0.0025. Confinement reinforcement comprises No. 4 hoops with crossties at 6 in (150 mm) spacing and 4.25 in (108 mm) longitudinal spacing. Length of confined region was $0.1l_w$, $0.1l_w$, $0.21l_w$, $0.34l_w$, and $0.5l_w$ for walls with $P/A_gf_c' = 0.05$, 0.1, 0.2, 0.3, and 0.4 and higher.

Curvature capacity of a wall can be increased by providing confining reinforcement in the wall boundaries. According to ACI 318 (2011 and earlier editions), the confinement reinforcement in each direction must satisfy $A_{sh}/b_c s \geq 0.09 f_c'/f_{yt}$, with maximum vertical spacing $b_w/3$, and must extend from the wall edge a distance not less than the greater of $c/2$ and $c - 0.5l_w$, where c is the depth to neutral axis from the extreme compression fiber calculated for nominal material properties with $\varepsilon_{cu} = 0.003$. Figure 6.43 plots calculated moment–curvature relations for walls without confinement and for walls satisfying these confinement requirements. In all cases, the boundary element longitudinal reinforcement ratio is 0.01. Walls with low axial load ratio show strain-hardening behavior, whereas walls with higher axial load show strain-softening behavior. Strain-softening behavior should be avoided in design because it would result in localization of inelastic curvatures and low displacement capacity. Strain-hardening of the cross section can be increased by increasing the longitudinal reinforcement ratio or increasing the amount of boundary element confinement.

Results plotted in Figure 6.43 were calculated assuming that the wall boundary was stable. Relatively slender wall boundaries, however, may be susceptible to lateral buckling (Figure 6.44). As discussed in Chapter 5, tensile yielding can reduce the stability of slender members. Spalling of cover concrete around a confined boundary element can also reduce the effective cross section, thereby

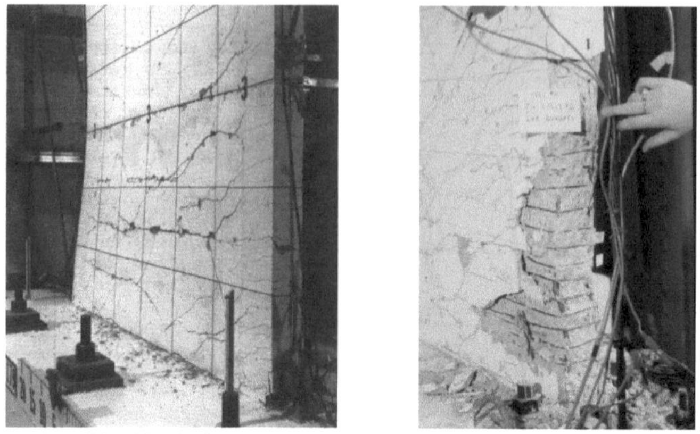

(a) Wall buckled before spalling (b) Wall buckled after spalling

Figure 6.44 Photographs of wall lateral buckling observed in laboratory tests. [*(a) After Corley et al., 1981, courtesy of American Concrete Institute; (b) after Thomsen and Wallace, 2004, used with permission from ASCE.*]

making it more susceptible to buckling. Conditions for lateral instability of prismatic members were analyzed in Section 5.7.2. Applications to slender structural walls are discussed in Section 13.4.3.

6.9.3 Nominal, Probable, and Design Strengths

The terms *nominal*, *probable*, and *design strengths* apply to structural walls in the same way as they apply to columns. Refer to Section 6.8.4 for discussion.

6.9.4 Reinforcement Limits

Building codes may prescribe limits for both the distributed reinforcement and the boundary element reinforcement. In ACI 318, special structural walls for seismic resistance require both vertical and horizontal distributed reinforcement with steel ratio not less than 0.0025 or as required for shear. Smaller steel ratio may be permitted under some conditions. Additionally, the distributed reinforcement may be required to be placed in two curtains (layers). See Chapter 13 for additional discussion. U.S. building codes do not place additional limits on the boundary element vertical reinforcement. As discussed in Section 6.9.2, to ensure distributed cracking it may be advisable to use at least $\rho_l = 0.01$ as the local reinforcement ratio at boundaries. An upper limit of $\rho_l = 0.06$ is recommended for constructability.

6.10 Flanged Sections

6.10.1 Beams

For beams cast monolithically with a flange, the flange increases beam stiffness and strength. Portions of the flange more distant from the web are less effective than those closer to the web (Figure 6.45*a*). Common design practice is to define an effective flange width (Figure 6.45*b*) that is assumed to act monolithically with the beam web without shear lag effects, and which develops the same total flange

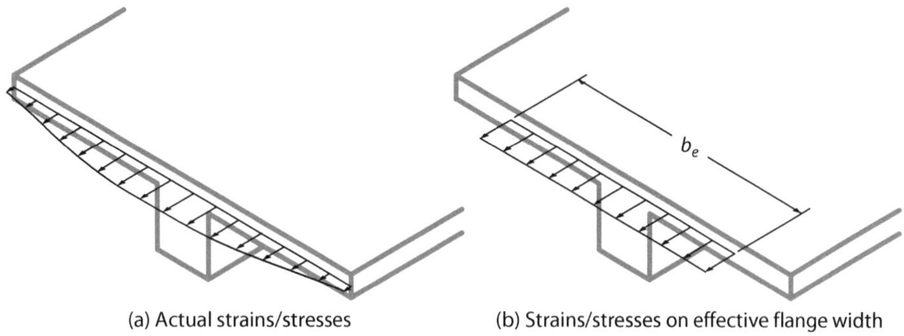

FIGURE 6.45 Flexural contribution of a flange.

force as is developed in the actual flange. In some building codes, the flange is defined only as a compression element, but a flange also can act as a tension element.

Whether the flange acts in tension or compression depends on the loading. Figure 6.46 depicts a floor system comprising a beam with monolithic floor slab framing between columns. Moment diagrams under idealized gravity and lateral loads are shown. Under gravity loads, moment puts the top surface of the beam in compression at section A-A and in tension at section B-B. This action determines whether the flange acts in tension or compression, as shown in Figure 6.46*c*. Actions under lateral loads can be determined similarly.

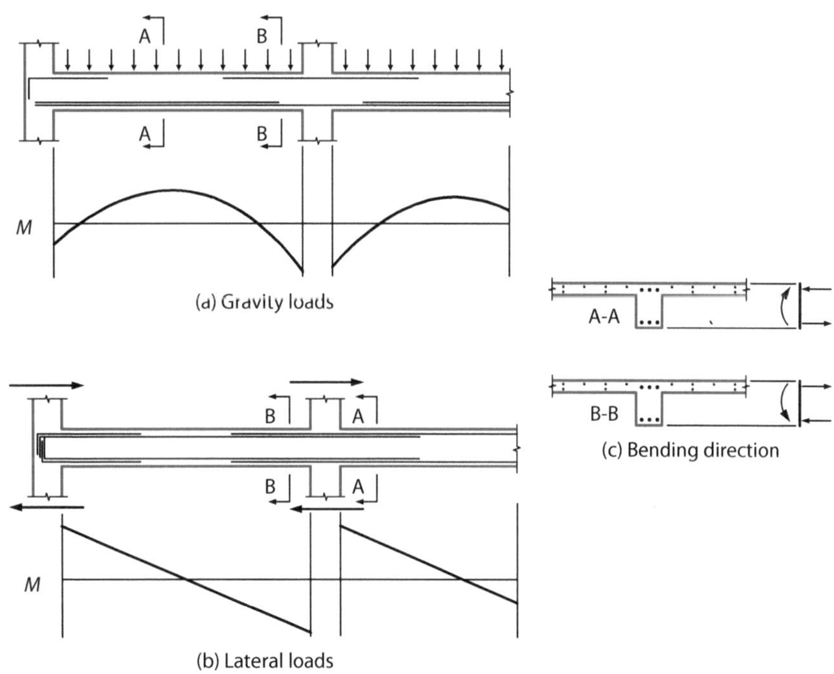

FIGURE 6.46 Beam-slab-column floor system under idealized vertical and lateral loading.

ACI 318 defines an effective flange width for the case of the flange in compression. For beams with slab on both sides of the web, the effective flange width is taken as one-quarter of the span length of the beam, except the effective overhanging flange width on each side of the web is not to exceed $8h_s$ nor half the clear distance to adjacent webs. For beams with a slab on one side only, the effective overhanging flange width is limited to the least of $l/12$, $6h_s$, and half the clear distance to adjacent webs. In isolated beams, flange thickness h_s must not be less than $b_w/2$ and the effective flange width is limited to $4b_w$. The flange is required to have reinforcement perpendicular to the beam spaced not farther apart than $5h_s$ and 18 in (460 mm). That reinforcement must also be capable of resisting moments due to loads acting on the flange; in monolithic floor systems, the approach is to design the flange as if it is a cantilever having length equal to the effective overhanging flange dimension.

To better understand behavior of a flange, consider a cantilever beam loaded so that the flange is stressed in tension (Figure 6.47). The beams including flange are fixed at the left end with all other surfaces free. Under this loading, elongation of the flexural tension region of the beam web is transferred to the flange along interface **ad**, inducing shear stress along that interface (Figure 6.47*d*). Equilibrium of forces in the direction of the beam span is achieved through tensile stresses acting on face **ab** of the flange (Figure 6.47*d*). Moment equilibrium about point **a** requires tensile stresses along interface **ad**, which in turn requires shear stress along **ab**.

Shear stresses acting within the flange panel **abcd** cause cracks to bend away from the fixed edge (Figure 6.47*c*). Shear distortion, shown shaded in Figure 6.47*d*, relieves

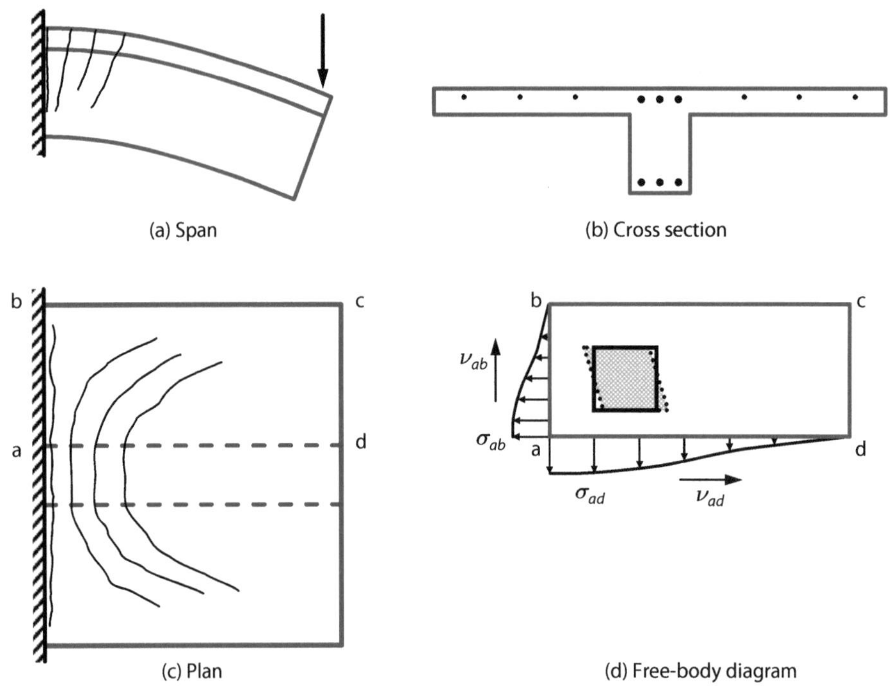

Figure 6.47 Deformations and stresses in a beam flange loaded in tension.

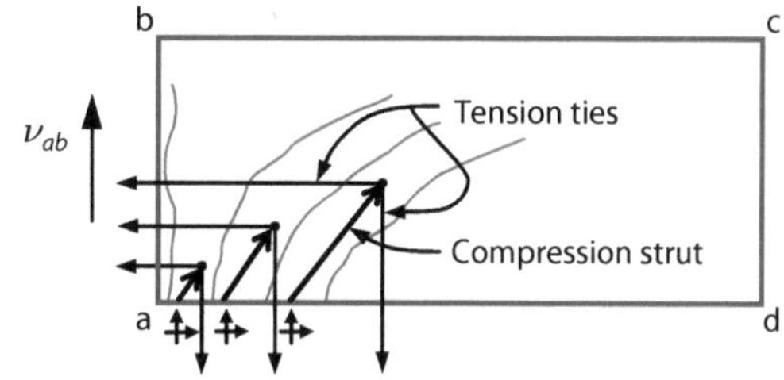

FIGURE 6.48 Strut-and-tie model of slab panel acting as a tension flange.

the tensile stresses with increasing distance along line **ab**, such that at sufficient distance the tensile stress drops essentially to zero. This action is known as *shear lag*.

Note that equilibrium requires flange tension in both the longitudinal and transverse directions (Figure 6.47*d*). Therefore, both longitudinal and transverse reinforcement are required in the flange.

Figure 6.48 shows an alternative idealization to represent the flow of forces within a tension flange. Cracks in the flange are considered to define isolated segments of concrete. Shear stresses along **ad** load these segments, inducing *diagonal compressive forces* acting along the axes of these segments, or *struts*. Normal stresses also are required along **ad** to equilibrate the diagonal compressive forces. Each diagonal compressive force is resolved at a *node* to which longitudinal and transverse flange reinforcement is assumed to be anchored. Flange reinforcement thus acts as a *tension tie* to equilibrate the diagonal compressive forces in the struts. This *strut-and-tie model* can be convenient for visualizing the flow of forces in concrete elements subjected to shear. Chapter 7 develops the concept of strut-and-tie models in greater depth. Note that, according to this model, concrete acts primarily in compression while reinforcement acts in tension.

Figure 6.49 depicts a test specimen used to assess seismic behavior of a slab-beam-column-joint assembly. Beams and columns extend to idealized inflection points at mid-length of the beam span and column height, respectively, with portions of the slab extending around the connection. Seismic loading is simulated by cycling the column back and forth while beam ends are supported on rollers or, alternatively, by holding the column ends in place while cycling the beam ends up

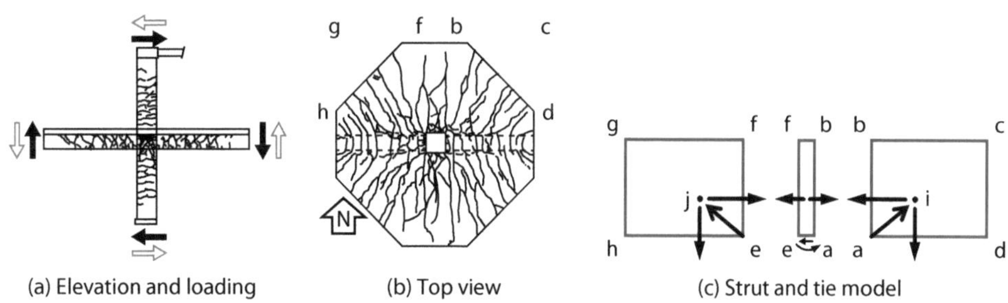

FIGURE 6.49 Actions on an isolated beam-slab-column connection under lateral loading. Test results for test specimen J1. (*After Kurose et al., 1988, courtesy of American Concrete Institute.*)

and down in opposite directions. Slab crack patterns observed in such tests (Figure 6.49*b*) are consistent with those shown in Figure 6.47*b*.

If the right-hand beam is sheared downward while the left-hand beam is sheared upward, then segment **abcd** in Figure 6.49*c* will act as the slab-in-tension side, similar to the case of an isolated cantilever. As discussed previously, a diagonal compression strut will develop along **ai**, being equilibrated by tension in longitudinal and transverse slab reinforcement. Longitudinal reinforcement is anchored through segment **abef** (which could include a transverse beam or the slab only) and into slab segment **efgh**. Although segment **abef** will experience in-plane shear, moment, and possibly torsion, the stiffer load path typically is into slab segment **efgh**. Anchorage of the slab reinforcement in segment **efgh** will induce diagonal compression (strut **ej**) and transverse tension. Consequently, the slab-in-compression side experiences a stress field somewhat similar to the stress field in the slab-in-tension side (French and Moehle, 1988). One interpretation is that the slab is in tension on both sides of the connection.

Figure 6.50*a* shows longitudinal strains measured in longitudinal slab reinforcement during tests of two slab-beam-column test specimens, one being the test specimen depicted in Figure 6.49. Strains increase as drift level increases. Strains also tend to be higher if there is a transverse beam, due to the added constraint it provides. At large drifts, it is not unusual for most (or all) of the slab reinforcement to reach the yield strain. The total force resisted by the flange in tension is $\sum A_{si} f_{si}$, where A_{si} and f_{si} are the area and stress of individual longitudinal slab bars. Assuming yielding of the flexural tension reinforcement, the effective flange width is thus $b_e = \sum A_{si} f_{si} / \rho_f h_s f_y + b_w$. Because strains and stresses increase with increasing drift, b_e increases with increasing drift, but at a decreasing rate as more bars yield, and with an upper bound based on total slab width and bar strength. Figure 6.50*b* shows the increase in b_e with increasing drift ratio for the two test specimens.

Tests have also been done on multi-bay frames. Rather than the inclined cracks observed in isolated subassembly tests, cracks tend to be more linear and oriented

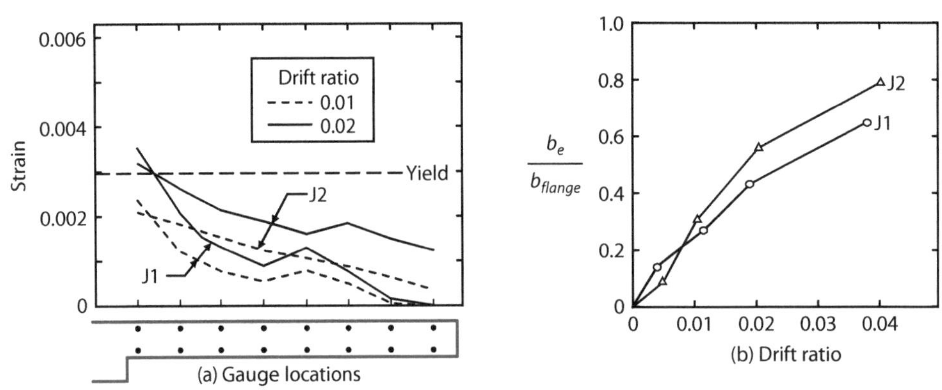

Figure 6.50 Test results for interior beam-column joints without (J1) and with (J2) transverse beams: (*a*) strain in top slab longitudinal reinforcement; (*b*) ratio of flange effective width to total flange width considering both top and bottom reinforcement. (*Test results after Kurose et al., 1988, used with permission from the University of Texas at Austin.*)

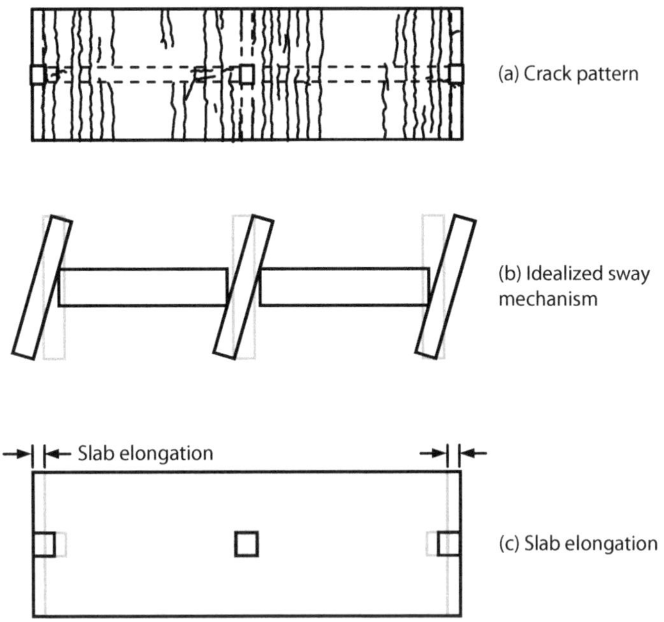

Figure 6.51 Slab mechanism in indeterminate frame. (*Crack pattern after Qi, 1986, used with permission from the University of California, Berkeley.*)

perpendicular to the framing direction (Figure 6.51*a*). This suggests a different resisting mechanism in which the slab is in tension along the entire length of the frame. As illustrated in Figure 6.51*b*, the neutral axis for beam flexure is located near the flexural compression side, such that beam flexure produces beam elongation that pushes the columns apart and induces elongation and tension in the slab (Qi and Pantazopoulou, 1991). This action produces compression in the beams that increases their moment strength. In laboratory tests on a full-scale, seven-story reinforced concrete test structure, yielding extended into the floor slab (Figure 6.52), resulting in significant increase in beam moment strength and, consequently, strength of the structural system.

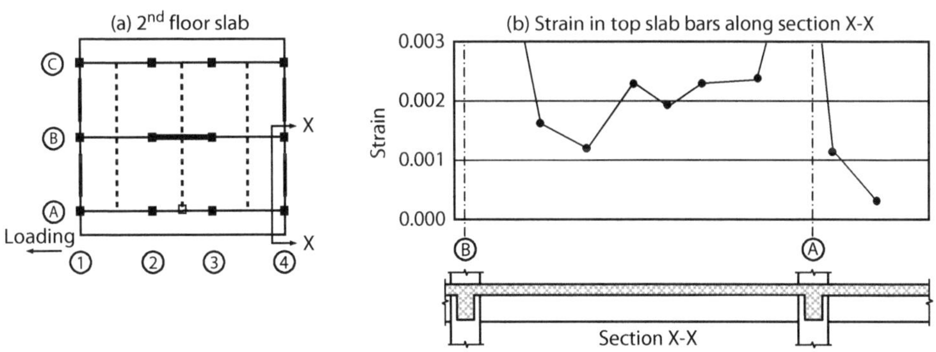

Figure 6.52 Test results from U.S.–Japan seven-story reinforced concrete test structure at drift ratio 1/68. (*After Yoshimura and Kurose, 1985, courtesy of American Concrete Institute.*)

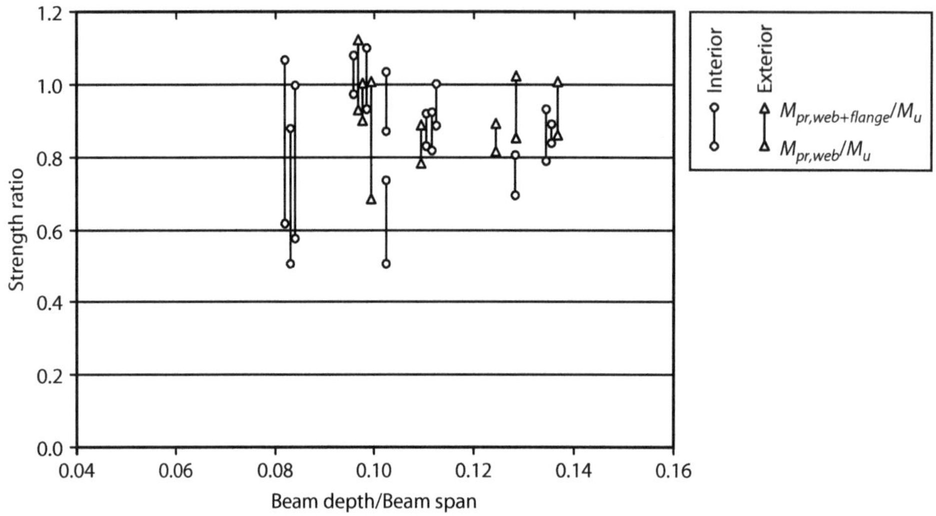

FIGURE 6.53 Ratio of calculated moment strength to moment resisted at drift ratio 0.02. (*Data after French and Moehle, 1991.*)

French and Moehle (1991) compared measured and calculated beam moment strengths from subassembly tests (similar to the one shown in Figure 6.49) for the case of the slab acting as a tension flange. Measured moment strength was defined as the moment recorded for subassembly drift ratio of 0.02. Probable moment strength was calculated according to the ACI 318, with two exceptions: (1) Concrete strength was taken equal to measured concrete compressive strength; (2) steel yield strength was taken equal to 1.25 times measured yield strength. Calculated moment strengths based on web longitudinal reinforcement generally fell well below measured strengths (Figure 6.53). Improved estimates of strength were obtained by including developed slab reinforcement within b_e, where b_e is the effective flange width according to ACI 318 for the slab-in-compression case.

Example 6.6 A cast-in-place floor system is supported by 18″ by 24″ beams, as shown in Figure 6.54. The 6″ thick floor slab is reinforced with top and bottom reinforcement, each

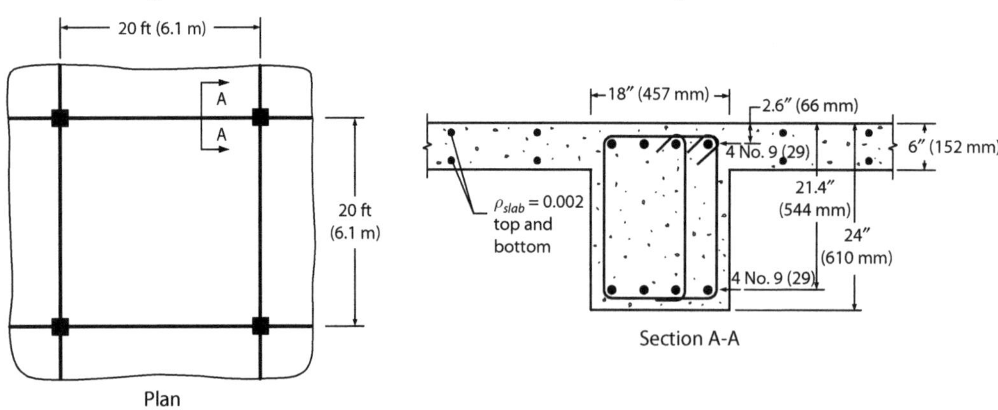

FIGURE 6.54 Cast-in-place floor system.

providing $\rho_{slab} = 0.002$; $f_c' = 4000$ psi (28 MPa); Grade 60 (420) reinforcement. For an interior beam, calculate M_{pr} for both positive moment (bottom in tension) and negative moment.

Solution

The effective flange width is $b_e = \min\,(l/4 = 240''/4 = 60''; b_w + 2 \times 8h_s = 18'' + 16 \times 6'' = 114''; 240'')$. Therefore, $b_e = 60''$.

For the case of positive moment, the effective flexural compression depth $\beta_1 c$ is assumed to fall within the 6″ slab thickness, in which case the beam flexural strength can be calculated as if it is a rectangular section of width b_e. Ignoring the slab reinforcement in compression, the areas of tension and compression reinforcement are $A_s = A_s' = 4$ in^2. Taking effective yield stress as $1.25f_y$, probable moment strength is calculated as $M_{pr} = 6270$ k-in (709 kN-m). Depth $\beta_1 c = 1.8''$ (45 mm), which falls within the flange depth as assumed.

For the case of negative moment, the area of slab reinforcement to be included as tension reinforcement is $2 \times 0.002\,(b_e - b_w)h_s = 2 \times 0.002 \times (60'' - 18'')(6'') = 1.01$ in^2. Thus, $A_s = 5.01$ in^2, $A_s' = 4$ in^2. From here, $M_{pr} = 7220$ k-in (819 kN-m).

A common construction form uses unbonded post-tensioned slabs cast monolithically with conventionally reinforced beams. Placing the unbonded strands outside the effective flange width does not mean those strands do not contribute to beam flexural strength. This is because, away from the slab edge, the post-tensioning produces a fairly uniform compressive stress field across the plate including the beam cross section. A reasonable approach is to calculate the average prestress acting on the combined slab-beam system, and then apply this prestress to the T-beam cross section to determine the effective axial compression on the T-beam (Moehle et al., 2008). This axial load, acting at the level of the slab, is used along with the beam longitudinal reinforcement to calculate the T-beam flexural strength. This recommendation applies only for interior connections that are far enough away from the slab edge so as to be fully stressed by the post-tensioning. It need not apply at an exterior connection close to the slab edge because post-tensioning located transversely away from the beam will not effectively compress the beam at that location. See Figure 6.55.

6.10.2 Walls

Flanged walls occur in L, T, C, tube, and other shapes. In a flanged wall, shear is resisted almost entirely within the plane of the wall (the wall web) while moment

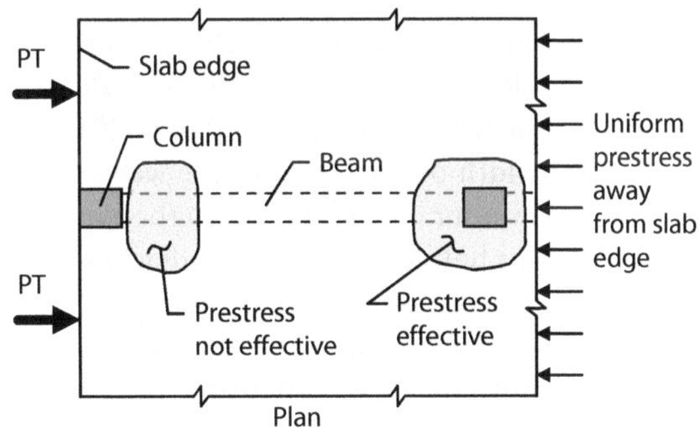

FIGURE 6.55 Effect of prestress on beam in unbonded, post-tensioned floor construction. (*After Moehle et al., 2008.*)

is resisted by both the web and the flanges. As with flanged beams, the flange in a wall acts both in tension and compression.

Figure 6.56 illustrates behavior of a T-wall tested in a laboratory under constant axial compression and lateral displacement reversals parallel to the stem of the T (Figure 6.56*a* and *b*). Flange strains are nearly uniform when the flange acts as a compression flange (Figure 6.56*c*). Shear lag is more apparent when acting as a tension flange (Figure 6.56*d*), but the entire flange width yields in tension before the end of the test.

Action of the flange as a tension or compression element can strongly affect the location of the neutral axis. When the flange of a T-wall acts in compression, the flange provides a wide compression zone such that the compression zone depth may be a relatively small fraction of the entire depth (Figure 6.56*e*). For this loading direction, the stem of the T sustains relatively large tensile strain. When the loading direction is reversed such that the flange acts in tension, the tensile force of the flexural tension zone including flange vertical reinforcement must be resisted by the narrow stem of the T, requiring a deeper compression zone (Figure 6.56*f*). Consequently, for a given curvature demand, the flexural compressive strain demand is larger for the flange-in-tension loading direction. If compressive strain capacities are the same at both ends of the wall, the curvature capacity will be lower for the flange-in-tension loading direction. Alternating tension and compression on the stem may make it more susceptible to instability (Figure 6.44*b*). Figure 6.57 illustrates the differences in strength and deformation capacity for the two loading directions.

Figure 6.56*e* and *f* shows both measured and calculated strain distributions for various drift ratios of the free end of the wall. Moment–curvature relations were calculated assuming plane sections remain plane, and results were integrated up the height of the wall using procedures described in Section 6.11. Correlation between measured and calculated results is typical of results obtained by these procedures, and demonstrates that calculated results are only an approximation of actual behavior. Building performance assessments done using these procedures should take their approximate nature into consideration.

The effective flange width varies with a number of parameters. Similar to flanged beams, the effective width increases with wall length and deformation level. The width also depends on whether the wall is acting in tension or compression. Hassan and El-Tawil (2003) report that effective flange width also increases with axial load and wall length (i.e., depth of the section).

Building codes specify wall effective flange widths based on limited laboratory test data and judgment. ACI 318 (2014) and Eurocode 8 (2004) define effective flange width on each side of the web equal to one-quarter of the height from the section under consideration to the top of the wall, but not greater than the actual width or half the distance to adjacent wall webs. NSZ 3101 defines the upper limit of the overhanging flange width as half the height from the section under consideration to the top of the wall for design flexural strength calculations, and twice that width for determination of the probable moment strength.

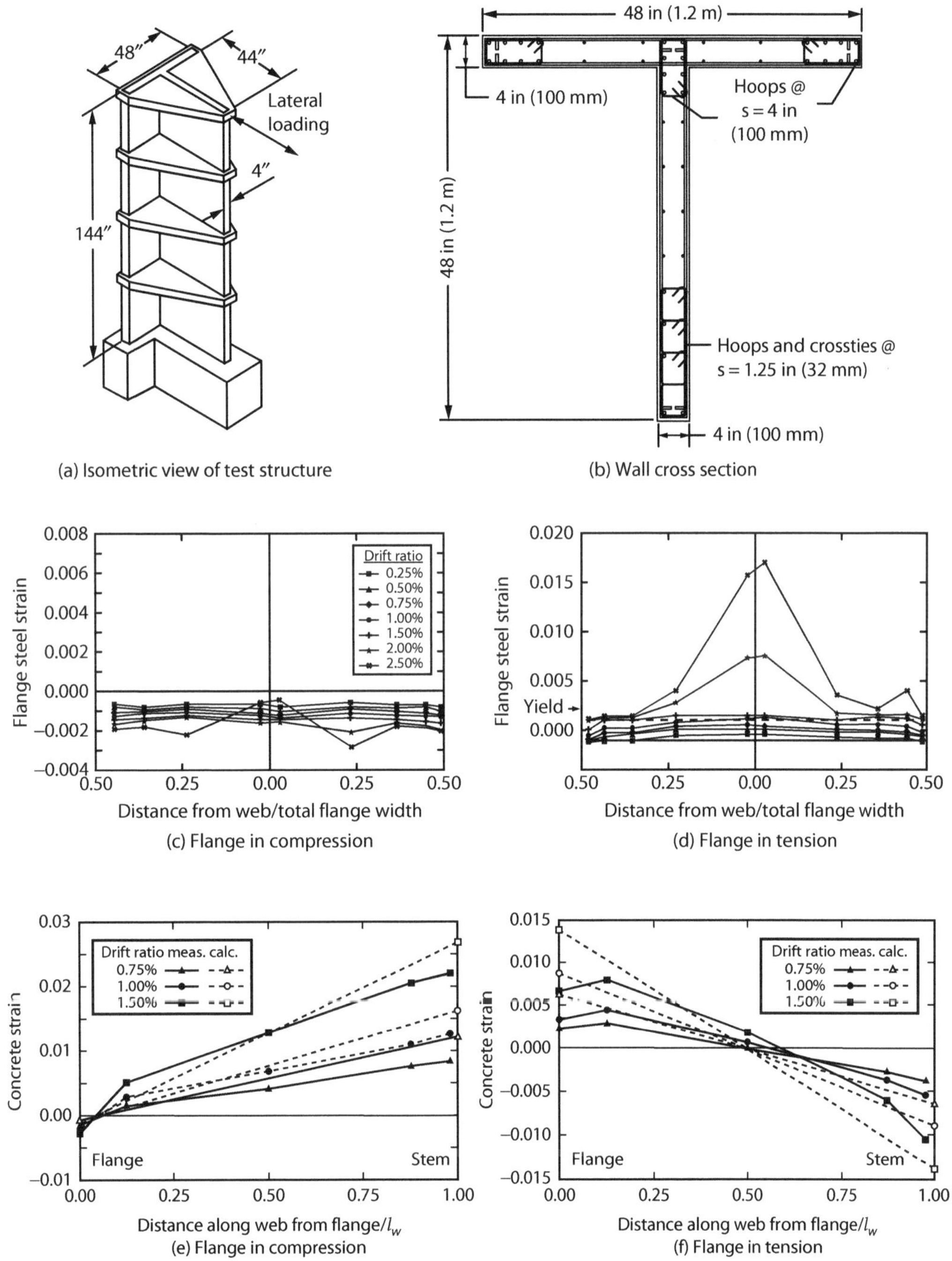

Figure 6.56 Measured behavior of T-wall. (*After Thomsen and Wallace, 1995 and 2004, used with permission from ASCE and Clarkson University.*)

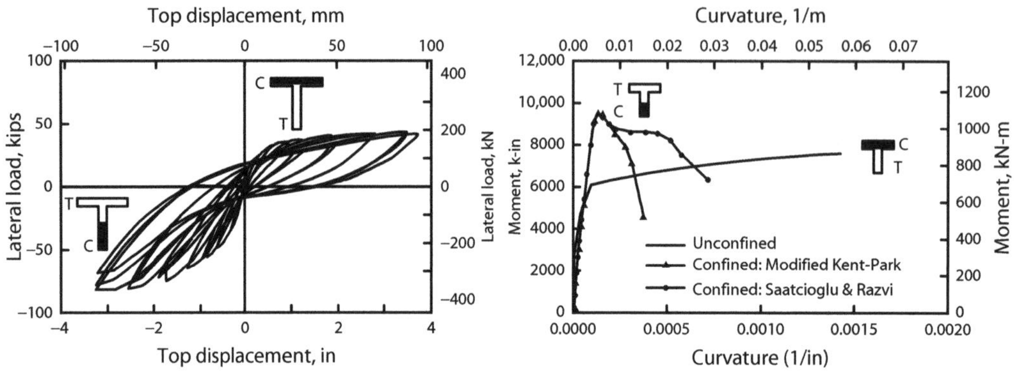

FIGURE 6.57 (*a*) Measured load–displacement relation and (*b*) calculated moment–curvature relations for a T-wall. (*After Thomsen and Wallace, 1995 and 2004.*)

6.11 Load-Deflection Calculations

The load-deflection behavior of relatively slender beams, columns, and walls can be calculated using conventional flexural theory that accounts for effects of concrete cracking and section yielding (e.g., Park and Paulay, 1975; Branson, 1977; ACI 435, 2003; Wight and MacGregor, 2012). The general procedure is to determine the curvature distribution along the member, and then integrate curvatures to determine rotations and displacements. For statically determinate members, the moment diagrams are readily determined, leading directly to the curvature distribution. For statically indeterminate structures, the internal moments depend on both the loading and the stiffness. One approach is to solve the statically indeterminate problem through a step-by-step analysis in which linear response is assumed in any given load step, with stiffness updates from one step to another. The reader is referred to texts on nonlinear structural analysis for further study. For discussion of long-term deflections, see ACI 435 (2003), ACI 318, and Branson (1977).

Flexural curvature may not adequately define the flexibility of a beam-column. More generally, it is necessary to account for shear deformations and for the effects of reinforcement slip from adjacent anchorage zones. For inelastic problems, interactions between flexural and shear deformations may need to be considered. The following sections introduce these effects along with analytical methods for estimating them, first considering linear response and then considering inelastic response.

6.11.1 Linear Response

Linear response refers to response before yielding of either the tension longitudinal reinforcement or the flexural compression zone. For response in the linear range, three components of flexibility should be considered (Figure 6.58). The flexural component refers to the displacements due to flexural curvatures. The shear component refers to conventional shear distortion. The slip component refers to rigid-body displacement that results from reinforcement slip from adjacent

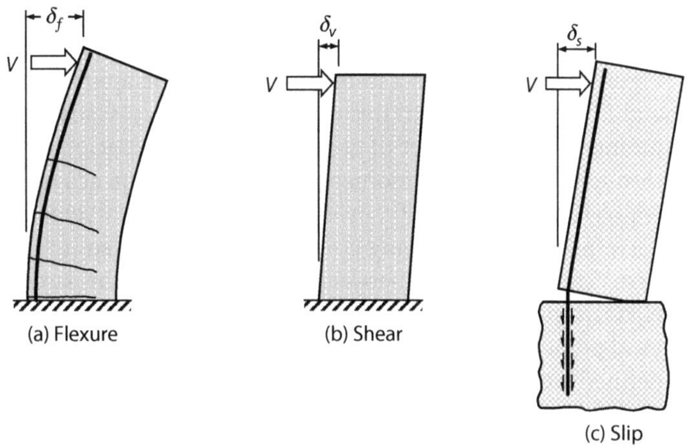

FIGURE 6.58 Components of flexibility in beam-columns.

anchorage zones. Total displacement is the sum of the three components, as expressed by Eq. (6.39).

$$\delta = \delta_f + \delta_v + \delta_s \tag{6.39}$$

The sections Flexural Deformation, Shear Deformation, and Slip Deformation consider these components separately.

Flexural Deformation

The first step in determining the flexural deformations is to define the moment–curvature relation. Although it is possible to use the relation exactly as calculated, it is usually preferred to idealize the relation. Figure 6.59*a* illustrates a trilinear approximation with break points corresponding to onset of cracking and onset of flexural yielding. The linear branch between cracking and yielding is stiffer than the calculated moment–curvature relation, which is appropriate as an approximation to the effects of tension-stiffening (see Sections 5.6 and 6.5.1). The

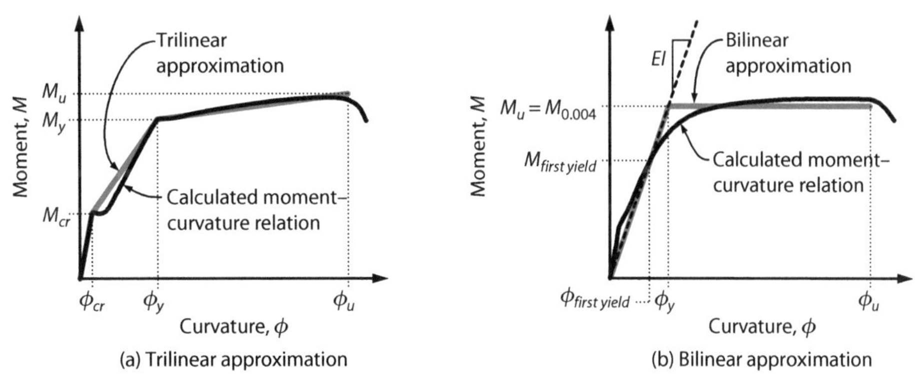

FIGURE 6.59 Idealized moment–curvature relations.

trilinear relation is useful where stiffness near the cracking load is an important consideration. In some applications, a bilinear approximation may be suitable. Figure 6.59*b* illustrates a bilinear, elasto-plastic approximation based on Elwood and Eberhard (2009). In this example, the effective strength is represented by the moment $M_{0.004}$ at onset of concrete spalling. A secant is drawn from the origin through the calculated moment–curvature relation at a moment equal to $M_{first\ yield}$, which is the moment at first yield of the longitudinal reinforcement or the moment corresponding to reaching concrete compressive strain 0.002, whichever occurs first. The intersection between the secant and the strength $M_{0.004}$ defines the effective yield curvature ϕ_y. This general approach to approximating the effective yield point is useful where the moment–curvature relation does not have a sharply defined yielding point. For sections that are strongly strain-hardening, however, definitions of the moment capacity other than $M_{0.004}$ may be more appropriate.

Given the moment–curvature relation, the flexural deformations of a beam-column can be calculated from first principles. For example, consider the reinforced concrete cantilever column of Figure 6.60 subjected to lateral force V at the free end, and developing moment M_y at the fixed end. Using the trilinear moment–curvature relation of Figure 6.59*a*, the corresponding curvature distribution is as shown in Figure 6.60*c*. Using the principle of virtual work, the deflection at the free end is obtained as the integral of the product of the virtual moments m_x due to a unit load at the free end and the curvatures due to the real loading, that is

$$\delta_f = \int_0^l m_x \phi_x dx \tag{6.40}$$

Where the curvature diagram is multi-linear, as in Figure 6.60*c*, Eq. (6.40) can be solved by integrating separately along each of the linear segments, and then summing the results.

If the cantilever column in Figure 6.60 is idealized with a linear effective stiffness, as in Figure 6.59*b*, Eq. (6.40) results in

$$\delta_{f,y} = \frac{\phi_y l^2}{3} \tag{6.41}$$

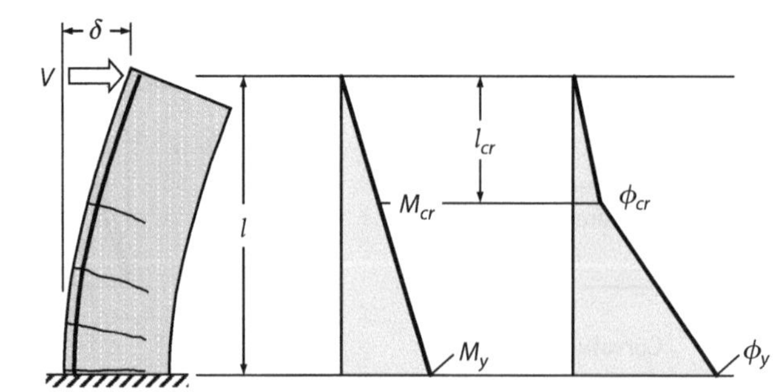

FIGURE 6.60 Flexural deflection of a cantilever column.

Shear Deformation

Shear deformations can be estimated by idealizing the beam-column as a homogeneous, isotropic material with constant shear modulus. For the cantilever shown in Figure 6.60, tip displacement due to shear can be calculated as

$$\delta_v = \frac{Vl}{A_v G_{eff}} \tag{6.42}$$

in which A_v = the effective shear area of the column cross section (5/6 A_g for rectangular sections and $0.85A_g$ for circular sections) and G_{eff} = effective shear modulus. Before shear cracking, a good approximation is $G_{eff} = 0.4E_c$. Effective shear modulus should decrease with increasing degree of concrete cracking. In this text, we adopt the recommendation of Elwood and Eberhard (2009) and use $G_{eff} = 0.2E_c$ for typical beams and columns. Shear stiffness is discussed in greater detail in Chapter 7.

Slip Deformation

For beams, columns, or walls having longitudinal reinforcement anchored in adjacent components, additional flexibility may be introduced due to rigid-body rotations that result from slip of the tensile longitudinal reinforcement from the anchorage (Figure 6.61). The reinforcement enters the anchorage in tension and transfers force through bond stress to the surrounding concrete. Consequently, the tension force gradually decreases until distance l_a into the anchorage, at which point the tension force reaches zero. The bar can be considered fixed at this point. Bar elongation from this point to the face of the anchorage can be calculated as the integral of the strain along the anchorage length. Assuming the concrete to be rigid, the reinforcement must slip from the surrounding concrete by an amount s_a equal to the total bar elongation. This slip results in a rigid-body rotation θ of the

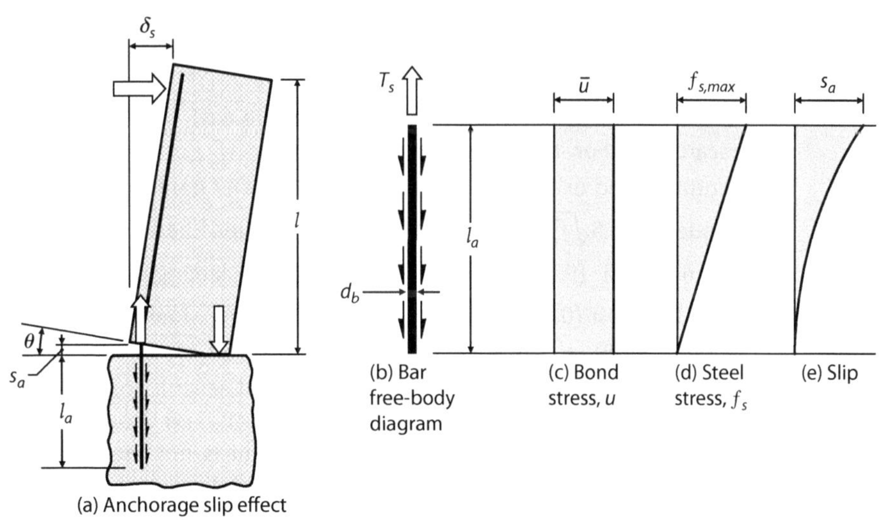

Figure 6.61 Bond-slip model for straight bar anchorage.

member that produces additional displacement; in the case of a cantilever the additional tip displacement is $\delta_s = \theta l$.

If we assume the bond stress is constant along the anchorage length at some average uniform bond stress $\bar{u}$, equilibrium of the anchored bar in Figure 6.61*b* requires

$$T_s = \pi \frac{d_b^2}{4} f_{s,max} = \pi d_b l_a \bar{u} \tag{6.43}$$

Solving for anchorage length l_a results in

$$l_a = \frac{f_{s,max} d_b}{4\bar{u}} \tag{6.44}$$

Slip at the face of the anchorage s_a is obtained by integrating bar strains along length l_a as

$$s_a = \int_0^{l_a} \varepsilon_s dx \tag{6.45}$$

For the case of linear bar response (i.e., for $f_{s,max} \le f_y$)

$$s_a = \frac{f_{s,max}}{2E_s} l_a = \frac{f_{s,max}^2 d_b}{8E_s \bar{u}} \tag{6.46}$$

The rotation θ is commonly assumed to occur about the neutral axis. Using c for compression zone depth, the rotation is

$$\theta = \frac{s_a}{d-c} = \frac{f_{s,max}^2 d_b}{8E_s \bar{u}} \frac{1}{d-c} \tag{6.47}$$

For the cantilever shown in Figure 6.61, the tip displacement due to anchorage slip is

$$\delta_s = \theta l \tag{6.48}$$

The uniform bond stress model of Figure 6.61 has been studied by several researchers. For the case of linear response ($f_{s,max} \le f_y$), Sozen et al. (1992) recommended using $\bar{u} = 6\sqrt{f_c'}$ psi ($0.5\sqrt{f_c'}$ MPa); Otani and Sozen (1972) recommended $\bar{u} = 6.5\sqrt{f_c'}$ psi ($0.54\sqrt{f_c'}$ MPa); Elwood and Eberhard (2009) recommended $\bar{u} = 9.6\sqrt{f_c'}$ psi ($0.80\sqrt{f_c'}$ MPa); Alsiwat and Saatcioglu (1992) recommended $\bar{u} = 10.5\sqrt{f_c'}$ psi ($0.87\sqrt{f_c'}$ MPa); and Lehman and Moehle (2000) and Sezen and Setzler (2008) recommended $\bar{u} = 12\sqrt{f_c'}$ psi ($1.0\sqrt{f_c'}$ MPa). The latter studies considered full-scale columns under simulated seismic loadings. Here we adopt the recommendation of $\bar{u} = 12\sqrt{f_c'}$ psi ($1.0\sqrt{f_c'}$ MPa) for the case of linear response.

The slip contribution to total displacement is not inconsequential in many practical cases. The ratio of slip to flexural yield displacements for a cantilever in which longitudinal reinforcement yields is obtained as the ratio of Eq. (6.48) to Eq. (6.41), resulting in

$$\frac{\delta_s}{\delta_y} = \frac{3 f_y d_b}{8 \bar{u} l} \tag{6.49}$$

Note that the relative contribution of slip is larger for larger bar diameter d_b and shorter member length l. For the case of f_y = 60,000 psi (414 MPa), No. 11 (36) longitudinal bar, $\bar{u} = 12\sqrt{f_c'}$ psi $(1.0\sqrt{f_c'}$ MPa), f_c' = 6000 psi (41 MPa), and l = length from column anchorage to inflection point for column bent in double curvature = 60 in (1.52 m), Eq. (6.49) produces $\delta_s/\delta_y = 0.56$.

As axial force increases, inelastic response of the flexural compression zone may limit the tensile stresses developed on the tension side of the member. For example, above the balanced point the tension reinforcement is not expected to yield at failure of an unconfined section. Thus, the contribution of slip to total displacement decreases as axial load increases (Elwood and Eberhard, 2009).

Total Deformation

Total displacement is the sum of the flexural, shear, and slip components [Eq. (6.39)]. Elwood and Eberhard (2009) compared measured and calculated stiffness at yield for 221 columns with rectangular cross sections with rectilinear transverse reinforcement and 108 columns with circular, octagonal, or square cross sections with spiral transverse reinforcement. The ratio of measured to calculated stiffness had mean and coefficient of variation equal to 0.97 and 0.27 for the rectangular reinforced columns and 1.04 and 0.21 for the spiral reinforced columns. Elwood and Eberhard (2009) also developed the following simplified expression for effective stiffness:

$$\frac{EI_e}{EI_g} = \frac{0.45 + 2.5P/A_g f_c'}{1 + 110\left(\dfrac{d_b}{h}\right)\left(\dfrac{h}{a}\right)} \tag{6.50}$$

in which h = diameter of circular column or overall thickness for rectangular column and a = shear span (distance from maximum moment to inflection point). It was suggested that d_b/h can be approximated as 1/18 for building columns and 1/25 for bridge columns. With these approximations, the ratio of measured to calculated effective stiffness had mean and coefficient of variation equal to 0.95 and 0.26 for the rectangular reinforced columns and 1.04 and 0.22 for the spiral reinforced columns.

Figure 6.62 compares results for a subset of the aforementioned data. For low axial loads, effective stiffness is overestimated if only flexural effects are considered. Results are significantly improved if effects of slip and (to a lesser degree) shear flexibilities are included. Elwood and Eberhard (2006) suggested the trilinear approximation shown in the figure. These results served as a basis for the effective stiffness values adopted by ASCE 41-13. ASCE 41 recommends using the effective stiffnesses listed in Table 6.1.

6.11.2 Nonlinear Inelastic Range

This section addresses load-deflection response beyond the yield point. As was done for linear response, the effects of flexure, shear, and bar slip should be considered. An additional complication arises in the interaction between flexure

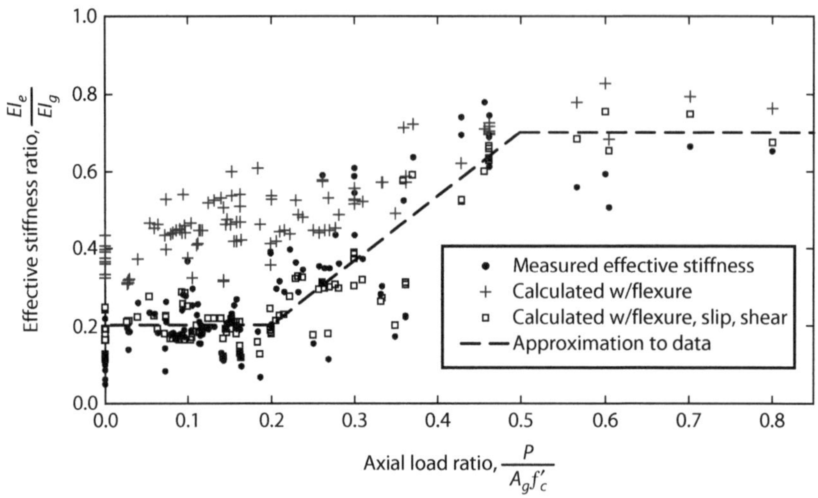

FIGURE 6.62 Comparison of effective stiffness ratio and axial load ratio. (*After Elwood and Eberhard, 2006, courtesy of American Concrete Institute.*)

$\frac{P}{A_g f'_c}$	$\frac{EI_e}{EI_g}$
≥0.5	0.7
Intermediate values	Interpolate linearly
≤0.2	0.3

TABLE 6.1 Effective Stiffness in Accordance with ASCE 41

and shear. The plastic hinge model will be introduced as a modeling tool to estimate flexural response beyond yielding.

Flexural Deformation

The procedures presented for the linear range of response apply similarly to the nonlinear inelastic range of response. Figure 6.63 depicts a cantilever column with

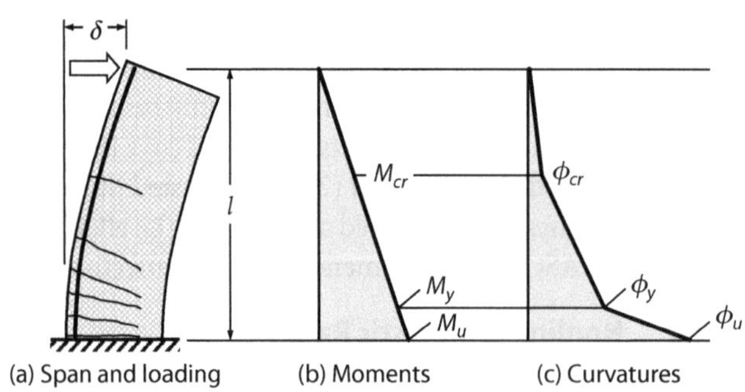

FIGURE 6.63 Deflection of a cantilever column in the inelastic range.

loading, moment diagram, and curvature diagram. Equation (6.40) can be used directly to determine the tip deflection, by integrating along each of the linear segments of the curvature diagram and summing the results.

Shear Deformation

Shear deformations can be estimated using the procedures for linear response described in Section 6.11.1.

Slip Deformation

If the longitudinal reinforcement yields in tension at the face of the anchorage, then yielding will penetrate some distance into the anchorage. Along the yielding length, bond stress will be reduced, mainly because of the increased local slip near the loaded end. Lehman and Moehle (2000) and Sezen and Setzler (2008) showed good correlation with test results using uniform bond stress of $\bar{u} = 6\sqrt{f'_c}$ psi $\left(0.5\sqrt{f'_c}\text{ MPa}\right)$ along the yielded embedment length and $\bar{u} = 12\sqrt{f'_c}$ psi $\left(1.0\sqrt{f'_c}\text{ MPa}\right)$ deeper within the anchorage along the elastic length. Alsiwat and Saatcioglu (1992) present an alternative approach for nonlinear response for straight and hooked bar anchorages, including the condition where the provided length is insufficient to fully develop bar stress.

Tension Shift Effect

For members subjected to inelastic deformations well beyond yield, and especially for members with relatively large shear force, the flexural curvatures shown in Figure 6.63*c* underestimate the spread of inelastic action (Park and Paulay, 1975). To understand the underlying cause, consider the member shown in Figure 6.64*a*. Figure 6.64*b* shows a free-body diagram bounded by the fixed end and by a diagonal crack. Shown on the right side of the free-body diagram are T_x, the flexural tension force in the longitudinal reinforcement at distance x from the fixed end; C, the flexural compression force; V_d, a dowel force acting between the longitudinal reinforcement and surrounding concrete, arising from vertical movement associated with rotation about the root of the shear crack; V_a, an inclined shear force associated with aggregate interlock, arising from the same vertical movement that engages V_d, assumed parallel to the inclined face; and V'_s, the force carried by stirrups crossing the inclined section. Note that V'_s is not necessarily equal to the full shear force assigned to the stirrups, because the angle swept out by the shear crack is not necessarily equal to the angle assumed for design of the stirrup reinforcement. Here we will assume $V'_s = V_s\, x/d$, where V_s is the total force resisted by the stirrups within a 45° diagonal (section D-D). Furthermore, we will assume that V_s is not the total shear force on the beam but some fraction η of that force, that is, $V_s = \eta V_u$. Summing moments about the root of the shear crack where the flexural compression force C is shown results in

$$-M_1 + T_x jd + V_d x + V'_s \frac{x}{2} = 0 \tag{6.51}$$

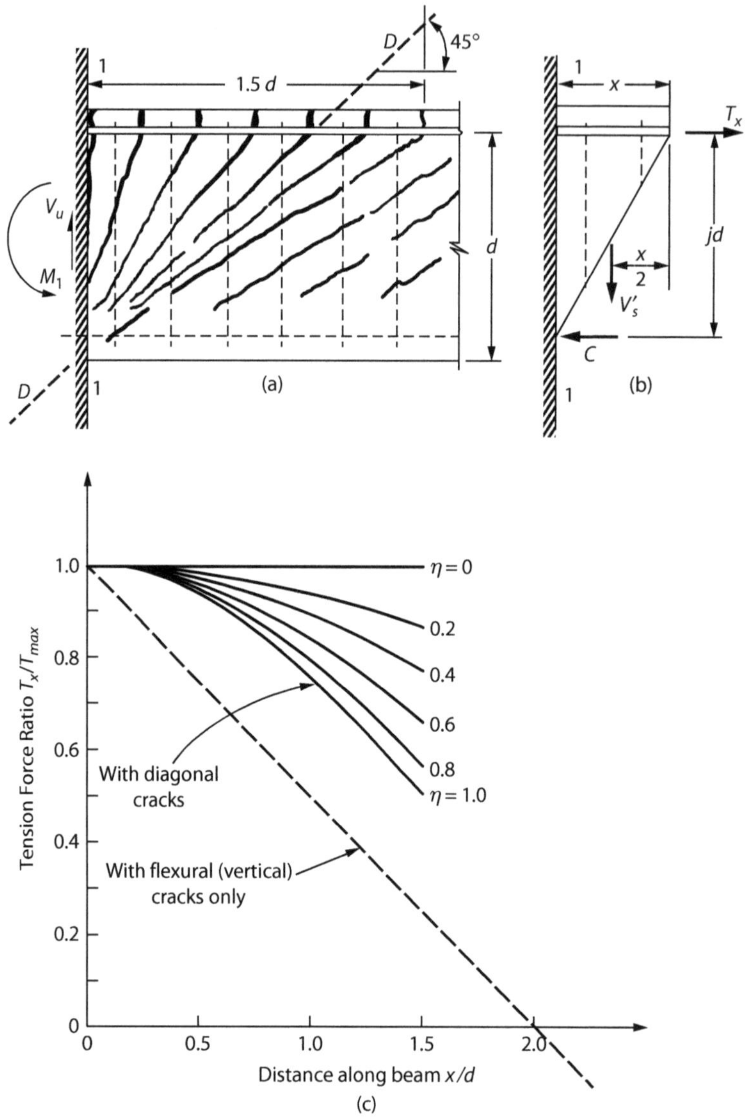

FIGURE 6.64 Distribution of longitudinal reinforcement forces at plastic hinge as affected by shear. (*After Park and Paulay, 1975, used with permission from Wiley Interscience.*)

At large deformations, it is reasonable to assume $V_d = 0$ because dowel action would produce horizontal cracks along the longitudinal reinforcement, thereby relieving the dowel force. Setting $V_d = 0$ and $V_s = \eta V_u$ results in

$$T_x = \frac{1}{jd}\left(M_1 - \frac{\eta x^2}{2d} V_u \right) \tag{6.52}$$

To illustrate the result, suppose $M_u/V_u d = 2$. Then

$$T_x = \frac{M_1}{jd}\left(1 - \frac{\eta x^2}{4d^2}\right) \tag{6.53}$$

For a beam without stirrups, Eq. (6.53) simplifies to

$$T_x = \frac{M_1}{jd} \tag{6.54}$$

Note that Eq. (6.54) states that the steel tension force at x is equal to the moment at section 1-1 divided by jd, rather than being the moment at section x divided by jd. Thus, the maximum tension force has shifted from the section of maximum moment. Figure 6.64*c* plots the tension shift for different values of η. Park and Paulay (1975) referred to this as the *tension shift* effect.

Tension shift has important implications. First, it results in steel stresses away from the critical section for flexure being higher than would be predicted by flexural theory; this effect needs to be considered when determining reinforcement development lengths. Second, by increasing the stress in the flexural tension reinforcement, additional flexural deformation occurs beyond that predicted by flexural theory alone. Third, as longitudinal reinforcement yields and the tension chord elongates, each of the inclined concrete segments rotates about the compression chord, leading to additional shearing deformation. This latter phenomenon will be discussed further in Chapter 7.

Plastic Hinge Model

Previous discussion identified components of nonlinear deformations, including curvature, shear deformation, slip from anchorages, and the tension shift effect including flexural and shearing components. Explicit consideration of each of these may be too complicated for routine application. In this section we introduce the plastic-hinge model as an approximate assessment and design tool that incorporates all of these effects (Figure 6.65).

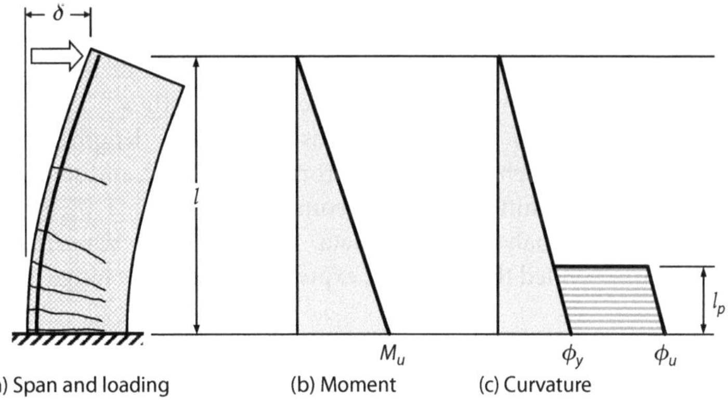

Figure 6.65 Plastic hinge model.

The plastic-hinge model uses an idealized elasto-plastic moment–curvature relation (Figure 6.59*b*). The linear-elastic segment approximates behavior before yield. Normally it is acceptable to use a secant passing through the calculated moment–curvature relation at a moment corresponding to either the onset of tension yielding or the onset of concrete compressive strain reaching 0.002, whichever occurs first. If the section is not strongly strain-hardening, the plastic portion can be defined by the moment strength corresponding to concrete compressive strain 0.004 (the nominal onset of spalling). For sections with greater strain-hardening, alternative definitions of the plastic strength may be more appropriate. With this approach, the intersection of the secant and the moment strength defines an effective yield curvature ϕ_y exceeding the curvature at first yield of the section.

The flexural member is assumed to develop elastic curvatures over height according to the elastic stiffness and applied moments. Inelastic curvature of magnitude $\phi_u - \phi'_y$ is assumed to extend along *plastic hinge length* l_p (Figure 6.65*c*). As originally conceived and as generally applied today, additional deformations associated with slip of reinforcement from anchorages and tension shift are accounted for indirectly in the plastic hinge length.

In the plastic-hinge model, plastic rotation is

$$\theta_p = (\phi_u - \phi_y) l_p \tag{6.55}$$

For the cantilever shown in Figure 6.65, displacement at the free end is

$$\delta_u = \delta_y + \theta_p \left(l - \frac{l_p}{2} \right) = \frac{\phi_y l^2}{3} + \left(\phi_u - \phi_y \right) l_p \left(l - \frac{l_p}{2} \right) \tag{6.56}$$

The plastic hinge length has been evaluated empirically by several researchers. The displacement δ_u corresponding to some performance point is measured in a laboratory experiment, curvatures ϕ_u and ϕ_y are either measured or calculated, and the plastic hinge length l_p required to satisfy Eq. (6.56) is determined. Based on various studies, the following expressions for plastic hinge length are of interest.

Priestley and Park (1987) studied a series of bridge columns subjected to simulated seismic loading. Plastic hinge length was assumed to be a function of column length (yielding extends over longer length for longer columns), bar diameter (which relates to slip from anchorages), and section dimension h (related to tension shift and local geometric constraints on spread of plasticity). In statistical analyses of the data, they found l_p was not sensitive to h. They recommended the following expression for plastic hinge length:

$$\begin{aligned} l_p &= 0.08l + 0.00015 d_b f_y, \text{ psi} \\ l_p &= 0.08l + 0.022 d_b f_y, \text{ MPa} \end{aligned} \tag{6.57}$$

Berry et al. (2008) considered a larger database including a wider range of pertinent variables. They recommended

$$l_p = 0.05l + 0.008 d_b f_y / \sqrt{f_c'}, \text{ psi}$$
$$l_p = 0.05l + 0.1 d_b f_y / \sqrt{f_c'}, \text{ MPa} \tag{6.58}$$

Berry et al. (2008) observed that accuracy of Eq. (6.58) was relatively insensitive to changes in the coefficients, suggesting that alternative forms of the equation could be similarly acceptable. For building columns having proportions typical of construction practice in regions of high seismicity in the United States, the following simple expression was reasonably accurate:

$$l_p = 0.5h \tag{6.59}$$

This expression is convenient for understanding design requirements, and will be used extensively in subsequent sections of this book.

Note that the length l_p is used solely for calculation of inelastic deflections. It does not reflect the length over which spalling will occur or over which confinement should be provided. This subject is discussed further by Watson et al. (1994) and Lehman et al. (2004).

Example 6.7. For the beam shown in Figure 6.66, calculate the displacement at the load point corresponding to onset of (a) cracking, (b) yielding, and (c) spalling. Plot the load–displacement relation. $f_c' = 4000$ psi (28 MPa), Grade 60 (420) reinforcement. Note that the beam cross section is identical to the cross section of Example 6.2. The moment–curvature relation from that example is applied here.

Solution

Cracking

The flexural contribution is

$$\delta_{f,cr} = \frac{\phi_{cr} l^2}{3} = \frac{(1.1 \cdot 10^{-5}\ \text{in}^{-1})(100'')^2}{3} = 0.037 \text{ in } (0.94 \text{ mm})$$

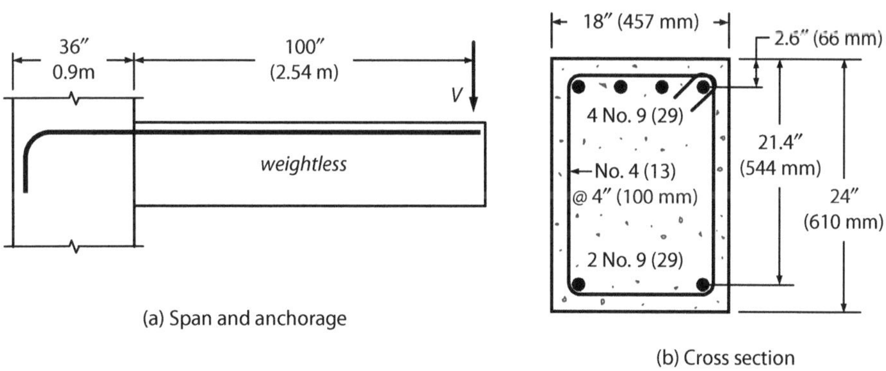

Figure 6.66 Beam span, anchorage, and cross section for Example 6.7.

For the shear contribution, use $G_{eff} = 0.4E_c = 0.4 \times 3600$ ksi $= 1440$ ksi. Thus, the shear contribution from Eq. (6.42) is

$$\delta_{v,cr} = \frac{Vl}{A_v G_{eff}} = \frac{M_{cr}}{A_v G_{eff}} = \frac{819 \text{ k-in}}{\frac{5}{6}(18'')(24'')(1440 \text{ ksi})} = 0.0016 \text{ in } (0.041 \text{ mm})$$

Prior to cracking, there is zero slip contribution. (The beam is likely to be cracked at the interface between the beam and its support due to differential volume change, but that effect is ignored here.)

Thus, the total deflection is $\delta_{cr} = 0.037 \text{ in} + 0.0016 \text{ in} = 0.04 \text{ in}$ (1.0 mm). The corresponding load is $V = M_{cr}/l = 8.19$ kips (36 kN).

Yielding

The flexural contribution is obtained by integrating deflection contributions from curvature along the length. For this purpose, the principle of virtual forces is employed. The real moment and curvature diagrams are presented in Figure 6.67*b* and *c*. A unit virtual force is applied at the top of the beam in the downward direction, resulting in the virtual moment diagram shown in Figure 6.67*d*. The integral of Eq. (6.40) is evaluated in parts, first from $x = 0''$ to 15.6″, and then from 15.6″ to 100″, using the expressions for m_x and ϕ_x given in Figure 6.67. The result is

$$\delta_{f,y} = \int_0^{100} m_x \phi_x dx = 0.510 \text{ in } (13.0 \text{ mm})$$

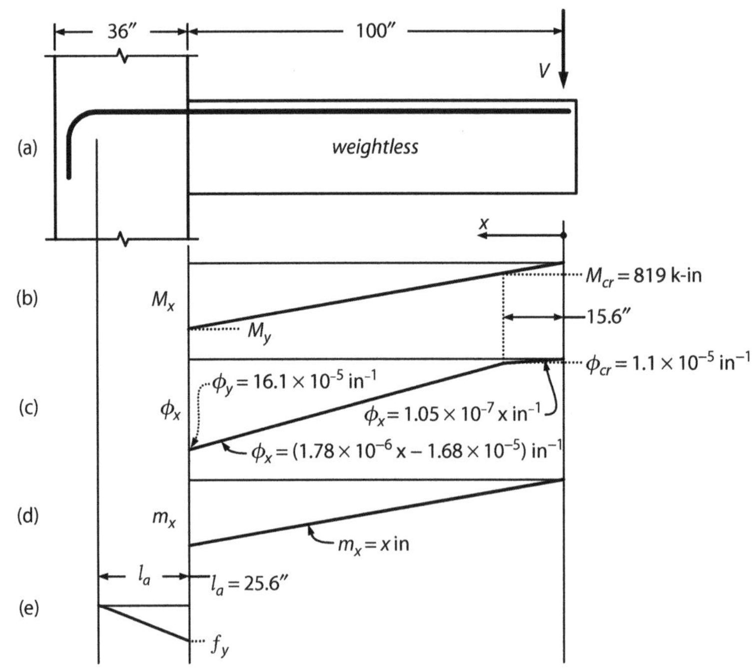

FIGURE 6.67 Parameters for calculation of yield displacement.

Note that Eq. (6.41), reproduced below, produces a close approximation with much less computational effort.

$$\delta_{f,y} = \frac{\phi_y l^2}{3} = \frac{(16.1 \cdot 10^{-5}\,\text{in}^{-1})(100'')^2}{3} = 0.537\,\text{in}\,(13.6\,\text{mm})$$

For the shear contribution, use $G_{eff} = 0.2E_c = 0.2 \times 3600$ ksi $= 720$ ksi. Thus, the shear contribution from Eq. (6.42) is

$$\delta_{v,y} = \frac{Vl}{A_v G_{eff}} = \frac{M_{cr}}{A_v G_{eff}} = \frac{5260\,\text{k-in}}{\frac{5}{6}(18'')(24'')(720\,\text{ksi})} = 0.0203\,\text{in}\,(0.52\,\text{mm})$$

For the slip contribution, assume uniform bond stress of $u = 12\sqrt{f'_c}$ psi $= 12\sqrt{4000} = 759$ psi. The length over which the bar is stressed is first calculated to determine whether sufficient length exists for uniform bond transfer, as

$$l_a = \frac{f_{s,max} d_b}{4u} = \frac{(69{,}000\,\text{psi})(1.128'')}{4 \times 759\,\text{psi}} = 25.6\,\text{in}\,(650\,\text{mm})$$

From Figure 6.67*a*, it appears that adequate straight development is available. The slip from the anchorage is

$$s_a = \frac{f_{s,max}^2 d_b}{8E_s u} = \frac{(69{,}000\,\text{psi})^2(1.128'')}{8(29{,}000{,}000\,\text{psi})(759\,\text{psi})} = 0.0305\,\text{in}\,(0.77\,\text{mm})$$

The fixed end rotation is assumed to occur about the cracked section neutral axis. Hence

$$\theta = \frac{s_a}{d-c} = \frac{0.0305''}{21.4'' - 6.74''} = 0.00208\,\text{rad}$$

Thus, the displacement at the fixed end is $\theta l = 0.00208 \times 100'' = 0.208$ in (5.3 mm).

Thus, the total deflection is $\delta_y = 0.510$ in $+ 0.0203$ in $+ 0.208$ in $= 0.74$ in (19 mm). The corresponding load is $V = M_y/l = 52.6$ kips (234 kN).

Spalling

The plastic-hinge model is used to estimate the displacement at spalling. This model includes effects of flexure, slip, and some effects of flexure–shear interaction in the plastic hinge zone. Elastic shear deformations are a relatively small part of the total for this beam, so shear deformations will otherwise be ignored.

The moment–curvature relation is idealized as an elasto-plastic relation (Figure 6.59*b*), with initial linear branch passing through the yield point of the calculated moment–curvature relation, and plastic strength taken equal to the strength at spalling. Thus, the effective yield curvature is

$$\phi_y = \frac{M_{spalling}}{M_{first\ yield}} \phi_{first\ yield} = \frac{5370\,\text{k-in}}{5260\,\text{k-in}}(16.1 \cdot 10^{-5}\,\text{in}^{-1}) = 16.4 \cdot 10^{-5}\,\text{in}^{-1}$$

The plastic-hinge length can be estimated from Eq. (6.58) or (6.59) as

$$l_p = 0.05l + 0.008 d_b \frac{f_y}{\sqrt{f'_c}}\,\text{psi} = (0.05)(100'') + (0.008)(1.128'')\frac{69{,}000\,\text{psi}}{\sqrt{4000\,\text{psi}}} = 14.8\,\text{in} \quad (6.58)$$

$$\text{or } l_p = h/2 = 24''/2 = 12\,\text{in} \quad (6.59)$$

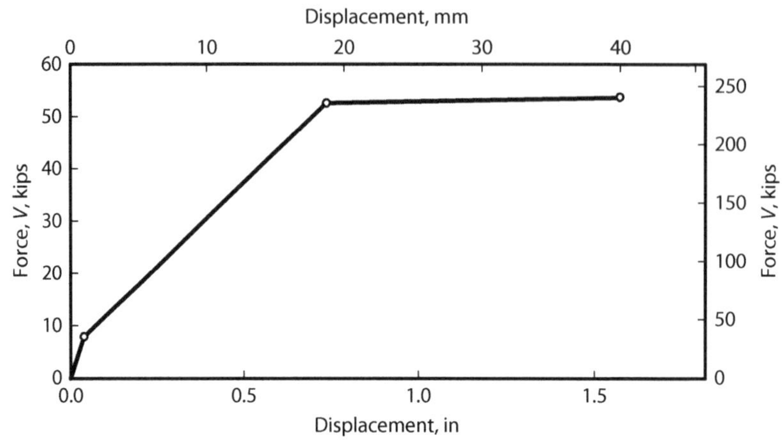

Figure 6.68 Calculated load–displacement relation.

Use $l_p = 12$ in. Thus, from Eq. (6.56), the spalling displacement is

$$\delta_u = \frac{\phi_y l^2}{3} + (\phi_u - \phi_y)(l_p)\left(l - \frac{l_p}{2}\right)$$

$$= \frac{(16.4 \cdot 10^{-5}\,\text{in}^{-1})(100'')^2}{3} + (108 - 16.4)(10^{-5}\,\text{in}^{-1})(12'')(100'' - 12''/2) = 1.6\ \text{in (40 mm)}$$

The corresponding load is $V = M/l = 53.7$ kips (239 kN).
The calculated load–displacement relation is plotted in Figure 6.68.

6.12 Reversed Cyclic Loading

Until now, the primary emphasis of this chapter has been flexural behavior under monotonic loading. Results for monotonic loading provide good insight into many aspects of behavior under reversed cyclic loading, but there are some distinct differences for reversed cyclic loading that merit discussion.

6.12.1 General Aspects of Response to Reversed Cyclic Loading

Consider the beam illustrated in Figure 6.69. Area of top longitudinal reinforcement exceeds area of bottom longitudinal reinforcement. The beam resists gravity loads at point **a**, and then undergoes deformation history **a-g**. The quarter cycle **ab** causes negative curvature ϕ (Figure 6.69*b* and *c*). Top reinforcement sustains tensile strain well beyond the yield strain, while bottom reinforcement sustains moderate compressive strain (Figure 6.69*d*).

Load reversal results in compression at the top and tension at the bottom of the beam. Flexural compression initially is resisted entirely by the top reinforcement until the previously formed cracks are closed. In the case considered here, the tensile force that can be developed in the bottom reinforcement is insufficient to yield the top reinforcement, so the cracks formed during half cycle **abc** remain

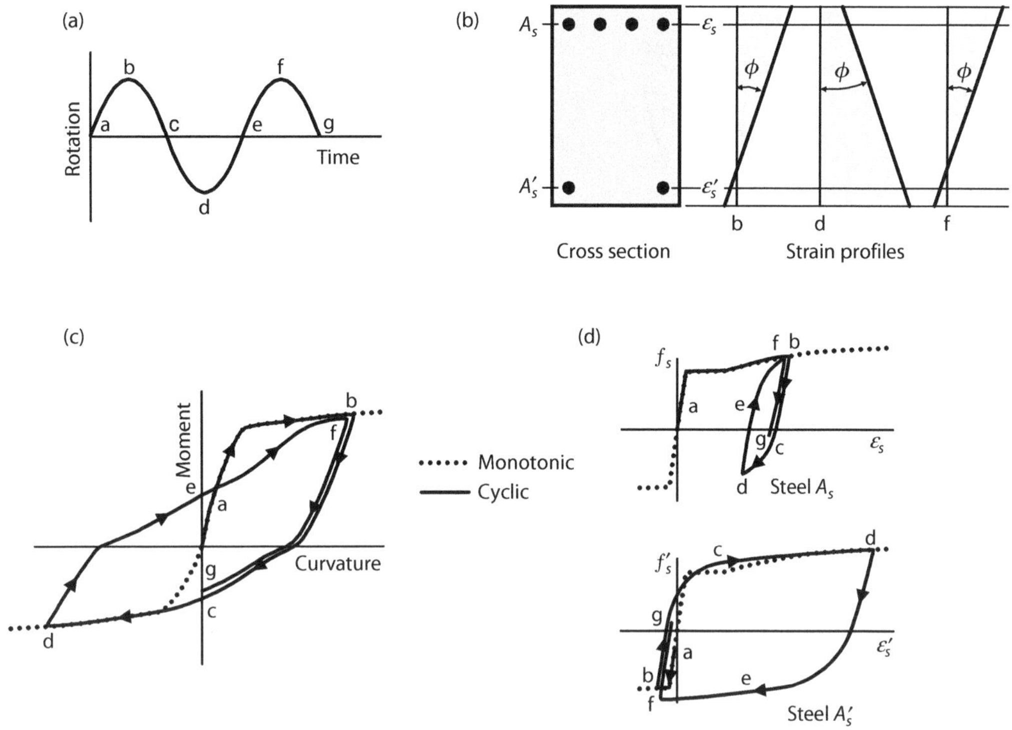

FIGURE 6.69 Idealized behavior of a flexural member under reversed cyclic loading.

open at loading point **d** (Figure 6.69*b*). The open cracks in the compression zone reduce flexural stiffness. Furthermore, because the compression zone does not fully compress, the tension side must reach higher tensile strain in order to reach the curvature ϕ. Half cycle **def** results in crack closure at the bottom of the beam because top reinforcement in tension is sufficient to yield the bottom reinforcement in compression. Stiffness and strength degradation due to cyclic loading may occur due to degradation of the constituent materials and the bond between them.

From the discussion in the preceding paragraph, several observations can be made. First, the strain conditions under reversed cyclic loading may not match those for monotonic loading. Second, stresses in materials are not uniquely related to strains. Consequently, notions such as balanced failure conditions calculated based on monotonic behavior may be inapplicable for reversed cyclic loading. Third, the moment at any particular curvature is not uniquely defined. However, it is useful to note that the monotonic moment–curvature envelope provides an approximate bound for the moment–curvature response under reversed cyclic loading.

6.12.2 Laboratory Tests

Figure 6.70 illustrates behavior of two beams with equal areas of top and bottom longitudinal reinforcement. The tests were run under displacement control, with multiple cycles at each of several progressively increasing peak displacement

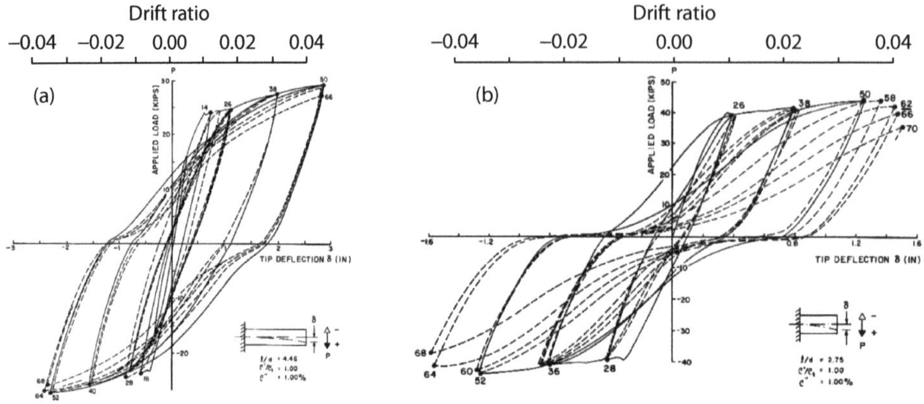

FIGURE 6.70 Measured behavior of two beams with equal areas of top and bottom longitudinal reinforcement. Beams had 9 in wide by 16 in deep (229 mm by 406 mm) rectangular cross sections, 4 No. 6 (19) longitudinal bars, restrained by No. 2 (6) four-legged closed hoops at 3½ in (89 mm) spacing. (*After Ma et al., 1976, used with permission from the University of California, Berkeley.*)

amplitudes. The load–displacement relation is approximately bounded by the load-deformation relation under monotonic loading. Yielding of longitudinal reinforcement is apparent in a sharp transition in the stiffness around drift ratio 0.01. Unloading slope is close to the secant to yield, but decreases with increasing peak displacement amplitude. The stiffness is relatively low immediately after load reversal; this occurs because flexural compression initially is resisted only by the longitudinal reinforcement spanning across previously formed concrete cracks. For relatively slender members with low shear (e.g., Figure 6.70*a*), the stiffness increases again after cracks close. For members with higher shear (e.g., Figure 6.70*b*), deformations along inclined cracks result in softer response with spindle-shaped load–displacement loops (this subject is covered in greater detail in Chapter 7).

Members with adequate, well-detailed transverse reinforcement and low to moderate shear stresses typically exhibit stable hysteresis to relatively large drift ratios, such that multiple cycles to constant drift amplitude may be possible without significant strength degradation. Strength degradation from one cycle to another typically (though not always) is an indication of impending failure. Such degradation is apparent after cycle 58 of test specimen R-5 (Figure 6.70*b*).

Beams with asymmetric longitudinal reinforcement exhibit asymmetric flexural strengths for loading in opposite directions. Flexural compression and flexural tension demands tend to be higher on the more lightly reinforced side, such that failure tends to initiate there, either as compressive failure (possibly involving longitudinal reinforcement buckling) or as tensile failure due to longitudinal reinforcement fracture. Figure 6.71 illustrates behavior from laboratory tests, in one case a rectangular cross section and in the other a flanged cross section.

Behavior of a symmetrically reinforced, confined column is illustrated in Figure 6.72. Crack closure usually is assured because of symmetric longitudinal reinforcement and axial compressive loading. Apparent strength degradation with increasing lateral displacement is partially attributable to the additional moment

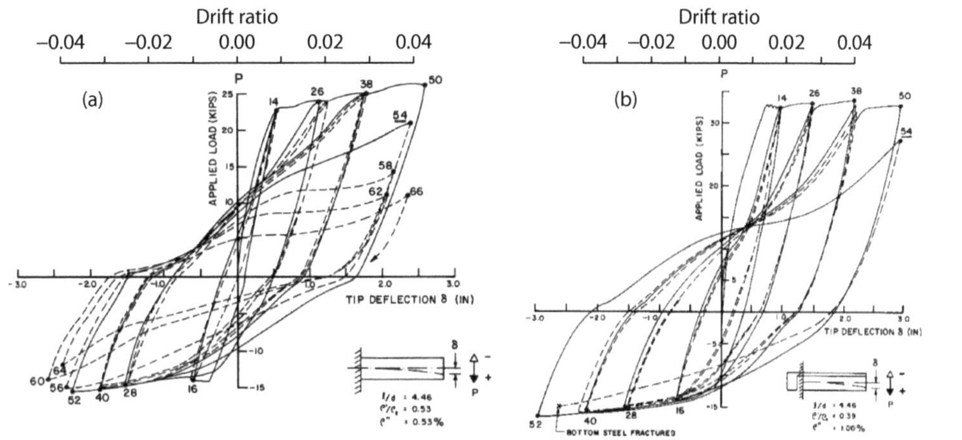

FIGURE 6.71 Measured behavior of two beams with top longitudinal steel area exceeding bottom longitudinal steel area: (*a*) rectangular cross section and (*b*) T cross section. Beams had 9 in wide by 16 in deep (229 mm by 406 mm) web cross sections, 4 No. 6 (19) top and 3 No. 5 (16) bottom longitudinal bars, restrained by No. 2 (6) four-legged closed hoops at 3½ in (89 mm) spacing. T-beam had two layers of 6 No. 2 (6) longitudinal bars in the slab. (*After Ma et al., 1976, used with permission from the University of California, Berkeley.*)

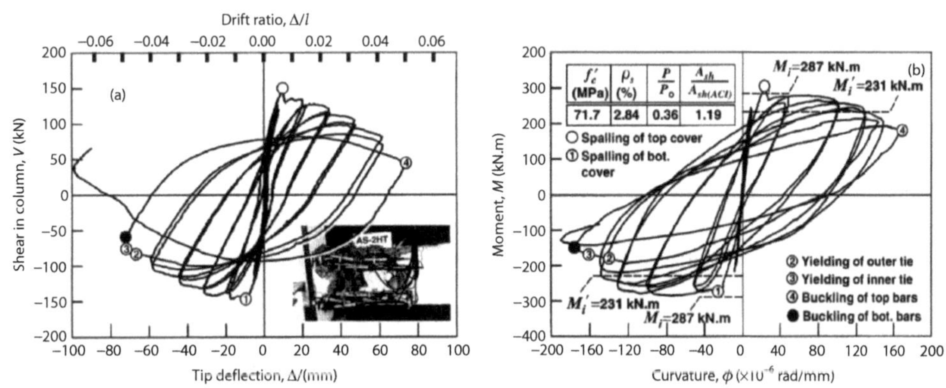

FIGURE 6.72 Measured behavior of symmetrically reinforced rectangular column supporting axial force near the balanced point: (*a*) column shear versus tip displacement; (*b*) base moment versus average curvature. (*After Bayrak and Sheikh, 1997, courtesy of American Concrete Institute.*)

caused by axial force acting through lateral displacement. Columns in frames may experience axial force variations as a result of overturning action, such that the column axial load increases for lateral load in one direction and decreases for lateral load in the opposite direction. Flexural stiffness increases with increasing axial load. Furthermore, for axial loads below the balanced point, flexural strength increases with increasing axial load. Figure 6.73 illustrates the effect of varying axial load on column behavior. The loading direction associated with increasing axial compression usually is the direction in which failure occurs.

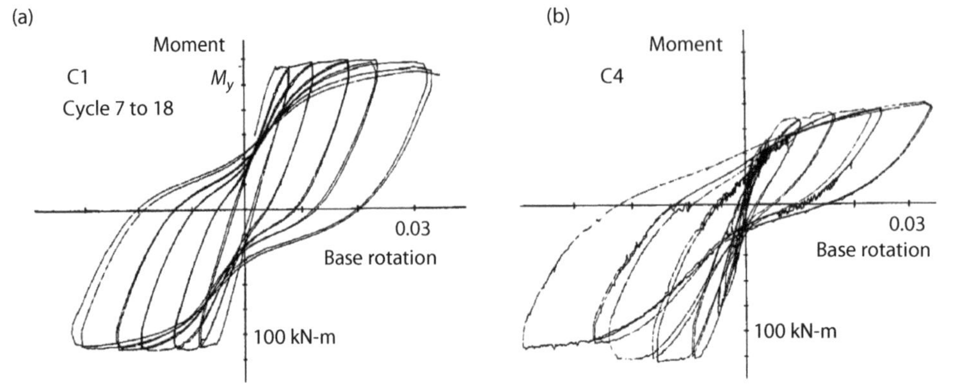

Figure 6.73 Effect of axial load on column behavior: (*a*) constant axial load; (*b*) variable axial load with axial compression increasing for negative moment and base rotation. Base rotation is measured over length *h* at base of column. (*Data after Abrams, 1987, courtesy of American Concrete Institute.*)

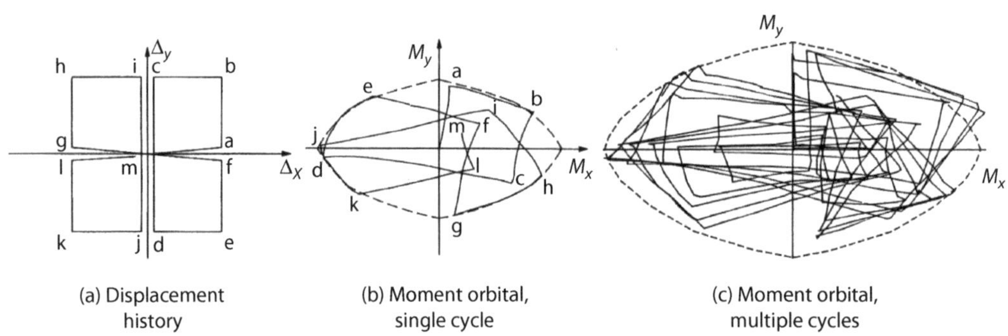

Figure 6.74 Biaxial response of a column with rectangular cross section: (*a*) cloverleaf displacement history; (*b*) example moment orbital for single cycle; (*c*) moment orbitals for multiple cycles at constant amplitude. (*Data after Low and Moehle, 1987, used with permission from the University of California, Berkeley.*)

When a column is loaded simultaneously in two horizontal directions, the lateral resistance is bounded by a biaxial yield surface (e.g., Figure 6.74*b* and *c*). In the example shown, column moment M_y increases as it is loaded along the *x* direction to point **a** (according to the sign convention, loading in the *x* direction causes moment M_y about the *y*-axis). The interaction surface requires that resistance in the *x* direction decrease as the column is subsequently loaded in the *y* direction to point **b**. Unloading in the *x* direction to point **c** not only reduces M_y but also causes relaxation of M_x. This complex interaction continues as the column traces out a complicated orbital in the M_x–M_y plane.

Numerically simulating member behavior under reversed cyclic loading generally requires the use of a computer. Cross-sectional behavior can be modeled using fiber models (Figure 6.23) in which the cyclic behavior of materials is simulated and integrated across the section. Cross-sectional behavior is then

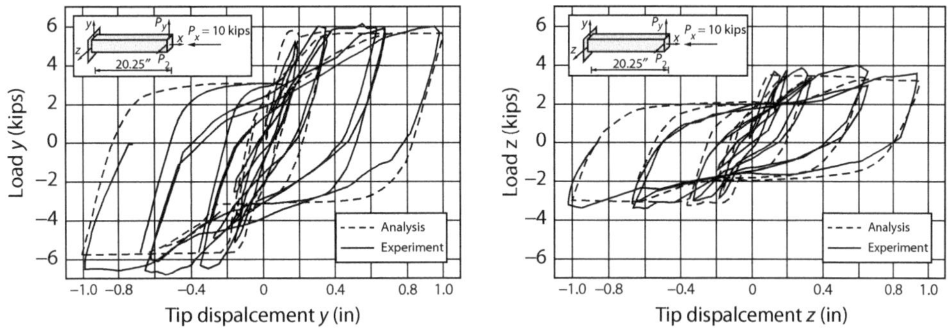

FIGURE 6.75 Measured versus calculated response. (*Test data after Low and Moehle, 1987, and analysis by Taucer et al., 1991, used with permission from the University of California, Berkeley.*)

integrated along the length using either the plastic-hinge model or fiber models at multiple locations along the member length. Figure 6.75 illustrates results of a simulation reported by Taucer et al. (1991).

Nonlinear flexural response also can be modeled using hysteresis rules that operate based on a calculated moment–rotation envelope (and possibly other parameters). Figure 6.76 shows moment–rotation hysteresis based on the well-known Takeda model (Takeda et al., 1970). This model shows stiffness degradation that mimics behavior observed for well-detailed flexural members deformed well within their deformation capacities. For less well detailed components or for components loaded near deformation capacities, strength-degrading hysteresis models are also available (e.g., Kunnath et al., 1990). For analysis of complete building systems under earthquake ground motions, these hysteresis models provide an efficient way to represent nonlinear response characteristics. Fiber models are computationally less efficient, but are advisable where axial force variations are significant or where bi-directional loading occurs.

Section 6.10.2 presented some results for structural walls subjected to reversed cyclic loading, which are not repeated here.

For a more detailed discussion of flexural modeling approaches, the interested reader is referred to ATC 72 (2010).

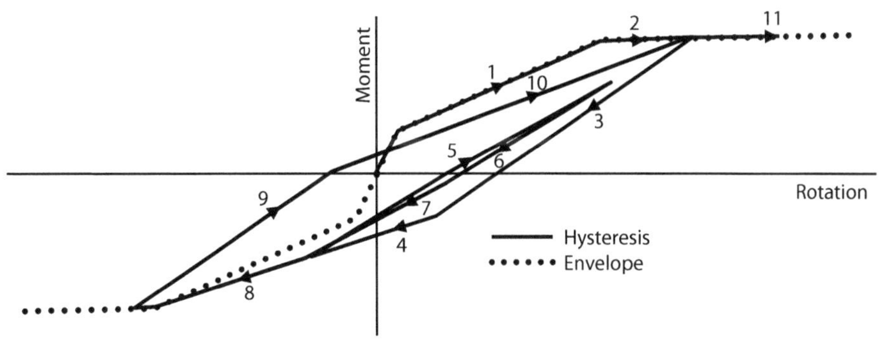

FIGURE 6.76 Takeda hysteresis model.

References

Abrams, D.P. (1987). "Influence of Axial Force Variations on Flexural Behavior of Reinforced Concrete Columns," *ACI Structural Journal*, Vol. 84, No. 3, pp. 246–254.

ACI 318 (2014). *Building Code Requirements for Structural Concrete (ACI 318-14) and Commentary*, American Concrete Institute, Farmington Hills, MI.

ACI 435 (2003). "Control of Deflection in Concrete Structures (ACI 435R-95)," *ACI Manual of Concrete Practice*, American Concrete Institute, Farmington Hills, MI, 89 pp.

ACI ITG-4.3R-07 (2007). *Report on Structural Design and Detailing for High-Strength Concrete in Moderate to High Seismic Applications*, American Concrete Institute, Farmington Hills, MI, 62 pp.

Alsiwat, J.M., and M. Saatcioglu (1992). "Reinforcement Anchorage Slip under Monotonic Loading," *Journal of Structural Engineering*, Vol. 118, No. 9, pp. 2421–2438.

ASCE 41 (2013). *Seismic Evaluation and Retrofit of Existing Buildings*, ASCE/SEI Standard 41–13, American Society of Civil Engineers, Reston, VA.

ATC 32 (1996). *Improved Seismic Design Criteria for California Bridges: Provisional Recommendations*, ATC-32, Applied Technology Council, Redwood City, CA.

ATC 40 (1996). *Seismic Evaluation and Retrofit of Concrete Buildings*, Vol. 1, Report No. SSC 96-01, Seismic Safety Commission, State of California.

ATC 72 (2010). *Modeling and Acceptance Criteria for Seismic Design and Analysis of Tall Buildings*, Applied Technology Council. Also available as PEER Report No. 2010/111 at http://peer.berkeley.edu/tbi/wp-content/uploads/2010/09/PEER-ATC-72-1_report.pdf, 222 pp.

Bae, S., and O. Bayrak (2003). "Stress Block Parameters for High-Strength Concrete Members," *ACI Structural Journal*, Vol. 100, No. 5, pp. 626–636.

Bayrak, O., and S.A. Sheikh (1997). "High-Strength Concrete Columns under Simulated Earthquake Loading," *ACI Structural Journal*, Vol. 94, No. 6, pp. 708–722.

Berry, M.P., D.E. Lehman, and L.N. Lowes (2008). "Lumped-Plasticity Models for Performance Simulation of Bridge Columns," *ACI Structural Journal*, Vol. 105, No. 3, pp. 270–279.

Bertero, V.V., and C. Felippa (1965). Discussion of "Ductility of Concrete," by H.E.H. Roy and M.A. Sozen, *Flexural Mechanics of Reinforced Concrete, SP-12*, American Concrete Institute, Farmington Hills, MI, pp. 227–234.

Bischoff, P.H. (2005). "Reevaluation of Deflection Prediction for Concrete Beams Reinforced with Steel and Fiber Reinforced Polymer Bars," *Journal of Structural Engineering*, Vol. 131, No. 5, pp. 752–767.

Bonelli, P., R. Tobar, and G. Leiva (1999). "Experimental Study on Failure of Reinforced Concrete Building," *ACI Structural Journal*, Vol. 96, No. 1, pp. 3–8.

Branson, D.E. (1963). *Instantaneous and Time-Dependent Deflections of Simple and Continuous Reinforced Concrete Beams*, HPR Report No. 7, Part 1, Alabama Highway Department, Bureau of Public Roads, AL, 78 pp.

Branson, D.E. (1977). *Deformation of Concrete Structures*, McGraw-Hill Book Company, New York, NY, 546 pp.

CSA (2004). *Design of Concrete Structures*, CSA A23.3-04, Canadian Standards Association, Mississauga, Canada.

Corley, W.G., A.E. Fiorato, and R.G. Oesterle (1981). "Structural Walls," *ACI Special Publication SP72-04*, American Concrete Institute, Farmington Hills, MI, pp. 77–131.

Elwood, K.J., and M.O. Eberhard (2006). "Effective Stiffness of Reinforced Concrete Columns," *PEER Research Digest No. 2006-1*, Pacific Earthquake Engineering Research Center, University of California, Berkeley, CA. http://peer.berkeley.edu/publications/research_digests/digests_2006/rd2006-1.pdf.

Elwood, K.J., and M.O. Eberhard (2009). "Effective Stiffness of Reinforced Concrete Columns," *ACI Structural Journal*, Vol. 106, No. 4, pp. 476–484.

Eurocode 8 (2004). *Eurocode 8: Design of Structures for Earthquake Resistance, Part 1, General Rules, Seismic Actions and Rules for Buildings*, Comité Européen de Normalisation, European Standard EN 1998-1:2004, Brussels, Belgium.

Fasching, C.J., and C.E. French (1998). "Effect of High-Strength Concrete (HSC) on Flexural Members," *High-Strength Concrete in Seismic Regions, SP-176*, C.W. French and M.E. Kreger, eds., American Concrete Institute, Farmington Hills, MI, pp. 137–178.

FEMA 306 (1999). *Evaluation of Earthquake Damaged Concrete and Masonry Wall Buildings*, Federal Emergency Management Agency, Washington, D.C., 249 pp.

French, C.W., and J.P. Moehle (1991). "Effect of Floor Slab on Behavior of Slab-Beam-Column Connections," *Design of Beam-Column Joints for Seismic Resistance, ACI SP 123*, 1991, American Concrete Institute, Farmington Hills, MI, pp. 225–258.

Hassan, M., S. El-Tawil (2003). "Tension Flange Effective Width in Reinforced Concrete Shear Walls," *ACI Structural Journal*, Vol. 100, No. 3, pp. 349–356.

Hognestad, E. (1951). *A Study of Combined Bending and Axial Load in Reinforced Concrete Members*, Bulletin Series No. 399, Engineering Experiment Station, University of Illinois, Urbana, IL, 128 pp.

Hognestad, E. (1952). "Fundamental Concepts in Ultimate Load Design of Reinforced Concrete Members," *Journal of the American Concrete Institute*, Vol. 48, No. 6, pp. 809–830.

Hognestad, E., N.W. Hanson, and D. McHenry (1955). "Concrete Stress-Distribution in Ultimate Strength Design," *Journal of the American Concrete Institute*, Vol. 52, No. 6, pp. 455–479.

Kunnath, S.K., A.M. Reinhorn, and Y.J. Park (1990). "Analytical Modeling of Inelastic Seismic Response of R/C Structures," *Journal of Structural Engineering*, Vol. 116, No. 4, pp. 996–1017.

Kurose, Y., G. Guimaraes, Z. Liu, M. Kreger, and J. Jirsa (1988). *Study of Reinforced Concrete Beam-Column Joints under Uniaxial and Biaxial Loading*, PMFSEL Report. No. 88-2, Phil M. Ferguson Structural Engineering Laboratory, University of Texas, Austin, TX, 146 pp.

Lehman, D., and J. Moehle (2000). *Seismic Performance of Well-Confined Concrete Bridge Columns*, Report No. PEER 1998/01, Pacific Earthquake Engineering Research Center, University of California, Berkeley, CA, 295 pp.

Lehman, D., J. Moehle, S. Mahin, A. Calderone, and L. Henry (2004). "Experimental Evaluation of the Seismic Performance of Reinforced Concrete Bridge Columns," *Journal of Structural Engineering*, Vol. 130, No. 6, pp. 869–879.

Low, S., and J.P. Moehle (1987). *Experimental Study of RC Columns Subjected to Multi-axial Cyclic Loading*, Report No. UCB/EERC 87/14, Earthquake Engineering Research Center, University of California, Berkeley, CA.

Ma, S-Y.M., V.V. Bertero, and E.P. Popov (1976). *Experimental and Analytical Studies on Hysteretic Behavior of Reinforced Concrete Rectangular and T-Beams*, Report No. UCB/EERC 76/2, University of California, Berkeley, CA, 241 pp.

Mattock, A.H., L.B. Kriz, and E. Hognestad (1961). "Rectangular Concrete Stress Distribution in Ultimate Strength Design," *Journal of the American Concrete Institute*, Vol. 57, No. 2, pp. 875–928.

Moehle, J.P., J.D. Hooper, and C.D. Lubke (2008). *Seismic Design of Reinforced Concrete Special Moment Frames: A Guide for Practicing Engineers*, NEHRP Seismic Design Technical Brief No. 1, NIST GCR 8-917-1, National Institute of Standards and Technology, Gaithersburg, MD.

NZS 3101 (2006). *Concrete Structures Standard—The Design of Concrete Structures*, Standards New Zealand, 696 pp.

Orakcal, K., and J.W. Wallace (2006). "Flexural Modeling of Reinforced Concrete Walls—Experimental Verification," *ACI Structural Journal*, Vol. 103, No. 2, pp. 196–206.

Otani, S., and M.A. Sozen (1972). *Behavior of Multistory Reinforced Concrete Frames during Earthquakes*, Structural Research Series No. 392, University of Illinois, Urbana, IL, 551 pp.

Panagiotou, M., T. Visnjic, G. Antonellis, P. Galanis, and J.P. Moehle (2013). *Effect of Hoop Reinforcement Spacing on the Cyclic Response of Large Reinforced Concrete Special Moment Frame Beams*, Report No. PEER 2013/16, Pacific Earthquake Engineering Research Center, University of California, Berkeley, CA, 92 pp.

Park, R., and T. Paulay (1975). *Reinforced Concrete Structures*, Wiley Interscience, New York, NY, 769 pp.

Priestley, M.J.N., and R. Park (1987). "Strength and Ductility of Concrete Bridge Columns under Seismic Loading," *ACI Structural Journal*, Vol. 84, No. 1, pp. 61–76.

Qi, X. (1986). *The Behavior of a R/C Slab-Beam-Column Subassemblage Under Lateral Load Reversals*, Report No. UCB/SESM CE 299 Report, University of California, Berkeley, CA, 67 pp.

Qi, X., and S.J. Pantazopoulou (1991). "Response of RC Frame under Lateral Loads," *Journal of Structural Engineering*, Vol. 117, No. 4, pp. 1167–1188.

Sezen, H., and E.J. Setzler (2008). "Reinforcement Slip in Reinforced Concrete Columns," *ACI Structural Journal*, Vol. 105, No. 3, pp. 280–289.

Sozen, M., P. Monteiro, J.P. Moehle, and H.T. Tang (1992). "Effects of Cracking and Age on Stiffness of Reinforced Concrete Walls Resisting In-Plane Shear," *Proceedings, Fourth Symposium on Current Issues Related to Nuclear Power Plant Structures, Equipment and Piping*, Orlando, FL.

Takeda, T., M.A. Sozen, and N.N. Neilsen (1970). "Reinforced Concrete Response to Simulated Earthquakes," *Journal of the Structural Division*, Vol. 96, No. 12, pp. 2557–2573.

Taucer, F., E. Spacone, and F.C. Filippou (1991). *A Fiber Beam-Column Element for Seismic Response Analysis of Reinforced Concrete Structures*, Report No. UCB/EERC 91/17, University of California, Berkeley, CA, 136 pp.

Thomas, K., and M.A. Sozen (1965). *A Study of the Inelastic Rotation Mechanism of Reinforced Concrete Connections*, Structural Research Series No. 301, Department of Civil Engineering, University of Illinois, Urbana-Champaign, IL, 121 pp.

Thomsen, J.H., and J.W. Wallace (1995). *Displacement-based Design of Reinforced Concrete Structural Walls: Experimental Studies of Walls with Rectangular and T-shaped Cross Sections*, Report No. CU/CEE-95/06, Department of Civil and Environmental Engineering, Clarkson University, Potsdam, NY, 353 pp.

Thomsen, J.H., and J.W. Wallace (2004). "Displacement-Based Design of Slender Reinforced Concrete Structural Walls—Experimental Verification," *Journal of Structural Engineering*, Vol. 130, No. 4, pp. 618–630.

Watson, S., F.A. Zahn, and R. Park (1994). "Confining Reinforcement for Concrete Columns," *Journal of Structural Engineering*, Vol. 120, No. 6, pp. 1798–1824.

Whitney, C.S. (1937). "Design of Reinforced Concrete Members under Flexure or Combined Flexure and Direct Compression," *Journal of the American Concrete Institute*, Vol. 33, No. 2, pp. 483–498.

Wight, J.K., and J.G. MacGregor (2012). *Reinforced Concrete Mechanics and Design*, 6th ed., Pearson Education, Upper Saddle River, NJ, 1157 pp.

Wood, S.L. (1989). "Minimum Tensile Reinforcement Requirements in Walls," *ACI Structural Journal*, Vol. 86, No. 5, pp. 582–591.

Yoshimura, M., and Y. Kurose (1985). "Inelastic Behavior of the Building," *Earthquake Effects on RC Structures: U. S. Japan Research, ACI SP 84-7*, American Concrete Institute, Farmington Hills, MI, pp. 163–201.

CHAPTER 7

Shear in Beams, Columns, and Walls

7.1 Preview

Structural framing is usually designed so that inelastic response, if it occurs during an earthquake, is predominantly in flexure, without significant inelastic response in shear. This objective, however, cannot be met for all structural elements. For example, inelastic response in shear may be accepted for some low-rise structural walls. Furthermore, shear and flexure are coupled such that apparent shear-yielding can occur even if the transverse reinforcement remains in the linear range. Thus, it is important to gain a deeper understanding of shear strength and deformability. This chapter considers analysis and design for shear in flexural members, with emphasis on beams, columns, and structural walls. This chapter also introduces strut-and-tie models for low-aspect-ratio members and discontinuity regions. Shear in beam-column connections, slab-column connections, structural wall connections, diaphragms, and foundations is covered in Chapters 9, 10, 13, 15, and 16.

7.2 Some Observations about Shear in Flexural Members

Structural members such as beams, columns, and structural walls resist earthquake effects through combinations of axial, flexural, and shearing actions. Figure 7.1 illustrates typical actions on a column of a moment-resisting frame. Note that body forces are ignored in this illustration, such that axial and shear forces are constant. The illustration also shows nominal shear and normal stresses expected if the column was made of a homogeneous, isotropic material, without stress due to volume change restraint. Prior to cracking, these stresses can be superimposed to determine directions of principal tensile stress. Cracks tend to form perpendicular to principal tensile stresses, suggesting crack orientations. In most earthquake-resisting structures, the largest principal tensile stresses are due to flexure and occur at the member ends. Hence, cracks tend to form first near member ends, approximately perpendicular to the longitudinal axis. As loading continues, additional flexural cracks initiate away from member ends, but shear causes these cracks to incline as they move deeper into the member. Figure 7.1 shows idealized

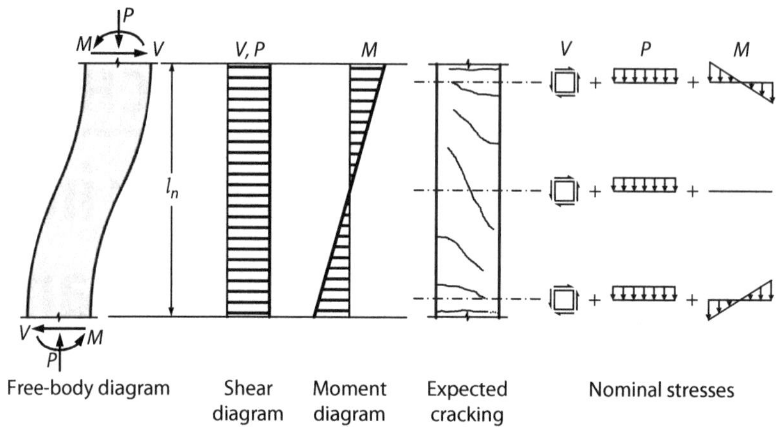

FIGURE 7.1 Axial, flexural, and shearing actions on a column.

crack orientations for loading in the direction shown. For loading in the opposite direction, shear and flexural (and possibly axial) stresses will reverse, producing new cracks that are reflected approximately about the column longitudinal axis.

Figure 7.2 shows details of a test to explore behavior of columns under lateral deformation reversals and constant axial force (Sezen and Moehle, 2006). The column is fixed at the base and connected to a stiff beam at the top. Two vertical actuators apply axial compression and maintain zero beam rotation. A horizontal actuator imposes lateral deformations. Boundary conditions in the test setup simulate those shown in Figure 7.1.

Key characteristics of the column are longitudinal reinforcement ratio $A_{st}/A_g = 0.025$, without splices, and transverse reinforcement with relatively wide spacing and poor configuration (hooks for perimeter hoops are not embedded in the core concrete). Concrete compressive strength is 3.06 ksi (21.1 MPa). Yield strengths of longitudinal and transverse reinforcement are 63 and 69 ksi (434 and 476 MPa), respectively. These details and materials are representative of those used in construction of the 1970s and earlier in highly seismic regions of the United States. Similar construction is used in many parts of the world today.

The column is tested by first applying axial compressive force of 150 kips (667 kN) (approximately $0.15f_c'A_g$, where f_c' = measured concrete compressive strength), and then subjecting it to multiple cycles of lateral displacements at progressively increasing amplitude. Figure 7.3 shows the progression of cracking and other damage in the column. Δ_y' is a nominal yield displacement corresponding approximately to effective yield in the force–displacement relation. At $0.5\Delta_y'$, cracks are approximately horizontal, suggesting predominant influence of flexural stresses. By $1\Delta_y'$, inclined cracks are evident, suggesting influence of shear. Cracking and damage increase as displacement amplitude increases.

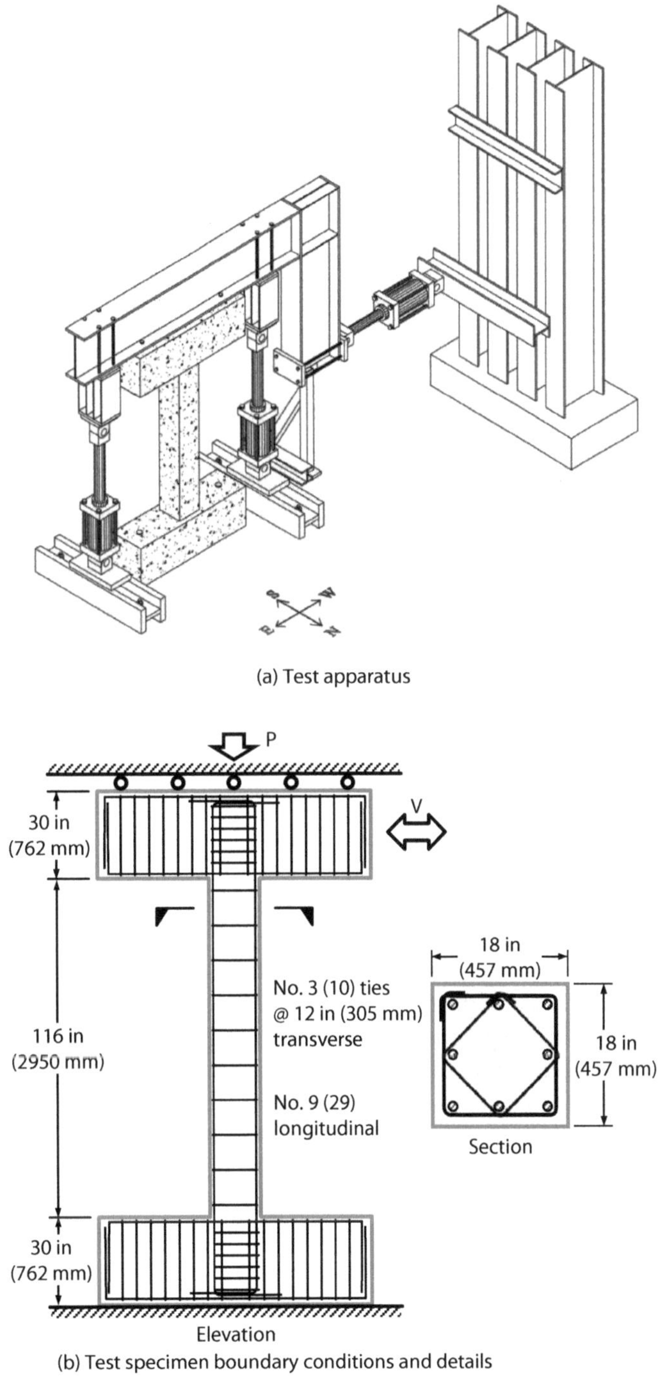

FIGURE 7.2 Laboratory test column. (*After Sezen and Moehle, 2006, courtesy of American Concrete Institute.*)

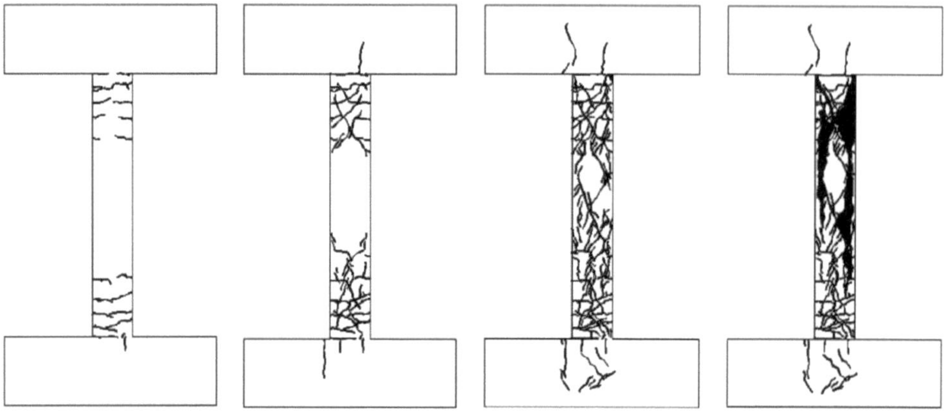

Figure 7.3 Development of cracking pattern during test, view from side. (*After Sezen and Moehle, 2006, courtesy of American Concrete Institute.*)

Figure 7.4*a* plots force–displacement relations for the column. Focusing first on the upper plot in Figure 7.4*a*, effective yielding (in flexure) occurs at around 1.5 in (40 mm) lateral displacement. Strength degradation (due to shear damage) is apparent at around 2.2 in (56 mm). Lateral strength drops below zero at large displacements due to the P-Δ effect. The lower plot in Figure 7.4*a* shows the relation between lateral and vertical displacements. Initially, the column elongates due to flexure because the neutral axis migrates toward the flexural-compression side, with the column centerline always in a tensile strain region. As damage progresses, the column begins to shorten. Eventually, the column is no longer able to support the constant applied axial force and sustains axial compression failure. Figure 7.4*b* shows the column at the end of test. Note the intersecting diagonal cracks. Note also the severe failure along the column longitudinal bars, especially in the upper half of the column. These latter failure patterns can be attributed to high bond stresses, which are common in members with high shear.

Behavior of the test column is somewhat typical for older existing construction with widely spaced transverse reinforcement and moderately high longitudinal reinforcement ratio. Where longitudinal reinforcement ratios and concurrent shear stresses are lower, flexure-dominated behavior can be expected without shear and axial failure. Where axial stresses are higher with light transverse reinforcement, axial failure becomes more likely (Elwood and Moehle, 2005a). Modern reinforced concrete construction may have higher transverse reinforcement ratios with better detailing, resulting in improved behavior. These subjects are discussed more fully in the remainder of this chapter.

7.3 Relations among Moment, Shear, and Bond

Our first goal is to review the relations among shear, moment, and bond between reinforcement and concrete. Consider a weightless cantilever beam with concentrated load at the tip (Figure 7.5).

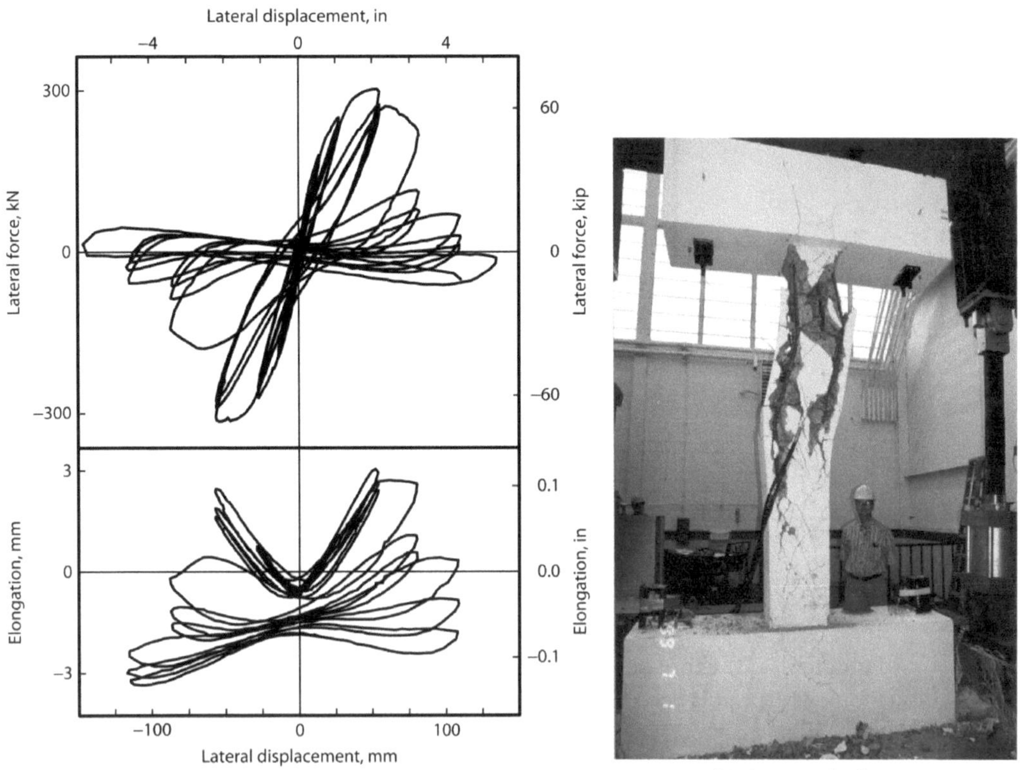

(a) Force–displacement relations

(b) Photograph after test

FIGURE 7.4 Force–displacement relations and final damage condition, photo from back side. (*After Sezen and Moehle, 2006, courtesy of American Concrete Institute.*)

From equilibrium, we know that

$$V = \frac{dM}{dx} \tag{7.1}$$

Substituting $M = T_s jd$ in Eq. (7.1) we find

$$V = \frac{dM}{dx} = \frac{d(T_s jd)}{dx} = \frac{dT_s}{dx} jd + \frac{d(jd)}{dx} T_s \tag{7.2}$$

The two terms on the right-hand side of Eq. (7.2) represent two mechanisms by which shear can be resisted in structural concrete members. Setting $\frac{d(jd)}{dx} T_s = 0$, Eq. (7.2) simplifies to

$$V = \frac{dT_s}{dx} jd \tag{7.3}$$

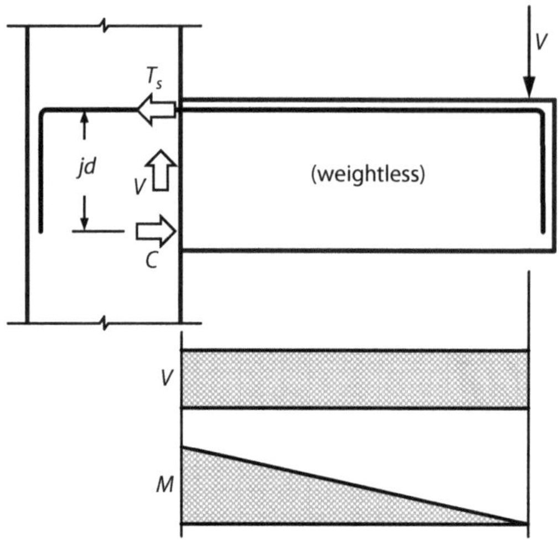

FIGURE 7.5 Cantilever beam with concentrated load.

Equation (7.3) expresses *beam action* for a fully cracked beam, that is, internal moment arm jd is constant and flexural tension force T_s varies linearly along the span. If, instead, we set $\frac{dT_s}{dx} jd = 0$, Eq. (7.2) simplifies to

$$V = \frac{d(jd)}{dx} T_s \tag{7.4}$$

Equation (7.4) expresses *arch (or truss) action*, that is, flexural tension force T_s is constant and internal moment arm jd varies linearly along the span. The physical implications of beam and arch action as represented by Eqs. (7.3) and (7.4) are examined more fully in Section 7.4.

Returning to Eq. (7.3), we see that beam action requires $\frac{dT_s}{dx}$ to be directly proportional to the shear force V. Thus, in a short length Δx of the beam, reinforcing steel would experience a change in tensile force equal to $\frac{dT_s}{dx}\Delta x$. Figure 7.6 shows the forces T_s and $\frac{dT_s}{dx}\Delta x$ acting on a reinforcing bar of length Δx. Horizontal force equilibrium requires that there be a bond stress u acting between the surface of the bar and the surrounding concrete. Horizontal force equilibrium requires

$$u\sum(\pi d_b)\Delta x = \frac{dT_s}{dx}\Delta x \tag{7.5}$$

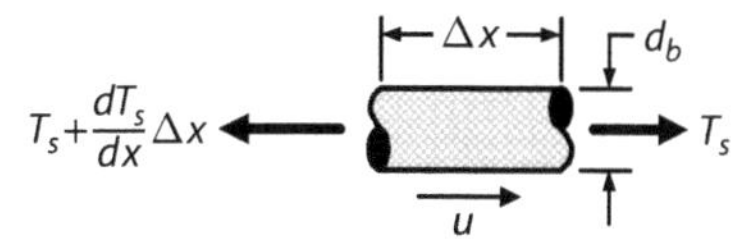

FIGURE 7.6 Free-body diagram of short length of reinforcing bar.

in which the summation is the sum of circumferences of all bars resisting force T_s. Thus, bond stress is

$$u = \frac{dT_s}{dx}\left(\frac{1}{\Sigma(\pi d_b)}\right) = \frac{V}{\Sigma(\pi d_b)\, jd} \tag{7.6}$$

Thus, we find that bond stress u between longitudinal reinforcement and surrounding concrete also is directly proportional to applied shear V. This partially explains why some members subjected to high shear and showing signs of shear failure may also show signs of bond failure, as in the vertical cracks along the longitudinal reinforcement shown in Figure 7.4*b*.

7.4 Beam Action and Arch Action

Equations (7.3) and (7.4) define mathematically the concepts of *beam action* and *arch action*, which are two distinct methods for resisting applied shear and moment. It is important to understand the physical meaning of these two equations. Figure 7.7*a* depicts beam action, by which shear and moment are resisted through a couple between tensile and compressive forces at

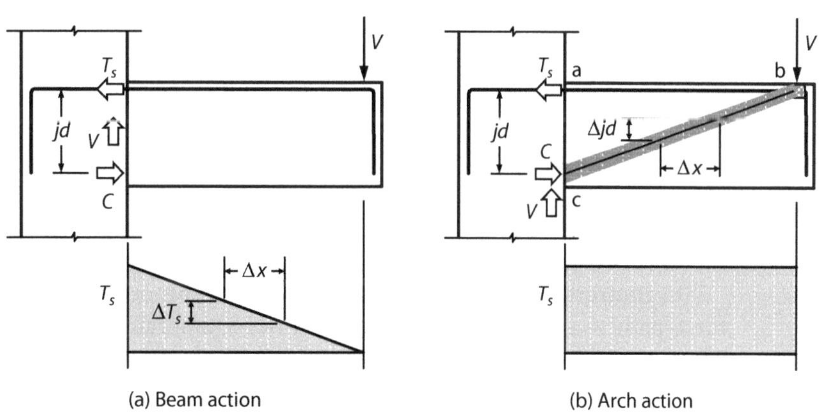

FIGURE 7.7 Theoretical beam action and arch action.

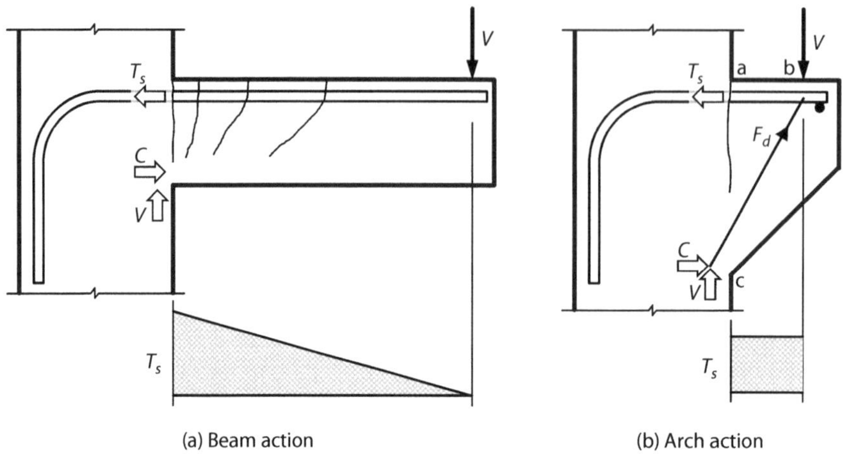

FIGURE 7.8 Practical conditions for beam action and arch action.

constant internal moment arm *jd*. Beam action results in flexural tension force T_s and flexural compression force C that vary in proportion with the applied moment. Basic flexural design (Chapter 6) is in accordance with the assumptions of beam action. Figure 7.7*b* depicts arch (or truss) action, by which shear and moment are resisted through constant tensile force T_s acting through varying internal moment arm *jd*. In essence, the beam is acting as pinned truss **abc**.

The natural question is: Which of the two mechanisms, beam action or arch action, is correct for reinforced concrete? The answer is that both mechanisms are at play in typical members, with one possibly more dominant than the other depending on member geometry.

In a relatively slender beam (Figure 7.8*a*), tests show that beam action predominates, with flexural tension force varying nearly linearly along the span. Two deviations commonly are noted. First, the flexural tension force will dip below the straight line where concrete is not cracked, as expected at uncracked sections. Second, at late stages of loading, opening of inclined cracks near the support may lead to tension shift, as discussed in Section 6.11.2. Tension shift is a manifestation of arch action occurring near the beam support. We will discuss this further in Section 7.6.

In contrast, in a deeper cantilever, as in the corbel in Figure 7.8*b*, the stiffest force path is a compression strut along **bc**, equilibrated by tension T_s along **ab**. Thus, this member resists shear primarily through arch or truss action. Note that arch or truss action in this component results theoretically in constant tensile force in the top reinforcement, such that the full tensile force T_s is acting at point **b**. As a result, anchorage design at the free end of the top reinforcement can be critical for such members.

7.5 Internal Forces in Members with Transverse Reinforcement

In members with transverse reinforcement subjected to shear, it is useful to develop an idealized force path to approximate internal forces. In this section, we develop the concept of reinforcement as *tension ties*, concrete as *diagonal compression struts*, and their intersections as *nodes*.

Consider the weightless cantilever shown in Figure 7.9*a*. We envision two idealized cracks, a vertical crack along **ac** and an inclined crack along **bc**. Figure 7.9*b* shows a free-body diagram of **abc**. Tension in top reinforcement is assumed to increase along **ba** by amount ΔT_s. The crack surface along **bc** can be presumed to be rough, such that any movement parallel to the crack produces aggregate interlock shear V_a. It is assumed that there is no tension perpendicular to the crack, which is not necessarily correct but which is a reasonable approximation for the present discussion. Crack opening along **bc** will also result in tensile force F_t in the transverse reinforcement. Vertical movement at point **b** will result in a dowel shear V_d acting on the longitudinal reinforcement. Near ultimate strength, the dowel shear is likely to cause concrete splitting cracks along the longitudinal reinforcement, the occurrence of which will reduce the dowel force. Therefore, for the present discussion, we assume $V_d \approx 0$. Finally, a portion of the total shear V' and a compressive force C' will act on the vertical face of segment **abc**. The line of action of force V_a is approximately concurrent with point **c**. Therefore, moment equilibrium about **c** is preserved mainly by the opposing moments due to forces F_t and ΔT_s.

Now consider the free-body diagram of Figure 7.9*c*, in which longitudinal and transverse reinforcement have been cut from the surrounding concrete, leaving a small concrete "node" at their intersection. The transverse reinforcement, being

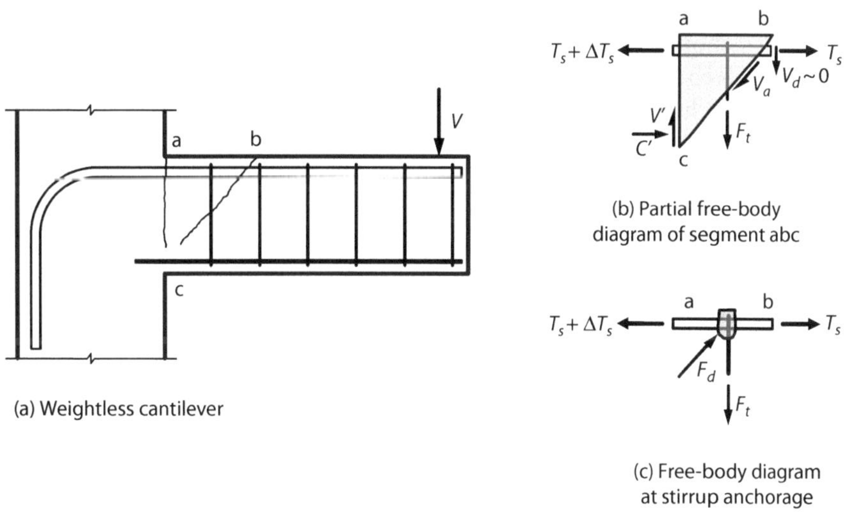

(a) Weightless cantilever

(b) Partial free-body diagram of segment abc

(c) Free-body diagram at stirrup anchorage

Figure 7.9 Internal forces in member with transverse reinforcement.

anchored around the longitudinal reinforcement, creates a downward reaction on the longitudinal reinforcement. Similarly, as a simplification, it is assumed that horizontal force ΔT_s is anchored entirely at the intersection of the longitudinal and transverse reinforcement. Consequently, a diagonal compression force F_d is required in the concrete to equilibrate the node.

The diagonal compression force F_d identified in Figure 7.9*c* is similar to the diagonal compression force F_d identified in Figure 7.8*b*. The concept of reinforcement acting as tension ties, concrete acting as struts, and their intersections acting as nodes is a powerful tool for analysis of reinforced concrete members. Section 7.6 further pursues this concept.

7.6 Strut-and-Tie Models

A *strut-and-tie model*, or *truss model*, is a structural idealization for analysis and design of reinforced concrete members subjected to shear, moment, axial force, or combinations of these and other actions. Figure 7.8*b* represents a strut-and-tie model for a corbel, comprising a tension tie at the top of the corbel and a diagonal compression strut extending diagonally downward to the flexural compression zone near the bottom of the corbel. Strut-and-tie models can be powerful tools for visualizing and analyzing stress flow in structural concrete members. This section develops the basic concept of strut-and-tie models and presents some examples. Proportioning requirements for strut-and-tie models are discussed in Section 7.7. For more detailed coverage see, for example, Marti (1985), ACI 445 (1999), ACI SP 273 (2010), and Wight and MacGregor (2012).

Truss models were introduced by Ritter (1899), who postulated that a beam after cracking could be modeled as a truss with parallel tension and compression chords, concrete diagonals at 45° with respect to the longitudinal axis, and vertical web members. Truss models received increased attention since the 1960s, including introduction of variable angle trusses, stress limits for concrete, and broader application to a range of analysis and design problems (ACI 445, 1999). Truss models also have been introduced in building codes, including the strut-and-tie method of ACI 318.

To assemble a truss model, we first envision the stress flow within a structural member. Experience with similar design problems, or results of linear structural analysis, can aid in identifying stress patterns. The directions of principal compressive stress suggest directions in which to orient compression struts. Struts can deviate moderately from principal stress directions of linear structural analysis because ductile reinforced concrete members can usually redistribute internal stresses to accommodate the deviations. In flexural members, reinforcement should be placed to resist flexural tension forces. Additional reinforcement is placed to equilibrate compression struts and form a complete and stable truss. In general, there is no unique truss model for a given design problem, and different solutions can produce acceptable results. It takes experience, and sometimes

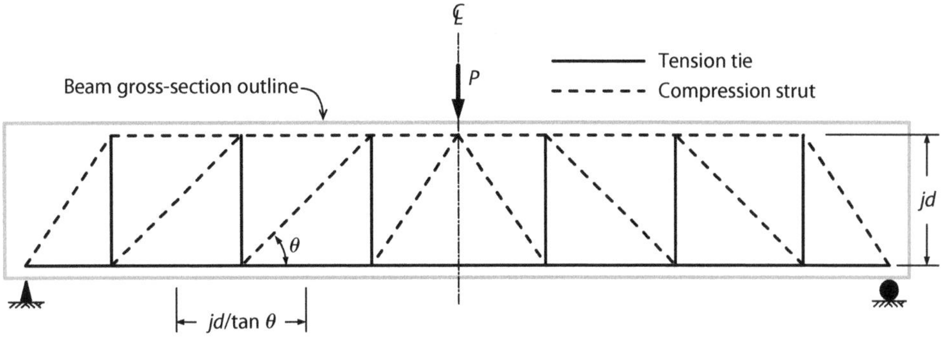

Figure 7.10 Truss model for simply supported beam.

design iterations, to develop the most efficient truss models for a particular design situation.

As an example, consider a weightless, simply supported beam subjected to midspan concentrated force P (Figure 7.10). A suitable truss model for a flexural member will have compression and tension chords at top and bottom. Shear will result in principal compressive stresses at mid-depth oriented approximately 45° with respect to the longitudinal axis. A truss is defined based on these considerations. Truss analysis is done by the usual procedures of structural analysis. Truss member forces can be used to determine requirements for concrete section, longitudinal reinforcement, and transverse reinforcement. Each of the vertical web members should be considered to represent transverse reinforcement uniformly spaced along length $jd/\tan\theta$, rather than a single, widely spaced stirrup. Concrete stress limits are described in Section 7.7.

7.6.1 Plastic Truss Analysis for Beams

A plastic truss model is one in which some or all tension ties are stressed to full tensile capacity. A plastic truss model can render an otherwise indeterminate truss into a determinate one, as forces in yielding ties are known. Such trusses are useful for understanding stress flow limits in beams.

Consider a weightless, simply supported beam subjected to concentrated force $P = 2V$ at midspan (Figure 7.11). Transverse reinforcement comprises stirrups at spacing $s = jd/2$. Each stirrup has strength $A_v f_{yt} = V/3$. We begin the analysis by assuming that a compression strut acts along the stiffest force path, that is, from node **r** to node **h** (Figure 7.11*b*). Stirrup **hq** resists the vertical component of this diagonal compression force with vertical force $V/3$ (i.e., the strength of that stirrup). The horizontal component of the strut is resisted by a change in tensile force in the longitudinal reinforcement at node **h**. Because a single stirrup has insufficient strength to resist the entire shear, it can be considered to soften upon yielding, increasing viability of an additional strut **rg** (Figure 7.11*c*). Similarly, an additional strut **rf** is required, by which the entire shear force V is now supported

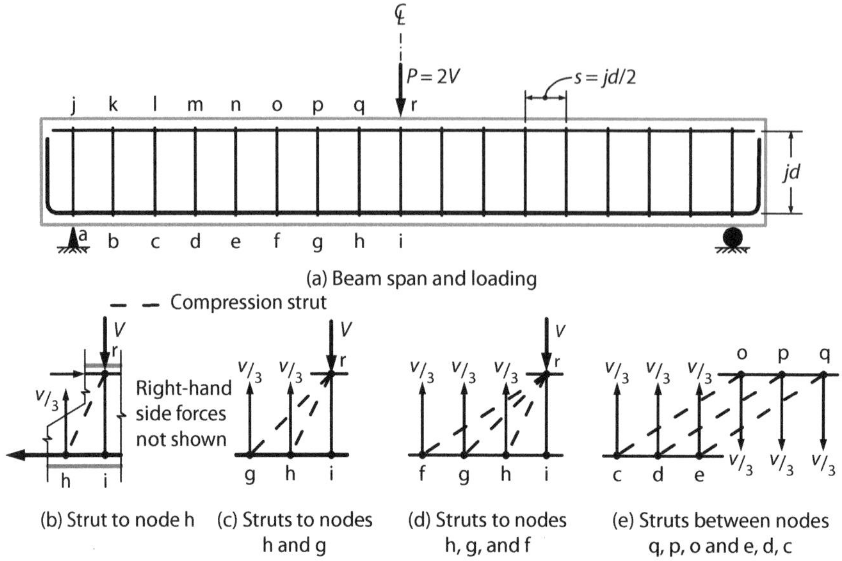

FIGURE 7.11 Plastic truss model. (Not all horizontal forces are shown in each diagram.)

(Figure 7.11*d*). The vertical forces from stirrups **hq**, **gp**, and **fo** are "hung" from the compression chord at nodes **q**, **p**, and **o** (Figure 7.11*e*). Continuing the analysis, the force at **q** requires strut **qe** (note that the strut cannot be resisted at nodes **f** or **g** because the stirrups at those points already are at capacity). The analysis continues until the end of the beam, where a series of struts converges above the reaction at point **a** (Figure 7.12).

Figure 7.12*a* shows the final layout of the plastic truss model. The truss layout suggests orientations of diagonal cracks, as these would be approximately parallel to the diagonal compression struts. Figure 7.12*b* shows cracks that were sketched based on the orientation of struts and experience in cracking patterns for reinforced concrete members. Exact crack locations are not known, although it is likely that the transverse reinforcement will act as crack initiators, which will influence crack locations.

Figure 7.12*c* shows free-body diagrams of nodes **a** and **k**. Knowing that each diagonal compression strut has vertical force of $V/3$, and knowing the strut angles, we can calculate the horizontal forces acting in the tension and compression chords at each of these points. At node **a**, the chord tensile force is $(1/2 + 2/2 + 3/2)V/3 = V$. At node **k**, the chord compressive force is $(1/2)V/3 = V/6$. Continuing the analysis using method of joints or sections reveals tensile and compressive chord forces along the span. These are plotted in Figure 7.12*d* along with the flexural tension or compression force obtained from $T_s = C = M/jd$. Note that the two different analysis models result in the same chord tensile force at point **i** (as required by statics). However, the truss model indicates a shift in the flexural tension force associated with action of the diagonal compression struts. This shift

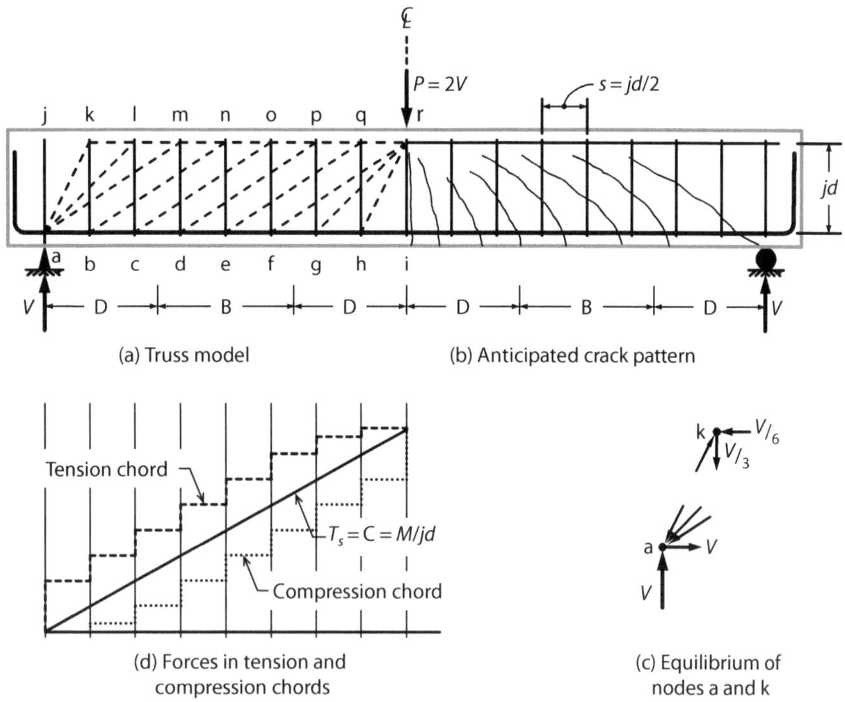

Figure 7.12 Solution of plastic truss model.

has been observed in tests on reinforced concrete beams, and is reflected in code requirements for development of longitudinal reinforcement (ACI 318). Note that tension shift will be larger for flatter orientation of the diagonal compression struts. Forces in the compression chord are generally smaller than those indicated by flexural theory. The results are sensible (e.g., one should not expect significant flexural compression force between nodes **j** and **k** as strutting action will carry shear into the support directly rather than by flexural action). The stepwise variation of chord forces is, of course, an outcome of model idealization, and should not be expected in actual structures.

The plastic truss model of Figure 7.12 suggests that cracks may fan out from discontinuity points such as the point of application of force P and the reaction at node **a**. This behavior is evident in tests of structural concrete members (Figure 7.13). Schlaich et al. (1987) introduced the concept of *D-regions and B-regions*, where D stands for discontinuity or disturbed, and B stands for Bernoulli or beam. In flexural members, D-regions extend approximately one member depth from any disturbance such as concentrated force or change in cross section, with B-regions composing the remaining beam length. B-regions and D-regions are identified beneath the beam sketch in Figure 7.12.

B-regions in slender beams can have an effectively uniform diagonal compression field associated with struts inclined at constant angle (Figure 7.12).

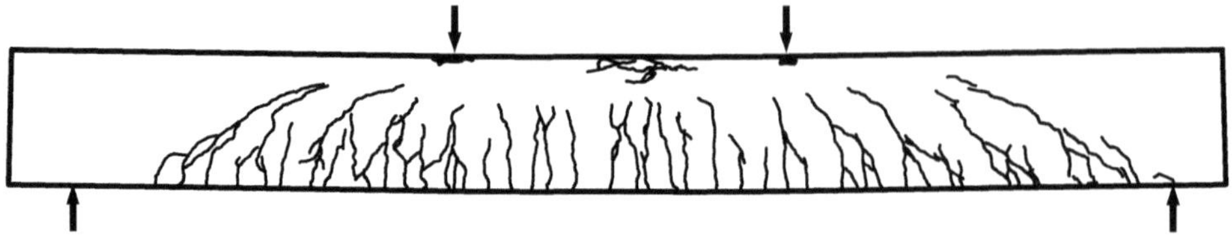

Figure 7.13 Crack orientation in a simply supported beam with two concentrated loads. (*After Leonhardt, 1965.*)

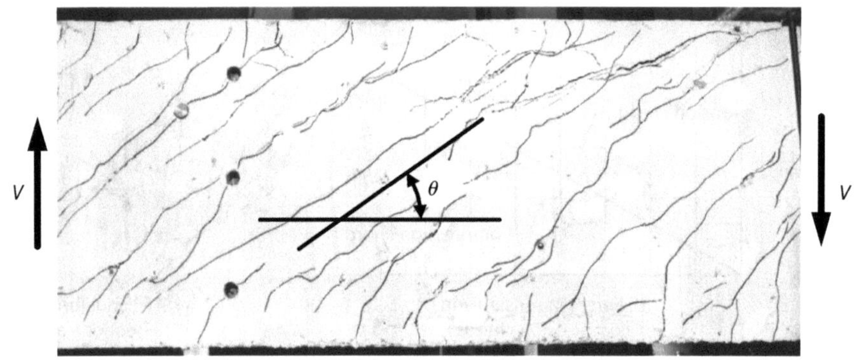

Figure 7.14 B-region observed during laboratory test. (*After Collins and Mitchell, 1987.*)

Figure 7.14 shows cracks forming approximately parallel to the uniform compression field at an angle $\theta \approx 35°$.

Shear cracks in a rectangular beam usually form as extensions of flexural cracks, initially propagating at approximately 45° with respect to the longitudinal axis. As loading progresses, redistribution of internal stresses enables formation of new cracks at shallower (or steeper) angle as necessary to equilibrate resistance provided by reinforcement. Figure 7.15 shows an example for a thin-webbed beam in which cracks in the B-region initially are approximately 45°, but are at shallower inclination for higher force levels. Angles approaching 30° are evident.

ACI 445 (1999) notes that strut inclinations θ in B-regions typically fall in the range 18° to 65° relative to the longitudinal axis. For design, ACI 318 limits the angle between the axes of any strut and any tie entering a single node to 25°. This book recommends $25° \leq \theta \leq 65°$.

In slender beams, diagonal compression struts fan out from points of reaction or concentrated force (Figure 7.12). These D-regions generally are stronger than adjacent B-regions, provided the cross section is constant and reinforcement is continuous. D-regions at supports, however, can be susceptible to anchorage failure of the tension chord because of tension shift effects. Good practice checks anchorage at supports. Hooks or headed reinforcement may be required.

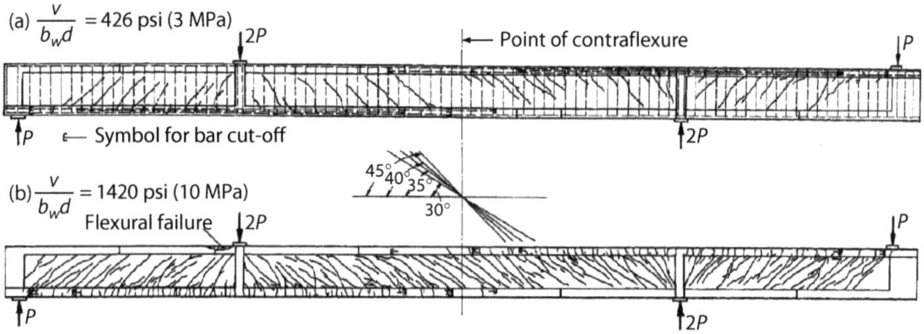

FIGURE 7.15 Crack formation in a thin-webbed beam. (*After Kupfer and Bauman, 1969.*)

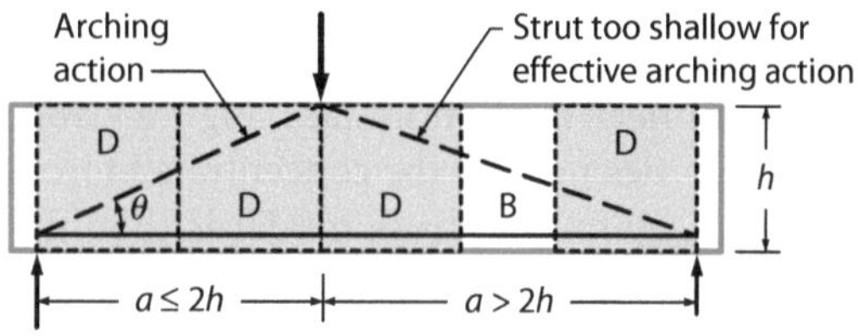

FIGURE 7.16 Description of deep and slender beams.

In deeper beams, two adjacent D-regions may intersect or overlap, as in the left-hand side of the beam in Figure 7.16. In this case, a compression strut can form directly between point of load application and the reaction, such that stirrups are not required to resist shear (regardless, reinforcement may be required for crack control). Where shear span $a > 2h$ such that D-regions do not overlap (right-hand side of Figure 7.16), the compression strut becomes too shallow to effectively transmit force directly to the reaction point. According to ACI 318, the angle θ between the axis of any strut and the tension tie restraining it is not permitted to be less than 25°, which corresponds approximately to shear span $a = 2h$. ACI 318 also uses this geometry to differentiate between deep and shallow beams. A *deep beam* is one having either (a) l_n not exceeding $4h$ or (b) a concentrated load within $2h$ from a support. Special design provisions apply to deep beams.

7.6.2 Example Strut-and-Tie Models

Development of efficient strut-and-tie models requires experience and may require design iterations. This section presents some examples to illustrate typical models.

Figure 7.17 depicts a dapped beam end, which is common in precast construction. Figure 7.17*a* illustrates one acceptable truss model (alternative models also have been shown to produce good performance). Diagonal compression strut **ac** supports vertical shear at the dapped end and requires tension ties **ae** and **cd**.

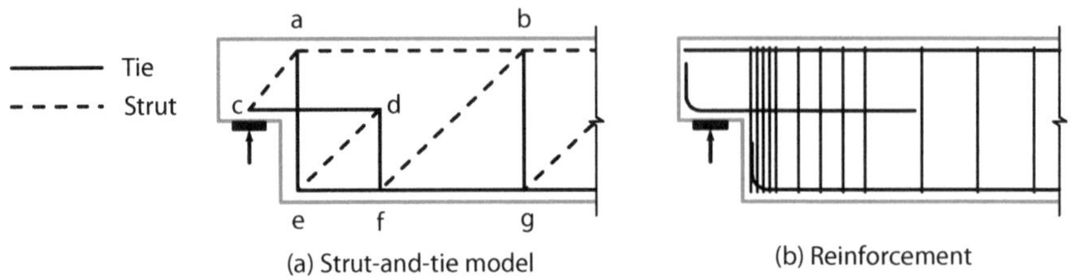

FIGURE 7.17 Dapped beam.

Note that anchorage of **cd** within the beam creates diagonal compression **de**, which in turn requires tension tie **df**. Figure 7.17*b* shows reinforcement satisfying requirements of the truss model.

Beams sometimes support concentrated forces near their bottom chord. A common example is where one beam frames into and is supported by a second beam—diagonal compression struts from the supported beam will tend to enter the supporting beam near the bottom face. Figure 7.18*a* identifies an acceptable truss model in which ties pick up the concentrated force and deposit the force at the top of the beam. A routine truss model is then used to carry the internal actions to the supports. Figure 7.18*b* shows hanger reinforcement used to pick up the concentrated force.

In precast construction, inverted T-beams are used to support orthogonal beams, resulting in the conditions shown in Figure 7.19*a*. In this case, an inclined strut transmits concentrated support forces to the hanger reinforcement of the supporting beam. At columns in precast construction, corbels are common (Figure 7.19*b*). In these two strut-and-tie models, struts of finite width react against nodal zones. Section 7.7 addresses proportioning of these struts and nodal zones.

Figure 7.19*c* depicts a strut-and-tie model for a deep beam. Note that the tension tie theoretically would have constant tension over the entire length; anchorage of the tie near the nodal zone may require a mechanical anchorage device.

Figure 7.19*d* depicts two concentrated forces supported by a structural wall. These forces spread out into the supporting wall. For design purposes it is common

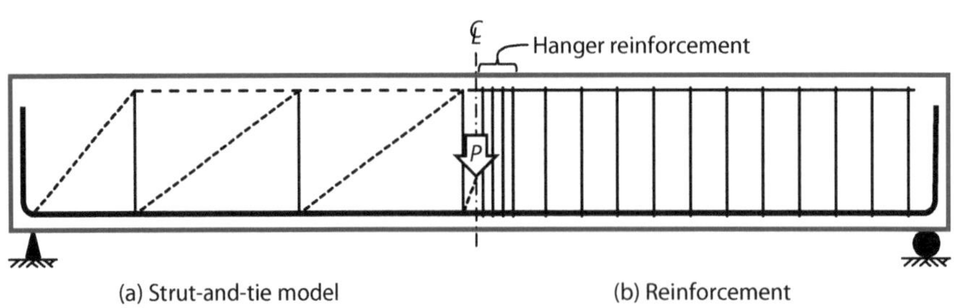

FIGURE 7.18 Beam supporting concentrated force near bottom chord.

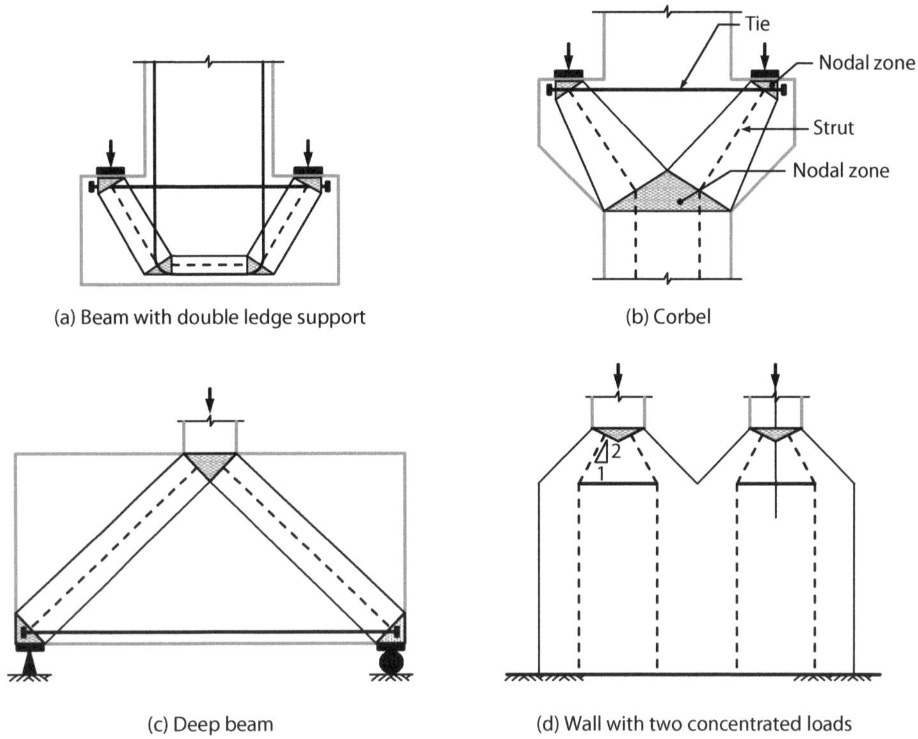

FIGURE 7.19 Various strut-and-tie models.

to assume the force spreads into the wider section at angle of 2:1 as shown. Note that tension ties are required to support the horizontal thrust created by spreading the force into the wall. Additional wall examples, including walls with openings, are described in Chapter 13.

7.7 Proportioning of Strut-and-Tie Models

Section 7.6 introduced strut-and-tie models as a way of visualizing and calculating the idealized internal forces within structural members. This section introduces additional detail about geometry of strut-and-tie models and presents the ACI 318 approach for proportioning struts, ties, and nodal zones for intended performance. This method was developed primarily for monotonic loading, but can be effective for seismic design applications as well. Von Ramin and Matamoros (2006) suggest modifications for cyclic loading that are not pursued here.

7.7.1 Overall Geometry

A strut-and-tie model is made up of struts and ties connected at nodes. A *strut* is a compression member representing the resultant of a parallel or a fan-shaped compression field. A *tie* is a tension member. A *node* is a point in a strut-and-tie

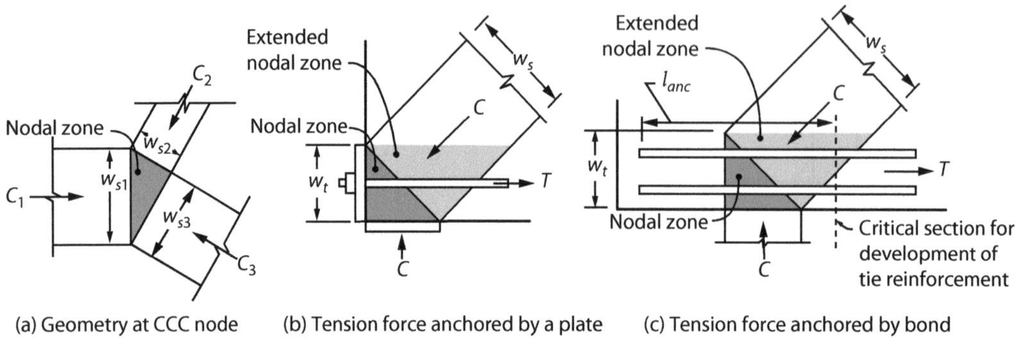

FIGURE 7.20 Definition of nodal zones and extended nodal zones. (*After ACI 318, 2014, courtesy of American Concrete Institute.*)

model corresponding to the joint at which axes of struts, ties, and concentrated forces intersect. Each strut or tie has a finite width and thickness; in a planar member such as a beam, the thickness corresponds to the dimension perpendicular to the plane of the member and width corresponds to the dimension in the plane of the member but perpendicular to the strut or tie. The intersection of struts and ties therefore occurs over a finite volume of concrete around the node known as the *nodal zone*. Finite dimensions of struts, ties, and nodal zones are shown in Figure 7.19. Nodal zones also are shown in Figure 7.20. An *extended nodal zone* occurs around the node connecting struts and ties, and is that portion of a member bounded by the intersection of the effective strut width, w_s, and the effective tie width, w_t. See Figure 7.20*b* and *c*. Extended nodal zones will be used to define anchorage lengths for ties.

As noted previously, ACI 318 limits the angle between the axes of any strut and any tie entering a single node to 25° or greater. This is to avoid incompatibility between shortening of struts and elongation of ties occurring in almost the same directions.

7.7.2 Design Strength

Structures designed by strut-and-tie models are required to have design strength at least equal to required strengths. Whereas traditional design methods refer to strength at sections, for strut-and-tie models we refer to strengths of the struts, ties, and nodal zones. The basic design equation is

$$\phi F_n \geq F_u \tag{7.7}$$

where $\phi = 0.75$; F_n = nominal strength of the strut, tie, nodal zone, or bearing area; and F_u = required strength based on governing load combinations. Sometimes the design strength will be in terms of stress, in which case force F in Eq. (7.8) is divided by area of the corresponding component. Where a strut is essentially a column, as in the case of a narrow pier in a perforated wall, the strength reduction factor for columns should apply instead.

7.7.3 Struts

In slender beams, the beam length is subdivided into B-regions and D-regions (Figure 7.12). In deep beams or other discontinuity regions, the structure may comprise only D-regions; for example, the strut-and-tie models shown in Figure 7.19 comprise D-regions only. Where B-regions and D-regions occur together in a slender beam, the strength of the shear span is controlled by the B-regions if the B-regions and D-regions have similar geometry and reinforcement. Normally, B-regions are designed by traditional shear design procedures such as those in Section 7.9, rather than a strut-and-tie model. Here, for completeness, we consider struts in both B-regions and D-regions.

B-Regions

We begin by referring to the slender beam shown in Figure 7.12. Strength in the B-region can be expressed in terms of transverse reinforcement using a free-body diagram cut along an inclined crack (Figure 7.21*a*).

The total number of stirrups crossing the crack is $n = jd/s \tan\theta$. Summing forces in the vertical direction results in

$$V = \sum A_v f_{yt} = \frac{A_v f_{yt} jd}{s \tan\theta} \tag{7.8}$$

Strength in the B-region also can be expressed in terms of concrete diagonal compressive stress using a free-body diagram cut along a zigzag between stirrups (Figure 7.21*b*). Summing forces in the vertical direction results in

$$V = f_{cd} b_w jd \cos\theta \sin\theta \tag{7.9}$$

Rearranging terms, the concrete compressive stress is

$$f_{cd} = \frac{V}{b_w jd \cos\theta \sin\theta} = \frac{V}{b_w jd}\left(\tan\theta + \frac{1}{\tan\theta}\right) \tag{7.10}$$

From Eq. (7.10), we see that, for a given shear V, the diagonal compressive stress increases with decreasing angle θ. In the limit, $f_{cd} \to \infty$ as $\theta \to 0$. As noted in Section 7.6.1, the angle θ between the axis of any strut and the tension tie restraining it should not be less than 25°.

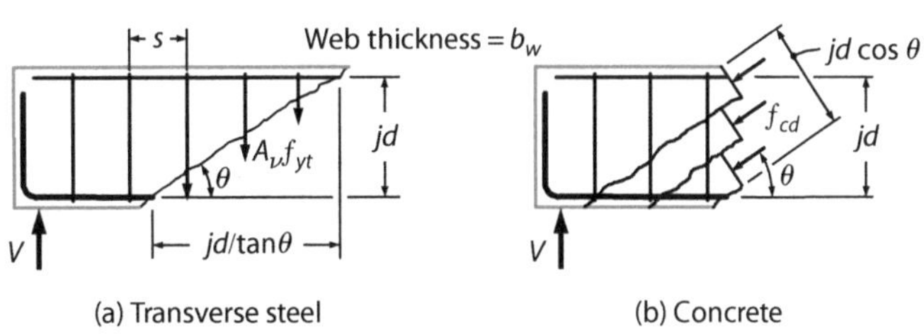

Figure 7.21 Internal actions in B-region.

The preceding results show that, according to the truss model, shear strength in the B-region is not the sum of steel and concrete contributions. Instead, the model requires both steel and concrete components if shear is to be resisted. Of course, it should be evident that a concrete beam without shear reinforcement has shear strength, at least until the shear reaches the concrete cracking strength. Thus, the truss model presented here is not a complete representation of shear resistance in beams. Nonetheless, it is a useful detailing and analysis tool.

We have seen that strut angle θ affects stress in concrete diagonals, tension shift in longitudinal reinforcement, and number of stirrups engaged to resist shear. In general, a shallower angle θ will result in greater economy because fewer stirrups are required to resist a given shear. This comes at the expense of increased tension shift and increased stress in concrete diagonals, both of which need to be considered in design.

In beams with relatively high shear and corresponding transverse reinforcement, diagonal compressive stresses can reach concrete compressive strength (Figure 7.22). Diagonal compression struts are crossed by transverse reinforcement strained in tension (Figure 7.21). As noted in Section 3.6, transverse tensile strain reduces concrete compressive strength. ACI 445 (1999) presents a range of recommendations for compressive strength considering these biaxial stress conditions. Here we adopt the stress limits of ACI 318. Specifically, for the case of a diagonal compression field in the B-region of a beam, nominal compressive strength is

$$f_{ce} = 0.85\beta_s f_c' = 0.85 \times 0.60\lambda f_c' = 0.51\lambda f_c' \quad (7.11)$$

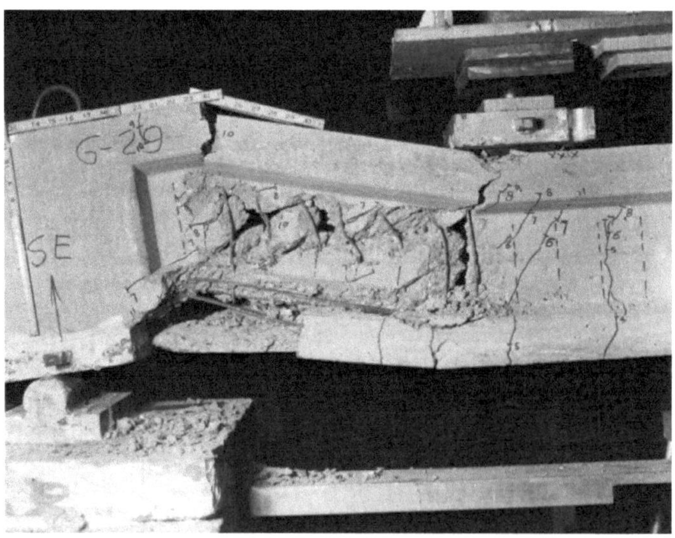

FIGURE 7.22 Web crushing of a beam. (*Courtesy of J.G MacGregor.*)

in which λ = modification factor for lightweight aggregate concrete = 0.75 for all lightweight concrete and 0.85 for sand lightweight concrete.

D-Regions

ACI 318 defines several different types of struts, and specifies different unit strength for each type. The different strut types are illustrated in Figure 7.23 and summarized in Table 7.1. For each of these struts, nominal compressive strength is defined by

$$F_{ns} = f_{ce} A_{cs} \tag{7.12}$$

in which A_{cs} = cross-sectional area at one end of the strut and f_{ce} = compressive strength, taken as the smaller of (a) the effective compressive strength of concrete in the strut and (b) the effective compressive strength of concrete in the nodal zone (Section 7.7.5).

The effective compressive strength of concrete in the strut is

$$f_{ce} = 0.85 \beta_s f_c' \tag{7.13}$$

in which β_s is defined according to Table 7.1.

Values of β_s in Table 7.1 warrant additional discussion. Case A applies to a strut equivalent to the rectangular stress block in a compression zone in a beam or column, as in strut **bc** in Figure 7.23*a*.

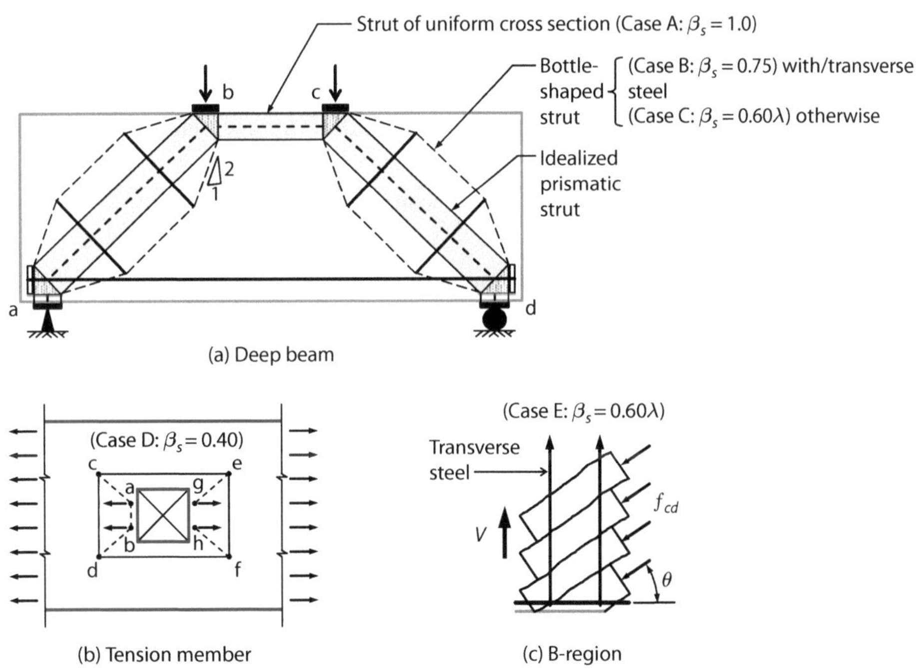

FIGURE 7.23 Example values of β_s.

Case	Illustration	Description	β_s
A	Figure 7.23*a*	For a strut of uniform cross-sectional area over its length	1.0
B	Figure 7.23*a*	Bottle-shaped struts with transverse reinforcement*	0.75
C	Figure 7.23*a*	Bottle-shaped struts otherwise	0.60λ
D	Figure 7.23*b*	For struts in tension members, or tension flanges of members	0.40
E	Figure 7.23*c*	For all other cases (applies to B-regions)	0.60λ

*Axis of strut crossed by transverse reinforcement proportioned to resist transverse tensile force resulting from the compressive force spreading in the strut.

TABLE 7.1 Values of β_s

Cases B and C apply to struts commonly known as *bottle-shaped struts*, as in struts **ab** and **cd** in Figure 7.23*a*. Although these struts are idealized as being prismatic, the presence of surrounding concrete means that compressive stress will spread transverse to the axis of the idealized strut. The internal spread leads to transverse tension that can split the strut. Reinforcement should be placed to resist this splitting force. Where sufficient reinforcement is placed, Case B in Table 7.1 applies. Otherwise, Case C applies.

The reinforcement crossing bottle-shaped struts can be in one direction provided the angle between the reinforcement and the axis of the strut is not less than 40°. Alternatively, it can be placed in two orthogonal directions. It is permitted to calculate the required amount of reinforcement from basic principles. Alternatively, for $f_c' \leq 6000$ psi (41 MPa), ACI 318 permits the reinforcement to be defined by

$$\sum_i \frac{A_{si}}{b_s s_i} \sin a_i \geq 0.003 \qquad (7.14)$$

in which b_s refers to strut thickness (in a two-dimensional structure, this is the dimension out of plane), and the other variables are defined in Figure 7.24.

Case D in Table 7.1 refers to struts in tension members or the tension flanges of members. Figure 7.23*b* shows an example of a tension member with an opening. Working our way from left to right in that figure, tensile force acting across the width of the opening generates struts **ac** and **bd**. Ties **ce** and **df** carry the tension past the opening, while tie **cd** provides equilibrium of struts **ac** and **bd**. The force path continues around the right side of the opening.

Case E in Table 7.1 applies in other cases not covered by A through D. The most common application is the B-region of a slender beam. Figure 7.23*c* illustrates the diagonal compression field of a B-region crossed by transverse reinforcement in tension.

It is permitted to increase the compressive strength by providing longitudinal reinforcement in a strut. Concrete confinement also is permitted to be considered,

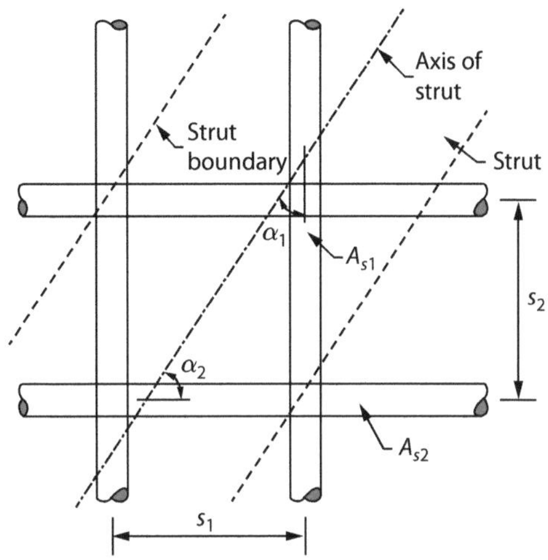

FIGURE 7.24 Reinforcement crossing strut. (*After ACI 318, 2014, courtesy of American Concrete Institute.*)

but it is likely that realization of higher strength due to confinement would require cover spalling, which might constitute unacceptable performance.

7.7.4 Ties

Ties can be made of deformed reinforcement, prestressed reinforcement, or both. The axis of reinforcement in a tie should coincide with the axis of the tie in the strut-and-tie model. Where tie reinforcement is in one layer, effective tie width w_t can be taken as twice the concrete cover, where concrete cover is measured from the centroid of the layer to the nearest concrete surface (Figure 7.25*a*). If, using this width, the stress on the nodal zone exceeds the stress limit for the nodal zone (Section 7.7.5), then the minimum width of the tension tie is defined by strength f_{ce} of the nodal zone, calculated as

$$w_t = F_{nt}/f_{ce}b_s \tag{7.15}$$

In this case, tie reinforcement should be distributed approximately uniformly over the width and thickness of the tie rather than being placed in a single layer (Figure 7.25*b*).

Nominal strength of a tie is

$$F_{nt} = A_{ts}f_y + A_{tp}(f_{se} + \Delta f_p) \tag{7.16}$$

where f_{se} = effective stress in prestressing steel, Δf_p = increase in stress in prestressing steel due to factored loads, which can be taken equal to 60,000 psi (414 MPa) for bonded prestressing steel or 10,000 psi (69 MPa) for unbonded prestressing steel, and $(f_{se} + \Delta f_p)$ should not exceed f_{py}.

Tie reinforcement can be anchored by mechanical devices, post-tensioning anchorage devices, standard hooks, or straight bar development. At nodal zones

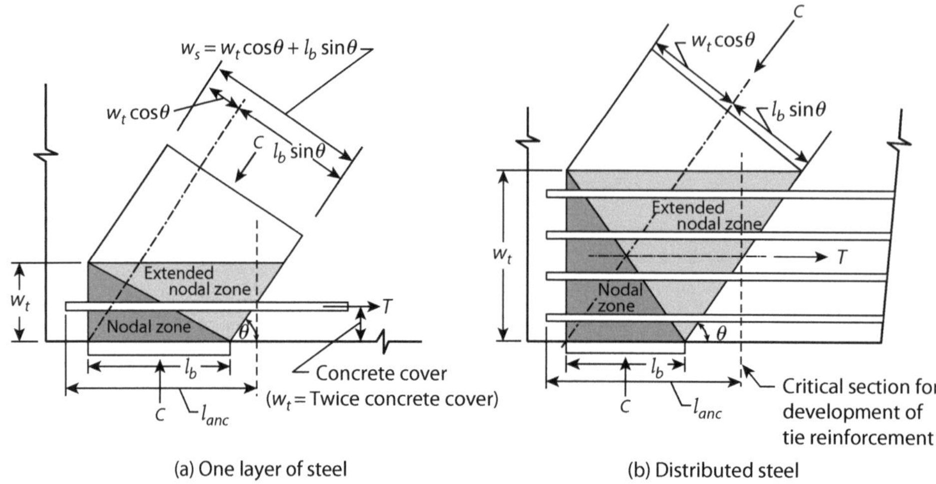

FIGURE 7.25 Placement and anchorage of tie reinforcement. *(After ACI 318, 2014, courtesy of American Concrete Institute.)*

anchoring one or more ties, the force in each tie must be developed at the point where the centroid of the reinforcement in the tie leaves the extended nodal zone and enters the span (Figures 7.20*c* and 7.25). Where a tie passes through a nodal zone, the nodal zone must develop the difference between the tie force on one side of the node and the tie force on the other side.

7.7.5 Nodal Zones

A strut-and-tie model idealizes the structure as a truss comprising struts and ties that intersect at nodes. In a planar structure, equilibrium requires at least three forces at each node. Nodes are classified according to the forces in the intersecting struts and ties. A C-C-C node resists compressive forces from three struts, a C-C-T node resists two struts and one tie, and so on. See Figure 7.26.

A nodal zone is the volume of concrete around a node that is assumed to transfer strut and tie forces through the node. A *hydrostatic nodal zone* is one that has loaded faces perpendicular to the axes of the struts and ties and has equal stresses on each face (nodal zones in Figure 7.20 are hydrostatic nodal zones). Setting up a strut-and-tie model to satisfy the requirements of a hydrostatic nodal zone can be tedious, and is not a requirement. Instead, it is more common to

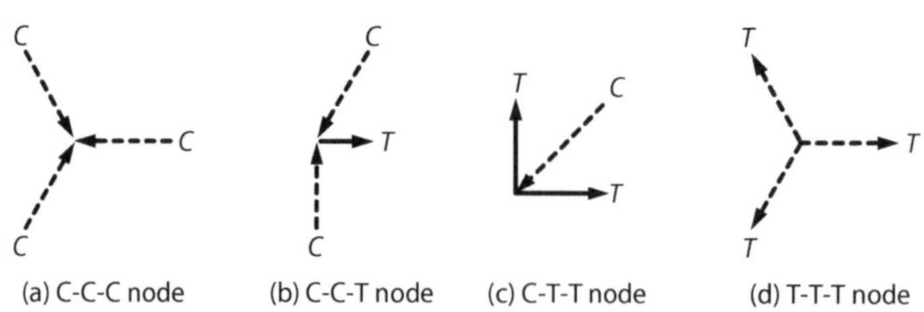

FIGURE 7.26 Classification of nodes.

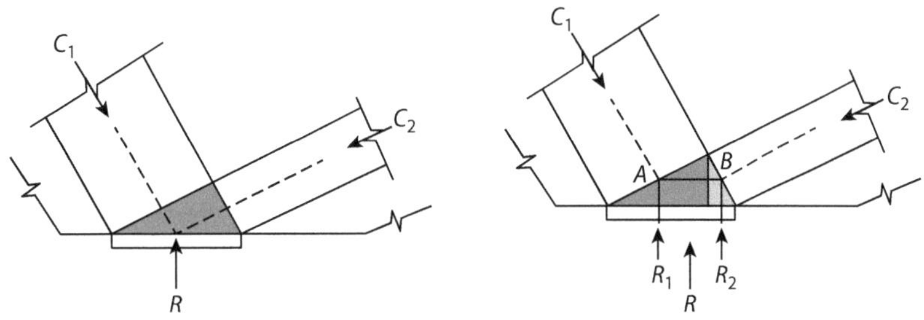

FIGURE 7.27 Subdivision of nodal zone. (*After ACI 318, 2014, courtesy of American Concrete Institute.*)

use an extended nodal zone, which is that portion of a member bounded by the intersection of effective strut and tie widths. In Figure 7.20*b* and *c*, the extended nodal zone includes both the darkly shaded and the lightly shaded volumes (i.e., the volumes marked as "nodal zone" and "extended nodal zone").

Nominal compressive strength of a nodal zone is

$$F_{nn} = f_{ce} A_{nz} \tag{7.17}$$

in which f_{ce} = effective compressive strength of the concrete in the nodal zone and A_{nz} = area of nodal zone on which f_{ce} is assumed to act. In Figure 7.27*a*, each of three faces of the nodal zone is perpendicular to the strut, tie, or reaction meeting at the node. In this case, the stress is checked on each of the three faces. Where the face of a nodal zone is not perpendicular to the strut, tie, or reaction (e.g., for the compression strut in Figure 7.25*b*), there is both normal stress and shear stress on the face. In this case, the stress is replaced by normal stress acting on the area A_{cs} of the strut.

Sometimes, calculations are easier if a nodal zone is subdivided, as in Figure 7.27*b*. Here the reaction is divided into R_1, which equilibrates the vertical component of C_1, and R_2, which equilibrates the vertical component of C_2. Note also that the subdivided nodal zones A and B have compressive force acting along line AB; stress acting on the vertical face where these two subdivided nodal zones intersect also has to be checked.

The effective compressive strength of concrete in a nodal zone is

$$f_{ce} = 0.85\beta_n f_c' \tag{7.18}$$

in which β_n is defined according to Table 7.2.

Case	Illustration	Description	β_n
C-C-C	Figure 7.26*a*	For a nodal zone bounded by struts or bearing areas, or both	1.0
C-C-T	Figure 7.26*b*	For a nodal zone anchoring one tie	0.80
C-T-T, T-T-T	Figure 7.26*c* and *d*	For a nodal zone anchoring two or more ties	0.60

TABLE 7.2 Values of β_n

7.8 Transverse Reinforcement Detailing

In the previous sections we have seen that transverse reinforcement in beams and other members acts as tension tie reinforcement. Details including anchorages need to be arranged such that the reinforcement can perform as intended. In beams, the transverse reinforcement equilibrates the vertical component of diagonal compression struts, and therefore needs to be distributed so as to effectively engage the full beam cross section. In a narrow beam (Figure 7.28*a*), two vertical stirrup legs may suffice to support the struts. As beam width increases (Figure 7.28*b*), additional vertical legs should be added to promote development of a more uniform diagonal compression field across the beam width. In flanged members, hooks on transverse reinforcement should be oriented to direct diagonal compression stresses into the beam web (Figure 7.28*c*) rather than away from it (Figure 7.28*d*). In more irregular shapes, such as I-beams, vertical legs of stirrups should be straight from the top of the beam to the bottom of the beam; bends in transverse reinforcement along the member depth should be avoided, because such bends result in directional changes in reinforcement tension and create bursting stresses in the concrete.

Building codes typically require that hooks engage longitudinal bars, both to stiffen the anchorage and to increase splitting resistance where the hooks bear on concrete. Precise prescriptions on placement of longitudinal bars, however, are lacking. Tests by Varney et al. (2011) suggest that exact placement is not critical for performance. Nonetheless, good detailing and construction practice achieves close tolerances with longitudinal bars held tightly within the bends of the transverse reinforcement.

Structural walls typically have relatively long, thin webs with boundary elements at the ends of the web where concentrations of longitudinal and transverse reinforcement are detailed. In special structural walls with special

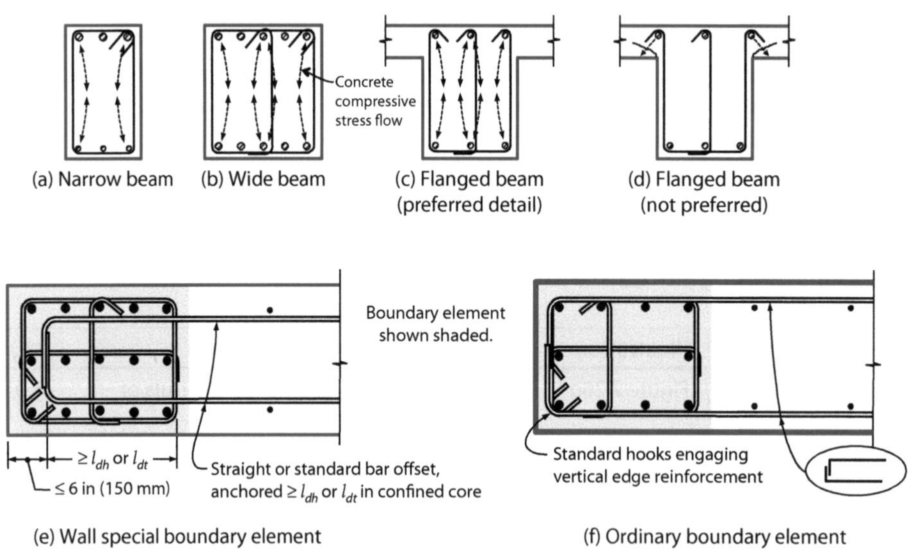

FIGURE 7.28 Transverse reinforcement details.

boundary elements (see Chapter 13), transverse reinforcement for shear should be effectively continuous from one edge of the wall to the opposite edge, and anchored within the confined boundary zone. In other structural walls, an accepted detail is to hook transverse reinforcement for shear around longitudinal bars at the wall edge. See Figure 7.28*e* and *f*.

7.9 Empirical Approach for Shear Strength of Beams and Columns

Advances in strut-and-tie models, especially since the 1980s, have greatly improved understanding of design requirements for shear, especially for D-regions where internal forces can be highly nonuniform. Prior to this development, designs were based largely on empirical expressions and ad hoc procedures. For B-regions, empirical expressions still provide an efficient and effective approach for design, and are used widely. This section presents test results demonstrating strength trends, followed by some of the empirical expressions and design requirements of ACI 318 for beam and column design.

7.9.1 Strength of Members without Transverse Reinforcement

Shear strength of beams without transverse reinforcement has been studied through hundreds of laboratory tests (ACI-ASCE 326, 1962; ACI 445, 1999; ASCE-ACI 426, 1973). From these tests we have gained an understanding of the mechanisms of shear resistance and the primary variables affecting shear strength.

We begin by considering a series of beam tests reported by Leonhardt (1962). The beams were simply supported and subjected to two point loads (Figure 7.29).

In the Leonhardt tests, behavior was strongly influenced by shear span ratio a/d, where a is span from point load to reaction and d is beam effective depth.

- Beams with largest a/d had shear strength exceeding the shear corresponding to flexural yielding (beams 10/1 and 9/1). They developed cracks mainly perpendicular to the beam longitudinal axis, with slight crack inclination due to shear. Failure was by crushing of the flexural compression zone, without shear failure.
- Beams with $2.5 \le a/d \le 6$ sustained shear failures associated with steeply inclined cracks (beams 8/1, 7/1, 6, 5, and 4). Initially, the beams were stable owing primarily to shear transfer through aggregate interlock across inclined cracks. Further loading increased crack opening, reducing aggregate interlock, and leading to *diagonal tension failure*. Failure is characterized by steeply inclined cracks extending along the bottom longitudinal bars, which are sheared off the bottom of the beam. Such failures can be especially brittle. The primary inclined cracks also may penetrate the flexural compression zone, leading to compression zone failure.
- Beams with $a/d < 2.5$ developed a diagonal compression strut that supported the concentrated loads through arch action (beams 1, 2, and 3). Failure can be by splitting of the diagonal compression strut, as occurs in a split cylinder test, or by failure of the compression chord at the top of the beam.

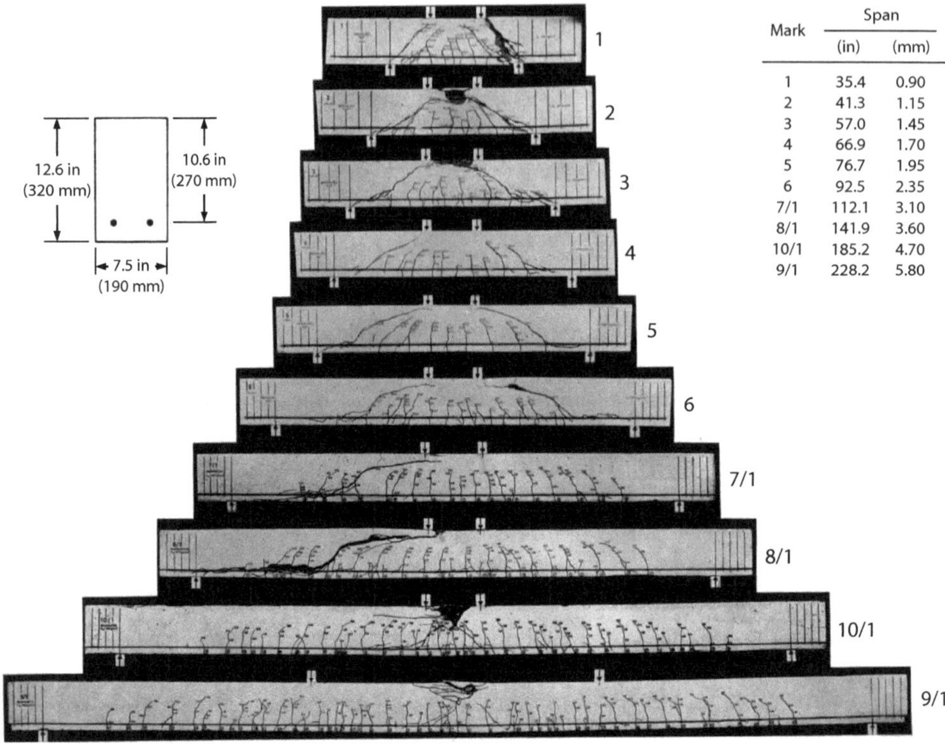

Mark	Span (in)	Span (mm)	$\frac{a}{d}$
1	35.4	0.90	1.0
2	41.3	1.15	1.5
3	57.0	1.45	2.0
4	66.9	1.70	2.5
5	76.7	1.95	3.0
6	92.5	2.35	4.0
7/1	112.1	3.10	5.0
8/1	141.9	3.60	6.0
10/1	185.2	4.70	8.0
9/1	228.2	5.80	7.0

Figure 7.29 Damage at failure in series of tests on beams without transverse reinforcement. (*After Leonhardt, 1962.*)

The different force-resisting mechanisms in beams with different aspect ratios result in different beam strengths. Figure 7.30 plots measured shear strength as function of aspect ratio a/d for the beams shown in Figure 7.29. Beams with aspect ratio $a/d = 7$ and 8 developed flexural strength prior to shear failure. For smaller a/d, the shear force corresponding to flexural failure increases (broken curve in Figure 7.30), such that shear failure occurred before flexural strength was reached. Shear strength was nearly constant for a/d to around 3, but increased rapidly for smaller a/d. Apparently, the arch mechanism that develops for $a/d \leq 2.5$ results in shear strength significantly exceeding the strength observed for larger values of a/d.

For a cracked reinforced concrete flexural member, flexural stresses and, hence, shear flow act over section depth extending from the extreme flexural compression fiber to the flexural tension reinforcement at depth d. Thus, it is conventional to define average nominal shear stress as $v = V/b_w d$, in which b_w is section width (or web width for flanged members) and d is effective depth to centroid of flexural tension reinforcement.

Studies show that nominal shear strength of beams increases with increasing concrete compressive strength f'_c, increasing longitudinal reinforcement ratio ρ_w, and decreasing a/d. The term a/d is applicable only to simply supported beams subjected to concentrated loads. To be applicable to more general loading cases,

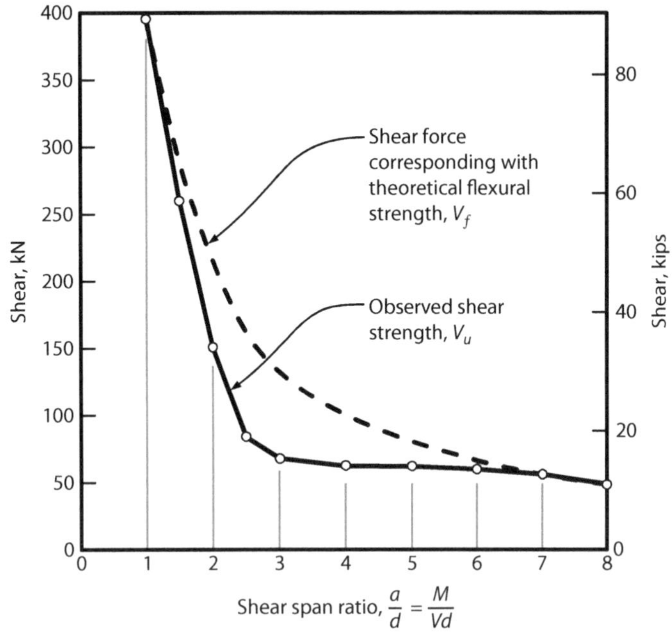

FIGURE 7.30 Shear at failure as function of aspect ratio for beams without transverse reinforcement. (*After Park and Paulay, 1975.*)

the term M/Vd can be substituted for a/d. Figure 7.31 plots nominal shear strength of beams without transverse reinforcement using a functional form presented by ACI-ASCE 326 (1962) as follows:

$$\begin{aligned} v_c &= 1.9\sqrt{f_c'} + 2500\rho_w Vd/M \le 3.5\sqrt{f_c'}, \text{psi} \\ &(0.16\sqrt{f_c'} + 17\rho_w Vd/M \le 0.29\sqrt{f_c'}, \text{MPa}) \end{aligned} \tag{7.19}$$

ACI ITG-4.3 (2007) reports data showing no bias in Eq. (7.19) up to compressive strength of 16,000 psi (110 MPa).

For beams without shear reinforcement, nominal shear stress capacity decreases as beam depth increases (Figure 7.32). One hypothesis is that deeper beams develop wider cracks, leading to reduction of unit shear strength. Fortunately, this effect is less pronounced in beams with transverse reinforcement. The effect for thick footings and foundation mats, where transverse reinforcement commonly is not used, is unknown.

Axial stress also affects shear strength. The effect can be understood by considering a planar element in the x-y plane subjected to normal stresses σ_x and σ_y (taken positive in tension) and shear stress τ_{xy}. From equilibrium, the principal stresses are

$$\sigma_1, \sigma_2 = \frac{\sigma_x + \sigma_y}{2} \pm \sqrt{\left(\frac{\sigma_x - \sigma_y}{2}\right)^2 + \tau_{xy}^2} \tag{7.20}$$

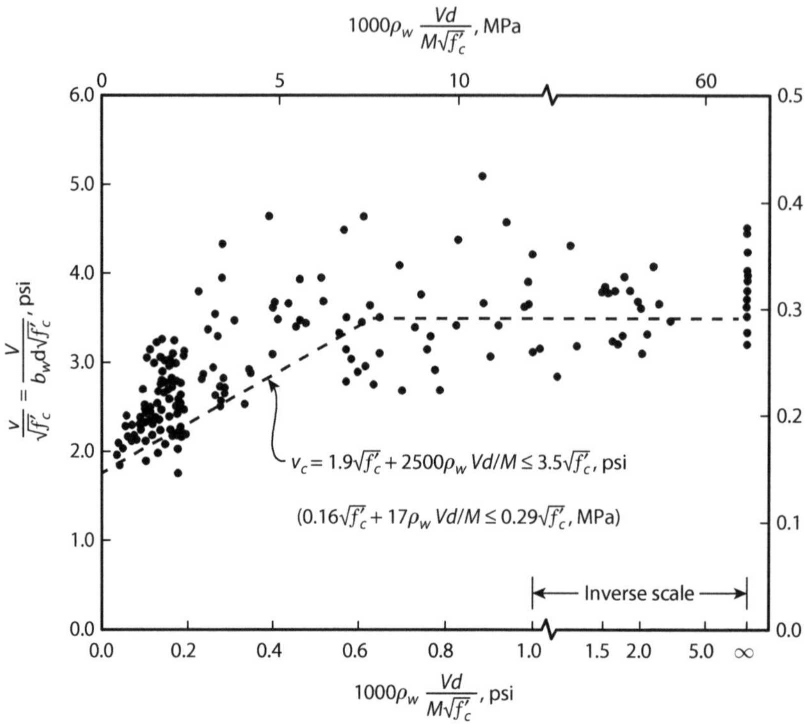

FIGURE 7.31 Nominal shear strength of beams without transverse reinforcement. (*After ACI-ASCE 326, 1962, courtesy of American Concrete Institute.*)

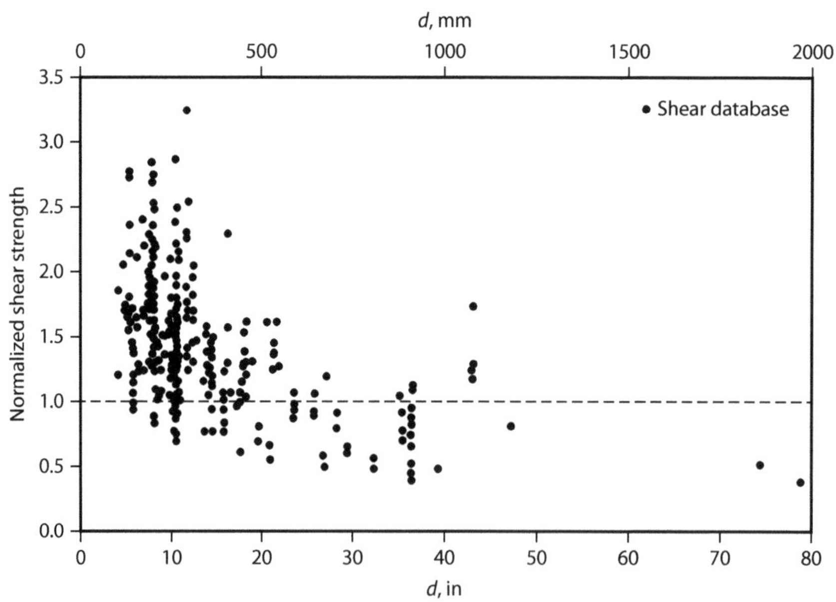

FIGURE 7.32 Effect of depth on strength of beams without transverse reinforcement. (*After Sneed and Ramirez, 2010, courtesy of American Concrete Institute.*)

Setting $\sigma_x = 0$ (a reasonable approximation for an axially loaded column), and solving for τ_{xy} in terms of σ_y and principal tensile stress σ_1 results in

$$\tau_{xy} = \sigma_1\sqrt{1 - \frac{\sigma_y}{\sigma_1}} \tag{7.21}$$

Onset of cracking in plain concrete is expected when σ_1 = concrete tensile strength f_t. Substituting into Eq. (7.21), we obtain

$$\tau_{xy} = f_t\sqrt{1 - \frac{\sigma_y}{f_t}} \tag{7.22}$$

Note that in Eq. (7.22) f_t is concrete tensile strength and σ_y is axial stress, both taken positive in tension. Thus, Eq. (7.22) indicates that shear stress at onset of cracking will increase under axial compression and decrease under axial tension. Figure 7.33 shows this effect as measured in laboratory tests.

7.9.2 Members with Transverse Reinforcement

Early truss models assumed compression diagonals inclined at 45° with respect to the beam longitudinal axis. With this assumption, and the approximation that $jd \approx d$, Eq. (7.8) simplifies to

$$V_s = \Sigma A_v f_{yt} = \frac{A_v f_{yt} d}{s} \tag{7.23}$$

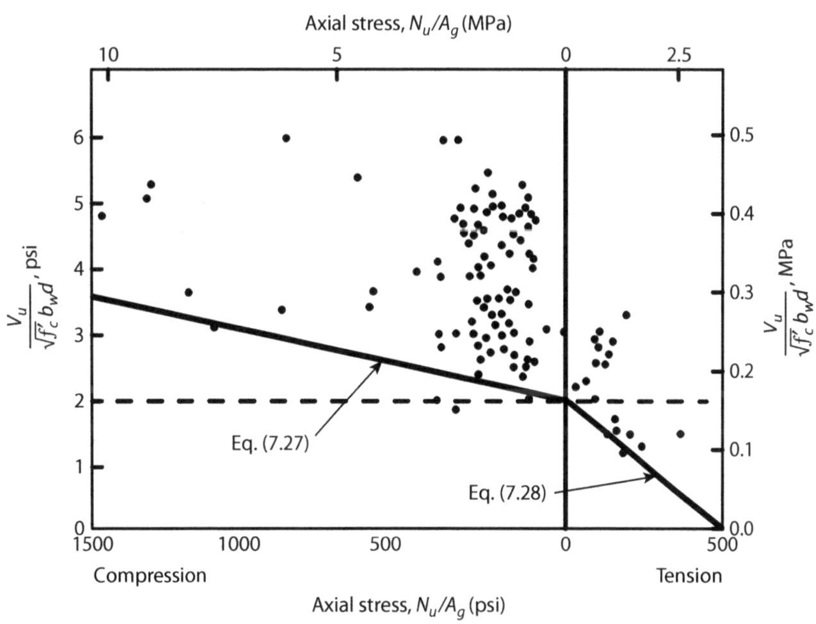

FIGURE 7.33 Effect of axial stress on shear strength of reinforced concrete members without transverse reinforcement. (*After Wight and MacGregor, 2012.*)

in which shear resisted by the transverse reinforcement is now denoted by V_s. Tests by Talbot (1909) confirmed that Eq. (7.23) was conservative compared with test results.

Given the conservatism of Eq. (7.23), and the observation that concrete beams resist shear without transverse reinforcement, a common approach has been to express shear strength as the sum of concrete and transverse reinforcement components, as in

$$V_n = V_c + V_s \tag{7.24}$$

in which $V_c = v_c b_w d$, with v_c defined by Eq. (7.19) and V_s is defined by Eq. (7.23). Figure 7.34 compares results of Eq. (7.24) with test data. In Figure 7.34, $v_s = V_s/b_w d$.

7.9.3 ACI 318 Design Equations and Requirements for Beams and Columns

ACI 318 contains design provisions for beams and columns that follow from the findings of the preceding sections. The strength design requirement is expressed by

$$\phi V_n \geq V_u \tag{7.25}$$

in which V_n = nominal strength, V_u = required strength based on load combinations defined in Chapter 1, and $\phi = 0.75$. (For diagonally reinforced coupling beams, an alternative design approach is used, including an alternative value for the strength reduction factor ϕ. See Section 7.11.)

Nominal shear strength V_n is defined by Eq. (7.24). For members without axial force, Eq. (7.19) is the basic expression for shear strength contribution of

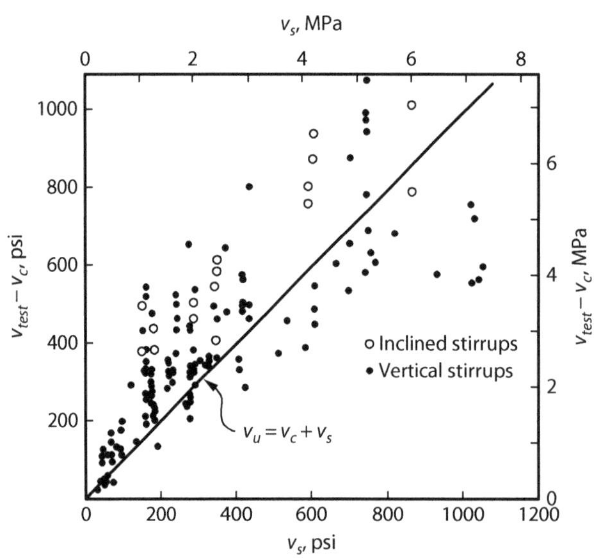

Figure 7.34 Comparison of measured and calculated strengths as function of v_s provided. *(After ACI-ASCE 326, 1962, courtesy of American Concrete Institute.)*

concrete. For most practical designs, however, design is based on the simpler expression

$$V_c = 2\lambda\sqrt{f_c'}b_w d, \text{ psi} \quad (0.17\lambda\sqrt{f_c'}b_w d, \text{MPa}) \tag{7.26}$$

For members subject to axial compression

$$V_c = 2\left(1+\frac{N_u}{2000A_g}\right)\lambda\sqrt{f_c'}b_w d, \text{ psi} \quad \left(0.17\left(1+\frac{N_u}{14A_g}\right)\lambda\sqrt{f_c'}b_w d, \text{MPa}\right) \tag{7.27}$$

where N_u is positive for compression.

For members subject to axial tension

$$V_c = 2\left(1+\frac{N_u}{500A_g}\right)\lambda\sqrt{f_c'}\,b_w d \geq 0, \text{ psi} \quad \left(0.17\left(1+\frac{0.29N_u}{A_g}\right)\lambda\sqrt{f_c'}\,b_w d \geq 0, \text{MPa}\right) \tag{7.28}$$

where N_u is negative for tension.

For high-strength concrete, shear cracks may propagate through the aggregates, leaving a smoother surface that is less capable of transferring shear through aggregate interlock (ACI 363, 2010). For this reason, $\sqrt{f_c'}$ is not permitted to exceed 100 psi (8.3 MPa) except higher values are permitted where transverse reinforcement exceeds minimum quantities discussed later in this section.

For circular members, the area used to compute V_c can be taken as the product of the diameter and effective depth of the concrete section. Alternatively, it is permitted to take d as 0.80 times the diameter of the concrete section. Although not specifically stated in ACI 318, it is common to take effective shear area for rectangular columns as $0.8A_g$. Use of $0.8A_g$ for rectangular columns is recognized by ASCE 41 (2013).

Shear strength provided by transverse reinforcement oriented perpendicular to the longitudinal axis is given by Eq. (7.23). Because of concerns about crack width under service loads, the value of f_{yt} used in design is not permitted to exceed 60,000 psi (414 MPa) for conventional reinforcement and 80,000 psi (552 MPa) for welded deformed wire reinforcement. Tests (Budek et al., 2002) have shown that higher strength transverse reinforcement is effective where crack width control is not a concern.

Based on the derivation by Ang et al. (1989), Caltrans (2010) expresses strength provided by spiral or circular-hoop reinforcement as

$$V_s = \frac{\pi}{2}\frac{A_{sp}f_{yt}D'}{s} \tag{7.29}$$

ACI 318 does not use this expression for circular sections, but instead uses Eq. (7.23). Collins et al. (2002) present data demonstrating the ACI 318 approach is appropriately conservative.

Transverse reinforcement in members of rectangular cross section should have straight legs that extend from extreme compressive fiber to extreme tensile fiber, with allowance for required cover, and should be developed at both ends. Circular hoops or spirals also are acceptable as shear reinforcement.

To ensure that inclined cracks cross multiple legs of transverse reinforcement, spacing is not permitted to exceed the smaller of $d/2$ and 24 in (610 mm). Where V_s exceeds $4\sqrt{f_c'}b_w d$, psi ($0.33\sqrt{f_c'}b_w d$, MPa), these values should be reduced by half. V_s is not permitted to exceed $8\sqrt{f_c'}b_w d$, psi ($0.67\sqrt{f_c'}b_w d$, MPa).

Where $V_u \geq 0.5\phi V_c$, transverse reinforcement is required to provide at least $A_{v,min}$.

$$A_{v,min} = 0.75\sqrt{f_c'}\frac{b_w s}{f_{yt}} \geq 50\frac{b_w s}{f_{yt}}, \text{ psi}$$
$$\left(0.062\sqrt{f_c'}\frac{b_w s}{f_{yt}} \geq 0.35\frac{b_w s}{f_{yt}}, \text{ MPa}\right) \tag{7.30}$$

Exceptions are made for footings and solid slabs; certain hollow-core units; concrete joist construction; beams with h not greater than 10 in (254 mm); beams integral with slabs with h not greater than 24 in (610 mm) and not greater than the larger of 2.5 times thickness of flange and 0.5 times width of web; and certain beams constructed of steel fiber-reinforced concrete.

7.9.4 Comparison of ACI 318 and Truss Models

For B-regions of beams it is permissible to use either the ACI 318 code empirical equations or the truss model for design. It is of interest to compare results of the two methods. Figure 7.35 compares results for a particular beam cross section using 4000 psi (28 MPa) normalweight aggregate concrete. The truss model is solved for shallow angle of diagonal compression struts (25°), which generally produces the most efficient design. For the typical range of interest, the two models produce similar results. For higher transverse reinforcement ratios, the truss model results in higher nominal strength than the ACI 318 empirical equations.

Returning to the angle of the diagonal compression struts, the designer has some flexibility in the selected angle. In general, however, the angle should be in the range 25° to 65°. Smaller angles result in more efficient use of the transverse

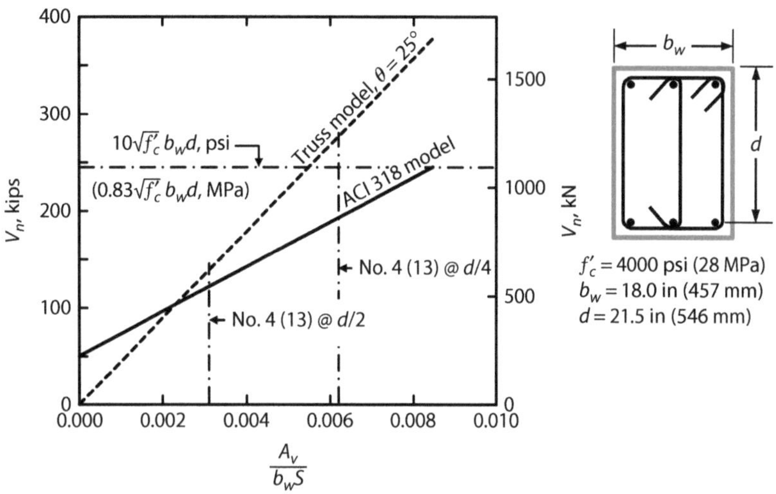

FIGURE 7.35 Comparison of ACI 318 empirical equations and truss model for shear design.

reinforcement, and therefore a smaller angle is usually selected for design when using the strut-and-tie method. Note, however, that a smaller strut angle results in a greater tension shift for the longitudinal reinforcement, and therefore will require longer bar extensions than are required for steeper strut angles.

7.10 Effects of Inelastic Cyclic Loading

Yielding of slender flexural members can lead to degradation of the shear-resisting mechanisms, and can lead to eventual shear failure. For this reason, it usually is good design practice to maintain a margin against shear failure in slender flexural members. In this section we discuss some of the effects that contribute to shear strength degradation and present some models that consider the effects.

Consider a cantilever beam idealized using truss models (Figure 7.36). Note that two different truss models are required, one for loading in the downward direction and another for loading upward. If the member is proportioned with insufficient transverse reinforcement, that reinforcement will yield in tension for loading in each direction. Thus, tensile strains in transverse reinforcement will accumulate with each loading cycle, leading to dilation of the member with continued cycling (Figure 7.36*d*). This is in contrast with flexural behavior at reversing plastic hinges where the longitudinal reinforcement is subjected to alternating tensile and compressive forces due to flexure (see Section 6.12).

In cases where shear-yielding dominates, reversed cyclic loading will produce a set of crisscrossed diagonal cracks that effectively subdivide the member into a series of concrete segments separated by inclined cracks (Figures 7.36*d* and 7.37). For shear applied in one direction, corresponding diagonal compression struts must form, closing any cracks that cross the struts. Shear force reversal requires

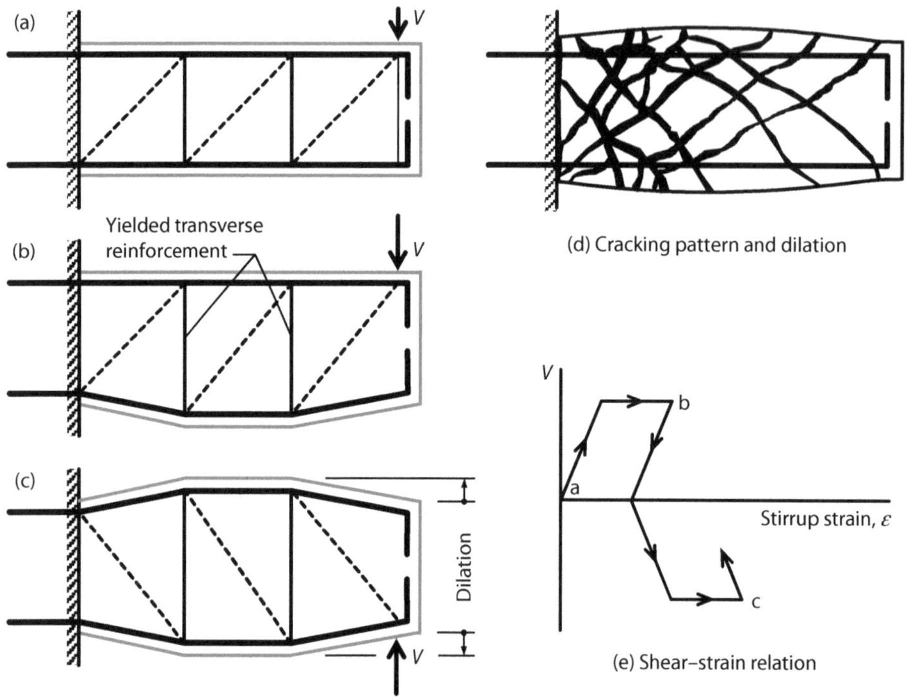

FIGURE 7.36 Idealized behavior for yielding transverse reinforcement.

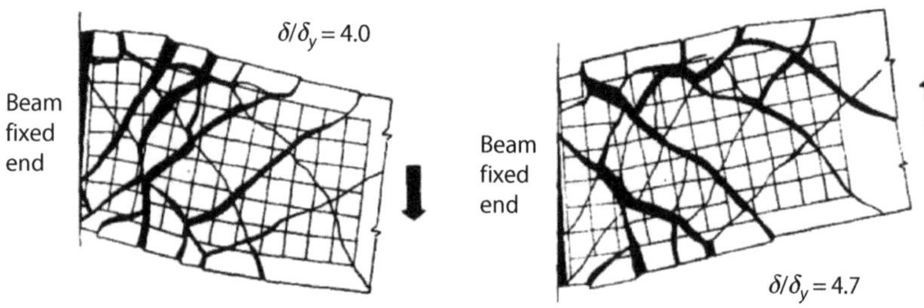

FIGURE 7.37 Surface deformations and cracks of beam in advanced stages of cyclic loading. Crack widths and deformations are amplified by factor 5.0. (*After Popov, 1984, courtesy of American Concrete Institute.*)

that diagonal compression struts develop in the opposing direction. This requires the segments to shift orientation, a process that usually occurs with limited shear resistance. Significant lateral force resistance can be achieved only after the segments have shifted to a position that enables formation of diagonal compression struts effective for the loading direction. In a member with well-developed set of diagonal cracks, considerable lateral displacement may be required to open and close the opposing cracks. This can result in "pinched" hysteresis loops in shear-damaged members.

Figure 7.38 illustrates shear–displacement relations measured in laboratory tests of two fixed-ended columns. Figure 7.38*a* shows response dominated by flexural behavior. For this condition, shear–displacement loops are relatively full without strength degradation to large displacements. Figure 7.38*b* shows results for a nearly identical column proportioned so that shear failure occurs shortly following flexural yielding. Rapid strength loss associated with shear failure is apparent. Additionally, the shear–displacement loops are "pinched" for reasons described in the preceding paragraph.

Although traditional design methods treat shear and flexure as independent mechanisms, truss models suggest instead that they are coupled. Members with low axial loads gradually elongate as a result of accumulated plastic strain in longitudinal reinforcement (Figure 7.39*a*). This elongation leads to rotation of inclined compression struts, which reduces shear resistance mechanisms. Longitudinal strain also

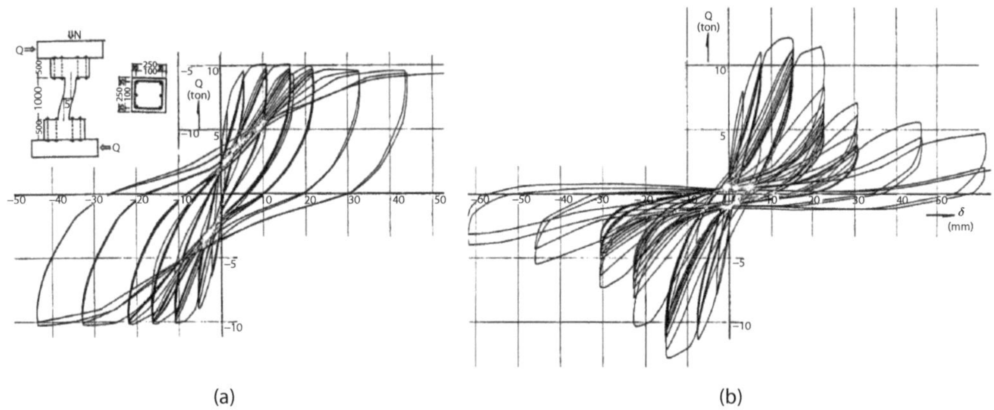

Figure 7.38 Shear–displacement relations for (*a*) flexure-controlled column and (*b*) shear-controlled column. (*After Higashi and Hirosawa, 1974.*)

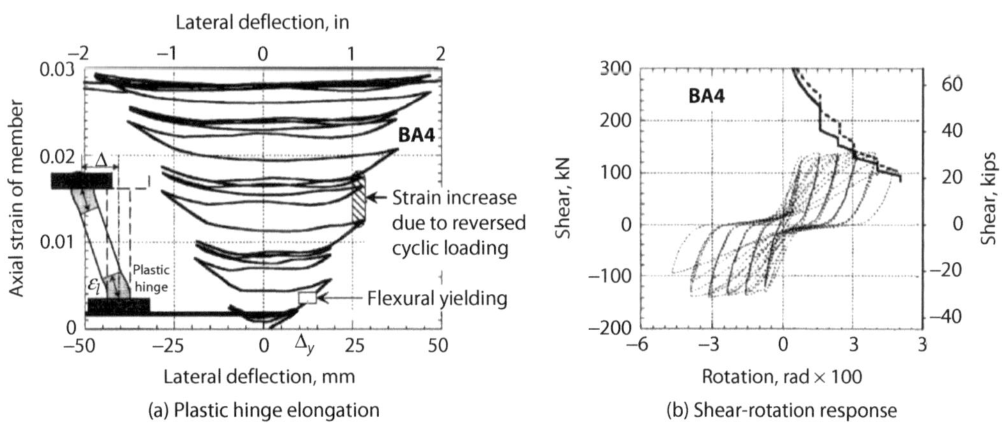

Figure 7.39 Elongation of flexural plastic hinge with cycling. (*After Lee and Watanabe, 2003, courtesy of American Concrete Institute.*)

may reduce strength of diagonal compression struts, similar to the strength reduction noted in Section 3.6.2, further contributing to shear strength degradation. Lee and Watanabe (2003) have presented empirical models that show good correlation between inelastic flexure, axial elongation, and shear strength degradation (Figure 7.39*b*). Degradation effects may be less pronounced for members with moderate axial loads because axial elongation is suppressed in such members. Effects of axial force above the balanced point are not extensively studied, though it seems likely that crushing of the flexural compression zone would reduce diagonal compression strength and, hence, result in accelerated shear failure. This effect was evident in tests reported by Sezen and Moehle (2006).

Several empirical models for shear strength degradation of flexural members have been developed. Most models express shear strength as $V_n = V_c + V_s$ or $V_n = V_c + V_p + V_s$, in which V_c represents contribution of concrete, V_p represents contribution of axial force, and V_s represents contribution of transverse reinforcement. In most of these models, based on an interpretation that concrete diagonal compressive strength and aggregate interlock are degrading, only the concrete contribution V_c degrades with increasing flexural deformation demand (ATC 12, 1983; Ang et al., 1989; Aschheim and Moehle, 1992; Ichinose, 1992; Priestley et al., 1994; Lee and Watanabe, 2003; Biskinis et al., 2004). Other models degrade both V_c and V_s, based on an interpretation that the overall shear-resisting mechanism is degrading (Biskinis et al., 2004; Sezen and Moehle, 2004).

ASCE 41 for seismic rehabilitation of buildings adopts the model by Sezen and Moehle (2004). According to this model, column shear strength is defined by

$$V_n = k\left(V_s + V_c\right) \tag{7.31}$$

in which V_s is defined by Eq. (7.23) and V_c is defined by

$$V_c = \left(\frac{6\lambda\sqrt{f_c'}}{M/Vd}\sqrt{1+\frac{N_u}{6\sqrt{f_c'}A_g}}\right)0.8A_g,\ \text{psi}$$

$$\left(\frac{0.5\lambda\sqrt{f_c'}}{M/Vd}\sqrt{1+\frac{N_u}{0.5\sqrt{f_c'}A_g}}\right)0.8A_g,\ \text{MPa} \tag{7.32}$$

in which $\lambda = 0.75$ for lightweight aggregate concrete and 1.0 for normalweight aggregate concrete; N_u = axial compressive force, taken equal to 0 for axial tension; M/Vd = largest ratio of moment to shear times effective depth under design loadings for the column but not less than 2, nor greater than 4; and it is permitted to take $d = 0.8h$. In Eq. (7.32), k is a parameter to represent shear strength degradation, defined by Figure 7.40*a*. Figure 7.40*b* compares test and calculated shear strengths according to Eq. (7.31), with an emphasis on older-type columns with relatively wide hoop spacing.

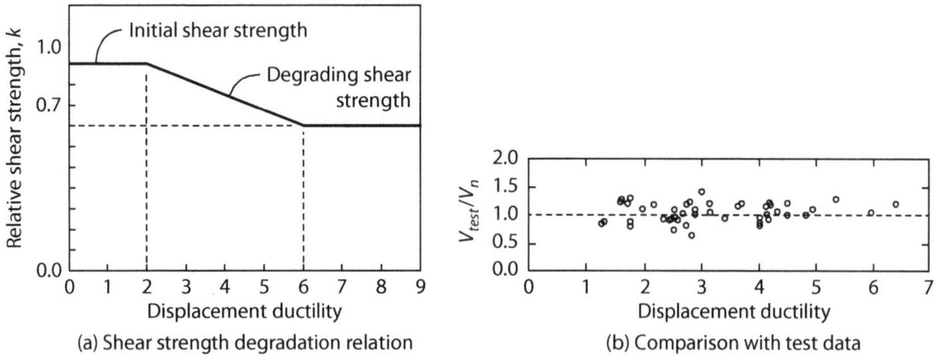

FIGURE 7.40 Shear strength degradation model. (*After Sezen and Moehle, 2004, courtesy of ASCE.*)

Elwood and Moehle (2005b) and Pujol et al. (2006) present models for deformation at onset of shear failure following flexural yielding. These are not pursued further here.

Building configuration sometimes results in lateral shears oriented primarily along one principal axis of a beam or column. More typically, however, column shears under earthquake loading involve bi-directional lateral loading. Woodward and Jirsa (1984), Maruyama et al. (1984), and Umehara and Jirsa (1984) present results of a series of column tests under multi-directional loading. The principal findings were as follows:

- Strength in any direction was not affected by previous loading unless previous loading had been to deflection that would have resulted in development of shear strength under unidirectional loading. Previous loading did, however, result in stiffness reduction.
- Shear strength under bi-directional loading could be estimated by circular interaction for square or circular columns or by elliptical interaction for rectangular columns.
- Complex loading histories involving loading in multiple directions do not much affect maximum shear strength but do result in accelerated strength deterioration following initiation of failure.
- Strength degradation was more rapid for higher axial loads. Behavior under alternating tensile and compressive axial load was similar to behavior under the maximum compressive axial load.

7.11 Diagonally Reinforced Beams

Flexural members in reinforced concrete frames usually have aspect ratio $l_n/h \geq 4$. Such members readily can be designed so that inelastic response is predominantly in a ductile flexural mode. For members with smaller aspect ratio, it can be more challenging to achieve ductile flexural response using conventional reinforcement details. Alternative reinforcement arrangements involving inclined reinforcement can be more effective for such members. This type of detail is common for coupling beams of special structural walls (see Chapter 13), but they also find application in other structural members of low aspect ratio.

Figure 7.41 illustrates a low-aspect-ratio coupling beam with equal but opposite moments applied at the ends. For a beam with conventional longitudinal reinforcement (Figure 7.41*a*), inclined cracking results in significant tension shift; for such low aspect ratios it is not unusual for both top and bottom longitudinal reinforcement to be in tension across the full beam span. Because both top and bottom bars are in tension, equilibrium of the cross section at the beam ends requires deeper flexural compression zone than would be calculated by the usual assumptions involving plane sections, and this in turn leads to reduced moment strength. More importantly, longitudinal reinforcement may be yielding in tension for loading in either direction, such that tensile strains accumulate, the beam elongates, and cracks open progressively with cycling. Shear transfer across open cracks relies on roughness of the crack surface, but this degrades as cracks open and as the interface grinds back and forth (Paulay and Binney, 1974). Longitudinal bars crossing cracks act as dowel bars, providing limited shear resistance until kinking of the bars brings the adjacent crack surfaces in contact, at which point shear resistance is restored at least partially. The result can be a severely pinched load-deformation behavior. Figure 7.42*a* illustrates the pinched hysteresis observed in tests reported in Barney et al. (1978).

Recognizing potential performance deficiencies of conventionally reinforced deep coupling beams, Paulay and Binney (1974) introduced groups of diagonally

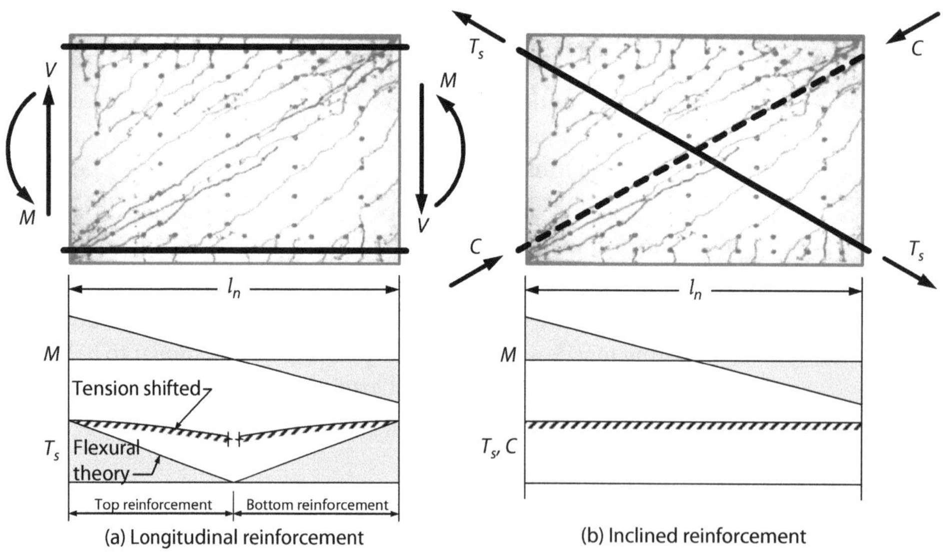

FIGURE 7.41 Idealized internal forces in beams with (*a*) longitudinal reinforcement and (*b*) inclined reinforcement. (*Inclined crack pattern from Paulay, 1974.*)

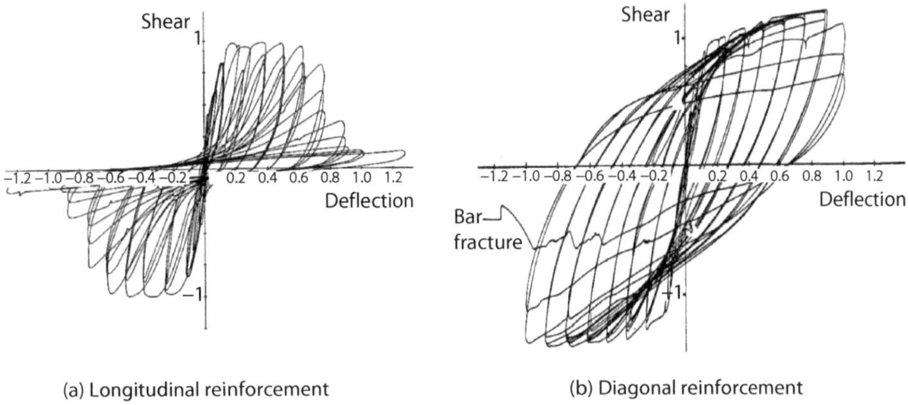

FIGURE 7.42 Normalized shear-deflection relations of beams with aspect ratio $l_n/h = 2.5$. (*After Barney et al., 1978.*)

oriented bars crossing at midspan. The aim was to provide an entirely different internal resisting system simulating behavior of an x-braced frame. As shown in Figure 7.41*b*, one set of diagonal bars is placed to resist compressive force C along the principal corner-to-corner diagonal compression strut and another set of diagonal bars is placed to resist tensile force T_s along the diagonal between the other corners. Axial forces in individual diagonals alternate from tensile to compressive as the beam is subjected to deformation reversals, such that plastic strains are fully reversed under reversed-cyclic loading. Additionally, the direct force path in tension and compression results in more stable force resistance. Figure 7.42*b* illustrates the full hysteresis loops commonly observed for diagonally reinforced beams.

A diagonally reinforced coupling beam can be idealized as a truss with tension and compression diagonals along the axes of the diagonally placed reinforcement (Figure 7.43). Vertical equilibrium of the truss defines shear strength V_n as

$$V_n = 2A_{vd} f_y \sin\alpha \tag{7.33}$$

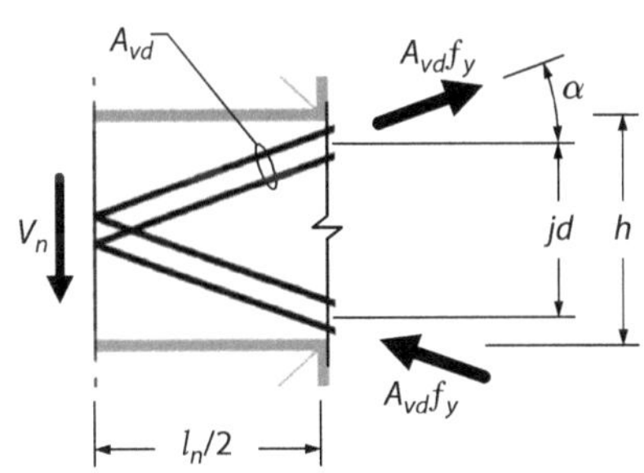

FIGURE 7.43 Free-body diagram of half-span of diagonally reinforced coupling beam. Gravity loads not shown.

According to ACI 318, nominal shear strength given by Eq. (7.33) is not to exceed $10\sqrt{f'_c}A_{cw}$, psi ($0.83\sqrt{f'_c}A_{cw}$, MPa), based on concerns for web crushing for higher nominal shear stresses.

The strength design requirement for diagonally reinforced coupling beams is given by Eq. (7.25). Available test data support the use of strength reduction factor $\phi = 0.85$ in ACI 318, rather than the value 0.75 used for conventionally reinforced beams.

An important aspect of a diagonally reinforced beam is that shear and moment are resisted by the same mechanism (the diagonal truss) such that shear and moment resistance are always in balance and capacity design requirements for shear are satisfied automatically. We can demonstrate this by comparing theoretical shear and moment strengths at the wall interface. Nominal shear strength is given by Eq. (7.33). Nominal moment strength is obtained by taking moments of horizontal components of the diagonal bar force $A_{vd}f_y$, resulting in

$$M_n = A_{vd} f_y jd \cos\alpha \tag{7.34}$$

From geometry of the coupling beam, $l_n = jd/\tan\alpha$. Equilibrium of the free-body diagram in Figure 7.43 requires $V_n = M_n/(l_n/2)$. Substituting the relations for M_n and l_n results in $V_n = (A_{vd} f_y jd \cos\alpha)(2\tan\alpha/jd) = 2A_{vd} f_y \sin\alpha$, which is the same result expressed in Eq. (7.33). Importantly, this means the shear strength provided by the diagonal bars is always exactly in equilibrium with the moment demand that they produce. Additional strength is provided by the concrete gross cross section and by any transverse reinforcement around the beam core, ensuring that shear failure by diagonal tension failure will not occur.

In U.S. practice, two options for transverse reinforcement are common, one in which individual diagonals are confined and another in which the entire beam cross section is confined (Figure 7.44). For additional discussion on required details and construction considerations, see Chapter 13.

Naish et al. (2012) report tests comparing performance of diagonally reinforced beams using the two alternative details. Strength and deformation capacity were slightly larger for the option using full beam section confinement (Figure 7.45). Coupling beams had $l_n/h = 2.4$ to 3.33, which is in the typical range for current tall building construction. For these aspect ratios, most damage was concentrated at the beam–wall interface in the form of slip of diagonal reinforcement from adjacent walls. Beams not detailed with full section confinement experienced more damage at large rotations. The test program also included beams cast monolithically with conventionally reinforced slabs and unbonded post-tensioned slabs. Conventionally reinforced slabs increased beam strength 15% to 20%, while post-tensioning increased strength an additional 10%. Fortney et al. (2008) report a test of a diagonally reinforced beam in which groups of diagonal bars were individually confined along their free length and partial core confinement was used along the length where the bars intersected. The beam reached chord rotation of 0.10 radians without apparent degradation.

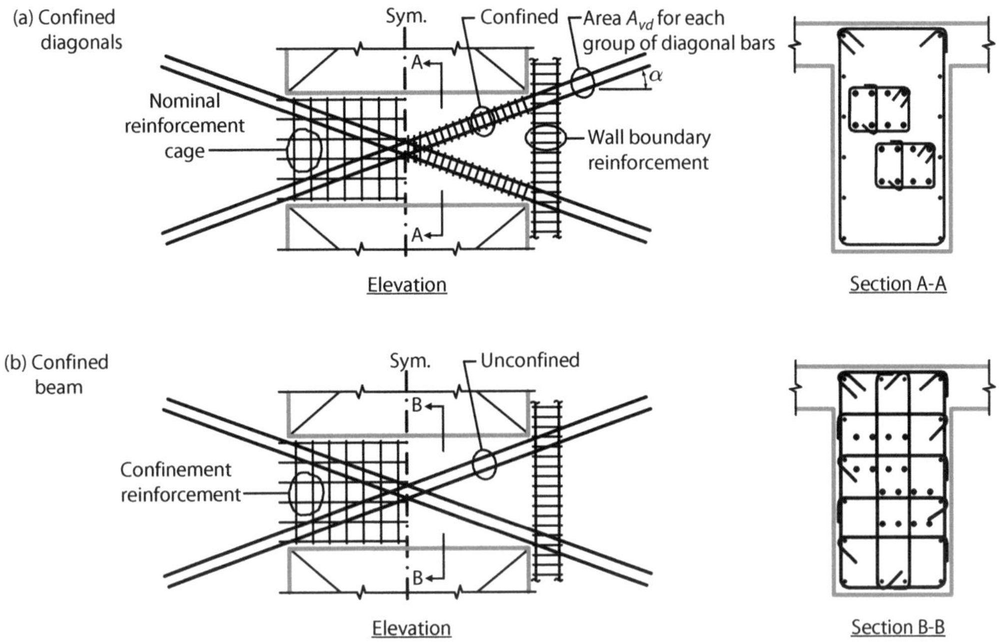

FIGURE 7.44 Reinforcement options for diagonally reinforced beams.

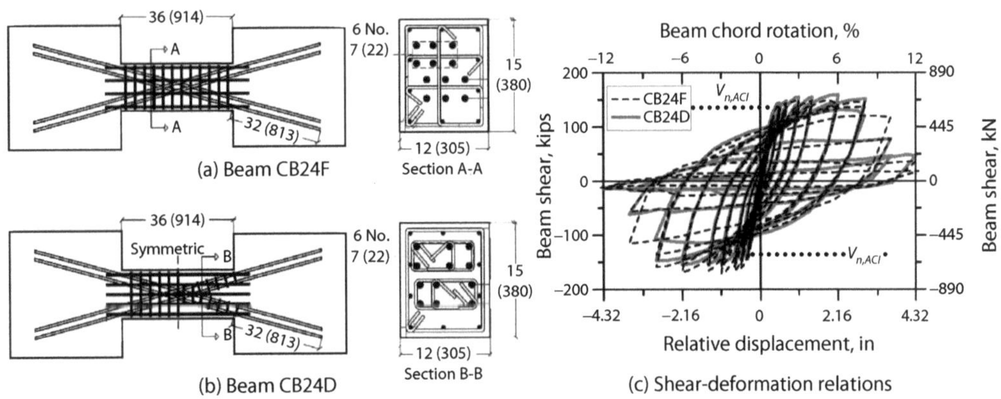

FIGURE 7.45 Shear-deformation relations for beams with diagonal group confinement and full section confinement. Dimensions in inches (millimeter). (*After Naish et al., 2012, courtesy of American Concrete Institute.*)

Various alternative reinforcement details involving diagonal reinforcement have been proposed. One hybrid solution includes both diagonal and conventional horizontal reinforcement (Figure 7.46*a*). For this solution, the diagonal reinforcement provides balanced shear and moment resistance; additional transverse reinforcement is required along the beam span to provide shear strength necessary to resist M_{pr} of the conventional (horizontal) flexural reinforcement.

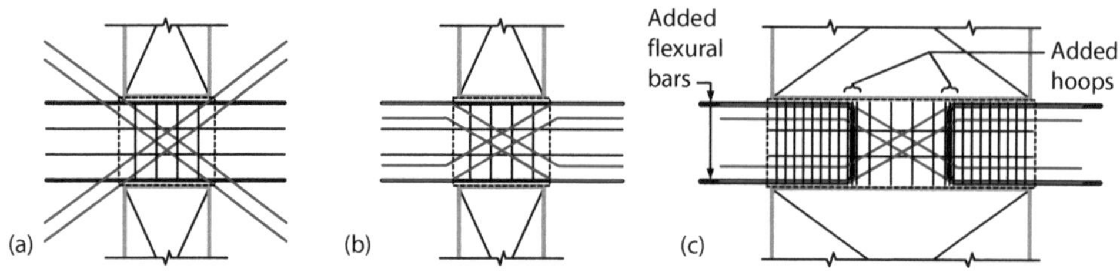

FIGURE 7.46 Alternative diagonal reinforcement details. Dotted lines show possible outlines for precast components.

In a similar approach, Canbolat et al. (2005) and Lequesne et al. (2011) used high-performance fiber-reinforced concrete (HPFRC) to resist shear and confine the concrete core. This permitted reduction of required diagonal reinforcement and required addition of horizontal flexural bars to balance the shear resisted by the HPFRC. One detail used conventional diagonal bars as in Figure 7.46*a*. Because the amount of diagonal reinforcement was significantly reduced, it also was possible to use smaller diameter bars and bend them to horizontal (Figure 7.46*b*). This detail has potential to simplify construction, as it avoids the difficulty of threading diagonal bars into adjacent framing members or wall boundaries. It also opens the possibility of precasting the coupling beam; precast concrete portions of the beams are suggested by the broken lines in Figure 7.46.

Figure 7.46*c* shows a concept where diagonal reinforcement placed in the center of the span is intended to act as a ductile "shear fuse." This detail requires added flexural bars in the end regions, shown here as U bars, to increase beam flexural strength in the end regions such that the midspan is the weak section. Under reversed cyclic deformations, the bend in the diagonal bars creates alternating bearing and bursting forces on the concrete. Added hoops are required to resist these forces. It is recommended that the ties be designed to resist in the linear range a force equal to the vertical component of the overstrength force in the diagonal bars. Using the ACI strut-and-tie model with required $\phi = 0.75$ and minimum overstrength factor 1.25, this requirement is $\phi A_{ts} f_y \geq 1.25 A_{vd} f_y \sin\alpha$, or for equal values of f_y, $A_{ts} \geq 1.67 A_{vd} \sin\alpha$. Note that ACI 318 requires factor 1.5 (rather than 1.67) for design of hoops to restrain forces from column offset bars; the lower factor of ACI 318 is appropriate considering the offset column bars are not expected to experience design loads causing longitudinal reinforcement yielding and strain-hardening.

Another factor to consider for the concept shown in Figure 7.46*c* is out-of-plane bursting forces. These arise because the diagonal bars and the added flexural bars are in different planes, such that force transfer between these bars along their non-contact lap splice results in horizontal strut forces that must be restrained by transverse reinforcement. When this added component is taken into consideration, this beam detail can work effectively (Restrepo et al., 1995).

7.12 Shear in Structural Walls

Shear behavior of structural walls can be interpreted using many of the same concepts introduced previously for beams and columns. The main differences arise because of the larger dimensions and the larger thickness-to-depth ratio of structural walls. Additionally, it is not unusual for some structural walls to be long (measured in the horizontal direction) compared with their height, more so than is typical for beams and columns. In this section we review the internal forces and reinforcement requirements, strength, and performance limits for shear in structural walls. Flexural response is reviewed in Chapter 6. Applications to design of buildings, including applicable strength reduction factors ϕ, are covered in detail in Chapter 13.

7.12.1 Wall Classification Based on Slenderness

Structural walls conventionally are classified by aspect ratio l_w/h_w (Figure 7.47). The classification relates partly to the primary force-resisting mechanisms under lateral load and, consequently, the required reinforcement. The term *slender wall* usually is reserved for walls having $h_w/l_w \geq 2$. A slender wall is not able to resist horizontal shear efficiently through a simple diagonal compression strut from point of load application to point of support, but instead requires horizontal tension ties (or more simply, distributed horizontal reinforcement) to resist shear (Figure 7.47*a*). The term *squat wall* usually is reserved for walls having $h_w/l_w < 2$. In contrast with slender walls, squat walls are able to transmit shear efficiently through diagonal compression struts direction to the foundation (Figure 7.47*b*). In very squat walls, shear can be resisted by multiple diagonal compression struts that are equilibrated by distributed vertical reinforcement, as shown in the right side of Figure 7.47*b*. Behavior also can be interpreted in terms of B-regions and D-regions, introduced in Section 7.6.1. Slender walls have B-regions and D-regions, whereas squat walls consist only of D-regions.

The classification also relates to design and performance expectations. Slender walls usually can be designed such that inelastic response is primarily in flexure. Achieving a flexural yielding mechanism can be more difficult in squat walls, especially for $h_w/l_w \leq 1$. This subject is discussed more fully in Chapter 13.

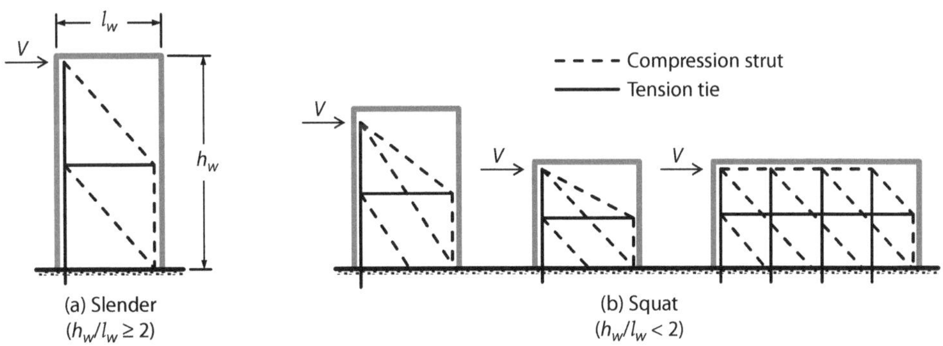

FIGURE 7.47 Classification of wall slenderness.

7.12.2 Slender Structural Walls

Observations on Wall Behavior

Similar to beams and columns, behavior of structural walls is affected by cyclic loading. Figure 7.48 illustrates load–displacement relations measured for three walls with confined boundary elements (Corley et al., 1981). Figure 7.49 shows damage patterns near failure. Key aspects of response include the following:

- Wall B4 was loaded monotonically, developing maximum shear of $3.9A_{cv}\sqrt{f_c'}$, psi ($0.33A_{cv}\sqrt{f_c'}$, MPa) and maximum drift ratio around 0.07. Concrete cracks near failure created a fan pattern within the yielding region, indicating influence of shear. Failure mainly was due to fracture of longitudinal reinforcement.
- Wall B3 was nominally identical to wall B4 but was loaded under deformation reversals. Maximum shear was $3.1A_{cv}\sqrt{f_c'}$, psi ($0.26A_{cv}\sqrt{f_c'}$, MPa). Drift

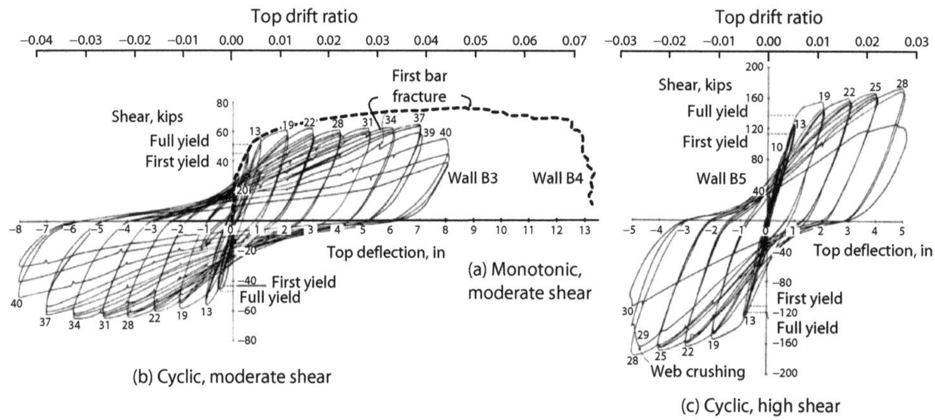

FIGURE 7.48 Load–displacement relations for structural walls subjected to (*a*) monotonic loading under moderate shear, (*b*) cyclic loading under moderate shear, and (*c*) cyclic loading under high shear. (*After Corley et al., 1981, courtesy of American Concrete Institute.*)

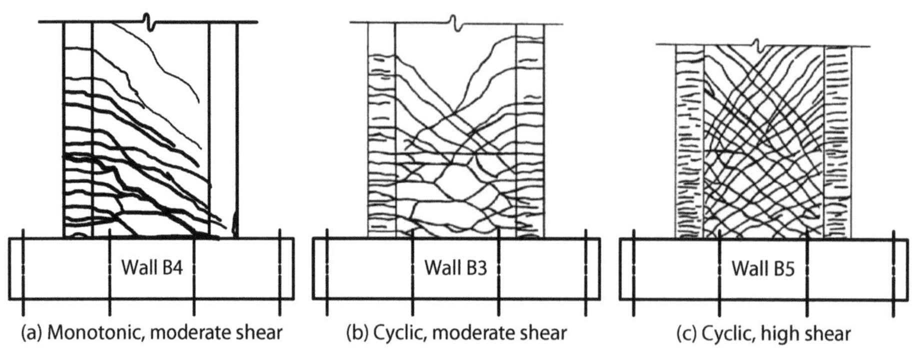

FIGURE 7.49 Cracking patterns for structural walls subjected to (*a*) monotonic loading under moderate shear, (*b*) cyclic loading under moderate shear, and (*c*) cyclic loading under high shear. (*After Oesterle et al., 1980, courtesy of American Concrete Institute.*)

ratio at failure (~0.045) was markedly reduced compared with the monotonically loaded wall. Deformation reversals resulted in significantly more damage to the wall web, including crisscrossed crack patterns and shear distortions leading to moderately pinched load-deformation relations.

- Wall B5 was reinforced with additional boundary element longitudinal reinforcement, leading to maximum shear $8.8A_{cv}\sqrt{f_c'}$, psi ($0.73A_{cv}\sqrt{f_c'}$, MPa). The wall developed the design moment strength but failed by crushing of diagonal compression struts in the wall web. More extensive diagonal cracking also was apparent due to the higher shear stress. Wall drift ratio capacity was reduced to approximately 0.028.

These observations demonstrate that deformation capacity and failure mode are sensitive to loading history and shear stress level.

Shear Strength

As with slender beams, shear strength of slender structural walls can be estimated using shear strength equations developed for the B-region. According to ACI 318, shear strength is the sum of concrete and steel contributions, as given by Eq. (7.35).

$$V_n = V_c + V_s = \alpha_c \lambda A_{cv}\sqrt{f_c'} + \frac{A_v f_{yt} l_w}{s} = A_{cv}\left(\alpha_c \lambda \sqrt{f_c'} + \rho_t f_{yt}\right) \qquad (7.35)$$

in which A_{cv} = web area = $l_w b_w$, α_c (psi units) = 2 for $h_w/l_w \geq 2$, = 3 for $h_w/l_w \leq 1.5$, and varies linearly for intermediate h_w/l_w (in MPa units, $\alpha_c = 0.17$ for $h_w/l_w \geq 2$, and = 0.25 for $h_w/l_w \leq 1.5$), and ρ_t = steel ratio of distributed web horizontal reinforcement. Equation (7.35) is identical to the combination of Eqs. (7.23), (7.24), and (7.26) for beams except for the variable multiplier α_c [which was taken as constant 2.0 psi or 0.17 MPa in Eq. (7.26)] and the use of the full wall length l_w instead of effective depth d. In ACI 318, V_n is limited to $10A_{cv}\sqrt{f_c'}$, psi ($0.83A_{cv}\sqrt{f_c'}$, MPa) for individual walls and $8A_{cv}\sqrt{f_c'}$, psi ($0.66A_{cv}\sqrt{f_c'}$, MPa) for all walls in a building resisting a common force. This limit is discussed in greater detail under the subheading *web crushing limit*. Its application in buildings is covered in Chapter 13.

Figure 7.50 plots ratio of measured to calculated shear strength [Eq. (7.35)] as function of displacement ductility. Data are limited to slender walls ($h_w/l_w \geq 2$) without openings, tested as cantilevers under reversed cyclic lateral loading, with axial load ranging from 0 to $0.13A_g f_c'$, and failing in shear. Failure is defined when strength loss reaches 20% of maximum strength. Similar to columns (Figure 7.40), wall shear strength degrades to around 70% of V_n as displacement demands increase beyond the linear range. (Similar trends are observed if wall shear strengths are plotted versus drift ratio.) Data in Figure 7.50 are sorted by maximum nominal shear stress. Strength and degradation patterns are similar for walls with

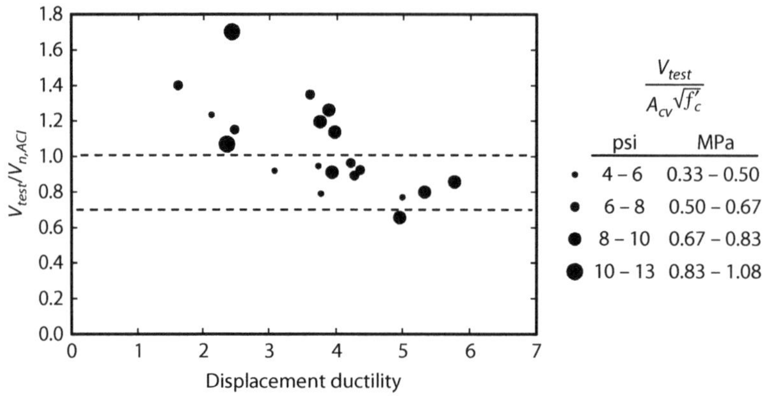

FIGURE 7.50 Ratio of measured to calculated shear strength of slender cantilever walls as function of displacement ductility at failure. (*Data from Vallenas et al., 1979; Iliya and Bertero, 1980; Corley et al., 1981; Aktan and Bertero, 1985; and Pilakoutas and Elnashai, 1995.*)

different nominal shear stress. The data in Figure 7.50 suggest that the shear strength equations of ACI 318 are conservative for walls with low ductility.

Web Crushing Limit

Shear strength of walls can be limited by diagonal tension strength (see preceding subsection), by interface shear strength at construction joints (see Section 7.13), or by web crushing strength (Figure 7.51). Web crushing occurs when compressive

FIGURE 7.51 Web crushing in flanged wall. (*After Oesterle et al., 1984.*)

stress on diagonal compression struts reaches the compressive strength. Web crushing sets an upper bound on shear strength for walls with given cross section and concrete compressive strength.

Web crushing strength is affected by the degree of inelastic flexural response in the wall (Oesterle et al., 1984; Hines and Seible, 2004). During early stages of lateral loading, while flexural cracks are relatively narrow at the wall base, shear is resisted along the entire wall length. This is evident in the orientation of diagonal cracks in Figure 7.52*a*. As flexural deformations increase into the inelastic range, with widening cracks in the flexural tension zone, the majority of shear must be resisted near the flexural compression zone. This requires re-alignment of inclined cracks, with concentration of diagonal compressive stress near the compressive toe of the web (Figure 7.52*b*). The concentration of diagonal compressive stress, coupled with inelastic straining of vertical web reinforcement, results in reduced web crushing strength with increasing lateral displacement.

Oesterle et al. (1984) studied web crushing for flanged and bar-bell cross sections with confined boundary elements. Shear was assumed to result in diagonal compressive stress f_{cd} in accordance with the truss model [Eq. (7.10)]. Strength of diagonal compression struts was assumed to decrease due to increasing transverse tensile strain (Figure 3.13), and this in turn was related to drift ratio. Setting an upper limit on story drift ratio of 0.02, the following expression for web crushing strength was defined:

$$V_n = 0.14 f_c' + \frac{N_u}{2A_g} \le 0.18 f_c' \tag{7.36}$$

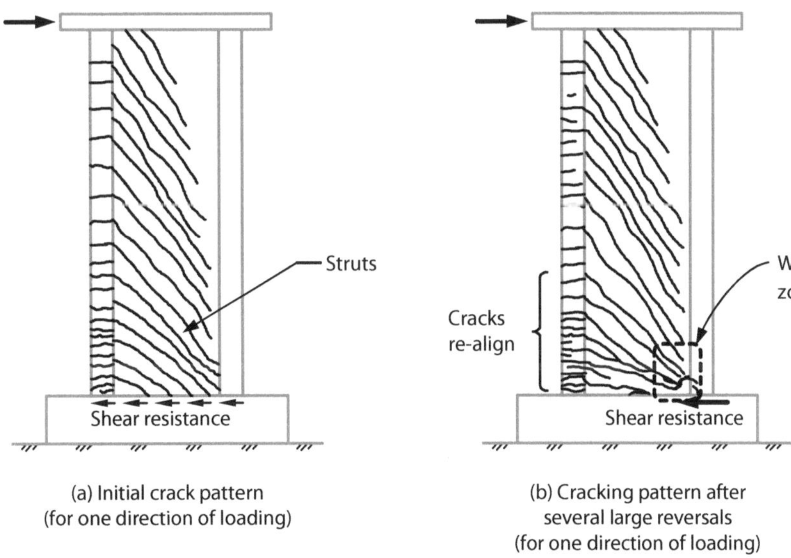

FIGURE 7.52 Re-alignment of cracks with increasing deformation demand. (*After Oesterle et al., 1984, courtesy of American Concrete Institute.*)

ACI 318, based on consideration of web crushing strength, sets the upper limit on shear strength as

$$V_{n,max} = 10A_{cv}\sqrt{f'_c}, \text{ psi } (0.83A_{cv}\sqrt{f'_c}, \text{ MPa}) \tag{7.37}$$

Figure 7.53 plots results of Eqs. (7.36) and (7.37) as function of axial stress level. It can be seen that the ACI 318 limit is conservative except for low compressive strength and low axial compression. Hines and Seible (2004) note that both equations may be overly conservative for walls in which the ratio of wall length to boundary element length is in the range $l_w/l_{be} \leq 6$.

According to ACI 318, the upper limit on shear expressed by Eq. (7.37) applies to all regions of structural walls. As described in this section, however, the limit was derived for web crushing near the flexural compression region of a flexural plastic hinge. This is a location where shear stress flow is constricted to a short length of the wall, as illustrated in Figure 7.52. Away from such regions, higher average shear stresses might be tolerated. See Section 7.12.4 for additional discussion.

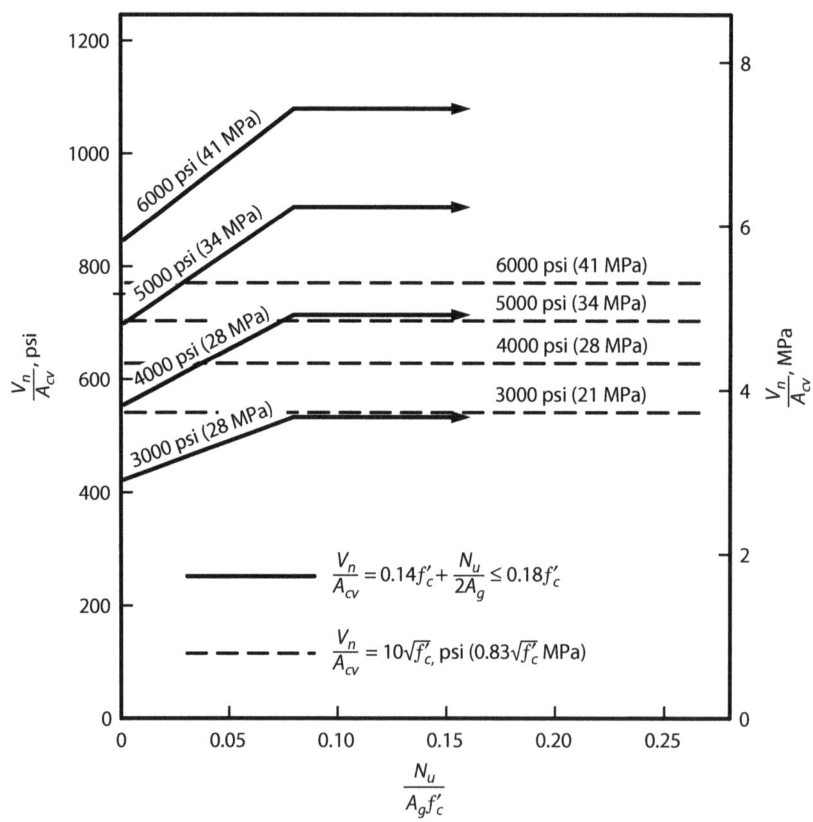

FIGURE 7.53 Maximum allowable nominal shear stress. (*After Oesterle et al., 1984, courtesy of American Concrete Institute.*)

7.12.3 Squat Structural Walls

Observations on Wall Behavior

Squat walls ($h_w/l_w < 2$) typically are capable of transferring a significant portion of applied shears through direct strut action between point of load application and point of reaction (Figure 7.47*b*). This force-resisting mechanism decreases the importance of horizontal distributed reinforcement and increases the importance of vertical distributed reinforcement to resist shear. Both horizontal and vertical distributed reinforcement, however, are important to shear resistance in squat walls.

The principal shear failure mechanisms in squat walls can be described in reference to Figure 7.54 as follows:

- *Diagonal tension failure:* When horizontal reinforcement is insufficient, a corner-to-corner diagonal failure plane can develop (Figure 7.54*a*). This mechanism is especially prominent in walls or wall segments having h_w/l_w approaching or exceeding 1.0 (Hidalgo et al., 2002). In more squat walls, diagonal tension failure also can occur along a steeper failure plane (Figure 7.54*b*), but viability of this mechanism depends on how shear is applied at the top of wall. In a building with floor slabs or tie beams at top of wall, the slab/beam would participate in the force-resisting mechanism, transferring forces back into the rest of the wall, thereby possibly avoiding the steeper diagonal tension failure.
- *Vertical yielding failure:* Where horizontal reinforcement is sufficient, vertical reinforcement, acting as the vertical ties in a strut-and-tie system, can become the weak link (Figure 7.54*b* and *c*). Yielding of vertical bars can provide limited ductility. Large rotations of the diagonal compression struts

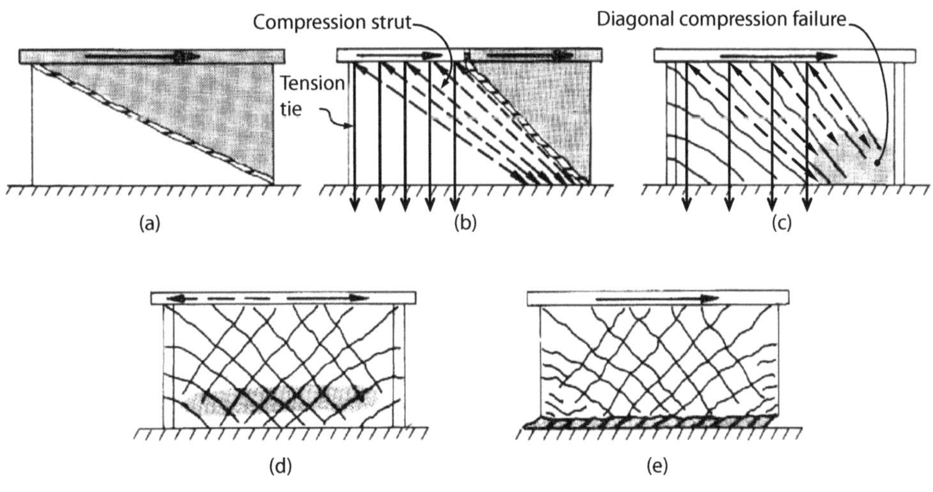

FIGURE 7.54 Shear failure modes in squat walls. (*Modified from Paulay et al., 1982, courtesy of American Concrete Institute.*)

may lead to diagonal compression failure, or crack opening under multiple displacement cycles may lead to sliding shear failure. These mechanisms are described next.

- *Diagonal compression failure:* Where provided horizontal and vertical reinforcement are sufficient to resist high nominal shear stress, failure can occur by crushing the diagonal compression struts (Figure 7.54*c*). Under reversed cyclic loading, failure can progress from one compression strut to another, leading to crushing spreading across the entire wall length (Figure 7.54*d*) (Barda et al., 1977).
- *Sliding shear failure:* If vertical reinforcement yields under lateral loading, residual tensile strain in the reinforcement can result in open cracks along the wall–foundation interface (or other construction joints). Upon force reversal, vertical reinforcement across these residual cracks must yield in compression before the cracks can fully close. If axial forces are small, as is typical in squat walls, these residual cracks may remain open. This can create a weak interface along which sliding deformations can occur, resisted mainly by kinking action of the vertical reinforcement extending across the crack opening (Paulay et al., 1982).

In this section we address the failure types illustrated in Figure 7.54*a* through *d*. Interface shear transfer resulting in sliding shear failure is addressed in Section 7.13.

Squat walls typically sustain inclined cracking associated with shear. Movement along these cracks can result in pinched hysteresis loops similar to those observed for beams or columns with high shear (Figure 7.38). Figure 7.55 shows shear–displacement relations for a series of walls of rectangular cross section fixed against rotation at top and bottom and subjected to shear reversals. Each wall has distributed vertical and horizontal reinforcement with $\rho_l = \rho_t = 0.0025$, the minimum permitted in ACI 318. The most slender wall develops flexural yielding followed by shear failure. Deformation capacity decreases as aspect ratio decreases.

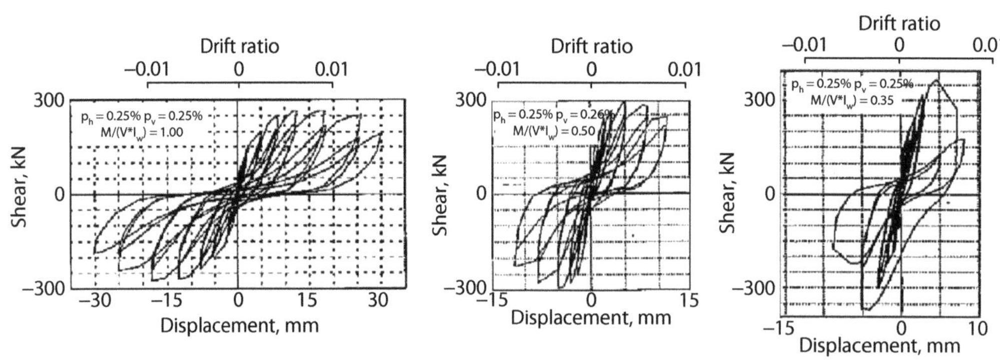

(a) $M/Vl_w = 1.00$ (b) $M/Vl_w = 0.50$ (c) $M/Vl_w = 0.35$

Figure 7.55 Shear–displacement relations for wall piers with minimum distributed vertical and horizontal reinforcement. (*After Hidalgo et al., 2002.*)

As illustrated in Figure 7.47, squat walls have multiple force paths. Some force paths require horizontal web reinforcement and others require vertical web reinforcement. To illustrate importance of vertical web reinforcement in very squat walls, consider the idealized shear transfer mechanism illustrated in Figure 7.56. Diagonal cracking is assumed to occur along a constant angle θ with respect to horizontal. Although shear and normal stresses can be resisted by concrete across inclined cracks, in this solution we assume that force transfer across cracks occurs only through reinforcement. We further simplify the problem by assuming all reinforcement is stressed to f_y. Considering the free-body diagram of Figure 7.56*b*, equilibrium of horizontal forces requires that shear $v_n x b_w$ applied at the top of free body **ab** be equilibrated by equal shear $v_n x b_w$ at the bottom. Similarly, considering vertical force equilibrium, axial force $n_u x b_w$ applied at the top is resisted by $n_u x b_w$ at the bottom.

Summing moments about point **b** at the bottom of the free-body diagram of Figure 7.56*b* results in

$$v_n x b_w h_w - n_u x b_w \frac{h_w}{\tan\theta} - A_{sy} f_y \frac{x}{s_x} \frac{h_w}{\tan\theta} = 0 \tag{7.38}$$

Assuming the reinforcement arrangement enables uniform stress v_n to act across the entire wall length l_w, we obtain the following expression for wall shear strength:

$$V_n = v_n l_w b_w = \left(n_u l_w b_w - \frac{A_{sy} f_y l_w}{s_x} \right) \frac{1}{\tan\theta} = (N_u + \rho_l f_y A_{cv}) \frac{1}{\tan\theta} \tag{7.39}$$

in which ρ_l = steel ratio for distributed vertical reinforcement. Note that the term in parentheses is the normal force due to externally applied axial force plus normal force due to yielding of the distributed vertical reinforcement. Thus, the right-hand side of Eq. (7.39) can be interpreted as the product of an effective normal force and a friction coefficient $1/\tan\theta$. This same concept is used to define interface shear strength in Section 7.13.

Notably, Eq. (7.39) states that shear strength is a function of the distributed vertical reinforcement, not distributed horizontal reinforcement. Tests by Barda et al. (1977) confirm this finding for very squat walls. Distributed horizontal

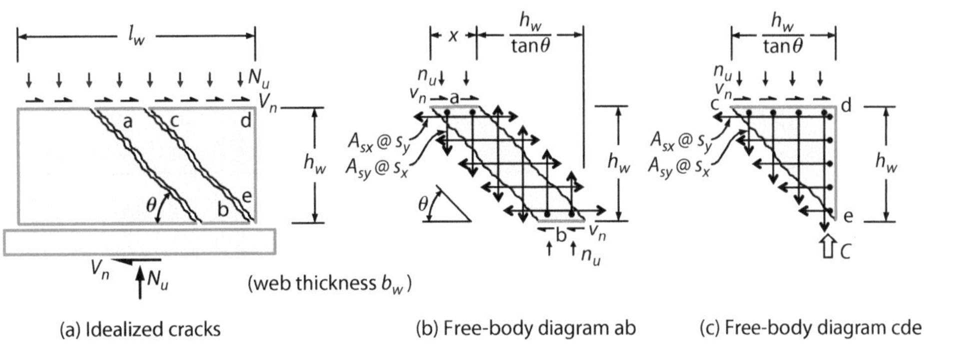

FIGURE 7.56 Idealized actions on squat wall.

reinforcement is also important for inhibiting crack opening and thereby improving overall response of the wall.

Distributed horizontal reinforcement also is important for equilibrium of the free-body diagram of Figure 7.56c. Specifically, from equilibrium of horizontal forces we have $v_n b_w h_w / \tan\theta = A_{sx} f_y h_w / s_y$. Moment equilibrium about point **e** suggests that distributed vertical reinforcement also is required. Finally, moment equilibrium of a point halfway between **c** and **e** requires a compressive force C located near point d and acting parallel to **de**. Thus, the idealization of Figure 7.56 breaks down in that concrete near the toe of the wall (point **e**) must be resisting compression (and possibly shear) across the idealized concrete crack.

Shear Strength

Several studies have reported models for shear strength of squat structural walls. Barda et al. (1977) present a shear strength expression for squat walls with inclined cracking and web crushing characteristic of shear failure modes. Cardenas et al. (1980) apply ACI 318 design equations of that era to squat walls experiencing a range of failure modes. Wood (1990) presents shear strength equations for squat walls that were reported to have failed in shear. Hwang et al. (2001) present a softened truss model for strength of squat walls reported to have failed in web shear. Gulec et al. (2008) evaluate shear strength equations for walls having measured strength less than the strength corresponding to expected flexural strength, without identifying actual observed failure mode.

Figure 7.57 plots ratio of measured to calculated wall strength as function of wall aspect ratio. Data are a subset of the data presented in Gulec (2009) for

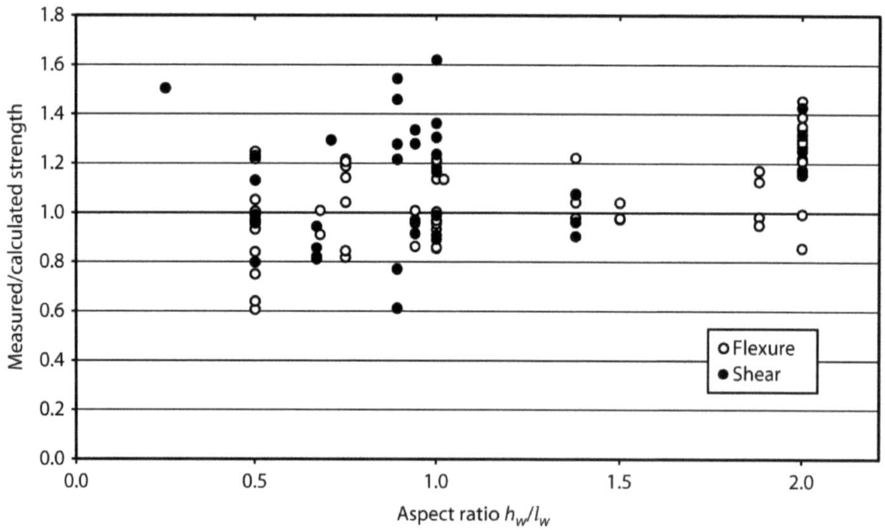

FIGURE 7.57 Ratio of measured to calculated strength as function of wall aspect ratio. "Shear" is for walls having calculated shear strength less than the shear corresponding to development of calculated flexure strength. "Flexure" is for walls having calculated shear strength greater than the shear corresponding to development of calculated flexure strength.

squat walls with rectangular cross section; having distributed horizontal web reinforcement $\rho_t \geq 0.0025$ and distributed vertical web reinforcement ρ_l not less than the greater of 0.0025 and ρ_t (i.e., steel ratios as required by ACI 318); and subjected to force reversals, applied either slowly or dynamically. In the figure, "Shear" is for walls having calculated shear strength less than the shear corresponding to development of calculated flexure strength. "Flexure" is for walls having calculated shear strength greater than the shear corresponding to development of calculated flexure strength. Calculated flexure strength is expected strength determined by conventional cross section analysis using measured material properties. Calculated shear strength is based on Eq. (7.35) using measured material properties. The designations "Shear" and "Flexure" do not intend to identify the failure mode, but instead serve only as indices for comparison with quantities commonly computed in design. For the entire data set, ratio of measured to calculated strength has mean of 1.05 with coefficient of variation of 0.20. For the data marked "Shear," the corresponding values are 1.12 and 0.22. Concrete compressive strengths for the data set ranged from 2000 to 7300 psi (14 to 50 MPa). Wallace (1998) reports somewhat more conservative results for low-rise walls with compressive strengths to 20,000 psi (138 MPa).

7.12.4 Shear in Panel Zones

The term *panel zone* refers to a region of a structural wall (or other element) subjected to relatively uniform shear, with or without normal stresses. Figure 7.58*a* illustrates a panel zone in a structural wall system. Several other examples are in Chapter 13. The nominal shear stress in such panel zones may be much greater than the nominal shear stress in the adjacent wall segments.

Shear resistance in a panel zone can be idealized in different ways. One idealization is that chords along the panel zone boundary deliver uniform shear around the perimeter of the panel zone (Figure 7.58*b*). According to this idealization, the panel zone is subjected to uniform shear. An alternative idealization is that shear is resisted by a diagonal compression strut along the direction of principal compressive stress in the panel zone (Figure 7.58*c*). According to this idealization,

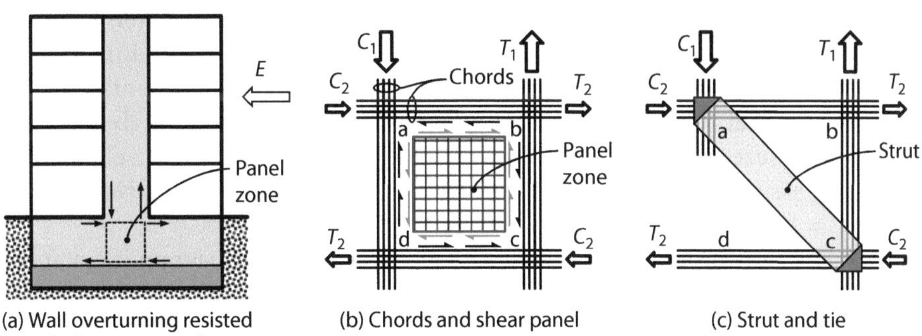

(a) Wall overturning resisted by basement wall

(b) Chords and shear panel

(c) Strut and tie

FIGURE 7.58 Shear panel zone in structural wall, and concepts for internal force resistance.

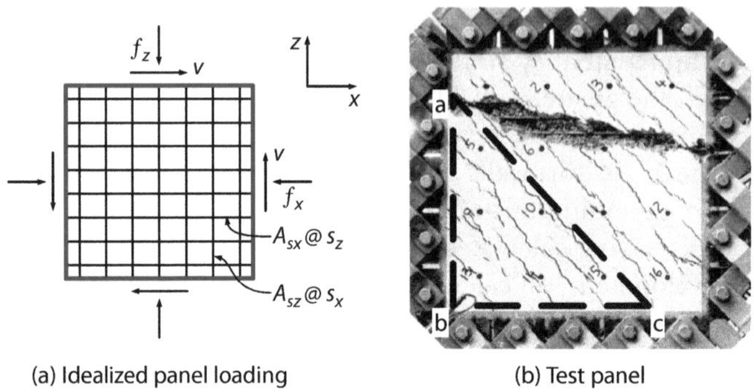

FIGURE 7.59 Shear panel idealization and test specimen. (*Photograph courtesy of Michael Collins.*)

the chords deliver forces to the nodal zones. In either idealization, vertical and horizontal chords are required to equilibrate the internal actions of the panel zone. Although both mechanisms can properly resolve the internal forces and can be used for design, the following paragraphs consider only the resistance mechanism shown in Figure 7.58*b*.

Special testing devices have been developed to impose shear and normal forces around the perimeter of an isolated panel zone in a laboratory (Figure 7.59). More than 100 such tests have been reported (Bentz et al., 2006). An important outcome has been the development of models for the stress–strain response of membrane elements subjected to axial and shear stresses. Two such models are the Modified Compression Field Theory (Vecchio and Collins, 1986) and the Softened Truss Model (Hsu, 1988). These models are capable of simulating the complete nonlinear stress–strain response of biaxially loaded membranes, and are therefore useful for modeling load-deformation behavior of panel zones. Our main focus here, however, is to identify a simple model for estimating the strength of a panel zone.

In a membrane test, a series of cracks typically will form consistent with the orientation of the stress field. These cracks can be considered to divide the panel into a series of diagonal compression struts located between and running parallel to the cracks. In a heavily reinforced panel, failure of the membrane can occur by crushing of the diagonal compression struts before the panel reinforcement yields. Under such conditions, Bentz et al. (2006) report that the unit shear strength can be taken conservatively as $v_n = 0.25 f'_c$. For failures occurring below this stress level, similar to Bentz et al. (2006), we can assume that the panel reinforcement reaches the yield stress, that is, $f_{sx} = f_{sz} = f_y$.

To establish the equilibrium equations, a free-body diagram is cut around the boundaries and along an inclined crack of a test panel (**abc** in Figures 7.59 and 7.60). Horizontal stress f_x is taken equal to zero, consistent with the typical stress condition of a structural wall. To further simplify the equilibrium statement, it is

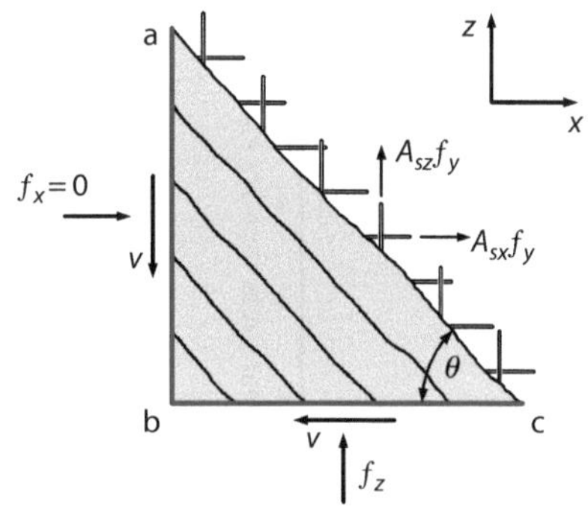

FIGURE 7.60 Free-body diagram of a portion of a shear panel.

assumed that crack surface along ac is free of shear and normal stresses. With these assumptions, the requirements for force equilibrium in the x and z directions result in Eqs. (7.40) and (7.41), respectively.

$$v - \rho_{sx} f_y \tan\theta = 0 \tag{7.40}$$

$$v \tan\theta - \rho_{sz} f_y - f_z = 0 \tag{7.41}$$

These expressions can be solved for shear stress v. Defining this shear stress as the nominal shear stress capacity v_n, and including the limit of $0.25\, f_c'$, results in the nominal strength equation

$$v_n = \sqrt{\rho_{sx} f_y (\rho_{sz} f_y + f_z)} \le 0.25\, f_c' \tag{7.42}$$

Note that axial compressive stress f_z is identified as having the same effect as an equivalent stress provided by additional vertical reinforcement. This effect was identified previously in Mau and Hsu (1987).

An alternative empirical expression for the nominal strength is

$$\begin{aligned} v_n &= 3\sqrt{f_c'} + \rho_{smin} f_y \le 0.25 f_c', \text{psi} \\ &= 0.25\sqrt{f_c'} + \rho_{smin} f_y \le 0.25 f_c', \text{Mpa} \end{aligned} \tag{7.43}$$

in which ρ_{smin} = the lesser of ρ_{sx} and ρ_{sz}. Equation (7.43) has the advantage of being similar to the equation used in ACI 318 for shear strength of low-rise walls [Eq. (7.35)].

The results of Eqs. (7.42) and (7.43) are compared with measured strengths in Figure 7.61*a* and *b*, respectively. Measured data include the 73 tests reported in Bentz et al. (2006) having distributed reinforcement in both horizontal and vertical directions. The calculated and measured strengths compare favorably. Slightly improved correlation (not shown) was obtained by setting f_z = 0 in Eq. (7.42). Note that these results are for monotonically loaded panel zones.

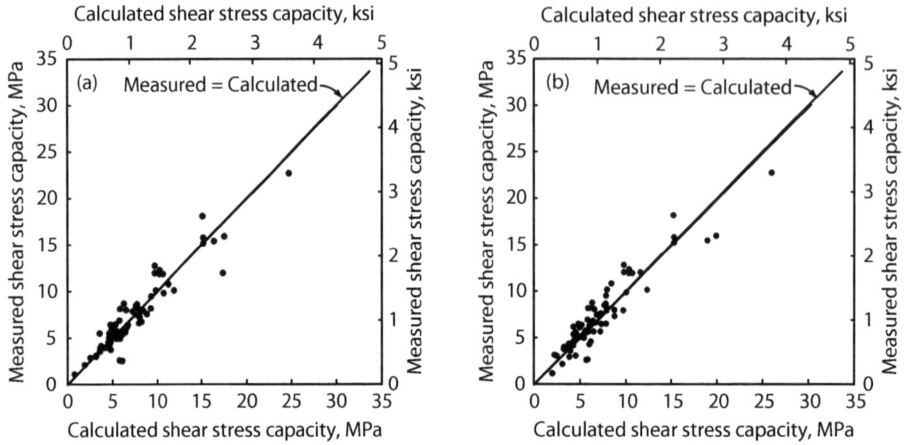

FIGURE 7.61 Measured and calculated strengths for membranes subjected to shear with or without axial stress: (*a*) Eq. (7.42); (*b*) Eq. (7.43).

7.13 Interface Shear Transfer

Interface shear transfer refers to the transfer of shear across the interface between two different volumes of concrete. One example is at construction joints where fresh concrete is placed against previously hardened concrete. In this case, a weak plane may exist at the joint or in concrete immediately adjacent to the joint. Another example is at the intersection between two dissimilar structural members, such as a structural wall and its foundation. As discussed in Section 6.11, flexural cracking and reinforcement strain penetration into the supporting element can lead to a relatively wide crack at the interface, creating a weakened plane for shear transfer.

Slip along construction joints has been observed following earthquakes and during laboratory tests (Figure 7.62). Once sliding begins, deformation reversals may grind the concrete back and forth, degrading concrete resistance and possibly leading to other failures. Slip reversals also can kink the longitudinal bars back and forth across the interface, leading to reinforcement failures.

Laboratory studies show two different mechanisms by which shear is resisted across a potential shear failure plane (Mattock and Hawkins, 1972). In previously uncracked interfaces, diagonal cracks form along the shear plane, creating a series of concrete struts (Figure 7.63). The applied shear creates a moment on the struts that is equilibrated by reinforcement crossing the interface plus externally applied normal force. When the shear is sufficient to induce yielding of the reinforcement, plastic straining of the reinforcement enables the struts to rotate, eventually leading to strut failure.

(a)

(b)

FIGURE 7.62 Distress at construction joints: (*a*) Cordova Building, Anchorage, 1964 (*photo by K. Steinbrugge, courtesy of the National Information Service for Earthquake Engineering, University of California, Berkeley*); (*b*) squat wall test (*after Paulay et al., 1982, courtesy of American Concrete Institute*).

If the shear plane is pre-cracked, applied shear results in slip along the shear plane. Because the crack faces are rough, they are forced to separate as slip occurs (Figure 7.63). This separation causes tension in reinforcement crossing the shear plane. Reinforcement tension is balanced by equal compression in the concrete across the crack. This creates frictional resistance to the shear force, sometimes

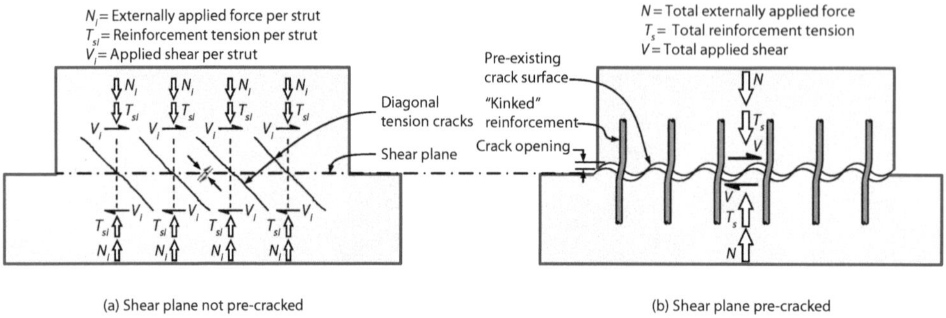

FIGURE 7.63 Interface shear force-resisting mechanisms.

referred to as *shear-friction*. Externally applied normal force can increase or decrease normal force on the crack (depending on whether it is compressive or tensile), and thereby directly affects frictional resistance. Note that reinforcement kinking occurs across the slip plane, leading to additional shear resistance through *dowel action*.

If the shear plane is pre-cracked, slip along the crack surface and crack opening is required before significant frictional resistance is provided by reinforcement. In contrast, if the shear plane is not pre-cracked, significant slip only occurs after cracking is induced. Consequently, a pre-cracked shear plane generally develops greater slip than an uncracked shear plane (Mattock and Hawkins, 1972). In cases of high external axial compression, the crack can "lock up," in which case the pre-cracked interface behaves as an uncracked interface (Mattock and Hawkins, 1972).

For low to moderate amounts of reinforcement and externally applied compression, strength of a pre-cracked shear plane is less than strength of an uncracked shear plane. Given uncertainty about whether a crack will form at an interface due to applied loads or volume change effects, in most cases the shear strength should be based on a shear-friction model.

According to ACI 318, where clamping reinforcement is perpendicular to the shear plane, the shear-friction strength is defined by

$$V_n = A_{vf} f_y \mu \le V_{n,max} \tag{7.44}$$

in which A_{vf} = area of reinforcement crossing the shear plane; f_y = its yield strength, not to be taken greater than 60,000 psi (414 MPa); μ is a friction coefficient defined by Table 7.3; and $V_{n,max}$ is an upper bound on shear-friction strength defined by Table 7.4.

Where concrete is placed against a previously formed surface, as between a floor slab (or diaphragm) and a previously cast structural wall, a smooth interface will limit shear transfer capacity. In such cases, where significant shear must be transferred, it may be necessary to roughen the previously formed surface through

Condition	μ
Concrete placed monolithically	1.4λ
Concrete placed against hardened concrete with surface intentionally roughened*	1.0λ
Concrete placed against hardened concrete not intentionally roughened†	0.6λ
Concrete anchored to as-rolled structural steel by headed studs or by reinforcing bars	0.7λ

*To use $\mu = 1.0\lambda$, the interface for shear transfer must be clean, free of laitance, and roughened to a full amplitude of approximately 1/4 in (6 mm).
†To use $\mu = 0.6\lambda$, the interface for shear transfer must be clean and free of laitance.

Table 7.3 Values of Shear-Friction Coefficient μ

Condition	$V_{n,max}$
For normalweight concrete either placed monolithically or placed against hardened concrete with surface intentionally roughened to full amplitude of approximately ¼ in (6 mm)	Least of (a), (b), and (c) (a) $0.2 f_c' A_c$ (b) $(480 + 0.08f_c')A_c$, psi $[(3.3 + 0.08f_c')A_c$, MPa$]$ (c) $1600A_c$, psi ($11A_c$, MPa)
For all other cases	Lesser of (d) and (e) (d) $0.2 f_c' A_c$ (e) $800A_c$, psi ($5.5A_c$, MPa)

Table 7.4 Values of $V_{n,max}$

chipping. Alternatively, shear keys can be cast at the interface to improve shear transfer capacity, thereby enabling use of $\mu = 1.0\lambda$.

Horizontal construction joints occurring between the top of one pour and the start of another require special attention. *Laitance*—a layer of weak material derived from cementitious material and aggregate fines carried by bleeding to the surface—should be removed before placing the fresh concrete. Studies by Paulay et al. (1974), Bass et al. (1989), and Djazmati and Pincheira (2004) have examined the degree of roughening required. Based on those studies, sandblasting [1/8-in (3-mm) roughness], "brooming" the surface, chipping the surface [1/4-in (6-mm) roughness], or shear keys effectively achieve the same result. Apparently, any reasonable degree of roughness will force the failure into the weak concrete that typically occurs at the very top of the first pour (Paulay et al., 1974). Regardless of the treatment, the joint should be dry, as wetting before casting the new concrete can reduce strength by as much as one-half (Djazmati and Pincheira, 2004).

The $V_{n,max}$ values effectively represent over-reinforced conditions (due to externally applied force, distributed reinforcement, or both) in which shear strength is limited by concrete capacity. The limit was revised upward from

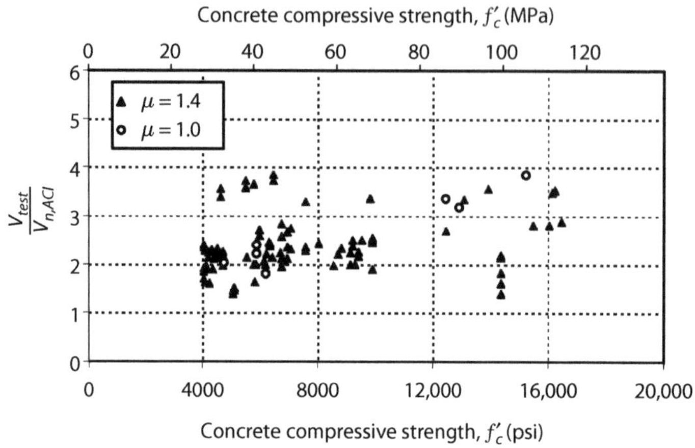

FIGURE 7.64 Comparison of measured and calculated shear-friction case for normalweight aggregate concrete. (*After ACI 318-CE002, 2006, courtesy of American Concrete Institute.*)

previous building codes based on test results for high-strength concrete reported by Mattock (2001) and Kahn and Mitchell (2002).

As shown in Figure 7.64, Eq. (7.44) is conservative over a range of concrete strengths. Hwang et al. (2000) present a softened-truss model that is a better predictor of interface shear strength.

Figure 7.65 shows a representative measured load–slip relation for reversed cyclic loading (Park and Paulay, 1975). The relation is characterized by the following: (a) high initial stiffness associated primarily with aggregate interlock; (b) maximum force capacity at slip around 0.2 in (5 mm); (c) reduction in shear transfer capacity for slip exceeding slip at maximum shear (shear strength degradation is especially noteworthy under the action of deformation reversals, as repeated and reversed loading results in gradual abrasion of the interlocking mechanism; at large deformations, dowel action becomes more prominent, and may be the primary source of shear resistance); (d) pinching of the hysteresis relation upon force reversal as the pre-damaged interface slides relatively easily until the previous maximum deformation in the opposite direction is reached.

Mattock et al. (1975) conducted tests to determine requirements where net tension is applied normal to the shear plane. They found that total reinforcement area should be the sum of areas required for tension and for interface shear.

Mattock et al. (1975) also studied combined moment and interface shear. In this case, flexural tension and flexural compression balance one another, such that there is no net tension, and therefore shear-friction strength is unaffected. The study recommends that shear-friction reinforcement be distributed through the flexural tension zone to be most effective. For earthquake engineering problems, where force and deformation reversals occur, it is common

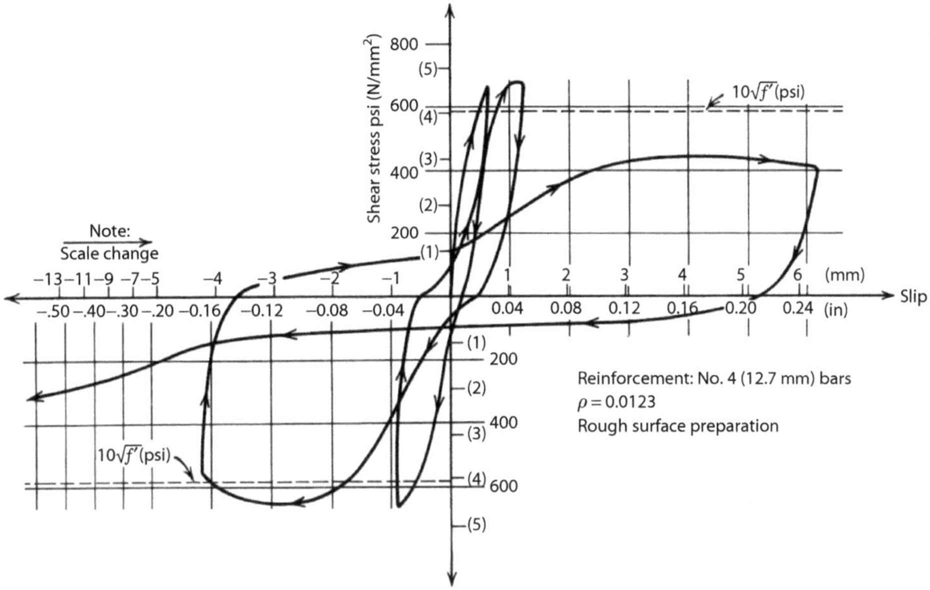

Figure 7.65 Load–slip relation for shear transfer across a construction joint. (*After Park and Paulay, 1975.*)

practice to distribute shear-friction reinforcement uniformly through the length of the wall web.

Note that shear-friction reinforcement need be developed only on both sides of the shear plane, whereas reinforcement for flexure may need to be extended farther to provide required moment resistance at other sections. Thus, a common practice is to provide longitudinal reinforcement as required for flexure as a first design step. If additional reinforcement is required for interface shear, shorter dowel bars are added.

The study by Mattock et al. (1975) does not address directly the case where inelastic moment reversals occur simultaneously with interface shear. If vertical reinforcement yields under lateral loading, residual tensile strain in the longitudinal reinforcement can result in open cracks along the wall–foundation interface (or other construction joints). Upon moment reversal, vertical reinforcement across these residual cracks must yield in compression before the cracks can fully close. If axial forces are small, as is typical in squat walls, these residual cracks may remain open. This can create a weak interface along which sliding deformations can occur, resisted mainly by kinking action of the vertical reinforcement extending across the crack opening (Figure 7.62*b*). Eventually, slip along the shear plane will kink the bars to the point that they pull the adjacent crack surfaces together, at which point the concrete once again participates in the force-resisting mechanism. This behavior can result in a severely pinched hysteresis (Figure 7.66). Paulay et al. (1982) recommend using inclined reinforcement to resist sliding in such cases. See Chapter 13 for additional discussion.

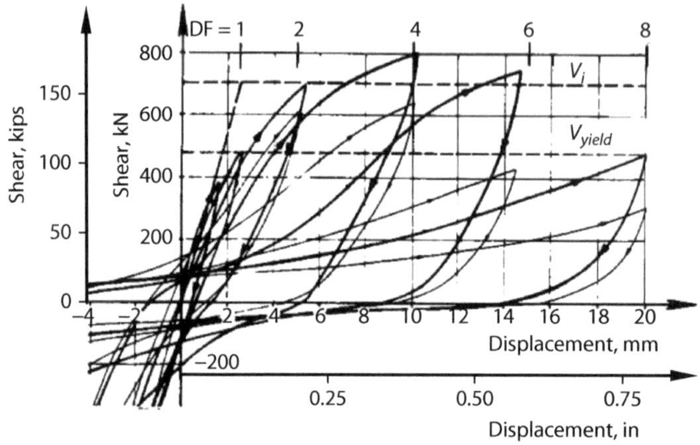

FIGURE 7.66 Shear–displacement relation of rectangular-section, squat wall. (*After Paulay et al., 1982, courtesy of American Concrete Institute.*)

ACI 318 covers the case where reinforcement is provided across the shear plane at an angle such that shear produces tension in the reinforcement. In this case, the bar both clamps the crack surfaces together and directly resists shear through tension. The resulting shear-friction strength is

$$V_n = A_{vf} f_y (\mu \sin \alpha + \cos \alpha) \le V_{n,max} \tag{7.45}$$

in which α is the angle between reinforcement and the shear plane.

ACI 318 does not cover the case where shear produces compression in inclined reinforcement. In this case, the reinforcement would tend to push the crack surfaces apart, thereby reducing shear-friction resistance. It seems intuitive to assume that if an orthogonal grid of reinforcement was placed at 45° relative to the shear plane, one bar would be in tension and one would be in compression, such that the tendency of one to clamp and the other to separate the crack faces would cancel. The resulting shear strength would thus be the vector sum of the forces in the reinforcement, that is, $V_n = 0.5A_{vf} f_y(\sin 45° + \cos 45°) = 0.71\, A_{vf} f_y$, where A_{vf} is total area of reinforcement across the shear plane (i.e., area of all the interface bars at +45° and at −45°). Monotonic tests and detailed analyses, however, show that the result is closer to $V_n = 0.5A_{vf} f_y$ (Mattock, 1974). Such reinforcement should be effective at resisting sliding across an interface under reversed cyclic loading.

7.14 Shear Stiffness

7.14.1 General Aspects

Shear stiffness is the shear that will cause a unit shear deflection in a member of unit length. For uncracked members, it is commonly taken as the gross-section stiffness considering the web only, defined by

$$K_{vg} = G_c A_{cv} / f = 0.4 E_c b_w h / f \tag{7.46}$$

in which G_c = shear modulus, taken as $0.4E_c$; A_{cv} = web area = $b_w h$; b_w = web width; h = section depth measured parallel to the applied shear V; and f = form factor accounting for nonuniform shear stress across the section depth, taken as 1.2 for rectangular sections.

For cracked sections, we can approximate shear stiffness using the truss idealization in B-regions (Figure 7.67). Section depth is approximated as d (rather than jd) and shear deformations are assumed to be due to shortening of the diagonal compression struts and elongation of the transverse reinforcement (top and bottom chord assumed infinitely stiff). Stress f_{cd} in diagonal compression struts is defined by Eq. (7.10). Stress f_s in transverse steel can be obtained from Eq. (7.8) as $f_s = Vs \tan\theta / A_v d$. A full length diagonal strut has length $d/\sin\theta$. Therefore, shortening of a full-length diagonal strut is $\Delta_c = f_{cd} d / E_c \sin\theta$, and its contribution to shear displacement is $\Delta_c/\sin\theta$. Elongation of transverse reinforcement is $\Delta_s = f_s d / E_s$. Combining terms, shear deflection along length $d/\tan\theta$ is

$$\Delta_{cr} = V\left(\frac{s \tan\theta}{E_s A_v} + \frac{1}{b_w E_c \cos\theta \sin^3\theta}\right) \tag{7.47}$$

Shear stiffness K_{vcr} is defined as force per unit deflection of a unit length of the truss, that is $K_{vcr} = V/(\Delta_{cr}/d \cot\theta)$. Defining modular ratio $n = E_s/E_c$ and transverse steel ratio $\rho_t = A_v/b_w s$, and manipulating terms, we can write the shear stiffness of the idealized truss as

$$K_{vcr} = \frac{nE_c b_w d\rho_t \cos^2\theta \sin^2\theta}{\sin^4\theta + \rho_t n} \tag{7.48}$$

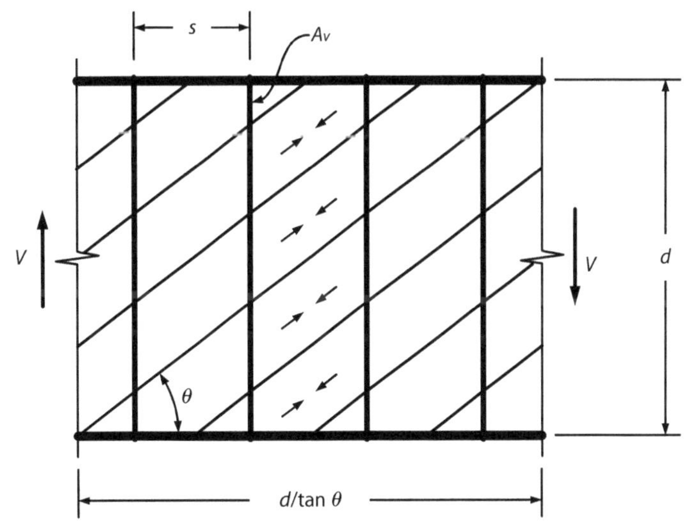

FIGURE 7.67 Idealized truss model for approximating cracked shear stiffness.

With the approximation $h \approx d$, the ratio of cracked shear stiffness [Eq. (7.48)] to uncracked stiffness [Eq. (7.46)] is given by

$$k = \frac{2.5\rho_t nf \cos^2\theta \sin^2\theta}{\sin^4\theta + \rho_t n} \tag{7.49}$$

For rectangular cross sections with $\theta = 45°$, Eq. (7.49) simplifies to

$$k = 2.5\rho_t n \tag{7.50}$$

Thus, shear stiffness for fully cracked members is a small fraction (on the order of one-tenth) of the uncracked shear stiffness.

7.14.2 Coupling Beams

Various standards and guidelines contain guidance on effective stiffness modeling of beams, which commonly has been considered applicable to coupling beams. ACI 318 provides several options for modeling flexural stiffness, including (a) 50% of stiffness values based on gross-section properties; (b) $E_cI_e = 0.35E_cI_g$; or (c) values based on more detailed analysis considering reduced stiffness under the loading conditions. ASCE 41 uses flexural rigidity $E_cI_e = 0.3E_cI_g$ and shear rigidity $0.4E_cA_{cw}$. These effective stiffness values are intended to provide an estimate of secant stiffness to the yield point.

The commentary to NZS 3101 (2006) contains more detailed guidance on effective stiffness modeling considering flexure, bond slip, and shear deformations. For conventionally reinforced beams, effective flexural rigidity at the yield point is given as $0.4E_cI_g$ for 44 ksi nominal yield (Grade 300) and $0.32E_cI_g$ for 73 ksi (Grade 500). Interpolating, flexural effective flexural rigidity is $0.37E_cI_g$ for Grade 60 (Grade 414) reinforcement. For typical reinforcement ratios, EI_e is nearly independent of concrete modulus but instead depends mainly on the reinforcement stiffness; consequently, these effective flexural rigidity values are intended to be used with constant E_c corresponding to $f_c' = 5800$ psi (40 MPa). Effective stiffness of short members, typical of coupling beams, is strongly affected by bar slip from anchorage zones. NZS 3101 recommends a modifier $R = \frac{l_n}{l_n + 0.8l_d}$. For practical purposes, we can take $l_d \approx 50d_b$. Finally, NZS 3101 adds modifier α_c to approximate effect of shear cracking. Where nominal shear stress is less than or equal to $3\sqrt{f_c'}$, psi ($0.25\sqrt{f_c'}$, MPa), shear cracking effect can be considered minimal, thus, $\alpha_c = 1.0$. Otherwise, α_c varies with aspect ratio $2M/Vd \approx l_n/h$. Combining effects of flexural rigidity, bar slip, and shear cracking, effective flexural rigidity for Grade 60 (414 MPa) reinforcement near yield point is as shown in Figure 7.68*a*.

For a diagonally reinforced beam, deformation is primarily in shear. Following the approach in NZS 3101, we can use a truss model to estimate the deformation near yield. The free length of a diagonal bar is $l_n/\cos\alpha$, where α is the angle between diagonal reinforcement and beam longitudinal axis. Adding an additional

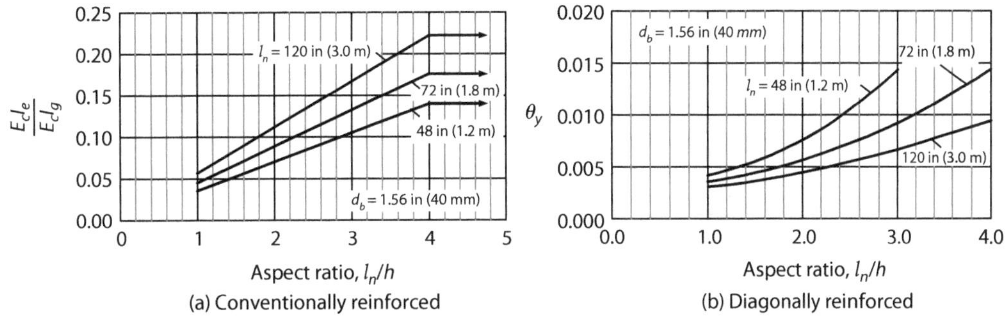

FIGURE 7.68 Effective stiffness of (*a*) conventionally reinforced and (*b*) diagonally reinforced coupling beams. *Note:* To use the chart in (a), use E_c based on 5800 psi (40 MPa) concrete compressive strength, regardless of the compressive strength in the beam.

effective length of $f_y d_b/2200$, psi ($f_y d_b/15$, MPa) to account for strain penetration on both ends, total elongation of a tension diagonal is $\delta_y = \left(\frac{l_n}{\cos\alpha} + \frac{f_y d_b}{2200}\right)\frac{f_y}{E_s}$, psi $\left[\left(\frac{l_n}{\cos\alpha} + \frac{f_y d_b}{15}\right)\frac{f_y}{E_s}, \text{MPa}\right]$. Further assuming that the deformation of the compression diagonal is 20% of δ_y, total vertical displacement at yield is $\delta_v = \frac{1.2\delta_y}{2\sin\alpha}$. Chord rotation at yield is thus

$$\theta_y = \frac{\delta_v}{l_n} = 0.6\varepsilon_y\left(\frac{1}{\cos\alpha} + \frac{f_y d_b}{2200 l_n}\right)\frac{1}{\sin\alpha}, \text{psi}$$
$$= 0.6\varepsilon_y\left(\frac{1}{\cos\alpha} + \frac{f_y d_b}{15 l_n}\right)\frac{1}{\sin\alpha}, \text{MPa} \tag{7.51}$$

The results are plotted in Figure 7.68*b*.

Results for diagonally reinforced beams also can be reported as ratio GA_e/GA, expressed as

$$\frac{GA_e}{GA/f} = 10\rho_{vd}\, n \frac{\sin^2\alpha}{\frac{1}{\cos\alpha} + \frac{f_y d_b}{2200\, l_n}}, \text{psi}$$
$$= 10\rho_{vd} n \frac{\sin^2\alpha}{\frac{1}{\cos\alpha} + \frac{f_y d_b}{15\, l_n}}, \text{MPa} \tag{7.52}$$

in which $\rho_{vd} = A_{vd}/b_w h$. Values from Eq. (7.52) vary widely with steel ratio, aspect ratio, and span. For l_n/h = 2, ρ_{vd} = 0.015, l_n = 72 in (1.8 m), and d_b = 1.56 in (40 mm), the stiffness ratio from Eq. (7.52) is approximately 0.1.

ATC 72 (2010) reviews results of laboratory tests and concludes that coupling beams with $l_n/h \geq 2$ are dominated by flexure. For such beams, it recommends $E_cI_e \approx 0.15E_cI_g$ and $G_c = 0.4E_c$. For beams with $l_n/h \leq 1.4$, elastic deformations due to flexure and shear are about equal, and nonlinear behavior is dominated by shear deformations. For such beams, it recommends $E_cI_e \approx 0.15E_cI_g$ and $G_c = 0.1E_c$. ATC 72 recommends linear interpolation of effective stiffness values for $1.4 < l_n/h < 2.0$.

7.14.3 Slender Walls

Shear stiffness of slender walls in the linear range of response can be estimated using the same methods used for slender beams (Section 7.14.1), including stiffness before cracking [Eq. (7.46)] and effective stiffness near yielding [Eqs. (7.47) through (7.50)]. Figure 7.69 presents measured flexural and shear deformations along a height equal to the wall length ($h = l_w$) for a slender wall. As shown in Figure 7.69*b*, the cracked stiffness model of Eq. (7.50) provides a close estimate of the secant shear stiffness near yielding.

For walls that have been proportioned with shear strength exceeding the shear corresponding to flexural strength, the common practice is to assume yielding occurs only in flexure while shear retains its original cracked-section stiffness. If this approach is implemented in a computer model, calculated shear deformations will reach their peak near the yield point without further increase as the wall flexural deformations increase. Test results, however, show that flexural and shearing deformations are coupled, such that flexural yielding causes additional shear deformations even if the wall transverse reinforcement is in the linear range and concrete struts are not crushing. This effect is illustrated in Figure 7.69, where it can be seen that apparent yielding in flexure and shear occur simultaneously. In general, if yielding occurs in flexure, the ratio of shear to flexural displacements in the hinge region remains approximately constant with increasing displacement demand (Oesterle et al., 1980; Beyer et al., 2011). On the other hand, if shear failure occurs without flexural yielding, the ratio of shear to flexural displacements increases with increasing displacement demand (Beyer et al., 2011).

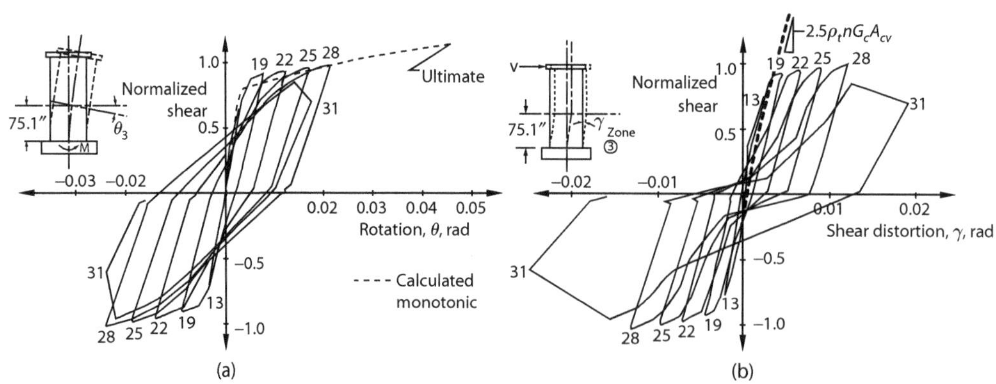

FIGURE 7.69 Measured flexural and shear force deformation relations. (*After Oesterle et al., 1980, courtesy of American Concrete Institute.*)

Oesterle et al. (1980) explained the coupling between flexural and shear yielding in terms of simple kinematics. If a wall has sufficient shear to develop inclined cracks, then yielding of the vertical (flexural) reinforcement leads to rotation of the concrete struts, which is manifest both as flexural rotation yielding and as shear distortion yielding. This coupling results in large effective shear distortions within the flexural hinging zone, but not in higher elevations where flexural hinging does not occur.

7.14.4 Squat Walls

Sozen et al. (1992) report a study of stiffness of squat cantilever walls resisting in-plane shear. According to their findings, the shear stiffness defined by Eq. (7.46) adequately represents initial shear stiffness of the wall web. Shear cracking strength is well represented by the following expression:

$$V_{cr} = 4A_{cv}\sqrt{f_c'}\sqrt{\left(1+\frac{P}{4\sqrt{f_c'}A_g}\right)}, \text{ psi}$$
$$= 0.33A_{cv}\sqrt{f_c'}\sqrt{\left(1+\frac{P}{0.33\sqrt{f_c'}A_g}\right)}, \text{ MPa} \tag{7.53}$$

The load-deformation relation following cracking is defined by a secant from the cracking point through a point corresponding to V_{uc} and Δ_{uc}, defined by

$$V_{uc} = V_{cr} + \rho_w f_y A_{cv} \tag{7.54}$$

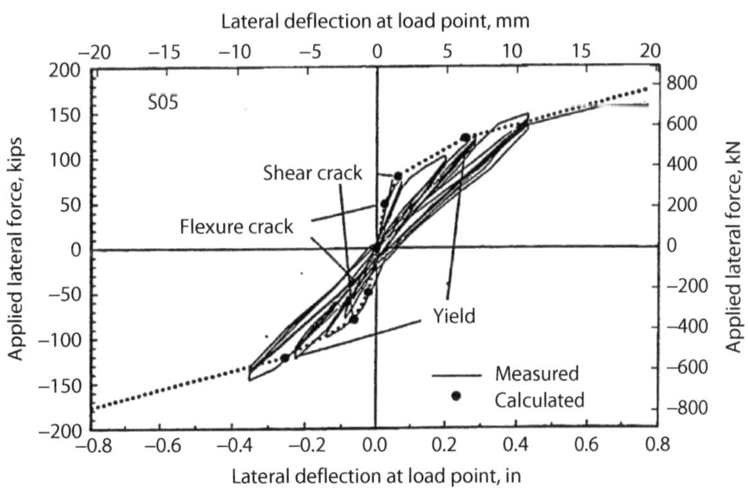

FIGURE 7.70 Comparison of measured shear–displacement relation and calculated envelope. (*After Sozen et al., 1992.*)

in which ρ_w = lesser of the web reinforcement ratios in the two directions, and

$$\Delta_{uc} = \frac{V_{uc} h_w}{\rho_w n G_c A_{cv} / f} \tag{7.55}$$

Figure 7.70 compares measured and calculated shear-displacement relations for a squat wall as reported by Sozen et al. (1992).

References

ACI 318-CE002 (2006). *ACI 318 Change Submittal CE002*, Revised.

ACI 318 (2014). *Building Code Requirements for Structural Concrete (ACI 318-14) and Commentary*, American Concrete Institute, Farmington Hills, MI.

ACI-ASCE 326 (1962). "Shear and Diagonal Tension," *Journal of the American Concrete Institute*, Vol. 59, No. 2, pp. 277–333.

ACI 363 (2010). *Report on High-Strength Concrete (363R-10)*, American Concrete Institute, Farmington Hills, MI, 65 pp.

ACI 445 (1999). *Recent Approaches to Shear Design of Structural Concrete (ACI 445R-99)*, American Concrete Institute, Farmington Hills, MI, 55 pp.

ACI ITG-4.3 (2007). *Report on Structural Design and Detailing for High-Strength Concrete in Moderate to High Seismic Applications (ACI ITG-4.3R-07)*, American Concrete Institute, Farmington Hills, MI, 62 pp.

ACI SP 273 (2010). *Further Examples for the Design of Structural Concrete with Strut-and-Tie Models*, Special Publication 273, American Concrete Institute, Farmington Hills, MI.

Aktan, A.E., and V.V. Bertero (1985). "RC Structural Walls: Seismic Design for Shear," *Journal of Structural Engineering*, Vol. 111, No. 8, pp. 1775–1791.

Ang, B.G., M.J.N. Priestley, and T. Paulay (1989). "Seismic Shear Strength of Circular Reinforced Concrete Columns," *ACI Structural Journal*, Vol. 86, No. 1, pp. 45–59.

ASCE 41 (2013). *Seismic Evaluation and Retrofit of Existing Buildings*, American Society of Civil Engineers, Reston, VA.

ASCE-ACI 426 (1973). "The Shear Strength of Reinforced Concrete Members," *Journal of the Structural Division*, Vol. 99, No. ST6, pp. 1091–1187.

Aschheim, A.M., and J.P. Moehle (1992). *Shear Strength and Deformability of RC Bridge Columns Subjected to Inelastic Cyclic Displacements*, Report No. UCB/EERC 92/04, University of California, Berkeley, CA, 100 pp.

ATC 12 (1983). *Seismic Retrofitting Guidelines for Highway Bridges*, ATC-12/FHWA/RD-83-007, Applied Technology Council/Federal Highway Administration, Washington, D.C., 105 pp.

ATC 72 (2010). *Modeling and Acceptance Criteria for Seismic Design and Analysis of Tall Buildings*, Applied Technology Council , Redwood City, CA. Also available as PEER Report No. 2010/111 at http://peer.berkeley.edu/tbi/wp-content/uploads/2010/09/PEER-ATC-72-1_report.pdf, 222 pp.

Barda, F., J.M. Hanson, and W.G. Corley (1977). "Shear Strength of Low-Rise Walls with Boundary Elements," Special Publication 53, American Concrete Institute, Farmington Hills, MI, pp. 149–201.

Barney, G.B., K.N. Shiu, B.G. Rabbat, A.E. Fiorato, H.G. Russell, and W.G. Corley (1978). *Earthquake Resistant Structural Walls—Tests of Coupling Beams*, R/D Series 1583, Portland Cement Association, Skokie, IL, 146 pp.

Bass, R.A., R.L. Carrasquillo, and J.O. Jirsa (1989). "Shear Transfer across New and Existing Concrete Interfaces," *ACI Structural Journal*, Vol. 86, No. 4, pp. 383–393.

Bentz, E.C., F.J. Vecchio, and M.P. Collins (2006). "Simplified Modified Compression Field Theory for Calculating Shear Strength of Reinforced Concrete Elements," *ACI Structural Journal*, Vol. 103, No. 4, pp. 614–624.

Beyer, K., A. Dazio, and M.J.N. Priestley (2011). "Shear Deformations of Slender Reinforced Concrete Walls under Seismic Loading," *ACI Structural Journal*, Vol. 108, No. 2, pp. 167–177.

Biskinis, D.E., G.K. Roupakias, and M.N. Fardis (2004). "Degradation of Shear Strength of Reinforced Concrete Members with Inelastic Cyclic Displacements," *ACI Structural Journal*, Vol. 101, No. 6, pp. 773–783.

Budek, A.M., M.J.N.Priestley, and C.O. Lee (2002). "Seismic Design of Columns with High-Strength Wire and Strand as Spiral Reinforcement," *ACI Structural Journal*, Vol. 99, No. 5, pp. 660–670.

Caltrans (2010). *Caltrans Seismic Design Criteria, Version 1.6*, California Department of Transportation, Sacramento, 160 pp. Available at http://www.dot.ca.gov/hq/esc/earthquake_engineering/SDC_site/2010-11-17_SDC_1.6_Full_Version_OEE_Release.pdf.

Canbolat, B.A., G. Parra-Montesinos, and J.K. Wight (2005). "Experimental Study on Seismic Behavior of High-Performance Fiber-Reinforced Cement Composite Coupling Beams," *ACI Structural Journal*, Vol. 102, No. 1, pp. 159–166.

Cardenas, A.E., H.G. Russell, and W.G. Corley (1980). "Strength of Low-Rise Structural Walls," Special Publication 63, American Concrete Institute, Farmington Hills, MI, pp. 221–241.

Collins, M.P., E.C. Bentz, and Y.J. Kim (2002). "Shear Strength of Circular Reinforced Concrete Columns," Special Publication 197, American Concrete Institute, Farmington Hills, MI, pp. 45–85.

Collins, M.P., and D. Mitchell (1987). *Prestressed Concrete Basics*, Canadian Precast Prestressed Concrete Institute, Ottawa, Ontario, Canada, 614 pp.

Corley, W.G., A.E. Fiorato, and R.G. Oesterle (1981). "Structural Walls," Special Publication 72, American Concrete Institute, Farmington Hills, MI, pp. 77–131.

Djazmati, B., and J.A. Pincheira (2004). "Shear Stiffness and Strength of Horizontal Construction Joints," *ACI Structural Journal*, Vol. 101, No. 4, pp. 484–493.

Elwood, K.J., and J.P. Moehle (2005a). "Axial Capacity Model for Shear-Damaged Columns," *ACI Structural Journal*, Vol. 102, No. 4, pp. 578–587.

Elwood, K.J., and J.P. Moehle (2005b). "Drift Capacity of Reinforced Concrete Columns with Light Transverse Reinforcement," *Earthquake Spectra*, Vol. 21, No. 1, pp. 71–89.

Fortney, P.J, G.A. Rassati, and B.M. Shahrooz (2008). "Investigation on Effect of Transverse Reinforcement on Performance of Diagonally Reinforced Coupling Beams," *ACI Structural Journal*, Vol. 105, No. 6, pp. 781–788.

Gulec, C.K. (2009). *Performance-based Assessment and Design of Squat Reinforced Concrete Shear Walls*, Doctoral Dissertation, State University of New York at Buffalo, NY, 719 pp.

Gulec, C.K., A.S. Whittaker, and B. Stojadinovic (2008). "Shear Strength of Squat Rectangular Reinforced Concrete Walls," *ACI Structural Journal*, Vol. 105, No. 4, pp. 488–497.

Hidalgo, P.A., C.A. Ledezma, and R.M. Jordan (2002). "Seismic Behavior of Squat Reinforced Concrete Shear Walls," *Earthquake Spectra*, Vol. 18, No. 2, pp. 287–308.

Hines, E.M., and F. Seible (2004). "Web Crushing Capacity of Hollow Rectangular Bridge Piers," *ACI Structural Journal*, Vol. 101, No. 4, pp. 569–579.

Higashi, Y., and M. Hirosawa (1974). "Experimental Research on Ductility of R/C Short Columns under Cyclic Lateral Loads," Preliminary report, IABSE Symposium, Quebec, International Association of Bridge and Structural Engineering, Zurich, Switzerland, pp. 277-296.

Hsu, T.T.C. (1988). "Softened Truss Model Theory for Shear and Torsion," *ACI Structural Journal*, Vol. 85, No. 6, pp. 624–635.

Hwang, S.-J., W.-H. Fang, H.-J. Lee, and H.-W. Yu (2001). "Analytical Model for Predicting Shear Strength of Squat Walls," *Journal of Structural Engineering*, Vol. 127, No. 1, pp. 43–50.

Hwang, S.-J., H.-W. Yu, and H.-J. Lee (2000). "Theory of Interface Shear Capacity of Reinforced Concrete," *Journal of Structural Engineering*, Vol. 126, No. 6, pp. 700–707.

Ichinose, T. (1992). "A Shear Design Equation for Ductile R/C Members," *Earthquake Engineering and Structural Dynamics*, Vol. 21, pp. 197–214.

Iliya, R., and V.V. Bertero (1980). *Effects of Amount and Arrangement of Wall-Panel Reinforcement on Hysteretic Behavior of Reinforced Concrete Walls*, Report No. UCB/EERC-80/04, Earthquake Engineering Research Center, University of California, Berkeley, CA, 188 pp.

Kahn, L.F., and Mitchell, A.D. (2002). "Shear Friction Tests with High-Strength Concrete," *ACI Structural Journal*, Vol. 99, No. 1, pp. 98–103.

Kupfer, H., and T. Baumann (1969). "Staffelung der Biegezugbewehrung bei hohen Schubspannungen in schlanken Stahlbetonträgern mit I-Querschnitt," *Beton-und Stahlbetonbau*, Vol. 64, No. 12, pp. 278–283.

Lee, J.Y., and F. Watanabe (2003). "Shear Deterioration of Reinforced Concrete Beams Subjected to Reversed Cyclic Loading," *ACI Structural Journal*, Vol. 100, No. 4, pp. 480–489.

Leonhardt, F., and R. Walther (1962). *Schubversuche an einfeldrigen Stahlbetonbalken mit und ohne Schubbewehrung*, Deutscher Ausschuss für Stahlbeton, Ernst & Sohn, Berlin, Vol. 151, 83 pp.

Leonhardt, F. (1965). "Reducing the Shear Reinforcement in Reinforced Concrete Beams and Slabs," *Magazine of Concrete Research*, Vol. 17, No. 53, pp. 187–198.

Lequesne, R., M. Setkit, C. Kopczynski, J. Ferzli, M.-Y. Cheng, G. Parra-Montesinos, and J.K. Wight (2011). "Implementation of High-Performance Fiber Reinforced Concrete Coupling Beams in High-Rise Core-Wall Structures," Special Publication 280, American Concrete Institute, Farmington Hills, MI, pp. 7-1–7-12.

Marti, P. (1985, December). "Truss Models in Detailing," *Concrete International*, American Concrete Institute, Farmington Hills, MI, pp. 66–73.

Maruyama, K., H. Ramirez, and J.O. Jirsa (1984). "Short RC Columns under Bilateral Load Histories," *Journal of Structural Engineering*, Vol. 110, No. 1, pp. 120–137.

Mattock, A.H. (1974). "Shear Transfer in Concrete Having Reinforcement at an Angle to the Shear Plane," Special Publication 42, American Concrete Institute, Farmington Hills, MI, pp. 17–26.

Mattock, A.H. (2001). "Shear Friction and High-Strength Concrete," *ACI Structural Journal*, Vol. 98, No. 1, pp. 50–59.

Mattock, A.H., and N.M. Hawkins (1972). "Shear Transfer in Reinforced Concrete—Recent Research," *PCI Journal*, Vol. 17, No. 2, pp. 55–75.

Mattock, A.H., L. Johal, and H.C. Chow (1975). "Shear Transfer in Reinforced Concrete with Moment or Tension Acting Across the Shear Plane," *PCI Journal*, Vol. 20, No. 4, pp. 76–93.

Mau, S.T., and T.T.C. Hsu (1987). "Shear Behavior of Reinforced Concrete Framed Wall Panels with Vertical Loads," *ACI Structural Journal*, Vol. 84, No. 3, pp. 228–234.

Naish, D., A. Fry, R. Klemencic, and J. Wallace (2012). "Reinforced Concrete Coupling Beams—Part I: Testing," *ACI Structural Journal*, Vol. 110, No. 6, pp. 1057–1066.

NZS 3101 (2006). *Concrete Design Standard, NZS3101:2006, Part 1* and *Commentary on the Concrete Design Standard*; *NZS 3101:2006, Part 2*, Standards Association of New Zealand, Wellington, New Zealand.

Oesterle, R.G., J.D. Aristizabal-Ochoa, K.N. Shiu, and W.G. Corley (1984). "Web Crushing of Reinforced Concrete Structural Walls," *ACI Journal*, Vol. 81, No. 3, pp. 231–241.

Oesterle, R.G., A.E. Fiorato, J.D. Aristizabal-Ochoa, and W.G. Corley (1980). "Hysteretic Response of Reinforced Concrete Structural Walls," Special Publication 63, American Concrete Institute, Farmington Hills, MI, pp. 243–273.

Park, R., and T. Paulay (1975). *Reinforced Concrete Structures*, Wiley-Interscience, New York, NY, 769 pp.

Paulay, T. (1974). "Coupling Beams of Reinforced Concrete Shear Walls," *Proceedings, Workshop on Earthquake-Resistant Reinforced Concrete Building Construction*, University of California, Berkeley, CA, pp. 1452–1460.

Paulay, T., and J.R. Binney (1974). "Diagonally Reinforced Coupling Beams of Shear Walls," ACI Special Publication 42, American Concrete Institute, Farmington Hills, MI, pp. 579–598.

Paulay, T., R. Park, and M.H. Phillips (1974). "Horizontal Construction Joints in Cast-in-Place Reinforced Concrete," Special Publication 42, American Concrete Institute, Detroit, MI, pp. 599–616.

Paulay, T., M.J.N. Priestley, and A.J. Synge (1982). "Ductility in Earthquake Resisting Squat Shearwalls," *ACI Journal*, Vol. 79, No. 4, pp. 257–269.

Pilakoutas, K., and A. Elnashai (1995). "Cyclic Behavior of Reinforced Concrete Cantilever Walls, Part 1: Experimental Results," *ACI Structural Journal*, Vol. 92, No. 3, pp. 271–281.

Popov, E.P. (1984). "Bond and Anchorage of Reinforcing Bars under Cyclic Loading," *ACI Journal*, Vol. 81, No. 4, pp. 340–349.

Priestley, M.J.N., R. Verma, and Y. Xiao (1994). "Seismic Shear Strength of Reinforced Concrete Columns," *Journal of Structural Engineering*, Vol. 120, No. 8, pp. 2310–2329.

Pujol, S., J.A. Ramirez, and M.A. Sozen (2006). "Displacement History Effects on Drift Capacity of Reinforced Concrete Columns," *ACI Structural Journal*, Vol. 103, No. 2, pp. 253–262.

Restrepo, J.I., R. Park, and A.H. Buchanan (1995). "Tests on Connections of Earthquake Resisting Precast Reinforced Concrete Perimeter Frames of Buildings," *PCI Journal*, Vol. 40, No. 5, pp. 68–80.

Ritter, W. (1899). "Die bauweise hennebique," *Schweizerische Bauzei-tung*, Vol. 33, No. 7, pp. 59–61.

Schlaich, J., I. Schäfer, and M. Jennewein (1987). "Towards a Consistent Design of Structural Concrete," *Journal of the Prestressed Concrete Institute*, Vol. 32, No. 3, pp. 74–150.

Sezen, H., and J.P. Moehle (2004). "Shear Strength Model for Lightly Reinforced Concrete Columns," *Journal of Structural Engineering*, Vol. 130, No. 11, pp. 1692–1703.

Sezen, H., and J.P. Moehle (2006). "Seismic Tests of Concrete Columns with Light Transverse Reinforcement," *ACI Structural Journal*, Vol. 103, No. 6, pp. 842–849.

Sneed, L.H., and J.A. Ramirez (2010). "Influence of Effective Depth on Shear Strength of Concrete Beams—Experimental Study," *ACI Structural Journal*, Vol. 107, No. 5, pp. 554–562.

Sozen, M.A., P. Monteiro, J.P. Moehle, and H.T. Tang (1992). "Effects of Cracking and Age on Stiffness of Reinforced Concrete Walls Resisting In-Plane Shear," *Proceedings, Fourth Symposium on Current Issues Related to Nuclear Power Plant Structures, Equipment and Piping*, Orlando, FL.

Talbot, A.N. (1909). *Tests of Reinforced Concrete Beams: Resistance to Web Stresses Series of 1907 and 1908*, Bulletin 29, University of Illinois Engineering Experiment Station, Urbana, IL, 85 pp.

Umehara, H., and J.O. Jirsa (1984). "Short Rectangular RC Columns under Bidirectional Loadings," *Journal of Structural Engineering*, Vol. 110, No. 3, pp. 605–618.

Vallenas, J.M., V.V. Bertero, and E.P. Popov (1979). *Hysteretic Behavior of Reinforced Concrete Structural Walls*, Report No. UCB/EERC-79/20, Earthquake Engineering Research Center, University of California, Berkeley, CA, 266 pp.

Varney, J.C., M.D. Brown, O. Bayrak, and R.W. Poston (2011). "Effect of Stirrup Anchorage on Shear Strength of Reinforced Concrete Beams," *ACI Structural Journal*, Vol. 108, No. 4, pp. 469–478.

Vecchio, F.J., and M.P. Collins (1986). "The Modified Compression Field Theory for Reinforced Concrete Elements Subjected to Shear," *ACI Journal*, Vol. 83, No. 2, pp. 219–231.

Von Ramin, M., and A.B. Matamoros (2006). "Effect of Cyclic Loading on the Shear Strength of RC Members," Special Publication 236, American Concrete Institute, Farmington Hills, MI, pp. 103–126.

Wallace, J.W. (1998). "Behavior and Design of High-Strength RC Walls," Special Publication 176, American Concrete Institute, Farmington Hills, MI, pp. 259–279.

Wight, J.K., and J.G. MacGregor (2012). *Reinforced Concrete: Mechanics and Design*, 6th ed., Pearson Education, Upper Saddle River, NJ, 1157 pp.

Wood, S.L. (1990). "Shear Strength of Low-Rise Reinforced Concrete Walls," *ACI Structural Journal*, Vol. 87, No. 1, pp. 99–107.

Woodward, K.A., and J.O. Jirsa (1984). "Influence of Reinforcement on RC Short Column Lateral Resistance," *Journal of Structural Engineering*, Vol. 11, No. 1, pp. 90–104.

CHAPTER 8

Development and Anchorage

8.1 Preview

The mechanics of structural concrete requires that forces be transferred between reinforcement and concrete. In conventional reinforced concrete, this occurs by bond stress that transfers forces gradually or by mechanical devices that transfer forces at essentially point locations. Proper bond and anchorage of reinforcement ensures that reinforcement does not pull out of anchorages, enables the development of moment gradients and shear, and contributes to serviceability through provision of stiffness and crack width control. This chapter considers the nature of bond stress in reinforced concrete as well as requirements for reinforcement development and anchorage.

8.2 Some Observations about Bond and Anchorage

Bond stress between reinforcement and concrete arises from a variety of sources. The three principal sources we are concerned with are illustrated in the idealized cantilever beam of Figure 8.1.

Bond stress develops within the anchorage zone along **ac** as required to anchor the flexural tension force at point **c**. Here, elongation and slip of the reinforcement relative to the surrounding concrete result in local bond stresses along all or part of the anchorage length, leading to reduction in the reinforcement tensile force and, ideally, full anchorage of the reinforcement. Actual variation of bond stress along the anchorage length is a complex function of local bond-slip characteristics (see Chapter 5). A common simplification, illustrated here, is that bond stress is constant within the anchorage until all of the tension force is delivered from the reinforcement to the surrounding concrete. This approach was discussed in Section 6.11.

Bond stress also occurs along shear span **cf**, where variation of moment results in variation of the flexural tension force. The tensile force can vary only if there is bond between reinforcement and concrete. In Figure 8.1 we assume the member is fully cracked such that concrete does not resist tension. Thus, flexural tension force varies linearly along the span. Required bond stress would thus be constant along the span, except at the lap splice.

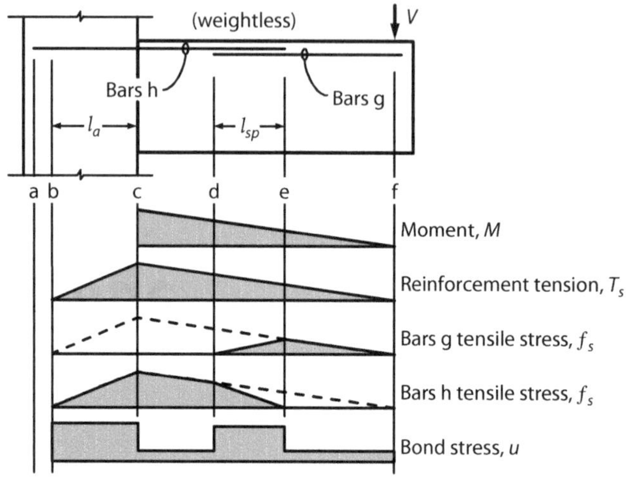

Figure 8.1 Idealized moment, flexural tension force, bar stress, and bond stress in flexural member. The bar stress diagrams assume bars g and h have equal cross-sectional area.

A lap splice of beam longitudinal reinforcement requires that tensile force in one set of reinforcing bars be transferred to the adjacent bars. This occurs through bond stress between the lap-spliced bars and the surrounding concrete along the lap length. In Figure 8.1, we idealize the stress transfer by showing uniform bond stress along lap splice length **de**.

High bond stress in flexural members commonly is manifest in a series of splitting cracks observed on the concrete surface, running along the longitudinal bars generating the bond stress. At the ultimate condition, excessive bond stress may burst the concrete cover, thereby freeing the bars to slip relative to the concrete (Figure 8.2). Failure can be rather sudden and brittle if there is no

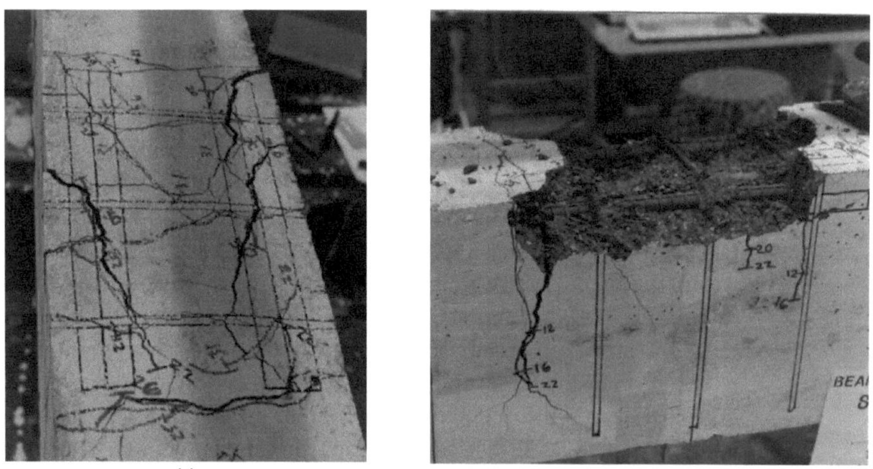

(a) (b)

Figure 8.2 Bond splitting failure of lap splice in a beam. (*After DeVries and Moehle, 1991.*)

transverse reinforcement confining the developed or spliced reinforcement. Transverse reinforcement confining a set of developed or spliced bars can increase strength and ductility. Where bars are anchored in an adjacent anchorage zone, such as a beam-column joint, either straight bars, hooked bars, headed bars, or alternative arrangements can be used, each with its own requirements and potential failure modes. These subjects are discussed more fully in the remainder of this chapter.

8.3 Relations among Bond Stress and External Forces

In this section we develop relations between bond stress and external forces. Some of these were derived in previous chapters. We limit the discussion to straight reinforcing bars, either anchored, developed, or spliced as shown in Figure 8.1.

Within anchorage zone ***ac*** of Figure 8.1, bond stress is required to anchor the embedded reinforcing bars and prevent pullout failure. Assuming uniform bond stress $\bar{u}$ along the anchorage length l_a, the total bond force per bar is $l_a \pi d_b \bar{u}$. This must develop the force per bar $T_{si} = f_{s,max} \pi d_b^2/4$. Equating and solving for l_a, the required anchorage length is

$$l_a = \frac{f_{s,max} d_b}{4\bar{u}} \tag{8.1}$$

Alternatively, given development length l_a, the required bond stress can be solved as

$$\bar{u} = \frac{f_{s,max} d_b}{4 l_a} \tag{8.2}$$

Equation (8.1) was first introduced in Section 6.11.

Along shear span ***cf***, bond stress enables the flexural tension force to vary as required by the varying internal moment. Assuming the member to be fully cracked such that concrete does not resist tension, the flexural tension force in longitudinal reinforcement varies in proportion with the moment. In the example of Figure 8.1, the tensile force varies linearly along the span; thus, the bond stress must be constant (except at the lap splice). In Chapter 7 we used the equilibrium requirement $V = \frac{dM}{dx}$ to derive the following expression:

$$V = \frac{dM}{dx} = \frac{d(T_s jd)}{dx} = \frac{dT_s}{dx} jd \tag{8.3}$$

This solution, in which jd is assumed constant, describes what is referred to as *beam action*. See Chapter 7 for *arch action* and additional details.

In a short length Δx of the beam shear span, reinforcing steel would experience change in tensile force equal to $\frac{dT_s}{dx}\Delta x$, as shown in Figure 8.3. Horizontal force equilibrium requires bond stress u satisfying the relation

$$u \sum (\pi d_b) \Delta x = \frac{dT_s}{dx} \Delta x \tag{8.4}$$

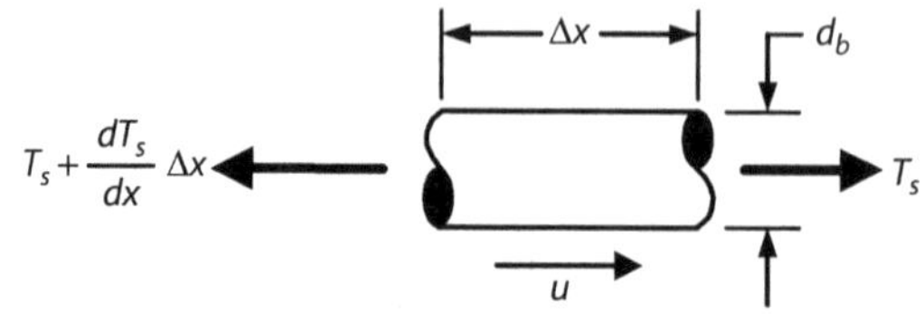

FIGURE 8.3 Free-body diagram of short length of reinforcing bar.

in which $\Sigma(\pi d_b)$ refers to the sum of circumferences of all bars resisting force T_s. Solving for u results in

$$u = \frac{dT_s}{dx}\left(\frac{1}{\Sigma(\pi d_b)}\right) = \frac{V}{\Sigma(\pi d_b)jd} \tag{8.5}$$

Thus, bond stress u between longitudinal reinforcement and surrounding concrete is directly proportional to shear V.

Along lap splice **de**, the tensile force in one set of bars must be transferred to the other set through the surrounding concrete. Assuming constant bond stress along the lap length, and linear moment diagram as occurs in this example, bar stress varies linearly from maximum value at the start of the lap to zero at the free end. If the lap splice occurs along a shear span, then tensile forces at opposite ends of the splice are not equal. In this case, the more heavily stressed bars have higher bond stress (assuming bars are of equal number and size). In Figure 8.1, bars **h** are more heavily loaded and develop higher bond stress. Thus, the critical bond stress is

$$\bar{u} = \frac{f_{s,max} d_b}{4 l_{sp}} \tag{8.6}$$

8.4 Bond Mechanics

As described by ACI 408 (2003), transfer of force between deformed reinforcement and the surrounding concrete occurs by

- chemical adhesion between the bar and concrete;
- frictional forces arising from roughness of the interface, forces transverse to the bar surface, and relative slip between the bar and surrounding concrete; and
- mechanical anchorage or bearing of bar deformations (ribs) against surrounding concrete.

See Figure 8.4. Chemical adhesion is relatively weak; after it is broken, it provides no appreciable resistance to slip. Once slip initiates, bearing forces on the ribs and friction forces on the ribs and barrel of the bar are mobilized. Of these, bearing and friction forces acting on the ribs provide most of the resistance.

Figure 8.5 depicts a deformed reinforcing bar being pulled from a concrete section. Tensile strain and elongation of the bar result in slip of the bar relative to the surrounding concrete, which generates bearing at the ribs. The action of bearing on the concrete, and resulting shear stresses in the concrete, may be

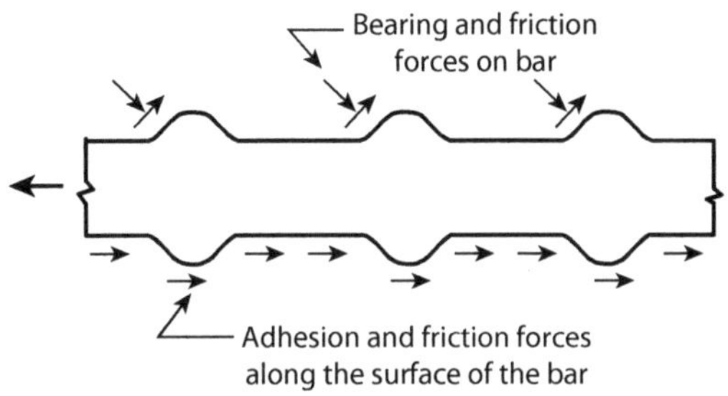

FIGURE 8.4 Force transfer between deformed reinforcement and surrounding concrete. (*After ACI 408, 2003, courtesy of American Concrete Institute.*)

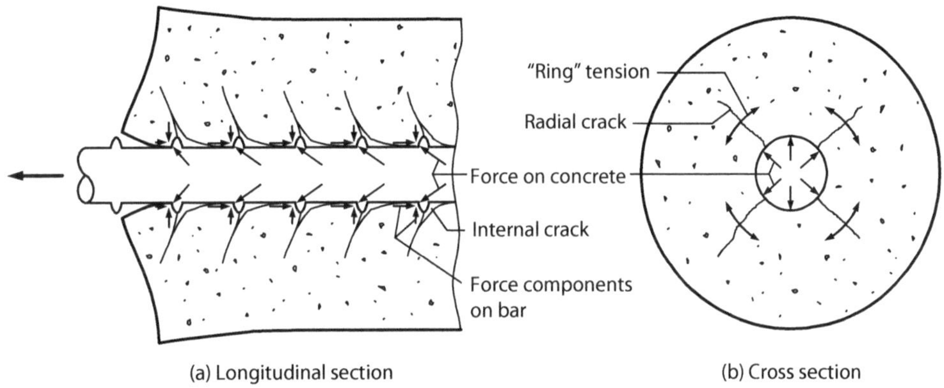

FIGURE 8.5 Internal stresses and cracks in zone of high bond stress for deformed reinforcement. (*Longitudinal section after Goto, 1971, courtesy of American Concrete Institute.*)

sufficient to cause internal cracking, as shown in Figure 8.5*a* (Goto, 1971). Slip is largest at the loaded end. Also, concrete shear deformations tend to peel the concrete away from the bar near the loaded end (Lutz and Gergely, 1967). As a result, at advanced stages of loading, bond stresses may be reduced at the loaded end, with higher bond stresses occurring within the anchorage.

Figure 8.5*b* illustrates a cross section of the concrete in which the centrally located reinforcement has been "cut" from the section and replaced by the internal bearing force that reinforcement produces on the surrounding concrete. Only the radial component of the bearing force is shown. The radial force produces circumferential (or "ring") tension in the surrounding concrete, which may result in radial cracks. These cracks propagate both toward the exterior of the concrete and along the length of the reinforcement.

The relative magnitude of the bearing and frictional forces in Figure 8.4, and the orientation of the force vector acting on the surrounding concrete in Figure 8.5, depend on inclination of the ribs and surface conditions of the reinforcement. Epoxy-coated reinforcement has lower frictional resistance and therefore develops larger radial or bursting force for a given longitudinal force on the reinforcing bar. Thus, epoxy-coated bars require longer development length than uncoated bars.

In members with cover on the order of two bar diameters or less, radial cracks (Figure 8.5*b*) may propagate to the concrete free surface and extend along the embedded bar. Figure 8.6 illustrates the types of splitting cracks that may occur in a beam or column with high bond stress. Where bars are widely spaced (Figure 8.6*a* and *b*), a critical condition is reached when two or more cracks reach the concrete surface, loosening a wedge of concrete, and freeing the bar to slip with significantly reduced resistance. Where bars are more closely spaced (Figure 8.6*c*), cracks may propagate both between individual bars and between outer bars and the concrete free surface, in which case a single failure surface occurs that is roughly aligned with the plane of the embedded reinforcement. Transverse reinforcement crossing a failure surface clamps the crack surfaces together and may increase bond strength (Figure 8.6*d*). Transverse reinforcement may also improve ductility, as the reinforcement will continue to clamp the embedded bars even after failure initiates.

From these cases, we can anticipate that anchorage strength will depend on the quality of concrete surrounding the anchored bars (typically measured by concrete tensile strength), the concrete cover and spacing between bars, and the quantity of transverse reinforcement. We can also anticipate that top-cast bars (Figure 8.6*a*) will be weaker than otherwise identical bottom-cast bars (Figure 8.6*b*) because of plastic settlement of concrete away from the bars as well as increased water–cement ratio due to bleeding, both of which result in a weak lens of concrete beneath top-cast bars. Finally, epoxy-coated bars will develop less friction and consequently less bond strength than otherwise equivalent uncoated bars.

In members having generous amounts of cover, bar spacing, or transverse reinforcement, the splitting strength can be increased to the point that an alternative failure mode becomes more likely. In such members, failure may occur by shearing the concrete along a surface bounded by the outer circumference of the reinforcement ribs (Figure 8.7). Provision of additional cover or transverse reinforcement generally does not increase strength associated with this failure mode.

An objective in design of reinforced concrete members is to provide sufficient cover, spacing, transverse reinforcement, and embedded length such that embedded bars are able to develop required strength without bond failure. In design for gravity and wind loads, it may be sufficient that the bar be capable of reliably

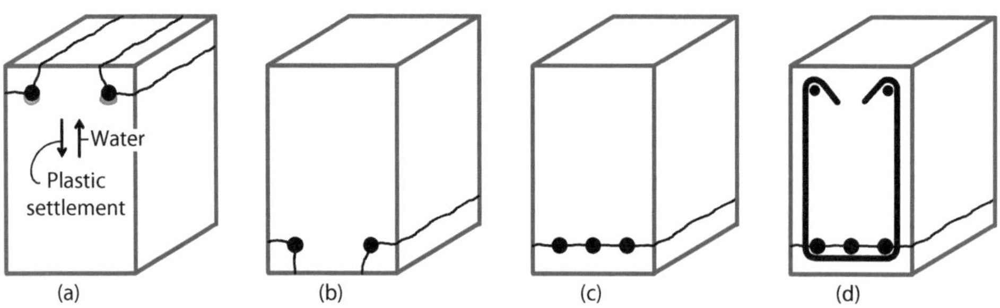

Figure 8.6 Splitting failures in rectangular sections.

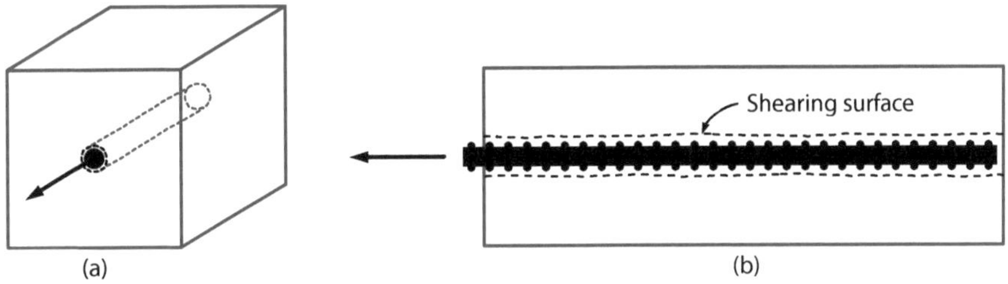

FIGURE 8.7 Shearing failure in member with large cover or heavy transverse reinforcement.

developing the yield strength of the reinforcement. In seismic design applications, it may be necessary to develop the full bar strength and associated inelastic strain demands under reversed loading cycles. In the immediately following sections we will address requirements for the former design conditions, that is, requirements for developing bar yield strength. Subsequent sections will consider additional requirements for seismic design.

8.5 Bond Strength of Deformed Reinforcement

8.5.1 Empirical Relations

Tests for bond strength and development length in flexural members typically are done using beam tests of the type shown in Figure 8.8. In such tests, one or more bars are developed along a shear span or are lap spliced in a region of constant moment or a shear span. Generally the test specimen is designed such that the developed or lapped bar fails in bond prior to yielding. See Orangun et al. (1977), Sozen and Moehle (1990), and ACI 408 (2003) for summaries of these types of tests.

Assuming uniform bond stress $\bar{u}$ along the development or lap length, the bond strength from a test is

$$\bar{u} = \frac{f_{s,max} d_b}{4 l_a} \tag{8.7}$$

Equation (8.7) repeats Eq. (8.2). To calculate $\bar{u}$, stress $f_{s,max}$ is taken equal to the bar stress at which bond failure occurs and either l_d or l_{sp} is substituted for l_a, depending on whether it is a development test or a splice test.

A usual approach is to gather bond strength data from large numbers of tests with variable embedment conditions, from which to derive empirical relations for bond strength. Tepfers (1973) advanced a bond strength model in which strength

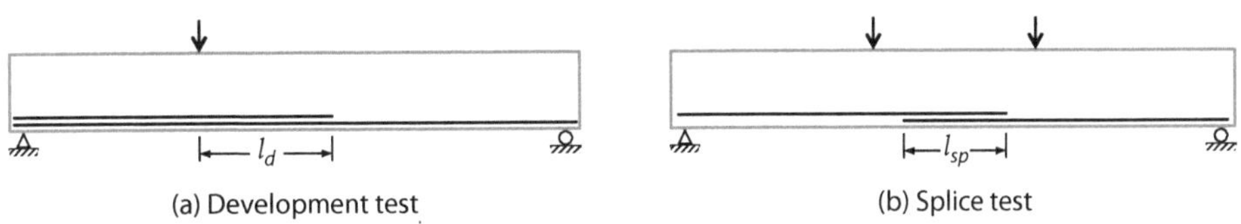

FIGURE 8.8 Typical test configurations for bond tests.

was determined by the weakest splitting plane, considering side cover, top (or bottom) cover, bar clear spacing, and transverse reinforcement, as suggested by Figure 8.6. Orangun et al. (1977) adopted this model and conducted a statistical analysis of available test data to define a best-fit empirical bond strength model. Their model is expressed as

$$\bar{u} = \left[1.2 + \frac{3C}{d_b} + \frac{50d_b}{l_a} + \frac{A_{tr} f_{yt}}{500\, nsd_b}\right]\sqrt{f'_c},\ \text{psi}$$

$$\bar{u} = \left[1.2 + \frac{3C}{d_b} + \frac{50d_b}{l_a} + \frac{A_{tr} f_{yt}}{3.45\, nsd_b}\right]\frac{\sqrt{f'_c}}{12},\ \text{MPa} \tag{8.8}$$

The following definitions and conditions apply to the terms in Eq. (8.8):

- The term C represents the effective clear cover or clear spacing, and is taken as the smallest of c_{c1}, c_{c2}, and $c_{c3}/2$. Figure 8.9 defines the terms. For the example shown, if bottom bars are being developed, it would appear from the figure that the smallest of these three quantities is $c_{c3}/2$. Thus, the weakest section for splitting failure is the clear spacing between the developed bottom bars, indicating that a horizontal splitting plane determines the strength. If top bars are being developed, it would appear from the figure that the smallest dimension is either c_{c1} or c_{c2}, indicating that splitting of the concrete cover determines the strength.
- For large values of the ratio C/d_b, shearing failure becomes more likely than splitting failure. Based on the available test data, Orangun et al. proposed the limit $C/d_b \le 2.5$.
- Bond strength varies with ratio d_b/l_a. This can be understood by considering the bond-slip relation (see, e.g., Figure 5.20). Bond stress reaches peak value for slip on the order of 0.01 in (0.2 mm), then decreases for increasing slip. A very short bonded length is capable of developing the peak bond strength

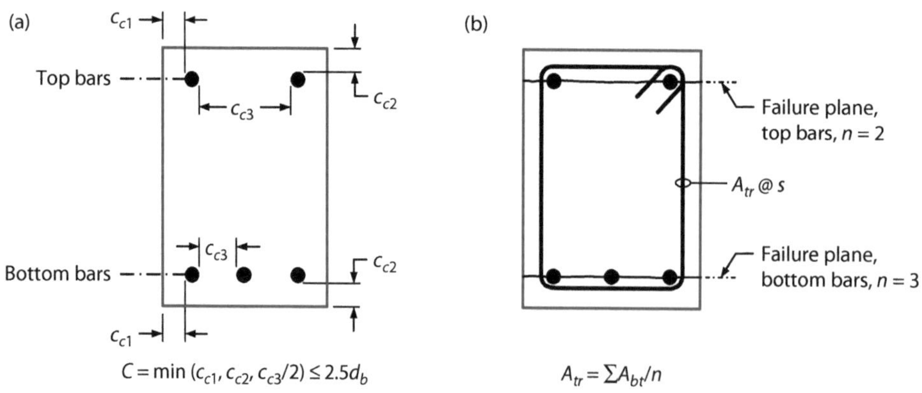

FIGURE 8.9 Definitions of terms defining failure mode and bond strength for Orangun et al. (1977).

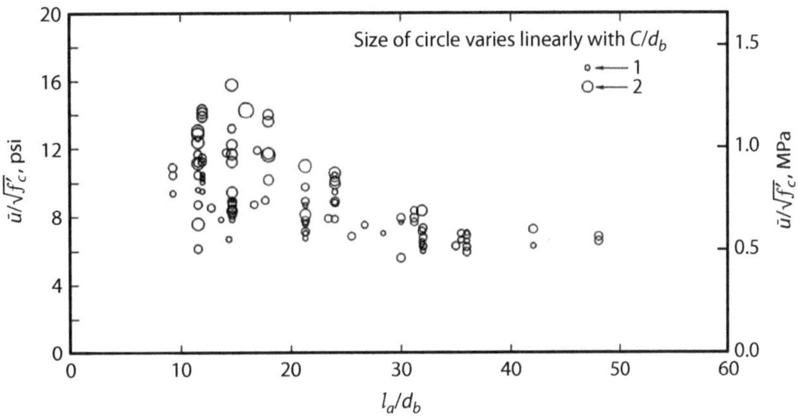

FIGURE 8.10 Variation of average bond stress with l_a/d_b. *(After Sozen and Moehle, 1990.)*

over its entire length. A longer bonded length, however, develops appreciable slip at the loaded end before reaching peak bar force. This results in reduced bond stress at the loaded end. Consequently, the average bond stress along the length is less than the peak bond stress for large values of d_b/l_a. This trend is evident in Figure 8.10.

- A_{tr} is the total cross-sectional area of all transverse reinforcement within spacing s that crosses the potential plane of splitting through the reinforcement being developed and n is the number of reinforcing bars being developed or lap spliced. Thus, A_{tr}/n represents an average area of transverse reinforcement clamping each developed or spliced bar. In reference to the beam shown in Figure 8.9*b*, $n = 2$ for top bars and 3 for bottom bars. Thus, $A_{tr}/n = 2A_{bt}/2 = A_{bt}$ for top bars, and $A_{tr}/n = 2A_{bt}/3$ for bottom bars. The quantity $A_{tr}f_{yt}/nd_bs$ can be defined as a transverse reinforcement index that describes the average clamping force per unit length divided by bar diameter of the clamped bar, assuming transverse reinforcement yields. Figure 8.11 plots the relation between average bond strength and the transverse reinforcement index. Transverse reinforcement is most effective in small quantities, reaching maximum effectiveness for $A_{tr}f_{yt}/nd_bs \approx 1500$ psi (≈ 10 MPa). Orangun et al. (1977) recommend this as a maximum value of the transverse reinforcement index. In addition to its effect on strength, transverse reinforcement improves the ductility of anchorage zones that are stressed to their capacity. Note that high bond and high shear commonly occur together. Examination of test data demonstrates that stirrups stressed in shear can be effective in resisting bond splitting as well.

Figure 8.12 compares results of Eq. (8.8) with test results for both development tests and lap splice tests. Note that the fit is similar for development length and for splice length, suggesting that developed bars and lap spliced bars develop the same strength (on average) for a given length.

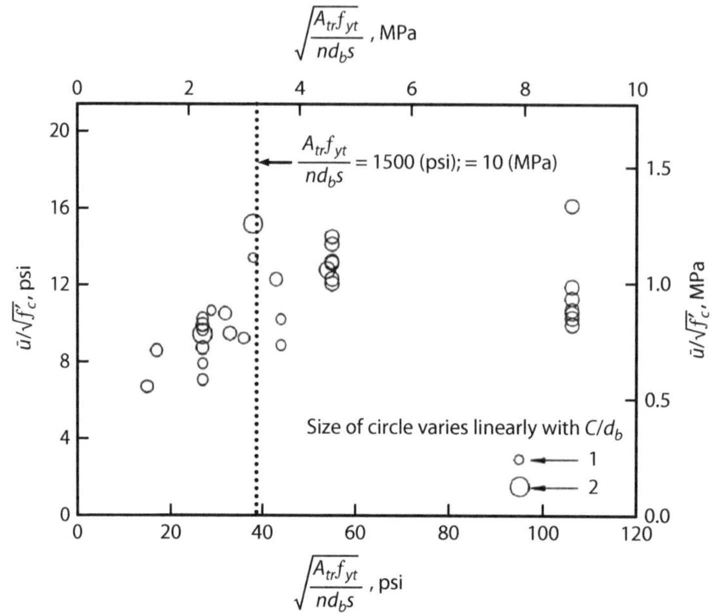

Figure 8.11 Variation of average bond stress with transverse reinforcement index. (*After Sozen and Moehle, 1990.*)

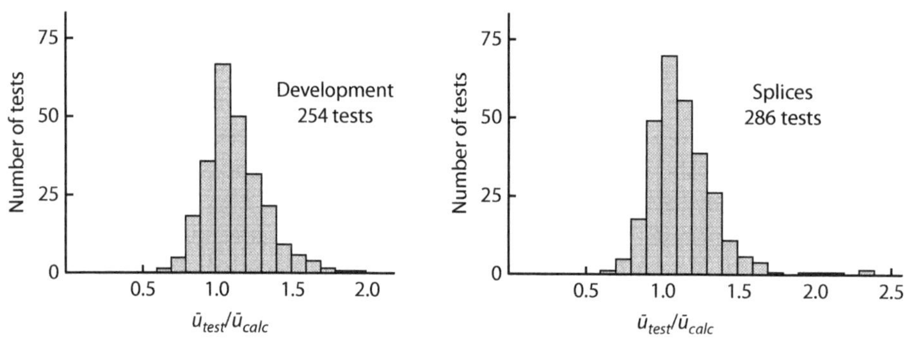

Figure 8.12 Comparison of measured and calculated average bond strengths at failure. (*After Jirsa et al., 1979, courtesy of American Concrete Institute.*)

8.5.2 ACI 318 Provisions for Development of Deformed Bars and Deformed Wires in Tension

ACI 318 (2014) prescribes two alternative approaches for determining required development length for non-seismic applications. The first is derived from the bond strength expression of Orangun et al. (1977) as presented in Eq. (8.8). A simplified approach is also prescribed as a conservative alternative. These two approaches are presented in the following text.

The first approach was developed in Jirsa et al. (1979). It involves combining Eqs. (8.8) and (8.1), adding a safety factor of 1.25, and simplifying the final design equations through a series of approximations. Deliberations of ACI Committee 318 resulted in further modifications. The final form of the development length provisions is as follows.

For deformed bars or deformed wire in tension, required development length l_d is

$$l_d = \left(\frac{3}{40} \frac{f_y}{\lambda\sqrt{f_c'}} \frac{\psi_t \psi_e \psi_s}{\left(\frac{c_b + K_{tr}}{d_b} \right)} \right) d_b, \text{ psi}$$

$$= \left(\frac{9}{10} \frac{f_y}{\lambda\sqrt{f_c'}} \frac{\psi_t \psi_e \psi_s}{\left(\frac{c_b + K_{tr}}{d_b} \right)} \right) d_b, \text{ MPa} \tag{8.9}$$

in which confinement term $\frac{c_b + K_{tr}}{d_b}$ is not taken greater than 2.5 and

$$K_{tr} = \frac{40 A_{tr}}{sn} \tag{8.10}$$

in which n is the number of bars or wires being spliced or developed along the potential plane of splitting. All variables are as defined previously, except c_b = smaller of (a) distance from center of a bar or wire to nearest concrete surface and (b) one-half the center-to-center spacing of bars or wires being developed. The length obtained by Eq. 8.9 is to be multiplied by the various modification factors listed in Table 8.1.

The second approach permitted by ACI 318 calculates the basic development length using the applicable expression in Table 8.2. These expressions are based on a re-evaluation of bond strength data considering typical cover, spacing, and transverse reinforcement conditions (Sozen and Moehle, 1990). The length obtained by the applicable expression in Table 8.2 is to be multiplied by the various modification factors listed in Table 8.1.

ACI 318 also permits development length to be shortened where excess reinforcement area is provided, because in such cases the bars will not be fully stressed to f_y. For this purpose, the development length calculated by the preceding methods is permitted to be reduced by multiplying it by the ratio $(A_{s,required})/(A_{s,provided})$. This modification factor should not be used where lateral deformations under earthquake loading may result in reinforcement yielding.

8.5.3 ACI 318 Provisions for Development of Deformed Bars and Deformed Wire in Compression

Development of bars in compression produces a more favorable stress field and enables shorter length than required in tension. According to ACI 318, development

Modification Factor	Condition	Value of Factor
Lightweight λ	Lightweight concrete	0.75
	Lightweight concrete, where f_{ct} is specified	See ACI 318
	Normalweight concrete	1.0
Epoxy* ψ_ε	Epoxy-coated or zinc and epoxy dual-coated reinforcement with clear cover less than $3d_b$ or clear spacing less than $6d_b$	1.5
	Epoxy-coated or zinc and epoxy dual-coated reinforcement for all other conditions	1.2
	Uncoated or zinc-coated (galvanized) reinforcement	1.0
Size ψ_σ	No. 7 (22) and larger bars	1.0
	No. 6 (19) and smaller bars and deformed wires	0.8
Casting position* ψ_τ	More than 12 in (305 mm) of fresh concrete placed below horizontal reinforcement	1.3
	Other	1.0

*The product $\psi_\tau \psi_\varepsilon$ need not exceed 1.7.

TABLE 8.1 Modification Factors for Development of Deformed Bars and Deformed Wires in Tension

Spacing and Cover	No. 6 (19) and Smaller Bars and Deformed Wires	No. 7 (22) and Larger Bars
Clear spacing of bars or wires being developed or spliced not less than d_b, clear cover not less than d_b, and stirrups or ties throughout l_d not less than the Code minimum or Clear spacing of bars or wires being developed or spliced not less than $2d_b$ and clear cover not less than d_b	$l_d = \left(\frac{f_y \psi_t \psi_e}{25\lambda\sqrt{f'_c}}\right) d_b$, psi $= \left(\frac{12 f_y \psi_t \psi_e}{25\lambda\sqrt{f'_c}}\right) d_b$, MPa (8.11) (a)	$l_d = \left(\frac{f_y \psi_t \psi_e}{20\lambda\sqrt{f'_c}}\right) d_b$, psi $= \left(\frac{3 f_y \psi_t \psi_e}{5\lambda\sqrt{f'_c}}\right) d_b$, MPa (8.11) (b)
Other cases	$l_d = \left(\frac{3 f_y \psi_t \psi_e}{50\lambda\sqrt{f'_c}}\right) d_b$, psi $= \left(\frac{18 f_y \psi_t \psi_e}{25\lambda\sqrt{f'_c}}\right) d_b$, MPa (8.11) (c)	$l_d = \left(\frac{3 f_y \psi_t \psi_e}{40\lambda\sqrt{f'_c}}\right) d_b$, psi $= \left(\frac{9 f_y \psi_t \psi_e}{10\lambda\sqrt{f'_c}}\right) d_b$, MPa (8.11) (d)

TABLE 8.2 Alternative Development Length Requirements for Deformed Bars or Deformed Wire in Tension

Modification Factor	Condition	Value of Factor
Lightweight λ	Lightweight concrete	0.75
	Lightweight concrete, where f_{ct} is specified	See ACI 318
	Normalweight concrete	1.0
Confining reinforcement ψ_r	Reinforcement enclosed within any of the following: 1. a spiral 2. a circular continuously wound tie with $d_b \geq 0.25$ in (6 mm) and pitch ≤ 4 in (100 mm) 3. No. 4 (13) ties as required for columns and spaced ≤ 4 in (100 mm) on center	0.75
	Other	1.0

Table 8.3 Modification Factors for Development of Deformed Bars and Deformed Wires in Compression

length, l_{dc}, for bars or wires in compression is the greater of (1) and (2), multiplied by the modification factors in Table 8.3, but not less than 8 in (203 mm).

1. $\left(\dfrac{f_y \psi_r}{50\lambda\sqrt{f_c'}}\right)d_b$, psi $\left[\left(\dfrac{6 f_y \psi_r}{25\lambda\sqrt{f_c'}}\right)d_b, \text{MPa}\right]$
2. $0.0003 f_y \psi_r d_b$, psi ($0.043 f_y \psi_r d_b$, MPa)

8.6 Lap Splices

Maximum lengths of reinforcing bars generally are limited by transportation restrictions; in the United States the maximum length generally available is 60 ft (18 m). Bars can be spliced to achieve greater length, or to otherwise provide reinforcement continuity. Lap splices can be used where sufficient length is available, where congestion is not a problem, and where permitted by the contract documents. Alternatively, mechanical splices can be used (Section 8.7).

Lap splices are formed by overlapping two or more bars by a length that is sufficient to enable force transfer between the bars. A *contact lap splice* is one in which the bars are in contact with one another, and a *non-contact lap splice* (also called a *spaced lap splice*) is one in which the bars are not in contact with one another. Some earlier building codes, out of concern that concrete would not completely surround bars if they were in contact, required lapped bars to be spaced. Today, most codes permit both spaced and contact lap splices, and treat them equally.

8.6.1 Tension Lap Splices

In a lap splice, force from one bar is transferred to the surrounding concrete and from there to the adjacent bar. As suggested by Figure 8.13, in the absence of transverse confining reinforcement, the radial component of the bond stress must be resisted entirely by concrete tensile stresses acting perpendicular to the plane of the splice. Failure commonly is by splitting along this plane (Sagan et al., 1991).

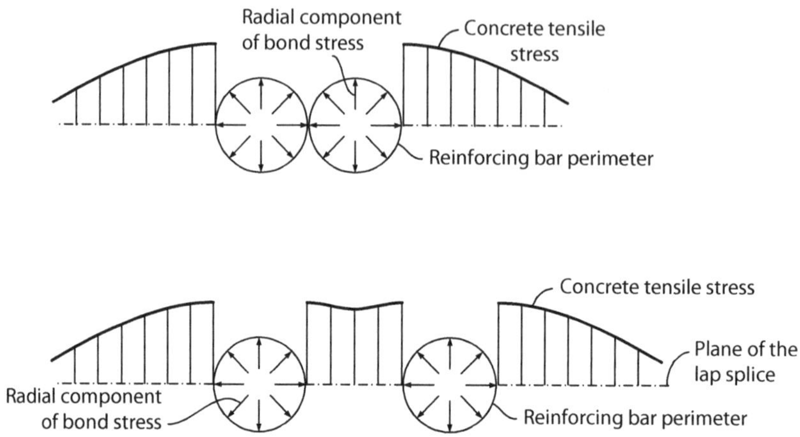

FIGURE 8.13 Radial component of bond stress and resulting concrete tensile stresses acting perpendicular to the plane of the lap splice.

The force path between two bars in a non-contact lap splice can be identified by cracking patterns observed during tests (Figure 8.14). The cracks suggest development of a well-defined truss mechanism in which concrete compression diagonals transfer forces between the bars. The transverse component of the diagonal compression struts must be resisted by concrete in tension, with additional tension resisted by reinforcement (if any) placed perpendicular to the lap splice. Lap splices also typically develop a major crack at the ends of the lapped bars. The crack width increases with bar size, leading many codes to place a limit on the maximum bar size permitted to be spliced by lapping. ACI 318 limits the bar size to No. 11 (36) and smaller.

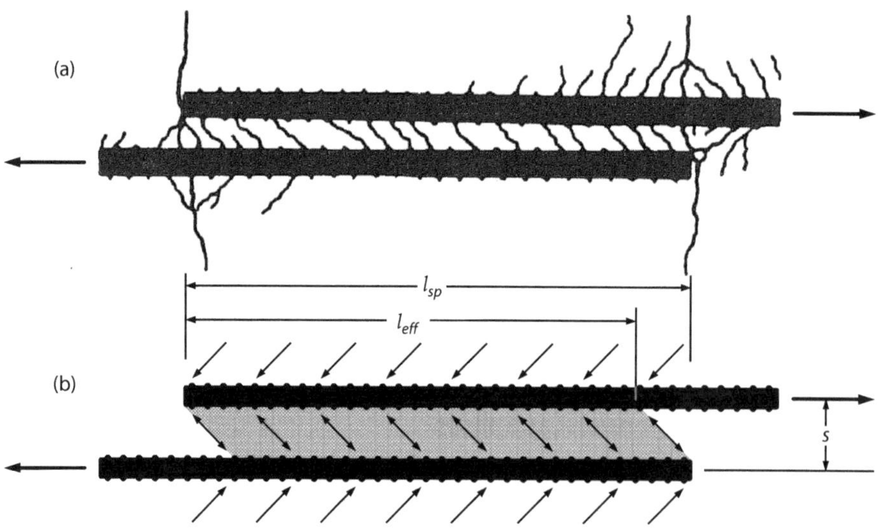

FIGURE 8.14 Non-contact lap splice: (*a*) crack pattern (*after Goto and Otsuka, 1979*); (*b*) force transfer mechanism.

Assuming a 45° inclination for the diagonal compression struts in Figure 8.14*b*, the effective length of a lap splice is reduced to approximately $l_{eff} = l_{sp} - s$, suggesting there should be a limit on the spacing between lapped bars. In ACI 318, the maximum spacing between lapped bars is the lesser of one-fifth of the required lap splice length and 6 in (150 mm). Tests summarized by ACI 408 (2003) show that splice strength is relatively insensitive to spacing within these limits.

As noted in Section 8.5.1, lap-spliced bars and developed bars achieve essentially the same bond strength under monotonic loading, all other factors being constant. Most building codes, however, require additional length for lapped bars. According to ACI 318, required lap splice length for bars in tension is given in Table 8.4.

Although not specifically addressed by ACI 318, lap splices can also be made using headed reinforcement. See Section 8.10.

8.6.2 Compression Lap Splices

In contrast with tension lap splices, compression lap splices develop bearing against the end of the bar, which strengthens the splice (ACI 408, 1966; Cairns and Arthur, 1979). Cairns and Arthur (1979) note that location of ties within the splice has a strong influence on strength, and recommend that ties always be provided close to the splice ends. In addition to end bearing, the compressive stress field associated with compression lap splices increases unit bond strength in comparison with tension lap splices. Thus, compression lap splices can be shorter than tension lap splices.

According to ACI 318, lap splice length, l_{sc}, of No. 11 (36) and smaller deformed bars in compression is according to (1) or (2), but not less than (3), and subject to (4):

1. For $f_y \leq 60{,}000$ psi (414 MPa), $l_{sc} = 0.0005 f_y d_b$ psi ($0.071 f_y d_b$ MPa)
2. For $f_y > 60{,}000$ psi (414 MPa), $l_{sc} = (0.0009 f_y - 24) d_b$ psi [$(0.13 f_y - 24) d_b$ MPa]
3. l_{sc} shall not be less than 12 in (300 mm)
4. For $f'_c < 3000$ psi (21 MPa), the length of lap obtained from (1), (2), or (3) shall be increased by one-third

$A_{s,provided}/A_{s,required}$* Over Length of Splice	Maximum Percent of A_s Spliced within Required Lap Length	Splice Type	Required Splice Length, l_{st}	
≥ 2.0	50	Class A	Greater of:	$1.0 l_d$ and 12 in (305 mm)
≥ 2.0	100	Class B	Greater of:	$1.3 l_d$ and 12 in (305 mm)
< 2.0	All cases	Class B	Greater of:	$1.3 l_d$ and 12 in (305 mm)

*Ratio of area of reinforcement provided to area of reinforcement required by analysis at splice location.

Table 8.4 Lap Splice Lengths, l_{st}, of Deformed Bars and Deformed Wires in Tension

ACI 318 does not permit compression lap splices of bars larger than No. 11 (36) except where they are spliced with smaller-diameter bars.

8.7 Mechanical Splices

Mechanical splices are an alternative to lap splices. They may be preferred where available lengths are relatively short (as in connections between precast elements or between new and existing construction) or where reinforcement congestion is high. Mechanical splices may be required by the building code in certain conditions, including splices of No. 14 (43) and No. 18 (57) bars; where spacing of reinforcing bars is insufficient to permit lap splicing; where tension lap splices are long and consequently intrude into regions where lap splices are not permitted; in tension tie members; and within yielding regions of special moment frame members.

There are two primary types of mechanical splices in the United States. Type 1 mechanical splices are required to develop in tension or compression, as required, at least 125% of the specified yield strength of the bar. These splices are used in elements where there is little concern for inelastic deformations and elevated tensile stresses associated with response to earthquakes. Type 2 splices are required to develop the specified tensile strength of the bars being spliced. Type 2 splices are intended for use where inelastic response associated with seismic loading is anticipated.

Figure 8.15 compares stress–strain relations for A615 and A706 Grade 60 reinforcing bars versus various stress limits applicable to mechanical splices. The

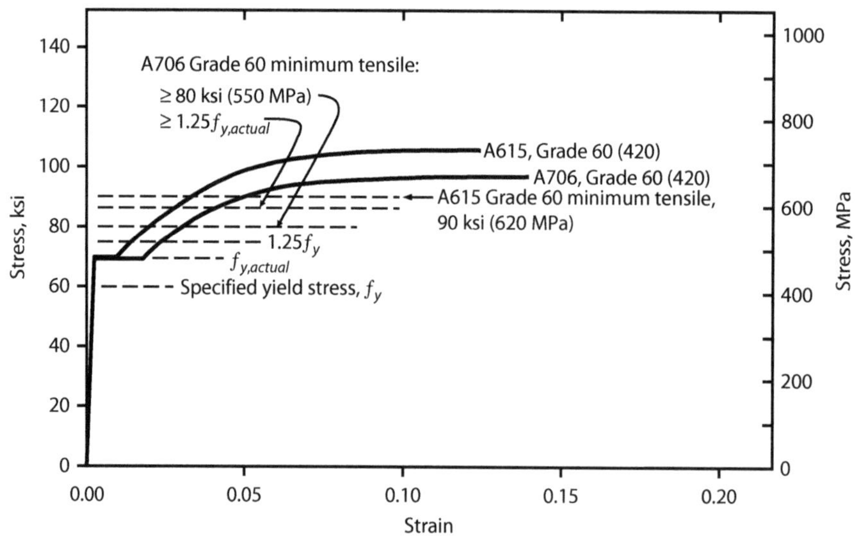

FIGURE 8.15 Stress–strain relations for A615 and A706 Grade 60 reinforcement based on mean properties from mill tests compared with various minimum strength requirements relevant to mechanical splices.

stress–strain relations represent average values from a large number of mill tests as presented in Chapter 2. Type 1 mechanical splices are only required to develop $1.25f_y$. Therefore, Type 1 bars are likely to develop the yield strength for average bar properties. There is no upper limit on f_y for A615 bars, and for A706 Grade 60 (420) bars [and A615 bars substituted for A706 Grade 60 (420) bars] the upper limit is $f_y = 78{,}000$ psi (540 MPa). Therefore, some bars may have strength exceeding the minimum required strength of a Type 1 splice. Type 2 mechanical splices are required to develop at least the specified tensile strength of the bar [90,000 psi (620 MPa) for A615 and 80,000 psi (550 MPa) but not less than $1.25f_{y,actual}$ for A706]. Therefore, some strain ductility is assured. The available ductility and the soundness of the splice under multiple strain reversals, however, may not be sufficient for some applications. The engineer should specify only splices that meet the performance requirements for the design, and may consider using only those splices that have been demonstrated to break the bar at full tensile strength. This information can be obtained by consulting with the splice manufacturer.

Many different types of mechanical splices are available from a range of manufacturers. Figure 8.16 depicts various generic splice types that manufacturers have claimed are capable of meeting Type 2 performance requirements. Each splice type has particular physical and installation features. Splice performance, components and available options, spacing and cover requirements, matching of

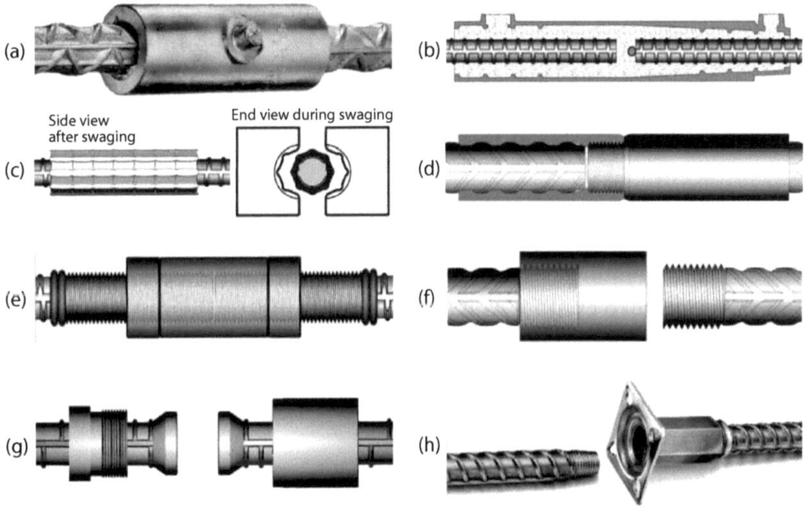

FIGURE 8.16 Various mechanical splice types: (*a*) steel-filled coupling sleeve; (*b*) grout-filled coupling sleeve (section); (*c*) cold-swaged steel coupling sleeve; (*d*) extruded steel coupler with parallel threaded ends (partial section); (*e*) friction-welded bar position coupler with parallel threaded ends; (*f*) threaded coupler with upsized bar threads, cold-forged; (*g*) upset bar and coupling sleeve with straight threads; (*h*) coupler with taper threads and mounting plate (to splice bars across construction joints after removal of formwork). (*After ACI 439.3R-07, 2007, courtesy of American Concrete Institute.*)

bar end alignments, bar coatings, installation requirements, and installed cost should be considered in the design process. Additional information is available in ACI 439.3R-07 (2007) or from the splice manufacturer.

8.8 Welded Splices

Welded splices primarily are intended for large bars [No. 6 (19) or larger]. The usual requirement is to develop at least $1.25f_y$. ACI 318 provides requirements and commentary on the welding process. The AWS D1.4 Welding Code (2011) covers aspects of welding reinforcing bars, including criteria to qualify welding procedures.

When welding of reinforcing bars is required, the weldability of the steel and compatible welding procedures must be considered. Weldability of the steel is based on its chemical composition or carbon equivalent (CE). The Welding Code establishes preheat and interpass temperatures for a range of carbon equivalents and reinforcing bar sizes. ASTM A706 covers low-alloy steel reinforcing bars intended for applications requiring controlled tensile properties or welding. Weldability is accomplished in ASTM A706 by limits or controls on chemical composition and on carbon equivalent. ASTM A706 requires the producer to report chemical composition and carbon equivalent. For bars other than ASTM A706, the producer does not routinely provide the chemical analysis required to calculate carbon equivalent. Where bars other than A706 are to be welded, the design drawings or project specifications should specifically require results of the chemical analysis to be furnished. Where no mill test report is available (e.g., welding to bars in an existing building), it may be possible to conduct a chemical analysis on representative bars. If the chemical composition is not known or obtained, the Welding Code requires a minimum preheat.

Welding of bars should be performed in accordance with AWS D1.4. It should also be determined if additional precautions are in order, based on other considerations such as stress level in the bars, consequences of failure, and heat damage to existing concrete due to welding operations.

8.9 Hooked Anchorages

8.9.1 Standard Hooks

Building codes use the term *standard hook* to define a specific reinforcement detail comprising a straight lead in length, a bend, and a tail. In ACI 318, standard hooks have a straight lead in length and one of the following:

1. 90° bend plus $12d_b$ extension at the free end, or
2. 180° bend plus $4d_b$ extension, but not less than 2.5 in (64 mm) at the free end

Standard hooks also have minimum bend diameters. Figure 8.17 shows details of these standard hooks.

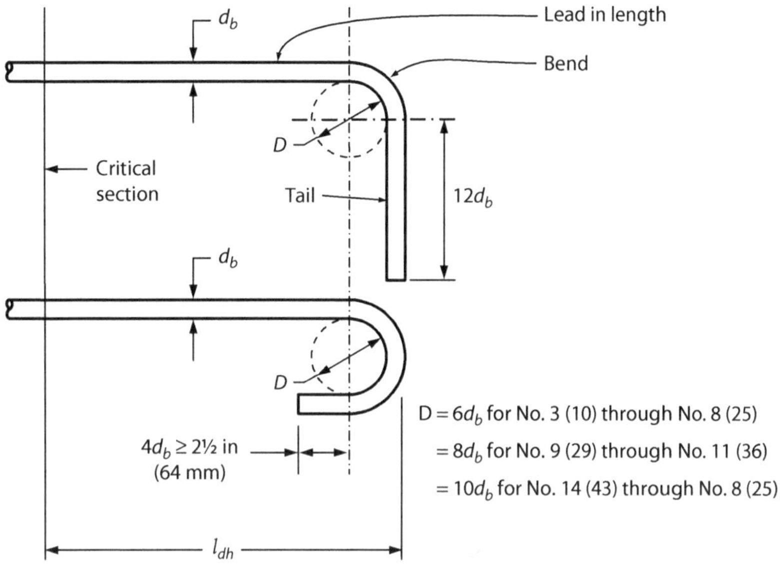

FIGURE 8.17 Hooked bar details for development of standard hooks.

ACI 318 also defines stirrup and tie hooks as follows:

1. No. 5 (16) bar and smaller: 90° bend plus $6d_b$ extension at free end.
2. No. 6 (19), No. 7 (22), and No. 8 (25) bar: 90° bend plus $12d_b$ extension at free end.
3. No. 8 (25) bar and smaller: 135° bend plus $6d_b$ extension at free end.
4. *Seismic hook:* A hook on a stirrup or crosstie having bend not less than 135°, except that circular hoops shall have a bend not less than 90°; a hook shall have $6d_b$ [but not less than 3 in (76 mm)] extension that engages the longitudinal reinforcement and projects into the interior of the stirrup or hoop.

For stirrup and tie hooks, diameter of bend measured on the inside of the bar is as shown in Figure 8.17, except for No. 3 (10) through No. 5 (16) the diameter of bend is permitted to be $4d_b$.

8.9.2 Force Transfer Mechanism

When a hooked anchorage is subjected to tension, some bond force is transferred along the straight lead in length, with the remainder transferred by the bend. To study the force transfer mechanism at the bend, Minor and Jirsa (1975) conducted tests of hooked bars in which the lead in length was unbonded (Figure 8.18). The hook was embedded in a block of concrete restrained by concentrated forces similar to those that develop in a beam-column joint. Width of the block was sufficient to prevent splitting failure of the block.

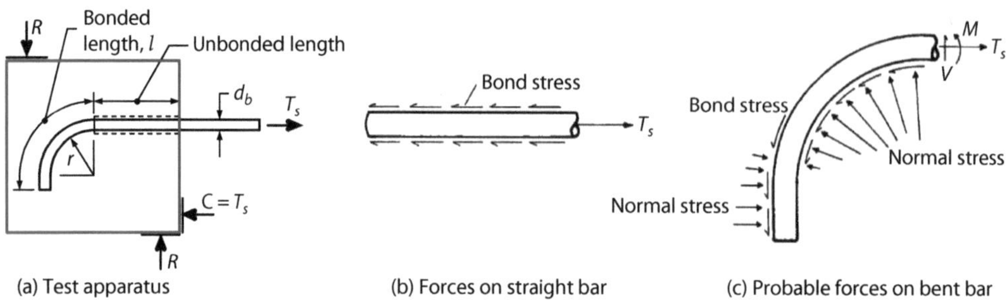

FIGURE 8.18 Tests to understand force transfer mechanism in hooked bars. (*After Minor and Jirsa, 1975, courtesy of American Concrete Institute.*)

The tests had the following primary conclusions:

- In a straight bar anchorage without hook, bond stress is resisted along the entire length of the anchored bar. In a hooked anchorage, however, most of the force is resisted by bearing along a relatively short length on the inside of the bend. The tail does not contribute appreciably to anchorage resistance. Instead, the tail tends to straighten out, leading to normal stresses on the outside portion of the tail acting in the direction of the applied tension (Figure 8.18*c*).
- The concentrated bearing stresses in a hooked anchorage result in it being more flexible than an otherwise equivalent straight anchorage having the same bonded length (Figure 8.19*a*). The slip is larger for 180° bends than for 90° bends, and increases with decreasing bend radius *r*.

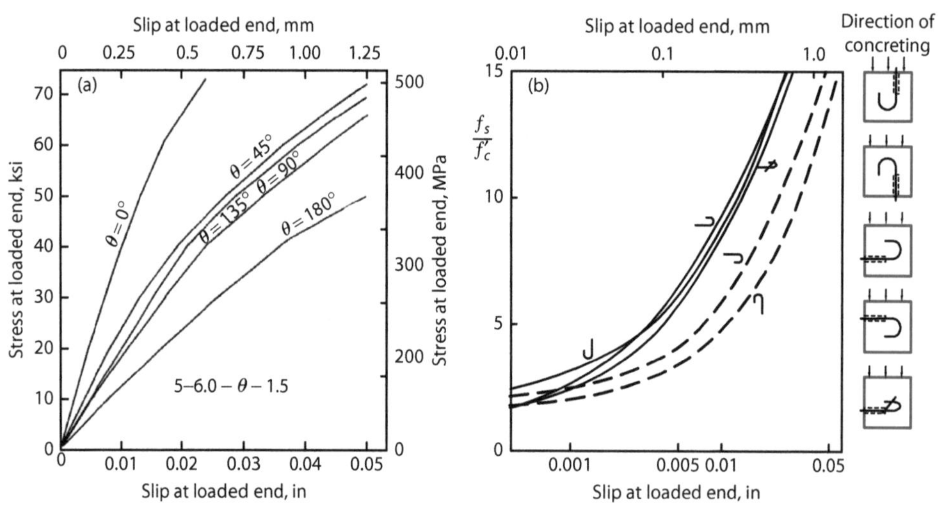

FIGURE 8.19 Load–slip relations. (*a*) Results from Minor and Jirsa (1975), *courtesy of American Concrete Institute*: for given load, slip increases going from included angle 0° (straight) to 180°. (*b*) Results from Rehm (1969): for given load, hooks are more flexible when the inside of bend faces downward, presumably because of plastic settlement and bleeding effects on the critical bearing surface of the bend.

Figure 8.19*b* shows that hooked anchorages have largest slip when they are cast with the inside of the hook oriented downward, which is consistent with accumulation of a soft lens of concrete inside the bend for these casting positions.

Marques and Jirsa (1975) report tests on hooked bar anchorages using either 90° or 180° hooks conforming to the geometries in Figure 8.17, fully bonded along bar length, with side cover and transverse steel variations similar to those occurring in typical construction. In these tests, there did not appear to be significant difference in stiffness of anchorages with 90° or 180° hooks. In almost all tests, high bearing stresses on the inside of the bend caused side face splitting, with sudden and complete failure (Figure 8.20). Strength was improved by increasing the side face cover to the hooked bar, by adding closely spaced transverse reinforcement, or both. Because failure was by side face splitting, column axial compression had no apparent effect on anchorage strength and could, conceivably, be detrimental. Tests by Hamad et al. (1993) showed that epoxy-coated hooked bars consistently developed lower anchorage strengths and stiffnesses than companion uncoated hooked bars.

8.9.3 ACI 318 Provisions

According to ACI 318, development length l_{dh} for deformed bars in tension terminating in a standard hook is

$$l_{dh} = \left(\frac{f_y \psi_e \psi_c \psi_r}{50\lambda\sqrt{f_c'}} \right) d_b \geq 8d_b \text{ and } 6 \text{ in, psi}$$

$$= \left(\frac{6}{25} \frac{f_y \psi_e \psi_c \psi_r}{\lambda\sqrt{f_c'}} \right) d_b \geq 8d_b \text{ and } 152 \text{ mm, MPa} \tag{8.12}$$

Applicable modification factors are listed in Table 8.5, except it is permitted to take either or both of ψ_c and ψ_r equal to 1.0.

Figure 8.20 Side face failure of hooked anchorages in beam-column joint. (*After Marques and Jirsa, 1975, courtesy of American Concrete Institute.*)

Modification Factor	Condition	Value of Factor
Lightweight λ	Lightweight concrete	0.75
	Normalweight concrete	1.0
Epoxy ψ_e	Epoxy-coated or zinc and epoxy dual-coated reinforcement	1.2
	Uncoated or zinc-coated (galvanized) reinforcement	1.0
Cover ψ_c	For No. 11 (36) bar and smaller hooks with side cover (normal to plane of hook) ≥ 2½ in (64 mm) and for 90° hook with cover on bar extension beyond hook ≥ 2 in (50 mm)	0.7
	Other	1.0
Confining reinforcement ψ_r^*	For 90° hooks of No. 11 (36) and smaller bars having ties or stirrups as shown in Figure 8.21*a* or *b*	0.8
	For 180° hooks of No. 11 (36) and smaller bars having ties or stirrups as shown in Figure 8.21*a* (i.e., ties or stirrups perpendicular to l_{dh})	
	Other	1.0

*The first tie or stirrup shall enclose the bent portion of the hook within $2d_b$ of the outside of the bend, where d_b is the nominal diameter of the hooked bar.

Table 8.5 Modification Factors for Development of Hooked Bars in Tension

Furthermore, for bars being developed by a standard hook at discontinuous ends of members (including ends of simply supported beams, free ends of cantilevers, and ends of members framing into a joint), if both side cover and top (or bottom) cover over hook are less than 2½ in (64 mm), ties or stirrups are required as shown in Figure 8.22. In this case, $\psi_r = 1.0$. The requirements of this paragraph do not apply for hooked bars at discontinuous ends of slabs with confinement provided by the slab continuous on both sides normal to the plane of the hook.

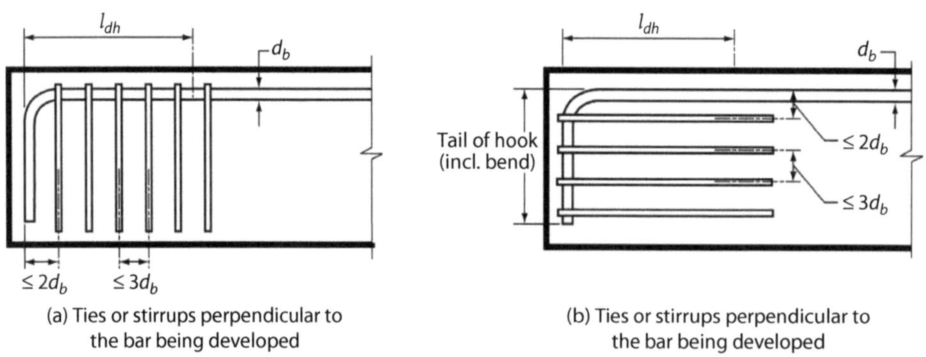

Figure 8.21 Options for placement of transverse reinforcement in cases where side cover ≥ 2.5 in (64 mm) and tail cover over 90° hooks ≥ 2 in (50 mm). (*After ACI 318, 2014, courtesy of American Concrete Institute.*)

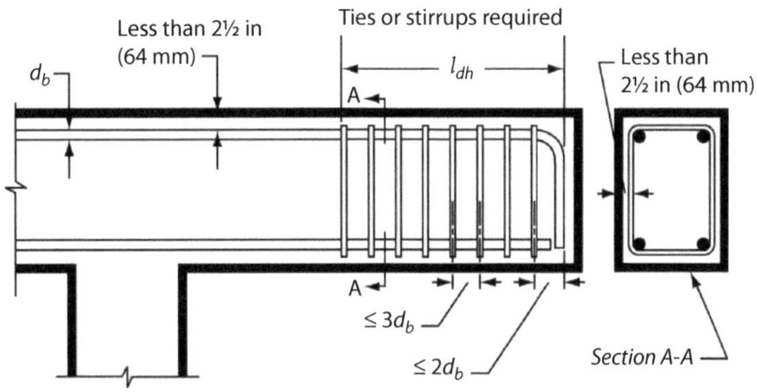

FIGURE 8.22 Required orientation and placement of transverse reinforcement where top (bottom) and side cover over hook ≤ 2.5 in (64 mm) at discontinuous ends of members. *(After ACI 318, 2014, courtesy of American Concrete Institute.)*

According to ACI 318, hooks are not to be considered effective in developing bars in compression. Instead, the lead in length must be sufficient to develop the bar in compression assuming the hook is not present.

8.10 Headed Reinforcement

Headed reinforcement refers to deformed or plain steel reinforcing bars with a head attached to one or both ends. Attachment can be through welding, hot forging, thread connection, or a separate threaded nut to secure the head to the bar. Heads typically have flat bearing faces of circular or rectangular shape. Net bearing area, A_{brg}, is defined as area of the head minus area of the body of the bar including any protrusions caused by the attachment of the head. ACI 318 contains provisions for heads having $A_{brg} \geq 4A_b$. See ASTM A970 (2009) for additional information. Figure 2.3 illustrates an example of a welded head.

8.10.1 Force Transfer Mechanism

Anchorage of headed bars is mobilized in two stages (Figure 8.23). During initial loading stages, most of the bar force is transferred to concrete through bond along the developed length, with maximum bond stress developed before fully developing bearing at the head. In later stages, as bond begins to deteriorate, bar force is transferred to the head. Thus, maximum strength typically is determined by bearing strength of the head acting together with some reduced bond stress (Thompson et al., 2005; Chun et al., 2009).

Bearing failure has been idealized as a driving wedge in front of the bearing surfaces that causes lateral splitting stress (Thompson et al., 2006a). Where bars are located close to and parallel to an exterior face, the side face can burst outward perpendicular to the line of the anchorage in what is known as a *side face blowout* failure. Transverse reinforcement placed perpendicular to headed bars has not been found to increase anchorage strength significantly, although some increase in ductility is observed (DeVries et al., 1999; Thompson et al., 2006b).

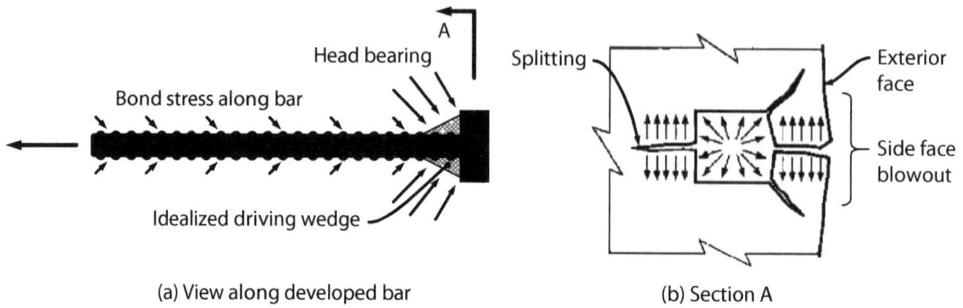

FIGURE 8.23 Force transfer mechanism for headed reinforcement. (*Modified from Thompson et al., 2006a, courtesy of American Concrete Institute.*)

8.10.2 ACI 318 Provisions

According to ACI 318, use of heads to develop deformed bars in tension is limited to conditions (1) through (7):

1. Bar shall conform to ASTM A970 including Annex A1 Requirements for Class HA Head Dimensions.
2. Bar $f_y \le 60{,}000$ psi (414 MPa).
3. Bar size ≤ No. 11 (36).
4. Net bearing area of head $A_{brg} \ge 4A_b$.
5. Normalweight concrete.
6. Clear cover for developed bar $\ge 2d_b$.
7. Clear spacing between developed bars $\ge 4d_b$. (See further discussion below, including an exception for beam-column joints in special moment frames.)

For such headed deformed bars, development length, l_{dt}, is

$$l_{dt} = \left(\frac{0.016 f_y \psi_e}{\sqrt{f'_c}} \right) d_b \ge 8d_b \text{ and } 6 \text{ in, psi}$$

$$= \left(\frac{0.19 f_y \psi_e}{\sqrt{f'_c}} \right) d_b \ge 8d_b \text{ and } 152 \text{ mm, MPa} \tag{8.13}$$

In Eq. (8.13), the value of f'_c is not to be taken greater than 6000 psi (41 MPa), and modification factor ψ_e is 1.2 for epoxy-coated or zinc and epoxy dual coated bars and 1.0 for uncoated or zinc-coated (galvanized) bars. Development length l_{dt} is measured from the critical section to the inside face of the head (Figure 8.24).

The requirement that clear spacing between developed bars be at least $4d_b$ is based on the parameters of those tests that were used to develop the ACI 318 provisions, none of which had multiple closely spaced bars. Kang et al. (2009) review available data for beam-column joints typical of those used in seismic

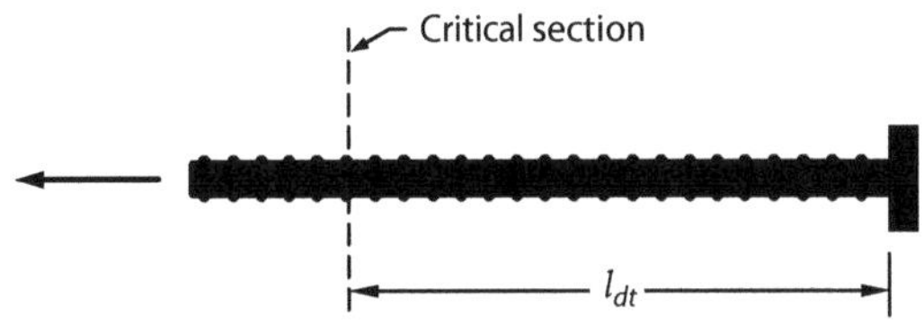

Figure 8.24 Development length for headed deformed bars.

applications. That study concludes that clear bar spacing can be reduced to $2d_b$ in beam-column joints of special moment frames. Following a review of the Kang et al. (2009) database, ACI Committee 318 concluded that the data set justified use of headed bars with clear spacing as close as $3d_b$ in beam-column joints of special moment frames, but no closer. This is the current limit for special moment frames in ACI 318.

Although not specifically addressed by ACI 318, lap splices can also be made using headed reinforcement. Thompson et al. (2006b) present data from a test series on lap splices using headed deformed reinforcement and reinforcement without heads. Bars were tested so as to fail the splice before yielding the bars. Substituting the measured stress f_s for f_y in Eq. (8.13) leads to an equivalent ACI 318 length required to develop f_s. Figure 8.25 plots ratio of required length to length provided in the test, considering only those headed bars that conform to ACI 318 requirements. The ACI required length is conservative in all cases.

Headed reinforcement is used for reinforcement development in tension. Applications in compression have not received much attention and are unknown.

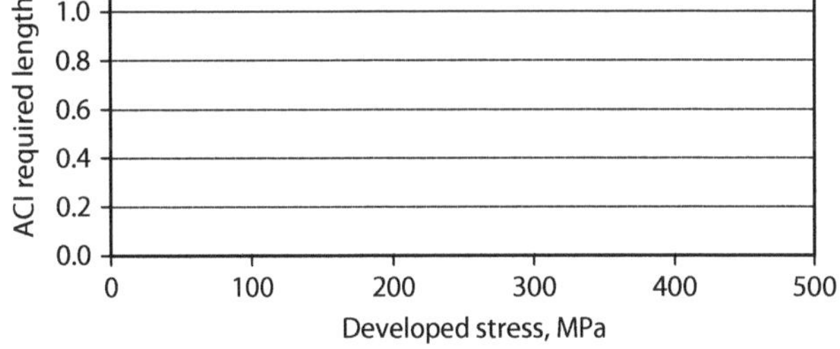

Figure 8.25 Test results for lap splices of headed deformed bars. Vertical axis is the ratio of length required by ACI 318 [substituting f_s for f_y in Eq. (8.13)] to the length provided. Values exceeding 1.0 indicate conservative estimation by ACI 318 provisions. (*Data from Thompson et al., 2006b.*)

8.11 Effects of Inelastic Cyclic Loading

8.11.1 Straight Bar Anchorages

Section 8.4 describes the mechanics of bond under monotonic loading, including the development of cracking around lateral bar deformations (ribs). With bar force reversal, new cracks having reversed inclination may intersect previously developed cracks, resulting in a weakened matrix of cracked and fractured concrete surrounding the developed bar (Figure 8.26). For a bar with relatively small amounts of cover, spacing, and transverse reinforcement, reversed cyclic loading can gradually degrade the bond, effectively "unzipping" the anchorage. This behavior is known qualitatively, but detailed models are lacking (ACI 408.2R-92, 1992).

Bond forces identified in the preceding paragraph have been known to result in bond splitting failures, especially in columns (Figure 8.27). According to Ichinose (1995), full yielding of longitudinal reinforcement within the plastic hinge zone leaves residual cracks that remain open until longitudinal reinforcement yields in compression during a deformation reversal. Thus, longitudinal reinforcement must resist C_s and T_s, both at the yield force level or greater, through bond along the bonded length. Because crack closure within the plastic hinge zone requires reinforcement yielding, the effective bonded length is reduced at each end by the plastic length. Ichinose estimates the plastic length is approximately equal to $0.7d$, which can be approximated as $0.6h$. Thus, bonded length is approximately $l_u - (2 \times 0.6h) = l_u - 1.2h$. This length should be approximately twice the development length l_d of the bars (because the bar is resisting $C_s + T_s$). Thus, the calculated development length l_d must satisfy

$$l_d \le 0.5l_u - 0.6h \tag{8.14}$$

in which l_n is clear span or clear height, and l_d can be calculated according to Eq. (8.9). Alternatively, l_d can be determined according to Table 8.2, although lengths by those equations usually will exceed available length. This aspect usually is important only in stocky members such as columns or conventionally reinforced coupling beams.

Where cover, spacing, and especially transverse reinforcement are sufficient to prevent bond splitting failure, a bond shearing mechanism can develop. Eligehausen

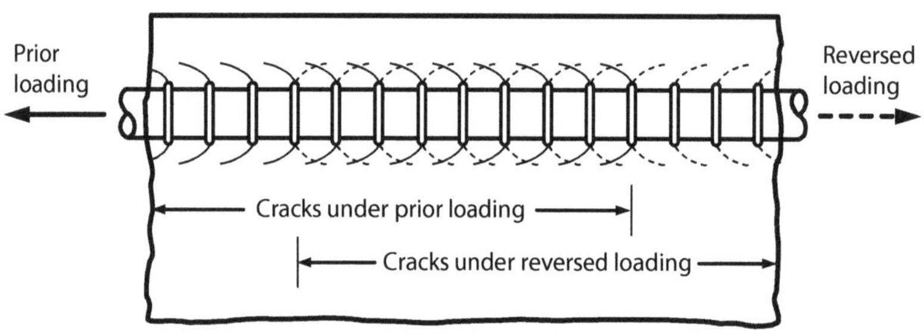

Figure 8.26 Radial cracking under reversed loading. The developed bar is considered to be loaded first in tension from right to left, followed by reversed tension loading from left to right. (*After Hassan and Hawkins, 1977, courtesy of American Concrete Institute.*)

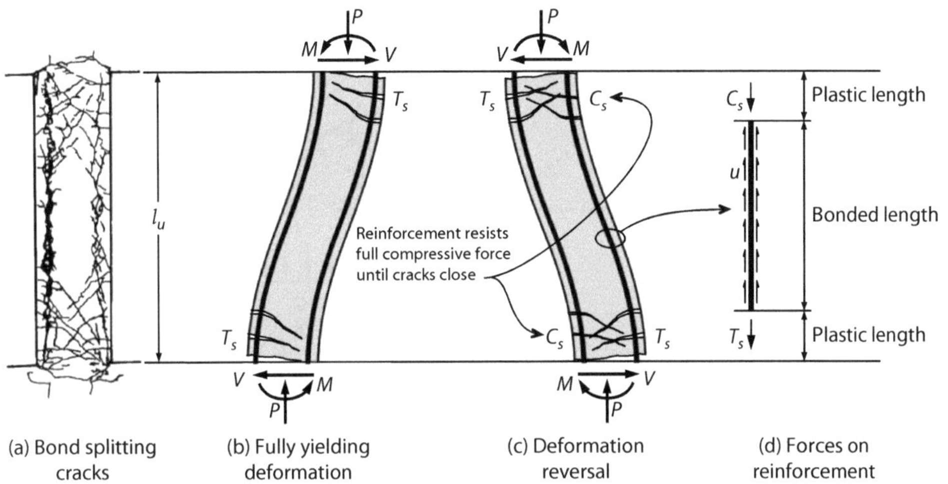

FIGURE 8.27 Actions leading to bond splitting cracks. (*After Ichinose, 1995, courtesy of American Concrete Institute.*)

et al. (1983) describe this behavior and present detailed analytical models. Under monotonic loading, inclined cracks initiate at relatively low bond stress during essentially linear response (Figure 8.28*a*). With increasing slip, concrete at the leading edge of the ribs will become increasingly crushed (Figure 8.28*b* and *c*), leading to development of bond strength and bond shear failure. If, instead of being loaded monotonically to failure, the force is reversed after reaching point **A**, the unloading branch will be very stiff as essentially elastic concrete unloads. Deformation reversal may involve some initial slip as the ribs slide past previously damaged concrete, followed by increasing bond for negative slip values (Figure 8.28*a′*). If prior loading has resulted in greater amount of crushing, including initiation of shear cracks in front of ribs, then additional sliding occurs, and strength may begin to degrade (Figure 8.28*b′*). Multiple reversed cycles at large slip can lead to relatively low resistance because concrete along the cylindrical shear failure surface has been smoothed during previous slip cycles (Figure 8.28*c′*).

Shearing failures will be a prominent consideration in beam-column joints where confinement usually prevents splitting failures but lengths may not be sufficient to prevent shearing failures.

8.11.2 Lap Splices

When a lap splice is subjected to tensile loading, the highest bond stresses initially occur near the two ends of the splice. Where concrete cover is limited to around three bar diameters or less, and where tensile forces approach yield, concrete cracking and splitting may initiate near the splice ends. Loading in the reverse direction may cause cracks to form in the opposing direction, leading to crisscrossed cracks that further weaken the surrounding concrete. Repeated force reversals thus can lead to progressively damaged concrete around the splice, possibly leading to concrete splitting that advances along the splice length, essentially "unzipping" the splice. Unlike concrete cover, which can split and spall away from the splice, closely spaced transverse reinforcement can be effective by

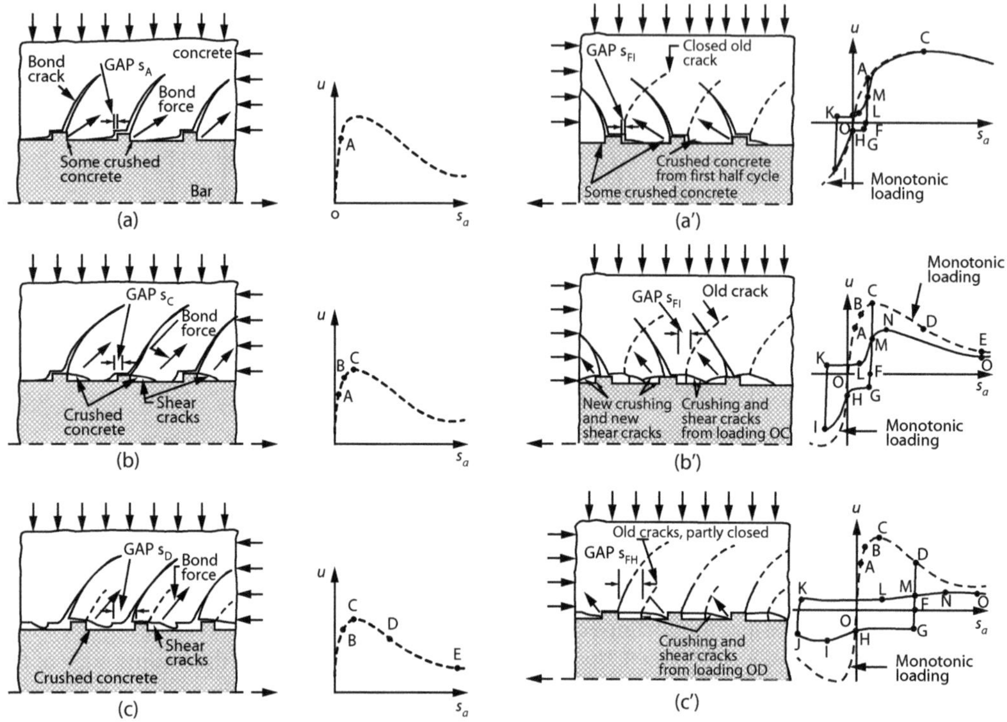

FIGURE 8.28 Mechanism of bond resistance in confined concrete. *(a, b, c)* for monotonic loading. *(a′, b′, c′)* for reversed cyclic loading. *(After Eligehausen et al., 1983, courtesy of University of California, Berkeley.)*

providing a continuous, uniform clamping force along the splice length. Properly proportioned and detailed, transverse reinforcement can enable a splice to develop the yield strength of the spliced bars under multiple inelastic loading cycles.

Alternative models have been described for lap splices confined by transverse reinforcement. Paulay (1982) proposed a truss model in which the lapped bars are the chords, concrete forms diagonal compression struts between the chords, and transverse reinforcement acts as transverse web members or ties equilibrating the transverse component of the diagonal compression struts (Figure 8.29). Assuming diagonal compression struts act at 45° to the longitudinal axis, the total force developed in the transverse reinforcement is equal to the longitudinal tensile force transferred by the splice. Assuming the longitudinal reinforcement is stressed to yield, we have

$$\frac{A_{tr} f_{st} l_{sp}}{s} = T_s = A_{bl} f_y \tag{8.15}$$

Paulay assumes transverse reinforcement develops yield stress. With this assumption, the required spacing of transverse reinforcement along the splice is

$$s \le \frac{A_{tr}}{A_{bl}} \frac{f_{yt}}{f_y} l_{sp} \tag{8.16}$$

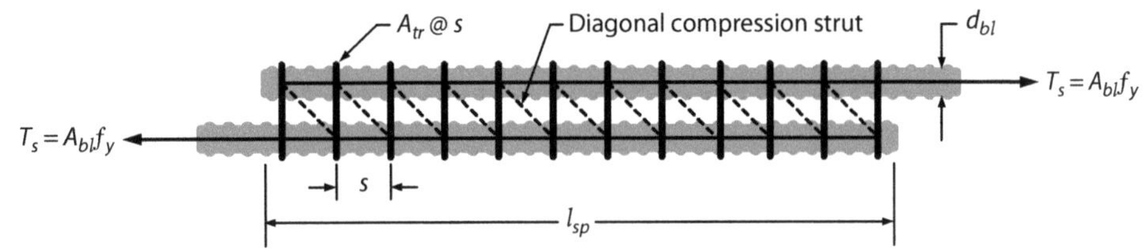

FIGURE 8.29 Strut-and-tie model of lap splice.

Paulay recommends splice length $l_{sp} \geq 30d_{bl}$. Paulay also proposes a stress limit for concrete, which is unlikely to control for typical concrete strengths and Grade 60 (420) reinforcement.

Sivakumar et al. (1983) considered the bursting stress from spliced reinforcement to derive required transverse reinforcement. Considering the free-body diagram of Figure 8.30*b*, the clamping force of transverse reinforcement per unit length must equilibrate bursting force $F_1 + F_2$, where F_1 and F_2 are radial components of forces from bond stresses acting transverse to the splice plane per unit length. Assuming bond stresses act at 45° relative to the splice axis, it can be shown that $F_1 = F_2 = d_{bl}^2 f_y / 4l_{sp}$. In tests reported by Lukose et al. (1982), yield of longitudinal reinforcement penetrated about $0.2l_{sp}$ along the splice, leaving an effective splice length of $0.8l_{sp}$ to resist the full bar yield stress. Combining results and solving for transverse reinforcement spacing gives

$$s \leq 1.6 \frac{A_{tr}}{d_{bl}^2} \frac{f_{st}}{f_y} l_{sp} \approx \frac{5}{4} \frac{A_{tr}}{A_{bl}} \frac{f_{st}}{f_y} l_{sp} \tag{8.17}$$

Note the similarity in formulations of Eqs. (8.16) and (8.17).

Sivakumar et al. (1983) noted that transverse reinforcement did not always yield along the splice length. They also noted improved performance of splices confined with closely spaced small diameter transverse reinforcement as compared with larger diameter hoops at wider spacing. Based on these observations, they proposed an effective limiting stress for transverse reinforcement as

$$f_{st} = 40{,}000/\pi\sqrt{A_{tr}}, \text{ psi } (7000/\pi\sqrt{A_{tr}}, \text{ MPa}) \tag{8.18}$$

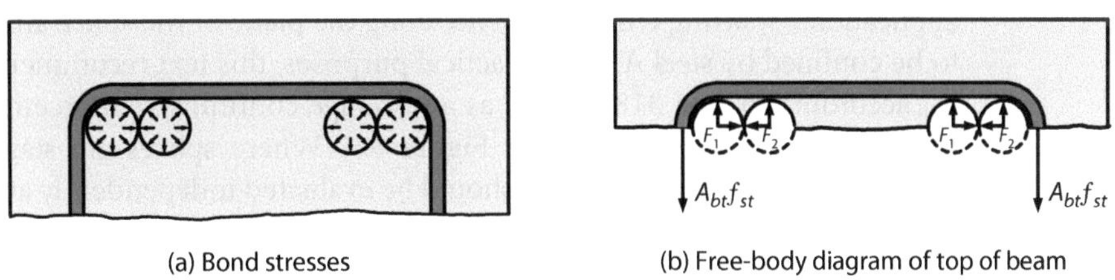

FIGURE 8.30 Confinement of lap splices by transverse reinforcement. (*After Sivakumar et al., 1983, courtesy of American Concrete Institute.*)

Assuming A_{tr} comprises a single reinforcing bar having diameter d_{bt}, such that $A_{tr} = \pi d_{bt}^2/4$, the limiting stress for transverse reinforcement is $f_{st} = 14{,}000/d_{bt}$, psi ($2500/d_{bt}$, MPa). Combining terms, Eq. (8.17) can be rewritten

$$s \le k_{tr} \frac{A_{tr}}{A_{bl}} \frac{36{,}000}{f_y} l_{sp}, \text{ psi} \left(k_{tr} \frac{A_{tr}}{A_{bl}} \frac{250}{f_y} l_{sp}, \text{ MPa} \right) \tag{8.19}$$

in which k is a modifier for diameter of transverse reinforcement expressed as

$$k_{tr} = \frac{0.5}{d_{bt}}, \text{ psi} \left(\frac{13}{d_{bt}}, \text{ MPa} \right) \tag{8.20}$$

Sivakumar et al. recommend maximum spacing of 6 in (152 mm) for transverse reinforcement.[1]

Sivakumar et al. (1983) also noted that a splice along a section with moment gradient will be more resistant to degradation because bond deterioration tends to propagate from one end only rather than from both ends. For this effect, they recommend another modifier k_M for transverse reinforcement spacing as

$$k_M = \frac{1}{1 - \frac{l_{sp}}{2a}} \text{ but not less than 1.0, nor greater than 2.0} \tag{8.21}$$

Sivakumar et al. (1983) also recommend that splice length for bottom-cast, Grade 60 (420) reinforcement satisfy Eq. (8.22).

$$l_{sp} \ge \frac{1860}{\sqrt{f_c'}}, \text{ psi} \left(\frac{3900}{\sqrt{f_c'}}, \text{ MPa} \right) \ge 20 d_{bl} \tag{8.22}$$

ACI 408.2R-92 (1992) adopts the recommendations of Sivakumar et al. (1983). It is noted that for $f_c' \ge 3800$ psi (26 MPa), l_{sp} conservatively may be taken as $30d_{bl}$. It is also recommended that all other modification factors in ACI 318 (presumably for top-cast bars, epoxy-coated bars, and lightweight concrete) be applied where applicable.

In the derivations by Sivakumar et al. (1983), the transverse reinforcement term A_{tr} is evaluated at each spliced bar. Where clear spacing between spliced bars equals or exceeds $4d_{bl}$, the corner bars are deemed most critical and A_{tr} applies only to those bars. Where clear spacing is less, which is most common in seismic applications, splitting tends to extend along the plane of the splice and all bars are to be confined by steel A_{tr}. For practical purposes, this text recommends to define A_{tr} according to ACI 318, that is, as an average confining reinforcement area per longitudinal bar as described in Figure 8.9. Where splices are staggered, as is sometimes done, the value of A_{tr} should be evaluated independently at each pair of spliced bars.

---

[1]Note that the form of equations presented here differs from that in Sivakumar et al. (1983), but the result is identical.

Lukose et al. (1982), Paulay (1982), and Aristizabal-Ochoa (1983) report results of tests on confined lap splices in rectangular beam-column cross sections. The principal findings include the following:

- Load cycles have little effect if load level is below about 75% of monotonic capacity. Lap strength deterioration accelerates as bar stress approaches yield. Reversed cyclic loading is more detrimental than repeated loading.
- Transverse reinforcement is most effective for pre-yielding excursions if located near the splice ends, but for inelastic cycles yield penetrates into the splice past the end stirrups, accelerating degradation if confinement is poor along the splice length. Transverse reinforcement uniformly spaced along the lap splice is preferred. Offset column bars, discussed subsequently, may require additional transverse reinforcement at the offset.
- For a given amount of total transverse reinforcement area, smaller diameter transverse reinforcement at smaller spacing is more effective than larger ties at wider spacing.

For lap splices confined by spirals or circular hoops, two splitting planes are identified (Figure 8.31). Paulay (1982) suggests that radial splitting applies for bars spliced circumferentially while circumferential splitting applies for radial splicing. It seems more likely, however, that splitting would occur in the weakest direction, regardless of the placement of the spliced bars. For radial splitting, the circular hoop or spiral steel is perpendicular to the splitting failure plane and should be fully effective as confinement reinforcement, providing clamping force per unit length equal to $N/s = A_{tr}f_{st}/s$ similar to a rectangular section. For circumferential splitting, no reinforcement crosses the splitting crack. The hoop force in the spiral or circular hoop nonetheless resists crack opening and thereby clamps the spliced bars. Assuming development of a normal force N per hoop spacing s, and cutting the hoop midway between spliced bars, the clamping force per unit length in the direction

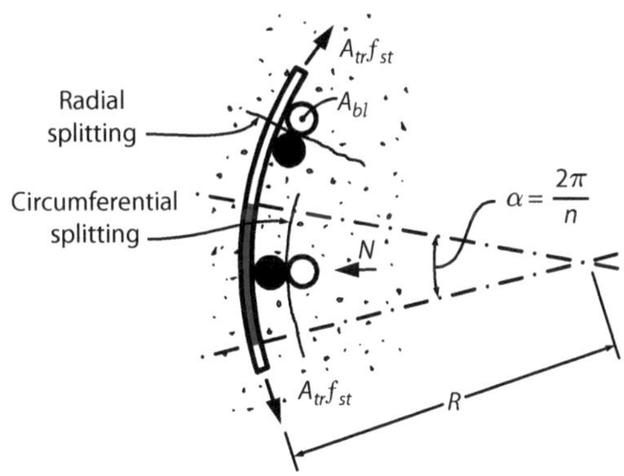

FIGURE 8.31 Splice confinement by circular hoops or spirals. (*After Paulay, 1982, courtesy of American Concrete Institute.*)

opposite N is $N/s = (2A_{tr} f_{st} \sin\frac{\alpha}{2})/s$. Alternatively, the hydrostatic pressure from circular hoop reinforcement can be assumed to act over arc αR, resulting in simpler expression $N/s = \alpha A_{st} f_{st}/s$. Multiplying by splice length to obtain total clamping force, and setting the total clamping force equal to the splice tensile force, leads to

$$s \leq \frac{A_{tr}}{A_{bl}} \frac{f_{st}}{f_y} l_{sp} \tag{8.23}$$

$$s \leq \frac{2\pi}{n} \frac{A_{tr}}{A_{bl}} \frac{f_{st}}{f_y} l_{sp} \tag{8.24}$$

Equation (8.23) corresponds to the radial crack while Eq. (8.24) corresponds to the circumferential crack. Selected spacing should be the smaller of the values obtained from the two expressions. The value of f_{st} should not exceed the value given by Eq. (8.19). Spacing can be modified by Eq. (8.21) if applicable. According to Paulay (1982), a test on a spirally reinforced column with lap splice satisfying these recommendations performed satisfactorily whereas another column having spiral pitch 1.4 times the recommended value failed at moderate ductility demand.

In tests reported by Paulay (1982), the lap splices were located immediately adjacent to the fixed end, corresponding to a maximum moment section. In effect, the available longitudinal reinforcement transitioned from single bars at the interface with the fixed end to double bars along the splice length. This resulted in strain concentration and early fracture in the longitudinal bars at the fixed end. Where lap splices are used, they should not be located immediately adjacent to maximum moment sections, but instead should be shifted away from those sections, thereby providing a length over which plastic hinging can occur outside the lap splice (see Figure 8.32).

Lap splices sometimes require offset bars (Figure 8.32). According to ACI 318, the slope of the offset should not exceed 1 transverse to 6 longitudinal. Transverse reinforcement should be provided that is capable of resisting 1.5 times the transverse force component produced by the offset bars. Lukose et al. (1982) and Paulay (1982) report good performance in lap splices satisfying these requirements.

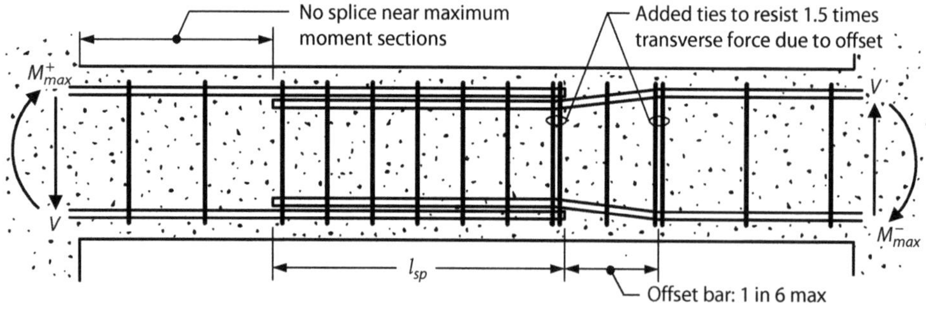

FIGURE 8.32 Detailing of splice location and offset bars.

Aristizabal-Ochoa (1983) reports excess strain in transverse reinforcement along offset bars not satisfying this requirement.

Panahshahi et al. (1992) conducted tests on beam-columns with compression lap splices. For Grade 60 (420) reinforcement, they recommend splice length of $35d_b$ with clear cover of at least $1.5d_b$ and minimum concrete strength of 4000 psi (28 MPa). Closely spaced, uniformly distributed transverse reinforcement should be provided along the entire splice length plus a distance equal to d beyond the splice ends. Based on considerations of radial bursting stresses, they recommended transverse reinforcement spacing

$$s \leq 0.62\left(\frac{A_{tr} l_{sp}}{d_{bl}^2}\right) \leq 6d_{bl} \tag{8.25}$$

Spacing also should not exceed requirements considering reinforcement buckling and concrete confinement. They concluded that, by following these recommendations, compression lap splices can be designed to sustain multiple cycles of inelastic loading.

The available test data indicate that tension and compression lap splices designed by these recommendations should be able to maintain strength under a multiple inelastic cycles with low to moderate ductility demands. In general, lap splices are not recommended at locations where many cycles of large inelastic demands are expected, nor in the end region of beams or columns where plastic hinges can be expected during design earthquake loading.

8.11.3 Hooked Bars

Hassan and Hawkins (1977), Eligehausen et al. (1982), and ACI 408.2R-92 (1992) report results of reversed cyclic tests on hooked bar anchorages, with the following principal observations:

- Reversed cyclic loading induces progressive deterioration of bond between concrete and reinforcing bars. Bond is reduced considerably if the current slip is smaller than previous maximum values, but resistance increases for slip larger than the previous maximum.
- Where demands in compression are considerably lower than demands in tension, tension behavior largely follows the monotonic loading envelope. Where ductility demands on the reinforcement are comparable in tension and compression, peak load and deformation capacity are reduced relative to the monotonic envelope.
- In well-confined anchorages, behavior of hooks loaded in compression is similar to behavior in tension.

ACI 408 (1992) recommends length of hooked bar anchorages for Grade 60 (420) reinforcement as

$$l_{dh} \geq \frac{1800d_b}{\sqrt{f_c'}},\ \text{psi}\left(\frac{150d_b}{\sqrt{f_c'}},\ \text{MPa}\right) \tag{8.26}$$

For beam-column joints in structural frames, however, ACI 408 recommends following the recommendations of ACI 352. For connections resisting seismic actions and confined by transverse reinforcement typical of that occurring in earthquake-resisting beam-column joints, ACI 352 (2002) recommends that bars should be anchored using a 90° standard hook with development length

$$l_{dh} \geq \frac{\alpha f_y d_b}{75\sqrt{f_c'}},\ \text{psi}\left(\frac{\alpha f_y d_b}{6.2\sqrt{f_c'}},\ \text{MPa}\right) \geq 8d_b \text{ and } 6 \text{ in } (152 \text{ mm}) \tag{8.27}$$

in which α = stress multiplier at the beam-column joint interface, typically taken as $\alpha = 1.25$.

For beam-column joints of special moment frames, ACI 318 requires development length of hooked bars in accordance with

$$l_{dh} \geq \frac{f_y d_b}{65\sqrt{f_c'}},\ \text{psi}\left(\frac{\alpha f_y d_b}{5.4\sqrt{f_c'}},\ \text{MPa}\right) \geq 8d_b \text{ and } 6 \text{ in } (152 \text{ mm}) \tag{8.28}$$

This expression is slightly less conservative than Eq. (8.27).

8.11.4 Headed Reinforcement

Kang et al. (2009) report data from reversed cyclic load tests of beam-column joints using headed reinforcement anchored in the joints. They note that most of the force resistance initially is due to bond along the deformed bar with little contribution from the head, but as bond deteriorates with cycles, anchorage is primarily attributable to head bearing. They conclude that the ACI 352 recommendations for headed reinforcement in earthquake-resistant beam-column joints are appropriate. According to ACI 352 (2002), development length for this condition is

$$l_{dh} \geq \frac{\alpha f_y d_b}{100\sqrt{f_c'}},\ \text{psi}\left(\frac{\alpha f_y d_b}{8.3\sqrt{f_c'}},\ \text{MPa}\right) \geq 8d_b \text{ and } 6 \text{ in } (150 \text{ mm}) \tag{8.29}$$

References

ACI 318 (2014). *Building Code Requirements for Structural Concrete (ACI 318-14) and Commentary*, American Concrete Institute, Farmington Hills, MI.

ACI 352 (2002). *Recommendations for Design of Beam-Column Connections in Monolithic Reinforced Concrete Structures (ACI 352R-02)*, American Concrete Institute, Farmington Hills, MI, 37 pp.

ACI 408 (1966). "Bond Stress—The State of the Art," *Journal of the American Concrete Institute*, Vol. 63, No. 11, pp. 1161–1190.

ACI 408 (2003). *Bond and Development of Straight Reinforcing Bars in Tension (ACI 408R-03)*, American Concrete Institute, Farmington Hills, MI, 49 pp.

ACI 408.2R-92 (1992). *Bond under Cyclic Loads (ACI 408.2R-92)*, Reapproved 2005, American Concrete Institute, Farmington Hills, MI, 26 pp.

ACI 439.3R-07 (2007). "Types of Mechanical Splices for Reinforcing Bars," *ACI Manual of Concrete Practice,* 2008 edition, American Concrete Institute, Farmington Hills, MI.

Aristizabal-Ochoa, J.D. (1983). "Earthquake Resistant Tensile Lap Splices," *Journal of Structural Engineering,* Vol. 109, No. 4, pp. 843–858.

ASTM A970 (2009). *Standard Specification for Headed Steel Bars for Concrete Reinforcement (A970/A970M-09),* American Society for Testing and Materials, West Conshohocken, PA, 7 pp.

AWS D1.4 (2011). *Structural Welding Code—Reinforcing Steel (D1.4/D1.4M:2011),* American Welding Society, Miami, FL, 85 pp.

Cairns, J., and P.D. Arthur (1979). "Strength of Lapped Splices in Reinforced Concrete Columns," *ACI Journal,* Vol. 76, No. 2, pp. 277–296.

Chun, S.-C., B. Oh, S.-H. Lee, and C.J. Naito (2009). "Anchorage Strength and Behavior of Headed Bars in Exterior Beam-Column Joints," *ACI Structural Journal,* Vol. 106, No. 5, pp. 579–590.

DeVries, R.A., J.O. Jirsa, and T. Bashandy (1999). "Anchorage Capacity in Concrete of Headed Reinforcement with Shallow Embedments," *ACI Structural Journal,* Vol. 96, No. 5, pp. 728–736.

DeVries, R.A., and J.P. Moehle (1991). *Lap Splice Strength of Plain and Epoxy-Coated Reinforcements: An Experimental Study Considering Concrete Strength, Casting Position, and Anti-Bleeding Additives,* Report No. UCB/SEMM-91-02, Department of Civil Engineering, University of California, Berkeley, CA, 86 pp.

Eligehausen, R., V.V. Bertero, and E.P. Popov (1982). "Hysteretic Behavior of Reinforcing Deformed Hooked Bars in RC Joints," *Proceedings, Seventh European Conference on Earthquake Engineering,* Athens, Vol. 4, pp. 171–178.

Eligehausen, R., E.P. Popov, and V.V. Bertero (1983). *Local Bond Stress-Slip Relationships of Deformed Bars under Generalized Excitations,* Report No. UCB/EERC-83/23, Earthquake Engineering Research Center, University of California, Berkeley, CA, 162 pp.

Goto, Y.M. (1971). "Cracks Formed in Concrete around Deformed Tension Bars," *ACI Journal,* Vol. 68, No. 4, pp. 244–251.

Goto, Y.M., and K. Otsuka (1979). *Experimental Studies on Cracks formed in Concrete around Deformed Tension Bars,* The Technology Reports of the Tohoku University, Vol. 44, No. 1, pp. 49–83.

Hamad, B.S., J.O. Jirsa, and N.I. D'Abreu de Paulo (1993). "Anchorage Strength of Epoxy-Coated Hooked Bars," *ACI Structural Journal,* Vol. 90, No. 2, pp. 210–217.

Hassan, F.M., and N.M. Hawkins (1977). "Anchorage of Reinforcing Bars for Seismic Forces," Special Publication 53, American Concrete Institute, Farmington Hills, MI, pp. 387–416.

Ichinose, T. (1995). "Splitting Bond Failure of Columns under Seismic Action," *ACI Structural Journal,* Vol. 92, No. 5, pp. 535–541.

Jirsa, J.O., L.A. Lutz, and P. Gergely (1979). "Rationale for Suggested Development, Splice, and Standard Hook Provisions for Deformed Bars in Tension," *Concrete International,* Vol. 1, No. 7, pp. 47–61.

Kang, T.H.-K., M. Shin, N. Mitra, and J.F. Bonacci (2009). "Seismic Design of Reinforced Concrete Beam-Column Joints with Headed Bars," *ACI Structural Journal,* Vol. 106, No. 6, pp. 868–877.

Lukose, K., P. Gergely, and R.N. White (1982). "Behavior of Reinforced Concrete Lapped Splices for Inelastic Cyclic Loading," *ACI Journal,* Vol. 79, No. 5, pp. 355–365.

Lutz, L.A., and P. Gergely (1967). "Mechanics of Bond and Slip of Deformed Bars in Concrete," *ACI Journal,* Vol. 64, No. 11, pp. 711–721.

Marques, J.L.G., and J.O. Jirsa (1975). "A Study of Hooked Bar Anchorages in Beam-Column Joints," *ACI Journal,* Vol. 72, No. 5, pp. 198–209.

Minor, J., and J.O. Jirsa (1975). "Behavior of Bent Bar Anchorages," *ACI Journal,* Vol. 72, No. 4, pp. 141–149.

Orangun, C.O., J.O. Jirsa, and J.E. Breen (1977). "A Reevaluation of Test Data on Development Length and Splices," *ACI Journal*, Vol. 74, No. 3, pp. 114–122.

Panahshahi, N., R.N. White, and P. Gergely (1992). "Reinforced Concrete Compression Lap Splices under Inelastic Cyclic Loading," *ACI Structural Journal*, Vol. 89, No. 2, pp. 164–175.

Paulay, T. (1982). "Lapped Splices in Earthquake-Resisting Columns," *ACI Journal*, Vol. 79, No. 6, pp. 458–469.

Rehm, G. (1969). "Kriterien zur Beurteilung von Bewehrungsstäben mit hochwertigen Verbund," *Stahlbetonbau, Berichte aus Forschung und Praxis, Festschrift Rüsch*, Wilhelm Ernst & Sohn, Berlin, pp. 79–96.

Sagan, V.E., P. Gergely, and R.N. White (1991). "Behavior and Design of Noncontact Lap Splices Subjected to Repeated Inelastic Tensile Loading," *ACI Structural Journal*, Vol. 88, No. 4, pp. 420–431.

Sivakumar, B., P. Gergely, and R.N. White (1983). "Suggestions for the Design of R/C Lapped Splices for Seismic Loading," *Concrete International*, Vol. 5, No. 2, pp. 46–50.

Sozen, M.A., and J.P. Moehle (1990). *Development and Lap-Splice Lengths for Deformed Reinforcing Bars in Concrete*, A report to the Portland Cement Association and the Concrete Reinforcing Institute, Schaumburg, IL, 112 pp.

Tepfers, R. (1973). *A Theory of Bond Applied to Overlapped Tensile Reinforcement Splices for Deformed Bars*, Publication 72:2, Chalmers University of Technology, Division of Concrete Structures, Göteborg, Sweden, 328 pp.

Thompson, M.K., J.O. Jirsa, and J.E. Breen (2006a). "CCT Nodes Anchored by Headed Bars—Part 2: Capacity of Nodes," *ACI Structural Journal*, Vol. 103, No. 1, pp. 65–73.

Thompson, M.K., A. Ledesma, J.O. Jirsa, and J.E. Breen (2006b). "Lap Splices Anchored by Headed Bars," *ACI Structural Journal*, Vol. 103, No. 2, pp. 271–279.

Thompson, M.K., M.J. Ziehl, J.O. Jirsa, and J.E. Breen (2005). "CCT Nodes Anchored by Headed Bars—Part 1: Behavior of Nodes," *ACI Structural Journal*, Vol. 102, No. 6, pp. 808–815.

CHAPTER 9

Beam-Column Connections

9.1 Preview

Moment-resisting frames comprise beams, columns, and beam-column joints (Figure 9.1). A *beam-column joint* is defined as that portion of the column within the depth of the deepest beam that frames into the column. The *beam-column connection* comprises the joint plus portions of the columns, beams, and slab immediately adjacent to the joint. Beam-column connections enable forces and moments to be transmitted among the beams and columns. In earthquake-resisting frames, a primary design goal is for joints to remain essentially elastic while plastic hinges form in adjacent framing members. Framing action can result in relatively large shear forces being transferred across the joint. In many buildings, joint design requirements will control the sizes and reinforcement of the adjacent framing members. This chapter describes the design forces, internal resisting mechanisms, and proportioning and detailing requirements for beam-column connections. It includes an introduction to behavior of beam-column connections without transverse reinforcement under ordinary loading. The main emphasis, however, is on connections required to resist inelastic deformation reversals associated with strong earthquake shaking.

9.2 Forces in Beam-Column Connections

9.2.1 Connection Forces from Gravity and Lateral Loading

Beam-column connections are important to framing action for gravity loads, lateral loads, and combinations of both. Usually, though, it is lateral loads that place the largest demands on beam-column joints. This can be understood by considering actions on a single floor level cut from a multi-level frame (Figure 9.2).

Uniform gravity loads typically produce moments similar to those shown in Figure 9.2*a*. These moments are balanced or nearly balanced at interior connections. Thus the main role of an interior joint is to transmit column axial force and beam flexural tension and compression forces through the joint. If concrete compressive strength within the joint is similar to that of adjacent beams and columns, and if reinforcement is continuous through the joint, an interior

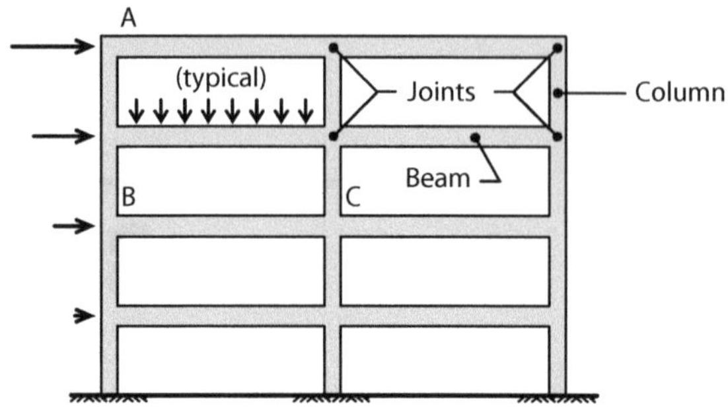

FIGURE 9.1 Reference frame subjected to gravity and lateral loading.

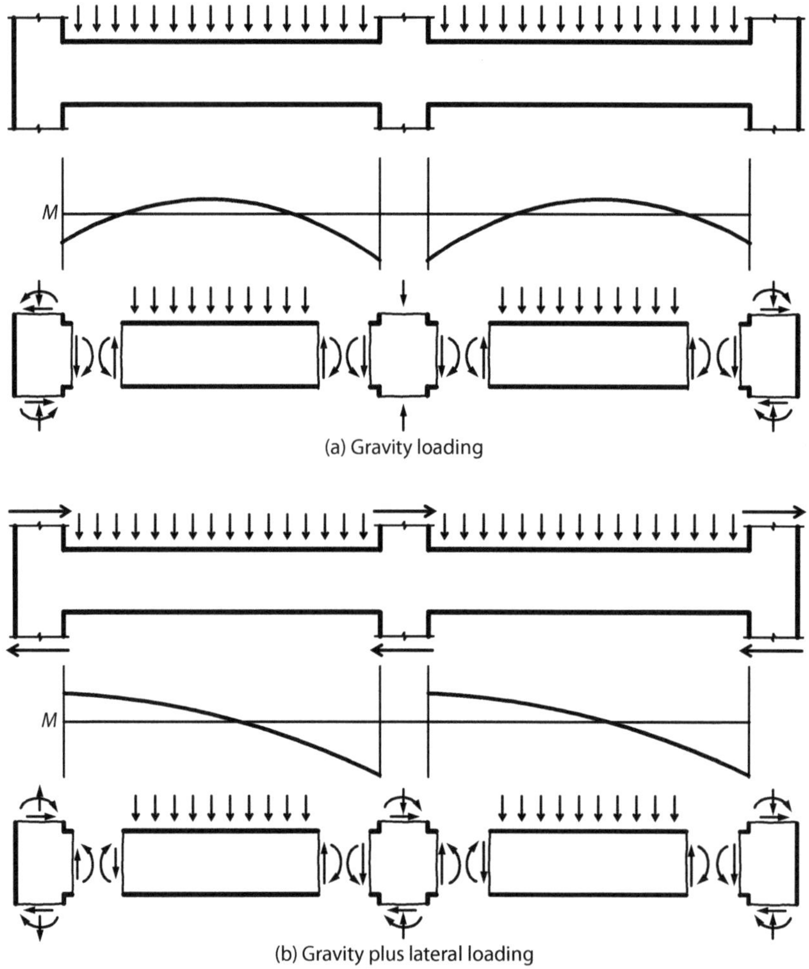

FIGURE 9.2 Frame elevations, moment diagrams, and internal force diagrams for one level of a moment-resisting frame under (*a*) gravity loading and (*b*) gravity plus lateral loading. In (b), the left-hand column may be in either tension or compression.

joint will be capable of transmitting these forces without further attention. At an exterior joint, beam moment usually is transmitted to adjacent columns through the joint. These moments typically are relatively small and can be accommodated by a joint with nominal joint reinforcement. If the joint becomes overloaded and has insufficient moment transfer capacity, then the exterior beam moment can be redistributed safely to the adjacent interior connection. Usually, the most significant performance consideration is exterior joint cracking under service loads. In special cases, however, greater attention to joints is warranted.

In contrast, lateral loads produce moments that must be transferred between beams and columns at all joints (Figure 9.2*b*). As will be demonstrated subsequently, these moment transfers can produce very high shear stresses in beam-column joints. Thus, attention to the design of beam-column joints is routinely required for frames subjected to lateral loads.

9.2.2 Calculation of Joint Shear

Joint shears can be calculated by resolving the flexural, axial, and shear forces acting at the joint boundaries. For a frame responding in the linear-elastic range, member actions at the joints can be calculated by conventional methods of linear structural analysis. For frames responding inelastically, as can occur under seismic loading, joint demands can be determined by methods of limit analysis (Chapter 1). Normally we will design earthquake-resisting frames such that yielding occurs predominantly in the beams, in which case joint demands are defined by the beam moment strengths.

General Procedure

To illustrate the general procedure for calculating joint shear demands in a frame with beams yielding in flexure, consider the portion of a frame shown in Figure 9.3*a*. Beams are presumed to reach their flexural strengths at the faces of the columns.

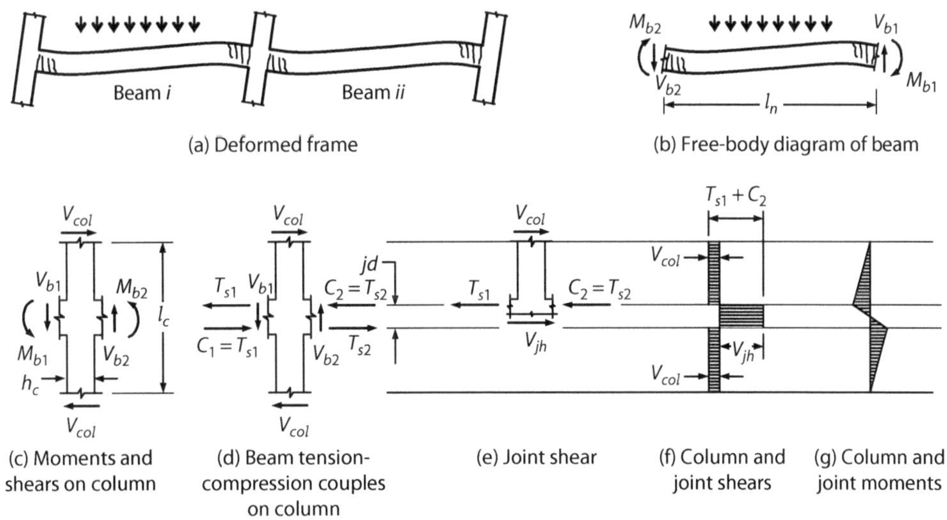

FIGURE 9.3 Equilibrium requirements of a beam-column connection in a frame with yielding beams. Column axial force does not enter the equilibrium equations, and is not shown.

Figure 9.3*b* shows a free-body diagram obtained by cutting through the beam at the faces of the columns. The bending moments at these sections are known (because the beam has reached its flexural strength at these sections), rendering the free body statically determinate. Shears V_{b1} and V_{b2} are obtained by summing moments about opposite ends.

Beam shears and moments obtained in the previous step are next applied to the beam-column joint, in this case shown in Figure 9.3*c* for an interior joint. The portion of column selected for this free-body diagram extends from an inflection point above the joint to an inflection point below the joint. In general, locations of column inflection points are unknown. In intermediate levels of a building, it may be acceptable to assume that the column inflection points are located at column mid-height. Near the base of the building, near the roof, and at other levels where discontinuities occur, the column moment distribution should be estimated based on analysis of the frame, and the column shear should be calculated from first principles. For the free-body diagram of Figure 9.3*c*, moment equilibrium requires

$$V_{col} = \frac{1}{l_c}\left[M_{b1} + M_{b2} + (V_{b1} + V_{b2})\frac{h_c}{2} \right] \tag{9.1}$$

Assuming the beams carry no net axial force, flexural tension and compression force resultants are defined by the familiar relation $C = T_s = M_b/jd$. In the more general case where a beam resists net axial force, alternative relations are required to obtain C and T_s. Note that T_s is tension force in beam longitudinal reinforcement when the beam develops its flexural strength. This tension force can be expressed as $\alpha A_s f_y$, where α is 1.0 or greater to account for material overstrength and strain-hardening. Figure 9.3*d* shows the free-body diagram including the flexural tension and compression force resultants C and T_s.

In Figure 9.3*e*, a horizontal cut through the joint exposes the horizontal joint shear V_{jh}. Summing horizontal forces and solving for V_{jh} result in

$$V_{jh} = T_{s1} + C_2 - V_{col} \tag{9.2}$$

in which T_{s1}, C_2, and V_{col} have directions shown in the figure.

Figure 9.3*f* plots the shear diagram for the column extending through the joint. Note that the shear diagram implies that the joint shear is opposite and much larger than the column shear. This is consistent with the moment diagram shown in Figure 9.3*g*, which shows opposite and steep slope through the joint depth. In the next section, an example illustrates the relative magnitudes of the column and joint shears, showing that the diagrams of Figure 9.3*f* and *g* are qualitatively correct.

Relative Magnitude of Column and Joint Shears

To illustrate the relative magnitude of column and joint shears, consider the case in which beams have equal top and bottom longitudinal reinforcement resulting in equal positive and negative moment strengths, and gravity loads are small relative to other forces acting on the frame and can be ignored. Inflection points are assumed to be at beam midspans and column mid-heights. Figure 9.4*a* shows

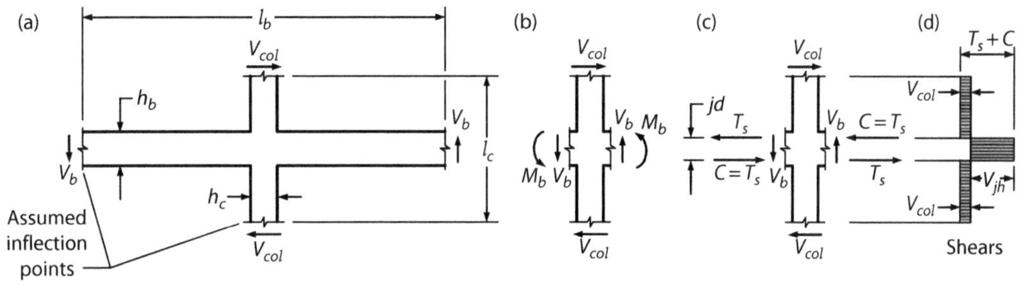

FIGURE 9.4 Free-body diagrams to calculate joint shear.

a free-body diagram of an interior joint cut through the inflection points of the beams and columns. Moment equilibrium of that free body requires

$$V_{col} = \frac{l_b}{l_c} V_b \tag{9.3}$$

Moment equilibrium of each beam requires $M_b = \frac{l_b - h_c}{2} V_b$, which can be solved for beam shear as $V_b = \frac{2M_b}{l_b - h_c}$. Figure 9.4*b* shows V_b and M_b acting at the joint faces. Substituting for V_b in Eq. (9.3) results in

$$V_{col} = 2\frac{l_b}{l_c}\frac{M_b}{(l_b - h_c)} \tag{9.4}$$

For a beam with zero axial force, $M_b = T_s jd = Cjd$. C and T_s are shown acting on the joint in Figure 9.4*c*. Substituting for M_b in Eq. (9.4) results in

$$V_{col} = 2\frac{jd}{l_c}\frac{l_b}{(l_b - h_c)}T_s \tag{9.5}$$

The shear diagram of Figure 9.4*d* shows that horizontal joint shear is

$$V_{jh} = T_s + C - V_{col} = 2T_s - V_{col} \tag{9.6}$$

Substituting V_{col} from Eq. (9.5) in Eq. (9.6) leads to

$$V_{jh} = 2\left[1 - \frac{jd}{l_c}\frac{l_b}{(l_b - h_c)}\right]T_s \tag{9.7}$$

Dividing Eq. (9.7) by Eq. (9.5) results in

$$\frac{V_{jh}}{V_{col}} = \frac{\left(\frac{l_c}{h_b}\frac{h_b}{jd} - \frac{1}{1 - h_c/l_b}\right)}{\left(\frac{1}{1 - h_c/l_b}\right)} \tag{9.8}$$

For typical values of h_b/jd, l_c/h_b, and h_c/l_b, Eq. (9.8) results in V_{jh}/V_{col} ranging from approximately 3 to 5. For exterior connections in frames of the same proportions, values of V_{jh}/V_{col} are about half those obtained for interior connections. These relatively high values of V_{jh}/V_{col} suggest that joint shear will be an important consideration in design of lateral-force-resisting frames.

Orientation of Shear Stress and Shear Cracking in Various Joints

The preceding discussion examined horizontal joint shear in an interior joint and found that joint shear was opposite and much larger than the shear in the adjacent columns. The joint shears result mainly from the large flexural tension and compression forces acting on the joint boundaries. Figure 9.5 illustrates the shears and moments acting on three typical joints, identifies the orientation of joint shear, and sketches the orientation of expected joint cracking if cracks occur. Note that the orientation of potential joint shear cracks is always opposite the orientation of potential shear cracks in the adjacent framing members.

So far we have emphasized horizontal joint shear, but equilibrium also requires vertical joint shear. Figure 9.6 illustrates this for a joint subjected to general loading within one plane. Column longitudinal reinforcement is assumed to be lumped near the edges rather than distributed along the face. Beams are assumed to have axial forces. Moments and axial forces acting around the joint are replaced by stress resultants in Figure 9.6*b*. As before, a horizontal cut through the joint exposes horizontal joint shear (Figure 9.6*c*). Similarly, a vertical cut through the joint exposes vertical joint shear V_{jv} (Figure 9.6*d*). From equilibrium of vertical forces

$$V_{jv} = T_{sc1} + C_{c2} - V_{b1} \tag{9.9}$$

Assuming the joint boundary is defined by the effective flexural moment arms jd_b in the vertical direction and jd_c in the horizontal direction, with V_{jh} and V_{jv}

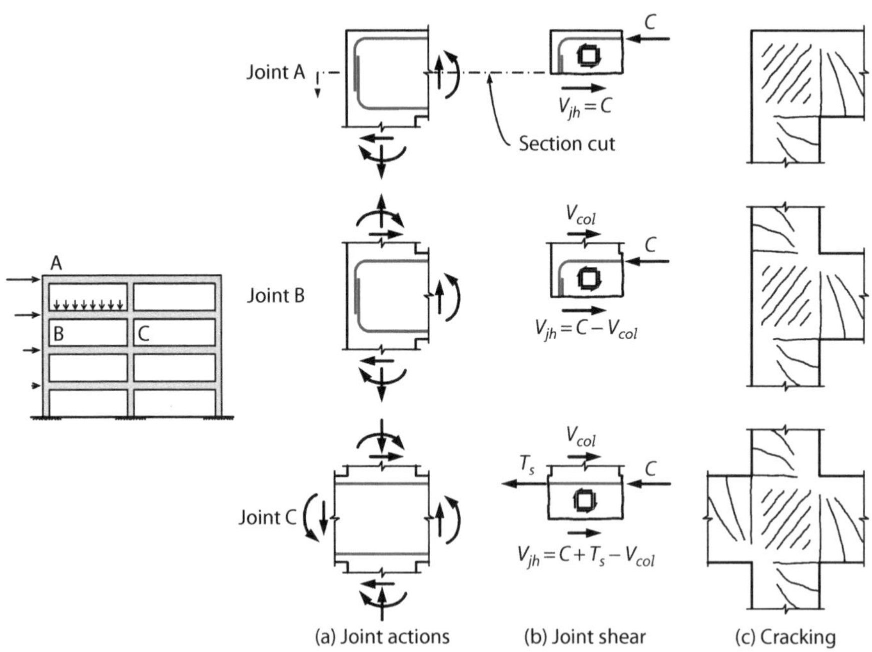

FIGURE 9.5 Resolution of internal forces at beam-column connections. Orientations of some of the beam shears and column axial forces might differ from orientations shown. For clarity, only horizontal forces are shown in (*b*).

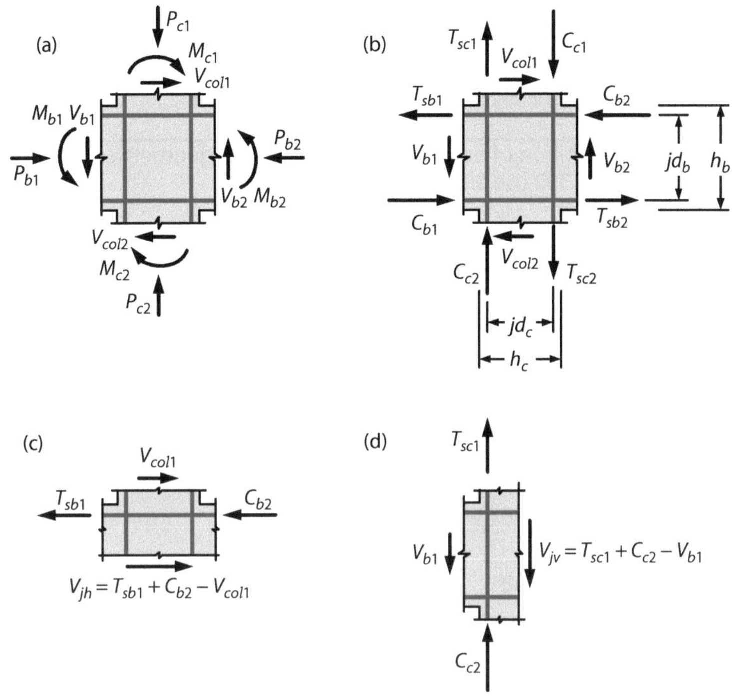

FIGURE 9.6 Horizontal and vertical joint shear. Note that (*c*) and (*d*) show only horizontal forces and vertical forces, respectively, acting on the free-body diagram.

acting on respective horizontal and vertical faces, the sum of moments on the joint requires

$$\frac{V_{jv}}{V_{jh}} = \frac{jd_b}{jd_c} \approx \frac{h_b}{h_c} \tag{9.10}$$

Joint equilibrium can also be considered in terms of average joint stresses. Defining average horizontal joint stress as $v_{jh} = V_{jh}/jd_c b_c$ and average vertical joint stress as $v_{jv} = V_{jv}/jd_b b_c$, we can use Eq. (9.10) to demonstrate that $v_{jh} = v_{jv}$, as is generally required for moment equilibrium. Paulay (1989) presents additional discussion on equilibrium requirements for joints.

Finally, it is worth noting that the discussion thus far has been restricted to consideration of joints resisting forces in a single plane. The state of stress and force-resisting mechanisms are more complicated for joints resisting forces in three dimensions.

9.3 Joint Classifications

Joint types are classified according to connection geometry, loading type, and joint reinforcement, as follows.

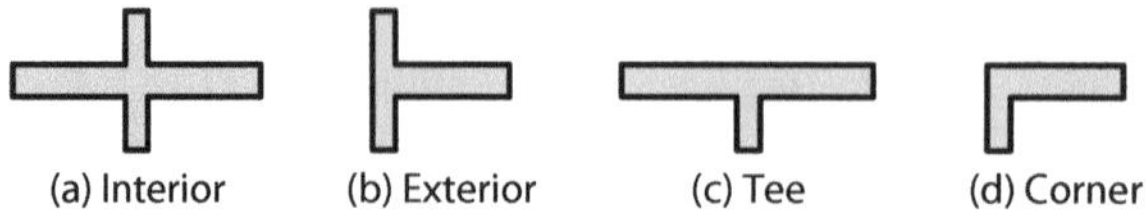

FIGURE 9.7 Definition of beam-column connection geometries. Note that ASCE 41, ACI 318, and ACI 352 use alternative designations.

9.3.1 Connection Geometry

Beam-column joints occur in several different configurations (Figure 9.1). This chapter generally refers to these as interior, exterior, tee, and corner connections, as shown in Figure 9.7. Only planar joints are shown, although beams sometimes frame into joints out of plane. Note that ASCE 41, ACI 318, and ACI 352[1] use alternative designations when specifying joint strengths and design requirements. Those alternative designations will be introduced where the ASCE 41, ACI 318, and ACI 352 provisions are presented.

9.3.2 Loading Type

Joint behavior depends on the type of loading. Adopting the ACI 352 classification, a *Type 1* connection is one designed to resist loading without significant inelastic deformation. A *Type 2* connection is one designed to sustain deformation reversals into the inelastic range. Although this book is mainly concerned with earthquake-resisting framing, which would have Type 2 connections, a brief description of Type 1 connections is also included.

9.3.3 Joint Reinforcement

In regions of high seismicity, beam-column joints in modern seismic force-resisting systems are designed with transverse (lateral) reinforcement and well-distributed column vertical reinforcement. This reinforcement serves many roles, including confinement of joint concrete, clamping of longitudinal reinforcement to improve bond and restrain buckling, and resistance of joint shear. In this text we refer to these as *reinforced joints*. Joints in older construction, and joints in parts of modern buildings not designed to resist seismic forces, commonly do not contain transverse reinforcement, or contain it in lesser quantities. We refer to these as *unreinforced joints*. Both unreinforced and reinforced joints are considered in this chapter.

9.4 Beam-Column Joints without Transverse Reinforcement

This section addresses behavior of beam-column connections without joint transverse reinforcement, sometimes referred to as unreinforced joints. It includes reversed cyclic loading typical of seismic loading where data are available. Where

[1]Unless otherwise noted, ACI 352 refers to the 2002 edition.

such data are lacking, it includes only behavior under monotonic loading. Interior, exterior, tee, and corner connections are presented sequentially.

9.4.1 Interior Connections

Figure 9.8*a* illustrates the internal forces from beams and columns acting on the boundary of an interior beam-column joint. These forces include tensile and compressive forces T_s and C_s in longitudinal reinforcement and compressive stresses having resultants C_c in concrete of the flexural compression zones. Concrete stresses are shown in the linear range, although this is not a requirement. The distribution of column and beam shears through the depth of the columns and beams is unknown in general. Where flexural cracks around the joint are relatively narrow, shear stresses may be distributed approximately uniformly across each joint face. Where wide flexural cracks occur at the joint boundary, it is reasonable to assume the shear forces enter the joint predominantly through the flexural compression zones.

Onset of joint cracking can be estimated based on average stresses within the joint. As described in Section 9.2.2, horizontal joint shear is

$$V_{jh} = T_{sb2} + C_{cb1} + C_{sb1} - V_{col1} \tag{9.11}$$

Average horizontal shear stress is

$$v_{jh} = v_j = V_{jh}/A_j \tag{9.12}$$

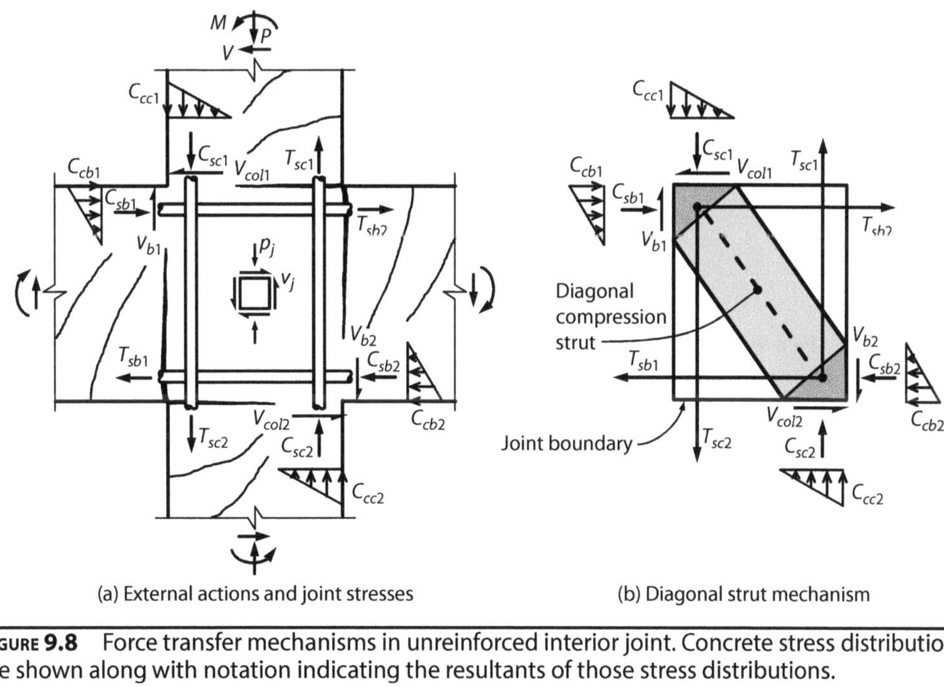

FIGURE 9.8 Force transfer mechanisms in unreinforced interior joint. Concrete stress distributions are shown along with notation indicating the resultants of those stress distributions.

in which A_j = cross-sectional area of the joint, commonly taken as $h_c b_c$, in which h_c = column dimension parallel to the beam longitudinal reinforcement generating the joint shear and b_c = column cross-sectional dimension in the orthogonal direction. (Alternative dimensions for b_c are sometimes used where beam and column widths differ significantly or where longitudinal axes do not align.) Average joint axial stress is

$$p_j = P/A_j \tag{9.13}$$

in which P can be taken as the average axial force for columns above and below the joint. Where beams are prestressed, or where lateral forces result in beam axial force, axial stress along the beam axis should also be considered. However, this case is not covered here.

The joint shear at onset of diagonal cracking of the joint panel can be estimated by equating nominal principal tensile stress and concrete tensile strength. Principal tensile stress is obtained from joint shear and normal stresses using the usual stress transformations (see Eq. 7.20) as

$$\sigma_1 = -\frac{p_j}{2} + \sqrt{\left(\frac{p_j}{2}\right)^2 + v_j^2} \tag{9.14}$$

in which p_j is joint axial stress taken positive in compression. Setting $\sigma_1 = f_t$ = tensile strength of concrete and solving for v_j result in

$$v_j = f_t\sqrt{1 + \frac{p_j}{f_t}} \tag{9.15}$$

Tensile strength of concrete in pure tension or in split cylinder tests can be approximated as $f_t = 6\sqrt{f'_c}$, psi ($0.5\sqrt{f'_c}$, MPa). Given the highly non-uniform stress field likely within a beam-column joint panel, a reduced effective tensile strength of $4\sqrt{f'_c}$, psi ($0.33\sqrt{f'_c}$, MPa) can be assumed. Therefore, shear stress at joint cracking can be approximated by

$$v_{cr} = 4\sqrt{f'_c}\sqrt{1 + \frac{p_j}{4\sqrt{f'_c}}}, \text{ psi} \left(0.33\sqrt{f'_c}\sqrt{1 + \frac{p_j}{0.33\sqrt{f'_c}}}, \text{ MPa}\right) \tag{9.16}$$

Past editions of ACI 352 (1976) also used the vehicle of diagonal tension cracking to estimate the contribution of concrete to joint shear strength, with additional strength considered to be provided by transverse reinforcement. According to ACI 352 (1976), the contribution of concrete to joint shear strength is given by

$$v_c = 3.5\sqrt{f'_c}\beta\eta\sqrt{1 + \frac{p_j}{500}}, \text{ psi} \left(0.29\sqrt{f'_c}\beta\eta\sqrt{1 + \frac{p_j}{3.45}}, \text{ MPa}\right) \tag{9.17}$$

in which β = 1.4 for monotonic loading and 1.0 for reversed cyclic loading, and η = 1.0 except η = 1.4 if the joint is confined perpendicular to the direction of shear considered using beams covering at least three-quarters of the joint width and three-quarters of the joint depth. The similarity between Eqs. (9.16) and (9.17) is noted.

The shear at onset of shear cracking does not necessarily correspond to the joint shear strength. This is because alternative force paths exist in beam-column joints. The primary alternative force path for interior joints is a diagonal compression strut from one corner of the joint to the opposite corner (Figure 9.8*b*). Dimensions of the nodal zones and struts are determined by dimensions of the flexural compression zones. The presence of flexural compression improves bond capacity for longitudinal bars passing through the joint. Thus, significant portions of the longitudinal reinforcement axial force can be developed within the nodal zones. Where development length within the nodal zone is insufficient, the mechanism allows the bar force to be developed beyond the joint within the flexural compression zone of the adjacent beam or column. In this case, the force resisted by concrete in the flexural compression zone is increased, with corresponding increase in the depth of the flexural compression zone and width of the diagonal compression strut. Development of the longitudinal reinforcement at the far side of the joint or in the framing member beyond the joint results in additional slip of the reinforcing bars, which reduces stiffness of the connection.

Figure 9.9 compares results of Eqs. (9.16) and (9.17) with test data. Test specimens were full-scale interior beam-column connections without transverse beams, subjected to constant axial force and lateral deformation reversals. Equation (9.17) is a fairly good estimator of shear at onset of visible joint cracking,

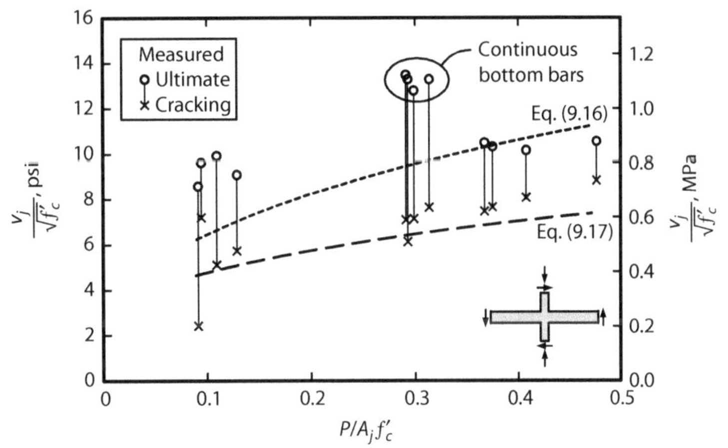

FIGURE 9.9 Shear strength of unreinforced interior joints. Curves associated with Eqs. (9.16) and (9.17) are best fit power curves for data points obtained using those equations. Test specimens with continuous beam bottom bars are noted. All other test specimens have discontinuous bottom bars embedded in the joint. (*Data from Pessiki et al., 1990, and Beres et al., 1992.*)

whereas Eq. (9.16) tends to overestimate cracking strength. In all cases, ultimate strength exceeds cracking strength by a considerable margin.

9.4.2 Exterior Connections

Similar considerations apply to exterior beam-column joints (Figure 9.10). In this case, however, effective joint depth is determined by the depth of the beam reinforcement anchorage; in Figure 9.10*b*, effective joint depth is defined by the bend in the hooked anchorage, typically taken to the outside edge of the standard hook. Some analysis approaches, however, ignore this detail and take the joint depth as the entire depth of the column parallel to the beam bars generating the joint shear.

Figure 9.11 plots joint shear strength versus joint aspect ratio h_b/h_c, where h_b = beam depth and h_c = column depth, for exterior beam-column joints subjected to reversed cyclic loading. Data are organized according to axial stress level and failure mechanism. Best fit trend lines are also shown. Salient observations include the following:

- Measured joint shear strength is highest if beams and columns do not yield in flexure. Where beams or columns yield prior to joint failure, joint shear demands are limited by the capacities of the beams or columns. Joint failure may still occur, but at reduced joint shear force. The reduced strength probably is associated with strain penetration of yielding reinforcement and reduced confinement by the yielding framing member.

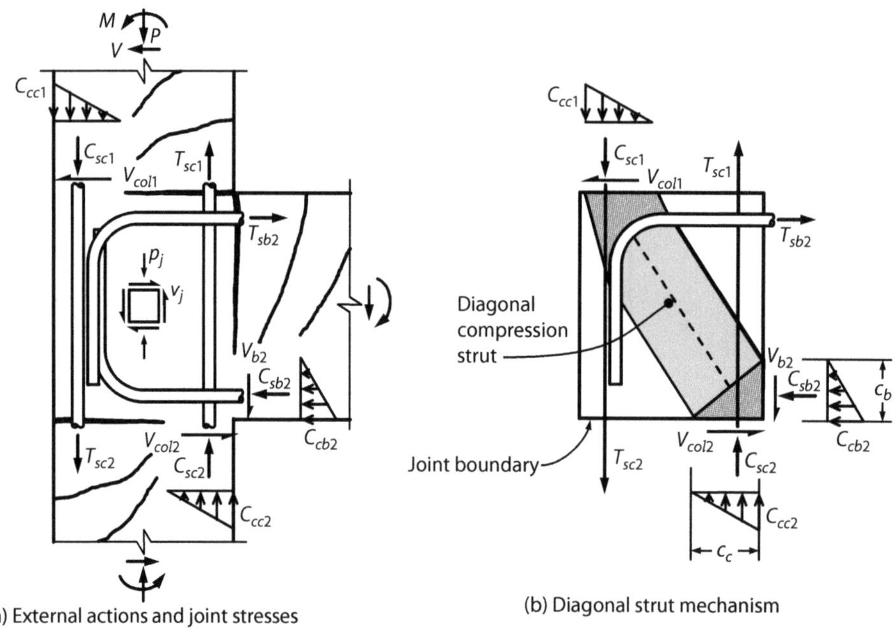

(a) External actions and joint stresses

(b) Diagonal strut mechanism

FIGURE 9.10 Force transfer mechanisms in unreinforced exterior joint. Concrete stress distributions are shown along with notation indicating the resultants of those stress distributions.

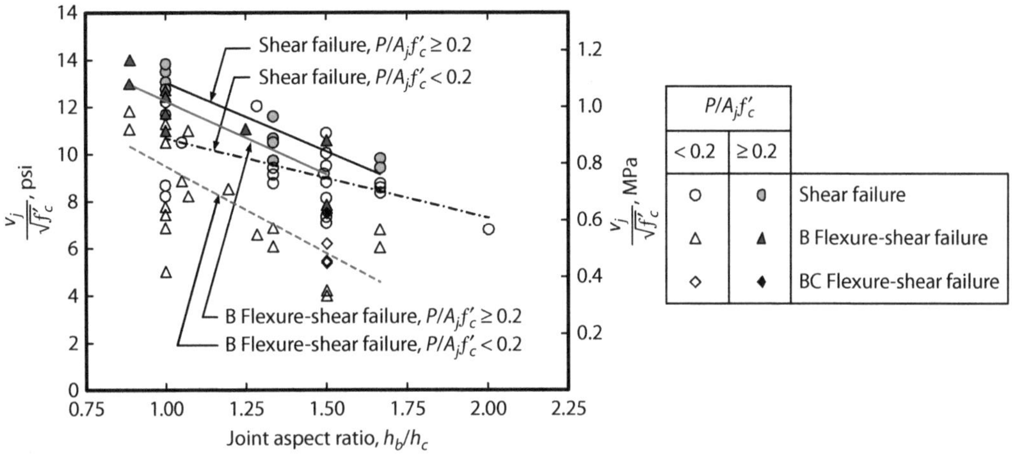

FIGURE 9.11 Joint shear strength as function of joint aspect ratio, axial stress level, and failure mode. "Shear Failure" refers to joint shear failure without flexural yielding; "B Flexure-Shear Failure" refers to beam flexural yielding followed by joint shear failure; "BC Flexure-Shear Failure" refers to combined beam and column yielding followed by joint shear failure. Best fit linear trend lines shown for "Shear Failure" and "B Flexure-Shear Failures." *(After Hassan, 2011, used with permission from W. Hassan.)*

- Joint shear strength is higher for higher column axial force P. One explanation is that higher axial force increases width of the diagonal compression strut shown in Figure 9.10*b*, thereby increasing joint shear strength.
- Joint shear strength reduces with increasing joint aspect ratio h_b/h_c. One explanation (Hassan, 2011) is that a joint with a higher joint aspect ratio requires a steeper joint strut (Figure 9.10*b*), which reduces joint strength. Hassan (2011) shows that joint shear strength can be modeled with good accuracy using strut-and-tie models.

9.4.3 Tee Connections

Tee connections are essentially exterior connections rotated through 90° (Figure 9.7). Consequently, many of the requirements and findings for exterior connections apply equally to tee connections. Nilsson and Losberg (1976) report monotonic load tests of tee connections with various details, some of which are shown in Figure 9.12. Details A and B show column longitudinal reinforcement bent into the joint core as required to develop a diagonal compression strut mechanism across the joint. In contrast, Detail C shows column longitudinal reinforcement bent away from the joint core. The graph plots joint efficiency as a function of column longitudinal reinforcement ratio. Joint efficiency is defined as the percentage of nominal column flexural strength that is developed by the joint. (100% joint efficiency indicates that the joint was capable of developing the full flexural strength of the column, whereas less than 100% means the joint was not capable of developing the column flexural strength.) Joints with Details A and B are much more efficient than joints with Detail C.

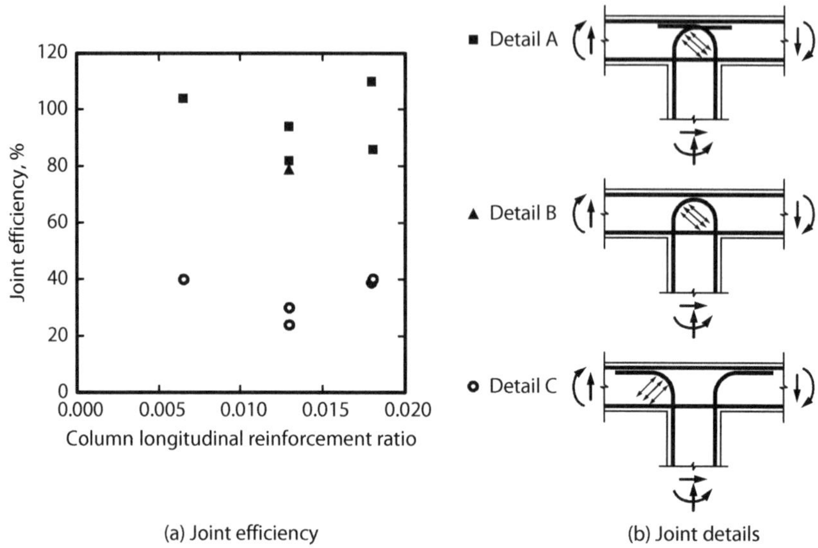

FIGURE 9.12 Joint efficiency obtained for tee connections having different reinforcement details. (*Data from Nilsson and Losberg, 1976.*)

These results suggest that Detail A or B should be used where it is necessary to develop column strength at the top of a building. These same details should also be used where a column frames into a footing or grade beam. Detail A is generally recommended because it is easier to construct and results in a stiffer connection. One test by Nilsson and Losberg (1976) using Detail B failed to develop adequate strength because of inadequate anchorage of the looped reinforcement.

9.4.4 Corner Connections

Corner connections (or joints) are sometimes referred to as *knee joints*. In corner connections we refer to closing moments and opening moments. *Closing moments* are moments that tend to reduce the 90° angle between the connecting members. *Opening moments* are moments that tend to increase the 90° angle.

Under closing moments, corner connections usually can develop flexural strengths of the adjacent framing members if reasonable reinforcement details are provided. Figure 9.13 shows typical details that have been shown to develop strengths of the framing members (Mayfield et al., 1971). These details work because the bend of the flexural tension reinforcement is appropriately oriented to develop the required diagonal compression strut.

Moment transfer is inherently less efficient under opening moments than under closing moments. Test results are reported by Mayfield et al. (1971), Mayfield et al. (1972), Nilsson and Losberg (1976), and Skettrup et al. (1984). Figure 9.14 illustrates the force-transfer problem. As shown, flexural compression forces tend to push off the corner of the joint, creating diagonal tensile stresses and leading to diagonal cracking that reduces joint strength. Nilsson and Losberg (1976) assumed

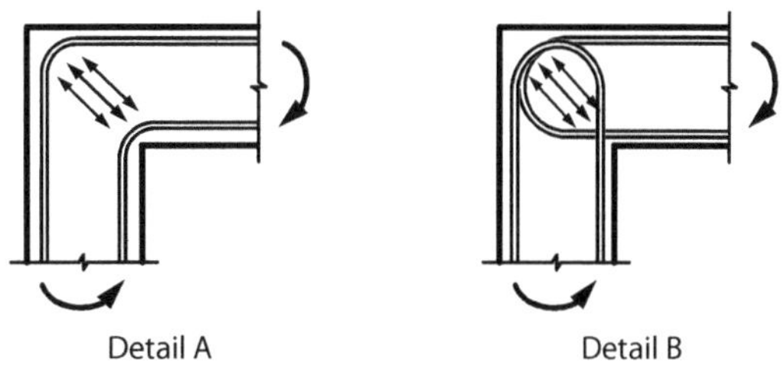

FIGURE 9.13 Acceptable details for closing moments on corner connections.

that a parabolic distribution of tensile stress developed in the joint (Figure 9.14*b*). Considering the free-body diagram of Figure 9.14*c*, the total tensile force resisted by the parabolic tensile stress distribution is $T = \frac{2}{3} f_t l_{dc} b_c$. Assuming 45° orientation of the stress field, the sum of forces on the free-body diagram of Figure 9.14*c* requires $C = T/\sqrt{2} = \frac{\sqrt{2}}{3} f_t l_{dc} b_c$. Noting that joint shear for a corner joint is simply $V_j = C$, the joint shear strength is $V_n = \frac{\sqrt{2}}{3} f_t l_{dc} b_c$. Nilsson and Losberg (1976) measured f_t and took l_{dc} as the length of the tensile stress distribution before cracking according to the theory of elasticity. These assumptions resulted in very good agreement with tests. Here, we take $l_{dc} \approx \sqrt{2} h_c$ and $f_t = 6\sqrt{f_c'}$, psi $(0.5\sqrt{f_c'}$, MPa). Thus, the joint nominal shear strength is

$$V_n = 4\sqrt{f_c'} h_c b_c, \text{ psi } (0.33\sqrt{f_c'} h_c b_c, \text{ MPa}) \tag{9.18}$$

Strength according to Eq. (9.18) applies to a plain concrete joint, in particular, Detail D shown in Figure 9.15. Behavior can be improved by detailing the longitudinal reinforcement from the beam and column such that it reinforces the corner of the joint. As shown in Figure 9.15, Details A, B, and C provide better joint efficiency compared with Detail D. Joint efficiency, as defined previously, is the percentage of nominal column (or beam) flexural strength that is developed by the joint.

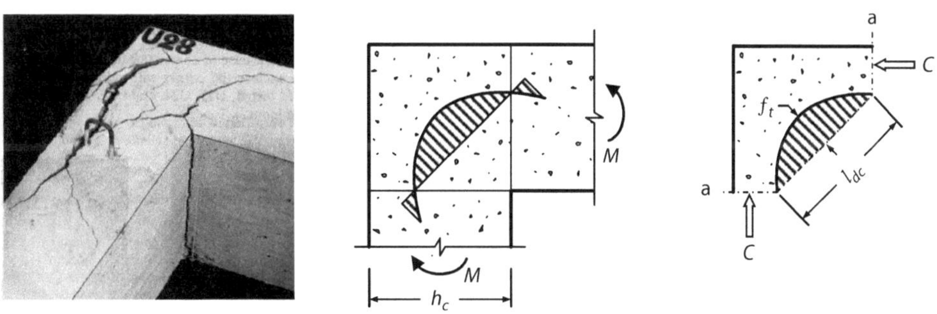

(a) Corner joint cracking (b) Opening moments and stresses (c) Free-body diagram

FIGURE 9.14 Failure of corner joint under opening moments. Beam and column shears and axial forces not shown. (*After Nilsson and Losberg, 1976, with permission of ASCE.*)

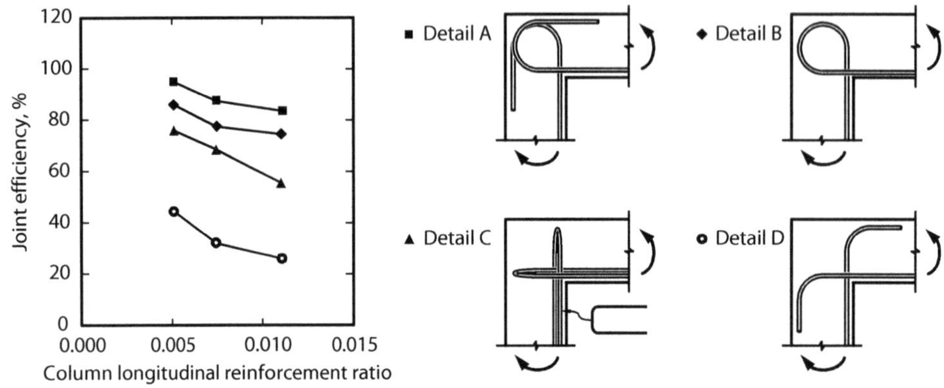

(a) Joint efficiency

(b) Joint details

Figure 9.15 Joint efficiency obtained for corner connections under opening moments and having different reinforcement details. (*Data from Nilsson and Losberg, 1976.*) *Note:* In Detail C, the bars are bent in a hairpin shape with the hook perpendicular to the direction of shear (i.e., the hook is perpendicular to the page).

9.4.5 ASCE 41 Joint Strength

ASCE 41 (2013) defines joint shear strength for unreinforced joints as

$$V_n = \lambda\gamma\sqrt{f_c'}A_j, \text{ psi } (0.083\lambda\gamma\sqrt{f_c'}A_j, \text{ MPa}) \tag{9.19}$$

in which $\lambda = 0.75$ for lightweight aggregate concrete and 1.0 for normalweight aggregate concrete, γ is defined in Table 9.1, and A_j is defined by a joint depth equal to the column dimension in the direction of framing and a joint width equal to the least of (a) the column width, (b) the beam width plus the joint depth, and (c) twice the smaller perpendicular distance from the longitudinal axis of the beam to the column side.

Connection Geometry				
Interior joint with transverse beams	Interior joint without transverse beams	Exterior joint with transverse beams	Exterior joint without transverse beams	Corner (knee) joint with or without transverse beams
	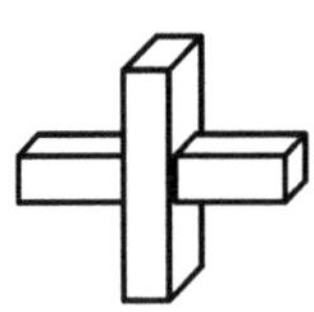		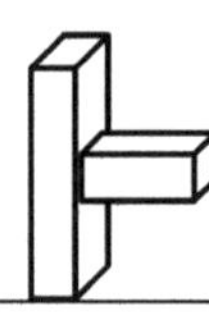	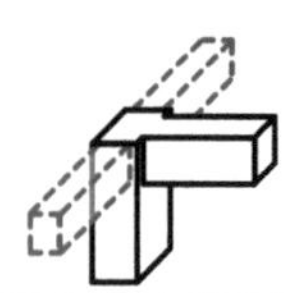
$\gamma = 12$	10	8	6	4

*Joint loading is in the plane of the paper.

Table 9.1 Values of γ for Joint Strength Calculation—Joints without Transverse Reinforcement (ASCE 41)*

Compared with measured strengths (see Sections 9.4.1 through 9.4.4), ASCE 41 joint strength values tend to be conservative.

9.5 Beam-Column Joints with Transverse Reinforcement

This section addresses behavior of beam-column connections with joint reinforcement subjected to inelastic deformation cycles representative of design-level earthquake loading. It is assumed that inelastic response is due to beam flexural yielding, either at the joint face or away from it, and that the joint and column respond in the essentially linear elastic range, as is the intent of most building codes for earthquake-resisting frames.

Joint reinforcement generally comprises horizontal transverse reinforcement in the form of hoops and crossties plus distributed column longitudinal reinforcement. Figure 9.16 shows an example of an interior connection with joint reinforcement that may be suitable for earthquake resistance.

9.5.1 Interior Connections

Joint Forces and Internal Resisting Mechanisms

Figure 9.17*a* illustrates the internal forces from beams and columns acting on the boundary of an interior beam-column joint. These forces include tensile and compressive forces T_s and C_s in longitudinal reinforcement and compression stress resultants C_c in concrete of the flexural compression zone. Concrete stresses are represented by a rectangular stress block, although alternative stress distributions can be considered. Assuming development of wide flexural cracks in beams and columns at the joint boundary, it is reasonable to assume that shear forces V_b in

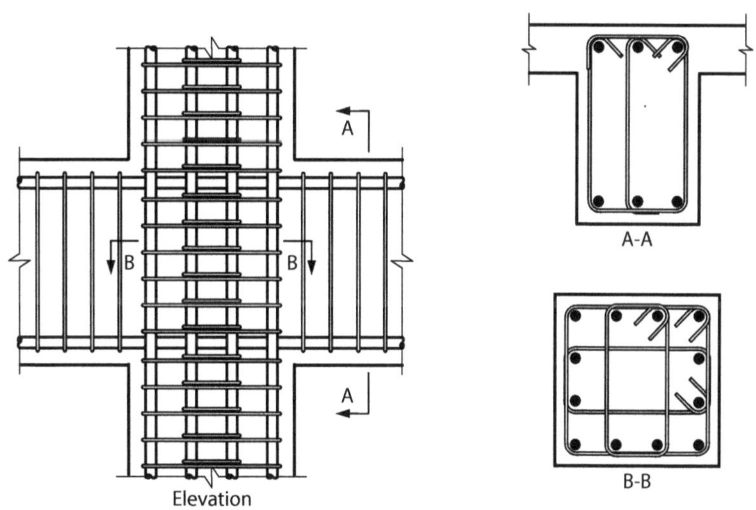

Figure 9.16 Typical reinforcement for an interior beam-column connection with joint transverse reinforcement.

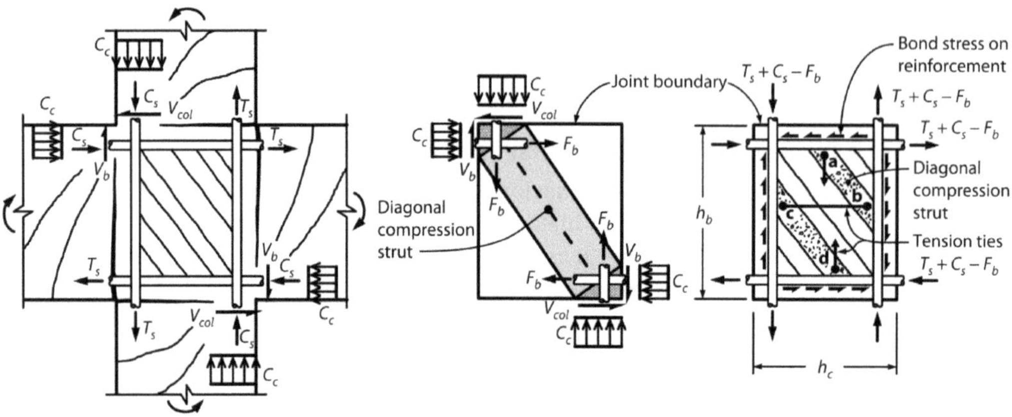

Figure 9.17 Force transfer mechanisms in interior beam-column joint. Values of flexural tension and compression forces T_s, C_s, and C_c, and values of the bond force F_b, may vary in the individual members, but are shown generically here. Concrete stress distributions are shown along with notation indicating the resultants of those stress distributions.

beams and V_{col} in columns enter the joint predominantly through the flexural compression zones.

Prior to cracking of the joint core, behavior of reinforced joints is similar to that of unreinforced joints. Meinheit and Jirsa (1981) report cracking strengths ranging from $3.6\sqrt{f'_c}$, psi ($0.30\sqrt{f'_c}$, MPa) for $P = 0.03P_o$ to $11\sqrt{f'_c}$, psi ($0.92\sqrt{f'_c}$, MPa) for $P = 0.36P_o$. In the same test series, joint ultimate shear strength generally ranged from $20\sqrt{f'_c}$ to $30\sqrt{f'_c}$, psi ($1.7\sqrt{f'_c}$ to $2.5\sqrt{f'_c}$, MPa) under monotonic loading, with strength decay occurring for multiple inelastic deformation cycles. Joint ultimate strength was relatively insensitive to column axial force. This could be partly because beam yielding limited the value of joint shears in the tests. Kitayama et al. (1991) and Bonacci and Pantazopoulou (1993) evaluated databases of interior joint tests and also did not find a discernible effect of axial force on joint strength.

Park and Paulay (1975), Meinheit and Jirsa (1981), and others have reported that joint cracking tends to develop from one corner of the joint to the opposite corner, at least within the range of joint aspect ratios tested. Figure 9.17*a* idealizes a series of well-developed diagonal cracks forming within the joint core with cracks oriented along the diagonal.

After cracking, two force transfer mechanisms commonly are considered, as presented by Park and Paulay (1975) and further elaborated by Paulay et al. (1978). These are a *diagonal strut mechanism* (Figure 9.17*b*) and a *truss mechanism* (Figure 9.17*c*). Total joint shear resistance is the sum of resistances provided by these two mechanisms.

Similar to an unreinforced joint, the *diagonal strut mechanism* (Figure 9.17*b*) resolves the forces acting in concrete at the joint boundaries, that is, forces C_c, V_{col}, and V_b, plus a portion of longitudinal reinforcement force F_b associated with bond stress within the nodal zone. Flexural compressive stresses from the adjacent beams and columns create ideal bond stress conditions for the longitudinal reinforcement within the nodal zones. Thus, force F_b can be a significant fraction of the total longitudinal reinforcement force. The diagonal compression strut does not require joint transverse reinforcement for equilibrium, as the strut force is fully equilibrated by forces acting on the nodal zones. Transverse reinforcement, however, can confine joint concrete, thereby potentially increasing compressive strength and ductility of the diagonal compression strut.

The *truss mechanism* (Figure 9.17*c*) is assumed to resist the remainder of the longitudinal reinforcement force, that is, $T_s + C_s - F_b$, through shear in the joint panel. To simplify the model, we can idealize the bond stress as acting uniformly around the joint panel, as shown. Bond stress is then equilibrated by a series of diagonal compression struts and tension ties. For example, in Figure 9.17*c*, the bond stress resultant near point **a** is balanced by the horizontal component of diagonal compression strut **ab**. Strut **ab** also requires a vertical tension tie at point **a**. At point **b**, forces in strut **ab** are resolved by bond stresses at **b** and by tension tie **bc**. Anchorage of tie **bc** at point **c** is resolved by strut **cd** and by vertical bond stress near **c**. Thus, the truss mechanism requires a series of vertical and horizontal tension ties to equilibrate bond forces acting on the column and beam longitudinal reinforcement. In the horizontal direction, the tension ties are provided by joint transverse reinforcement (Figure 9.16). Vertically oriented transverse reinforcement generally would be too difficult to place during construction. Instead, the column longitudinal reinforcement, which is intended to remain in the linear-elastic range, can act as the vertical ties if the longitudinal bars are well distributed across the joint depth.

The amount of horizontal and vertical reinforcement required to sustain the truss mechanism is readily obtained. Assuming a corner-to-corner diagonal crack, with no shear resisted by concrete across diagonal cracks, the total horizontal reinforcement is determined from horizontal force equilibrium as

$$nA_{jh}f_{yt} = C_s + T_s - F_b \tag{9.20}$$

in which n = total number of hoop sets distributed vertically within the joint core, A_{jh} = cross-sectional area of each hoop set in the direction parallel to the joint shear, and f_{yt} = yield strength of the joint shear reinforcement. The amount of vertically distributed reinforcement is similarly determined.

Effects of Inelastic Displacement Reversals

Under monotonic loading, the diagonal strut mechanism and the truss mechanism each may be capable of resisting a significant portion of the total joint shear. Reversed cyclic loading may alter the relative contributions of the two mechanisms, depending on how the resistance mechanisms decay with inelastic cycling. Two arguments concerning which mechanism dominates, one in favor of the truss mechanism and the other in favor of the diagonal strut mechanism, are presented in the following paragraphs.

The argument in favor of the truss mechanism as the predominant force transfer mechanism begins with consideration of how moment is transferred into the joint. If beam top reinforcement yields on one side of the joint for loading in one direction, and then the load is released, a residual crack will be present where the yielded beam intersects the joint face. Under load reversal, the entire flexural compression force will be resisted by that longitudinal reinforcement until the flexural crack closes. Usually the bottom reinforcement area is equal to or less than the top reinforcement area, such that top reinforcement is unlikely to develop significant yield strains in compression, leaving a residual crack width. Paulay et al. (1978) argue that, because of this effect, all of the joint shear must be introduced to the joint by bond between beam longitudinal bars and the joint concrete. This results in an increase in the participation of the truss mechanism and a decrease in the participation of the diagonal strut mechanism.

The average bond stresses associated with developing T_s on one side of the joint and C_s on the other side can be very high. Equating bond force to the change in reinforcement force across the joint results in $\bar{u}\,\Sigma(\pi d_b)h_c = C_s + T_s$, in which $\Sigma(\pi d_b)$ = sum of the perimeters of longitudinal reinforcing bars. For a single bar, this can be expressed as $\bar{u}\,\pi d_b h_c = \Delta f_s(\pi d_b^2/4)$, in which Δf_s = change in bar stress along the length h_c. Thus, the average bond stress is

$$\bar{u} = \frac{\Delta f_s}{4}\left(\frac{d_b}{h_c}\right) \tag{9.21}$$

To get a sense of the magnitude of bond stress, consider the case of a beam-column joint having h_c = 28 in (710 mm) and beams with equal numbers of top and bottom Grade 60 (420) No. 11 (35) bars, resulting in d_b/h_c = 1.41/28 = 1/20. Assuming the bars develop overstress of $f_s = 1.25f_y$, the resulting average bond stress from Eq. (9.21) is $\bar{u} = \frac{2\times 1.25 f_y}{4}\left(\frac{d_b}{h_c}\right)$ = 1900 psi (13 MPa). Tests reported by Viwathanatepa et al. (1979) and others show that a cone failure usually occurs on the tension side of the joint (Figure 9.18), further reducing the effective bonded length and increasing the local bond stress.

Leon (1991) reports that local bond stresses as high as 3500 psi (24 MPa) can be obtained over short bar lengths under monotonic loading, but average bond strength is significantly reduced for longer lengths of reinforcement subjected to multiple reversed cycles including bar yielding. For such cases, with anchorage lengths in the

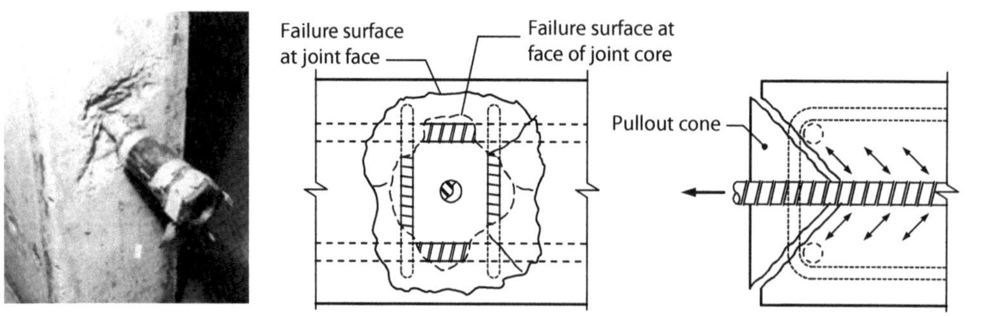

(a) Test specimen photograph (b) View along axis of embedded bar (c) Section through embedded bar

FIGURE 9.18 Pullout cone observed in tests of bars embedded in confined concrete blocks representing confined beam-column joints. (*After Viwathanatepa et al., 1979, used with permission from University of California, Berkeley.*)

range of $20d_b$ to $24d_b$, sustained bond strengths typically range from 700 to 900 psi (5 to 6 MPa) for concrete with compressive strength in the range from 3000 to 5000 psi (21 to 34 MPa), or about $12\sqrt{f'_c}$, psi ($1\sqrt{f'_c}$, MPa). This bond strength is a fraction of the bond demand in a typical joint. Thus, significant bar slip should be anticipated.

Slip of bars within the joint may result in crack closure, reactivating compression of concrete in the beam flexural compression zone, and shifting the joint shear force-resisting mechanism from the truss mechanism to the diagonal strut mechanism. On the basis of this argument, several references (e.g., Otani, 1991; Jirsa, 1995; ACI 352) have suggested that the diagonal strut mechanism, rather than the truss mechanism, is the primary shear force-resisting mechanism for beam-column joints subjected to inelastic deformation cycles. The diagonal compression strut inevitably will be weakened by reversed cyclic loading and by dilation transverse to the strut. Thus, according to this interpretation, the principal role of transverse reinforcement is to strengthen and toughen the diagonal compression strut by confining the joint core concrete.

A reasonable conclusion is that the truss mechanism is viable as the primary mechanism where there is sufficient bonded length within the joint and sufficient transverse reinforcement to support the truss action induced by bond. Where the bonded length is insufficient, where the transverse reinforcement is insufficient, or both, a diagonal strut mechanism will dominate. Tests by Goto et al. (1987, 1988) and analyses by Ichinose (1991) demonstrate the interplay between bond, transverse reinforcement, and the two mechanisms.

Figure 9.19 shows test results for two interior beam-column joints designed by two different methods as part of a collaborative research program including U.S. and New Zealand researchers. Yielding of beams was the primary source of inelastic action, with beam strength limiting the joint shear demands. Figure 9.19*a* shows results for a joint designed with sufficient joint depth and sufficient transverse reinforcement so that the truss mechanism could resist essentially the entire joint shear. Ratio d_b/h_c for Grade 40 (275) beam longitudinal bars was

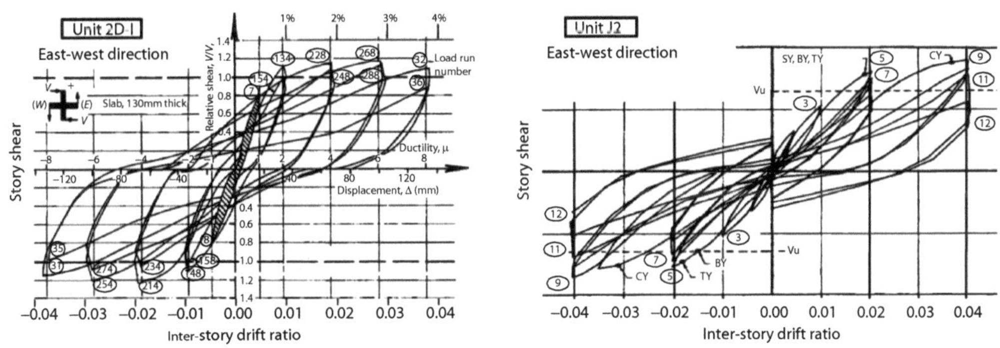

(a) Joint designed by truss mechanism

(b) Joint designed by diagonal strut mechanism

FIGURE 9.19 Column shear versus story drift ratio for interior beam column joints subjected to bi-directional lateral loading. Only east-west test results are shown. [*(a) After Cheung et al., 1991; (b) after Kurose et al., 1991.*] *(Courtesy of American Concrete Institute.)*

1/25, requiring relative large columns. Maximum joint shear stress resisted by the joint during the test was 619 psi (4.27 MPa), corresponding to $8.4\sqrt{f'_c}$, psi ($0.70\sqrt{f'_c}$, MPa). Figure 9.19*b* shows results for a joint designed with less transverse reinforcement and shorter embedment length through the column, such that the diagonal strut mechanism would have to resist a greater portion of the total joint shear. Ratio d_b/h_c for Grade 60 (420) beam longitudinal bars was 1/20, leading to smaller columns. Maximum joint shear stress resisted by the joint during the test was 1640 psi (11.3 MPa), corresponding to $26\sqrt{f'_c}$, psi ($2.2\sqrt{f'_c}$, MPa). Both joints developed the beam flexural strengths as intended, and both resisted design forces to story drift ratios of 0.04. The main difference in performance is that the joint designed with more transverse reinforcement, greater development length, and lower joint shear stress (Figure 9.19*a*) has wider hysteresis loops indicative of better bond and less shear damage than the joint designed with less transverse reinforcement, less development length, and higher joint shear stress (Figure 9.19*b*). The latter joint eventually failed in shear.

Effects of Bond Deterioration

The previous section noted the interplay between bond, transverse reinforcement, and the joint shear-resisting mechanism. Joints with excessive bond demand or with inadequate transverse reinforcement develop bar slip, which affects overall behavior of the beam-column connection. Figure 9.20 illustrates behavior of three beam-column joints with different values of h_c/d_b. Note that not only embedment lengths but also joint shear stresses varied from one specimen to another because joint area increased with increasing dimension h_c. The following principal observations are made:

1. Joints with smaller values of h_c/d_b have greater bar slip within the joint.
2. Greater bar slip results in reduced connection stiffness.
3. Greater bar slip results in hysteresis relations that are more pinched, thereby reducing energy dissipation capacity.
4. Joints with smaller values of h_c/d_b may have reduced overall deformation capacity at failure. Because the development length within the joint is relatively short, the tension reinforcement cannot be developed entirely within the joint but instead must be developed within the beam on the opposite side of the joint. This results in increased flexural compression in

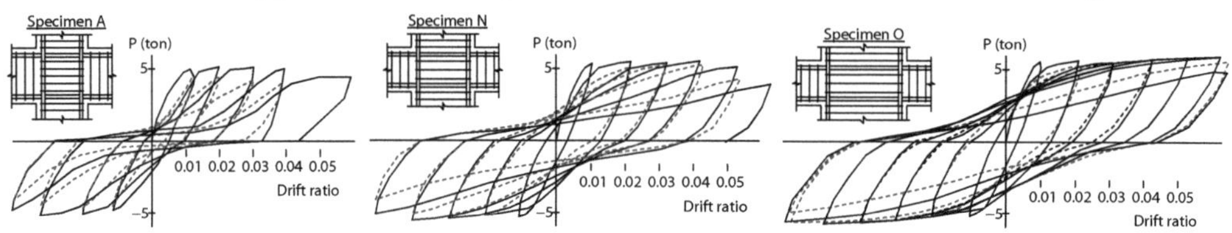

(a) $\frac{h_c}{d_b} = 18.8$ (b) $\frac{h_c}{d_b} = 28.0$ (c) $\frac{h_c}{d_b} = 37.5$

FIGURE 9.20 Column shear versus story displacement for interior beam-column joints with different values of h_c/d_b. (*After Kaku and Asakusa, 1991, Courtesy of American Concrete Institute.*)

that beam and decreased flexural deformation capacity. Loss of deformation capacity may also arise because joints with smaller h_c/d_b have higher joint shear stresses.

5. Severe damage to the interface between reinforcement and concrete enables bars to slip almost freely, resulting in a joint that deforms excessively under service loads. This damage is difficult to repair.

Because of the aforementioned factors, most guidelines and codes for earthquake-resisting frames prescribe limits to the ratio d_b/h_c for interior beam-column joints. In U.S. practice, the ACI 318 limit is $d_b/h_c \le 1/20$, applicable for any permitted reinforcement grade. Some other practices are more conservative. For example, in NZS 3101 (2006), the requirement for Grade 40 (300) reinforcement is $d_b/h_c \le 40\alpha_f\alpha_d\sqrt{f_c'}/1.25f_y$, psi ($3.3\alpha_f\alpha_d\sqrt{f_c'}/1.25f_y$, MPa), in which $\alpha_f = 0.85$ where beams pass through a joint in two directions, as in two-way frames, or $\alpha_f = 1.0$ for one-way frames, and $\alpha_d = 1.0$ for ductile plastic regions, $\alpha_d = 1.2$ for limited ductile regions. For higher strength reinforcement [typically Grade 72 (500) in New Zealand], the same equation may apply except lower values may be permitted where drift levels are sufficiently lower than the code maximum values. For Grade 60 (420) reinforcement and $f_c' = 4000$ psi (28 MPa), with $\alpha_f = \alpha_d = 1.0$, this results in $d_b/h_c \le 1/30$. AIJ (1999) requires $d_b/h_c \le \frac{15}{1+\alpha_b}\left(1+\frac{P}{A_g f_c'}\right)\frac{(f_c')^{2/3}}{f_y}$, psi $\left[\le \frac{2.8}{1+\alpha_b}\left(1+\frac{P}{A_g f_c'}\right)\frac{(f_c')^{2/3}}{f_y}, \text{MPa}\right]$, in which α_b = ratio of areas of beam tension reinforcement to compression reinforcement, but not more than 1.0. For Grade 60 (420) reinforcement, $f_c' = 4000$ psi (28 MPa), $\alpha_b = 1$, and $\frac{P}{A_g f_c'} = 0.2$, this results in $d_b/h_c \le 1/26$. Somewhat more complicated expressions are given as alternatives in NZS 3101, as well as in Eurocode 8 (2004).

Joint Strengths

Strength of the diagonal strut mechanism depends mainly on compressive strength of the diagonal strut. In contrast, strength of the truss mechanism depends mainly on bond strength and joint shear strength provided by joint transverse reinforcement (as well as distributed vertical joint reinforcement). By studying trends of joint shear strength for many joint tests, we can learn which parameters most affect joint strength, and better understand the predominant mechanisms of joint shear transfer.

Figure 9.21 plots measured shear strength as function of joint transverse reinforcement ratio for planar beam-column tests from Japan and the United States (Kitayama et al., 1991). Test results are sorted by those that sustained joint failure before beam yielding, joint failure after beam yielding, or beam flexural failure without joint failure. Joint strength was relatively insensitive to amount of joint transverse reinforcement. Joints were capable of sustaining joint shear stress of $0.25f_c'$ without failure for drift ratios up to 1/25 if joint transverse reinforcement ratio was at least 0.4%. These results suggest that the diagonal strut mechanism was the predominant mechanism in these tests; if the truss mechanism predominated, shear strength would increase approximately in proportion with

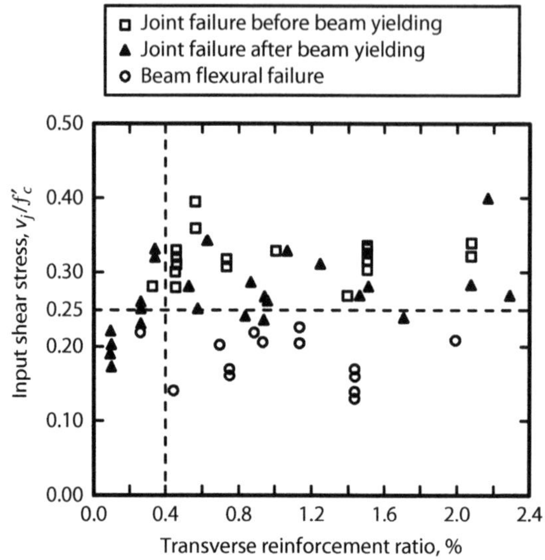

Figure 9.21 Interior joint strength as function of transverse reinforcement. (*After Kitayama et al., 1991, courtesy of American Concrete Institute.*)

the amount of transverse reinforcement. Similar results have been reported by Kim and LaFave (2008).

Figure 9.21 suggests that a safe limit for average joint shear stress is $v_j = 0.25 f'_c$ in interior joints with at least 0.4% transverse reinforcement. Relating joint shear strength to f'_c seems appropriate if joint strength is controlled by compressive strength of a diagonal strut. For higher strength concrete, however, the rate of increase in joint strength with increasing concrete strength is reduced (Figure 9.22). The linear relation $v_j = 0.25 f'_c$ becomes unconservative for higher concrete strengths. Two other functions using $\sqrt{f'_c}$ correlate better with observed strength. Data reported by Kurose et al. (1988) show that, for interior joints without transverse

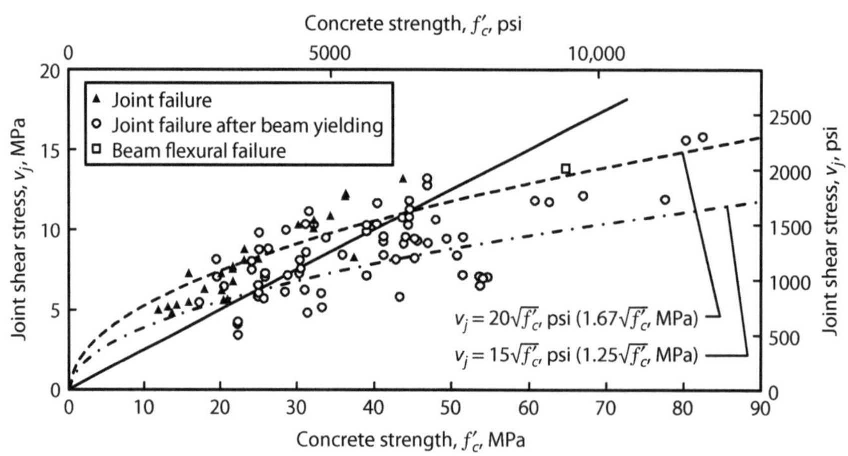

Figure 9.22 Interior joint shear strengths. (*Data from Sugano et al., 1991, courtesy of American Concrete Institute.*)

beams, $v_j = 15\sqrt{f'_c}$, psi ($1.25\sqrt{f'_c}$, MPa) fits the boundary between joints failing in shear without beam yielding and those failing in shear after beam yielding. For interior joints with transverse beams on both sides of the joint, $v_j = 20\sqrt{f'_c}$, psi ($1.67\sqrt{f'_c}$, MPa) is a better fit. Apparently, unloaded transverse beams confine the joint and increase joint strength. Whether transverse beams loaded by seismic actions provide effective joint confinement has been a matter of debate (e.g., Paulay, 1989; Kitayama et al., 1991).

Saqan and Kreger (1998) report additional tests for concrete having compressive strength up to 15 ksi (100 MPa). The results follow the trends shown in Figure 9.22.

Two-way frames with beams along both principal axes can have simultaneous yielding in both directions, increasing joint shear demands and column flexural demands. Leon and Jirsa (1986) note the importance of designing the column to be capable of resisting bi-axial moments with overstrength such that the distributed column reinforcement can serve as vertical joint reinforcement. In their tests, bi-directional joint shear strengths averaged $v_j = 27\sqrt{f'_c}$, psi ($2.25\sqrt{f'_c}$, MPa), corresponding to $v_j = 19\sqrt{f'_c}$, psi ($1.6\sqrt{f'_c}$, MPa) simultaneously in each direction. Note that this value is very close to values obtained in uni-directional tests. Bi-directional tests are also reported by Cheung et al. (1991) and Kurose et al. (1991).

Most tests reported in the literature have considered joints with $h_b/h_c \approx 1$. Such joints develop a diagonal compression strut at an efficient inclination of approximately 1:1 as shown in Figure 9.23*a*. Joints with larger values of h_b/h_c (e.g., Figure 9.23*b*) require a steeper strut, which would both increase strut demands and decrease strut capacity (see Chapter 7 for additional discussion of strut-and-tie models). Reduced joint strength should be anticipated for such joints. ACI 318 recognizes this effect by limiting the joint aspect ratio to $h_b/h_c \leq 2$. An alternative approach is to apply a strut-and-tie model to the joint design, maintaining efficient inclinations for the diagonal compression struts (Figure 9.23*c*). Hwang and Lee

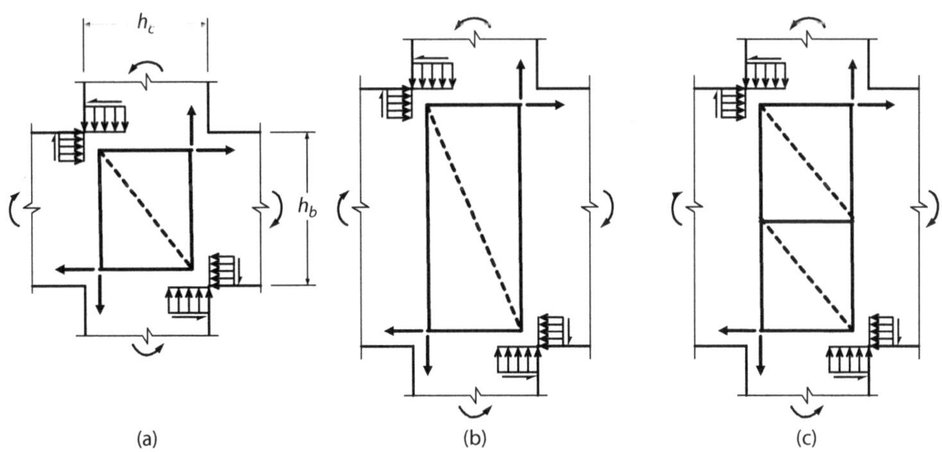

FIGURE 9.23 Strut-and-tie models for joints.

(1999, 2000) and Hwang et al. (2005) have developed a strut-and-tie modeling approach that shows good agreement with test results for both interior and exterior joints.

Wong et al. (1990) and Li and Tran (2009) report tests of interior beam-column joints with vertically distributed beam longitudinal reinforcement. Vertically distributed beam longitudinal reinforcement is found to be nearly as efficient as conventionally placed top and bottom longitudinal reinforcement insofar as beam flexural strength and ductility are concerned. Similar joint shear demands are also obtained. Where the beam yields at the joint face, the vertically distributed reinforcement located near mid-depth will be yielding in tension on both sides of the joint. This action tends to dilate the joint, potentially reducing strength of the diagonal strut. Horizontal joint reinforcement in the form of hoops is required to compensate for this effect. Full capabilities and requirements for such joints have not been established.

In a different design approach, Abdel-Fattah and Wight (1987) introduced supplementary beam longitudinal reinforcement near mid-depth. The reinforcement terminated a small distance from the joint face such that beam plastic hinges were located away from the joint. Joint performance was improved by keeping beam longitudinal reinforcement in the linear range at the joint boundary. Beam shears and plastic rotation demands, however, may be increased by this design approach.

Parra-Montesinos et al. (2005) report tests on beam-column joints with high-performance fiber-reinforced cement composites in the joint region. Similar to the joints reported by Abdel-Fattah and Wight (1987), the joints were reinforced with short lengths of beam longitudinal reinforcement that were located at mid-depth and that extended a short distance into the beams. The joints otherwise had no horizontal joint reinforcement. The connections showed excellent performance with reduced joint damage compared with conventional construction. The high-performance concrete also seemingly improved bond characteristics for beam longitudinal bars passing through the joint.

9.5.2 Exterior Connections

In the conventional detail for exterior connections, beam bars terminate in standard hooks within the joint core (Figure 9.24*a*). The hooked bars extend to near the back face of the joint core, but not less than the necessary development length into the joint. By extending the bars to the back face of the joint core, the maximum joint depth is engaged for joint shear resistance. In congested joints, it may be preferred to hold the tails of the hooks slightly back from the column longitudinal bars at the far face of the joint, thereby avoiding a "wall" of reinforcement that impedes concrete placement. Congestion can be reduced by eliminating the hook from the joint core. One option is to use headed reinforcement that extends to the far side of the joint (Figure 9.24*b*). Another option is to anchor the bent beam bars in a beam stub extending past the joint (Park, 1986) (Figure 9.24*c*). Formwork costs and architectural considerations may preclude use of this last detail. It is not considered further in this text.

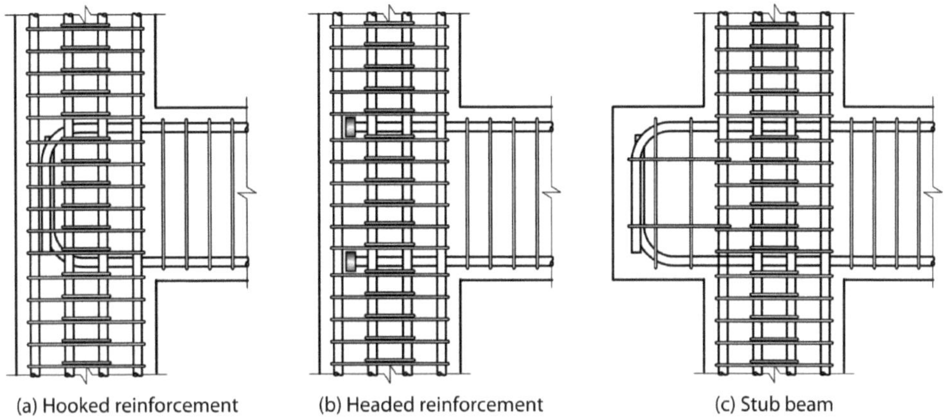

Figure 9.24 Elevation views of typical reinforcement for exterior beam-column connections with joint transverse reinforcement.

The applied forces and internal resisting mechanisms for exterior connections are similar to those for interior connections (Figure 9.25). For exterior connections, however, the horizontal component of the diagonal strut mechanism is actuated at the far side of the joint by bearing on the inside radius of the bent bars (Figure 9.25*b*). Effectiveness of the diagonal strut in resisting shear will be improved by setting the hook as close to the far face of the joint as practicable, but still within the joint core. Note that the shear and flexural compression from the upper column could create a failure surface along the outside of the hooked beam bars. Column and joint hoops in this region are important to reinforce the potential failure plane. These hoops also help "pull" the column flexural compression force from the outer edge of the column into the joint where it provides the vertical component of the diagonal strut. As

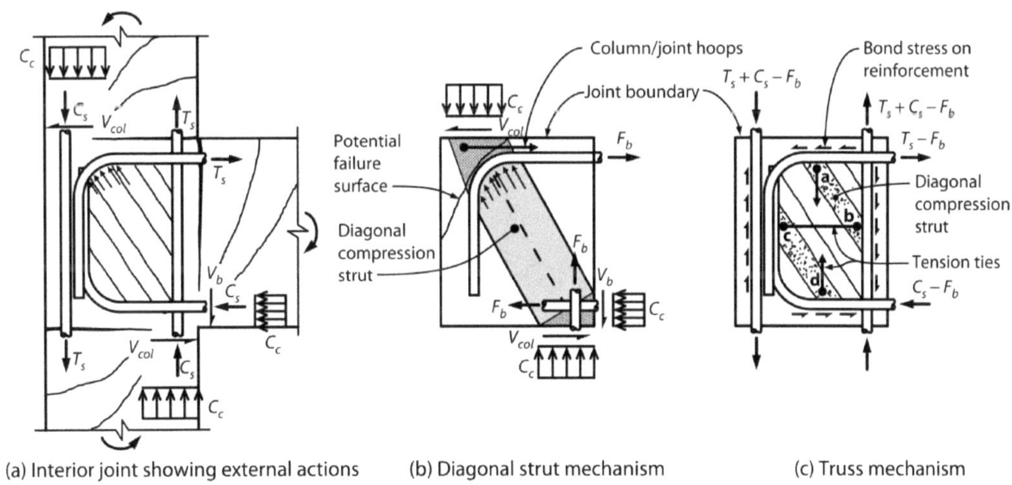

Figure 9.25 Force transfer mechanism in exterior beam-column joint. Concrete stress distributions are shown along with notation indicating the resultants of those stress distributions.

with interior joints, transverse reinforcement can confine joint concrete, thereby potentially increasing compressive strength and ductility of the diagonal compression strut. A truss mechanism can also resist joint shear forces (Figure 9.25*c*).

Similar to interior joints, inelastic displacement reversals may result in tensile yielding of longitudinal reinforcement followed by compressive loading that imposes large demands on the bar in compression. Anchorage conditions are less severe than for similarly sized interior connections because beam longitudinal reinforcement needs to be developed from only one side of the joint instead of both sides. Hawkins et al. (1987) report tests of confined joints showing that standard hooks having l_{dh} as required for tension are capable of developing f_y of the bar in compression. Thus, crack closure through bar yielding can be expected where beams have equal areas of top and bottom longitudinal reinforcement. Where reinforcement areas differ, crack closure is likely for only one direction of loading.

Figure 9.26 compares measured exterior joint strengths with two reference curves.

Headed reinforcement resists tension and compression in the joint through both bond and head bearing. Where bars are in compression, head bearing has been observed to spall the concrete cover off the far face of the joint. The situation is improved by holding the heads back a small distance from the far side of the joint core. Kang et al. (2009) review available data, concluding that headed bars are effective where clear spacing of bars within a layer or between layers is at least $2d_b$. Joints having normalweight concrete with strength as high as 18 ksi (120 MPa) and reinforcement yield strength as high as 100 ksi (690 MPa) have performed successfully. ACI 318, however, concluded that the available data do not support use of headed bars with clear spacing closer than $3d_b$ in special moment frames. See Section 9.6.4.

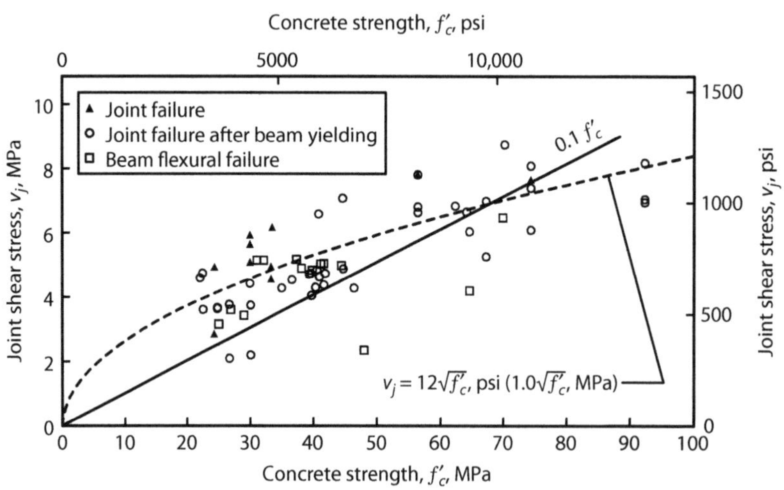

FIGURE 9.26 Exterior joint shear strengths. (*Data from Hwang and Lee, 1999, supplemented by personal communication from Hwang in 2012.*)

9.5.3 Tee (Roof) Connections

Tee connections in buildings occur at roof and mezzanine levels. The traditional detail was to terminate column bars without hooks or with 90° bends bent outward, away from the joint center. Although this detail was relatively easy to construct, tests have shown that joint efficiency is low (Nilsson and Losberg, 1976; Lowes and Moehle, 1999). Joint efficiency can be improved by either bending the column hooks inward or by using headed reinforcement (Figure 9.27). Where column bars are hooked, the hooks commonly will be located below the top beam reinforcement. In this case, an option is to add "U" bars to confine the top surface of concrete and to improve bond conditions for the top bars. An option common in bridge construction is to use column longitudinal bars (without hooks) lap-spliced with vertical joint reinforcement (Naito et al., 2002). Where headed reinforcement is used, beam top reinforcement can be confined by the reinforcement heads.

Relatively few tests have been reported on tee connections. Ishiwata et al. (2002), Ishibashi et al. (2003), Ishibashi and Inokuchi (2004), and Shimizu et al. (2005) report tests of building-type tee joints with headed reinforcement. Naito et al. (2002) report tests of bridge tee connections.

9.5.4 Corner Connections

In the conventional detail for corner connections, horizontal hoop reinforcement is provided through the joint height, and beam and column bars terminate in standard hooks with tails projecting into the joint core (Figure 9.28*a*). As with exterior connections, the hooked bars should extend almost to the far face of the joint core with at least the necessary development length into the joint. Tests on joints with this detail (Mazzoni et al., 1991; Kramer and Shahrooz, 1994) have performed poorly, mainly because of inadequate joint confinement in the vertical direction. Mazzoni et al. (1991) and Wallace et al. (1998) showed that behavior can

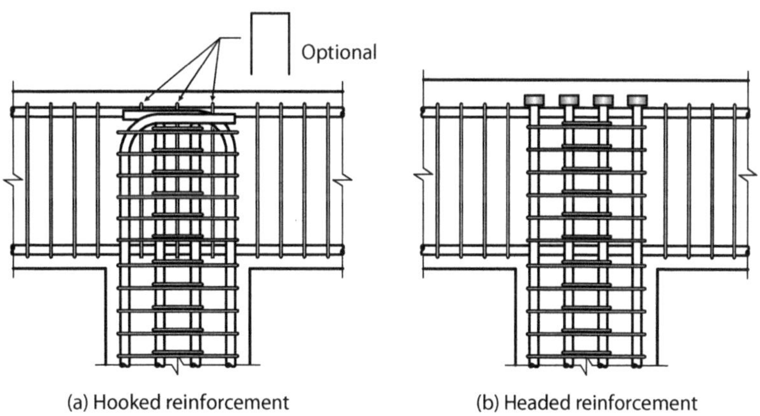

FIGURE 9.27 Elevation views of typical reinforcement for tee beam-column connections with joint transverse reinforcement.

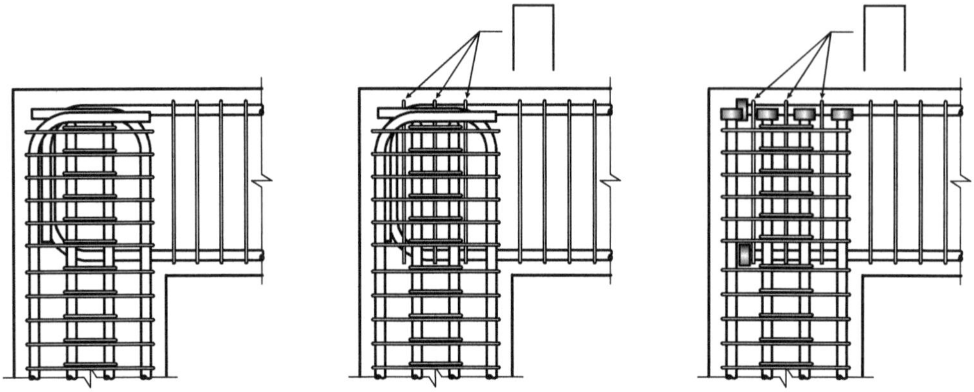

FIGURE 9.28 Elevation views of typical reinforcement for corner beam-column connections with joint transverse reinforcement.

be improved by inserting "U" bars from the top of the joint. The legs of the "U" bars should extend through the full joint depth but not less than a development length from the top beam bars (Figure 9.28*b*). This detail both confines the joint core vertically and clamps the beam longitudinal reinforcement to improve its anchorage.

Reinforcement congestion is a common problem in corner joints with hooked longitudinal reinforcement. Congestion can be reduced by using headed reinforcement (Figure 9.28*c*). Tests by Wallace et al. (1998) demonstrate that use of headed reinforcement improves joint confinement and longitudinal bar development, although these effects may vary with the size of the head relative to the bar size. Figure 9.29 shows a photograph of a corner joint after testing, showing a typical configuration of headed reinforcement and inserted "U" bars. Use of the inserted "U" bars is especially important in corner joints with headed bars because, without the "U" bars, the top headed bars would be effectively disconnected from the joint once the cover spalled off the top of the joint. Where a corner joint uses headed reinforcement for the top beam bars, ACI 318 requires that either (a) the column shall extend above the top of the joint a distance at least the depth h of the joint or (b) the beam reinforcement shall be enclosed by additional vertical joint reinforcement providing equivalent confinement to the top face of the joint.

Maximum joint shear strength of corner joints is reduced compared with interior and exterior joints. Kim and Lafave (2008) report corner joint strengths are approximately 0.4 times the strength of equivalent interior joints.

9.5.5 Predictive Models for Joint Shear Strength

Hwang and Lee (1999, 2000), Hwang et al. (2005), and Hong et al. (2011) have presented strut-and-tie models for joint shear strength that show excellent

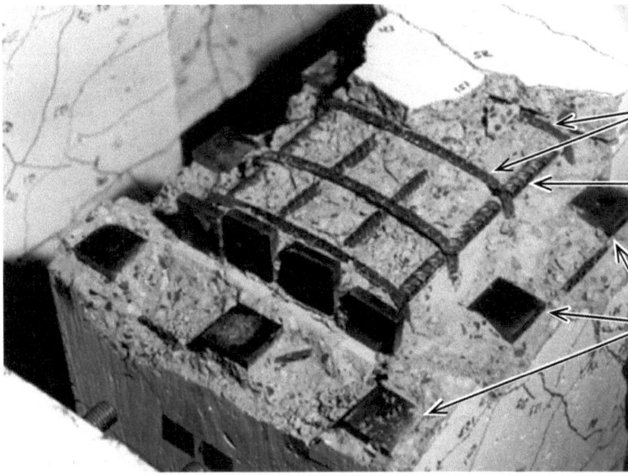

FIGURE 9.29 Photograph of corner joint showing "U" bars, beam reinforcement, and column reinforcement heads at end of test. (*After Wallace et al., 1998, courtesy of American Concrete Institute.*)

agreement with test results. These models, however, can be somewhat difficult to apply in routine analyses and designs due to complexity associated with softened models for concrete struts.

Kim and LaFave (2008) analyzed a large database of beam-column connections with joint transverse reinforcement approximating current design requirements. They found strength to be relatively insensitive to axial load, joint aspect ratio h_b/h_c, hoop spacing s, and amount of transverse reinforcement A_{sh}. Retaining the most important variables, they proposed joint shear strength as

$$v_n = \kappa \gamma_1 \gamma_2 \gamma_e (JI)^{0.07} (BI)^{0.25} (f_c')^{0.75} \tag{9.22}$$

in which $\kappa = 3.54$ for psi units and 1.02 for MPa units; γ_1 = parameter describing joint geometry = 1.0, 0.7, and 0.4 for interior, exterior, and corner (knee) joints; γ_2 = parameter describing out-of-plane geometry = 1.0 if one or no transverse beams and 1.18 if transverse beams on both sides of the joint; $\gamma_e = (1 - e/b_c)^{0.77}$ describes eccentricity of the beam and joint; JI = joint transverse reinforcement index = $\rho_j f_{yt}/f_c'$; BI = beam reinforcement index = $\rho f_y/f_c'$; f_c' = concrete compressive strength; e = eccentricity between beam and column centerlines; b_c = column width; ρ_j = volume of joint transverse reinforcement in direction parallel to joint shear divided by joint volume; f_{yt} = yield stress of joint transverse reinforcement; ρ = beam longitudinal reinforcement ratio; and f_y = yield stress of beam longitudinal reinforcement. Figure 9.30 shows that Eq. (9.22) is a fairly good predictor of joint strength.

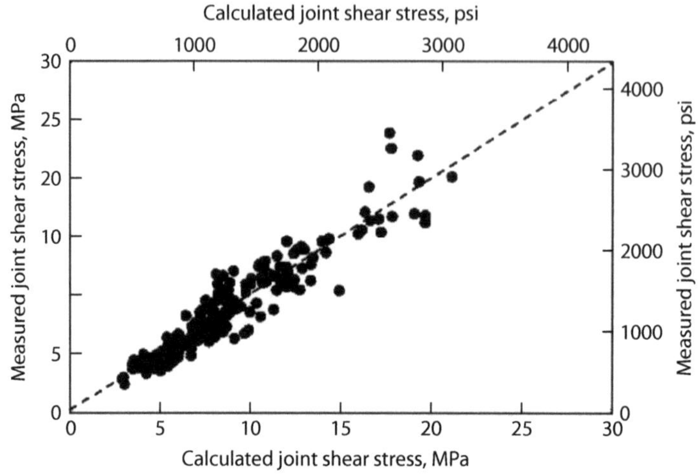

FIGURE 9.30 Measured versus calculated joint shear strength based on Eq. (9.22). *(After Kim and LaFave, 2008, courtesy of American Concrete Institute.)*

9.6 Recommended Design Procedure for Beam-Column Joints

Professional associations in several different countries have developed design recommendations or requirements for beam-column connections. These differ from one another depending on underlying philosophies on joint resistance mechanisms as well as differences in connection performance expectations. Notable recommendations/requirements include ACI 318 (2014), ACI 352, AIJ (1999), Eurocode 8 (2004), and NZS 3101 (2006). This text recommends following the procedures of ACI 318 and ACI 352, which are based primarily on the diagonal strut mechanism. ACI 318 contains *minimum requirements* for beam-column joints in buildings located in regions of highest seismicity; those requirements are adopted in most highly seismic jurisdictions of the United States. A related document ACI 352 (2002) provides *recommended practice* that generally equals or exceeds requirements of ACI 318. ACI 352 also includes connections for both highly seismic regions and less demanding regions. This text mainly follows the recommendations of ACI 352, with reference to the requirements of ACI 318 where they differ substantially.

The design procedure can be described in a series of six steps as follows:

1. Classify joints according to loading conditions and geometry.
2. Determine joint shear demands.
3. Size the connection for joint shear demands.
4. Develop beam and column longitudinal reinforcement.
5. Provide joint confinement.
6. Provide adequate strength and detailing in columns.

These steps are further detailed in the following sections.

9.6.1 Classify Joints According to Loading Conditions and Geometry

Connections are classified into two types based on the loading conditions for the connection and the anticipated deformations of the connected frame members when resisting lateral loads. A *Type 1* connection is intended for loading conditions without significant inelastic deformation. A *Type 2* connection is designed to have sustained strength under deformation reversals into the inelastic range. Type 2 connections are required for special moment frames used in regions of highest seismicity.

Connections are also classified as *interior*, *exterior*, or *corner* connections (Figure 9.31). Beams are assumed to have rectangular cross section with web width b_w, except the web can be integral with a slab, thereby forming a flanged beam.

Connections for which the beam web is wider than the column are classified as *wide-beam connections*. For such connections, beam width b_w cannot exceed the lesser of $3b_c$ and $(b_c + 1.5h_c)$, where b_c and h_c are the column width and depth, respectively. If the beam and column are concentric, it is permitted for some of the beam longitudinal reinforcement to pass outside the column core. Otherwise, all beam longitudinal reinforcement is required to be anchored in or pass through the column core. This latter requirement is because of lack of research data covering eccentric wide-beam connections. Note that this is not intended to exclude longitudinal reinforcement in beam flanges that passes outside the column core.

ACI 352 defines joint shear strength based in part on the number and position of beams that frame into and thereby effectively confine a joint face. For this purpose, a beam is considered to effectively confine a joint face if the beam web width is at least three quarters of the width of the column face and the beam depth is at least three quarters of the depth of the deepest beam framing into the joint. In Figure 9.31 dashed lines indicate beams that either do not exist or do not satisfy the aforementioned dimensional requirement.

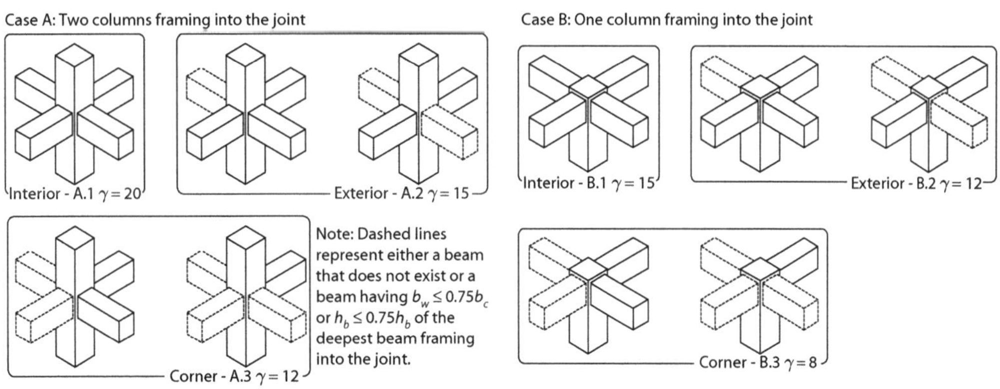

FIGURE 9.31 Beam-column connection geometries. Slabs not shown for clarity. Wide-beam cases not shown. Values of γ shown are for Type 2 connections only. (*After ACI 352, 2002, courtesy of American Concrete Institute.*)

9.6.2 Determine Joint Shear Demands

Joint shear demands are defined by horizontal joint shear V_{jh}. As indicated in Figure 9.3, V_{jh} is calculated from beam flexural tension and compression stress resultants (T_s and C) and column shear (V_{col}). Where yielding is not anticipated under design load combinations, T_s, C, and V_{col} can be calculated from moments and shears obtained from the design load combinations. Where inelastic response is anticipated, T_s, C, and V_{col} can be calculated from the inelastic mechanism as follows.

For Type 2 connections in special moment frames, joint forces are determined from the intended inelastic mechanism, which typically is flexural yielding of the beams at the joint faces (Figure 9.32*a*). A statically determinate free-body diagram of the beams is obtained by cutting through the beam plastic hinges (Figure 9.32*b*). The moments at the section cuts are set equal to the probable moments M_{pr}. Moment equilibrium then establishes the beam shears. Beam moments and shears are applied to a free-body diagram of the column including the joint, as shown in Figure 9.32*c*. Although the location of the column inflection points is generally not known, an acceptable approximation of the column shear V_{col} can be obtained by assuming the inflection points are at the center of the column clear height (an exception is at the base of the building where the column may be "pinned" at the footings). V_{col} is then calculated to satisfy moment equilibrium requirements. For the interior column shown, summing moments about the joint centroid leads readily to

$$V_{col} = \frac{1}{l_c}\left((M_{pr1} + M_{pr2}) + (V_{b1} + V_{b2})\frac{h_c}{2} \right) \tag{9.23}$$

Probable moment strength is obtained following the usual calculations for nominal moment strength, except reinforcement is assumed to have elasto-plastic stress–strain relation with effective yield strength of $1.25f_y$, where f_y = specified yield strength.

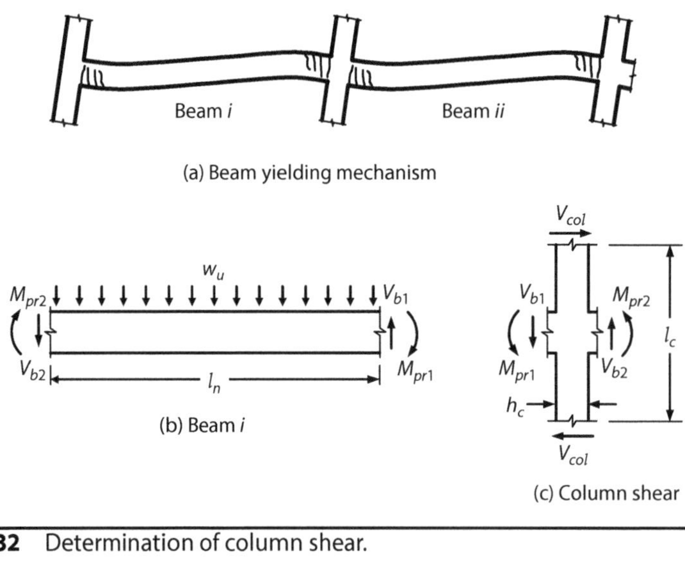

FIGURE 9.32 Determination of column shear.

For Type 2 connections with integrally cast slabs, the effect of the slab on beam moment strength should be included. The effect is especially important when the slab is located in the flexural tension zone. As discussed in Section 6.10, developed slab reinforcement within the effective width b_e should be included. Effective widths are as follows:

- For beams that are continuous through a beam-column joint, or for discontinuous beams at joints with transverse beams, the effective flange width b_e is defined as b_w plus overhanging flange width on each side of the web equal to the least of $8h_s$, $s_w/2$, and $l_n/8$ for beams with flanges on two sides of the web, and b_w plus the least of $6h_s$, $s_w/2$, and $l_n/12$ for beams with flange on one side of the web. Additionally, the effective flange width should not be taken less than $2b_w$. The relevant definitions are h_s = slab or flange thickness, s_w = clear distance to adjacent web, and l_n = beam clear span.
- For discontinuous beams at joints without transverse beams, the effective width b_e is defined by a 45° line drawn from the inside face of the column to the slab edge but not beyond the outside edge of the column (Figure 9.33).

Joint horizontal shear V_{jh} is next determined by substituting flexural tension and compression forces T_s and C (Figure 9.34*b*), then making a horizontal cut through the joint (Figure 9.34*c*), and finally summing forces in the horizontal direction to determine V_{jh}. From the free-body diagram for the interior joint shown in Figure 9.34*c*, the joint shear is given by Eq. (9.24). For Type 2 joints, the tensile force in beam flexural tension reinforcement is assumed to be αf_y, where $\alpha \geq 1.25$. In practice, a value $\alpha = 1.25$ is almost always used. Although ACI 318 does not require inclusion of slab reinforcement in determining the flexural

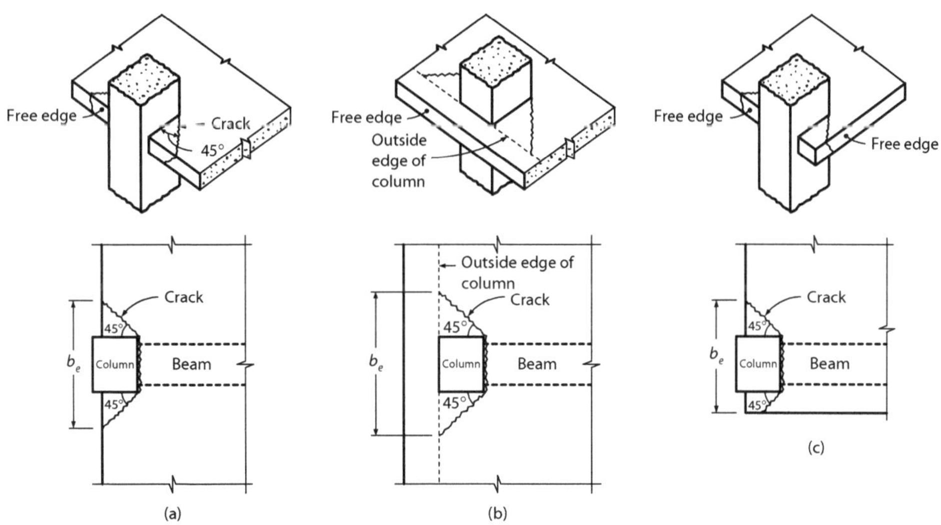

FIGURE 9.33 Effective flange width b_e for exterior connections without transverse beams.

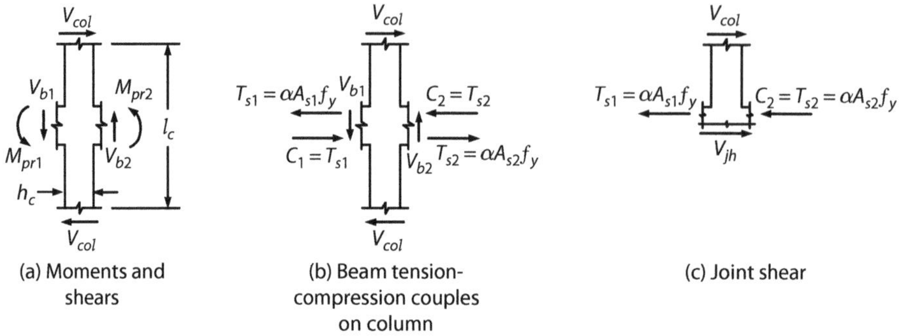

FIGURE 9.34 Determination of joint horizontal shear V_{jh}.

tension force, ACI 352 (and this book) recommends including developed reinforcement within the effective flange width b_e.

$$V_{jh} = T_{s1} + T_{s2} - V_{col} \tag{9.24}$$

9.6.3 Size the Connection for Joint Shear Demands

The strength design requirement is

$$\phi V_n \geq V_{jh} \tag{9.25}$$

where $\phi = 0.85$ and nominal joint shear strength V_n is

$$\begin{aligned} V_n &= \gamma\sqrt{f_c'}b_j h_c, \text{ psi} \\ &= 0.083\gamma\sqrt{f_c'}b_j h_c, \text{ MPa} \end{aligned} \tag{9.26}$$

where b_j = effective joint width defined by Eq. (9.27), and h_c = column depth in the direction of joint shear being considered. Where column depth changes at a joint, h_c is taken as the smaller value. If the column does not have a rectangular cross section or if the sides of the rectangle are not parallel to the spans, it should be treated as a square column having the same cross-sectional area.

According to ACI 352, the effective joint width b_j should not exceed the least value obtained using Eq. (9.27).

$$\begin{aligned} b_j &= \frac{b_w + b_c}{2} \\ &\leq b_w + \sum\frac{mh_c}{2} \\ &\leq b_c \end{aligned} \tag{9.27}$$

b_w is web width of the beam generating joint shear. If there is only one beam in the direction of loading, b_w is the width of that beam. Where beams of different width frame into opposite sides of the column in the direction of loading, b_w is the

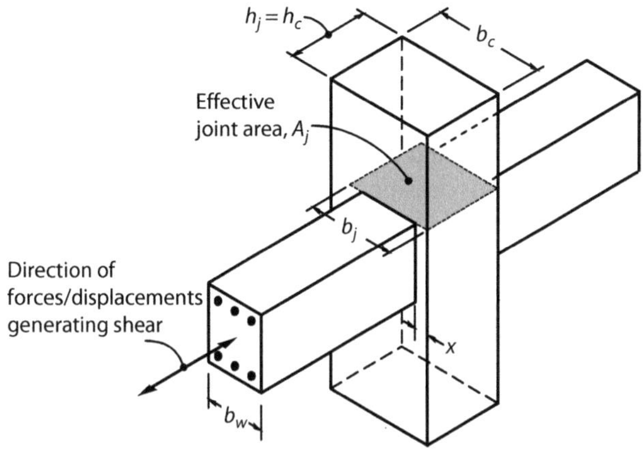

Effective joint width, b_j	
ACI 352	ACI 318
$b_j = \frac{b_w + b_c}{2}$ $\le b_w + \sum \frac{mh_c}{2}$ $\le b_c$	$b_j = b_w + h_c$ $\le b_w + 2x$

Notes:
1. For column wider than beam, x = the smaller of the dimensions from face of beam to edge of column.
2. Effective area of joint for forces in each direction of framing is to be considered separately.

FIGURE 9.35 Effective joint area in accordance with ACI 318 and ACI 352.

average of the two widths. For joints where the eccentricity between the beam centerline and the column centroid exceeds $b_c/8$, $m = 0.3$; for all other cases, $m = 0.5$. The summation term is applied on each side of the joint for which the edge of the column extends beyond the edge of the beam. The value of $mh_c/2$ should not be taken greater than the extension of the column beyond the edge of the beam. Note that ACI 318 uses a different definition of joint width, which governs in jurisdictions where ACI 318 is adopted.

Figure 9.35 illustrates the various terms and definitions.

Values for the constant γ in Eq. (9.26) are listed in Table 9.2. The values depend on the connection loading conditions and geometry. Figure 9.31 illustrates the connection geometries.

Connection Geometry*	Loading Conditions	
	Type 1	**Type 2**
A. Joints with columns framing into bottom and top of joint		
A.1 Joints effectively confined on all four vertical faces	24	20
A.2 Joints effectively confined on three vertical faces or on two opposite faces	20	15
A.3 Other cases	15	12
B. Joints with column framing into bottom of joint only		
B.1 Joints effectively confined on all four vertical faces	20	15
B.2 Joints effectively confined on three vertical faces or on two opposite faces	15	12
B.3 Other cases	12	8

*ACI 318 does not differentiate between Type A and Type B connections, but instead uses the Type A γ values for both continuous and discontinuous columns.

TABLE 9.2 Values of γ for Beam-Column Connections

For calculating joint shear strength, connections are classified according to the number of vertical sides of the joint confined by beams framing into the joint, and by whether the column is continuous or discontinuous. For a joint side to be considered effectively confined by a beam, the beam should cover at least 3/4 of the width of the column, and total depth of the beam should be at least 3/4 of the total depth of the deepest beam framing into the joint. This classification is valid for joints with unloaded beam or column stubs if they extend at least one effective depth beyond the joint face and meet the dimensional requirements for full frame members. See Figure 9.31.

9.6.4 Develop Beam and Column Longitudinal Reinforcement

Different reinforcement requirements apply depending on whether (a) the longitudinal reinforcement passes through the joint and continues as reinforcement in the adjacent member, as occurs in interior connections, or (b) the longitudinal reinforcement terminates at the joint, as occurs in connections around the perimeter of the building.

Interior Connections (Longitudinal Reinforcement Passing through the Joint)

For Type 1 connections with longitudinal reinforcement passing through the joint, there are no requirements. For Type 2 connections, concerns about excessive

Applicability	Parameter	Minimum Requirement (psi)	Minimum Requirement (MPa)
(a) Longitudinal reinforcement passing through the joint			
No requirements			
(b) Longitudinal reinforcement terminating at the joint			
Hooked reinforcement	l_{dh}*	$=\dfrac{\psi_e f_y d_b}{50\lambda\sqrt{f'_c}}$† but not less than $8d_b$ and 6 in	$=\dfrac{\psi_e f_y d_b}{4.2\lambda\sqrt{f'_c}}$† but not less than $8d_b$ and 150 mm
Headed reinforcement	l_{dt}‡	$0.75\,l_{dh}$ but not less than $8d_b$ and 6 in	$0.75\,l_{dh}$ but not less than $8d_b$ and 150 mm

*For No. 11 (36) and smaller bars, if side cover normal to the plane of the hook is at least 2.5 in (65 mm), and cover on the bar extension beyond the hook is at least 2 in (50 mm), l_{dh} as given in the table can be multiplied by 0.7. For No. 11 (36) and smaller bars, if the hook is enclosed vertically or horizontally within ties or stirrup ties that are provided along the full development length at spacing not greater than $3d_b$, where d_b is diameter of the anchored bar, then l_{dh} as given in the table can be multiplied by 0.8. Both factors can be applied if both conditions are satisfied.

†$\psi_e = 1.2$ for epoxy-coated bars, $= 1.0$ otherwise. $\lambda = 0.75$ for lightweight concrete, $= 1.0$ otherwise.

‡ACI 318 requires use of alternative expressions for l_{dt}. See Section 8.10.2. Note that ACI 318 requires clear spacing of at least $4d_b$ for headed bars in Type 1 beam-column connections. Although not specifically addressed in ACI 352, spacing as close as $2d_b$ is considered acceptable.

Table 9.3 Requirements for Development of Longitudinal Reinforcement at Type 1 Beam-Column Joints

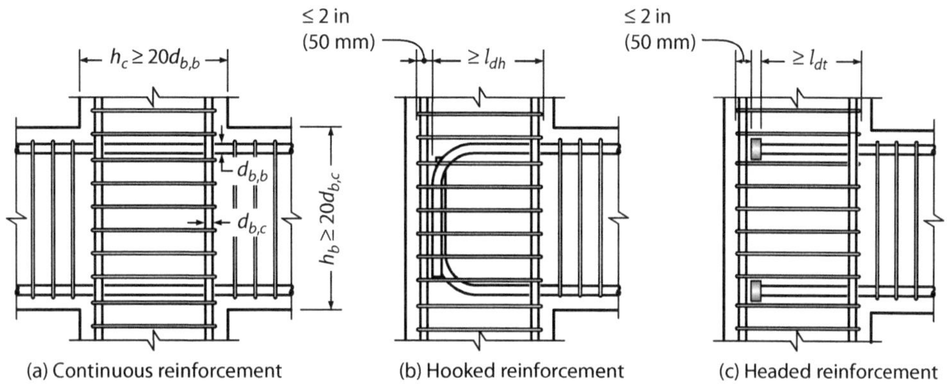

FIGURE 9.36 Location and dimensional requirements for straight bars, hooks, and headed bars in Type 2 connections.

bar slip have led to requirements for the ratio of a member cross-sectional dimension to the diameter of a longitudinal bar crossing it. See Figure 9.36*a* for the case where a column is wider than the beams framing into it. See Table 9.4(a) for more general case.

Exterior Connections (Longitudinal Reinforcement Terminating in the Joint)

See Tables 9.3 and 9.4 for development length requirements. Although ACI 352 measures l_{dt} to the back side of the head, here it is recommended to measure to the bearing face of the head, consistent with ACI 318, and as shown in Figure 9.36*c*.

For Type 1 connections, the critical section for development of reinforcement is the face of the column. For Type 2 connections, recognizing the pullout cone that commonly forms on the tension side (Figure 9.18), the critical section for beam reinforcement is taken at the outside edge of the transverse reinforcement confining the joint core. For column reinforcement in Type 2 connections, the critical section is taken at the outside edge of the beam longitudinal reinforcement that passes into the joint. (Note that in ACI 318, the critical sections are taken at the face of the column for beam reinforcement and at the face of the beam for column reinforcement.)

For hooked or headed reinforcement, the hook or head should extend at least a development length l_{dh} into the joint, and should extend at least to within 2 in (50 mm) of the far side of the confined core. The tail extension of hooks should project toward the mid-height of the joint so that the intended diagonal compression strut can develop across the joint. See Figure 9.36*b*. For multiple layers of reinforcement, the bars in each layer should follow the preceding requirements, except it is permitted for hooks of inner bars to exceed the 2-in (50-mm) limit of Figure 9.36 within reason.

ACI 352 recommends transverse reinforcement to resist bursting forces from heads near exposed faces of joints. Based on a review of a broad data set,

Applicability	Parameter	Minimum Requirement (psi)	Minimum Requirement (MPa)
(a) Longitudinal reinforcement passing through the joint			
Column bars or beam bars within the joint core	$\frac{h_c}{d_{b,b}}$ and $\frac{h_b}{d_{b,c}}$*	$20\frac{f_y}{60{,}000}$ but not less than 20	$20\frac{f_y}{420}$ but not less than 20
Wide beams, for main beam bars outside the joint core	$\frac{h_c}{d_{b,b}}$	$24\frac{f_y}{60{,}000}$ but not less than 24	$24\frac{f_y}{420}$ but not less than 24
(b) Longitudinal reinforcement terminating at the joint			
Hooked reinforcement	l_{dh}	$\frac{\alpha f_y d_b}{75\sqrt{f'_c}}$† but not less than $8d_b$ and 6 in	$=\frac{\alpha f_y d_b}{6.2\sqrt{f'_c}}$† but not less than $8d_b$ and 150 mm
Headed reinforcement	l_{dt}‡	$0.75\,l_{dh}$ but not less than $8d_b$ and 6 in	$0.75\,l_{dh}$ but not less than $8d_b$ and 150 mm

*In ACI 318, it is not required to check the ratio of beam depth to column longitudinal bar diameter.

†α is the stress multiplier for longitudinal reinforcement at the beam-to-joint interface for Type 2 connections (typically taken as $\alpha = 1.25$). If transverse joint reinforcement is provided at spacing $s \le 3d_b$ of the bar being developed, l_{dh} as given in the table can be multiplied by 0.8. At exterior connections, beam longitudinal reinforcement that passes outside the column core should be anchored in the core of the transverse beam following the requirements listed in the table including appropriate modification factors. The critical section for development of such reinforcement is the outside edge of the beam core. ACI 318 uses a similar expression, specifically, $l_{dh} = \frac{f_y d_b}{65\sqrt{f'_c}}\left(\frac{f_y d_b}{5.4\sqrt{f'_c}}\right)$, with the same limits on minimum length. ACI 318 also contains provisions for lightweight concrete that are not repeated here.

‡ACI 318 requires use of alternative expressions for l_{dt}. See Section 8.10.2. Note that ACI 318 requires clear spacing of at least $3d_b$ for headed bars in Type 2 beam-column connections. Although not specifically addressed in ACI 352, spacing as close as $2d_b$ is considered acceptable.

Table 9.4 Requirements for Development of Longitudinal Reinforcement at Type 2 Beam-Column Joints

Kang et al. (2009) recommend instead that clear side cover not be less than $2d_b$, in which case restraining reinforcement is not necessary. ACI 318 requires minimum clear spacing not less than $4d_b$. The review by Kang et al. (2009) justifies reducing the requirement to $2d_b$ between bars in a layer or between bars in adjacent layers.

The development length for a straight bar terminating in a Type 1 connection should be in accordance with ACI 318 development length requirements. The bar should pass within the core of the joint. Any portion of the required straight embedment length extending outside the confined core should be increased by

30%. In general, straight bars should not be terminated in Type 2 connections, although they can be developed outside the connection. ACI 318 provides requirements.

9.6.5 Provide Joint Confinement

Transmission of column axial force and joint shear through the joint requires adequate confinement of the joint core concrete by transverse reinforcement, transverse members, or both. Transverse reinforcement should satisfy general requirements of the governing building code, in addition to requirements for seismic resistance outlined here.

Type 1 Connections

At least two layers of transverse reinforcement should be provided between the top and bottom levels of beam longitudinal reinforcement of the deepest beam framing into the joint. The center-to-center tie spacing or spiral pitch should not exceed 12 in (300 mm). If the joint is part of the primary system for resisting non-seismic lateral loads, the center-to-center spacing or pitch of the transverse reinforcement should not exceed 6 in (150 mm). To facilitate placement of transverse reinforcement in Type 1 joints, cap or split ties may be used, provided the lap length is sufficient to develop the tie yield strength. Where beams frame into all four sides of the joint and where each beam width is at least 3/4 of the column width and does not leave more than 4 in (100 mm) of the column width uncovered on either side of the beams, joint ties are not needed within the depth of the shallowest beam.

Where beams frame into two opposite sides of a joint, and where each beam width is at least three quarters of the column width, leaving no more than 4 in (100 mm) of the column width on either side of the beam, transverse reinforcement perpendicular to those two covered faces is not required, but transverse reinforcement should be provided in the perpendicular direction.

For joints with a free horizontal face at the top of a discontinuous column, and for which discontinuous beam reinforcement is the nearest longitudinal reinforcement to the free horizontal face, vertical transverse reinforcement should be provided through the full height of the joint. At least two layers of vertical transverse reinforcement should be provided between the outermost longitudinal column bars. To ease placement of vertical transverse reinforcement, inverted "U"-shaped stirrups without 135° hooks may be used, provided straight bar length beyond the outermost layer of discontinued beam longitudinal reinforcement is at least l_d.

Type 2 Connections

Where spiral transverse reinforcement is used, the volumetric ratio ρ_s should be at least

$$\rho_s = 0.12\frac{f_c'}{f_{yh}} \tag{9.28}$$

and not less than

$$\rho_s = 0.45\left(\frac{A_g}{A_{ch}} - 1\right)\frac{f'_c}{f_{yh}} \tag{9.29}$$

ACI 352 limits f_{yh} to 60,000 psi (420 MPa), but ACI 318 permits up to 100,000 psi (69 MPa).

Where rectangular hoop and crosstie horizontal transverse reinforcement is used, the total cross-sectional area of transverse reinforcement in each direction should be at least

$$A_{sh} = 0.09 s b_{ch} \frac{f'_c}{f_{yh}} \tag{9.30}$$

and not less than

$$A_{sh} = 0.3 s b_{ch}\left(\frac{A_g}{A_{ch}} - 1\right)\frac{f'_c}{f_{yh}} \tag{9.31}$$

For connections that are part of the seismic-force-resisting system, center-to-center spacing between layers of horizontal transverse reinforcement (hoops or hoops and crossties) should not exceed the least of 1/4 of the minimum column dimension, $6d_b$ of the column longitudinal bars, and 6 in (150 mm). Crossties, where used, should be provided at each layer of horizontal transverse reinforcement. The lateral center-to-center spacing between crossties or legs of overlapping hoops should not exceed 12 in (300 mm), and each end of a crosstie should engage a peripheral longitudinal bar.

For joints confined on all sides by beams having width at least $0.75b_c$ and depth at least 0.75 times depth of the deepest beam framing into the joint, it is permitted to use half the amount of reinforcement specified by Eqs. (9.28) through (9.31). Spacing limits of the preceding paragraph, however, still apply to such joints.

All hoops should be closed with seismic hooks. Where crossties are used, the 90° bends of adjacent crossties should be alternated on opposite faces of the joint, except for exterior and corner connections where the 90° crosstie bends always should be used at the interior face of the joint.

Horizontal transverse reinforcement should be continued into columns adjacent to the joint, unless a greater amount of transverse reinforcement is required in the column for confinement or shear.

At roof connections, beam top longitudinal reinforcement terminating in the joint should be enclosed in vertical stirrups, as shown in Figure 9.28*b* and *c*. The stirrups should extend through the full height of the joint. The area of vertical stirrup legs should satisfy Eq. (9.30). Center-to-center spacing of stirrups should not exceed the smallest of $b_w/4$, $6d_b$ of the beam longitudinal bars, and 6 in (150 mm). Each corner and alternate beam longitudinal bar in the outermost layer should be enclosed in a 90° stirrup corner. To facilitate placement of vertical transverse reinforcement, inverted "U"-shaped stirrups without 135° hooks may be used provided the anchorage length is sufficient to develop the stirrup yield strength. The critical section for anchorage of this reinforcement is taken as the centerline of the beam longitudinal reinforcement nearest to the top face of the joint.

9.6.6 Provide Adequate Strength and Detailing in Columns

For Type 1 connections, it is acceptable for the columns to be weaker than the beams. For Type 2 connections, column moment strengths should exceed beam strengths, for two main reasons. First, frames with yielding columns are susceptible to developing story mechanisms, which may result in very large deformation demands. This subject is discussed more fully in Chapter 12. Second, the column longitudinal bars should be in the linear range of response such that they can serve the additional role of confining the joint in the vertical direction.

For Type 2 connections that are part of the seismic-force-resisting system, the specific strength requirement of ACI 352 and ACI 318 is that the sum of the nominal moment strengths of the column sections above and below the joint, calculated using the factored axial force that results in the minimum column moment strength, should be at least 1.2 times the sum of the nominal moment strengths of the beam sections at the joint. For connections with beams framing in from two perpendicular directions, this provision can be checked independently in each direction. This verification is not required in connections at the roof level of buildings.

For Type 1 connections, longitudinal column bars may be offset within the joint. The usual provisions for offset bars should be followed.

For Type 2 connections, longitudinal column bars extending through the joint should be distributed around the perimeter of the column core. The center-to-center spacing between adjacent column longitudinal bars should not exceed the greater of 8 in (200 mm) and 1/3 of the column cross-sectional dimension (or diameter) in the direction that the spacing is being considered, but never greater than 12 in (300 mm). Longitudinal column bars may be offset within the joint if extra ties are provided to resist 1.5 times the horizontal thrust due to the offset.

9.7 Beam-Column Joint Deformations

As we have seen, framing action under earthquake loading can result in relatively high shear and bond stresses in a joint. These stresses lead to local joint deformations that increase the overall flexibility of the structural framing. These deformations are especially important to recognize in older buildings with weak joints, as nonlinear response can be concentrated in the joints (Bayhan et al., 2014). Even in buildings with strong joints, though, joint deformations can appreciably affect overall building response and should be considered in structural analysis models.

Joint deformations are usually considered to arise from two effects:

1. *Bar slip:* Anchorage of beam and column longitudinal bars in the joint produces high bond stresses. Slip of the reinforcement relative to the joint concrete results in rigid-body rotations of the beams and columns relative

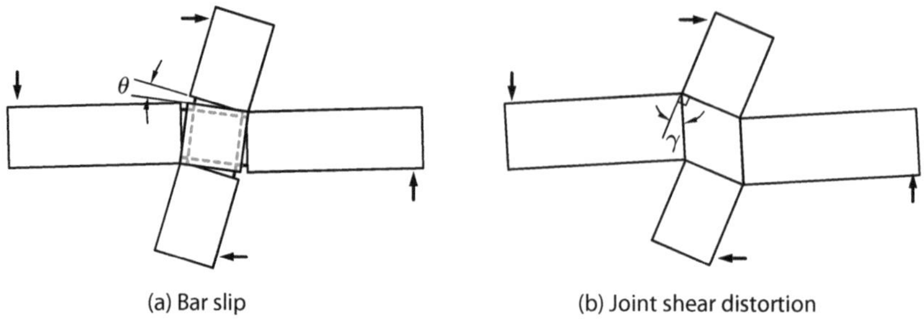

FIGURE 9.37 Sources of joint deformation contributing to overall framing flexibility.

to the joint boundaries (Figure 9.37*a*). This effect was considered in Section 6.11.

2. *Joint shear:* Joints are subjected to large shear stresses, as described in Section 9.2. This results in joint shear deformations and beam and column rigid-body rotations of the type illustrated in Figure 9.37*b*.

Linear structural analysis is commonly used to model effective stiffness of moment frames for strength design, as well as to simulate response at serviceability limit states. Some engineers model the joints as completely rigid. Others, recognizing joint deformations illustrated in Figure 9.37, model the joint as having zero dimensions and extend the beam and column flexibilities to the joint centerline. Yet others use a partially rigid joint. These options are illustrated in Figure 9.38. ASCE 41 recommends using a partially rigid model, with rigid joint dimensions depending on relative moment strengths of the columns and beams. Specifically, according to ASCE 41, for $\Sigma M_{nc}/\Sigma M_{nb} \geq 1.2$, $\beta_c = 0.0$ and $\beta_b = 1.0$; for $\Sigma M_{nc}/\Sigma M_{nb} \leq 0.8$, $\beta_c = 1.0$ and $\beta_b = 0.0$; and for intermediate cases, $\beta_c = \beta_b = 0.5$.

Birely et al. (2012) report statistical analyses of laboratory test results for 45 beam-column connections subjected to reversed cyclic loading. The connections

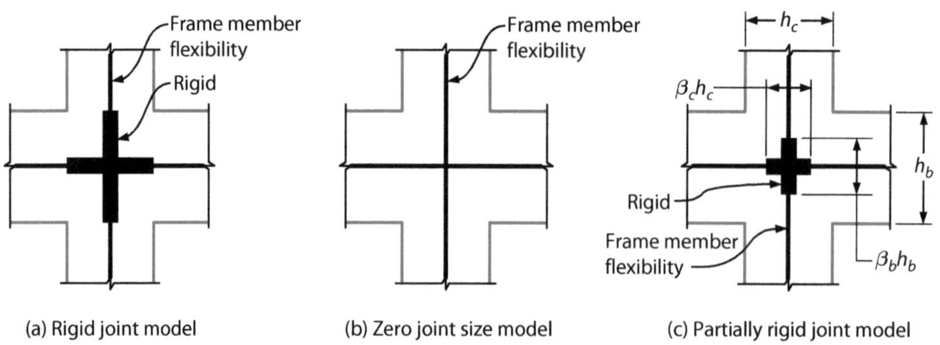

FIGURE 9.38 Joint modeling approaches: (*a*) rigid joint model; (*b*) zero joint size model; (*c*) partially rigid model.

all had columns and beams extending on both sides of the joint in one plane without transverse beams and without slabs. Concrete was normalweight. Good estimates of stiffness to the yield point were obtained by (a) modeling the beams and columns in accordance with the ASCE 41 recommendation (Table 6.1) and (b) modeling the joint as partially rigid (Figure 9.38*c*). The recommended coefficients for the partially rigid joint are $\beta_c = \beta_b = 0.6$ for beam-column connections that are compliant with the ACI 318/352 design requirements, and $\beta_c = \beta_b = 0.0$ for non-compliant connections.

In frames designed to have strong joints (i.e., joints proportioned to resist strengths of adjacent beams), some inelastic response of the joint nonetheless occurs because of joint cracking, inelastic strain penetration of longitudinal bars extending into the joint, and redistribution of internal effects as the adjacent members respond inelastically. Figure 9.39 plots the relative contributions of beams, columns, and joints to the total drift ratio as a function of the peak drift ratio, for the two test specimens reported in Figure 9.19. Unit 2D-I had comparatively low joint nominal shear stress and reinforcement bond stress. For that test specimen, joint deformations constituted approximately 25% of the total drift ratio. In contrast, Unit J2 had comparatively high joint nominal shear stress and reinforcement bond stress. Initially, joint deformations constituted approximately 25% of the total drift ratio, but as drifts increased, progressive damage to the joint resulted in an ever-increasing contribution to the total drift ratio.

For connections with high joint shears or high bond stresses, joint deterioration can reduce the connection ductility capacity. Birely et al. (2012) report the ductility capacity of 45 planar test connections having continuous beams and columns without slabs or transverse beams, subjected to cyclic loads within the plane of the test specimens. Displacement ductility was defined as the displacement at which

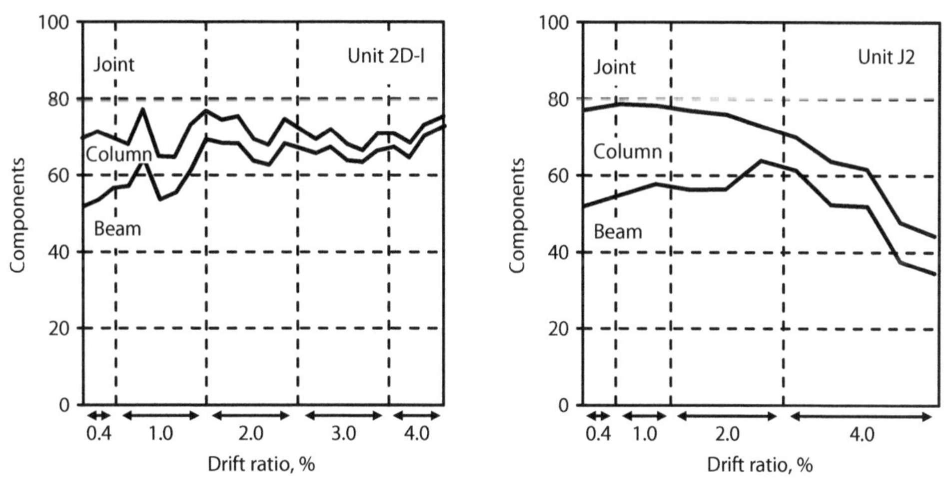

Figure 9.39 Relative contributions of beams, columns, and joints to total connection drift: (*a*) Unit 2D-I; (*b*) Unit J2. See Figure 9.19.

10% strength loss occurred divided by the displacement at yield. Test specimens were classified as brittle, ductile, or limited ductility:

- *Brittle:* Specimens for which maximum strength was less than the strength required to develop the beam yield moment, where the yield moment is defined by initial yielding of beam longitudinal reinforcement in tension.
- *Ductile:* Specimens not classified as brittle and with displacement ductility $\mu_\delta \geq 4$.
- *Limited ductility:* Specimens not classified as brittle or ductile.

The study found that ductility capacity could be gauged based on the joint nominal shear stress v_{jh} and joint nominal bond stress u, defined by

$$v_{jh} = \frac{1}{h_c b_j}[\alpha f_y (A_{s1} + A_{s2}) - V_{col}] \tag{9.32}$$

$$u = \frac{\alpha f_y d_b}{2h_c} \tag{9.33}$$

In Eq. (9.32), the term in square brackets is the nominal horizontal joint shear force assuming that the beam top and bottom longitudinal reinforcement reaches stress αf_y, with $\alpha = 1.25/1.1$ and f_y = measured yield strength of the longitudinal reinforcement. [The measured yield stress was assumed to be 1.1 times the specified value. Thus, applying $\alpha = 1.25/1.1$ in Eq. (9.32) to the measured yield stress is approximately equivalent to applying factor 1.25 to the specified yield strength, as is commonly done in design.] In Eq. (9.33), u is the average bond stress between a single reinforcing bar and the joint concrete assuming the bar yields in compression and tension on the two sides of the joint. As with Eq. (9.32), $\alpha = 1.25/1.1$ and f_y = measured yield strength of the longitudinal reinforcement.

Figure 9.40 plots the nominal joint shear stress demand versus the nominal bond stress demand for the test specimens, with markers indicating the ductility classification. The solid lines in Figure 9.40 indicate the proposed division between ductile and brittle specimens. Specifically, a connection having this geometry, and having $v_{jh}/\sqrt{f'_c} \leq 20$ psi (1.67 MPa) and $u/\sqrt{f'_c} \leq 35$ psi (2.9 MPa), can be expected to exhibit ductile response or limited ductility.

For this connection geometry, ACI 352 and ACI 318 limit the design shear stress to $v_{jh}/\sqrt{f'_c} = 15\phi$ psi $= 12.75$ psi (1.06 MPa). Note that this limit applies to a design loading that considers beam longitudinal reinforcement to be at a stress αf_y, where $\alpha = 1.25$ and f_y = specified yield strength. Similarly, the $20d_b$ limit for longitudinal bars passing through the joint corresponds to bond stress $u = \alpha f_y d_b/(40 d_b)$, which for $f_y = 60$ ksi (410 MPa), $f'_c = 5300$ psi (37 MPa), and $\alpha =$ 1.25 results in $u/\sqrt{f'_c} = 1.25 \times 60000/(40\sqrt{5300}) = 25.8$ psi (2.15 MPa). These ACI limits are indicated by the broken lines in Figure 9.40. All data points within the ACI limits correspond to "ductile" tests.

Connections defined as "limited ductility" or "brittle" tend to sustain greater joint damage and joint inelasticity than connections defined as "ductile." A variety

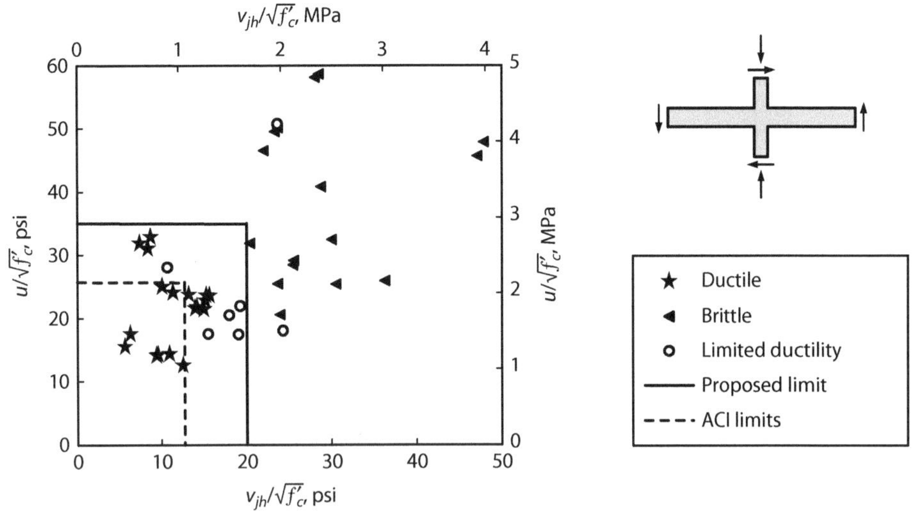

Figure 9.40 Ductility classification as a function of the nominal horizontal joint shear stress and nominal beam longitudinal bar bond demand. Median concrete compressive strength = 5300 psi (37 MPa). (*After Birely et al., 2012, courtesy of American Concrete Institute.*)

of approaches have been proposed for modeling frames with weak joints. Kunnath et al. (1995) represented effects of inadequate joint shear strength by reducing the moment strengths of the beams and columns framing into the joint. Alath and Kunnath (1995) modeled joint shear deformation through a rotational spring model, with degrading hysteresis determined empirically. Biddah and Ghobarah (1999) and Youssef and Ghobarah (2001) modeled joint shear and bond-slip deformations through multiple springs. Other models of interest include those of Lowes and Altoontash (2003) and Shin and LaFave (2004). Celik and Ellingwood (2009) review previous joint models and calibrate parameters to represent joint shear stress–strain relationships using results of full-scale beam-column joint tests.

References

Abdel-Fattah, B., and J.K. Wight (1987). "Study of Moving Beam Hinging Zones for Earthquake-Resistant Design of R/C Buildings," *ACI Structural Journal*, Vol. 84, No. 1, pp. 31–39.

ACI 318 (2014). *Building Code Requirements for Structural Concrete (ACI 318-14) and Commentary*, American Concrete Institute, Farmington Hills, MI.

ACI 352 (1976). "Recommendations for Design of Beam-Column Joints in Monolithic Reinforced Concrete Structures," *Journal of the American Concrete Institute*, Vol. 73, No. 7, pp. 375–393.

ACI 352 (2002). *Recommendations for Design of Beam-Column Connections in Monolithic Reinforced Concrete Structures (ACI 352R-02)*, American Concrete Institute, Farmington Hills, MI, 37 pp.

AIJ (1999). *Design Guidelines for Earthquake Resistant Reinforced Concrete Buildings Based on Inelastic Displacement Concept*, Architectural Institute of Japan, Tokyo, Chapter 8 translation by H. Shiohara.

Alath, S., and S.K. Kunnath (1995). "Modeling Inelastic Shear Deformations in RC Beam-Column Joints," *Proceedings, 10th Engineering Mechanics Conference*, ASCE, pp. 822–825.

ASCE 41 (2013). *Seismic Evaluation and Retrofit of Existing Buildings*, American Society of Civil Engineers, Reston, VA.

Bayhan, B., J.P. Moehle, S. Yavari, K.J. Elwood, S.H. Lin, C.L. Wu, and S.J. Hwang (2014). "Seismic Response of a Concrete Frame with Weak Beam-Column Joints," *Earthquake Spectra*, Earthquake Engineering Research Institute, doi: http://dx.doi.org/10.1193/071811EQS179M.

Beres, A., R.N. White, and P. Gergely (1992). *Seismic Performance of Interior and Exterior Beam-to-Column Joints Related to Lightly Reinforced Concrete Frame Buildings: Detailed Experimental Results*, Report NCEER-92-7, National Center for Earthquake Engineering Research, State University of New York at Buffalo, NY, 207 pp.

Biddah, A., and A. Ghobarah (1999). "Modelling of Shear Deformation and Bond Slip in Reinforced Concrete Joints," *Structural Engineering and Mechanics*, Vol. 7, No. 4, pp. 413–432.

Birely, A.C., L.N. Lowes, and D.E. Lehman (2012). "Linear Analysis of Concrete Frames Considering Joint Flexibility," *ACI Structural Journal*, Vol. 109, No. 3, pp. 381–391 plus appendixes.

Bonacci, J., and S. Pantazopoulou (1993). "Parametric Investigation of Joint Mechanics," *ACI Structural Journal*, Vol. 90, No. 1, pp. 61–71.

Celik, O.C., and B.R. Ellingwood (2009). "Seismic Risk Assessment of Gravity Load Designed Reinforced Concrete Frames Subjected to Mid-America Ground Motions," *Journal of Structural Engineering*, Vol. 135, No. 4, pp. 414–424.

Cheung, P.C., T. Paulay, and R. Park (1991). "New Zealand Tests on Full-Scale Reinforced Concrete Beam-Column-Slab Subassemblages Designed for Earthquake Resistance," Special Publication 123, American Concrete Institute, Farmington Hills, MI, pp. 1–37.

Eurocode 8 (2004). *Eurocode 8: Design of Structures for Earthquake Resistance, Part 1, General Rules, Seismic Actions and Rules for Buildings*, Comité Européen de Normalisation, European Standard EN 1998-1:2004, Brussels, Belgium.

Goto, Y., O. Joh, and T. Shibata (1987). "Influence of Lateral Reinforcement in Beams and Joints on the Behavior of R/C Interior Beam-Column Subassemblages," *Proceedings of the Japan Concrete Institute*, Vol. 9, No. 2, pp. 187–192.

Goto, Y., O. Joh, and T. Shibata (1988). "Influence of Transverse Reinforcement in Beam Ends and Joints on the Behavior of R/C Beam-Column Subassemblages," *Proceedings, 9th World Conference on Earthquake Engineering*, Japan, Vol. IV, pp. 585–590.

Hassan, W. (2011). *Analytical and Experimental Assessment of Seismic Vulnerability of Beam-Column Joints without Transverse Reinforcement in Concrete Buildings*, Doctoral Dissertation, University of California, Berkeley, CA, 471 pp.

Hawkins, N.M., I. Lin, and T. Ueda (1987). "Anchorage of Reinforcing Bars for Seismic Forces," *ACI Structural Journal*, Vol. 84, No. 5, pp. 407–418.

Hong, S.-G., S.-G. Lee, and T.H.-K. Kang (2011). "Deformation-Based Strut-and-Tie Model for Interior Joints of Frames Subject to Load Reversal," *ACI Structural Journal*, Vol. 108, No. 4, pp. 423–433.

Hwang, S.-J., and H.-J. Lee (1999). "Analytical Model for Predicting Shear Strengths of Exterior Reinforced Concrete Beam-Column Joints for Seismic Resistance," *ACI Structural Journal*, Vol. 96, No. 5, pp. 846–858.

Hwang, S.-J., and H.-J. Lee (2000). "Analytical Model for Predicting Shear Strengths of Interior Reinforced Concrete Beam-Column Joints for Seismic Resistance," *ACI Structural Journal*, Vol. 97, No. 1, pp. 35–44.

Hwang, S.-J., H.-J. Lee, T.-F. Liao, K.-C. Wang, and H.-H. Tsai (2005). "Role of Hoops on Shear Strength of Reinforced Concrete Beam-Column Joints," *ACI Structural Journal*, Vol. 102, No. 3, pp. 445–453.

Ichinose, T. (1991). "Interaction between Bond at Beam Bars and Shear Reinforcement in RC Interior Joints," Special Publication 123, American Concrete Institute, Farmington Hills, MI, pp. 379–399.

Ishibashi, K., and R. Inokuchi (2004). "Experimental Study on T-shaped Beam-Column Joints with Anchor-heads on Column's Rebars: Part 3," *Proceedings of the Architectural Institute of Japan*, pp. 819, 820.

Ishibashi, K., R. Inokuchi, H. Ono, and K. Masuo (2003). "Experimental Study on T-shaped Beam-Column Joints with Anchor-heads on Column's Rebars: Part 1," *Proceedings of the Architectural Institute of Japan*, pp. 533, 534.

Ishiwata, Y., S. Hattori, M. Ichikawa, H. Takeuchi, K. Nakamura, and K. Hayakawa (2002). "Development of Mechanical Anchorage Used Circular Anchor Plate: Part 5," *Proceedings of Architectural Institute of Japan*, pp. 563, 564.

Jirsa, J.O. (1995). "Shear in Joints and Other Places," Special Publication 157, American Concrete Institute, Farmington Hills, MI, pp. 59–74.

Kaku, T., and H. Asakusa (1991). "Bond and Anchorage of Bars in Reinforced Concrete Beam-Column Joints," Special Publication 123, American Concrete Institute, Farmington Hills, MI, pp. 401–423.

Kang, T.H.-K., M. Shin, N. Mitra, and J.F. Bonacci (2009). "Seismic Design of Reinforced Concrete Beam-Column Joints with Headed Bars," *ACI Structural Journal*, Vol. 106, No. 6, pp. 868–877.

Kim, J., and J. LaFave (2008). "Probabilistic Joint Shear Strength Models for Design of RC Beam-Column Connections," *ACI Structural Journal*, Vol. 105, No. 6, pp. 770–780.

Kitayama, K., S. Otani, and H. Aoyama (1991). "Development of Design Criteria for RC Interior Beam-Column Joints," Special Publication 123, American Concrete Institute, Farmington Hills, MI, pp. 97–123.

Kramer, D.A., and B.M. Shahrooz (1994). "Seismic Response of Beam-Column Knee Connections," *ACI Structural Journal*, Vol. 91. No. 3, pp. 251–260.

Kunnath, S.K., G. Hoffmann, A.M. Reinhorn, and J.B. Mander (1995). "Gravity-Load-Designed Reinforced Concrete Buildings—Part I: Seismic Evaluation of Existing Construction, *ACI Structural Journal*, Vol. 92, No. 3, pp. 343–354.

Kurose, Y., G.N. Guimaraes, Z. Liu, M.E. Kreger, and J.O. Jirsa (1988). *Study of Reinforced Concrete Beam-Column Joints under Uniaxial and Biaxial Loading*, PMFSEL Report No. 88-2, Phil M. Ferguson Structural Engineering Laboratory, The University of Texas at Austin, Austin, TX, 146 pp.

Kurose, Y., G.N. Guimaraes, Z. Liu, M.E. Kreger, and J.O. Jirsa (1991). "Evaluation of Slab-Beam-Column Connections Subjected to Bidirectional Loading," Special Publication 123, American Concrete Institute, Farmington Hills, MI, pp. 39–67.

Leon, R.T. (1991). "Towards New Bond and Anchorage Provisions for Interior Joints," Special Publication 123, American Concrete Institute, Farmington Hills, MI, pp. 425–442.

Leon, R., and J.O. Jirsa (1986). "Bidirectional Loading of R.C. Beam-Column Joints," *Earthquake Spectra*, Vol. 2, No. 3, pp. 537–564.

Li, B., and C.T.N. Tran (2009). "Seismic Behavior of Reinforced Concrete Beam-Column Joints with Vertically Distributed Reinforcement," *ACI Structural Journal*, Vol. 106, No. 6, pp. 790–799.

Lowes, L.N., and A. Altoontash (2003). "Modeling Reinforced-Concrete Beam-Column Joints Subjected to Cyclic Loading," *Journal of Structural Engineering*, Vol. 129, No. 12, pp. 1686–1697.

Lowes, L.N., and J.P. Moehle (1999). "Evaluation and Retrofit of Beam-Column T-Joints in Older Reinforced Concrete Bridge Structures," *ACI Structural Journal*, Vol. 96, No. 4, pp. 519–533.

Mayfield, B., F.-K. Kong, and A. Bennison (1972). "Strength and Stiffness of Lightweight Concrete Corners," *ACI Journal*, Vol. 69, No. 7, pp. 420–427.

Mayfield, B., F.-K. Kong, A. Bennison, and J.C. Twiston Davies (1971). "Corner Joint Details in Structural Lightweight Concrete," *ACI Journal*, Vol. 68, No. 5, pp. 366–372.

Mazzoni, S., J.P. Moehle, and C.R. Thewalt (1991). *Cyclic Response of RC Beam-Column Knee Joints—Test and Retrofit*, Report No. UCB/EERC-91/14, Earthquake Engineering Research Center, University of California, Berkeley, CA, 18 pp.

Meinheit, D.F., and J.O. Jirsa (1981). "Shear Strength of R/C Beam-Column Connections," *Journal of the Structural Division*, Vol. 107, No. ST11, pp. 2227–2244.

Naito, C.J., J.P. Moehle, and K.M. Mosalam (2002). "Evaluation of Bridge Beam-Column Joints under Simulated Seismic Loading," *ACI Structural Journal*, Vol. 99, No. 1, pp. 62–71.

Nilsson, I.H.E., and A. Losberg (1976). "Reinforced Concrete Corners and Joints Subjected to Bending Moment," *Journal of the Structural Division*, Vol. 102, No. ST6, pp. 1229–1254.

NZS 3101 (2006). *Concrete Design Standard, NZS3101:2006, Part 1* and *Commentary on the Concrete Design Standard, NZS 3101:2006, Part 2*, Standards Association of New Zealand, Wellington, New Zealand.

Otani, S. (1991). "The Architectural Institute of Japan (AIJ) Proposal of Ultimate Strength Design Requirements for RC Buildings with Emphasis on Beam-Column Joints," Special Publication 123, American Concrete Institute, Farmington Hills, MI, pp. 125–144.

Park, R. (1986). "Ductile Design Approach for Reinforced Concrete Frames," *Earthquake Spectra*, Vol. 2, No. 3, pp. 565–619.

Park, R., and T. Paulay (1975). *Reinforced Concrete Structures*, Wiley Interscience, New York, NY, 769 pp.

Parra-Montesinos, G.J., S.W. Peterfreund, and S.-H. Chao (2005). "Highly Damage-Tolerant Beam-Column Joints through Use of High-Performance Fiber-Reinforced Cement Composites," *ACI Structural Journal*, Vol. 102, No. 3, pp. 487–495.

Paulay, T. (1989). "Equilibrium Criteria for Reinforced Concrete Beam-Column Joints," *ACI Structural Journal*, Vol. 86, No. 6, pp. 635–643.

Paulay, T., R. Park, and M.J.N. Priestley (1978). "Reinforced Concrete Beam Column Joint under Seismic Actions," *Journal of the American Concrete Institute*, Vol. 75, No. 11, pp. 585–593.

Pessiki, S.P., C.H. Conley, P. Gergely, and R.N. White (1990). *Seismic Behavior of Lightly-Reinforced Concrete Column and Beam-Column Joint Details*, Report NCEER-90-0014, National Center for Earthquake Engineering Research, State University of New York at Buffalo, NY, 209 pp.

Saqan, E.I., and M.E. Kreger (1998). "Evaluation of U.S. Shear Strength Provisions for Design of Beam-Column Connections Constructed with High-Strength Concrete," Special Publication 176, American Concrete Institute, Farmington Hills, MI, pp. 311–328.

Shimizu, Y., K. Ishibashi, and R. Inokuchi (2005). "Experimental Study on T-shaped Beam-Column Joints with Anchor-heads on Column's Rebars: Part 5," *Proceedings of the Architectural Institute of Japan*, pp. 281, 282.

Shin, M., and J.M. LaFave (2004). "Testing and Modeling for Cyclic Joint Shear Deformations in RC Beam-Column Connections," *Proceedings, Thirteenth World Conference on Earthquake Engineering*, Paper No. 0301, Vancouver.

Skettrup, E., J. Strabo, N.H. Andersen, and T. Brøndum-Nielsen (1984). "Concrete Frame Corners," *ACI Journal*, Vol. 81, No. 6, pp. 587–593.

Sugano, S., T. Nagashima, H. Kimura, and A. Ichikawa (1991). "Behavior of Beam-Column Joints Using High-Strength Materials," Special Publication 123, American Concrete Institute, Farmington Hills, MI, pp. 359–377.

Viwathanatepa, S., E.P. Popov, and V.V. Bertero (1979). *Effects of Generalized Loadings on Bond of Reinforcing Bars Embedded in Confined Concrete Blocks*, Report No. UCB/EERC-79/22, Earthquake Engineering Research Center, University of California, Berkeley, CA, 316 pp.

Wallace, J.W., S.W. McConnell, P. Gupta, and P.A. Cote (1998). "Use of Headed Reinforcement in Beam-Column Joints Subjected to Earthquake Loads," *ACI Structural Journal*, Vol. 95, No. 5, pp. 590–602.

Wong, P.K.C., M.J.N. Priestley, and R. Park (1990). "Seismic Resistance of Frames with Vertically Distributed Longitudinal Reinforcement in Beams," *ACI Structural Journal*, Vol. 87, No. 4, pp. 488–498.

Youssef, M., and A. Ghobarah (2001). "Modelling of RC Beam-Column Joints and Structural Walls," *Journal of Earthquake Engineering*, Vol. 5, No. 1, pp. 93–111.

CHAPTER 10

Slab-Column and Slab-Wall Connections

10.1 Preview

Slab-column framing and slab-wall framing generally are not used as part of the seismic-force-resisting system in regions of high seismicity. Such framing is, however, used to support gravity loads. As such, its design must ensure that it is capable of supporting the gravity loads as the building sways under earthquake motions. This latter subject is the main focus of this chapter. In regions of lower seismic risk, slab-column frames may be used to provide resistance to lateral forces, including forces due to wind and earthquake loading. This latter application is not considered directly in this book.

This chapter introduces fundamental aspects of the behavior and design of slab-column and slab-wall framing when subjected to gravity and lateral loads. The main emphasis, however, is on aspects that are most relevant for seismic design, including lateral load stiffness; shear and moment transfer at connections; deformation capacity of connections; and requirements for structural integrity. For slab-column framing, both conventionally reinforced construction and unbonded post-tensioned construction are considered. Emphasis is on slab-column and slab-wall framing without beams. The text includes basic discussion about slab deformations, shears, and moments under gravity and lateral loadings, but details of design for flexural strength and for serviceability are not included. The interested reader is referred to other texts for those subjects.

10.2 Some Observations on Seismic Behavior of Slab-Column Connections

The most common form of slab-column framing is the flat plate in which a slab of uniform thickness frames into the columns without beams. The slab-column connection, which refers to the region of slab immediately adjacent to the column, may be subjected to relatively large shear stresses due to applied gravity loads. Imposed

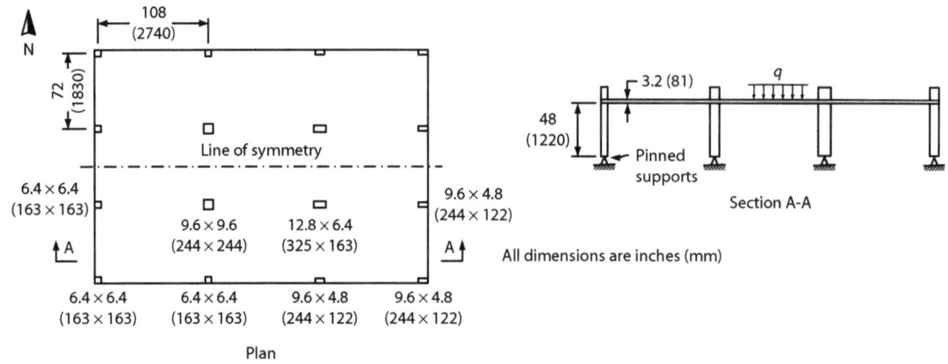

Figure 10.1 Test specimen. (*After Hwang and Moehle, 2000a.*)

lateral displacement during earthquake response produces additional stress in the connection, possibly leading to shear failure in the slab. Such failures usually occur suddenly and extend all the way around the column. This may enable the column to "punch" through the slab. If the slab has shear reinforcement or appropriately detailed flexural reinforcement, the failure may be localized at the punched connection. Otherwise, vertical movement at the punched connection can result in load being redistributed to adjacent connections, with the possibility of additional punching failures and even complete progressive collapse. The following discussion illustrates some of these aspects of behavior for a slab-column frame tested in a laboratory.

Figure 10.1 illustrates a 0.4-scale flat-plate test specimen designed to explore seismic resistance of slab-column framing (Hwang and Moehle, 2000a). A slab of uniform thickness spans between columns of various cross sections. The slab is reinforced with top and bottom reinforcement designed to resist slab moments in two orthogonal directions. The slab has no shear reinforcement. The supporting columns extend below the slab to pinned supports. Slab gravity load comprises self-weight plus uniformly distributed load applied on the slab surface. The total gravity load corresponds to a full-scale loading equal to slab self-weight plus 18 psf (0.86 kPa) superimposed load.

Reversed cyclic lateral displacements were applied to the slab in the NS and EW directions, with progressively increasing displacement amplitude. The slab developed nominal strengths according to ACI 318 (2014) at lateral drift ratio around 1/50 (Figure 10.2). The relatively large drift ratio required to develop nominal strength suggests that slab-column framing may be inadequate as part of the seismic-force-resisting system. The test specimen was, however, capable of safely supporting the imposed gravity loads to lateral drift ratios on the order of 1/25.

Punching failure first occurred for one of the interior connections during cycles to drift ratio of 1/25. Almost immediately, all the other interior connections sustained punching failure. The progressive failure suggests that punching at one connection caused load redistribution to the remaining intact connections, causing their failure. Punching at exterior connections also occurred later during these large drift cycles.

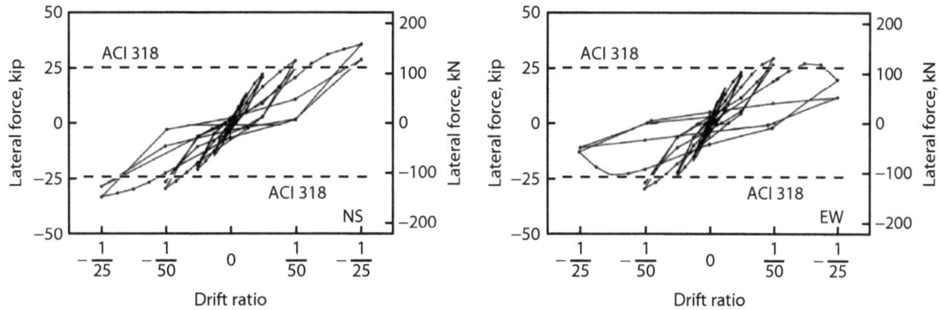

FIGURE 10.2 Measured relations between total lateral force and drift ratio in NS and EW directions. (*After Hwang and Moehle, 2000a, courtesy of from American Concrete Institute.*)

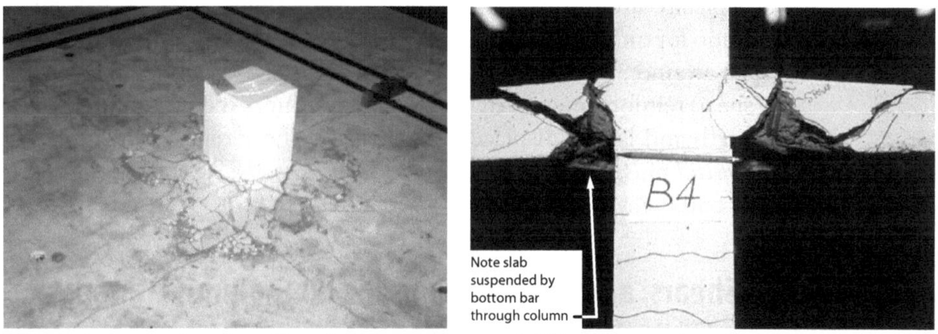

(a) Interior connection viewed from top at angle (b) Exterior connection viewed from exterior

FIGURE 10.3 Punching shear failures at (*a*) interior and (*b*) exterior connections. (*After Hwang and Moehle, 2000a.*)

The slabs had been reinforced according to structural integrity provisions of ACI 318. Those provisions require continuous bottom reinforcement extending over the columns that is intended to suspend the slab from the column in the event of punching shear failure. Figure 10.3 shows photographs of interior and exterior connections after several additional displacement cycles had been imposed on the already punched connections. Although the connections were heavily cracked and had displaced noticeably downward, the structural integrity reinforcement enabled them to maintain gravity load support.

10.3 Slab-Column Systems

Modern concrete building construction commonly uses two forms of slab-column framing (Figure 10.4). In the *flat plate*, a slab of uniform thickness is supported directly on columns. *Drop capitals*, sometimes referred to as *drop caps*, may be added to increase the thickness of the slab immediately around the supporting column, thereby improving shear strength at the connection. The thickness of the drop capital is selected considering shear at the perimeter of the column. The

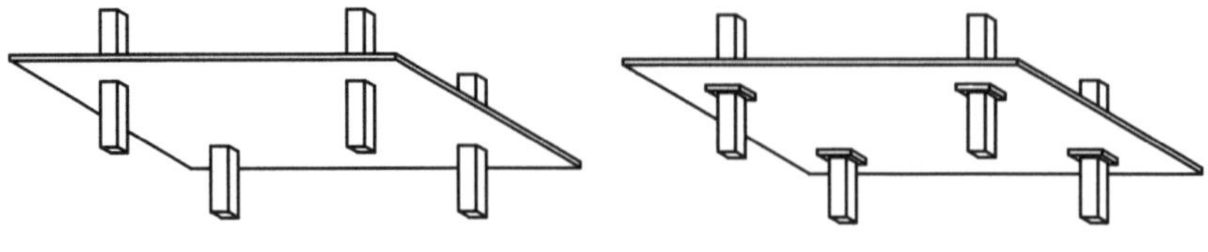

Figure 10.4 Typical slab-column framing configurations.

horizontal extent of the drop capital is determined such that adequate shear strength is provided at the critical section outside the perimeter of the drop capital.

Both conventionally reinforced and unbonded post-tensioned slab-column systems are common. Span-to-thickness ratio for conventionally reinforced flat plates is typically around 30 with typical span around 20 ft (6 m). Span-to-thickness ratio for unbonded post-tensioned flat plates is typically around 45, with typical span around 30 ft (9 m).

Slab shear reinforcement can be provided around the columns to improve shear strength and lateral displacement capacity. Shear reinforcement can be in the form of *stud rails*, *shearbands*, *stirrups*, or *shearheads*. Section 10.7.1 describes these items and their use.

10.4 Moments, Shears, and Deformations in Slab-Column Framing

Consider an idealized slab of infinite plan dimensions supported on point supports with uniform spacing l_1 and l_2 in two orthogonal directions (Figure 10.5*a*). The slab deflects downward under uniformly distributed downward load, q, with displacements as shown in Figure 10.5*b*. To visualize the displaced shape, consider first the grid lines **1**, **3**, **A**, and **C** passing over the supports. Displacements must be zero at the supports, and symmetry of the structure and the loading requires that rotations also must be zero at the supports. At points **a**, **b**, **d**, and **e**, these grid lines are displaced downward and symmetry again requires zero rotations. Next, consider grid lines **2** and **B**. At points **a**, **b**, **d**, and **e**, these grid lines also have zero

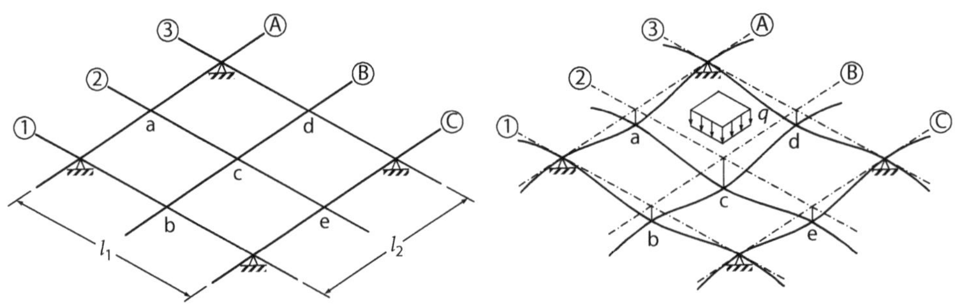

Figure 10.5 Deflected shape of point-supported flat plate under uniform load.

rotation. These grid lines develop additional displacements at point **c**, and again the rotations at this point must be zero.

The orientations of slab curvatures and corresponding moments can be extracted from the displaced shape shown in Figure 10.5*b*. The curvatures and corresponding moments are negative (top in tension, bottom in compression) at each of the support points. At points **a**, **b**, **d**, and **e**, the curvatures and corresponding moments are negative in one direction and positive in the orthogonal direction. Finally, at point **c**, the curvatures and corresponding moments are positive in each grid line direction.

For actual slabs of finite dimensions supported by columns or walls, slab displacements, shears, and moments under applied loads are commonly obtained from finite-element analyses. Alternatively, shears and moments for design can be obtained by idealizing the slab as a series of equivalent frames spanning between columns along one principal direction and bounded by midspans between columns in the orthogonal direction. For the shaded equivalent frame in Figure 10.6, subjected to factored load q_u, the uniform loading along the span of the equivalent frame is $q_u l_2$. Maximum shears for this equivalent frame can be approximated from tributary areas as $q_u l_1 l_2/2$. Panel moments are approximated as shown. Moment values for design can be obtained from the *equivalent frame method* or *direct design method*, which are commonly accepted design methods for two-way slab systems.

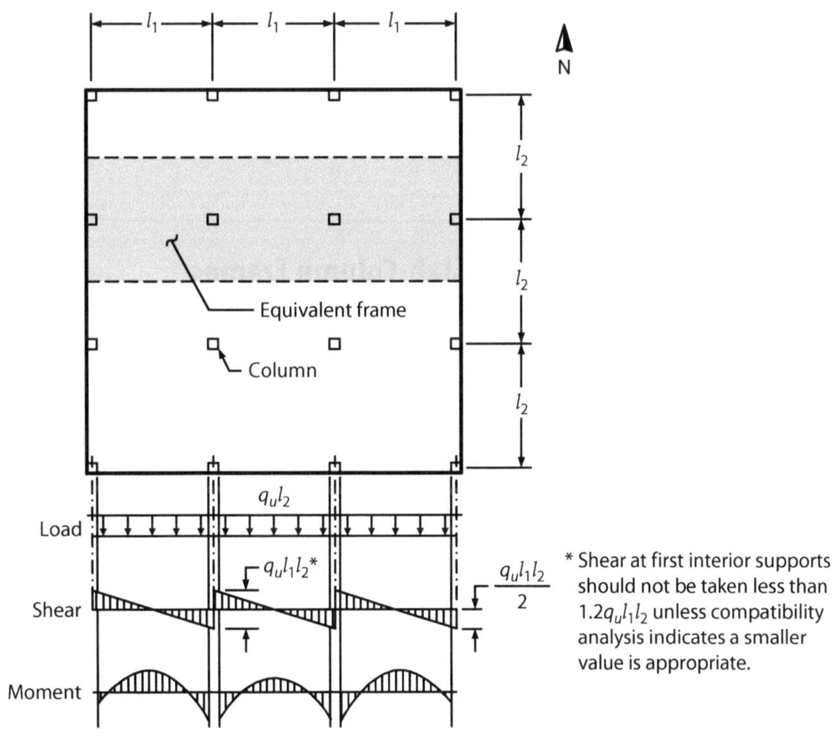

FIGURE 10.6 Panel load, shear, and moment diagrams for flat plate with uniform bay dimensions.

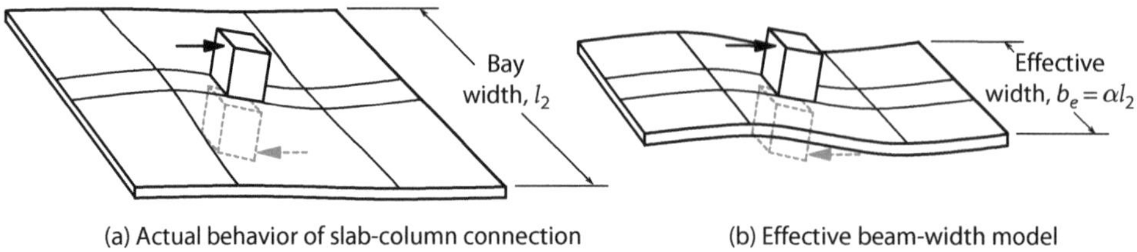

FIGURE 10.7 Deflected shape of interior slab-column connection subjected to lateral force.

These methods are adopted in some building codes, including ACI 318. For additional information, the interested reader is referred to general texts on reinforced concrete design.

From Figure 10.6 it is evident that the total shear transferred between the slab and an interior column is approximately $q_u l_1 l_2$ (there is a small reduction in shear force for loads applied within the slab critical section surrounding the column). For first interior connections (which would be all of the interior connections in Figure 10.6), continuity effects result in an increase in the shear force to approximately $1.2q_u l_1 l_2$. For exterior and corner connections without slab overhangs, the shear transferred between the slab and supporting columns is approximately $q_u l_1 l_2/2$ and $q_u l_1 l_2/4$, respectively.

Figure 10.7*a* illustrates an interior slab-column connection subjected to lateral force. As with gravity loads, the greatest slab curvatures (and moments) develop immediately adjacent to the column. Slab curvatures decrease with increasing transverse distance from the column. This action suggests that flexural reinforcement should be concentrated near the columns. It also suggests that the slab is not fully effective in resisting lateral forces. Section 10.6 provides additional discussion of lateral load stiffness.

10.5 Flexural Reinforcement in Slab-Column Frames

Although slab moments vary continuously along the span and across the bay width, it would be impractical to place reinforcement exactly according to the moment patterns. Instead, the moment patterns are simplified to accommodate simpler placement of reinforcement.

In conventionally reinforced slabs, the two-way slab is idealized as a series of frames spanning between columns in two principal directions. For example, the shaded panel in Figure 10.6 represents a frame spanning in the EW direction. Each frame is then subdivided into a column strip (running along the columns) and a middle strip (the remainder of the panel width) (Figure 10.8*a*). At critical negative and positive moment sections, the total moment across each column strip and middle strip is determined, and then reinforcement capable of resisting those moments is distributed uniformly across each strip. Figure 10.8*a* shows typical reinforcement patterns resulting from this approach. Additional reinforcement (not shown) is sometimes required immediately adjacent to the columns if large moments need to be transferred from the slab to the supporting columns.

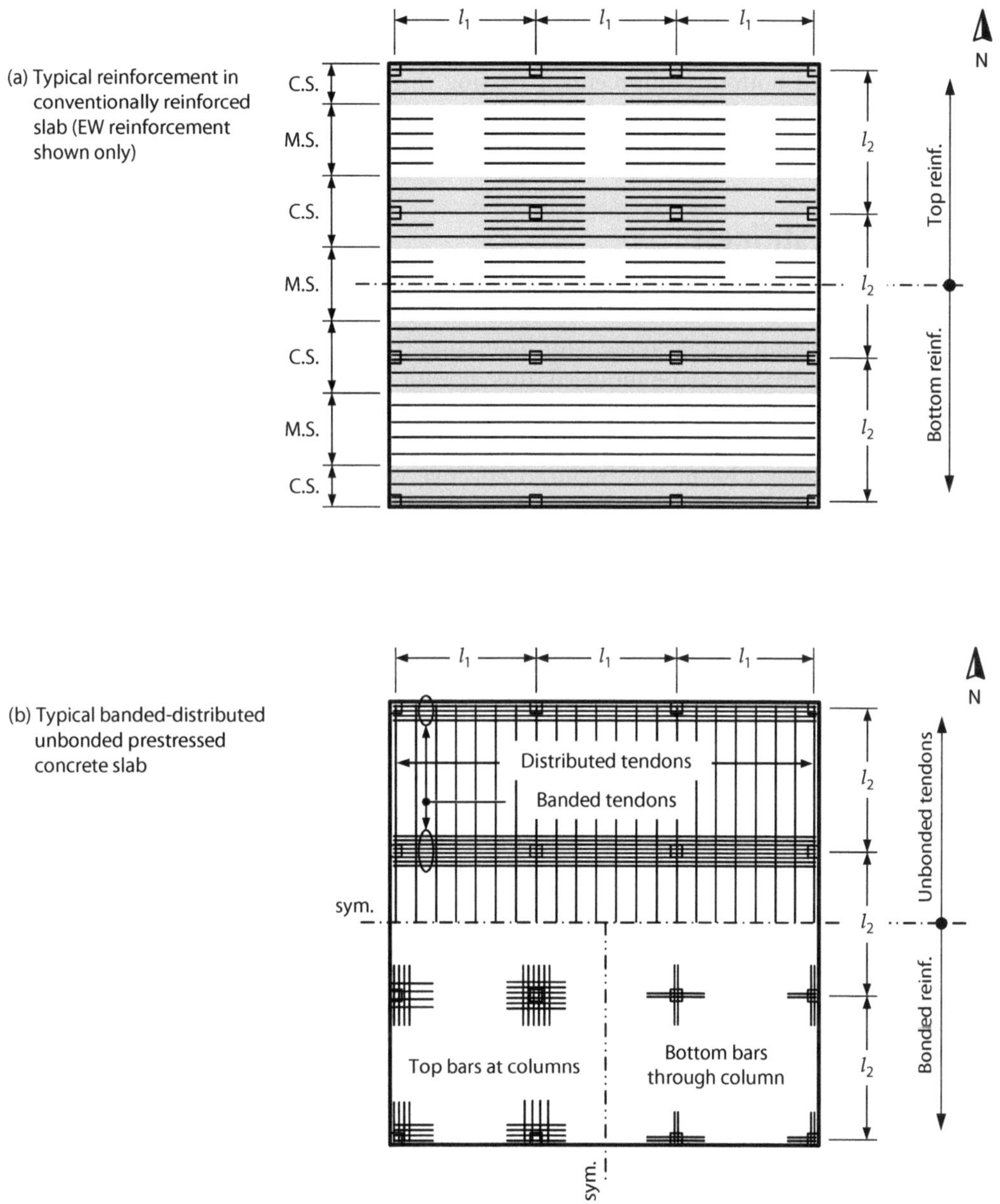

FIGURE 10.8 Characteristic reinforcement in (*a*) conventionally reinforced flat plates and (*b*) unbonded post-tensioned flat plates. C.S. refers to a column strip, and M.S. refers to a middle strip.

A different approach is generally used for unbonded post-tensioned slabs. The most common layout uses banded tendons spanning along column lines in one direction, with uniformly distributed tendons in the transverse direction. In effect, the system is treated as a one-way slab in one direction (the uniformly distributed direction) supported on beams (the banded cross sections) spanning between columns in the orthogonal direction. Figure 10.8*b* shows characteristic tendon placement. In addition to the unbonded tendons, bonded top reinforcement is usually required adjacent to and through the column core to control cracking. Bonded bottom reinforcement passing through the column core may also be required to support the slab in the event of punching shear failure. This last reinforcement is described more fully in Section 10.9.

10.6 Lateral Stiffness

As illustrated in Figure 10.7*a*, when a slab-column connection is subjected to lateral loading, slab flexural moments and curvatures are concentrated near the column and decrease with increasing transverse distance from the column. This behavior can be modeled directly with finite element models. Such models, however, can be computationally expensive and can be overly stiff if the effects of slab cracking are not accounted for. A practical alternative is to represent the slab using a beam with reduced effective width, as illustrated in Figure 10.7*b*. [Alternative models such as the equivalent frame model (Vanderbilt and Corley, 1983; Hwang and Moehle, 2000b) can also be used, though their use is less prevalent.]

In the effective beam-width model, the column is modeled directly and the slab is modeled using a slab-beam having an effective width that is a fraction of the actual slab width (Figure 10.7*b*). Hwang and Moehle (2000b) report studies of the effective beam-width model. For interior connections and for edge connections with bending perpendicular to the edge, the recommended effective width of an uncracked slab can be approximated as

$$\alpha l_2 = 2c_1 + \frac{l_1}{3} \tag{10.1}$$

in which αl_2 = effective width, l_1 and l_2 = center-to-center spans in the longitudinal and transverse directions, and c_1 = rectangular column cross-sectional dimension parallel to dimension l_1. For exterior connections with bending parallel to the edge, the recommended effective width is half the width obtained from Eq. (10.1). This effective width applies to models in which the slab-column joint is modeled as a rigid element. Where the joint is not modeled as a rigid joint but instead slab flexibility extends to the center of the joint, the effective width in Eq. (10.1) should be modified by the factor $(1 - c_1/l_1)^{-3}$.

Lateral stiffness of slab-column frames is further reduced by effects of slab cracking. Figure 10.9 illustrates the reduced stiffness observed in a two-story flat-plate frame subjected to vertical loads followed by lateral drifts due to earthquake shaking (Moehle and Diebold, 1992). Slab cracking may arise from restrained volume changes and externally applied loads. Construction loads applied early in the life of a flat plate can be especially damaging. Hwang and Moehle (2000b) report analytical and laboratory studies to evaluate the reduced effective stiffness of conventionally reinforced slab-column connections. The studies found that the maximum stiffness reduction can be expressed as a function of the service live load, slab geometry, and material properties. For a slab having $c_1 = c_2 = c$, $l_1 = l_2 = l$,

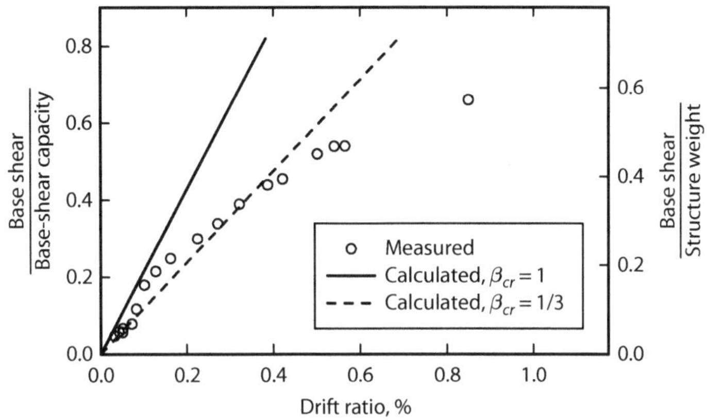

FIGURE 10.9 Measured stiffness versus stiffness calculated by effective beam-width model. (*After Hwang and Moehle, 2000b, courtesy of American Concrete Institute.*)

f'_c = 4000 psi (28 MPa), and f_y = 60 ksi (414 MPa), the slab stiffness reduction factor is given by

$$\beta_{cr} = 5\frac{c}{l} - \left(\frac{L}{40} - 1\right) \geq \frac{1}{3} \tag{10.2}$$

in which L = live load in psf (1 psf = 48 Pa), β_{cr} = ratio of the cracked to gross-section stiffnesses of the slab-beam in the effective beam width model, c = column cross-sectional dimension, and l = center-to-center span between columns. Equation (10.2) can be approximated by

$$\beta_{cr} = 4\frac{c}{l} \geq \frac{1}{3} \tag{10.3}$$

For prestressed slabs, a greater value of β_{cr} is appropriate because of reduced cracking due to prestressing. ASCE 41 (2013) recommends β_{cr} = 1/2 for unbonded post-tensioned slab-column construction based on studies reported by Kang and Wallace (2005) and Elwood et al. (2007).

10.7 Shear and Moment Transfer Strength at Slab-Column Connections

10.7.1 Connections Transferring Shear without Moment

Shear strength is governed by the more severe of two considerations:

1. Beam action, where critical sections are located a distance d from the face of the column, capital, or other discontinuity, and extend in a plane across the entire width of the slab. Shear strength for beam action is in accordance with the provisions of Chapter 7.
2. Two-way action, where critical sections are located a distance $d/2$ from the face of the column, capital, or other discontinuity, and extend around the column, capital, or other discontinuity.

Figure 10.10 Punching failure of slab-column connection viewed from below. (*After Dragosavić and van den Beukel, 1974.*)

Of the two shear failure modes, two-way action usually controls for floor systems. Therefore, two-way action is the focus of the discussion in this chapter.

Shear failure under two-way action typically involves a failure surface that extends around the column, and results in the column punching through the slab as shown in Figure 10.10. This failure mode is referred to as a *punching shear failure*. In a flat plate without shear reinforcement, the failure surface runs along the intersection between the slab and column at the bottom of the slab, extending through the slab at an angle of approximately 30° relative to horizontal. In a slab with a drop capital (or shear reinforcement), the failure can occur either around the perimeter of the column or at the edge of the drop capital (or shear reinforcement) away from the column.

Slab punching is complicated by the three-dimensional interaction between slab moments and shears in the vicinity of the column (Figure 10.11). Flexural

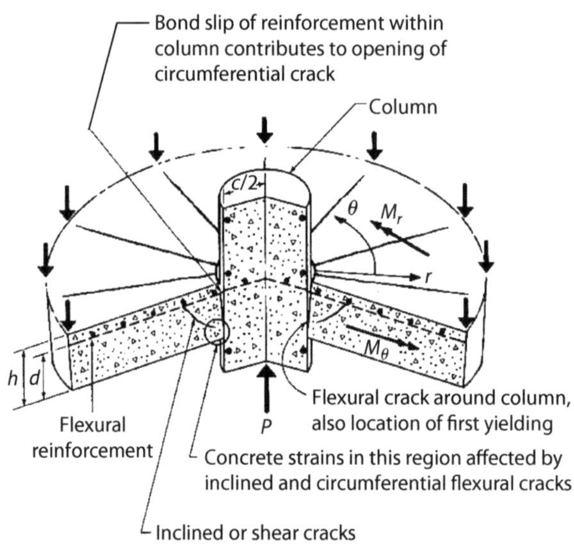

Figure 10.11 Crack formation around supporting column. (*After ASCE 426, 1974, with permission of ASCE.*)

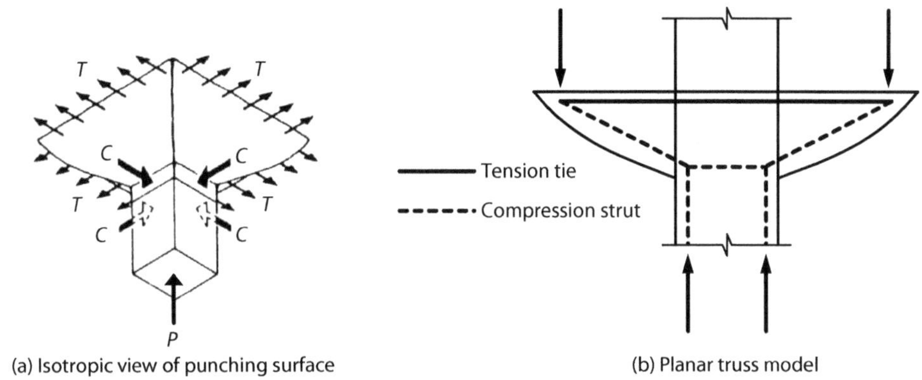

Figure 10.12 Flexural tension actions and idealized truss modle for slab-column connecation. [(*a*) *With permission of ASCE.*]

cracking usually initiates as circumferential cracks around the column perimeter, followed by cracks radiating from the column. Figure 10.12*a* illustrates the punching shear failure surface, showing flexural tension and compression forces. If the reinforcement is in the linear range of response, crack opening along the failure surface is constrained, enabling the failure surface to sustain a unit shear stress higher than the stress typically observed in reinforced concrete beams. If the reinforcement yields, however, widening cracks can lead to a reduction in aggregate interlock with concurrent reduction in punching shear strength. Alternatively, we can visualize shear resistance using a truss model, as shown in Figure 10.12*b*. According to this model, shear is resisted by a diagonal compression strut that is equilibrated by the slab flexural reinforcement. Yielding of the reinforcement limits the magnitude of the diagonal compression strut and, consequently, limits the shear strength of the connection.

Punching shear strength can be estimated using the design provisions of ACI 318 or the similar but more conservative provisions of ACI 352.1 (2011). A critical section for shear is defined around the column perimeter, located such that its perimeter b_0 is a minimum distance but no closer than $d/2$ from the column. In this context, d is defined as the average effective depth for flexural reinforcement in the two orthogonal directions. For columns with rectangular cross section, it is permitted to use a critical section composed of straight lines parallel to the column sides, resulting in a critical section having rectangular plan. Punching shear strength for connections with circular columns exceeds the punching shear strength for connections with square columns having the same perimeter (Vanderbilt, 1972); thus, it is conservative to represent circular columns by square columns having the same perimeter.

Figure 10.13 illustrates slab shear critical sections for several support geometries. Figure 10.13*a* and *b* illustrates critical sections for typical interior connections. Shear critical sections for exterior connections depend on the location of the slab edge relative to the column, as shown in Figure 10.13*c*. Two critical sections should be considered for connections with shear capitals or shear reinforcement because

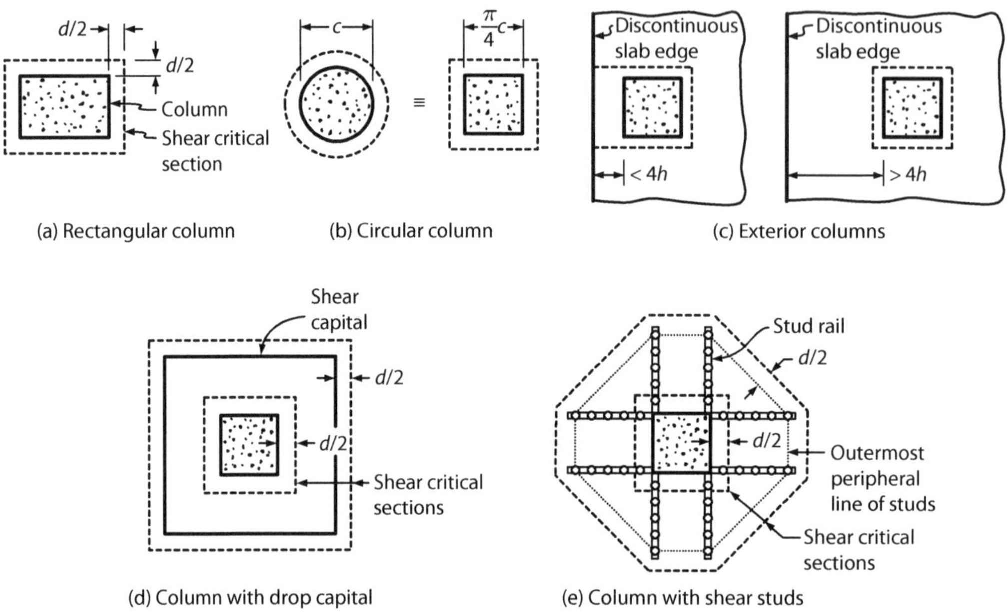

FIGURE 10.13 Examples of slab shear critical sections.

failure may occur either adjacent to the column or around the outside of the shear capital or shear reinforcement (Figure 10.13*d* and *e*). Effects of openings near a slab-column connection should also be considered in accordance with ACI 318 or ACI 352.1.

In U.S. design practice, for connections resisting shear in the absence of moment transfer, the design must satisfy the following requirement:

$$V_u \le \phi V_n \tag{10.4}$$

in which $\phi = 0.75$. Nominal shear strength V_n is given by[1]

$$V_n = C_v V_c + V_s \le 6\sqrt{f_c'} b_o d, \text{ psi } (0.5\sqrt{f_c'} b_o d, \text{ MPa}) \tag{10.5}$$

For connections without shear reinforcement, V_c is calculated in accordance with Eqs. (10.6) through (10.8) or Eq. (10.9), whichever applies. For connections with shear reinforcement, V_c in the region with shear reinforcement is calculated in accordance with Eq. (10.10). According to ACI 318, $C_v = 1.0$ in all cases. According to ACI 352.1, $C_v = 1.0$ for Type I connections (those not subjected to earthquake-induced inelastic deformations) and $C_v = 0.75$ for Type 2 connections (those required to sustain gravity loads under earthquake-induced inelastic deformations). The reduced value of C_v for Type 2 connections reflects the expected reduction in shear strength that occurs if the slab reinforcement yields under earthquake-induced deformations.

[1]The limit on maximum shear is recommended by ACI 352.1. ACI 318 permits nominal shear as high as $8\sqrt{f_c'} b_o d$, psi $(0.67\sqrt{f_c'} b_o d$, MPa) for slab-column connections reinforced with shear studs.

Nonprestressed Two-Way Slabs[2]

For nonprestressed two-way slabs, V_c is the least of values given by Eqs. (10.6) to (10.8).

$$V_c = \left(2+\frac{4}{\beta}\right)\lambda\sqrt{f_c'}b_o d,\ \text{psi}\left[\frac{1}{12}\left(2+\frac{4}{\beta}\right)\lambda\sqrt{f_c'}b_o d,\ \text{MPa}\right] \tag{10.6}$$

in which $\lambda = 1.0$ for normalweight concrete, 0.85 for sand-light weight concrete, and 0.75 for all-lightweight concrete; and β = ratio of long side to short side of the column or other reaction area. For shapes other than rectangular, β is taken to be the ratio of the longest overall dimension of the support to the longest overall perpendicular dimension of the support.

$$V_c = \left(\frac{\alpha_s d}{b_o}+2\right)\lambda\sqrt{f_c'}b_o d,\ \text{psi}\left[\frac{1}{12}\left(\frac{\alpha_s d}{b_o}+2\right)\lambda\sqrt{f_c'}b_o d,\ \text{MPa}\right] \tag{10.7}$$

in which α_s = 40 for interior columns, 30 for edge columns, and 20 for corner columns.

$$V_c = 4\lambda\sqrt{f_c'}b_o d,\ \text{psi}\ (0.33\lambda\sqrt{f_c'}b_o d,\ \text{MPa}) \tag{10.8}$$

Equation (10.8) represents the limiting strength for connections having supports with square or nearly square cross sections. For supports having cross section aspect ratios exceeding 2.0, Eq. (10.6) reduces the shear strength due to the reduced confinement of the failure surface as the connection geometry elongates. For large aspect ratios, Eq. (10.6) reduces to the shear strength commonly assigned for one-way shear. Equation (10.7) reduces the unit strength for slabs supported by large columns because of reduced constraint of the failure surface away from the corners of the supporting element.

Prestressed Two-Way Slabs

For prestressed two-way slabs with bonded reinforcement in the connection region, V_c is given by Eq. (10.9)[3]:

$$V_c = (\beta_p\lambda\sqrt{f_c'}+0.3f_{pc})\,b_o d + V_p,\ \text{psi}\left[\left(\frac{1}{12}\beta_p\lambda\sqrt{f_c'}+0.3f_{pc}\right)b_o d + V_p,\ \text{MPa}\right] \tag{10.9}$$

in which β_p is the lesser of 3.5 and $(\alpha_s d/b_o + 1.5)$, α_s is 40 for interior columns, 30 for edge columns, and 20 for corner columns, f_{pc} is the average value of prestress f_{pc} in the concrete for the two directions, and V_p is the vertical component of all effective prestress forces for prestressed reinforcement crossing the critical section. It is permitted to calculate V_c according to Eq. (10.9) only if all of the following conditions are satisfied:

1. No portion of the column cross section is closer to a discontinuous edge than four times the slab thickness.

---

[2]Where shear reinforcement is used, V_c is limited to the value given by Eq. (10.10).

[3]Where shear reinforcement is used, V_c is limited to the value given by Eq. (10.10).

2. The value of $\sqrt{f_c'}$ is not take greater than 70 psi (0.5 MPa).
3. The value of f_{pc} is at least 125 psi (0.86 MPa) in each direction.
4. The value of f_{pc} is not taken greater than 500 psi (3.4 MPa) in Eq. (10.9).

If any of these conditions is not satisfied, the connection should be treated as a nonprestressed connection, with V_c defined by the least of Eqs. (10.6), (10.7), and (10.8).

Stiff walls or other structural elements may restrain slab deformations due to shrinkage and creep, potentially resulting in partial or complete loss of post-tensioning in the connection region. Therefore, when computing f_{pc}, loss of prestress due to restraint of the slab by structural walls and other structural elements should be taken into account.

The term V_p corresponds to the vertical component of the prestressing force in reinforcement crossing the shear failure surface. In typical cases, the inclination (drape) of the prestressing reinforcement is relatively small, and it can be difficult to ensure that it is maintained at the proper angle during construction. Therefore, it is recommended to design for shear assuming $V_p = 0$.

Slab Shear Reinforcement

Slab shear reinforcement can be added to increase slab shear strength or to increase deformation capacity under earthquake-induced lateral displacements. Accepted types of slab shear reinforcement include steel stirrups, shearheads, and shear studs. These are described in the following:

Steel stirrups: These may be single- or multiple-leg stirrups, such as shown in Figure 10.14. Development of stirrups can be difficult to achieve in thin slabs. Therefore, ACI 318 restricts stirrups to slabs with $d \geq 6$ in (150 mm), but not less than 16 times the stirrup reinforcement bar diameter. Furthermore, effective anchorage of the stirrups requires that each stirrup vertical leg engage a longitudinal reinforcing bar at both the top and the bottom of the slab.

The spacing limitations of ACI 318 and ACI 352.1 vary, with the ACI 352.1 recommendations being more conservative than the ACI 318 requirements.

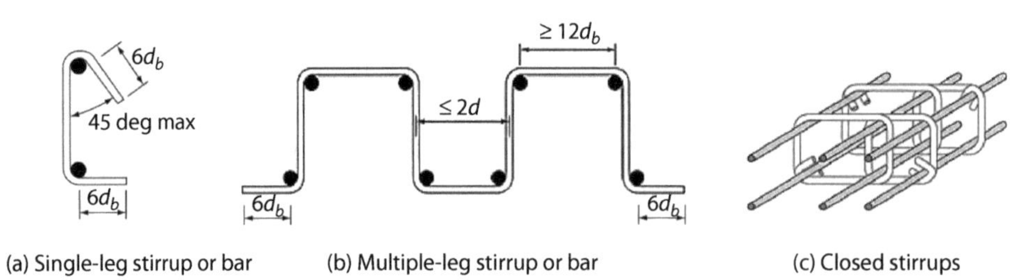

FIGURE 10.14 Stirrup-type slab shear reinforcement. (*After ACI 318, 2014, courtesy of American Concrete Institute.*)

According to ACI 318, the spacing between successive lines of shear reinforcement that surround the column should not exceed $s = d/2$ measured in a direction perpendicular to the column face. Also according to ACI 318, the spacing between adjacent stirrup legs in the first two lines of shear reinforcement should not exceed $2d$ measured in a direction parallel to the column face. Based on deterioration observed in dynamic tests (Kang and Wallace, 2005), ACI 352.1 recommends that the first line of stirrups be placed no farther than $s/2$ from the column face, where s is the typical spacing away from the column, with $s \leq d/2$. Also according to ACI 352.1, the shear reinforcement should be selected such that the distance between adjacent stirrup legs along the first and second peripheral lines does not exceed $2d$. Figure 10.15 illustrates the ACI 352.1 recommendations.

Shearheads: Shearheads are fabricated from structural steel sections that project from the column into the slab along principal framing directions. Shearheads are seldom used in modern construction. See ACI 318 for additional details.

Shear studs: Headed shear stud reinforcement, placed perpendicular to the plane of the slab, is widely used in modern construction. Shear studs are usually prefabricated as a *stud rail*, where a number of studs are attached to a single plate, anchoring one end of the stud, while the opposite end is anchored with individual anchorage plates attached to the studs (Figure 10.16). Effective anchorage is provided by a head with an area equal to 10 times the

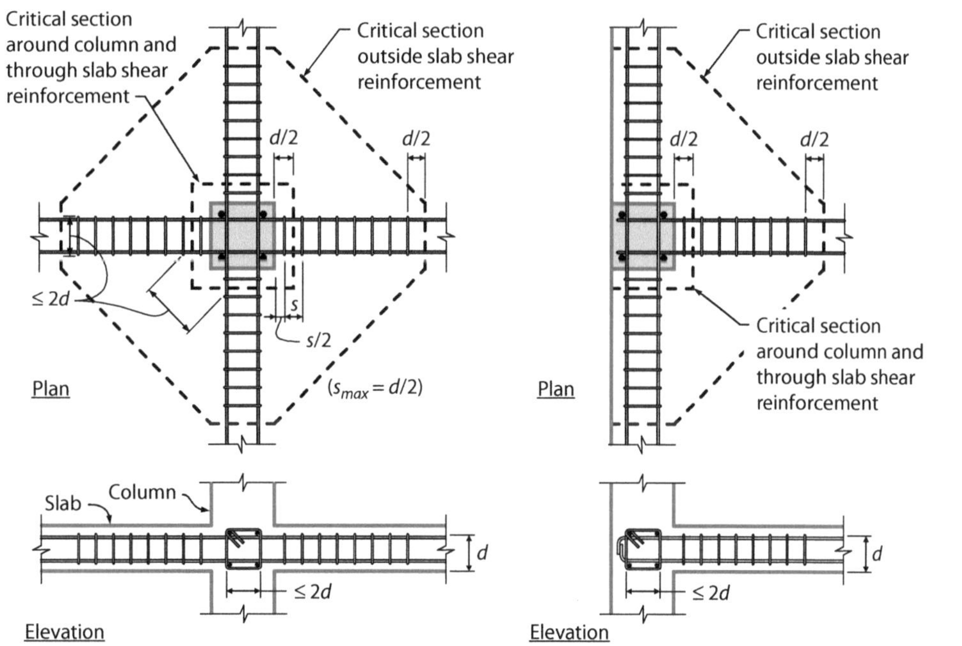

Figure 10.15 Arrangement of stirrup shear reinforcement. (*After ACI 318, 2014, and ACI 352.1, 2011, courtesy of American Concrete Institute.*)

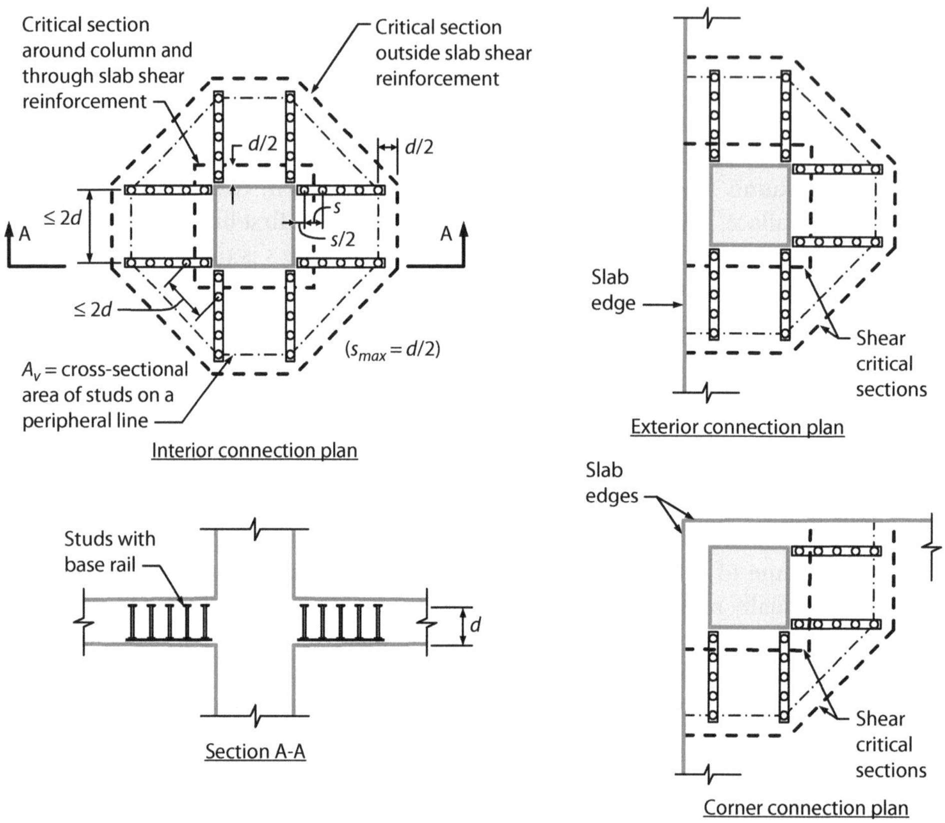

FIGURE 10.16 Typical arrangements of headed shear stud reinforcement and critical sections. (*After ACI 318, 2014, and ACI 352.1, 2011, courtesy of American Concrete Institute.*)

cross-sectional area of the stud (Dilger and Ghali, 1981; Megally and Ghali, 1994). The overall height of the shear stud assembly should not be less than the thickness of the slab minus the sum of (1) the concrete cover on the top flexural reinforcement; (2) the concrete cover on the base rail; and (3) one-half the bar diameter of the flexural tension reinforcement.

Figure 10.16 illustrates typical arrangements of headed shear stud reinforcement, as recommended by ACI 352.1. The spacing limits are based on the same considerations for stirrup reinforcement.[4]

Shearbands: Shearband reinforcement consists of thin steel strips with punched circular holes along the centerline (Figure 10.17). The strips can be bent into a variety of shapes. The pre-bent strips are placed over the uppermost layer of top slab bars around the slab-column joint. The strip thickness may range between 0.03 and 0.065 in (0.76 to 1.65 mm), and the strip width may range

[4]ACI 318 permits the maximum spacing of $s = 0.75d$ in some cases, but that spacing is not recommended here.

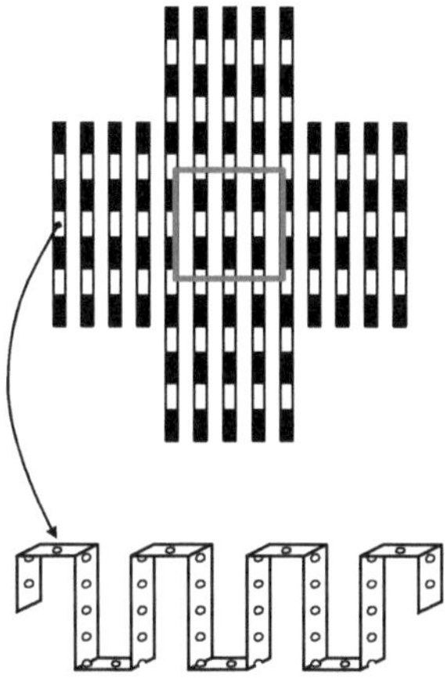

Figure 10.17 Shearband-type slab shear reinforcement. (*After ACI 352.1, 2011, courtesy of American Concrete Institute.*)

from 0.75 to 1.25 in (19 to 32 mm). Shearbands have been shown to be effective in improving shear strength and deformation capacity (Pilakoutas and Li, 2003; Kang and Wallace, 2008). Limitations on the shearband cross-sectional dimensions are based on the range of parameters considered in laboratory testing.

For connections using stirrups, shearheads, or shearbands, nominal strength V_n is according to Eq. (10.5). For both reinforced and post-tensioned slabs, in the region where shear reinforcement is used, the quantity V_c is limited to[5]

$$V_c = 2\lambda\sqrt{f'_c}b_o d, \text{ psi } (0.17\lambda\sqrt{f'_c}b_o d, \text{ MPa}) \tag{10.10}$$

The reduced value of V_c relative to values obtained from Eqs. (10.6) through (10.9) is attributed to the anticipated wider shear cracks when shear reinforcement resists a significant portion of the shear force (Park and Gamble, 1980). Outside the shear-reinforced region, V_c can be defined according to Eqs. (10.6) through (10.9).

Shear strength V_s provided by shear reinforcement oriented perpendicular to the plane of the slab is given by

$$V_s = \frac{A_v f_{yt} d}{s} \tag{10.11}$$

in which A_v = cross-sectional area of all vertical legs of shear reinforcement on one peripheral line that is geometrically similar to the perimeter of the column section; f_{yt} = yield stress of the shear reinforcement, not to be taken greater than 60,000 psi (420 MPa); d = average distance from the extreme compression fiber to the centroid of tension reinforcement in the two orthogonal directions; and s = spacing between adjacent peripheral lines of shear reinforcement.

[5]ACI 318 allows V_c to be 1.5 times the value given by Eq. (10.10) where shear studs are used. ACI352.1, however, recommends V_c according to Eq. (10.10) wherever shear reinforcement is used in two-way slabs.

Note that the upper limit of nominal shear strength according to Eq. (10.5) is $V_n = C_v V_c + V_s \le 6\sqrt{f'_c}b_o d$, psi ($0.5\sqrt{f'_c}b_o d$, MPa).

For slab-column connections with shear reinforcement, shear strength of the connection should be checked at multiple locations, including (a) $d/2$ from the column face within the shear-reinforced region [column critical section, in which Eq. (10.11) applies] and (b) $d/2$ outside the shear-reinforced region [outer critical section, in which Eqs. (10.6) through (10.9) apply]. Figures 10.15 and 10.16 show these critical sections.

Where shear reinforcement is provided to increase the shear strength of a connection or to enable a connection to withstand earthquake-induced drifts, V_s defined by Eq. (10.11) should not be less than $3.5\sqrt{f'_c}b_o d$, psi ($0.29\sqrt{f'_c}b_o d$, MPa).[6] The shear reinforcement should extend away from, and perpendicular to, the face of the support, drop panel, or column capital a distance at least $4h$ for reinforced concrete connections and $3h$ for prestressed concrete connections. Minimum distances for extending shear reinforcement are intended to reduce the likelihood of punching failure at the outer critical section before punching failure at the column critical section.

10.7.2 Connections Transferring Shear and Moment

Slab-column connections transfer shear and moment through a combination of slab flexure and eccentric shear stresses acting around the connection. Several models for shear and moment transfer have been proposed (Park and Gamble, 1980). Here we consider only the ACI 318 models, which are recommended also by ACI 352.1.

The moment to be transferred between the slab and the supporting column is designated M_{tr}.[7] Note that this is not necessarily the same as the total slab moment. Design for the total slab moment must be considered separately. A portion of the transfer moment equal to $\gamma_v M_{tr}$ is assumed to be resisted by slab shear stresses acting around a slab critical section, with the remainder of the moment equal to $\gamma_f M_{tr}$ assumed to be resisted by slab flexure (Figure 10.18).

The fraction of the transfer moment transferred by slab flexure is given by

$$\gamma_f = \frac{1}{1 + (2/3)\sqrt{b_1/b_2}} \tag{10.12}$$

in which b_1 = dimension of the critical section measured in the direction of the span for which moments are determined and b_2 = dimension of the critical section measured perpendicular to b_1. Figure 10.18 illustrates dimensions b_1 and b_2 of a

---

[6]This limit is recommended by ACI 352.1 and is required by ACI 318 for slab-column framing in earthquake-resisting construction wherever shear reinforcement is required based on considerations of shear and lateral drifts. See additional discussion in Section 10.8.

[7]The term transfer moment M_{tr} is used to represent the total moment transferred between the slab and the column, which is different from the total moment in the slab. ACI 318 and ACI 352.1 use the term unbalanced moment. That term is avoided here because the moment is, in fact, balanced by the supporting column.

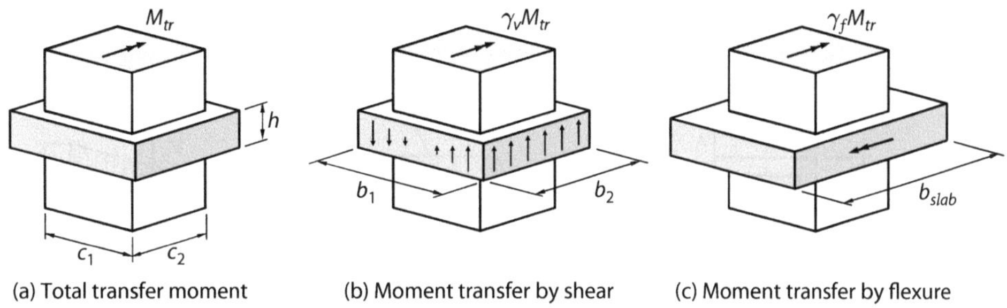

FIGURE 10.18 Moment transfer by slab shear and flexure.

slab critical section. For nonprestressed slabs, it is permitted to increase the value of γ_f in accordance with Table 10.1.

Flexural reinforcement to resist moment $\gamma_f M_{tr}$ should be placed within an effective slab width b_{slab} equal to the width of the column or capital plus 1.5h of the slab or drop panel on either side of the column or capital. Concentration of slab reinforcement to achieve this requirement is permitted. The reinforcement should be placed at the top of the slab if the moment field near the connection is dominated by gravity loads, divided equally between top and bottom faces where the moment field is dominated by moment transfer, or in some intermediate distribution for intermediate cases.

The fraction of the moment transferred by slab shear stress is

$$\gamma_v = 1 - \gamma_f \tag{10.13}$$

To check slab shear stress, a critical section is defined at distance $d/2$ surrounding the column perimeter, as described in Section 10.7.1 and illustrated in Figure 10.13. The nominal shear stress acting on the slab critical section is defined by

$$v_u = \frac{V_u}{b_o d} \pm \frac{\gamma_v M_{tr} c}{J_c} \tag{10.14}$$

in which V_u = shear force transferred from slab to column, J_c = property of slab critical section analogous to the polar moment of inertia, and c = distance from centroid of slab critical section to location where stress v_u acts. Figure 10.19

Column Location	Span Direction	V_u	ε_t (within b_{slab})	Maximum Value of γ_f
Corner	Either direction	$\leq 0.5\phi V_c$	≥ 0.004	1.0
Edge	Perpendicular to the edge	$\leq 0.75\phi V_c$	≥ 0.004	1.0
	Parallel to the edge	$\leq 0.4\phi V_c$	≥ 0.010	$\dfrac{1.25}{1+\left(\frac{2}{3}\right)\sqrt{\frac{b_1}{b_2}}} \leq 1.0$
Interior	Either direction	$\leq 0.4\phi V_c$	≥ 0.010	$\dfrac{1.25}{1+\left(\frac{2}{3}\right)\sqrt{\frac{b_1}{b_2}}} \leq 1.0$

TABLE 10.1 Permitted Values of γ_f for Nonprestressed Slabs

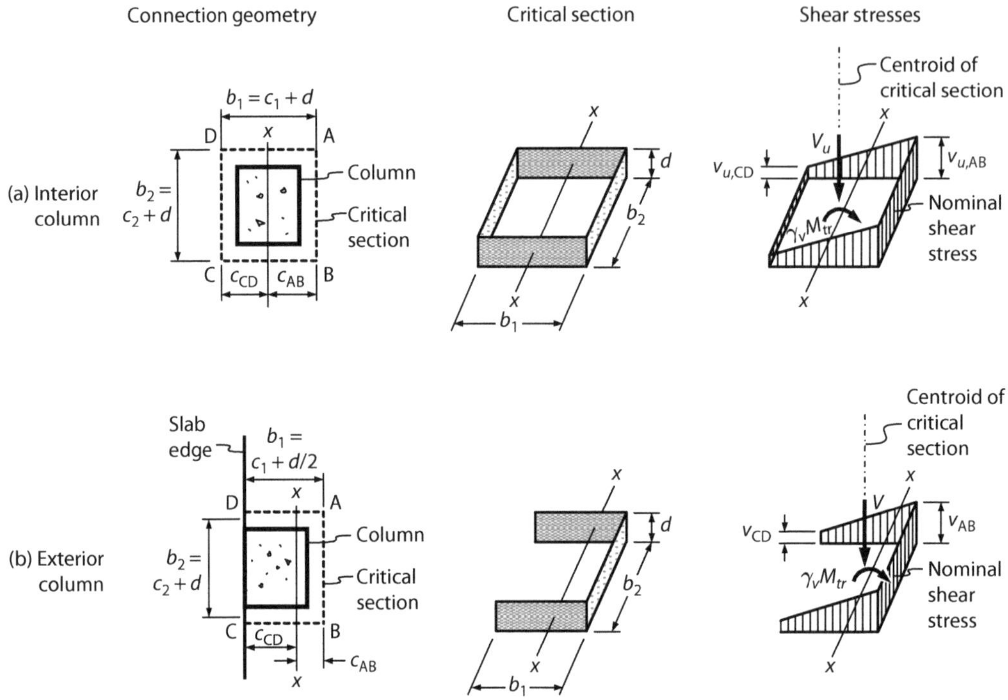

FIGURE 10.19 Variation of nominal shear stresses around critical section for (*a*) pure shear and (*b*) moment plus shear.

illustrates slab critical sections and calculated shear stress variations for rectangular cross section columns at interior and exterior locations.

For the critical section surrounding an interior rectangular column (Figure 10.19*a*), the centroid of the slab critical section coincides with the centroid of the column. In this case

$$J_c = 2\left[\frac{b_1^3 d}{12} + \frac{b_1 d^3}{12} + b_2 d\left(\frac{b_1}{2}\right)^2\right] \tag{10.15}$$

For the slab critical section surrounding an exterior rectangular column (Figure 10.19*b*), the centroid of the slab critical section is located at

$$c_{AB} = \frac{b_1^2 d}{2b_1 d + b_2 d} \tag{10.16}$$

and

$$J_c = 2\left[\frac{b_1^3 d}{12} + \frac{b_1 d^3}{12} + b_1 d\left(\frac{b_1}{2} - c_{AB}\right)^2\right] + b_2 d c_{AB}^2 \tag{10.17}$$

Similar expressions can be developed for other connection geometries.

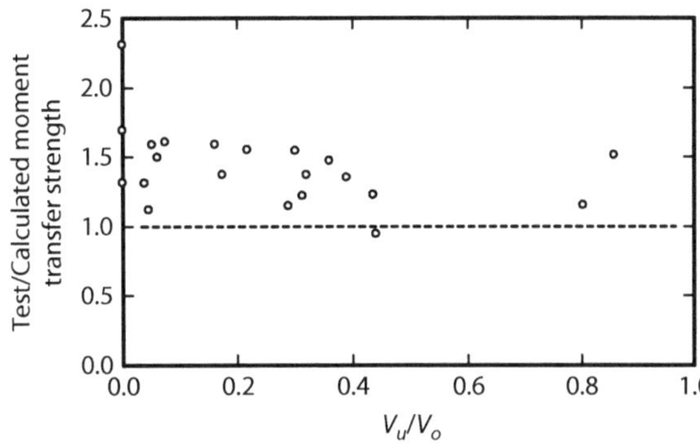

FIGURE 10.20 Comparison of measured and calculated moment transfer strength for reinforced concrete interior slab-column connections. (*After Moehle et al., 1988, courtesy of American Concrete Institute.*)

For slab-column connections designed to resist load combinations including gravity loads with or without lateral loads, maximum shear stress calculated according to Eq. (10.14) is not to exceed $\phi V_n/b_o d$, where V_n corresponds to the strength calculated in accordance with Eq. (10.5), and $\phi = 0.75$.

Tests on slab-column connections confirm that the design procedure of the preceding paragraphs is fairly effective for interior slab-column connections. Figure 10.20 plots ratios of measured to calculated moment transfer strengths as a function of the shear stress ratio V_u/V_o for reinforced concrete (nonprestressed) interior connections subjected to combined gravity and lateral loading. In the figure, calculated moment transfer strength is the lesser of (a) moment for which the calculated critical shear stress (Eq. 10.14) reaches the nominal stress limit and (b) moment strength of the slab within the effective transfer width $(c_2 + 3h)/\gamma_f$. Based on additional analyses of the test data, ASCE 41 (2013) recommends using an effective transfer width equal to $(c_2 + 5h)$, combined with the shear stress limit, to assess strength of existing construction.

The same design procedures are conservative for exterior slab-column connections transferring moment about an axis parallel to the slab edge, but the test strengths do not follow the calculated strengths very closely (Moehle et al., 1988). As an alternative (Moehle, 1988), for exterior slab-column connections transferring moment about an axis parallel to the edge and having $V_u \leq 0.75\phi V_c$, and for corner connections having $V_u \leq 0.5\phi V_c$, it is acceptable to set $\gamma_f = 1$ and $\gamma_v = 0$. Accordingly, moment transfer is assumed to not affect shear stresses, and the entire moment is resisted by slab reinforcement within an effective width. The effective width is defined by idealized yield lines along the inside face of the column and extending at 45° angle to the slab edge, but not to exceed $1.5h$ on either side of the column (Figure 10.21).

Figure 10.22 plots ratio of measured to calculated moment transfer strength as a function of the gravity shear ratio. Within the intended range of application, that is $V_u/V_o \leq 0.75$, the design procedure produces a close estimate of actual moment transfer strength.

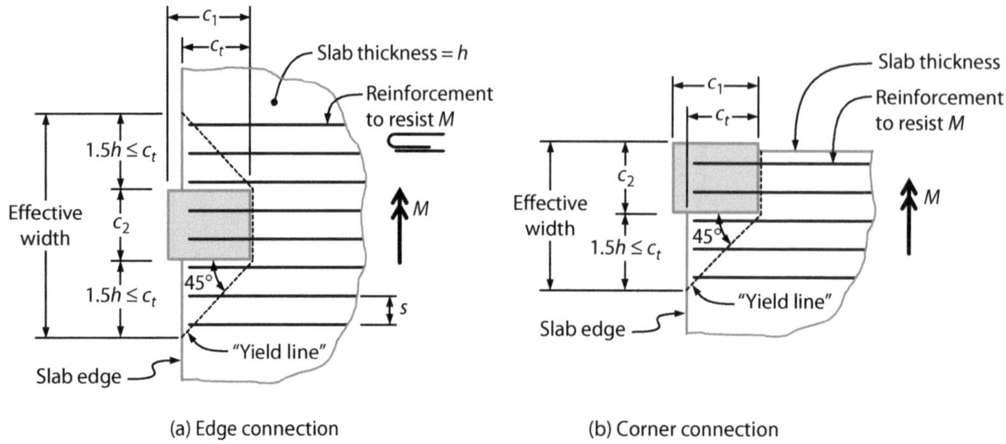

Figure 10.21 Effective transfer width for reinforcement placement in edge and corner slab-column connections.

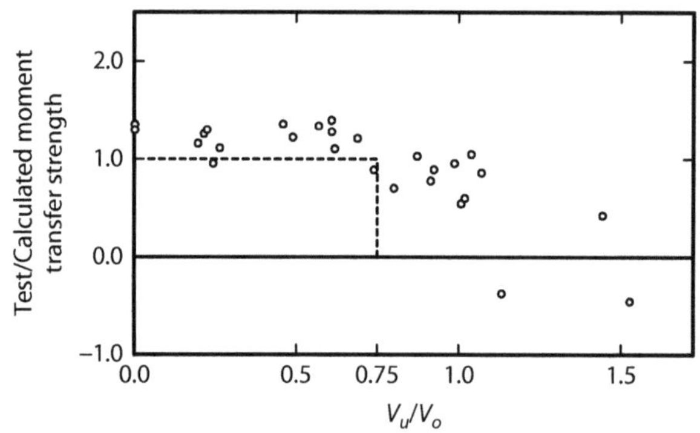

Figure 10.22 Comparison of measured and calculated moment transfer strength for reinforced concrete exterior slab-column connections. (*After Moehle, 1988, courtesy of American Concrete Institute.*)

For slab-column connections designed as gravity framing and therefore not considered part of the seismic-force-resisting system, it is not necessary to satisfy the strength requirements of the preceding paragraph under load combinations including seismic forces. Instead, the connection is checked for adequate strength under gravity loads, and for adequate drift capacity in accordance with Section 10.8 and Chapter 14.

10.8 Drift Capacity under Deformation Reversals

Lateral drift of a slab-column frame results in moment transfer at the connections. This has two deleterious effects on gravity shear transfer strength. First, moment transfer increases shear stresses on part of the slab critical section. Second, if moment transfer results in localized yielding of slab flexural reinforcement, the shear transfer strength may be reduced, as described in Section 10.7.1. Thus, moment transfer can induce punching shear failures in a slab that otherwise had adequate shear transfer capacity.

In this section we present test data to demonstrate the relation between the direct shear resisted by a connection and the drift ratio at onset of punching shear failure.

10.8.1 Slabs without Shear Reinforcement

Figure 10.23 plots measured relations between column lateral shear force and drift ratio for two interior, nonprestressed, slab-column connections without slab shear reinforcement. The plot on the left is for the case where the slab-column connection supports a gravity shear of $V_g = 0.88\sqrt{f_c'}b_o d$, psi $(0.073\sqrt{f_c'}b_o d$, MPa), whereas the plot on the right is for gravity shear of $V_g = 1.4\sqrt{f_c'}b_o d$, psi $(0.12\sqrt{f_c'}b_o d$, MPa). The reduction in drift ratio capacity for greater gravity shear is apparent.

The relation between drift ratio capacity and gravity shear ratio was first reported in Pan and Moehle (1989). Figure 10.24 plots drift ratio at failure versus gravity shear ratio, V_g/V_o, for several tests. V_g is the direct shear force resisted by the connection at failure, and V_o is the shear strength provided by concrete, V_c, in accordance with Eqs. (10.6) through (10.8). Apparent failure modes are indicated by the shape of the data symbol (flexure-shear failure is indicated where flexural yielding was followed by shear failure at a higher drift ratio). The drift ratio is defined as the drift ratio corresponding to the peak lateral force.[8] Reduction in drift ratio capacity with increasing values of V_g/V_o is apparent.

With few exceptions, the data in Figure 10.24 are from uni-directional tests. Bi-directional loading reduces stiffness, strength, and deformation capacity relative to uni-directional loading (Pan and Moehle, 1992). Specific recommendations for bi-directional lateral drift ratio capacity are not available in the literature. A reasonable design approach would be to limit the lateral drift ratio vector to the drift ratio capacity indicated in Figure 10.24.

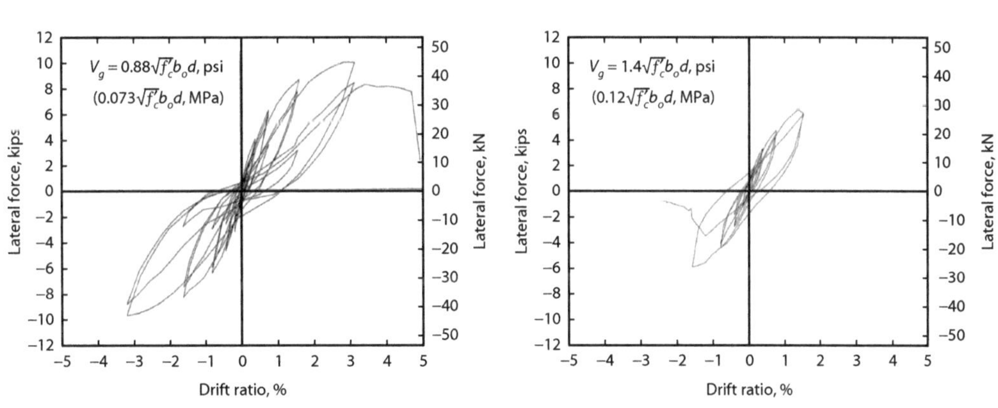

Figure 10.23 Measured relations between lateral force and story drift ratio. (*After Pan and Moehle, 1992, courtesy of American Concrete Institute.*)

[8] An alternative measure of drift capacity, such as drift at onset of punching or occurrence of some other damage state, might be preferred but was not available at the time of this writing.

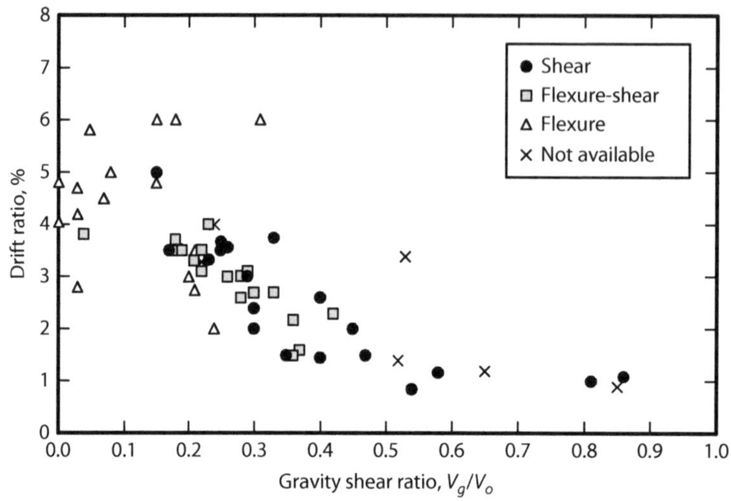

Figure 10.24 Drift ratio at failure for conventionally reinforced concrete slab-column connections without shear reinforcement. (*After Hueste et al., 2007, courtesy of American Concrete Institute.*)

Figure 10.25 compares results for conventionally reinforced and unbonded post-tensioned connections. For the post-tensioned connections, V_o is the shear strength provided by concrete in accordance with Eq. (10.9). The data show similar trends, but the post-tensioned connections tend to have higher drift ratio capacity than similarly loaded connections with conventional reinforcement. The improved behavior of the post-tensioned connections is believed to be partly

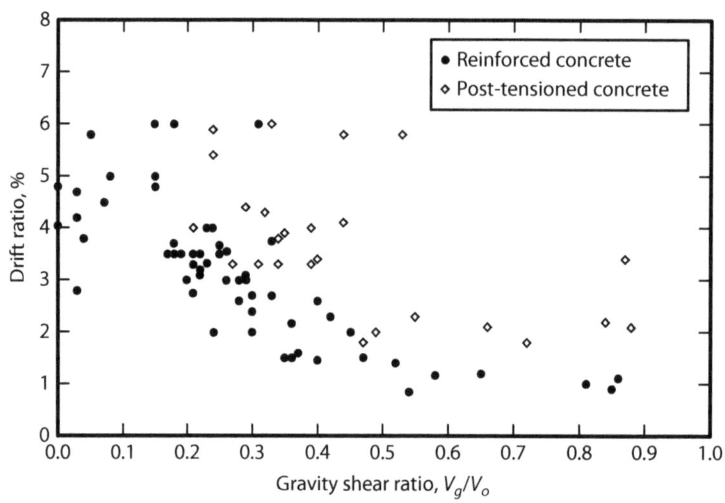

Figure 10.25 Drift ratio at failure for conventionally reinforced and post-tensioned slab-column connections without shear reinforcement. (*After ACI 352.1, 2011, courtesy of American Concrete Institute, based on data from Kang and Wallace, 2006; Hueste et al., 2007; Kang et al., 2008.*)

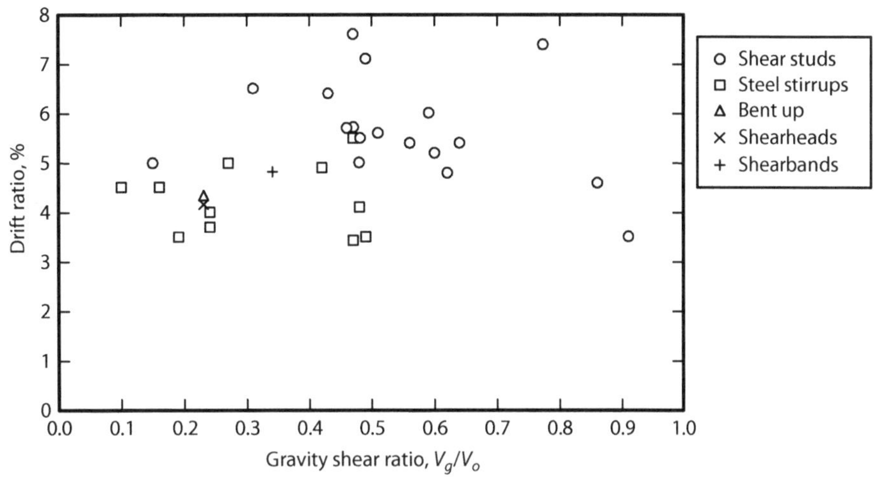

FIGURE 10.26 Drift ratio at failure for slab-column connections with shear reinforcement. *(After ACI 352.1, courtesy of American Concrete Institute, 2011, based on data from Hueste et al., 2007 and Kang and Wallace, 2008.)*

attributable to their higher span-to-thickness ratio compared with conventionally reinforced connections.

The trends observed in Figures 10.24 and 10.25 have been incorporated in ACI 318 requirements for slab-column gravity frames. This subject is covered in Chapter 14.

10.8.2 Slabs with Shear Reinforcement

Where drift ratio capacity of slab-column connections is found to be inadequate, behavior can be improved by adding slab shear reinforcement. Figure 10.26 plots drift ratio capacity versus gravity shear ratio for slab-column connections with a variety of shear reinforcement types. In all cases, the drift ratio capacity exceeded 0.03, which should be adequate for modern concrete buildings with well-designed seismic-force-resisting systems.

Matzke et al. (2012) report lesser drift ratio capacities (~0.02) in tests on connections with shear studs. The study suggests that improved behavior is obtained by using stud spacing in accordance with ACI 352.1 rather than ACI 318. Further study is required to better understand the requirements for design of slab shear reinforcement.

10.9 Post-Punching Behavior and Structural Integrity Reinforcement

The onset of punching shear failure can result in the sudden loss of vertical load-carrying capacity at a slab-column connection. If the slab at a failed connection drops vertically following local failure, then the vertical load will be redistributed to adjacent connections, which may fail under the increased vertical load. Thus, in some cases the failure of one connection can lead to progressive failure across several connections in a single floor level. If vertical load-carrying capacity at a level is effectively lost, the failed slab may drop onto the slab below, leading also to its failure. The failure can thus cascade into a collapse that spreads over many levels of a building. This type of failure is sometimes referred to as "building pancaking," because of the appearance of failed floor slabs piled atop one another at the base of the building.

Slab shear reinforcement may be effective in preventing progressive collapse in slab-column frames, as the shear reinforcement may provide some post-punching resistance (Kang and Wallace, 2005). A more effective solution is to provide slab longitudinal reinforcement, either conventional or prestressed, passing through the failure cone that surrounds the column. This reinforcement can resist vertical movement of a punched floor slab through catenary action (Figure 10.27). Top-placed conventional reinforcement is less effective than bottom reinforcement due to the potential to spall the thin top cover. Bottom-placed reinforcement passing through the column is much more effective. In tests reported by Hwang and Moehle (2000a), slab bottom reinforcement was able to safely support service loads while the framing system was subjected to lateral displacement cycles to drift ratio of 0.05.

Vertical load tests on slab-column connections (Moehle, 1990) have demonstrated that unbonded post-tensioned reinforcement can be effective in supporting a slab after initial punching, provided the strands pass through the failure cone surrounding the column. Significant vertical movement may be required, however, before the strands arrest the vertical movement through catenary action. Effectiveness of strands to support a slab during earthquake-induced lateral drift reversals is unknown.

On the basis of these observations, ACI 352.1 recommends *structural integrity*[9] reinforcement in the form of continuous bottom reinforcement passing through the column core at every slab-column connection. The amount of reinforcement

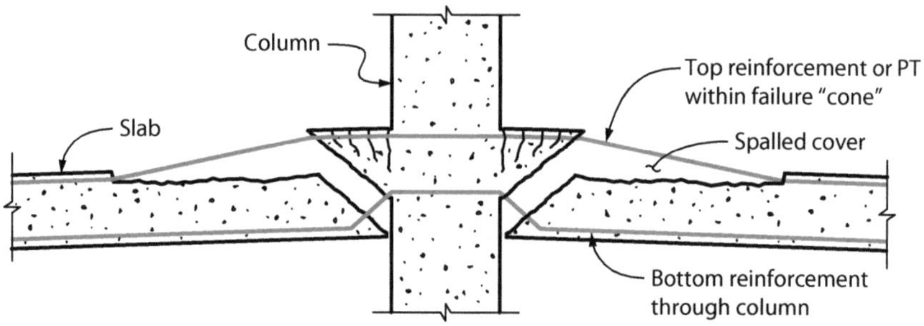

Figure 10.27 Disposition of various components following punching shear failure. *(After Pan and Moehle, 1992, courtesy of American Concrete Institute.)*

[9]Building codes and design guidelines (e.g., ACI 318 and ACI 352.1) refer to the continuous reinforcement through a connection as structural integrity reinforcement. Structural integrity reinforcement is intended to improve the redundancy and ductility in structures so that, in the event of damage to a major supporting element, the damage will be contained to a relatively small area and the structure will have an improved chance to maintain overall stability.

is calculated from equilibrium considerations. The total load to be resisted at an interior connection is taken equal to $q_u l_1 l_2$. Defining the area of reinforcement along each principal direction as A_{sm}, the total available steel area at an interior connection is $4A_{sm}$ (area A_{sm} enters each of four faces of the column). Assuming the catenary effective at an angle of 30° with respect to horizontal, the total resistance at yield stress is $2A_{sm}f_y$. Equating demand and capacity, and using a strength reduction factor of $\phi = 0.9$, the design recommendation is to provide continuous bottom slab reinforcement passing within the column core in each principal direction satisfying

$$A_{sm} \geq \frac{q_u l_1 l_2}{2\phi f_y} \tag{10.18}$$

For edge connections, the quantity of reinforcement A_{sm} in the direction perpendicular to the slab edge may be reduced to two-thirds of that given by Eq. (10.18); for corner connections, the quantity of reinforcement in each direction may be reduced to one-half of that given by Eq. (10.18).

ACI 352.1 recommends that q_u be taken equal to twice the slab dead load, based on studies indicating that the total load resisted by a connection during construction may be approximately twice the slab dead load. Where calculations and field monitoring of construction loads indicate lower loads, the design may be based on lower loads.

The requirements of ACI 318 differ from the recommendations of ACI 352.1. For conventionally reinforced slab-column connections, at least two of the column strip bottom bars or wires in each direction are required to pass within the region bounded by the longitudinal reinforcement of the column. At exterior supports, these bars are to be anchored to develop f_y.

For unbonded post-tensioned slab-column connections, ACI 318 requires a minimum of two 1/2 in (13 mm) diameter or larger, seven-wire post-tensioned strands in each direction at columns, either passing through or anchored within the region bounded by the longitudinal reinforcement of the column. Outside the face of the column or shear cap, these two structural integrity tendons are required to pass under any orthogonal tendons in adjacent spans. Where the two structural integrity tendons are anchored within the region bounded by the longitudinal reinforcement of the column, the anchorage must be located beyond the column centroid and away from the anchored span.

The "weaving" of post-tensioning strands as required in the previous paragraph can be difficult in construction. Alternatively, the connection can be provided with bottom reinforcement in each direction passing within the region bounded by the longitudinal reinforcement of the column and anchored at exterior supports. The area of bottom reinforcement in each direction must not be less than the greater of $\frac{4.5\sqrt{f'_c}}{f_y}bd$, psi ($\frac{3\sqrt{f'_c}}{8f_y}bd$, MPa) and $300bd/f_y$, psi ($2.1bd/f_y$, MPa), where b is the width of the column face through which the reinforcement passes. Minimum extension of these bars beyond the column or shear cap face must be equal to or greater than the bar development length.

10.10 Slab-Wall Connections

10.10.1 One-Way Slab-Wall Connections

The term "one-way slab-wall connection" refers to moment transfer between a slab and a wall loaded out-of-plane. Common examples include a slab framing perpendicular to a flange of a wall or a slab framing into a core wall. Although moment transfer capacity can be important in some cases, usually the most important consideration is drift capacity of the connection while the slab supports gravity loads acting out of the plane of the slab and diaphragm shears acting in the plane of the slab.

Klemencic et al. (2006) reports reversed cyclic tests of the connection between an unbonded post-tensioned slab and a wall segment representing a portion of the flange of a core wall (Figure 10.28). The wall was cast ahead of the slab, and the slab was then cast against the precast wall. Shear keys were provided to transfer out-of-plane gravity shears (simulated in the test) and diaphragm in-plane shears (not simulated in the test). Reinforcement continuity for out-of-plane moment and shear-friction was provided by nonprestressed dowel bars with a mechanical connection at the wall–slab interface. Anchorages for unbonded post-tensioned strands were held back from the wall face by a distance equal to the slab depth to provide a length h for inelastic flexure to occur.

Two test specimens were subjected to displacement reversals at progressively increasing peak displacement amplitude, with results shown in Figure 10.29. In an actual building, the flange of the wall uplifts for loading in one direction due to wall flexure. This causes different rotation demands to be placed on the slab-wall connection for loading to the same drift ratio in the two opposite directions. In the figure, this additional slab-wall rotation is factored into the results and presented as "building equivalent drift ratio." The tests demonstrate that the connections are capable of building drift ratio of at least 0.05 without failure. Significant crack width, likely requiring repair, is reported for the greater drift ratios (Klemencic et al., 2006).

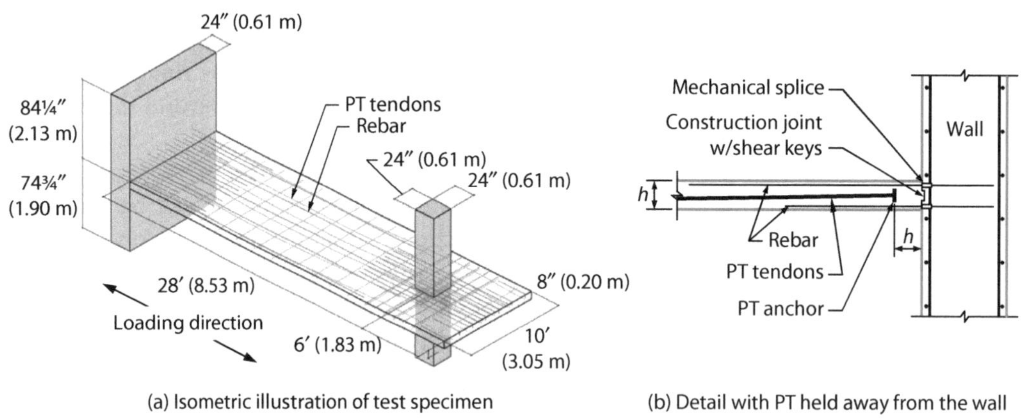

FIGURE 10.28 Test details for connection between unbonded post-tensioned slab and wall flange. (*After Klemencic et al., 2006, courtesy of the Post-Tensioning Institute.*)

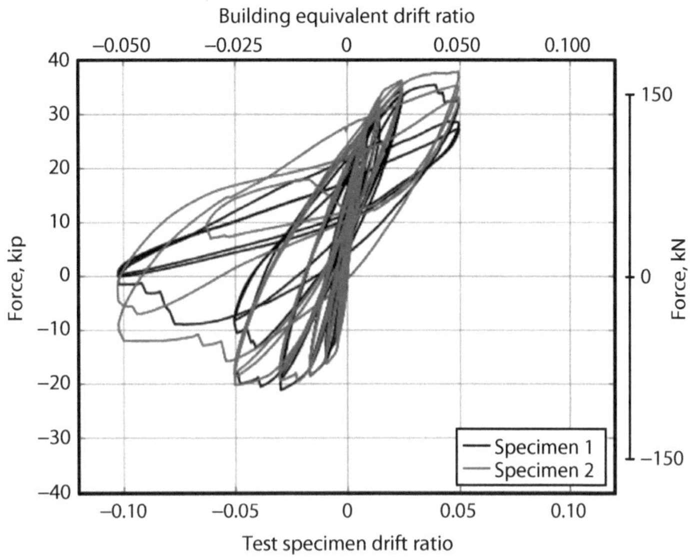

Figure 10.29 Force-drift ratio relations for two test specimens shown in Figure 10.28. (*After Klemencic et al., 2006, courtesy of the Post-Tensioning Institute.*)

Alternative details for slab-wall connections should be considered where construction methods allow. For nonprestressed slab-wall connections, top and bottom reinforcement should be hooked or otherwise developed into the wall flange or should be continuous into the slab on the other side of the flange. For prestressed slab-wall connections, tendons can pass through the flange to be anchored on the far side of the flange or should be continuous into the slab on the other side of the flange.

10.10.2 Slab-Wall Coupling

Slab-wall coupling refers to shear and moment transfer between a slab and two walls. Although slab-wall coupling can occur between slabs and flanged walls, this section focuses on coupling between slabs and walls with a rectangular cross section. The interior wall framing lines in Figure 10.30*a* depict a typical condition. The framing is characterized by rectangular wall and slab cross sections, small corridor width relative to length of the wall, small corridor width relative to transverse bay width, and large corridor width relative to wall thickness. Relevant studies are reported in Paulay and Taylor (1981), Qadeer and Smith (1969), and Schwaighofer and Collins (1972).

Lateral loading of walls coupled by thin slabs results in out-of-plane distortion of the slab, both along the framing direction and perpendicular to it (Figure 10.30*b*). Thus, the effective width of the slab is reduced from the total width. Paulay and Taylor (1981) and Schwaighofer and Collins (1972) report that the initial stiffness can be approximated by using an effective width b_e equal to one-half of the corridor opening width l_h (Figure 10.31*a*). As cracking progresses, the effective moment of inertia should be reduced to the cracked section moment of inertia. Furthermore, Schwaighofer and Collins (1972) recommend that the effective span should be increased to $l_h + b$ because of damage that progresses beyond the length of the corridor opening l_h. In this application, b is the width of the wall at the edge of the opening.

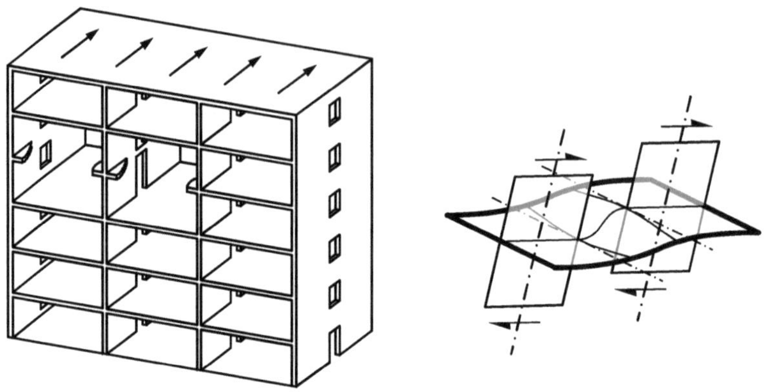

(a) Wall building (b) Distortion of coupling slab

Figure 10.30 Slab-wall coupling.

The contribution of longitudinal reinforcement to moment transfer strength decreases with transverse distance from the wall. At large lateral drift ratios, it may be possible to develop flexural strength of the full transverse width (Paulay and Taylor, 1981). The reinforcement located closest to the walls, however, is most effective in transferring moment. Schwaighofer and Collins (1972) recommend that reinforcement provided for moment transfer should be placed in width $l_h + b$ and length $3l_h$, centered on the corridor opening (Figure 10.31*b*).

Coupling action may be limited by punching shear strength at the wall boundary. Schwaighofer and Collins (1972) recommend defining a three-sided slab critical section having dimensions $b + d$ on each side (Figure 10.31*c*), resulting in critical section area equal to $3(b+d)d$. Shear transfer strength can be estimated by assuming a uniform limiting shear stress on the critical section, resulting in $V_n = 4\sqrt{f'_c}[3(b+d)d]$, psi $\{0.33\sqrt{f'_c}[3(b+d)d]$, MPa$\}$.

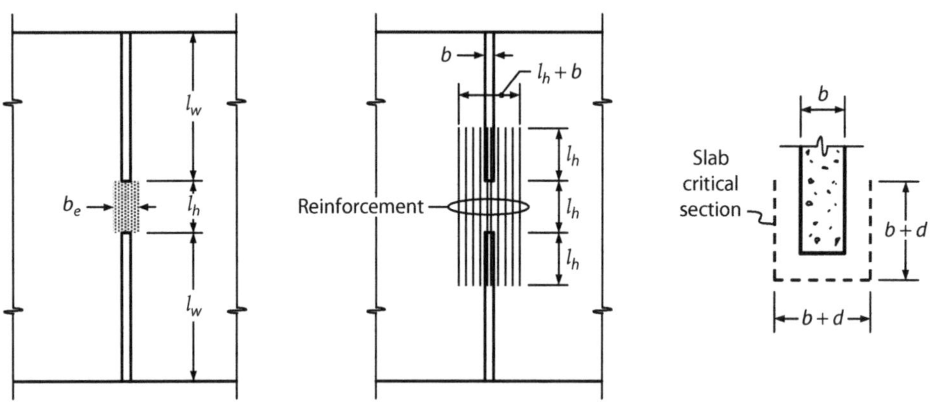

(a) Effective width for stiffness (b) Effective width for reinforcement (c) Slab critical section

Figure 10.31 Effective sections for design.

References

ACI 318 (2014). *Building Code Requirements for Structural Concrete (ACI 318-14) and Commentary*, American Concrete Institute, Farmington Hills, MI.

ACI 352.1 (2011). *Guide for Design of Slab-Column Connections in Monolithic Concrete Structures*, American Concrete Institute, Farmington Hills, MI, 28 pp.

ASCE 41 (2013). *Seismic Evaluation and Retrofit of Existing Buildings*, American Society of Civil Engineers, Reston, VA.

ASCE 426 (1974). "The Shear Strength of Reinforced Concrete Members—Slabs," Joint ASCE-ACI Task Committee 426 on Shear and Diagonal Tension of the Committee on Masonry and Reinforced Concrete of the Structural Division, *Journal of the Structural Division*, Vol. 100, No. ST8, pp. 1543–1591.

Dilger, W.H., and A. Ghali (1981). "Shear Reinforcement for Concrete Slabs," *Journal of Structural Engineering*, ASCE, Vol. 107, No. ST12, pp. 2403–2420.

Dragosavić, M., and A. van den Beukel (1974). "Punching Shear," *Heron*, Vol. 20, No. 2, 48 pp.

Elwood, K.J., A.B. Matamoros, J.W. Wallace, D.E. Lehman, J.A. Heintz, A.D. Mitchell et al. (2007). "Update to ASCE/SEI 41 Concrete Provisions," *Earthquake Spectra*, Vol. 23, No. 3, pp. 493–523.

Hueste, M.D., J. Browning, A. Lepage, and J.W. Wallace (2007). "Seismic Design Criteria for Slab-Column Connections," *ACI Structural Journal*, Vol. 104, No. 4, pp. 448–458.

Hwang, S.-J., and J.P. Moehle (2000a). "Vertical and Lateral Load Tests of a Nine-Panel Flat-Plate Frame," *ACI Structural Journal*, Vol. 97, No. 1, pp. 193–210.

Hwang, S.-J., and J.P. Moehle (2000b). "Models for Laterally Loaded Slab-Column Frames," *ACI Structural Journal*, Vol. 97, No. 2, pp. 345–353.

Kang, T.H.-K., I.N. Robertson, N.M. Hawkins, and J.M. LaFave (2008). "Recommendations for Design of Post-Tensioned Slab-Column Connections Subjected to Lateral Loading," *PTI Journal*, Vol. 6, No. 1, pp. 45–59.

Kang, T.H.-K., and J.W. Wallace (2005). "Dynamic Responses of Flat Plate Systems with Shear Reinforcement," *ACI Structural Journal*, Vol. 102, No. 5, pp. 763–773.

Kang, T.H.-K., and J.W. Wallace (2006). "Punching of Reinforced and Post-Tensioned Concrete Slab-Column Connections," *ACI Structural Journal*, Vol. 103, No. 4, pp. 531–540.

Kang, T.H.-K., and J.W. Wallace (2008). "Seismic Performance of Reinforced Concrete Slab-Column Connections with Thin Plate Stirrups," *ACI Structural Journal*, Vol. 105, No. 5, pp. 617–625.

Klemencic, R., J.A. Fry, G. Hurtado, and J.P. Moehle (2006). "Performance of Post-Tensioned Slab-Core Wall Connections," *PTI Journal*, Vol. 4, No. 6, pp. 7–23.

Matzke, E., C.K. Shield, G.J. Parra-Montesinos, and M.-Y. Cheng (2012). "Effect of Shear Stud Layout on the Seismic Behavior of Slab-Column Connections," *Proceedings, 15th World Conference on Earthquake Engineering, Lisbon*.

Megally, S.H., and A. Ghali (1994). "Design Considerations for Slab-Column Connections in Seismic Zones," *ACI Structural Journal*, Vol. 91, No. 3, pp. 303–314.

Moehle, J.P. (1988). "Strength of Slab-Column Edge Connections," *ACI Structural Journal*, Vol. 85, No. 1, pp. 89–98.

Moehle, J.P. (1990). Unreported tests of unbonded post-tensioned slab-column connections with drop capitals.

Moehle, J.P., and J.W. Diebold (1992). "Lateral Load Response of a Flat Plate Frame," *Journal of Structural Engineering*, Vol. 111, No. 10, pp. 2149–2164.

Moehle, J.P., M.E. Kreger, and R. Leon (1988). "Background to Recommendations for Design of RC Slab Column Connections," *ACI Structural Journal*, Vol. 85. No. 6, pp. 636–644.

Pan, A.D., and J.P. Moehle (1989). "Lateral Displacement Ductility of Reinforced Concrete Flat Plates," *ACI Structural Journal*, Vol. 86, No. 3, pp. 250–258.

Pan, A.D., and J.P. Moehle (1992). "An Experimental Study of Slab-Column Connections," *ACI Structural Journal*, Vol. 89, No. 6, pp. 626–638.

Park, R., and W.L. Gamble (1980). *Reinforced Concrete Slabs*, John Wiley & Sons, Inc., New York, NY, 618 pp.

Paulay, T., and R.G. Taylor (1981). "Slab Coupling of Earthquake-Resisting Shearwalls," *ACI Journal*, Vol. 78, No. 11, pp. 130–140.

Pilakoutas, K., and X. Li (2003). "Alternative Shear Reinforcement for Reinforced Concrete Flat Slabs," *Journal of Structural Engineering*, Vol. 129, No. 9, pp. 1164–1172.

Qadeer, A., and B.S. Smith (1969). "The Bending Stiffness of Slabs Connecting Shear Walls," *ACI Journal*, Vol. 66, No. 6, pp. 464–473.

Schwaighofer, J., and M.P. Collins (1972). "Experimental Study of the Behavior of Reinforced Concrete Coupling Slabs," *ACI Journal*, Vol. 74, No. 3, pp. 123–127.

Vanderbilt, M.D. (1972). "Shear Strength of Continuous Plates," *ASCE Proceedings*, Vol. 98, No. ST5, pp. 961–973.

Vanderbilt, M.D., and W.G. Corley (1983). "Frame Analysis of Concrete Building," *Concrete International*, Vol. 5, No. 12, pp. 33–43.

CHAPTER 11
Seismic Design Overview

11.1 Preview

Previous chapters of this book examined the behavior of structural materials and components under various loadings. This chapter and the following chapters will discuss how these materials and components are assembled to form complete building structures. Emphasis is on the design, behavior, and analysis of the structural framing including foundations, with limited consideration of requirements for nonstructural components and systems.

This chapter presents an overview of the seismic design process, and serves as a basis for the chapters that follow it. We begin with discussion of seismic hazard and how it is represented for use in building design. This leads to a review of seismic demands on building structures, emphasizing requirements for strength, deformability, and ductility. Seismic design approaches based on expected inelastic response of the structural framing are introduced. The chapter concludes with an overview of various additional considerations that affect the choice and proportioning of a seismic-force-resisting system.

11.2 Earthquakes and Engineering Representation of Seismic Hazard

11.2.1 Earthquakes and Earthquake Hazards

The term *earthquake* refers generally to any event that generates seismic waves. Earthquakes can be due to natural or man-made causes. From an earthquake engineering perspective, however, the most significant earthquakes are due to rupture along geologic faults. The vast majority of such earthquakes occur near boundaries of tectonic plates, and are the result of energy release and fault rupture as plate boundaries slip past one another along a zone of geologic faults. Intraplate earthquakes, occurring distant from the tectonic plate boundaries, can also be important, as in the New Madrid earthquake fault zone in the Central United States, or the volcanic-origin earthquakes of Hawaii.

Ground shaking is the main cause of earthquake damage to buildings. For this reason, earthquake ground shaking hazard is the main focus of most seismic designs and performance assessments. Other seismic hazards that may cause damage to buildings include surface fault rupture, liquefaction and associated settlement and lateral spreading, differential settlement of foundation material, landsliding, and tsunami. These latter effects should be included in assessment and design where they occur. For additional discussion of these effects, see Kramer (1996), ASCE 7 (2010), and Chock (2012).

Earthquake ground shaking has been recorded using strong motion instruments since the 1933 Long Beach, California earthquake. Instruments typically record the acceleration at a point in two horizontal directions and the vertical direction. Acceleration recordings can be processed to remove errors and noises, and then integrated to obtain velocity and displacement records. Figure 11.1 shows examples of recorded accelerations and derived velocities from several earthquakes.

The nature of ground shaking at a building site is a function of several factors, including earthquake magnitude, style of faulting, depth to top of fault rupture, source-to-site distance, site location on hanging wall or footwall of dipping faults, near-surface soil response, sedimentary basin depth/depth to basement rock, and other effects related to the three-dimensional wave propagation from the source to the site. The following paragraphs describe some of the main effects.

Earthquake magnitude is a measure of the energy released in an earthquake, and therefore relates to the rupture area and displacement along the fault. Larger magnitude earthquakes generally have longer fault rupture lengths. Because it takes time for rupture to extend along a fault, we should anticipate that larger magnitude earthquakes generally have a potential for longer shaking duration at a site. The larger energy release also creates a potential for higher shaking intensity at a site, especially longer period ground motions. A given fault can be capable of generating earthquakes with magnitudes ranging from the low end to some upper bound constrained by the length of the fault. Generally, smaller magnitude earthquakes occur more frequently, with larger magnitude earthquakes occurring less frequently. Some faults and fault segments, however, tend to repeatedly generate *characteristic earthquakes* of comparable magnitude.

Earthquake hazard experts develop and use empirical or simulation-based ground motion *attenuation models* (or ground motion prediction equations) to estimate how ground motion intensity varies with magnitude and distance from the fault. Figure 11.2 shows the median attenuation for peak horizontal ground acceleration from Campbell and Bozorgnia (2014). The model shows that peak ground acceleration at short distances is nearly independent of magnitude for moment magnitude greater than about M6.5, and that ground motion only slightly attenuates within approximately 5 km of the fault. For additional discussion on attenuation models, see Bozorgnia et al. (2014).

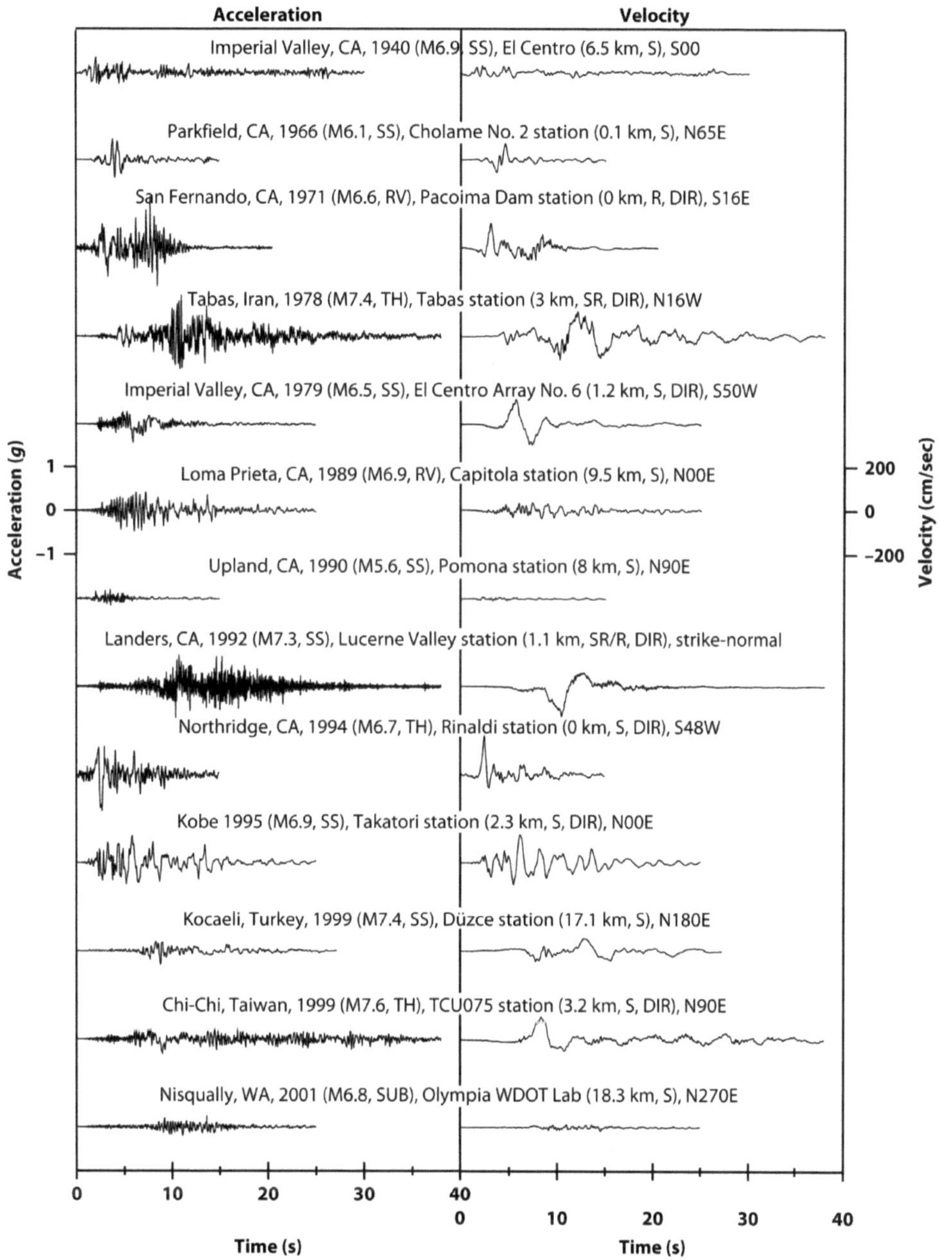

Figure 11.1 Selected recorded ground accelerations and corresponding ground velocities (plotted at the same scale). SS = strike-slip faulting; RV = reverse faulting; TH = thrust faulting; SUB = subduction intraslab earthquake; S = soil site; R = rock site; SR = soft rock site; DIR = record includes fault rupture directivity effects. Distance measure is from the recording site to surface projection of fault rupture plane (epicentral distance for the Nisqually earthquake). (*After Bozorgnia and Campbell, 2004, courtesy of CRC Press.*)

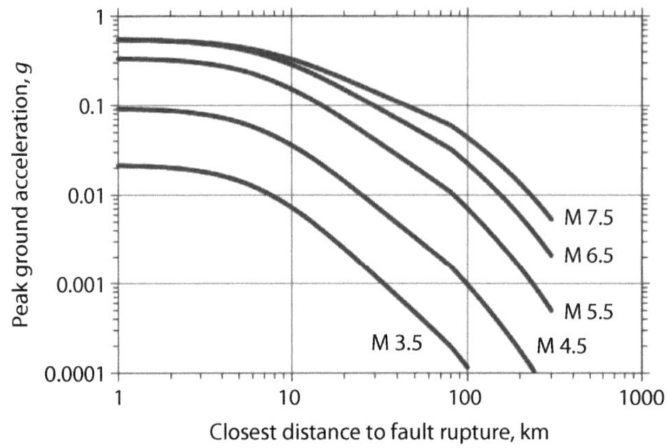

FIGURE 11.2 Expected attenuation of ground motion with rupture distance showing its dependence on moment magnitude (M). (*After Campbell and Bozorgnia, 2014, courtesy of Earthquake Engineering Research Institute.*)

Ground motions in close proximity to the seismic source (within approximately 10 km) can be significantly influenced by near-fault effects referred to as *rupture directivity*. Fault rupture releases energy in the form of waves that propagate from the rupture source. Because the rupture velocity and the resulting wave velocity typically are relatively close to each other, the progression of rupture along the fault results in a buildup of energy in the direction of rupture. Earthquake rupture toward a site tends to produce strong impulsive ground motions, an effect referred to as *forward directivity*. The impulsive motion may be especially strong in the fault-normal direction. *Neutral directivity* or *backward directivity* away from a site produces longer duration motion of relatively lower amplitude. Figure 11.3 illustrates the effect of rupture directivity on earthquake ground motion.

Ground shaking at a site also is affected by near-surface soil flexibility. Soft soil deposits tend to amplify earthquake ground motions, especially for longer periods. Complicating the problem, nonlinear response of soft sites may result in deamplification as the intensity of input motions increases. Building codes have developed site amplification factors based on a geotechnical site classification system and the intensity of input motions. As an alternative to building code site amplification factors, geotechnical engineers sometimes use one-dimensional and three-dimensional modeling procedures to estimate site amplification effects.

Ground shaking at a site is influenced by other geologic factors. Thrust faults, especially if the rupture does not reach the surface, generate higher than average ground motions especially at shorter periods. Furthermore, shaking tends to be higher on the hanging wall (i.e., the portion of the earth's crust above the fault plane) for thrust faults. As earthquake waves travel from the source to the site, complex geologic structures can reflect and refract earthquake ground motions, resulting in focusing of earthquake energy at some sites. On the other hand,

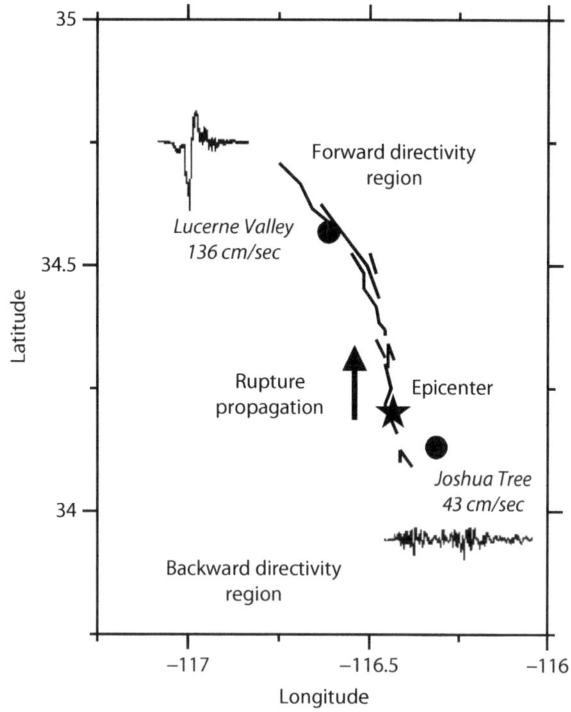

FIGURE 11.3 Velocity records from the 1979 Imperial Valley, California, earthquake at the Joshua Tree and Lucerne Valley strong ground motion recording sites. Note the shorter duration, impulsive motion at Lucerne Valley (forward directivity) and longer duration lower amplitude motion at Joshua Tree (backward directivity). (*After Bolt, 2004, courtesy of Y. Bozorgnia.*)

normal faults generally produce ground motions either comparable or less than those generated by strike-slip faults.

While some of these effects can be anticipated based on knowledge of the geologic setting, others depend on details of the faulting mechanism and source-to-site path, which cannot be known before an earthquake. Thus, quantification of uncertainty in forward estimation of ground shaking is an important topic in earthquake engineering, and earthquake ground motion commonly is described in probabilistic terms that enable general statements about the expected shaking and the variability about that expectation.

11.2.2 Engineering Characterization of Ground Motion

Earthquake ground motion can be characterized in terms of peak ground motion values (e.g., peak ground acceleration, or PGA), impulsive character, duration, energy and frequency contents, time variation of ground acceleration or velocity, or the response of a structure subjected to the motion. Among these, the most commonly used measure is the *elastic response spectrum*. The elastic response spectrum is a plot of the maximum response of a viscously damped linear-elastic single-degree-of-freedom (SDOF) oscillator as a function of its vibration period, considering a single earthquake ground motion record.

To construct an elastic response spectrum, the response history of a viscously damped linear-elastic SDOF oscillator having a specified vibration period, T, is calculated, and the absolute value of the maximum response is plotted as a function of the vibration period. The process is repeated for all vibration periods and damping values of interest. The resulting plot is the elastic response spectrum. The response quantity to be plotted versus period can be one of the following:

S_d = maximum displacement of the oscillator relative to the ground

SV = maximum velocity of the oscillator relative to the ground

SA = maximum absolute (total) acceleration of the oscillator

S_v (or PSV) = pseudo-velocity = ωS_d

S_a (or PSA) = pseudo-acceleration = $\omega^2 S_d$

The quantities S_v and S_a are usually preferred over SV and SA by earthquake engineers. To understand why this is the case, we first define some basic properties of the SDOF oscillator. The vibration period of the oscillator is

$$T = 2\pi\sqrt{M/K} \tag{11.1}$$

The relations between period, frequency, and circular frequency are $T = 1/f = 2\pi/\omega$. For a linear SDOF oscillator, the restoring force is equal to $V_e = KS_d$. Using Eq. (11.1), we can write

$$V_e = KS_d = M\omega^2 S_d = MS_a \tag{11.2}$$

Thus, the peak restoring force (equal to the base shear) is equal to the product of the mass and the pseudo-acceleration read from the response spectrum at the period T.

Another term commonly used by earthquake engineers is the *elastic seismic coefficient,* defined as

$$C_e = V_e/W = S_a/g \tag{11.3}$$

Figure 11.4 presents examples of S_a, S_v, and S_d elastic response spectra for an oscillator having damping equal to 5% of the critical value subjected to the three components of an earthquake ground motion record. From these response spectra, we can determine peak values of response to this earthquake record. For example, for a vibration period $T = 1$ s, the maximum response in the S48W direction is relative displacement of $S_d = 1.8$ in (45 cm), pseudo-acceleration of $S_a = 1.82g$, and peak base shear equal to $S_a W/g$.

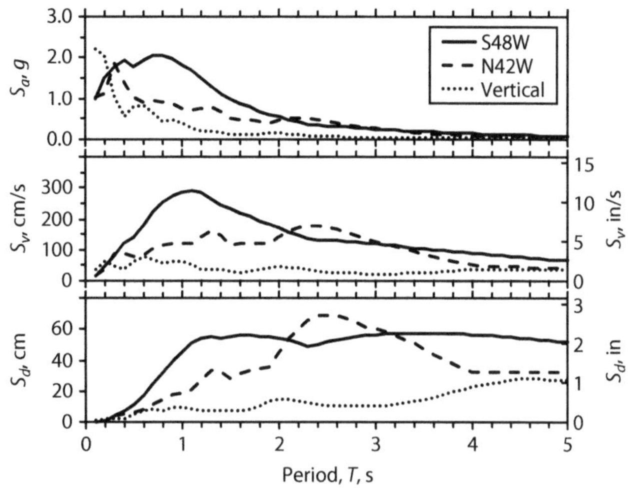

Figure 11.4 Linear-elastic pseudo-acceleration (S_a), pseudo-velocity (S_v), and relative displacement (S_d) response spectra for 5% damping for the ground motion recorded at the Rinaldi Receiving Station during the 1994 Northridge, California earthquake.

11.2.3 Site-Specific Seismic Hazard Evaluation

Probabilistic seismic hazard analysis (PSHA) is the most commonly used approach for assessing the seismic hazard at a site. This method is also implemented at the regional level to develop seismic hazard maps. Usually PSHA is done by a specialist in the field of engineering seismology. The outcome of the analysis is a quantification of the seismic hazard at the site in terms of rates at which response spectral ordinates will be exceeded, mean response spectral ordinates as function of the return period of the shaking hazard, definition of the faults and magnitudes that control the seismic hazard, and (possibly) specification of representative earthquake ground motions that can be used directly for response history analysis of the building.

The analysis begins with seismic source characterization. Usually, two types of earthquake sources are characterized. *Fault sources* are specifically identified fault surfaces along with details of their behavior. *Areal source zones* are regions where earthquakes are assumed to occur randomly, but are not associated with known or mapped faults. Figure 11.5 shows mapped fault sources for Los Angeles.

In a PSHA, the occurrence of earthquakes is modeled with a numerical procedure that accounts for the distribution of earthquakes in both time and space for the various faults, areal seismic source zones, and background seismicity. Another component of PSHA is a set of ground motion attenuation models such as the one depicted in Figure 11.2 to scale ground shaking at the building site as a function of various parameters including earthquake magnitude, distance from the source to the site, and site characteristics. The total annual frequency of exceedance for various ground motion parameters (e.g., PGA) at a site is then

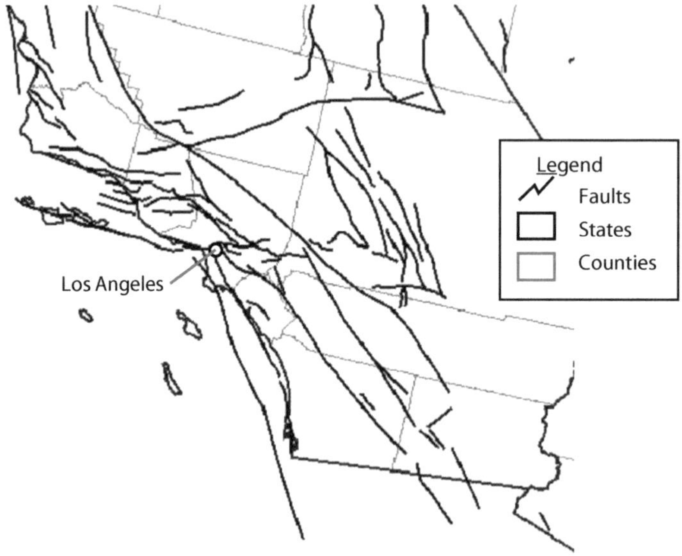

FIGURE 11.5 Fault sources in Los Angeles region. (*After USGS, 2014.*)

obtained by combining contributions from all seismic sources. For site-specific PSHA of important facilities, generally the probabilistic estimate of ground motion at a site is computed for a "rock" site condition and then the rock ground motion is modified for the specific soil characteristics through a separate site response analysis.

The PSHA produces seismic hazard curves that relate the amplitudes of PGA and spectral pseudo-accelerations to the annual probabilities of exceedance of those amplitudes. These results are combined to create response spectra corresponding to a return period or, alternatively, a probability of exceedance in a given time span. These response spectra are referred to as *equal-hazard spectra* (or *uniform-hazard spectra*) because the probability of exceedance of the response spectral ordinates is the same at every period. Figure 11.6 plots the equal-hazard spectra for a site in Los Angeles (Figure 11.5) for different probabilities of exceedance. Table 11.1 presents some commonly used equivalent expressions for return period and probability of exceedance in a given period.

A standard procedure is to deaggregate the PSHA results to identify which earthquake magnitudes and distances contribute most to the seismic hazard at the structural period(s) of interest. Figure 11.7 shows the geographic deaggregation for the same Los Angeles site identified in Figure 11.5. According to Figure 11.7, several different earthquake sources, with different magnitudes and distances, contribute to the hazard at the site. If earthquake ground motion records are to be used to analyze the response of the building, the records should have mechanisms, magnitudes, and distances representative of those earthquake sources that contribute most to the seismic hazard. The seismic hazard for long return periods,

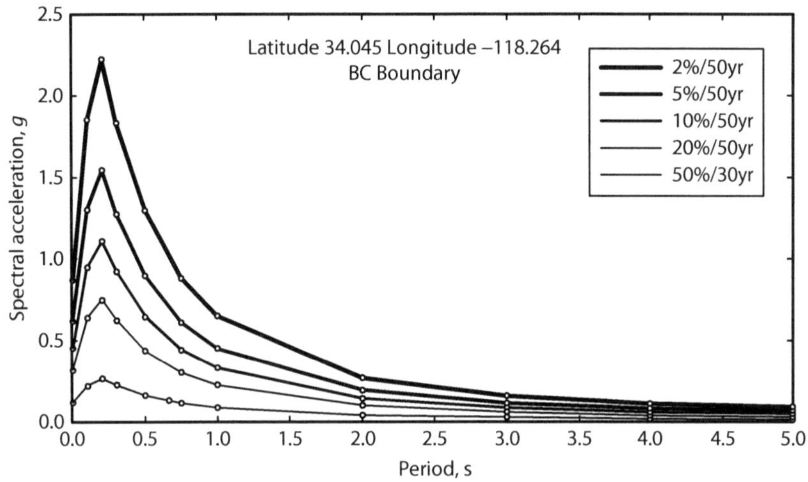

Figure 11.6 Equal-hazard response spectra for 5% damping for different exceedance levels for a site in Los Angeles with a "B/C" NEHRP soil category. (*After USGS, 2014.*)

Return Period (Years)	Probability of Exceedance
25	50% in 17 years
43	50% in 30 years
72	50% in 50 years
224	20% in 50 years
475	10% in 50 years
975	5% in 50 years
2475	2% in 50 years

Table 11.1 Equivalent Expressions of Return Period and Probability of Exceedance in a Given Time

that is, very rare shaking levels, may be controlled not only by large-magnitude earthquakes but also by unusually strong shaking given the magnitude–distance pair. USGS deaggregation resources (not shown here) also can quantify this effect, which can further aid in the selection of ground motion records.

Importantly, the equal-hazard spectrum has contributions at all periods from all earthquakes contributing to the seismic hazard. As such, the equal-hazard spectrum is not representative of the spectrum from any individual ground motion. A design-level earthquake ground motion may have spectral ordinates that reach the equal-hazard spectrum in one period range, but that same earthquake is likely to have spectral ordinates lower than the equal-hazard spectrum at other periods. As an alternative to uniform hazard spectra, one can select a vibration period of interest, and then use *scenario spectra* (*conditional*

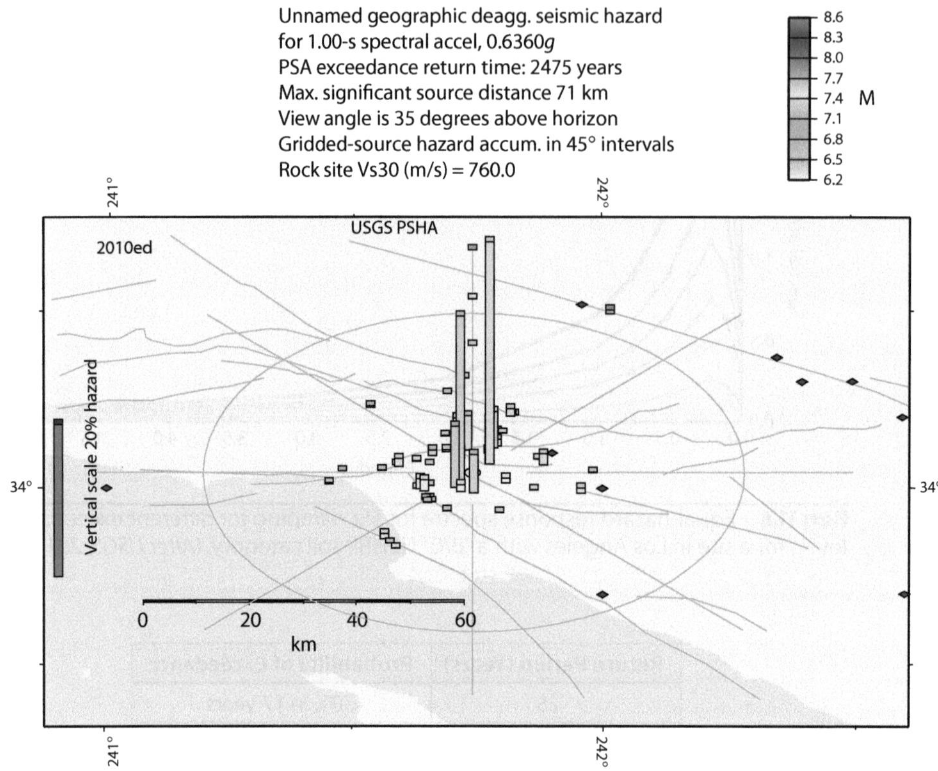

Figure 11.7 Geographic deaggregation of the seismic hazard for 1 s spectral acceleration for a site in Los Angeles for return period equal to 2475 years. Height of bars indicates relative contribution to the seismic hazard. Bar shading indicates earthquake magnitude. Jagged lines represent Quaternary fault locations. (*After USGS, 2014.*)

mean spectra) to represent the expected spectral ordinates at other periods conditioned on having the spectral ordinate at the target period. See Baker and Cornell (2006) for additional discussion.

Probabilistic seismic hazard analysis of sites within the United States can be done using the resources at the USGS (2014), as described in the previous paragraphs. The results are inherently approximate and conservative because they represent average site properties and interpolate between sites. For major projects, site-specific seismic hazard analysis can be done by a ground motion specialist using specialized software and data sets. This is usually the preferred approach as it takes into consideration the detailed geologic conditions at the site.

11.2.4 Design Response Spectra in U.S. Building Codes

Building codes in the United States use the USGS seismic hazard analysis resources along with site amplification factors and a standard response spectrum shape to determine seismic design spectra. For most of the United States, the spectral values are approximately equal to pseudo-acceleration response values having 2% probability of exceedance in 50 years. The spectral values were adjusted up or down so that the probability of collapse for an individual facility was equal to approximately 1% in 50 years. Near known active faults with significant slip rates and characteristic earthquakes with magnitudes in excess of about 6.0, the design values are limited by 1.8 times median response spectral values associated with a characteristic earthquake on the fault. The resulting spectral response values are referred to as the *Risk-Targeted Maximum Considered Earthquake,* designated MCE_R level. For design, the *Design Earthquake* (DE) level is set at DE = ⅔MCE_R, in anticipation of structural safety margin factor of 1.5 inherent in the design procedures. For additional information on the derivation of the design values, see Luco et al. (2007) and NEHRP (2009).

The specific procedure is as follows:

1. Use the USGS resources to determine the 5%-damped spectral response pseudo-accelerations at short-period, S_S, and 1-s period, S_1, at the *Risk-Targeted Maximum Considered Earthquake* (MCE_R) shaking level.
2. Adjust the values for effects of geotechnical site class, as $S_{MS} = F_a S_S$ and $S_{M1} = F_v S_1$. For this purpose, first determine the geotechnical site class in accordance with Table 11.2. Then use Tables 11.3 and 11.4 to obtain site coefficients F_a and F_v.

Site Class	Soil Shear Wave Velocity, $\bar{v}_s$, ft/s (m/s)	Standard Penetration Resistance, $\bar{N}$	Soil Undrained Shear Strength, $\bar{S}_u$, psf (Pa)
A. Hard rock	>5000 (1520)	N/A	N/A
B. Rock	2500 (760) to 5000 (1520)	N/A	N/A
C. Very dense soil and soft rock	1200 (370) to 2500 (760)	>50	>2000 (0.096)
D. Stiff soil	600 (180) to 1200 (370)	15 to 50	1000 (0.048) to 2000 (0.096)
E. Soft clay soil	<600 (180)	<15	<1000 (0.048)
	Any profile with more than 10 ft (3 m) of soil having the following characteristics: • Plasticity index PI > 20 • Moisture content $w \geq 40\%$ • Undrained shear strength $\bar{S}_u < 500$ psf (0.024 Pa)		
F. Soils requiring site response analysis	Soils vulnerable to failure or collapse under seismic loading, peats and/or highly organic clays, very high plasticity clays, or very thick soft/medium stiff clay*		

*Simplified description. See ASCE 7 for complete description.

Table 11.2 Geotechnical Site Class Definitions (*after ASCE 7, 2010, with permission from ASCE*)

	Mapped Maximum Considered Earthquake Spectral Response Acceleration Parameter at Short Period				
Site Class	$S_S \leq 0.25$	$S_S = 0.5$	$S_S = 0.75$	$S_S = 1.0$	$S_S \geq 1.25$
A	0.8	0.8	0.8	0.8	0.8
B	1.0	1.0	1.0	1.0	1.0
C	1.2	1.2	1.1	1.0	1.0
D	1.6	1.4	1.2	1.1	1.0
E	2.5	1.7	1.2	0.9	0.9
F	Site-specific analysis required				

Note: Use straight-line interpolation for intermediate values of S_S.

TABLE 11.3 Site Coefficient F_a to Modify S_S Values (*after ASCE 7, 2010, with permission from ASCE*)

	Mapped Maximum Considered Earthquake Spectral Response Acceleration Parameter at Short Period				
Site Class	$S_1 \leq 0.1$	$S_1 = 0.2$	$S_1 = 0.3$	$S_1 = 0.4$	$S_1 \geq 0.5$
A	0.8	0.8	0.8	0.8	0.8
B	1.0	1.0	1.0	1.0	1.0
C	1.7	1.6	1.5	1.4	1.3
D	2.4	2.0	1.8	1.6	1.5
E	3.5	3.2	2.8	2.4	2.4
F	Site-specific analysis required				

Note: Use straight-line interpolation for intermediate values of S_1.

TABLE 11.4 Site Coefficient F_v to Modify S_1 Values (*after ASCE 7, 2010, with permission from ASCE*)

3. Adjust the values to the design level using Eqs. (11.4) and (11.5).

$$S_{DS} = \frac{2}{3} S_{MS} = \frac{2}{3} F_a S_S \tag{11.4}$$

$$S_{D1} = \frac{2}{3} S_{M1} = \frac{2}{3} F_v S_1 \tag{11.5}$$

4. Given values of S_{DS} and S_{D1}, the design response spectrum is defined using the standard spectrum shape shown in Figure 11.8.

In this figure:

T = fundamental period of the structure, in seconds

$$T_S = \frac{S_{D1}}{S_{DS}}$$

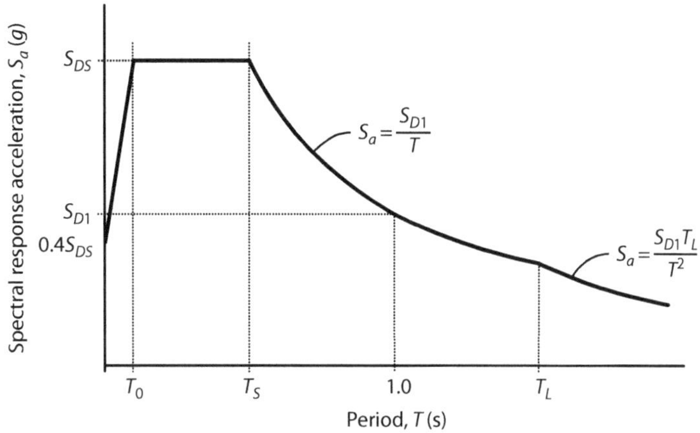

FIGURE 11.8 Design response spectrum. (*After ASCE 7, 2010.*)

$T_0 = 0.2\,T_S$

T_L = long-period transition period. Values for T_L are 4 s or greater. See ASCE 7 for details.

11.3 Earthquake Demands on Building Structures

11.3.1 Linear-Elastic Response

Earthquake demands on buildings will vary from earthquake to earthquake, and maximum expected demands will vary from region to region. For buildings located in regions of high seismicity, the maximum expected earthquake shaking levels may produce lateral displacements of several inches, with lateral forces for linear systems approaching or even exceeding the weight of the building. Except for very special structures, it will not be economically feasible to design buildings with conventional structural systems to respond linearly to such strong shaking. Some nonlinear response may have to be accepted.

We can demonstrate this for sites in the highly seismic western United States using the DE response spectrum presented in Section 11.2.4. We begin by selecting a specific site, in this case a site with latitude 34.045 and longitude −118.264 in the City of Los Angeles. The site is determined to be site class C. Using the USGS resources, $S_S = 2.320g$, $S_1 = 0.815g$, and $T_L = 8$ s. From Tables 11.3 and 11.4, F_a = 1.0 and $F_v = 1.3$. From Eqs. (11.4) and (11.5), $S_{DS} = 1.547g$ and $S_{D1} = 0.707g$. Using the standard spectrum shape in Figure 11.8, the pseudo-acceleration design response spectrum is as shown in Figure 11.9*a*. The corresponding displacement design response spectrum (Figure 11.9*b*) is derived from the pseudo-acceleration design response spectrum using the relation $S_d = S_a/\omega^2$.

Adopting the rule of thumb that a concrete frame building has a fundamental vibration period $T_1 \approx N/10$, where N = number of stories, we can estimate that a

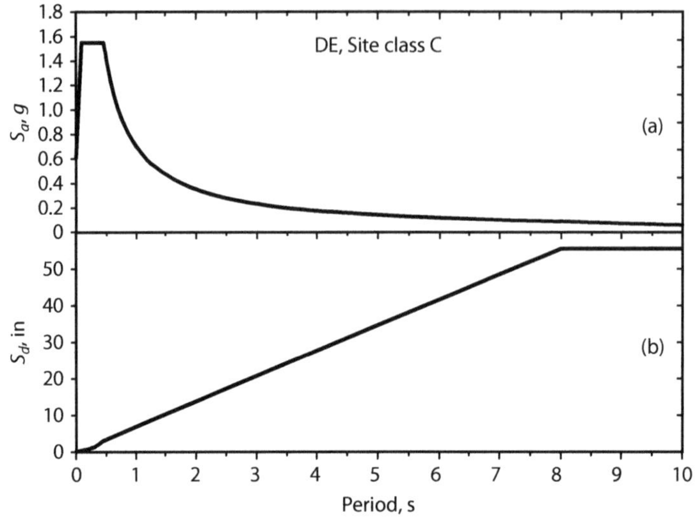

FIGURE 11.9 Pseudo-acceleration and displacement response spectra for the Design Earthquake (DE), site class C, for a site in the City of Los Angeles.

five-story frame building has a vibration period around 0.5 s. From Figure 11.9, for $T = 0.5$ s, $S_a = 1.4g$ and $S_d = 3.5$ in (89 mm). The corresponding base shear can be estimated to be $V_b = S_a W$, that is, 1.4 times the building weight. For a typical frame building, the roof displacement can be estimated to be 1.3 S_d, which in this case is approximately 4.6 in (120 mm). These are DE values. MCE_R values would be 1.5 times these values, or $V_b = 2.1W$ and roof displacement of 6.8 in (170 mm).

While it is possible to design a structure to remain linearly-elastic under the forces and deformations determined above, doing so would require considerable expense. It might also require use of a massive structural system that would interfere with the programmatic requirements of the building. Neither the expense nor the functional disruption can be justified in most building projects, especially considering the rarity of DE or MCE_R shaking levels.

In consideration of the preceding arguments, most building codes allow, either implicitly or explicitly, for nonlinear response of a building during strong shaking. As a minimum, we should design the structure so that anticipated inelastic response can occur without critical decay in resistance. For buildings with higher performance objectives, we might design the structure for reduced inelastic response. Regardless, it is important to understand the nature of inelastic response of buildings subjected to earthquake shaking.

11.3.2 Nonlinear Inelastic Response

Consider the single-degree-of-freedom oscillator shown in Figure 11.10*a*. The mass M (= W/g) is set equal to 1 kip-s^2/in (0.17 kN-s^2/mm) and the stiffness K is set equal to 39.5 kip/in (6.9 kN/mm), resulting in linear period $T = 1$ s. Damping is modeled as viscous with damping ratio equal to 5% of the critical value. The

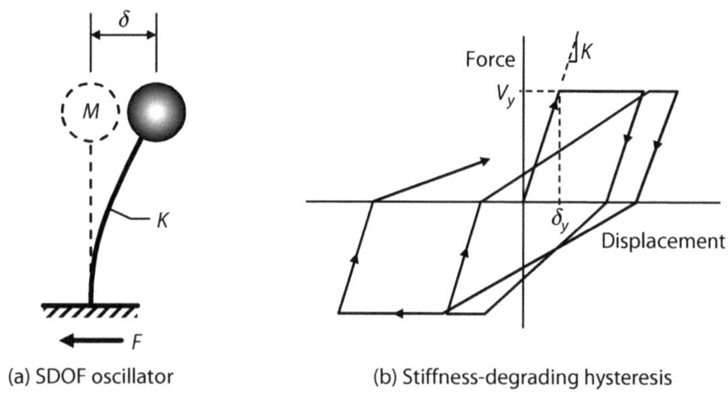

Figure 11.10 (*a*) Single-degree-of-freedom (SDOF) oscillator; (*b*) force-displacement response.

force-displacement relation (Figure 11.10*b*) has stiffness-degrading behavior that approximates behavior of reinforced concrete construction with low bond stress and low shear stress. By setting the yield force V_y sufficiently high, the response will be linear-elastic. For strength lower than the elastic force demand, hysteresis follows the behavior shown in Figure 11.10*b*.

Response of the oscillator is calculated using the software BISPEC (2009). Figure 11.11 plots the displacement response history for linear response and for moderately nonlinear response. For linear response (Figure 11.11*a*), the maximum displacement is 17.9 in (455 mm), consistent with the spectral displacement S_d that can be read from the linear response spectrum at $T = 1$ s (Figure 11.4). The maximum restoring force in the spring can be obtained from Eq. (11.2) as $V_e = KS_d$ = 39.5 kip/in × 17.9 in = 705k (3140 kN).

We here introduce the *response modification coefficient*, *R*, defined by

$$R = V_e / V_y \tag{11.6}$$

in which V_y is the yield force of the spring. Linear response requires $R = 1$ (or less). $R = 2$ corresponds to an oscillator having a yield base shear equal to half the elastic shear force for a linearly elastic oscillator. The response modification coefficient is widely used to define design strength requirements for seismic designs complying with ASCE 7, as will be discussed subsequently.

The nonlinear oscillator whose response is plotted in Figure 11.11*b* was defined to have $R = 2$, that is, yield strength equal to 705 kips/2 = 353 kips (1570 kN). The effects of yielding are apparent in two characteristics of the response history. First, the apparent vibration period is elongated relative to the initial period of $T = 1$ s; this is because yielding results in effective stiffness degradation in the load-displacement relation. Second, nonlinear response is apparent in the permanent offset of 4.9 in (120 mm).

An important observation from Figure 11.11 is that the peak displacements for linear response (17.9 in) and nonlinear response with $R = 2$ (18.6 in) are nearly

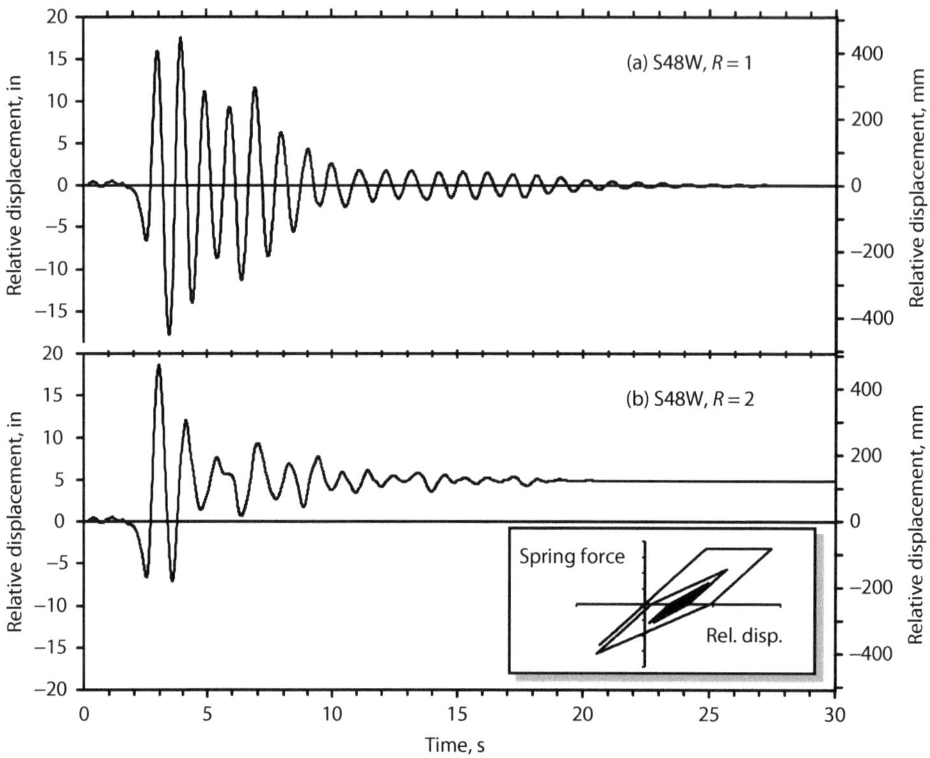

FIGURE 11.11 Calculated response of an oscillator having initial period $T = 1$ s and 5% damping subjected to the ground motion recorded at the Rinaldi Receiving Station during the 1994 Northridge, California earthquake. (*a*) $R = 1$ corresponds to linear-elastic response; (*b*) $R = 2$ corresponds to inelastic response for the oscillator having yield strength equal to half the elastic strength demand. The hysteretic relation between spring force and relative displacement is shown in the inset.

equal. If we were to further investigate this observation for $R = 3$, 4, and 6, we would find peak displacements of 21.0, 19.5, and 16.4 in, respectively. Apparently, for this structure and this ground motion record, the peak displacement is relatively insensitive to the strength within this range of strengths.

We can investigate whether this observation is generally correct. Figure 11.12 shows results from such an investigation. For this example, 5% damped nonlinear oscillators having bilinear hysteresis (i.e., not stiffness-degrading) with various periods and various strength reduction factors were subjected to 20 earthquake ground motions recorded at sites having site class C. The results suggest that, for oscillators having periods exceeding approximately $T = 1$ s and not having very large R values subjected to site class C motions, the mean displacement is relatively insensitive to strength. For shorter periods, peak displacement for nonlinear response is larger than for linear response, increasingly so as R increases. Studies for other oscillator restoring force properties and other earthquake ground motion characteristics have found similar (but not identical) results.

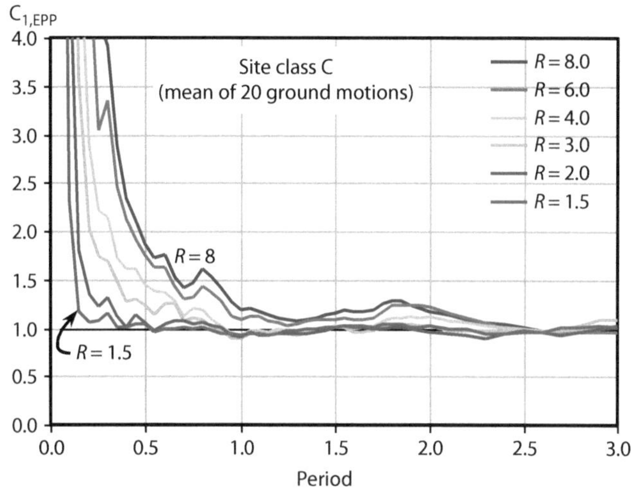

FIGURE 11.12 Ratio C_1 of maximum relative displacement of a yielding oscillator to the maximum relative displacement of a linear-elastic oscillator having the same initial period as a function of strength ratio R and initial vibration period T. The 5%-damped oscillator has bilinear load-displacement relation without strain-hardening. Results are the average values for 20 ground motions recorded at sites having site class C. (*After FEMA 440, 2005.*)

The observation that the peak displacement for nonlinear response is approximately equal to the peak displacement for linear response is known as the *equal displacement rule*. We should note that it is not a mathematically derived result, but instead is an empirical observation. Further, the observation is limited to the average of the results of several ground motions; for any one ground motion the peak displacement for inelastic response can vary appreciably from that for linear response. Furthermore, as shown in Figure 11.12, its applicability is limited to periods beyond about 1 s (or longer for certain types of ground motions) and for typical R values.

11.3.3 Drift and Ductility Demands

Despite all the caveats of the preceding paragraphs, the equal displacement rule provides a convenient approximate starting point for discussing expected earthquake demands for yielding systems. One measure of interest is the lateral *drift ratio*, that is, the lateral drift divided by height. For the response spectrum plotted in Figure 11.9, the spectral displacement is $S_d = 6.92T$ (in) for the period range 0.5 to 8 s. For a frame building, the roof drift ratio can be approximated as $1.3S_d = 9T$. If a building is designed to have a first-mode vibration period $T_1 = N/10$, with typical story height of 100 in (2.5 m), then the roof drift ratio for response in the first mode is approximately $9 \times (N/10)/(100N) = 0.009$, or 0.9% of total building height. If the building is made more flexible, with $T = N/5$, the roof drift ratio is increased to 1.8% of total building height. Clearly, damage to drift-sensitive components will be greater for a more flexible building.

Ductility demand may also be of interest. *Displacement ductility demand*, μ_δ, is the ratio of displacement demand to yield displacement. For the SDOF oscillator of Figure 11.10, displacement ductility demand is written as

$$\mu_\delta = \frac{|\delta_{max}|}{\delta_y} \tag{11.7}$$

If maximum displacements for a yielding oscillator are equal to those for a non-yielding oscillator of the same initial stiffness, then we can write

$$\mu_\delta = \frac{|\delta_{max}|}{\delta_y} = \frac{|V_e|}{V_y} = R \tag{11.8}$$

Thus, adopting the equal displacement rule, an oscillator designed for lateral strength $V_y = V_e/R$ is expected to develop displacement ductility $\mu_\delta = R$.

We may also be interested in plastic-hinge rotation ductility demand, μ_θ, which can be defined as the total chord rotation of a beam divided by the yield rotation. For a cantilever, as illustrated in Figure 11.13, chord rotation is the displacement of the free end relative to a tangent to the fixed end divided by the length l. Defining the plastic hinge to be concentrated at the interface between the beam and its support, the total rotation is $\theta_y + \theta_p$, and rotation ductility $1 + \theta_p/\theta_y$. Other definitions of plastic-hinge rotation ductility are possible, but are not used here.

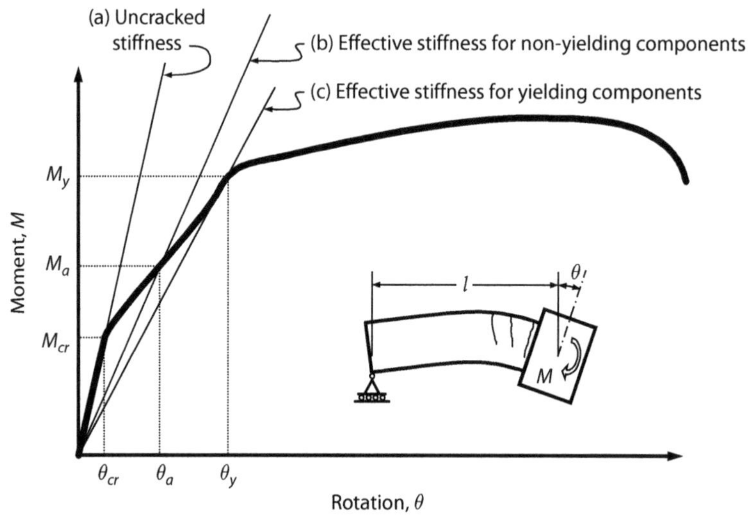

Figure 11.13 Definition of chord rotation, and effective flexural stiffness of concrete components accounting for cracking: (*a*) uncracked stiffness; (*b*) effective stiffness for component loaded to maximum moment M_a; (*c*) effective stiffness for yielding component.

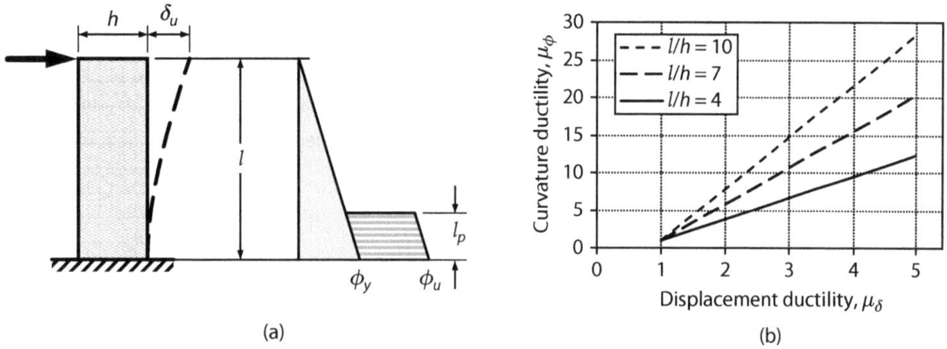

Figure 11.14 (*a*) Plastic-hinge model for a cantilever; (*b*) calculated relation between curvature ductility and displacement ductility for a cantilever.

Curvature ductility demand, μ_ϕ, is defined as the maximum curvature divided by the yield curvature. We can derive a relation between curvature ductility and displacement ductility using the plastic-hinge model of Figure 11.14*a* (see Section 6.11.2 for discussion of the plastic-hinge model). The tip displacement can be written

$$\delta_u = \delta_y + \theta_p\left(l - \frac{l_p}{2}\right) = \frac{\phi_y l^2}{3} + (\phi_u - \phi_y)l_p\left(l - \frac{l_p}{2}\right) \tag{11.9}$$

Using Eq. (11.9), the displacement ductility can be written

$$\mu_\delta = \frac{\delta_u}{\delta_y} = 1 + 3\left(\frac{\phi_u}{\phi_y} - 1\right)\frac{l_p}{l}\left(1 - \frac{l_p}{2l}\right) \tag{11.10}$$

Solving for curvature ductility results in

$$\mu_\phi = \frac{\phi_u}{\phi_y} = 1 + (\mu_\delta - 1)\left(3\frac{l_p}{l}\left(1 - \frac{l_p}{2l}\right)\right)^{-1} \tag{11.11}$$

Substituting $l_p = 0.5h$, the relation between μ_ϕ and μ_δ can be expressed as a function of member aspect ratio l/h. The result is plotted in Figure 11.14*b*. As shown, $\mu_\phi \geq \mu_\delta$ for all values of μ_δ and l/h. Note that μ_ϕ increases with increasing l/h. This can be understood by recognizing that the elastic curvatures are spreading over a larger member length while plastic hinge rotations are concentrating over the constant length $h/2$. Alternative relations between μ_ϕ and μ_δ can be obtained by selecting alternative expressions for l_p (see Chapter 6), but the end result is similar to the result shown in Figure 11.14*b*.

The results of Figure 11.14*b* do not translate directly to more complicated framing systems for two reasons. One is that the sources of linear-elastic flexibility change as the framing geometry changes. The other is that plastic hinges may form progressively in the structural system, such that no one relation can describe

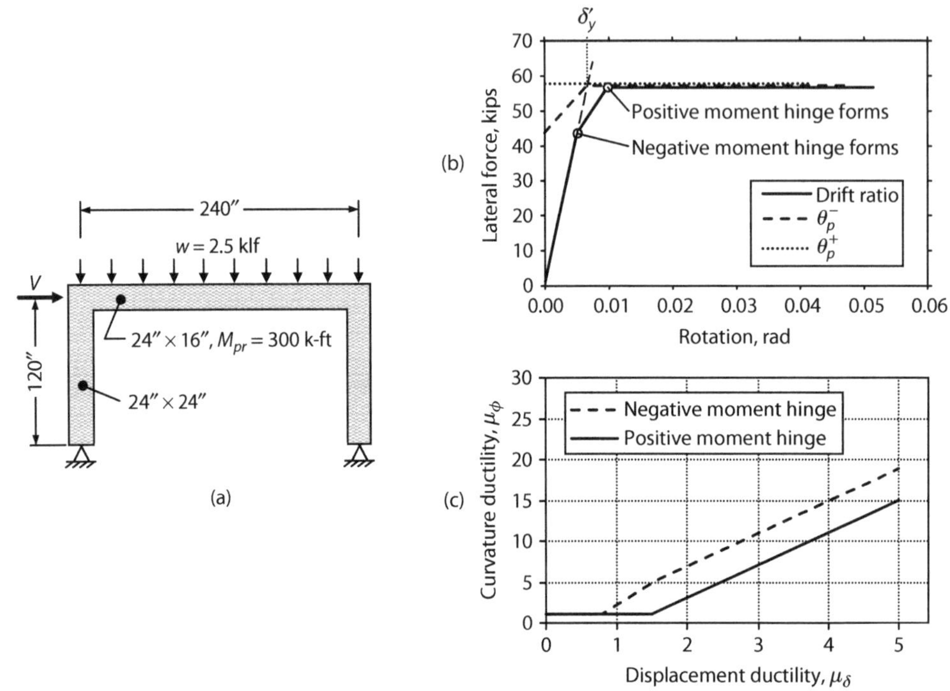

FIGURE 11.15 (*a*) Example frame; (*b*) relations between lateral force *V* and drift ratio (lateral displacement divided by column height), and between *V* and plastic hinge rotations; (*c*) relations between curvature ductility and displacement ductility for example frame.

ductility demand of each plastic hinge. Figure 11.15 illustrates this for a one-bay, one-story reinforced concrete frame. Flexural response of the frame was modeled by assigning flexural rigidity $E_cI = 0.5E_cI_g$ for all members. Vertical load *w* was applied first, followed by lateral load *V*. Plastic hinges were assumed to form at a joint centerline when the beam moment at that location reached the moment strength M_{pr}. The continuous curve in Figure 11.15*b* is the relation between lateral force *V* and drift ratio. A secant extended to the strength at $V = 57.3$ kips defines the effective yield displacement δ'_y. The negative moment plastic hinge rotation θ_p^- begins to accumulate after *V* reaches 43.7 kips (194 kN), and the positive moment plastic hinge rotation θ_p^+ begins to accumulate after *V* reaches 57.3 kips (255 kN).

Figure 11.15*c* plots the relations between plastic hinge curvature ductility and displacement ductility for the example frame. The derivation assumes the yield curvature at a beam plastic hinge is equal to $\varepsilon_y/(1-k)d$, that is, the reinforcement yield strain divided by the depth from the neutral axis for linear response (see Chapter 6 for additional discussion), and that plastic-hinge length is $l_p = 0.5h$. The results are similar to those for a cantilever (Figure 11.14*b*). However, the negative-moment and positive-moment hinges have different demands and, because of additional frame flexibility adding to the displacement at yield, the curvature ductility values trend higher than those shown in Figure 11.14*b*.

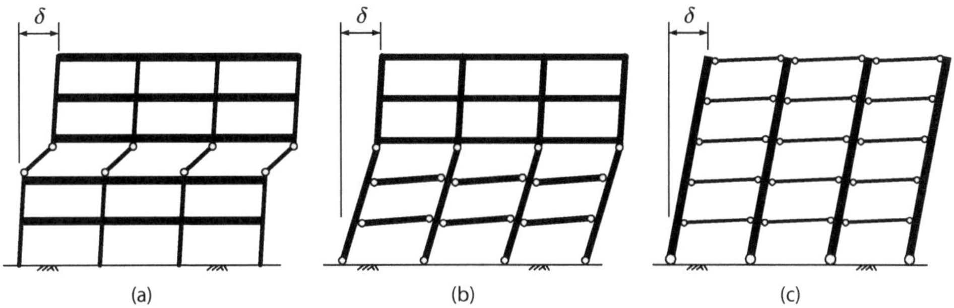

FIGURE 11.16 Idealized yielding mechanisms: (*a*) story mechanism; (*b*) intermediate mechanism; (*c*) beam mechanism.

As framing geometry and loadings become more complex, it becomes impractical to develop general relations between local curvature ductility and global seismic demands. For example, consider the multi-story beam-column moment frames shown in Figure 11.16, each displaced to roof displacement δ. Whether the frame develops a story mechanism, intermediate mechanism, or beam mechanism will strongly affect the local deformations required to achieve the displacement δ. It should be apparent that, for a given displacement δ, the inelastic demands will be greatest for the story mechanism and least for the beam mechanism. A primary aim in the design of special moment frames is to avoid the story mechanism (see Chapter 12).

Notwithstanding the complexities of building frame response, it can be informative to compare the design level of current codes with the expected overall system response. Here we follow the approach summarized in SEAOC (2007), which is applicable to design in accordance with ASCE 7.

Consider the load-displacement relations of Figure 11.17, including a linear system and a nonlinear building structure having the same initial effective stiffness. The building structure initially has stiffness corresponding to the uncracked member stiffness, but cracking, yielding, strain-hardening, and strength degradation occur progressively. The effective stiffness corresponds to a point on the load-displacement relation for which the shear is approximately two-thirds of the maximum value, at which point the structure is cracked and may be yielding locally.

For linear response to DE shaking, the base shear and corresponding displacement are designated as V_e and δ_e (Figure 11.17). According to the ASCE 7 design approach, the structure can be designed for a base shear equal to V_e/R. Coefficient R is the *response modification coefficient*, introduced earlier, and described in greater detail in Section 11.5.1. The base shear V_e/R is the minimum expected yield strength in the building. Actual first yield in the structural framing and actual expected strength of the building are expected to be higher than the design base shear. Several factors contribute to the overstrength, including material overstrength and strain-hardening, conservatism in design load combinations

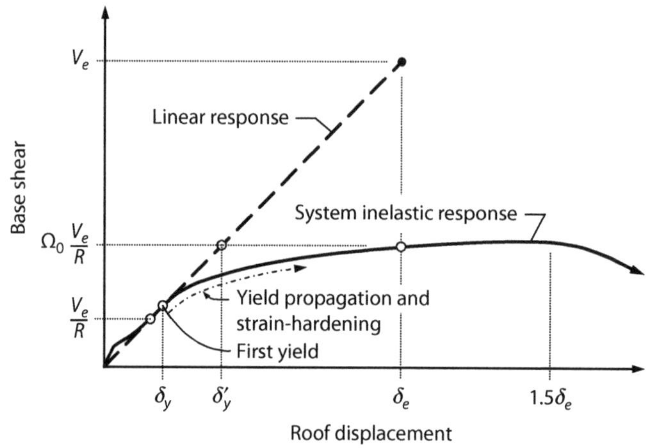

FIGURE 11.17 Idealized relations between linear-elastic response, design base shear, and expected behavior of complete structural system. (*After SEAOC, 2007.*)

and strength reduction factors, three-dimensional interactions among structural (and nonstructural) parts of the building that are not considered in analytical models used in design, and other factors such as drift limits that may control building proportions. The expected base-shear strength is designated as $\Omega_0 \frac{V_e}{R}$, in which Ω_0 is the *system overstrength factor*, that is

$$\Omega_0 = \text{expected building strength/minimum design building strength} \quad (11.12)$$

In Figure 11.17, the expected building strength is defined at displacement δ_e, but this is a somewhat arbitrary definition. The numerical value of Ω_0 will vary from one building to another depending on the configuration of the building and its constituent materials. ASCE 7 specifies values of Ω_0 to be used for the design of components that are sensitive to building overstrength. For reinforced concrete construction in regions of high seismicity, the specified value of Ω_0 is either 2.5 or 3 depending on the framing system. See Section 11.5.1 for additional discussion.

Engineers sometimes refer to an effective response modification coefficient, R_{eff}, defined as the elastic base shear divided by the expected base shear strength, that is

$$R_{eff} = \frac{V_e}{\Omega_0 \frac{V_e}{R}} = \frac{R}{\Omega_0} \quad (11.13)$$

The preferred way to determine this factor is to use nonlinear analysis of the as-designed building to estimate the expected building strength. It is also instructive, however, to estimate R_{eff} using the code-specified range of values for R (5 to 8) and Ω_0 (2.5 to 3). The resulting range is $1.7 \leq R_{eff} \leq 3.2$.

If we adopt the equal displacement rule as a means of approximating the expected displacement, then the expected peak displacement for the building responding inelastically to DE shaking is δ_e. For shaking at the MCE_R level, the peak displacement will be approximately $1.5\delta_e$. [The factor 1.5 is the reciprocal of the 2/3 factor used to scale the MCE_R hazard level to the DE level. See Eqs. (11.4) and (11.5).] These displacements can be used to check deformation capacities of structural and nonstructural components. Considering that the building code intent is to avoid collapse for MCE_R shaking, structural components required to support lateral and gravity loads should be able to survive deformations to approximately $1.5\delta_e$ without collapse.

We can also define a system ductility demand. First we estimate the effective system yield displacement, δ_y', as the abscissa at the intersection of the linear response line and the expected building strength, as shown in Figure 11.17. Knowing δ_y', the *system displacement ductility demand* is

$$\mu_\delta = \frac{\delta_e}{\delta_y'} = \frac{V_e}{\Omega_0 \dfrac{V_e}{R}} = R_{eff} \tag{11.14}$$

Thus, under the assumptions given, the expected system ductility demand is in the same range as R_{eff}, that is, $1.7 \le \mu_\delta \le 3.2$. Recall, however, that yielding of individual components may occur well before the building reaches the displacement δ_y'. Thus, local ductility demands may be much larger than the system ductility demand.

11.4 Earthquake-Resisting Buildings

Reinforced concrete building structures comprise a three-dimensional framework of structural elements designed to support loads, resist environmental effects, and provide enclosed spaces. In the case of earthquake-resisting structures, we are concerned primarily with the ability to support gravity loads while resisting earthquake shaking effects. Although the complete three-dimensional system acts integrally to resist lateral loads, it is common to conceive of the seismic-force-resisting system as being composed of vertical elements, horizontal elements, and the foundation.

Figure 11.18 illustrates a basic structural configuration for a building. Diaphragms make up the *horizontal elements* of the seismic-force-resisting system. These act to transmit inertial forces from the floor system to the vertical elements of the seismic-force-resisting system. They also tie the vertical elements together, and thereby transmit forces between these elements as may be required during earthquake shaking. The *vertical elements* of the seismic-force-resisting system are typically structural walls or moment-resisting frames. These elements transmit lateral and vertical forces from upper levels of the superstructure into the foundation. The *foundation* may include footings, foundation mats, pile caps, grade beams, structural slabs-on-grade, basement walls, and piles, piers, and caissons.

The preceding paragraph identifies elements designated as part of the seismic-force-resisting system. A building may have other parts of the structural framing

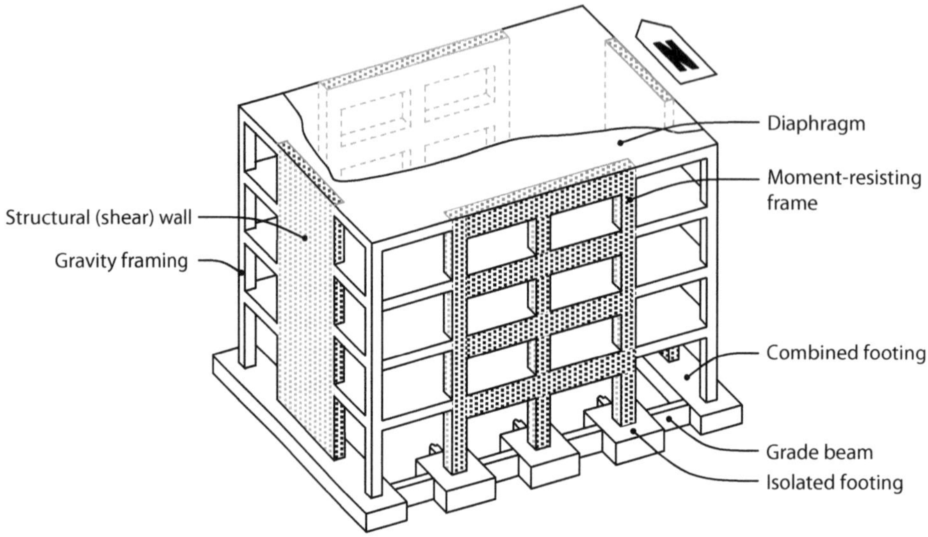

Figure 11.18 Sketch of a basic building system comprising horizontal spanning elements (diaphragms), vertical spanning elements (walls and frames), and foundation.

that are not designated to be part of the seismic-force-resisting system. The primary role of this other framing is to support gravity loads; hence, it is commonly referred to as *gravity framing*. In traditional moment-frame construction, moment-resisting frames were used along every framing line and bay of the building. The practice of designating a portion of a frame to be part of the seismic-force-resisting system, with the remainder designated as gravity framing, is an early development that became popular in the 1980s and is now widespread in the United States.

In this section we briefly review the different types of structural elements and their roles in supporting gravity loads and resisting earthquake shaking. Later chapters cover each of these elements in greater detail.

The layout of the seismic-force-resisting system is driven by consideration of functionality and economics, but also should be driven by consideration of how forces are generated and resisted during an earthquake. As the ground shakes, the motion is transferred to the foundation and into the superstructure. The resulting motion of the superstructure leads to inertial forces (product of mass and acceleration), which are distributed through the structure according to the distribution of mass and acceleration. The seismic-force-resisting system must be designed to provide a balanced and continuous load path from the source of the inertial forces back down to the foundation.

In a typical building the majority of the mass is in the floor system. An important consideration in laying out the seismic-force-resisting system is to locate the vertical elements so the center of resistance is close to the center of mass, thereby reducing torsional effects. The fundamental problem is illustrated in

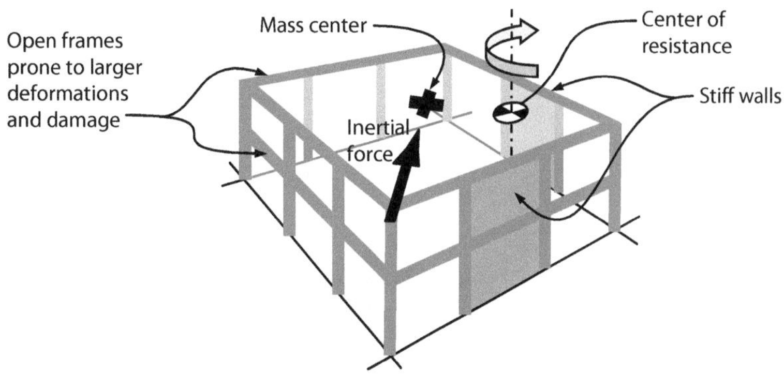

Figure 11.19 Building in which location of the vertical elements of the seismic-force-resisting system results in severe torsion problem.

Figure 11.19, in which severe eccentricity between the centers of mass and resistance results in significant torsion. In a building with rectangular plan, an ideal location for the vertical elements of the seismic-force-resisting system is around the building perimeter, such that the centers of mass and resistance coincide and high torsional resistance is provided. This location for vertical elements may not be ideal from the perspective of overturning resistance or building function.

Acceleration of the floor diaphragm (and all the components attached to it) results in inertial forces within the plane of the diaphragm that must be transmitted to the vertical elements of the seismic-force-resisting system. Consider the roof-level sketch of a building in Figure 11.20, with acceleration of the upper floor in the south direction. This results in inertial forces distributed along the diaphragm in the north direction (Figure 11.20), which must be resisted by the structural walls along the east and west ends of the building. The diaphragm must be designed for the shears and moments that result from spanning between the walls, the connections between the diaphragm and the walls must be designed to transfer

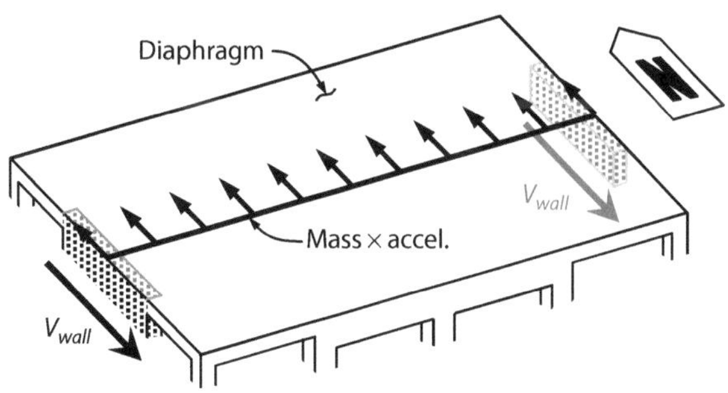

Figure 11.20 Inertial forces in a diaphragm and forces transferred from a diaphragm to the vertical elements of the seismic-force-resisting system.

FIGURE 11.21 Southwest garage at Northridge Fashion Center, Northridge earthquake, 1994. (*After EERI, 1996, courtesy of Earthquake Engineering Research Institute.*)

the forces at those locations, and the wall must be designed for the force transmitted from the diaphragm.

Excessive flexibility, inelastic response, or failure of inadequate diaphragm components contributed to the collapse of parking structures during the 1994 Northridge earthquake (EERI, 1996). Figure 11.21 shows a collapsed parking structure at the Northridge Fashion Center; collectors (elements transferring loads from the diaphragm to the walls) apparently yielded, resulting in large lateral displacements of gravity columns that subsequently contributed to their failure.

The vertical elements of the seismic-force-resisting system are required to transmit the accumulated seismic forces to the foundation system. Generally, it is preferable for the vertical elements to be continuous over height. Figure 11.22 illustrates two examples of discontinuous vertical elements. In Figure 11.22*a*, a

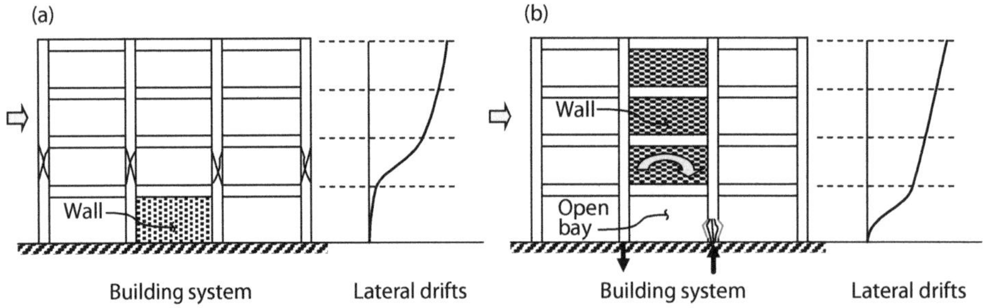

FIGURE 11.22 Buildings with discontinuous vertical elements of the seismic-force-resisting system.

(a) (b) (c)

Figure 11.23 (*a*) Upper story failure at point of discontinuity in the Kobe City Hall Annex, Kobe earthquake, Japan, 1995; (*b*) side-sway failure of the first story of the Olive View Medical Treatment Building, San Fernando earthquake, California, 1971; (*c*) crushed first-story columns of the Imperial County Services Building, Imperial Valley earthquake, California, 1979. (*Photographs used with permission from the National Information Service for Earthquake Engineering, University of California, Berkeley.*)

structural wall in the first story terminates at the first elevated level; the sudden discontinuity in stiffness and strength can result in distress in the story immediately above the wall cutoff. In Figure 11.22*b*, a wall in the upper stories is discontinuous in the first story. The resulting discontinuity in stiffness and strength can result in a weak first story that is highly vulnerable to earthquake effects. The condition is exacerbated by the overturning forces from the wall, which must be resisting in the first story by the supporting columns. Seismic building codes may prohibit the type of discontinuity shown in Figure 11.22*b* for buildings in highly seismic zones because it is known to result in poor performance.

Figure 11.23 illustrates failures of buildings with discontinuities of the type shown in Figure 11.22. In the Kobe City Hall Annex, steel reinforced concrete columns transitioned to reinforced concrete columns at the mid-story level that collapsed (Figure 11.23*a*). In the Olive View Medical Treatment Building, walls in the upper stories were discontinued in the first story, leaving a weak first story that nearly collapsed during the earthquake (Figure 11.23*b*). In the Imperial County Services Building, crushing failures of columns were attributed partly to overturning axial loads from discontinued structural walls (Figure 11.23*c*).

The foundation system supports the combined forces from earthquake and gravity load effects. Usually, the geotechnical engineer recommends a foundation type based on considerations of the site conditions and soil properties. In a structure founded on isolated or combined footings, such as illustrated in Figure 11.18, foundation elements are sized for the horizontal shears, axial loads, and moments from the structural elements framing into them. Structures with subterranean levels may have a more complicated load path (Figure 11.24). A slab at grade or at the top of a podium level acts in-plane as a stiff reaction point, resulting in large shears transferring from the vertical elements of the seismic-force-resisting system laterally to other stiff elements (to basement walls in the figure shown). The percentage of base shear transferring to the transfer

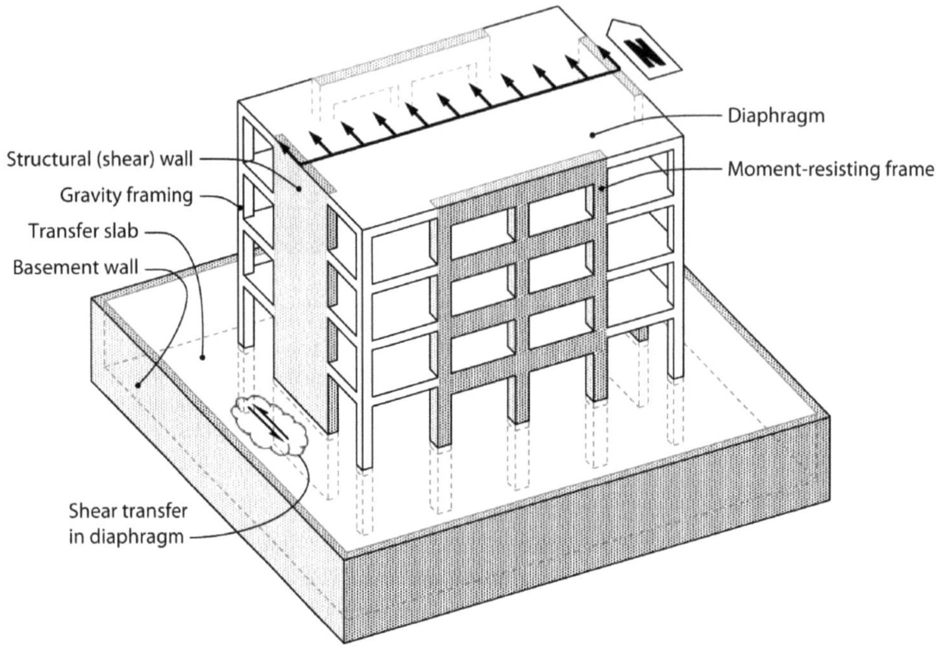

FIGURE 11.24 Sketch of the load path in a building with subterranean levels.

slab versus that remaining in the vertical element depends on relative stiffness in the load path to the foundation. Axial loads and overturning moments in the vertical elements almost always are transferred through the podium slab directly down to the foundation.

In conventional buildings, the intended inelastic response ideally is restricted to the vertical elements of the seismic-force-resisting system. This can be accomplished by first sizing the vertical elements for expected earthquake demands (reduced for anticipated inelastic response) and then designing the diaphragm and foundation elements to have sufficient strength to avoid significant inelastic response. This can be done using capacity design, using design forces that are amplified to account for expected overstrength using a factor Ω_0, or using nonlinear dynamic analysis to determine overstrength forces. For some components of the structural system, such as collector elements of diaphragms, the seismic code may require this approach to be taken. For other components, the use of amplified forces for the foundation and diaphragm elements is left to the discretion of the engineer.

Inelastic response also is permitted for elements not designated as part of the seismic-force-resisting system, such as the gravity framing shown in Figure 11.24, but it must be checked to be certain the deformation capacity is adequate. Inadequately detailed gravity framing was a primary cause of failure including

(a) (b)

Figure 11.25 California State University, Northridge, California earthquake, 1994: (*a*) interior failed gravity column; (*b*) extensive collapse of structural system due to failure of the gravity columns. [*Photograph (a) used with permission from the National Information Service for Earthquake Engineering, University of California, Berkeley; photograph (b) by P.C. Riddle, Los Angeles Times.*]

collapse of some concrete buildings during the 1994 Northridge, California earthquake (e.g., Figure 11.25).

11.5 Design Approach

Seismic design practices for buildings vary worldwide, and different practices may be applied depending on the complexity of the building, the seismic environment, and the performance objective. The following sections outline three seismic design approaches. Section 11.5.1 presents the strength-based design approach, with main emphasis on the provisions of ASCE 7. Those provisions emphasize required strength, with checks that lateral drifts do not exceed prescribed limits. Section 11.5.2 presents the displacement-based design approach. That approach emphasizes system proportioning to achieve a target displacement, with structural strength either an outcome of the design process or a secondary check. Section 11.5.3 presents the performance-based approach. There, the emphasis is on achieving target performance objectives that are defined in the design process.

11.5.1 Strength-Based Design in Accordance with ASCE 7

Risk Category and Seismic Design Category

Each building is assigned to a *Risk Category* in accordance with Section 1.4.3 of this book. Each building is also assigned to a *Seismic Design Category*. For typical buildings, the Seismic Design Category is the more severe of the categories identified in Tables 11.5 and 11.6. However, where the mapped spectral response acceleration parameter at 1 s period is $S_1 \geq 0.75$, Risk Category I, II, or III buildings are assigned to Seismic Design Category E, and Risk Category IV

Value of S_{DS}	Risk Category I or II	III	IV
$S_{DS} < 0.167$	A	A	A
$0.167 \le S_{DS} < 0.33$	B	B	C
$0.33 \le S_{DS} < 0.50$	C	C	D
$0.50 \le S_{DS} < 0.50$	D	D	D

Table 11.5 Seismic Design Category Based on Short-Period Response Acceleration Parameter (after ASCE 7, with permission from ASCE)

Value of S_{D1}	Risk Category I or II	III	IV
$S_{D1} < 0.067$	A	A	A
$0.067 \le S_{D1} < 0.133$	B	B	C
$0.133 \le S_{D1} < 0.20$	C	C	D
$0.20 \le S_{D1}$	D	D	D

Table 11.6 Seismic Design Category Based on 1 s Period Response Acceleration Parameter (after ASCE 7, with permission from ASCE)

buildings are assigned to Seismic Design Category F. Additional special cases are in ASCE 7.

Allowable Structural Systems

According to ASCE 7, the lateral and vertical seismic-force-resisting system must be consistent with the Risk Category and the Seismic Design Category to which the building is assigned. Table 11.7 lists the types of permitted vertical seismic-force-resisting systems composed of structural concrete, along with height limits (where applicable) and various design factors. The different structural systems are defined as follows:

- A *moment-resisting frame* is a vertical element in which beams, columns, and beam-column joints are connected to compose a rigid framework that resists lateral and vertical forces through moment, shear, and axial forces in the members. A *special moment frame* is one that is proportioned and detailed to enable it to meet the performance requirements in regions of highest seismicity. See Chapter 12.
- A *structural wall* is a wall that is designed to resist lateral (and perhaps vertical) loads within the plane of the wall. A structural wall is also known as a *shear wall*. A *special structural wall* is one that is proportioned and detailed to enable it to meet the performance requirements in regions of highest seismicity. See Chapter 13.

Seismic-Force-Resisting System	Response Modification Coefficient, R	System Overstrength Factor, Ω_0	Deflection Amplification Factor, C_d	Structural System Limitations and Building Height (ft) Limit† Seismic Design Category: B	C	D	E	F
A. Bearing wall systems								
Special reinforced concrete shear walls	5	2.5	5	NL	NL	160	160	100
Ordinary reinforced concrete shear walls	4	2.5	4	NL	NL	NP	NP	NP
B. Building Frame Systems								
Special reinforced concrete shear walls	6	2.5	5	NL	NL	160	160	100
Ordinary reinforced concrete shear walls	5	2.5	4.5	NL	NL	NP	NP	NP
C. Moment-resisting frame systems								
Special reinforced concrete moment frames	8	3	5.5	NL	NL	NL	NL	NL
D. Dual systems with special moment frames capable of resisting at least 25% of prescribed seismic forces								
Special reinforced concrete shear walls	7	2.5	5.5	NL	NL	NL	NL	NL
E. Dual systems with intermediate moment frames capable of resisting at least 25% of prescribed seismic forces								
Special reinforced concrete shear walls	6.5	2.5	5	NL	NL	160	100	100

*Partial presentation. See ASCE 7 for complete description.

†NL = Not Limited; NP = Not Permitted. 40 ft = 12.2 m, 100 ft = 30.5 m, 160 ft = 48.4 m. Heights are measured from the base of the structure.

Table 11.7 Design Coefficients and Factors for Seismic-Force-Resisting Systems Applicable to Regions of Highest Seismicity (after ASCE 7, with permission from ASCE)*

- A *dual system* is a combination of structural walls and moment-resisting frames, proportioned to resist the design earthquake loads in proportion with their rigidities, except the moment-resisting frames must be capable of resisting at least 25% of the prescribed earthquake forces.
- A *building frame system* is a structural system with an essentially complete space frame providing support for vertical loads. Seismic force resistance is provided by shear walls. See Chapter 13 for a more complete description.
- A *bearing wall system* is a structural system with structural walls providing support for all or major portions of the vertical loads, in addition to providing earthquake force resistance. See Chapter 13 for a more complete description.

Analysis to Determine Design Forces and Deformations

A structural analysis model is established to determine the design forces and deformations. The model should appropriately represent the effective mass/weight and stiffness of the building system.

The seismic weight includes the total dead load of the building, including all structural elements (both seismic-force-resisting and gravity elements), cladding, partitions, total operating weight of permanent equipment, plus a portion (usually 25%) of storage live loads, and snow loads.

The structural analysis model should represent the stiffness of all elements designated as part of the seismic-force-resisting system. Diaphragms of concrete slabs or concrete-filled metal deck with span-to-depth ratios of 3 or less in structures that have no horizontal irregularities generally can be modeled as rigid. Otherwise the diaphragm flexibility should be considered. For purposes of determining seismic forces, it is generally permitted to consider the structure to be fixed at the base; however, in special cases it may be important to model flexibility of the foundation-soil system. Where linear-elastic structural models are used, the stiffness properties of the concrete elements should be defined considering effects of cracking. Elements that are expected to respond inelastically should have stiffness approximating the secant to the effective yield point (Figure 11.13). Elements that are expected to respond essentially elastically should have stiffness consistent with the maximum loading.

Most designs are done using linear-elastic models, using either *equivalent lateral force (ELF)* analysis or *modal response spectrum analysis (MRSA)*. In both cases, the design forces are linked to the response spectrum of Figure 11.8. In recognition of force reductions that can be accommodated through nonlinear response, the spectral ordinates of the response spectrum of Figure 11.8 are divided by the factor R/I_e. Table 11.7 lists values of R. These values reflect, among other things, the judgments of code-writing committees about the inelastic response capabilities of different structural systems. Table 1.1 lists values of I_e. These values provide additional strength for structures in higher risk categories. In addition to these adjustments to the design strength, ASCE 7 also includes minimum base-shear strength requirements. See ASCE 7 for additional details.

The ELF analysis is the simplest of the linear analysis options and can be used effectively for basic low-rise structures. The base shear is calculated using an approximate fundamental period, T_a, defined by a simple equation in the building code. Alternatively, the fundamental period can be calculated using a structural analysis model of the building, which generally will produce a longer period and therefore a lower design base shear. In U.S. codes, ELF analysis is not permitted for long-period structures (fundamental period T greater than 3.5 s) or structures with certain horizontal or vertical irregularities.

MRSA is often preferred because it accounts for the overall dynamic behavior of the structure. It is not unusual for the modal base shear to be less than the design base shear calculated with the ELF analysis. When the modal base shear is less than 85% of the ELF base shear, U.S. codes require that the modal base shear be scaled up to 85% of the ELF base shear. Thus, the modal response spectrum procedure can result in cost savings relative to the ELF procedure. MRSA can use either a 2-D or 3-D computer model. A 3-D model can help identify any inherent torsion in the system as well as combined effects at corner conditions. It is common for the design load combination to include 100% of the seismic response in one direction plus 30% in the orthogonal direction, and then to switch the load combination directions to 30%/100%. Multiple load combinations are required to envelope the orthogonal effects in both directions. The design of each structural member is then based on an axial and biaxial flexural interaction for each load combination. See additional discussion on load combinations below.

Some members, and some actions in members, should remain essentially elastic while the overall structural system undergoes inelastic response. It may be feasible to use capacity design to determine the maximum forces that can be delivered to an element intended to remain essentially elastic. See Section 1.5.4 for a description of capacity design. An alternative is to use design forces amplified by the *system overstrength factor*, Ω_0, listed in Table 11.7, for those elements intended to remain essentially elastic. ASCE 7 requires the use of this factor to determine design forces in collectors of diaphragms and in vertical elements supporting discontinuous stiff elements such as structural walls. Alternatively, it may be permitted to derive alternative Ω_0 factors based on nonlinear dynamic analysis, but such permission generally is subject to approval by the building official.

Load Combinations

Design load combinations are determined using the Load and Resistance Factor Design (LRFD) method described in Section 1.5.3. According to ASCE 7, the seismic load effect E in the load combinations includes effects associated with horizontal shaking, E_h, and vertical shaking, E_v, as

$$E = E_h + E_v \tag{11.15}$$

The term E_h is the force effect obtained from the application of horizontal seismic forces to the structural model, modified by a factor that increases the effect for structural systems with low redundancy. For structural elements affected by loading in two orthogonal directions, x and y, E_h is generally defined by the orthogonal load combinations

$$E_{h1} = 1.0E_{hx} + 0.3E_{hy} \tag{11.16}$$

$$E_{h2} = 0.3E_{hx} + 1.0E_{hy} \tag{11.17}$$

The term E_v is determined as

$$E_v = 0.2S_{DS}D \tag{11.18}$$

in which D is the effect of dead load. Thus, the seismic load combinations from Chapter 1 become

$$(1.2 + 0.2S_{DS})D \pm 1.0(E_{h1} \text{ or } E_{h2}) + 1.0L + 0.2S \tag{11.19}$$

$$(0.9 - 0.2S_{DS})D \pm 1.0(E_{h1} \text{ or } E_{h2}) \tag{11.20}$$

(Note that the load factor on L is permitted to be 0.5 for all occupancies in which the live load is less than or equal to 100 psf (4.79 kN/m^2), with the exception of garages or areas occupied as places of public assembly.)

Lateral Drifts

The application of the factor R/I_e to the design response spectrum reduces not only design forces but also displacements. To compensate for this, ASCE 7 requires that the deflections at level x at the center of mass, δ_x, be determined in accordance with

$$\delta_x = \frac{C_d \delta_{xe}}{I_e} \tag{11.21}$$

where C_d is from Table 11.7, I_e is from Table 1.1, and δ_{xe} is the deflection determined by elastic analysis with seismic forces reduced by factor R/I_e. Using a value C_d equal to R would result in the displacement δ_x being equal to δ_{xe}, which would be consistent with the equal displacement rule. Empirical studies of single-degree-of-freedom oscillators show the average value of δ_x should be equal to or greater than δ_{xe} (Figure 11.12). For this reason, this book recommends against using values of C_d less than R.

Acceptance Criteria

Acceptance criteria are two-fold:

First, the force demands on individual components must not exceed reliable force capacities. For this purpose, the LRFD method is applied. See Section 1.5.3 for additional discussion on the LRFD method.

Structure	Risk Category		
	I or II	III	IV
Structures four stories or fewer with interior walls, partitions, ceilings, and exterior wall systems that have been designed to accommodate the story drifts	$0.025h_{sx}$[†]	$0.020h_{sx}$	$0.015h_{sx}$
All other structures	$0.020h_{sx}$	$0.015h_{sx}$	$0.010h_{sx}$

*Simplified presentation. See ASCE7 for complete information.
†h_{sx} is the story height below level x.

Table 11.8 Allowable Story Drift, Δ_a*

Second, story drift demands must not exceed allowable story drifts. The design story drift, Δ, is calculated as the difference of the deflections at the center of mass at the top and bottom of the story under consideration. According to ASCE 7, the design story drift Δ is not to exceed the allowable story drift, Δ_a, listed in Table 11.8 for any story. For special moment frames in Seismic Design Category D, E, or F, an additional requirement applies for structures with low redundancy.

Note that the design story drift Δ presented in the preceding paragraph is the drift for the Design Earthquake (DE). Using the equal displacement rule and the 1.5 factor implied by the reciprocal of the 2/3 factor in Eqs. (11.4) and (11.5), story drifts at the MCE_R level would be approximately 1.5Δ. Structural framing, including gravity framing, should be capable of supporting the applied loads as the structure is displaced to MCE_R drift levels.

Other Strength-Based Building Codes

Several other building codes follow similar practices. The *National Building Code of Canada* (2005) separates the response modification coefficient R into the product of two factors R_0 and R_d, where R_0 represents an estimate of the building overstrength and R_d can be considered a modification coefficient considering inelastic response. In the *Mexico Federal District Code* (2004) and *Eurocode* 8 (2004) the force reduction is period-dependent, decreasing with decreasing period [this trend partially offsets the displacement amplification observed in the short-period range (Figure 11.12)].

11.5.2 Displacement-Based Design

Overall Approach

Section 11.5.1 outlined the widely used design approach in which strength requirements are first determined followed by a check that drifts are within acceptable limits. An alternative approach is to select stiffness (and possibly other properties) required to achieve target drifts, and then to determine associated strength requirements. This approach is commonly known as displacement-based design. It is mainly suited to the design of buildings for which displacement response to earthquake loading is primarily in the first translational mode, which will include most low- and moderate-rise buildings without significant torsional response. With modifications, the method can be applied to buildings with other dynamic properties, but details are omitted here.

Different displacement-based designs have been used for seismic assessment of existing buildings (ASCE 41, 2013) and for seismic design of bridges (Caltrans, 2003). Similar procedures for buildings have not been widely codified. For this reason, the following text presents a very general overview of the concept of displacement-based design, without prescriptive procedures. For additional discussion, see Moehle (1992), Chopra and Goel (2001), and Priestley et al. (2007).

We begin by defining first-mode modal response quantities for the building. Although, strictly speaking, a structure responding nonlinearly does not have modal properties in the same way as does a linear structure, we here adopt the approximation that linear elastic modal properties apply for nonlinear response. Given a preliminary building design, structural analysis software can be used to generate the first-mode shape, or other approximate methods can be used. Figure 11.26 illustrates the terminology that describes the building and its mode shape, considering only a planar frame.

The story drift, Δ_j, for any story j can be calculated from the following:

$$\Delta_j = \Gamma(\phi_j - \phi_{j-1}) S_d \tag{11.22}$$

in which S_d is the design-level spectral displacement (e.g., Figure 11.9b) and the quantity Γ, sometimes referred to as the modal participation factor, is calculated as

$$\Gamma = \frac{\sum_{j=1}^{N} w_j \phi_j}{\sum_{j=1}^{N} w_j \phi_j^2} \tag{11.23}$$

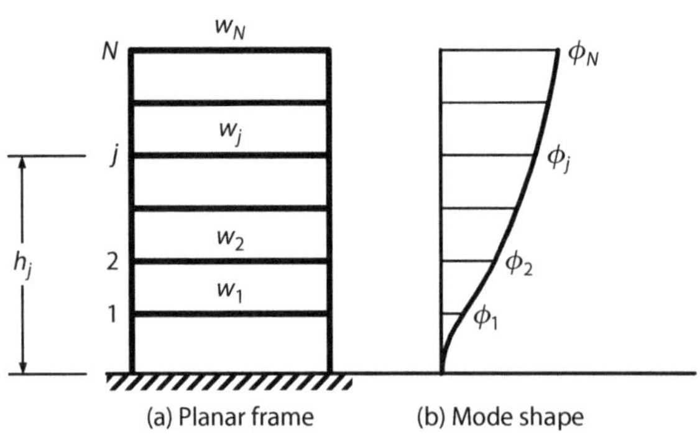

FIGURE 11.26 Approximate determination of first-mode displaced shape.

Note that Eq. (11.22) assumes that the drift of the inelastic system can be approximated as being equal to the drift of the elastic system, that is, the equal-displacement rule is being employed. Alternative assumptions or procedures can be adopted to relate the structural properties to expected nonlinear response.

The maximum calculated story drift from Eq. (11.22) should not exceed the allowable story drift Δ_a. Table 11.8, adopted from ASCE 7, provides a suitable definition of allowable story drift. For most buildings in Risk Category II, the allowable story drift will thus be $0.02h_{sx}$.

To solve the design problem, select a target story drift Δ_t not exceeding the allowable story drift Δ_a. Substituting Δ_t for Δ_j in Eq. (11.22) and defining $(\phi_j - \phi_{j-1})_{max}$ as the maximum value of $(\phi_j - \phi_{j-1})$, we can solve Eq. (11.22) for S_d as

$$S_d = \frac{\Delta_t}{\Gamma(\phi_j - \phi_{j-1})_{max}} \tag{11.24}$$

Using the displacement response spectrum, the corresponding vibration period, T, is identified. The building design must provide a period smaller than T if the drift target is to be achieved. Thus, the design problem becomes one of establishing a structural system having an appropriate fundamental vibration period.

Note that, according to the procedure outlined above, the required structure strength is not a direct outcome of the process. Instead, the strength is determined indirectly as an outcome of designing the structure to have the required stiffness. Alternatively, once a structural design with suitable period has been obtained, the strength-based procedure of Section 11.5.1 can be used to establish code-required strength or to establish strength as required for more frequent earthquakes.

Alternative approaches to identifying required properties have been proposed. For example, Chopra and Goel (2001) use inelastic response spectra instead of the elastic response spectrum. Priestley et al. (2007) use a secant stiffness approach.

Acceptance Criteria

Acceptance generally can be defined by three criteria:

1. The expected displacement of the structural system should be within the target value used to proportion the structural system. This should be a natural outcome of the process. Success of the process can be checked through nonlinear dynamic analysis.
2. Components should be detailed for the anticipated deformations, including anticipated inelastic deformations. Either nonlinear analysis can be done or simplified mechanism analysis can be used to explore inelastic deformation demands.
3. Components should be proportioned to have strength as required for the forces that develop as the structure is deflected to the target levels. The LRFD method can be used as a general guide to selecting load factors, load combinations, and strength reduction factors.

These criteria should be applied at appropriate hazard levels. Considering the dual objectives to have buildings that are both serviceable and safe, a suitable approach may be to apply criteria 1 and 2 at an appropriate serviceability level and criteria 1 through 3 at the MCE level.

11.5.3 Performance-Based Design

In its most basic form, performance-based design involves design to achieve target performance objectives rather than adherence to prescriptive strength or drift requirements. Although the performance metrics, their measurement, and the performance objectives traditionally have not been explicitly considered by the structural engineer, guidelines for implementation have been recently developed. Here we adopt the performance-based guidelines of the PEER Tall Buildings Initiative (TBI, 2010).

Design Approach

Performance-based approaches are not generally accepted by building officials. Thus, prior to embarking on a performance-based design, the structural engineer should ascertain that the building official is amenable to performance-based design alternatives. In addition, the structural engineer should assure that the development team is aware of and accepts the risks associated with the use of alternative design procedures, that the engineer has the appropriate knowledge and resources, and that construction quality will be adequate to ensure that the design is properly executed.

Performance Objectives

The structural engineer should work with the development team and the building official to establish the performance objectives. In general, these should be consistent with the building Risk Category (Table 1.1) and Seismic Design Category (Tables 11.5 and 11.6). Normally, the performance objectives will include at least two objectives: (1) essentially elastic response and limited damage under Service Level Earthquake (SLE) shaking and (2) stable response under MCE shaking without loss of gravity-load-carrying capacity, without inelastic straining of important lateral-force-resisting elements to a level that will severely degrade their strength, and without experiencing excessive permanent lateral drift or development of global structural instability. For Risk Category II structures, a typical selection is SLE shaking having a return period of 43 years (50% exceedance probability in 30 years; see Table 11.1), and MCE shaking consistent with the MCE_R definition in ASCE 7. If enhanced performance objectives are desired, the engineer should develop a formal Design Criteria document that defines the enhanced objectives.

Seismic Input

Seismic input is in the form of earthquake base motions. The motions should be selected and scaled to be compatible with the seismic hazard and geologic conditions at the site. Motions are sometimes scaled such that the average response spectral ordinates approximate target response spectra. In other projects, the motions will be spectrum matched to the target spectrum. The target spectrum can be either a uniform hazard spectrum or a conditional mean (or scenario) response spectrum. Decisions on these aspects depend on the nature of the design problem. In current practice, the earthquake base motions usually consider only the horizontal components, but vertical components should be considered under special circumstances.

Conceptual Design

In this step, the engineer selects the structural systems and materials, their approximate proportions, and the intended primary mechanisms of inelastic behavior. The engineer should use capacity design principles to establish the target inelastic mechanisms. Where possible, the structural configuration should follow a form that simplifies analysis and increases confidence in analysis results. Simple, continuous load paths should be achieved where possible.

Design Criteria

The structural engineer of record should develop a formal Design Criteria document that describes the structural systems and materials of construction, the anticipated mechanisms of inelastic response and behavior, the design performance objectives, the specific design and analysis measures to be conducted to demonstrate acceptable performance capability, and all exceptions to the prescriptive provisions of the building code. This Design Criteria document should be submitted to and approved by the authority having jurisdiction and third-party reviewers prior to undertaking substantial design effort.

Preliminary Design

Dynamic structural analysis is used to confirm that building designs are capable of meeting the intended performance objectives. To perform a meaningful analysis, the engineer must develop the building design to a sufficient level of detail to allow determination of the distribution of its stiffness, strength, and mass, as well as the hysteretic properties of elements that will undergo inelastic straining in response to strong ground shaking. ATC 72 (2010) provides useful guidance for defining component properties.

Service Level Evaluation

The Service Level Evaluation is intended to demonstrate that the building will be capable of withstanding relatively frequent, moderate-intensity shaking with limited structural damage. Usually, the damage should (a) not interrupt occupancy and function; (b) not require repair, except perhaps for cosmetic reasons; and (c) not impair capacity to resist future SLE or MCE shaking with satisfactory performance. This evaluation can be done using either linear-elastic analysis or using nonlinear analysis with limited inelastic response permitted in ductile components.

Maximum Considered Earthquake Evaluation

The Maximum Considered Earthquake Evaluation is intended to demonstrate that the building will be capable of withstanding MCE shaking without collapse. Nonlinear dynamic analysis is required for this evaluation. Current software does not permit reliable simulation of structural collapse for most structural systems. Therefore, the analysis is carried out to demonstrate stable response with some margin relatively to collapse. See TBI (2010) for accepted approaches.

Final Design

The final design should be documented by the construction documents, including detailed drawings and specifications, supported by extensive calculations and analyses. The design process may generate large amounts of data. These should be organized in a manner that facilitates review by building departments and third-party reviewers.

Peer Review

Independent, third-party review should include the design criteria; seismic hazards analysis; selection and scaling of ground motions; proportioning, layout, and detailing of the structure; and modeling, analysis, and interpretation of results.

11.6 Chapter Review

This chapter reviewed the nature of earthquakes and the general requirements for the design of buildings to resist earthquake effects. We first noted that earthquake ground shaking is the main cause of earthquake damage and, therefore, is the main focus of design provisions in building codes. For buildings located in regions of highest seismicity, the design-level shaking intensity exceeds the linear-elastic resistance of conventionally designed structures. One option, which is widely adopted, is to admit inelastic response as part of the design approach. The inelastic demands on a structure so designed can be estimated based on considerations of the stiffness, strength, and configuration of the structural framing. Structural behavior concepts and structural design procedures based on these considerations were presented in this chapter. The next chapters will extend the general information of this chapter to develop specific design requirements for structural elements of earthquake-resistant buildings.

References

ASCE 7 (2010). *Minimum Design Loads for Buildings and Other Structures,* American Society of Civil Engineers, Reston, VA.

ASCE 41 (2013). *Seismic Evaluation and Retrofit of Existing Buildings,* American Society of Civil Engineers, Reston, VA.

ATC 72 (2010). *Modeling and Acceptance Criteria for Seismic Design and Analysis of Tall Buildings,* Applied Technology Council. Also available as PEER Report No. 2010/111 at http://peer.berkeley.edu/tbi/wp-content/uploads/2010/09/PEER-ATC-72-1_report.pdf, 222 pp.

Baker, J.W., and C.A. Cornell (2006). "Spectral Shape, Epsilon and Record Selection," *Earthquake Engineering & Structural Dynamics*, Vol. 35, No. 9. pp. 1077–1095.

BISPEC (2009). Mahmoud M. Hachem, http://www.ce.berkeley.edu/~hachem/bispec. Version 1.62.

Bolt, B.A. (2004). "Engineering Seismology," in *Earthquake Engineering: From Engineering Seismology to Performance-based Engineering,* Y. Bozorgnia and V. Bertero, eds. CRC Press, pp. 24–74.

Bozorgnia, Y., N.A. Abrahamson, L. Al Atik, T.D. Ancheta, G.M. Atkinson, J.W. Baker, A. Baltay, et al. (2014). "NGA-West2 Research Project," *Earthquake Spectra*, Vol. 30.

Bozorgnia, Y., and K.W. Campbell (2004). "Engineering Characterization of Ground Motion," in *Earthquake Engineering: From Engineering Seismology to Performance-based Engineering*, Y. Bozorgnia and V. Bertero, eds. CRC Press, pp. 215–315.

Caltrans (2003). *Bridge Design Specifications, Section 4—Foundations*, California Department of Transportation, 70 pp.

Campbell, K.W., and Y. Bozorgnia (2014). "NGA-West2 Ground Motion Model for the Average Horizontal Components of PGA, PGV, and 5%-Damped Linear Acceleration Response Spectra," *Earthquake Spectra*, Vol. 30.

Chock, G. (2012). "ASCE 7 and the Development of a Tsunami Building Code for the U.S.," *Proceedings, 14th U.S.-Japan Workshop on the Improvement of Structural Design and Construction Practices*, Applied Technology Council, Redwood City, California.

Chopra, A.K. (2012). *Dynamics of Structures: Theory and Applications to Earthquake Engineering*, 4th ed., Prentice Hall, Englewood Cliffs, New Jersey, 992 pp.

Chopra, A.K., and R.K. Goel (2001). "Direct Displacement-Based Design: Use of Inelastic Design Spectra Versus Elastic Design Spectra," *Earthquake Spectra*, Vol. 17, No. 1, pp. 47–64.

EERI (1996). "Northridge Earthquake of January 17, 1994 Reconnaissance Report, Volume 2," *Earthquake Spectra*, Suppl C to Vol. 11, January 1996.

Eurocode 8 (2004). *Eurocode 8: Design of Structures for Earthquake Resistance, Part 1, General Rules, Seismic Actions and Rules for Buildings*, Comité Européen de Normalisation, European Standard EN 1998-1:2004, Brussels, Belgium.

FEMA 440 (2005). *Improvement of Nonlinear Static Seismic Analysis Procedures*, Federal Emergency Management Agency, Washington, D.C., 392 pp.

Kramer, S.L. (1996). *Geotechnical Earthquake Engineering*, Prentice-Hall, Englewood Cliffs, New Jersey, 653 pp.

Luco, N., B.R. Ellingwood, R.O. Hamburger, J.D. Hooper, J.K. Kimball, and C.A. Kircher (2007). "Risk-Targeted versus Current Seismic Design Maps for the Conterminous United States," *Proceedings, SEAOC 2007 Convention, Structural Engineers Association of California*, 13 pp.

Mexico Federal District Code (2004). "Complimentary Technical Norms for Seismic Design," Government of the Federal District, Official Gazette of the Federal District, October 6, 2004, Mexico.

Moehle, J.P. (1992). "Displacement-Based Design of RC Structures Subjected to Earthquakes," *Earthquake Spectra*, Vol. 8, No. 3, pp. 403–428.

National Building Code of Canada (2005). *The National Building Code of Canada*, Canadian Commission on Building and Fire Codes, National Research Council, Ottawa.

NEHRP (2009). *NEHRP Recommended Provisions (and Commentary)*, National Earthquake Hazard Reduction Program, FEMA P-750, Federal Emergency Management Agency, Washington, D.C., 388 pp.

Priestley, M.J.N., G.M. Calvi, and M.J. Kowalsky (2007). *Displacement-Based Seismic Design of Structures*, IUSS Press, Pavia, Italy, 721 pp.

SEAOC (2007). "Development of System Factors," *The SEAOC Blue Book: Seismic Design Recommendations*, Structural Engineers Association of California, Sacramento, CA. Available at http://www.seaoc.org/bluebook/index.html

TBI (2010). *Guidelines for Performance-Based Seismic Design of Tall Buildings*, Report No. 2010/05, Pacific Earthquake Engineering Research Center, University of California, Berkeley, CA, 84 pp.

USGS (2014). "Custom Mapping and Analysis Tools," US Geological Survey. Available at http://earthquake.usgs.gov/hazards/apps/.

CHAPTER 12
Special Moment Frames

12.1 Preview

Buildings resist earthquake effects through a combination of structural diaphragms, vertical framing elements, and the foundation. In reinforced concrete buildings, the vertical elements are usually either moment-resisting frames or structural walls. This chapter addresses behavior and design requirements for moment-resisting frames intended for use in regions of highest seismicity. In the United States, these are referred to as *special moment frames.*

Special moment frames comprise beams (with or without slabs), columns, and beam-column joints. These are proportioned and detailed to resist combinations of shear, moment, and axial force that occur as a building responds to strong earthquake ground shaking. A special moment frame can be reinforced with conventional nonprestressed reinforcement, prestressed reinforcement, or a combination of both. In this book only the conventionally reinforced frame is considered. *Intermediate moment frames* are also defined in the building codes, but these are not permitted in regions of highest seismicity, except in some lower-rise buildings where they may be permitted in combination with special structural walls. That application is not widespread, and is not covered here.

The chapter begins with a review of how special moment frames are used in modern buildings. It then describes the behavior of moment frames under strong earthquake shaking, and sets forth a set of principles for their design. Latter portions of the chapter describe in detail the analysis, design, and construction of special moment frames. Although the design requirements are drawn mainly from ACI 318, provisions of other building codes are introduced where applicable.[1]

[1]Content in some sections is derived from Moehle et al. (2008).

12.2 The Use of Special Moment Frames

Reinforced concrete special moment frames are used as part of the seismic-force-resisting system in buildings designed to resist earthquakes. Beams, columns, and beam-column joints are proportioned and detailed to resist flexural, axial, and shearing actions that result as a building sways through multiple displacement cycles during strong earthquake ground shaking. Special requirements for materials, member proportioning, detailing, and construction and inspection result in a frame capable of resisting strong earthquake shaking without critical loss of stiffness or strength.

Most special moment frames use cast-in-place, normalweight concrete having rectilinear cross sections without prestressing. This form of special moment frame will be considered the general case, and is the main emphasis of this chapter. It is also possible, however, to construct special moment frames using lightweight concrete, prestressed beams, spiral-reinforced columns, and precast concrete. These latter subjects are special cases, with only limited discussion in this chapter.

12.2.1 Historic Development

Reinforced concrete special moment frame concepts were introduced in the United States starting around 1960 (Blume et al., 1961). Their use at that time was essentially at the discretion of the designer, as it was not until 1973 that the Uniform Building Code (UBC, 1973) first required use of the special frame details in regions of highest seismicity. The earliest detailing requirements have many similarities to those in use today, though there are notable differences.

In most early applications, where special moment frames were used in a building they were used in all framing lines of the building. A trend developing in the United States through the 1990s and continuing today is to use special moment frames in fewer framing lines of the building, with the remainder comprising gravity-only framing that is not designated as part of the seismic-force-resisting system. Some of these gravity-only frames did not perform well in the 1994 Northridge earthquake, leading to more stringent requirements for proportioning and detailing of those frames. Analysis and design requirements for members not designated as part of the seismic-force-resisting system are presented in Chapter 14. The detailing requirements for the gravity-only elements may approach those for the special moment frame, such that it may be more economical to include those components in the seismic-force-resisting system if they can be made to satisfy all the applicable requirements.

Special moment frames have also found use in dual systems that combine special moment frames with shear walls or braced frames. In current U.S. codes, if a seismic-force-resisting system is designated as a dual system it is required that the moment frame be capable of resisting at least 25% of the design seismic forces, while the total seismic resistance is provided by the combination of the moment frame and the shear walls or braced frames in proportion with their relative stiffnesses. The dual system is not widely used in current practice, although there are some special situations where the dual system is used. See Section 12.2.2.

12.2.2 When to Use Special Moment Frames

Moment frames are generally selected as the seismic-force-resisting system when architectural program flexibility is desired. When concrete moment frames are selected for buildings assigned to the highest seismic design categories (Seismic Design Categories D, E, or F in the United States), they are required to be detailed as special moment frames. The special proportioning and detailing requirements for a special moment frame will enable the frame to safely undergo extensive inelastic deformations that are anticipated in these seismic design categories. Special moment frames may be used in lower seismic design categories (Seismic Design Categories A, B, and C in the United States), thereby allowing use of lower seismic design forces. This option is seldom used because it usually does not lead to the most economical design. Regardless of where they are used, if special moment frames are selected as the seismic-force-resisting system, all requirements for the frames must be satisfied to promote ductile behavior.

Dual systems combining walls or braced frames with special moment frames are seldom used, although there are some exceptions. The first is for tall buildings. Some building codes (including ASCE 7 in the United States) limit the height of certain seismic-force-resisting systems such as special reinforced concrete shear walls when such systems provide the entire seismic force resistance (see Table 11.7). These height limits do not apply when special moment frames are added to create a dual system. The second use of dual systems is where buildings are constructed on poor soils requiring expensive foundations. By using a dual system rather than a special shear wall without frames, the design forces may be reduced (in ASCE 7 the response modification coefficient R is 7 for a dual system compared with 5 or 6 for the shear wall system). The resulting savings in the foundation costs may more than offset the extra cost of the added frames.

12.2.3 Frame Layout and Proportioning

Typical economical beam spans for special moment frames are in the 20 to 30 ft (6 to 9 m) range. This range will generally result in beam depths that will support typical gravity loads and the requisite seismic forces without overloading the adjacent beam-column joints and columns. Special moment frames with story heights up to 20 ft (6 m) are not uncommon. For buildings with these relatively tall stories, it is important to make sure that soft and/or weak stories are not created.

The building code may also have limits on aspect ratios of the framing members (Figure 12.1). In U.S. codes, the clear span of a beam must be at least four times its effective depth; this requirement promotes stable flexural response with conventional details. Minimum beam width is $0.3h_b$, but not less than 10 in (250 mm). Beams can be wider than the supporting columns, within limits related to moment transfer at the beam-column connection, but constructability considerations normally dictate that beam width not exceed the column width. Slab-column framing (without beams) is generally not permitted as part of the seismic-force-resisting system because of excessive flexibility and problematic shear and moment transfer at connections.

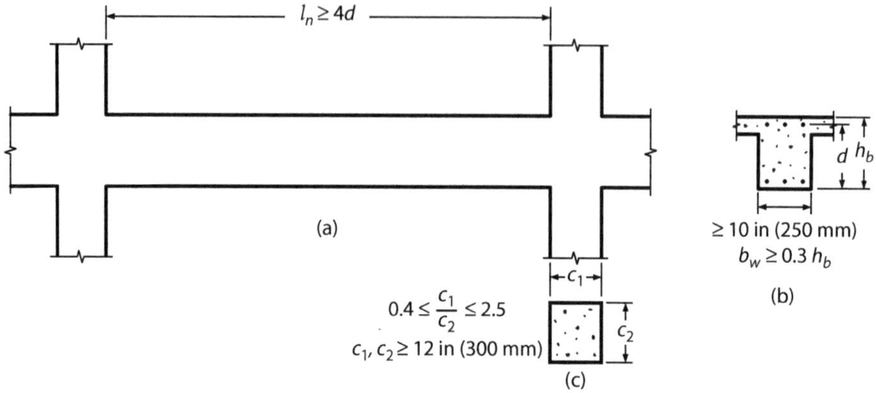

Figure 12.1 Dimensional limits of beams and columns of special moment frames according to ACI 318: (*a*) beam span; (*b*) beam cross section; (*c*) column cross section. The building code permits beams wider than columns within some limits, not shown.

In the United States, the ratio of the cross-sectional dimensions for columns is limited to $0.4 \leq c_1/c_2 \leq 2.5$. This limits columns to members with compact cross sections; columns having dimensions outside this range have to be detailed as walls or wall piers (see Chapter 13). The U.S. code minimum column dimension is 12 in (300 mm), which is often not practical to construct. A 16-in (400-mm) minimum dimension is suggested except for unusual cases or for low-rise buildings.

Both strength and stiffness need to be considered for the design of a special moment frame. In the United States, special moment frames are allowed to be designed for a force reduction factor of $R = 8$. That is, they are allowed to be designed for a base shear equal to one-eighth of the value obtained from an elastic response analysis. Therefore, where strength is controlled by this limit, extensive inelastic response might be anticipated for design-level earthquake shaking.

Some building codes impose a minimum base-shear strength that controls the design of buildings with longer vibration periods such as tall moment frame buildings. Wind loads must also be checked and may govern the strength requirements of special moment frames. Regardless of whether gravity, wind, or seismic forces are the largest, the proportioning and detailing provisions for special moment frames apply wherever special moment frames are used.

Additionally, the stiffness of the frame must be sufficient to control the story drifts to acceptable limits. In the United States, allowable drifts under required seismic design forces are specified in the building code (see Table 11.8). Wind drift limits are not specified in U.S. codes; therefore, engineering judgment is required to determine the appropriate limit. Consideration should be given to the attachment of the cladding and other elements, and to the comfort of the occupants.

12.3 Principles for Design of Special Moment Frames

We previously have noted that special moment frames are typically designed for base-shear strength that is considerably less than the strength required for fully elastic response under design-level ground shaking. Thus, we should anticipate that a building frame will respond to that ground shaking with response well into the inelastic range. Experience from past earthquakes, laboratory testing, and analytical studies suggests that we must apply certain design principles to ensure that the frame can sustain the anticipated inelastic response without critical decay in resistance. These principles are as follows:

1. Design a strong-column/weak-beam system.
2. Detail beams and columns for ductile flexural response.
3. Avoid more brittle failure modes such as shear, axial, connection, and splice failures.
4. Avoid interaction with nonstructural components.

These four principles are discussed more fully in the following sections.

12.3.1 Design a Strong-Column/Weak-Beam System

When a building sways during an earthquake, the distribution of damage over height depends on the distribution of lateral drift. If the building has weak columns, drift tends to concentrate in one or few stories (Figure 12.2*a*), and may exceed the drift capacity of the columns. On the other hand, if columns provide a stiff and strong spine over the building height, drift will be more uniformly distributed (Figure 12.2*c*), and localized damage will be reduced. Additionally, it is important to recognize that the columns in a given story support the weight of the entire building above those columns, whereas the beams only support the gravity loads of the floor of which they form a part; therefore, failure of a column is of greater consequence than failure of a beam. Recognizing this behavior, building codes specify that the columns be stronger than the beams that frame into them. This *strong-column/weak-beam principle*

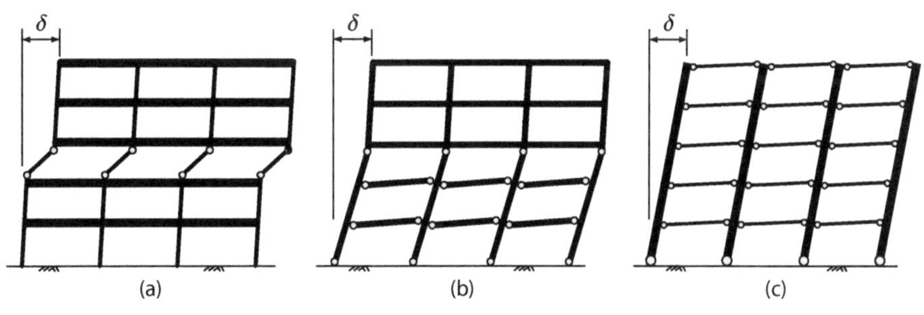

FIGURE 12.2 Idealized beam-yielding mechanisms: (*a*) story mechanism; (*b*) intermediate mechanism; (*c*) beam mechanism.

is fundamental to achieving safe behavior of frames during strong earthquake ground shaking.

Studies show that if the column moment strength exceeds the beam moment strength at every connection by a small margin, it is likely that the building will develop yielding mechanisms that involve several stories (Figure 12.2*b*) rather than a single story (Figure 12.2*a*). Achieving a complete beam mechanism (Figure 12.2*c*) may require column moment strengths several times beam moment strengths, increasingly so for taller buildings, which may prove uneconomical. Therefore, some yielding of the columns has to be anticipated, and reinforcement details consistent with this anticipated behavior must be provided. This subject is discussed further in Section 12.4.2.

Many older buildings (roughly prior to 1980 in the United States) in highly seismic zones were built without adhering to the strong-column/weak-beam design concept, and these may be vulnerable to weak story collapse during strong ground shaking. Figure 12.3 illustrates collapse of weak columns in a mid-story of a mid-rise building in which columns were weaker than adjacent beams.

12.3.2 Detail Beams and Columns for Ductile Flexural Response

The intended yield mechanism involves yielding of the beams throughout the height of the structure plus the columns at the base. Realistically, however, some column yielding along the height of the structure also has to be anticipated unless the columns are much stronger than the beams (see Section 12.4.2). Therefore, we

Figure 12.3 Weak-story failure in the Hyogo-ken Police Department Building, Kobe earthquake, 1995. Note complete collapse of third story, partial collapse of fourth story. (*Photo by T. Kabeyasawa.*)

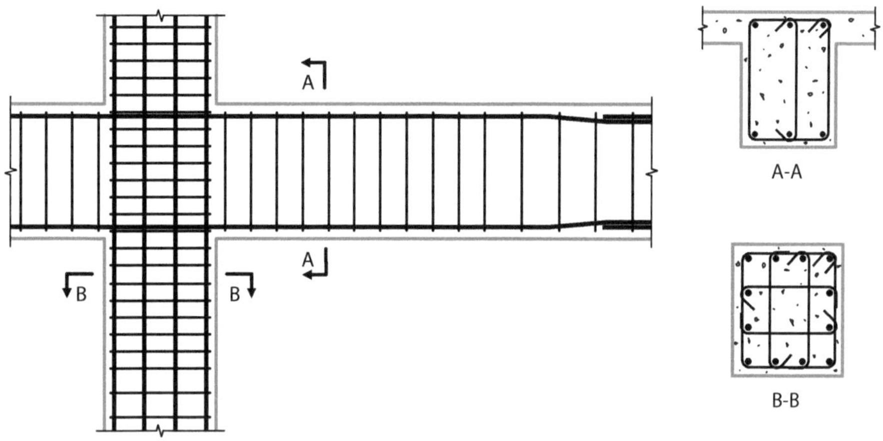

Figure 12.4 Confinement of potential yielding regions at ends of beams and columns.

should detail the end regions of the beams and columns at every beam-column joint so these regions can undergo inelastic flexural response without critical strength decay. Beam and column longitudinal reinforcement should be continuous through the joints without splices, unless such splices are proven to be capable of sustaining multiple post-yielding cycles. Transverse reinforcement should confine the core concrete and provide restraint against buckling of the longitudinal reinforcement. This transverse reinforcement should extend from the joint face along a length that will envelope the likely yielding region at the end of beams and columns. Figure 12.4 illustrates the typical required details near a beam-column connection.

12.3.3 Avoid Nonductile Failure Modes

The design should include proportions (member sizes and corresponding strengths) and reinforcement details to promote the formation of the intended flexural yielding mechanisms without failure in less ductile modes such as shear, axial, and connection failures. Normally, we will determine required proportions using a capacity design approach (Section 1.5.4). By this procedure, we first identify the flexural yielding mechanism, design-yielding cross sections for code-required moment strengths, and then calculate design actions on capacity-protected members assuming the flexural yielding regions develop probable moment strengths. The probable moment strength is calculated using procedures that produce a high estimate of the moment strength of the as-designed cross section. The following paragraphs describe some of the more important design considerations.

Column and Beam Shear

Shear failure is relatively brittle and can lead to rapid loss of shear resistance and, in the case of columns, axial load capacity. Strength degradation associated with shear failure of one or more columns in a story can lead to a story mechanism, further exacerbating the problem. Column shear failure is the most frequently cited cause of concrete building failure and collapse in earthquakes (Figure 12.5).

(a) (b)

Figure 12.5 Building damage associated with column shear failure: (*a*) shear and axial failure of middle column of multi-story building, Izmit, Turkey earthquake, August 17, 1999. (*Photograph by H. Sezen, used with permission from the National Information Service for Earthquake Engineering, University of California, Berkeley.*) (*b*) Interior view of exterior column showing shear damage and initiation of axial failure, Van Nuys Holiday Inn, Northridge, California earthquake, January 17, 1994. (*Used with permission from Earthquake Engineering Research Institute.*)

Chapter 7 discussed shear strength degradation under inelastic deformation reversals. Design procedures should take this degradation into account. In U.S. design practice, this is accomplished by assuming the contribution of concrete to shear resistance is negligible in the design equations for beams and columns with low axial load. The effect of assuming $V_c = 0$ in these members is that additional shear reinforcement is required to resist the entire shear force.

Sections 12.6.3 and 12.6.4 discuss the shear design approach for beams and columns more thoroughly.

Column Axial Load

Column axial failure can trigger progressive collapse in which axial loads from the overloaded column are transferred to adjacent columns, potentially overloading them in turn and leading to collapse of an entire story or building. We can estimate column axial loads if we know the building weight and the lateral yielding mechanism of the building (Figure 12.6*a*). The problem of estimating column axial loads is complicated by time variations in the lateral force distribution and

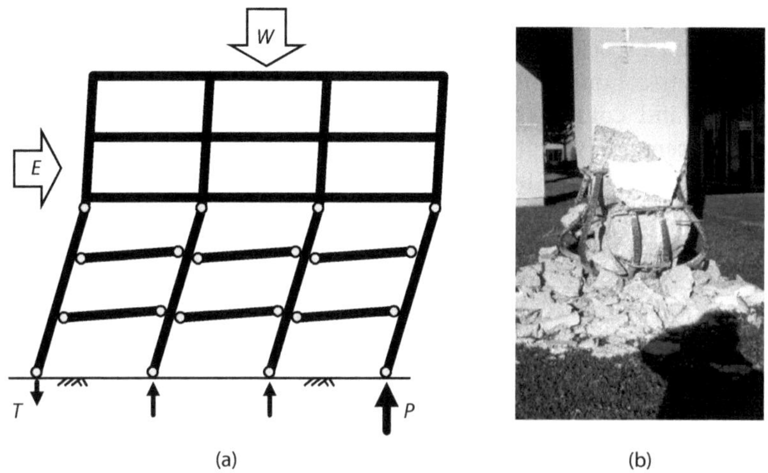

Figure 12.6 Column axial loads: (*a*) development of column axial loads under the frame yielding mechanism; (*b*) compressive failure of column supporting stiff wall in one direction and frame overturning in the orthogonal direction, Imperial County Services Building, Imperial County, California earthquake, 1979. (*Photograph by K. Steinbrugge, used with permission from the National Information Service for Earthquake Engineering, University of California, Berkeley.*)

yielding mechanism, as well as the effects of multi-directional earthquake input motion. The problem is discussed more fully in Section 12.4.3.

Column axial load failures may be associated with other actions, especially shear overloading (Figure 12.5). They are not uncommon in columns supporting discontinuous walls, including both reinforced concrete walls and walls formed by infilling bays of moment frames above the overloaded column (Figure 12.6*b*).

Connections

Seismic-force-resisting systems must provide continuous load paths through structural members and their interconnections. In reinforced concrete special moment frame construction, we are concerned with connections between horizontal and vertical elements of the seismic-force-resisting system (Chapter 15), as well as splices, anchors, and beam-column joints within the special moment frame.

Reinforcement splices include lap splices, mechanical splices, and welded splices. Lap splices rely partly on the confinement of cover concrete to develop their strength, and therefore they may be susceptible to concrete cover spalling associated with inelastic behavior of a member. Lap splices and mechanical splices have cross sections exceeding the cross section of the spliced reinforcement; this results in disruption of the reinforcement stresses within the splice length and can cause strain concentrations in the spliced reinforcement. Furthermore, some types of splices are sensitive to effects of inelastic cyclic loading. For these reasons, building codes may have restrictions on the use of splices within intended yielding regions of a special moment frame.

Beam-column joints in special moment frames transfer forces between beams and columns and serve to anchor beam and column longitudinal reinforcement (Chapter 9). They generally are regions of high stress under earthquake loading. Beam-column joints are especially vulnerable to failure at the perimeter of buildings because exterior faces are not confined by adjacent concrete framing members. Transverse reinforcement is required in special moment frame joints to confine the joint concrete and participate in the resistance of joint forces. Building codes typically require joint transverse reinforcement similar to that required in the end regions of the columns (Figure 12.4).

Figure 12.7 illustrates two examples where connection failures contributed to damage and collapse of buildings during earthquakes.

12.3.4 Avoid Interaction with Nonstructural Components

The building frame has to be designed and constructed so that it is free to sway without interference from stiff nonstructural elements. This can be especially problematic for building frames because of their relative flexibility. Infill panels, stairways, and ramps in parking structures are examples of nonstructural components that commonly interfere with the free movement of a moment frame. Figure 12.8*a* shows an example

(a) (b)

Figure 12.7 Connection failures in building frames: (*a*) apparent failure of joints and splices in the collapsed Juarez Hospital, Mexico City, Mexico earthquake, 1985 (*photograph by M. Sozen*); (*b*) joint failures in the partially collapsed Kaiser Permanente office building, Granada Hills, Northridge, California earthquake, 1994 (*used with permission from the National Information Service for Earthquake Engineering, University of California, Berkeley*).

FIGURE 12.8 Interference between nonstructural components and the structural frame: (*a*) column failure due to interference with stair landing at its mid-height, converting it into a short column with shear twice that computed neglecting the stairway, Medical Clinic, El Asnam, Algeria earthquake, 1980; (*b*) captive column restrained by partial-height masonry infill, Managua, Nicaragua earthquake, 1972. Note that the column shear in the example shown would be the sum of the moment strengths at the column ends divided by the effective clear height of the column. This shear may be much higher than the shear anticipated in the absence of the masonry infill. (*Photographs by V. Bertero, used with permission from the National Information Service for Earthquake Engineering, University of California, Berkeley.*)

where a stair landing supported by frame columns reduced the column clear height by half, thereby making the column more vulnerable to shear failure.

Concrete frame buildings sometimes have partial-height masonry infills or partial-height concrete walls to form window openings or passageways. If the infill is not adequately separated from the concrete framing, it can restrain movement of a column, thereby reducing its effective length. Figure 12.8*b* shows an example in which masonry infill restrained movement of a frame column. Columns restrained in this manner are known as *captive columns*. Captive columns have caused building damage and collapse in earthquakes worldwide, and should be avoided through appropriate separation of the nonstructural and structural framing elements.

12.4 Seismic Response of Special Moment Frames

Section 12.3 outlined the general principles for design of special moment frames, and showed earthquake reconnaissance examples where failure to follow those principles resulted in building damage and collapse. In this section we further examine the seismic response of special moment frames to gain a broader understanding of the design requirements.

12.4.1 Observations on Dynamic Response

Many buildings using well-designed special moment frames have been subjected to earthquakes, with generally positive performance observations. Models of special moment frames have also been studied analytically and in the laboratory. The following paragraphs illustrate some general aspects of the seismic response of special moment frames using results from a shaking table experiment (Shahrooz and Moehle, 1987).

The shaking table specimen was a one-quarter-scale model of a six-story special moment frame building with setback at mid-height (Figure 12.9). The

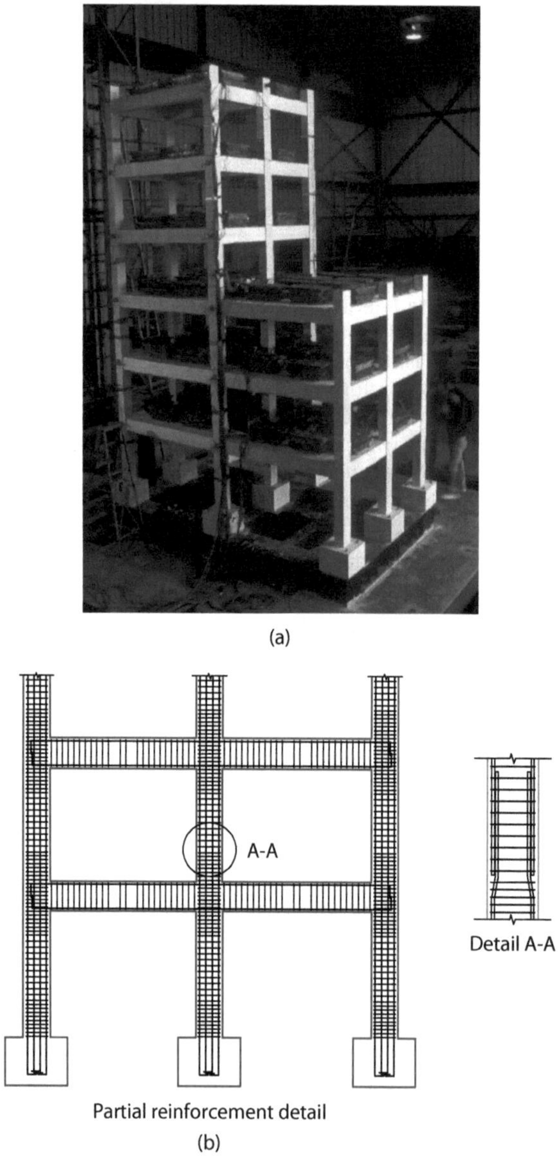

Figure 12.9 Shaking table test structure: (*a*) photograph of test frame on shaking table; (*b*) representative beam, column, and joint details.

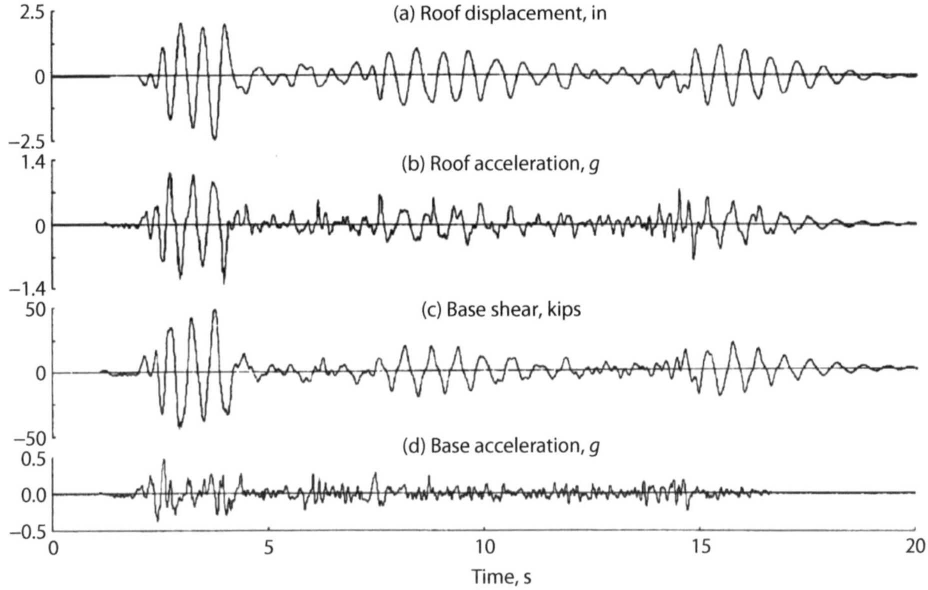

Figure 12.10 Test structure response histories: (*a*) roof relative displacement; (*b*) roof absolute acceleration; (*c*) base shear; (*d*) base acceleration.

prototype frame satisfied requirements of the Uniform Building Code (UBC, 1983) and ACI 318-83 (1983), including seismic design requirements. As was typical at that time, the contribution of the slab acting as a tension flange of the beam was not considered in calculating beam strengths. The test model was subjected to uni-directional scaled earthquake motions parallel to the setback direction, resulting in peak story drift ratio as large as 0.016. The model was subsequently subjected to uni-directional motion oriented 45° relative to the setback direction, resulting in peak story drift ratio in the setback direction as large as 0.033. This compares with maximum allowable drift ratio 0.02 in ASCE 7 for Design Earthquake (DE)-level motions.

Figure 12.10 plots the input base acceleration and resulting roof displacement, roof acceleration, and base-shear histories for the largest uni-directional shaking test parallel to the setback. The input base acceleration traces the NS motion recorded at El Centro in the 1940 Imperial Valley earthquake (Figure 11.1), but with peak base acceleration 0.49*g*. Characteristic of moderate-rise buildings, the roof displacement trace is dominated by apparent first-mode response, whereas roof acceleration and base-shear traces display apparent higher-mode contributions.

Figure 12.11*a* and *b* plots the displaced shape and floor-level accelerations at the time of the peak positive roof displacement. The displaced shape is dominated by the apparent fundamental mode, while the acceleration profile shows some apparent higher-mode contribution (the sign of accelerations changes over height).

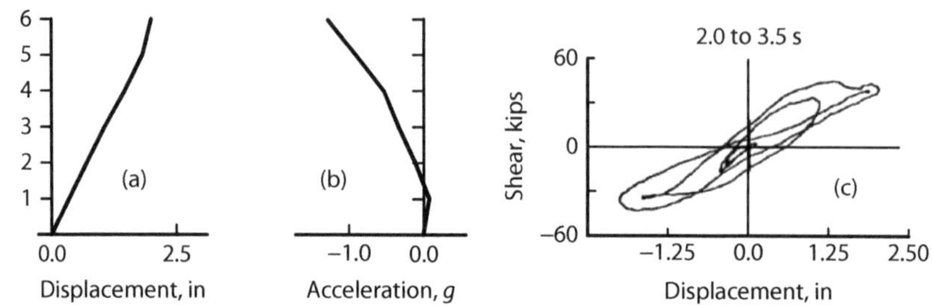

FIGURE 12.11 Test structure response: (*a*, *b*) relative displaced shape and absolute accelerations at $t = 3.98$ s; (*c*) relation between base shear and roof displacement from $t = 2.0$ s to $t = 3.5$ s.

Figure 12.11*c* plots the relation between base shear and roof displacement. The relation is complicated by contributions from higher-mode response, but clearly shows nonlinearity indicative of inelastic response. Inelastic response would be expected to result in elongation of the apparent fundamental vibration period. This can be seen in the roof displacement trace (Figure 12.10). (Vibration period can be approximated by the time between zero crossings, which is shorter around $t = 3$ s than it is at $t = 10$ s.)

Figure 12.12 plots strain histories at selected locations. Note that the gauges are located such that lateral load from left to right produces flexural tension simultaneously at gauge locations a, b, and c. At the time of peak strain in the beam bottom longitudinal reinforcement (gauge b), the strain in the beam top longitudinal reinforcement (gauge a) is relatively small, while that in the column

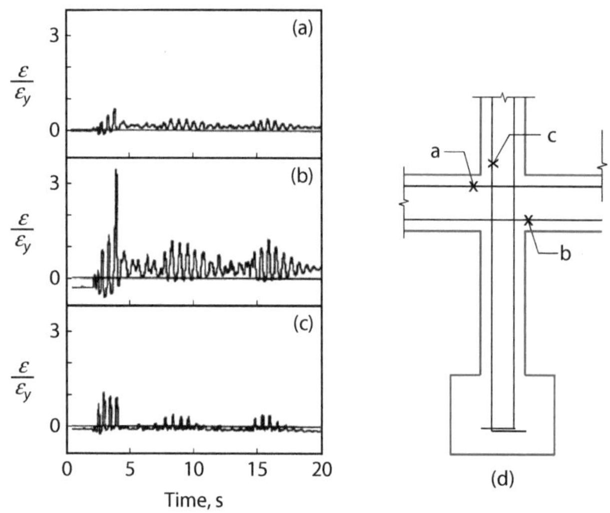

FIGURE 12.12 Test structure response: (*a*) strain history for top beam bar; (*b*) strain history for bottom beam bar; (*c*) strain history for column bar; (*d*) locations of strain gauges.

longitudinal reinforcement is slightly past the yield point (gauge c). Apparently, the beam reached its positive moment strength but not its negative moment strength during this test. The reason is that the slab reinforcement acted as a tension flange for the beam, significantly increasing the negative moment strength. The higher beam moments cause an increase in the column moment and yielding of the column longitudinal reinforcement.

In subsequent tests, the base motion intensity was increased, resulting in increased lateral drift and increased damage. Damage consisted of extensive concrete cracking; cover concrete spalling in beams, columns, and joints; and extensive yielding of column longitudinal reinforcement, suggesting that strong-column/weak-beam behavior was not fully realized. Nonetheless, overall behavior showed distributed yielding without signs of localized distress during even the strongest shaking tests.

12.4.2 Frame Yielding Mechanisms

In the previous section we observed column yielding in a special moment frame even though it had been designed following the strong-column/weak-beam design philosophy. This section identifies some of the reasons why this can occur and how this can affect seismic performance.

Beam Flexural Overstrength

Column yielding in a special moment frame can be partly attributed to beam flexural overstrength. Flexural overstrength arises due to (1) oversizing of the beam during the design process; (2) the contribution of developed slab reinforcement to be moment strength; and (3) material overstrength and strain-hardening. Normally, overstrength of the materials in the columns will be similar to that in the beams, partially offsetting this effect. However, some column yielding should be anticipated as the beam reinforcement is strained well into the strain-hardening range.

Plastic and Collapse Mechanisms

Even if beam overstrength was negligible (which generally is not the case), column yielding along the building height would be unavoidable unless the columns were made much stronger than the beams. We can demonstrate this using limit analysis of a frame under lateral loading. Consider the frame shown in Figure 12.13*a*. Impose a virtual displacement in a kinematically acceptable yield mechanism such as the one shown in Figure 12.13*b*. For this example, we will consider only virtual mechanisms extending from the base to some upper level; we could study other mechanisms with some additional effort but the conclusion will not be affected. Virtual work is done by the external forces acting through virtual lateral displacements and by internal moments acting through the hinge virtual rotations. The correct solution is the one corresponding to the minimum base shear. For this example, the correct solution is a mechanism extending from the base through level 6 (Figure 12.13*c*).

Although the correct mechanism was found to extend to level 6, we can see from Figure 12.13*c* that other mechanisms were nearly as critical. Slight changes

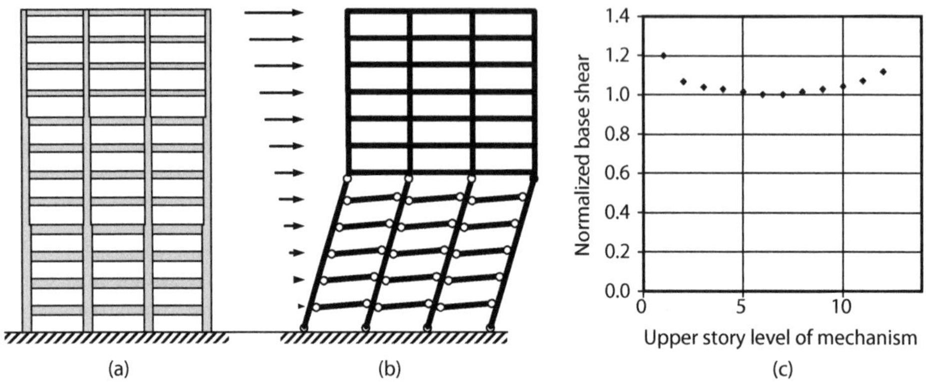

FIGURE 12.13 Calculated plastic mechanisms for a 12-story frame. (*a*) Frame proportions. Beam negative moment strength is 1.5 times beam positive moment strength at every joint, and the sum of column moment strengths is 1.2 times the sum of beam moment strengths at every joint except the roof. Beam relative strengths vary over height, being 1.0 for levels 1 to 4, 0.8 for 5 to 8, and 0.6 for 9 to 12. Beam span is twice the column height, joint width is 1/12 beam span and joint height is 1/6 column height. (*b*) Triangular lateral force distribution, with mechanisms assumed to extend from the base through some level above the base. (*c*) Normalized work associated with different mechanism heights.

in relative strengths within the frame would result in different mechanisms. Similarly, slight changes in the lateral force distribution could change the controlling yielding mechanism. Thus, we should expect that yield mechanisms during earthquake ground shaking will vary from one time to another because the lateral load profile is continuously changing with time.

We can investigate the effect of varying the relative strengths of columns and beams using the limit analysis approach. Figure 12.14 shows the results for relative

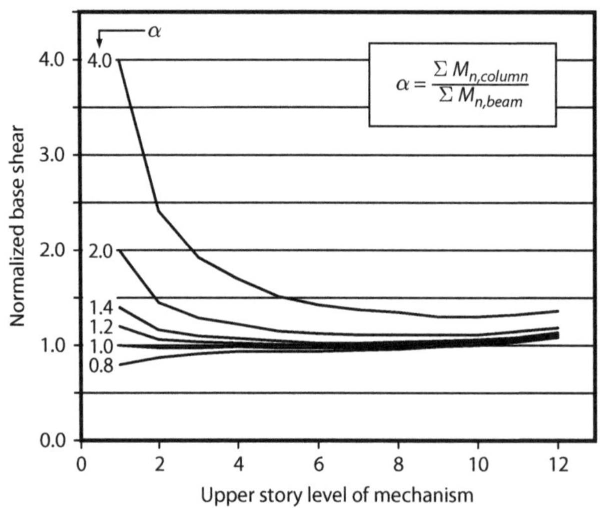

FIGURE 12.14 Yield mechanism solutions for frames designed with different ratios of column-to-beam strengths. Base shear is normalized to the value obtained for $\alpha = 1.2$.

strength ratios α ranging from 0.8 to 4.0. For $\alpha = 0.8$ or 1.0, the critical mechanism is a single-story mechanism. As α increases above 1.2, the base-shear strength increases and the controlling yield mechanism (corresponding to minimum base shear for that value of α) extends higher into the structure, but a full beam mechanism (Figure 12.2*c*) is never achieved.

Yield mechanisms during earthquake ground shaking will vary with time because of the sensitivity of the mechanism to lateral loading profile, but predominant collapse mechanisms can be identified. In one study (Haselton and Deierlein, 2008), series of frames representing the then current U.S. practice were subjected to ground motions whose intensity was scaled up until collapse occurred. Figure 12.15*a* shows results for buildings having moment frames in every bay. Although yielding occurred over most of the frame height, the apparent collapse mechanisms involved only the lower stories. The spread of yielding could be increased marginally by increasing the ratio of column-to-beam flexural strengths, but for taller buildings the predominant collapse mechanisms remained concentrated in the lower stories.

An alternative way of measuring the importance of the column-to-beam strength ratio is the probability of collapse given MCE-level shaking. Figure 12.15*b* plots calculated probability of collapse for a four-story space frame and a 12-story perimeter frame system as a function of the column-to-beam strength ratio. For the four-story space frame shaken in its plane, there is a significant benefit to the strength ratio being at least 1.0, with decreasing benefit for higher strength ratio. For the taller perimeter frame system, the results suggest that the design would be safer with strength ratio larger than 1.2. It is noted, however, that the results for the perimeter frame are based on the assumption that gravity framing does not contribute to lateral resistance. More likely, a typical gravity frame with substantial columns but weak horizontal framing members would improve behavior of a perimeter frame building. (The higher collapse probabilities for taller buildings, shown in Figure 12.15*b*, were deemed unacceptable by members of the ASCE 7 code committee. A minimum base-shear requirement, mainly applicable to tall buildings, was reinstated to reduce the collapse probabilities to acceptable levels. That minimum base-shear strength requirement remains in effect at the time of this writing.)

Higher-Mode Effects

The previous results suggest that, for practical column-to-beam strength ratios, it may not be feasible to prevent some column yielding as the structure forms a yield mechanism. Higher-mode response of multi-story frames also contributes to the tendency for column yielding. Frame distortions under higher-mode responses cause column moments to be unequally distributed above and below beam-column joints. For the example shown in Figure 12.16, at some locations and times (e.g., fourth level at $t = 5.48$ s), the beam inelastic moment demand at the beam-column joint is resisted almost entirely by one column above or below the joint while the column on the other side of the joint is only lightly stressed. For this condition, the moment strength of a single column (rather than the sum

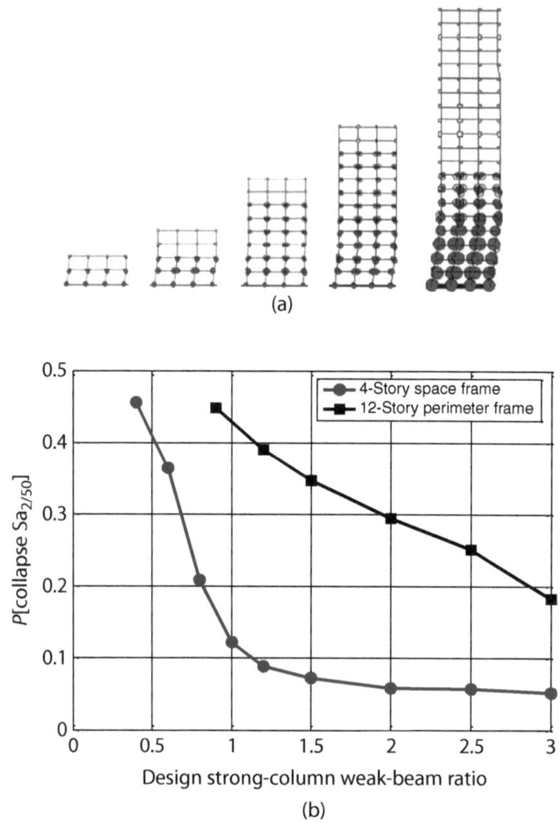

FIGURE 12.15 (*a*) Predominant collapse mechanisms calculated for space frames complying with current U.S. codes. (*b*) Probability of collapse given MCE shaking. (*After Haselton and Deierlein, 2008, used with permission from University of California, Berkeley.*)

of column moment strengths) would need to be approximately equal to the sum of the beam moment strengths if the objective was to prevent column yielding.

Biaxial Loading

The studies presented thus far have considered only loading within the plane of the frame. A space frame with moment frames in two orthogonal directions may have especially large column moment demands if the earthquake displaces the building along the diagonal (Figure 12.17). In this case, moment equilibrium requires

$$\sum M_c \approx \frac{M_{b1} + M_{b2} + M_{b3} + M_{b4}}{\sqrt{2}} \tag{12.1}$$

The equilibrium equation is approximate because the moments due to member shears acting at the faces of the joint have not been included. If we adopt the

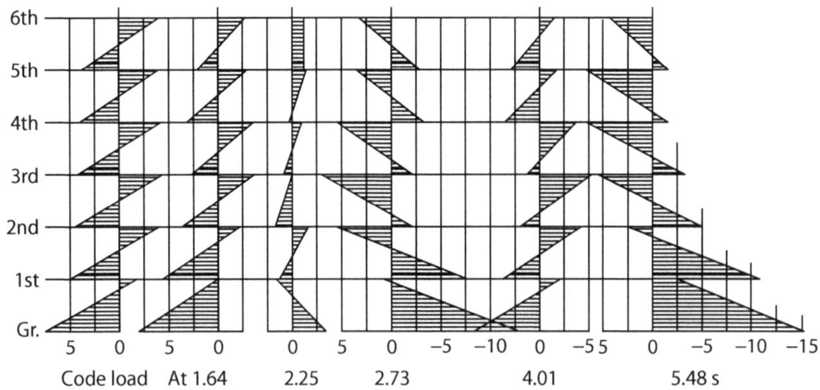

FIGURE 12.16 Column moment distributions calculated for a nonlinear analysis with inelastic beam elements and linear column elements. The results are for the lower 6 stories of a 12-story building. Plot shown left is for code static lateral loading. Remaining results are at times noted during dynamic response history analysis. (*After Kelly, 1974.*)

approximation that the column biaxial moment strength M_{uxy} is approximately equal to the uniaxial strength $M_{ux} = M_{uy}$, and set $M_{ub1} = M_{ub3}$ and $M_{ub2} = M_{ub4}$, then the equivalent strong-column/weak-beam requirement is

$$\sum M_{nc} \geq \frac{2}{\sqrt{2}} \sum M_{nb} \approx 1.4 \sum M_{nb} \tag{12.2}$$

The factor 1.4 in Eq. (12.2) would be in addition to mechanism amplification factor α (Figure 12.14) and dynamic amplification (Figure 12.16) discussed previously.

Beam-Yielding Mechanisms

We have assumed that beam yielding occurs at the face of the beam-column joint, but this is not necessarily the case. For long beam spans or heavy gravity loads,

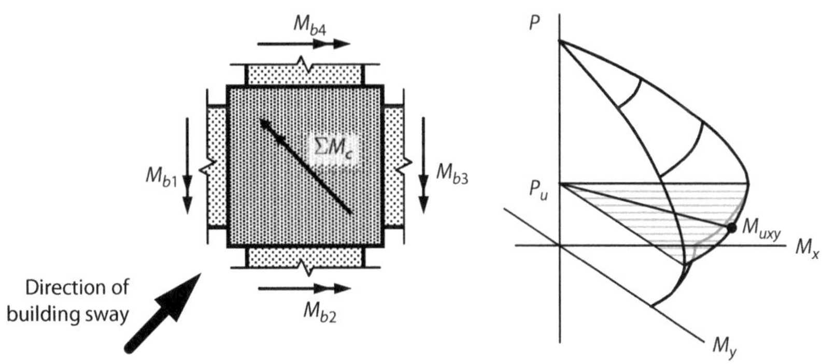

FIGURE 12.17 Biaxial loading of column: (*a*) plan view cutting through column; (*b*) axial load versus moment interaction diagram for biaxial bending.

yielding might occur away from the face of the beam-column joint. This is an undesirable behavior that should be avoided through design.

Ideally, beam yielding in a special moment frame will occur within the specially detailed lengths of the beams adjacent to the joint faces (Figure 12.18*a*). This type of yielding will occur if the beam is relatively short and the gravity loads are relatively low compared with seismic design effects. Where this occurs, the beam plastic hinges undergo yielding reversals as the building sways back and forth. This is the intended behavior.

In contrast, if gravity moments are relatively large compared with earthquake-induced moments, then a less desirable behavior can occur (Figure 12.18*b*). As the beam is deformed by the earthquake, the moments reach the plastic moment strengths in negative moment at the column face and in positive moment away from the column face. The deformed shape is shown. Upon reversal, the same situation occurs, but on opposite ends of the beam. Thus, beam plastic hinges do not reverse when the loading direction reverses. Under multiple yielding cycles, the plastic hinge rotations increase progressively. For long earthquake ground motions with multiple yielding cycles, the accumulated rotations can exceed

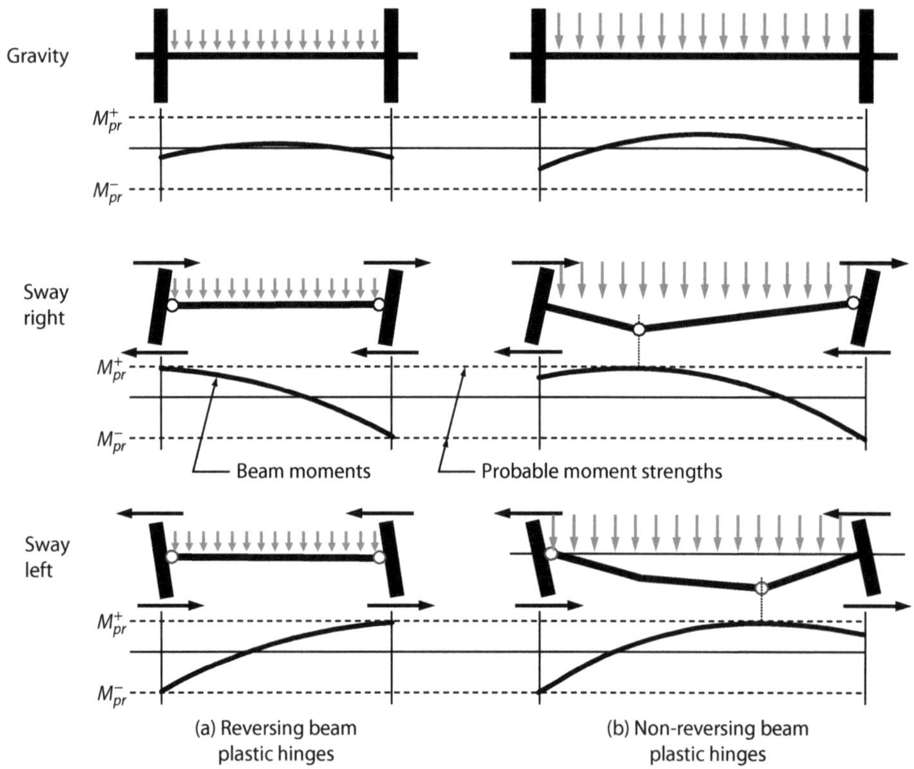

Figure 12.18 Beam-yielding mechanisms: (*a*) reversing beam plastic hinges; (*b*) non-reversing beam plastic hinges.

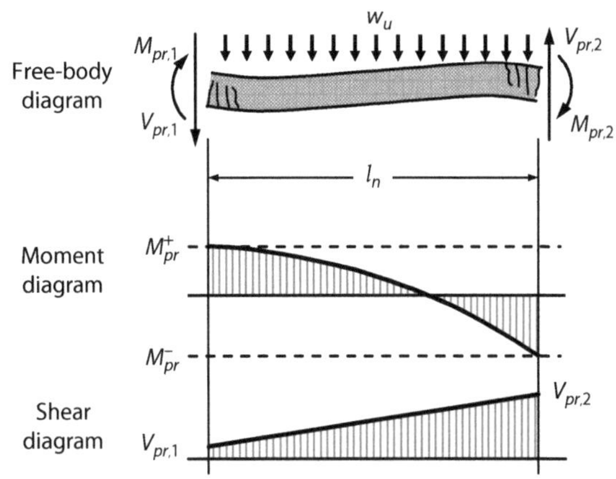

Figure 12.19 Free-body diagram of beam with plastic hinges at the face of the columns and corresponding moment and shear diagrams.

rotation capacities and vertical movement of the floor can exceed serviceable values.

To determine whether reversing or non-reversing plastic hinges occur, consider a free-body diagram of the beam (Figure 12.19). The left-hand plastic hinge (positive moment) will be located at the column face if the slope of the moment diagram is negative or zero, and will move from the column face if the slope is positive. Thus, the condition for the hinge located at the column face is that the shear $V_{pr,1}$ be in the direction shown, or zero. Thus, non-reversing plastic hinges can be avoided if

$$M^+_{pr} + M^-_{pr} \geq \frac{w_u l_n^2}{2} \tag{12.3}$$

This expression is valid for uniformly distributed load on beams for which moment strength does not change along the span. Requirements for other conditions can be derived using a similar approach.

Flexural yielding of a beam adjacent to a beam-column joint can lead to inelastic response within the joint itself, which may be detrimental to overall frame behavior. An alternative approach is to design the beam to have flexurally weak sections located a short distance from the face of the joint, thereby maintaining effectively linear response of the members at the beam-column joint (Abdel-Fattah and Wight, 1987). Figure 12.20 illustrates beam details to achieve this behavior. While this approach may improve behavior of the joint, there are some potential disadvantages to this approach. One disadvantage is that it results in a shorter beam span between yielding regions and therefore increases the plastic rotation demand on the beam plastic hinge. This detail also complicates steel placement during construction. It is seldom used in practice.

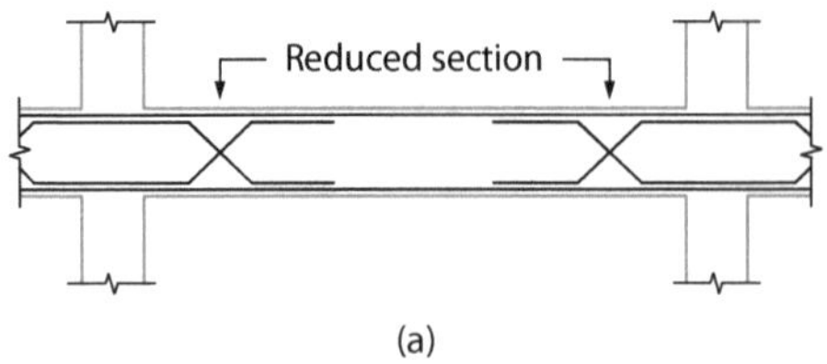

Figure 12.20 Beam design to locate reversing plastic hinges away from the face of the beam-column joint: (*a*) bent bars to create a reduced section; (*b*) straight bars terminated a short distance from the joint face.

12.4.3 Member Forces

Design in accordance with the ASCE 7 strength design procedures requires analysis of the structural framing under load combinations including gravity loads and seismic loads. Figure 12.21 illustrates the general nature of beam and column moments generated by gravity and lateral loads. The analysis results are combined using load combinations defined in Chapter 1 of this book to determine member design actions. Of primary interest are the beam flexural moments at the locations of the intended plastic hinges. Other quantities of interest include member axial forces and shears, although these quantities may be adjusted in consideration of the yielding mechanism of the frame. The following subsections discuss procedures for determining design forces in the various members.

Beam Moments

In the usual design approach, the intended primary yielding mechanism of a special moment frame is beam flexural yielding adjacent to the beam-column joints. Thus, the beam design moments at the joint faces are obtained from the controlling load combinations (Chapter 1) for both positive and negative moment.

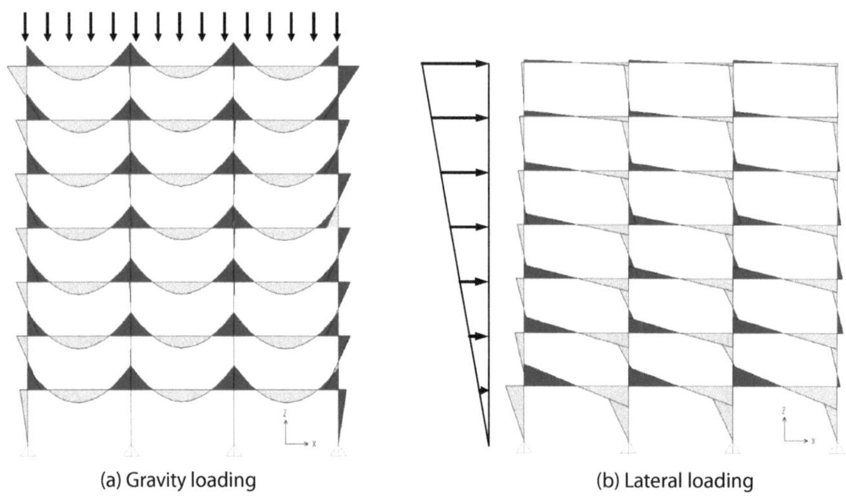

Figure 12.21 Moment diagrams under (*a*) gravity loading and (*b*) lateral loading.

At other locations along the span, beam moment strengths will be selected such that the provided moment strengths exceed the design moments.

Beam Shears

Beam shears are readily calculated using equilibrium of a free-body diagram of the beam. For the beam shown in Figure 12.19, a first analysis case considers beam moments acting clockwise at the ends of the free-body diagram, as shown. Summing moments about the right end yields $V_{pr,1}$, and summing moments about left end yields $V_{pr,2}$, as

$$V_{pr,1} = \frac{M_{pr,1} + M_{pr,2}}{l_n} - \frac{w_u l_n}{2} \text{ and } V_{pr,2} = \frac{M_{pr,1} + M_{pr,2}}{l_n} + \frac{w_u l_n}{2} \tag{12.4}$$

This equation assumes that the plastic hinges form at the joint faces. It also assumes that load w_u is due to gravity alone, without additional vertical acceleration effects associated with the earthquake. A second analysis case is defined for the beam moments reversed from those shown in Figure 12.19, and the design is based on the envelope of the two cases.

The quantity M_{pr} is the probable moment strength of the beam. In U.S. practice it is calculated for the as-built beam cross section using strength reduction factor $\phi = 1$, the rectangular stress block considering the specified concrete compressive strength, and elasto-plastic steel stress–strain relation with yield stress equal to αf_y. A value of $\alpha = 1.25$ is commonly assumed. See Section 6.7.2 for additional details.

The value of 1.25 for α may at first seem low considering that expected yield stress for typical Grade 60 reinforcement is approximately $1.15f_y$, and ultimate stress can be as high as $1.5 \times 1.15f_y$, or $1.7f_y$. A reasonable conclusion is that actual beam moment strength is probably larger than the value calculated based on $\alpha = 1.25$, and consequently the internal actions associated with development of beam plastic hinges are probably larger than actions based on M_{pr}. However, development of flexural plastic hinges in a beam is likely to be associated with spalling of cover concrete, which will reduce the effective section dimensions. Furthermore, any beam overstrength beyond M_{pr} is partially offset by material overstrengths and use of strength reduction factors, ϕ, that are less than 1.0 elsewhere in the structure. Therefore, the approach is deemed to be reasonably conservative for most applications.

Column Moments

Columns should be designed to resist at least the moments associated with development of the beam probable moment strengths. Considering the beam-column joint elevation shown in Figure 12.22, equilibrium of moments about the joint center results in

$$(M_{ct} + M_{cb}) + (V_{ct} + V_{cb})\frac{h_b}{2} = (M_{pr,1} + M_{pr,2}) + (V_{pr,1} + V_{pr,2})\frac{h_c}{2} \tag{12.5}$$

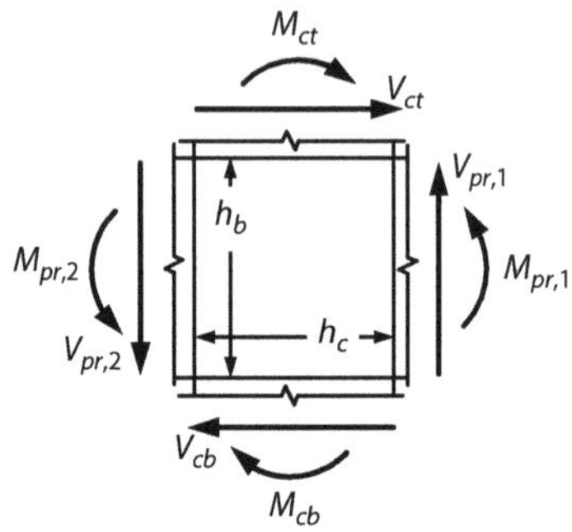

Figure 12.22 Elevation of beam-column joint.

If we consider the special (but common) case of a frame with equal beam spans and loads on either side of the joint, with essentially identical column moment and shear distributions above and below the joint, Eq. (12.5) can be expressed as

$$\Sigma M_c = \frac{l_u}{l_n}\frac{l_b}{l_c}\Sigma M_{pr} \tag{12.6}$$

in which ΣM_c is the sum of column moments at the faces of the joint, l_n and l_b are the clear and center-to-center span lengths of the beams, l_u and l_c are the clear and the center-to-center lengths of the columns, and ΣM_{pr} is the sum of the beam probable moment strengths at the faces of the joints. For typical dimensions, Eq. (12.6) can be approximated conservatively by

$$\Sigma M_c \approx \Sigma M_{pr} \tag{12.7}$$

If we assume that column moments above and below the joint have equal absolute value, then the column moment at the joint face is $\Sigma M_c/2$. Equation (12.7) serves as the basis for the required strength of a column as specified by ACI 318. Rather than compare moments at the plastic or probable level, however, ACI 318 uses nominal moment strengths for both the columns and beams. This is based on the assumption that if the nominal moment strengths are in the correct proportion, then the probable moment strengths will be similarly in proportion. See Section 12.6.4 for additional discussion.

The column moments may not be equally distributed above and below a joint, as assumed in the preceding paragraph. This is especially the case near the base and near the top of a building, where column fixity conditions change, or near other discontinuities. At such locations, the moment on one side of the column may be significantly greater than $\Sigma M_c/2$. This is illustrated in Figure 12.16. Thus, the usual practice is to design for at least the moments obtained from the analysis of the building frame under the design earthquake loading, in addition to the moments indicated by Eq. (12.7). An overstrength factor associated with beam flexural overstrength can also be considered, but this is not the usual practice.

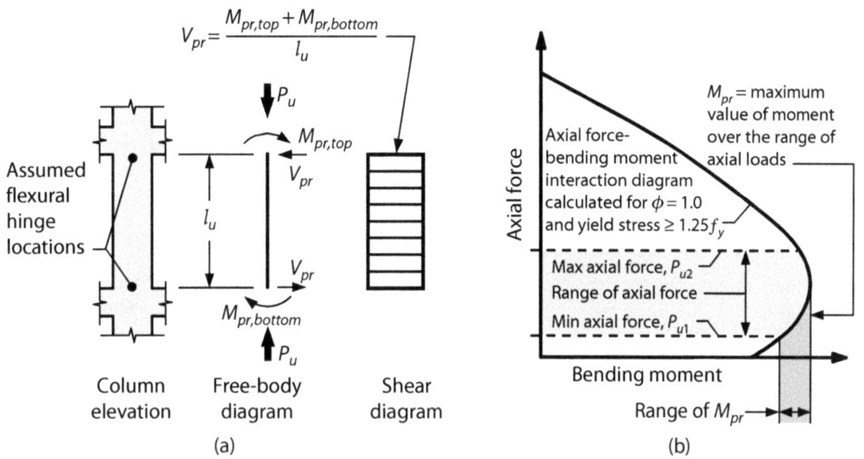

Figure 12.23 Column actions assuming development of probable moment strengths at both ends of the column: (*a*) column actions; (*b*) axial force-bending moment interaction diagram.

Column Shears

We can calculate an upper bound for the column shears using the same procedure used for beams, that is, by assuming the column develops its probable moment strengths at both ends and solving for the associated shears (Figure 12.23*a*). A complication is that the probable moment strength of a column varies with the axial force acting on the column. A suitable approach is to estimate the range of axial forces expected under combined earthquake and gravity loads and to select the value of M_{pr} that is highest within that range (Figure 12.23*b*).

Assuming that M_{pr} develops at both ends of the column simultaneously may be excessively conservative, especially where columns are considerably stronger than beams. An alternative sometimes used (and permitted by U.S. codes) is to assume that the frame develops its intended beam-yielding mechanism, and then calculate the column shear corresponding to development of M_{pr} of the beams framing into the joints (Figure 12.24). This approach is problematic because the distribution of column moments above and below any beam-column joint is indeterminate. A common practice is to assume that the input moment from the beams is distributed equally above and below the joint, or in proportion with the stiffness of the columns above and below the joint. However, as shown in Figure 12.16, the moment patterns can vary widely during seismic response, such that this approach is unreliable and not recommended. An alternative is to assign all of the moments from the beams to the column under consideration. This approach has been found to be overly conservative (Visnjic et al., 2014), and is not widely used.

The remaining alternative is to use dynamic analysis results to define the column design shear. The best option is to conduct nonlinear dynamic analysis under multiple ground motions, and select the column design shears based on statistical evaluation of the results (TBI, 2010). Where this option is not feasible,

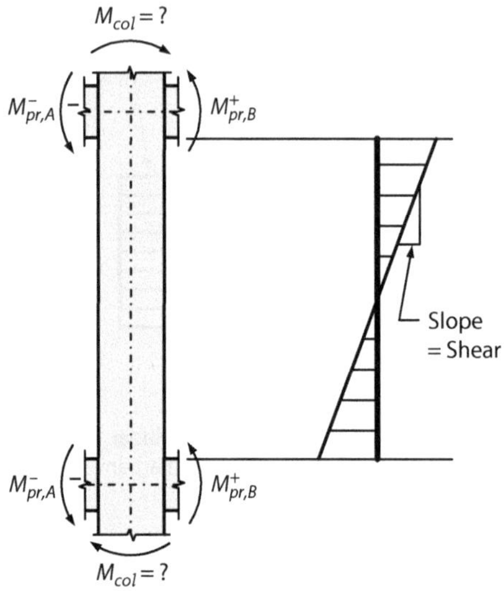

FIGURE 12.24 Column design shear based on probable moment strengths of the beams framing into the joints. Only moments are shown in the free-body diagram. Note that the column moments at the top and at the bottom of this free-body diagram are indeterminate, so the moments in the column and the corresponding shears are also indeterminate.

the column shears can be estimated based on results of linear response spectrum analysis. However, the shears calculated from linear analysis require adjustment to account for effects of anticipated nonlinear response of the building frame (Visnjic et al., 2014). Two effects should be considered, as described in the following text:

- Structural overstrength: As the building responds to design-level earthquake shaking, it will develop internal forces that are consistent with the provided building strength. Some overstrength relative to the design requirements should be anticipated, due to material overstrength, member oversizing, and various building code requirements including strength reduction factors and minimum strength requirements. Consequently, the internal forces, including column shears, will increase relative to the design values.
- Dynamic response: As the building responds to design-level earthquake shaking, the displacements and resulting distortions will be larger than those that are considered when determining the member shear forces. Especially in elements that are intended to respond in the linear-elastic range, the amplified deformations will result in larger internal member forces, including column shears.

Based on the preceding discussion, the design column shears can be expressed as

$$V_u = \omega \Omega_0 V_{MRSA} \tag{12.8}$$

In this equation, V_{MRSA} is the column shear as calculated in the modal response spectrum analysis. Factor ω is a dynamic amplification factor, which can be taken as $\omega = 1.3$. Factor Ω_0 measures the overstrength of the structural system, and is defined as

$$\Omega_0 = \frac{M_{b,\Omega}}{M_{b,MRSA}} \tag{12.9}$$

in which $M_{b,\Omega}$ = overturning moment strength of the structural system measured at the base and $M_{b,MRSA}$ = overturning moment demand on the structural system obtained from the modal response spectrum analysis. The quantity $M_{b,\Omega}$ can be calculated using nonlinear static analysis or limit analysis, as presented in Section 12.4.2. A simple approximation is $\Omega_0 = \sum M_{pr} / \sum M_{u,MRSA}$, in which $\sum M_{pr}$ = sum of probable moment strengths of all beam and column plastic hinges in a beam-yielding mechanism (Figure 12.2*c*), and $\sum M_{u,MRSA}$ = sum of the moments calculated from the modal response spectrum analysis at all beam and column plastic hinge locations of the same beam-yielding mechanism. Note that the moments M_{MRSA} do not include the moments due to design gravity loads, but instead include only the moments obtained from the modal response spectrum analysis.

The procedure of the previous paragraph is based on the work of Visnjic (2014). In that study, series of special moment frames were designed in accordance with ASCE 7-10 and ACI 318. Analytical models representing the inelastic properties of the frames were subjected to earthquake ground motions scaled to the DE shaking intensity, with the goal of identifying approximate methods to establish design forces. Figure 12.25 compares column shears calculated by various methods for an interior column of a 10-story frame.

a. The results from nonlinear dynamic analysis for 30 earthquake ground motions scaled to the DE level are shown with thin lines. There is some record-to-record variation in the results. Visnjic (2014) found that the results typically could be represented by a normal distribution with coefficient of variation of approximately 0.1 in lower stories, increasing to approximately 0.2 or greater in upper stories.

b. Shears calculated from the sum of column M_{pr} values divided by column clear height ($\sum M_{pr,col,i} / l_{u,i}$) are conservative, in some cases by what might be considered an excessive margin.

c. Shears calculated from the sum of beam M_{pr} values, distributed equally to columns above and below joints that the beams frame into, and divided by column clear height ($\sum M_{pr,beam,i} / 2l_{u,i}$) are unconservative in the critical lower stories, and conservative in the less critical upper stories.

d. Shears calculated from the modal response spectrum analysis, V_{MRSA}, with the response spectrum ordinates reduced by response modification coefficient R as permitted by ASCE 7, are grossly unconservative.

e. Shears calculated from Eq. (12.8), with $\omega = 1.3$ and Ω_0 calculated as the sum of member probable moment strengths divided by the sum of member

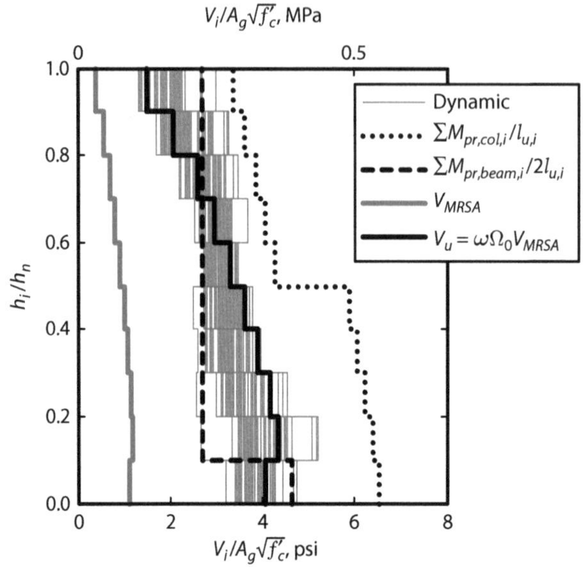

FIGURE 12.25 Column shears in a 10-story frame subjected to earthquake ground motions scaled to the DE intensity. (*After Visnjic, 2014.*)

moments from the modal response spectrum analysis, are appropriately conservative in the critical lower stories and slightly unconservative in the upper stories.

Local stiffness variations and local deformations may also affect the column shears (Pantazopoulou and Qi, 1991). Overturning increases the axial compressive force on some columns and decreases it on others, resulting in variations in effective stiffness that affect the distribution of story shears (Figure 12.26*a*). Visnjic (2014) recommends increasing the shear in exterior columns by factor 1.1 to account for this effect.

Flexural cracking of beams results in beam elongation (see Section 6.10), which pushes the columns outward. This can result in a large increase in first-story column shears if the base of the building is effectively fixed against lateral expansion (Figure 12.26*b*). Some building codes (e.g., NZS3101, 2006) require that this effect be considered by designing the first-story columns for the shear corresponding to development of M_{pr} at both column ends. Some analytical studies suggest, however, that this approach may be overly conservative (Visnjic et al., 2014).

Joint Shear

Joint shears can be estimated based on equilibrium of member forces around the beam-column joints. Normally we assume that the beams have reached their probable moment strengths as intended in the beam-yielding mechanism; this determines the beam moments and shears on the vertical faces of the joint. Unless the boundary conditions dictate an alternative assumption, we usually assume that the column inflection point is located at the column mid-height, such that the

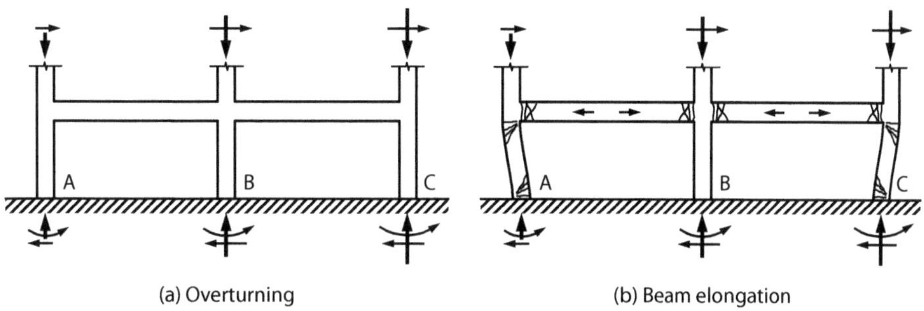

Figure 12.26 Effect of (*a*) overturning and (*b*) beam elongation on column shears. (*After Pantazopoulou and Qi, 1991.*)

column shear can be calculated from a free-body diagram of the beam-column subassemblage. Conventional practice is to carry out these calculations independently along each of the principal axes of the frames and to design for each separately. Design for biaxial loading generally is not considered. See Chapter 9 for details of the joint shear calculation procedures.

Column Axial Load

Column axial loads can be determined based on analysis of the building frame under appropriate load combinations. Where a nonlinear analysis model is subjected to multiple ground motions, the axial loads can be extracted directly from the analysis results and used without further modification. Linear structural analysis models using seismic forces reduced by response modification factor R will generally underestimate the variation of axial forces due to earthquake shaking, mainly because of overstrength built into the yielding members (mainly beams).

We can establish an upper bound to the column axial forces using limit analysis and the assumption that the frame has developed the intended beam-yielding mechanism (Figure 12.27). The beam probable shear forces, V_{pr}, known from equilibrium of the yielding beam are applied to the joint faces of the column over the building height. The sum of these shear forces plus the weight of the column and any other loads supported by the column (including beams framing in the transverse direction and cladding) are summed from the roof level to the level of interest to obtain the upper-bound column axial force. (Note that a somewhat higher force would be obtained if we considered vertical earthquake input motion, but that is not done in typical design-office practice.)

We can follow the same limit analysis procedure, but with beam yielding in two directions, to calculate upper-bound axial forces in corner columns (Figure 12.28). The relatively high vulnerability of corner columns, observed from damage in some past earthquakes, is partly due to this bi-directional loading effect.

The limit analysis approach outlined above provides a good estimate of likely axial loads for columns in low-rise buildings. It tends to overestimate axial load variations for taller frames because they are more strongly affected by

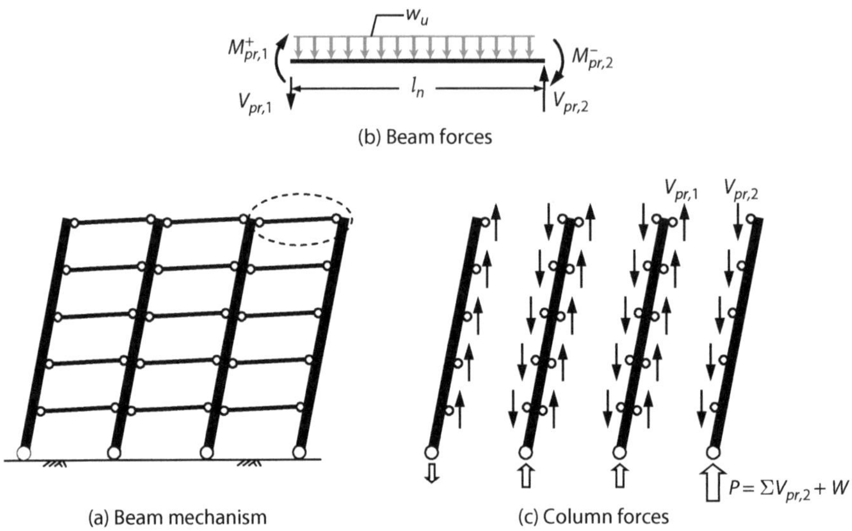

FIGURE 12.27 Column axial forces determined by limit analysis.

higher modes and also because they tend to develop partial-height mechanisms (Section 12.4.2). To recognize the development of a partial-height mechanism, the column axial load at level i can be defined by

$$P_{u,i} = P_{g,i} \pm \gamma_P \sum_{j=i}^{N} V_{pr,j} \tag{12.10}$$

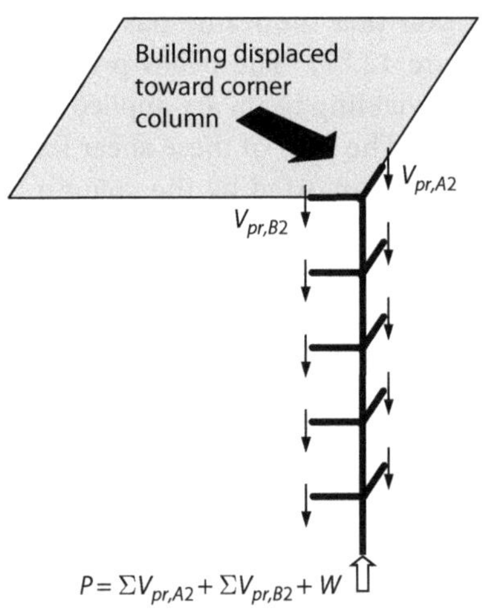

FIGURE 12.28 Limit analysis for corner column shared by intersecting moment frames.

in which $P_{u,i}$ = design axial load at level i, $P_{g,i}$ = design axial load at level i due to gravity loads only, $V_{pr,j}$ = shear due to development of M_{pr} at both levels of the beam at level j calculated for zero gravity load on the beam, γ_P = reduction factor to recognize that not all beams develop M_{pr} from levels i through N, i = level for which the axial load is being determined, and N = roof level. A study reported by Visnjic et al. (2014) supports using $\gamma = 0.8$. NZS3101 (2006) uses an approach similar to this.

12.4.4 Member Deformation Demands and Capacities

Special moment frame members should be capable of being deformed through multiple deformation cycles without critical loss of vertical or lateral force resistance. As a design problem, we are interested in estimating the deformation demands and determining the required detailing such that the member has the capacity for those demands.

Member Inelastic Deformations

Member deformation demands can be estimated either through detailed inelastic analysis of a building under anticipated earthquake ground motions or using simplified methods. The approach of inelastic dynamic analysis is favored for the performance-based design of important buildings. Analytical models should be calibrated using available laboratory test data so that the degree of modeling accuracy can be established. A representative suite of earthquake ground motions should be used to excite the model, resulting in a suite of response results. These results are then analyzed to determine a suitable estimate for design purposes. See TBI (2010) for additional details on using this approach.

Most building design projects do not have the resources required for nonlinear dynamic analysis studies. Thus, it is necessary to have a generic approach to estimating demands and capacities that can be used for prescriptive building designs. Figure 12.29 illustrates a simple and relatively conservative idealization for this purpose. The building frame is assumed to develop an inelastic mechanism, with all of the deformations associated with rotation of the plastic hinges. According to this idealization, the plastic hinge rotation at the base of the columns is equal to the story drift ratio. The plastic hinge rotations of the beams will be somewhat larger than the story drift ratio because of the rigid joint offsets between the column centerline and the center of the plastic hinge.

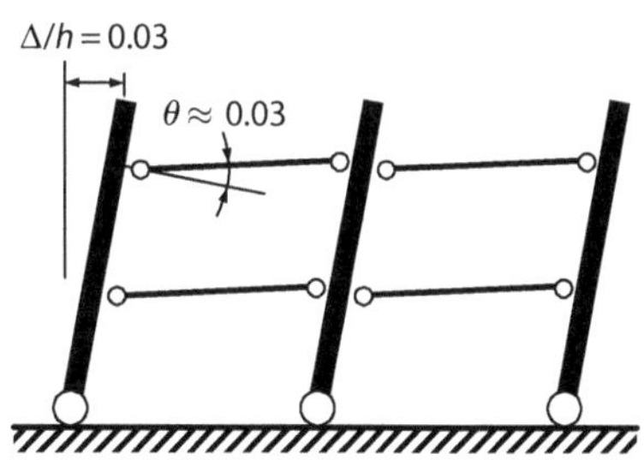

Figure 12.29 Idealized displaced shape of yielding frame.

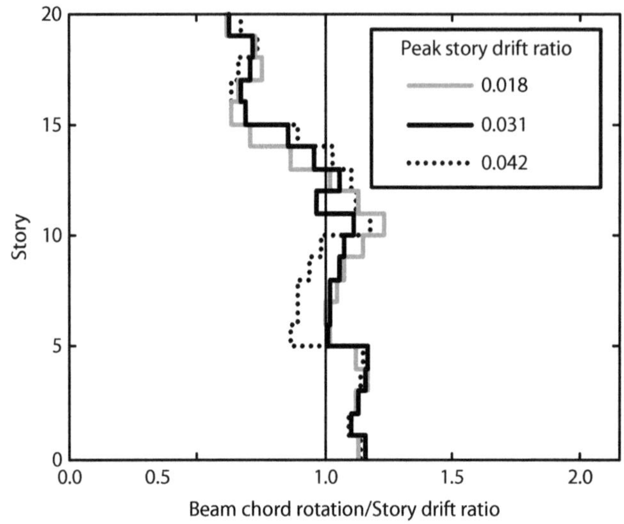

FIGURE 12.30 Calculated ratios of beam plastic hinge rotations to story drift ratios for a 20-story planar special moment frame subjected to earthquake ground shaking. (*After Visnjic, 2014.*)

However, not all of the frame drift is due to plastic hinge rotation, partially offsetting this effect. Thus, we can approximate the beam rotations as being equal to the story drift also.

We can check the approximations of the preceding paragraph using results of inelastic dynamic analysis of example building frames. Figure 12.30 shows sample results for a high-rise building frame subjected to ground motions that cause design-level or greater story drifts. As shown, the ratio of beam plastic rotation to story drift ratio ranges from approximately 0.7 to 1.3, with a central value near 1.0. Thus, the approximation of Figure 12.29 is reasonable, but not exact.

In U.S. design practice, the calculated story drift is not to exceed $0.02h_{sx}$ (ASCE 7, 2010) for Design Earthquake (DE) shaking. Given that Maximum Considered Earthquake (MCE) shaking is typically about 1.5 times DE shaking in the United States, the story drift limit for MCE shaking should be approximately $1.5 \times 0.02h_{sx} = 0.03h_{sx}$. This is the value adopted by performance-based design documents such as TBI (2010). Thus, for special moment frames designed in accordance with U.S. provisions, a reasonable estimate for the maximum hinge rotations for beams and columns is $\theta = 0.03$.

In some applications, it is useful to have an equivalent estimate of the curvature ductility demand. Assuming a plastic hinge length $l_p = h/2$, a total rotation of 0.03 requires total curvature within the plastic hinge length equal to $\phi_u = \theta_u/l_p = 0.03/(h/2) = 0.06/h$. Approximating the yield curvature $\phi_y \approx \frac{\varepsilon_y}{\frac{2}{3}h} \approx \frac{0.003}{h}$, the curvature ductility is on the order of $\mu_\phi = \frac{\phi_u}{\phi_y} \approx 20$.

Beam Rotation Capacities

Beam rotation capacity depends on the detailing of the cross section. This subject was reviewed in depth in Chapter 6. Two variables found to have important effect on rotation capacity were as follows:

1. The ratio of tension reinforcement area A_s to compression reinforcement area A_s'. Beam rotation capacity decreases as the ratio of A_s/A_s' increases.
2. The volume ratio and spacing of confining transverse reinforcement. Increasing the volume ratio and decreasing the spacing of individual hoop sets improves compressive strain capacity, supports longitudinal bars against buckling, and resists shear. Thus, beams with generous amounts of closely spaced transverse reinforcement have greater rotation capacity than equivalent beams without it. A good rule is that hoop spacing should not exceed $d/4$, $6d_b$, 6 in (150 mm), and the spacing required for shear.

Figure 12.31 illustrates several examples of beam cross sections and results of moment-rotation analyses using procedures described in Section 6.11.2. Each of

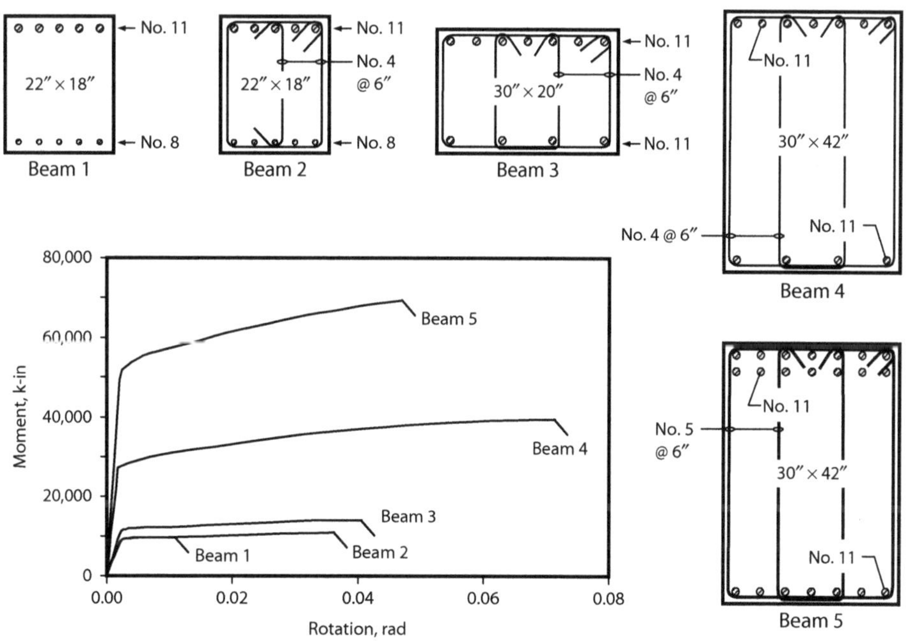

FIGURE 12.31 Examples of unconfined (Beam 1) and confined (Beams 2 through 5) beam cross sections and calculated moment rotation behavior. Typical A706 Grade 60 longitudinal reinforcement properties with $f_y = 70$ ksi (480 MPa) and $f_u = 100$ ksi (690 MPa), $f_c' = 6000$ psi (41 MPa), $f_{yt} = 60$ ksi (410 MPa). Concrete confinement was modeled using the procedures outlined in Chapter 6. Rotation was calculated as the product of the ultimate curvature capacity, ϕ_u, and plastic hinge length, $l_p = h/2$. Note this rotation includes total curvature (from zero to the ultimate value).

the cross sections has $A_s/A'_s \approx 2$, which is an upper limit permitted by some building codes. Beam 1 does not have transverse reinforcement. The other beams were sized considering the hoop spacing rule identified above. Beam 5 is considered to be an extreme example of a large beam; it would be difficult to design a beam-column joint capable of resisting the shear forces delivered by a beam with this reinforcement. With the exception of Beam 1, all of the beams shown have calculated rotation capacity exceeding the estimated MCE rotation demand of 0.03 rad.

Column Rotation Capacities

Like beams, column rotation capacity depends on the detailing of the cross section. It also depends on the axial force. This subject was reviewed in depth in Section 6.8. Defining the total rotation capacity of a column plastic hinge as the sum of the elastic and plastic rotations, we can write the total rotation capacity as $\theta_u \approx \phi_u l_p \approx \frac{\varepsilon_{cu}}{c}\frac{h}{2}$. According to this expression, as the flexural compression depth c increases with increasing axial force P, the strain capacity of the concrete ε_{cu} must increase if the rotation capacity θ_u is to be maintained. Thus, the amount of confinement reinforcement must be increased as axial force P increases.

This result is apparent in calculated moment–curvature relations such as those plotted in Figure 6.36. If the detailing is held constant, the curvature capacity decreases as axial load increases. Additional confinement would be required to maintain constant curvature capacity for increasing axial force. This result is also apparent in laboratory tests. Figure 12.32*a* plots drift ratio capacity for columns as a function of axial load ratio and confinement ratio. The confinement ratio is the ratio of A_{sh} provided in the test divided by the ratio A_{sh} required by Eqs. (5.22) and (5.23). Those expressions are repeated as expressions (a) and (b) in Table 12.2. As shown in Figure 12.32*a*, many columns satisfying these expressions fail to reach the

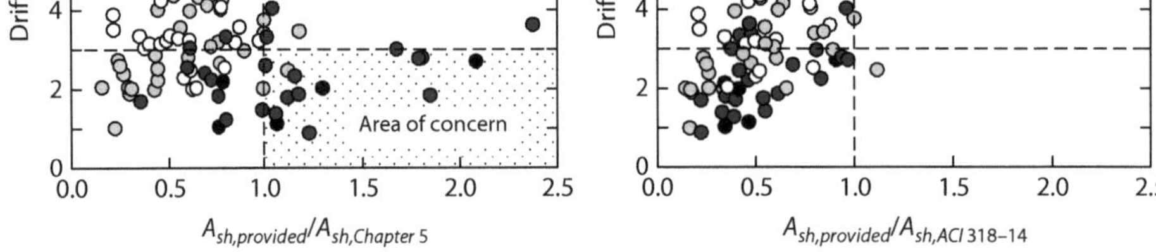

Figure 12.32 Drift ratio capacity as function of confinement ratio and axial load ratio. (*After Elwood, 2014.*)

target drift ratio of 0.03 for axial loads exceeding approximately $0.4A_g f'_c$. Prior to the 2014 edition of ACI 318, these were the only expressions defining the quantity of confinement reinforcement for rectilinear columns in special moment frames. ACI 318-14 introduced revisions to the confinement requirements to improve behavior of columns with higher axial loads. The requirements are summarized in Section 12.6.4.

Figure 12.33 plots calculated moment-rotation behaviors for columns designed to satisfy the ACI 318-14 confinement requirements. All of the calculated relations achieve rotation capacity of at least 0.03 as intended. Figure 12.32*b* plots the test data of Figure 12.32*a* but with different denominator on the horizontal axis. The results suggest that the ACI 318-14 provisions adequately achieve the target drift rotation of 0.03.

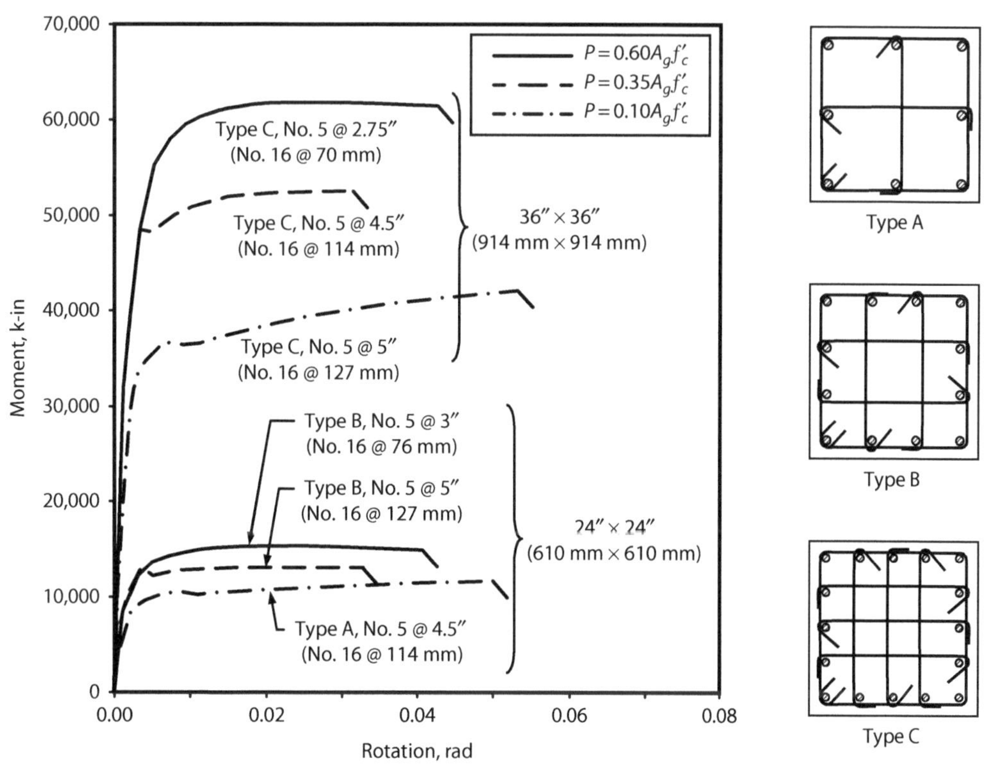

FIGURE 12.33 Calculated moment-rotation relations for various columns designed by ACI 318-14 confinement requirements. Longitudinal reinforcement ratio for all columns is 0.018 to 0.019. Transverse reinforcement varies as shown. Typical A706 Grade 60 (420) longitudinal reinforcement properties with $f_y = 70$ ksi (480 MPa) and $f_u = 100$ ksi (690 MPa), $f'_c = 6000$ psi (41 MPa), $f_{yt} = 60$ ksi (410 MPa). Concrete confinement was modeled using the procedures outlined in Chapter 6. Rotation was calculated as the product of the ultimate curvature capacity, ϕ_u, and plastic hinge length, $l_p = h/2$. Note this rotation includes total curvature (from zero to the ultimate value).

12.5 Modeling and Analysis

Chapter 11 described the overall seismic design requirements for buildings, including layout of the structural system, required strength, and drift limits. Once a schematic building design has been completed, the structural engineer can develop a structural analysis model to determine seismic design actions for the global structural system. The analysis model can be a linear elastic model or it can model the inelastic load-deformation behavior of the individual components. Linear elastic models, which are more commonly used, are the focus of discussion in this section.

12.5.1 Analysis Procedure

Many building codes provide three different linear analysis options for determining the seismic forces within the special moment frame. These are the *equivalent lateral force procedure*, *modal response spectrum analysis*, and *seismic response history analysis*. Chapter 11 provided a brief overview of the first two of these procedures. Seismic response history analysis is seldom used with linear models.

The structural analysis model should include all structural and nonstructural members that affect the linear and nonlinear response of the structure to earthquake motions. This can be especially important for special moment frames, which may be flexible in comparison with other parts of the building, including parts intended to be nonstructural in nature. Important examples include interactions with masonry infills (partial height or full height), architectural concrete walls, stairwells, cast-in-place stairways, and inclined parking ramps. If interactions with stiff nonstructural elements are permitted, the effects of localized failures of one or more of these elements must be considered. For example, the failure of a rigid architectural element in one story could lead to formation of a story mechanism. Generally, it is best to provide ample separation by jointing between the special moment frame and rigid elements assumed not to be part of the seismic-force-resisting system.

12.5.2 Stiffness Recommendations

When analyzing a special moment frame, it is important to model appropriately the cracked-section stiffness of the beams, columns, and joints, as this stiffness determines the resulting building periods, base shear, story drifts, and internal force distributions. Table 12.1 shows the range of values for the effective, cracked stiffness for each element based on the requirements of ACI 318. For beams cast monolithically with slabs, the gross moment of inertia I_g can be based on the combined beam web and effective flange width (see Section 6.10).

Element	I_e/I_g
Beam	0.35–0.50
Column	0.50–0.70

Table 12.1 Cracked Stiffness Modifiers

More detailed analysis may be used to calculate the reduced stiffness based on the applied loading conditions. For example, ASCE 41 (2013) recommends using $I_e/I_g = 0.7$ for $P/A_g f_c' \geq 0.5$, $I_e/I_g = 0.3$ for $P/A_g f_c' \leq 0.1$, with linear interpolation for intermediate axial loads. See Section 6.11.1 for additional discussion on effective stiffness. When considering serviceability under wind loading, it is common to assume gross-section properties for the beams, columns, and joints.

In a special moment frame, the beam-column joint is stiffer than the adjoining beams and columns, but it is not perfectly rigid. A partially rigid joint is recommended to account for joint flexibility. See Chapter 9 for additional discussion.

12.5.3 Foundation Modeling

The base restraint can have a significant effect on the behavior of a moment frame. Building codes commonly permit the base of the structure to be modeled as fixed for the purpose of determining seismic loads. Alternatively, foundation flexibility can be modeled. The engineer has to decide the most appropriate analysis assumptions for the frame considering its construction details. Figure 12.34 illustrates four types of base restraint conditions that may be considered.

Modeling pinned restraints at the base of the columns (Figure 12.34*a*) is typical for frames that do not extend through floors below grade. This assumption results in the most flexible column base restraint. The high flexibility will lengthen the period of the building, resulting in a lower calculated base shear but larger calculated drifts. Pinned restraints at the column bases will also simplify the design of the footing. Where pinned restraints have been modeled, the dowels connecting the column based to the foundation need to be capable of transferring the shear and axial forces to the foundation.

One drawback to the pinned column base is that the drift of the frame, especially the story drift in the lowest story, is more difficult to control within code-allowable limits. This problem is exacerbated because the first story is usually taller than typical stories. In addition, a pinned column base may lead to a soft or weak story, which should be avoided. If the drift of the structure is over the acceptable limits, then rotational restraint can be increased at the foundation by a variety of methods, as illustrated in Figure 12.34*b*–*d*. Regardless of which modeling technique is used, the base of the column and the supporting footing or grade beam must be designed

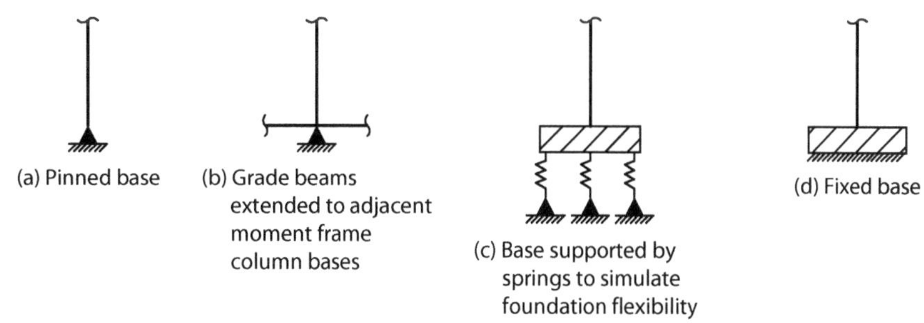

FIGURE 12.34 Column base restraint conditions.

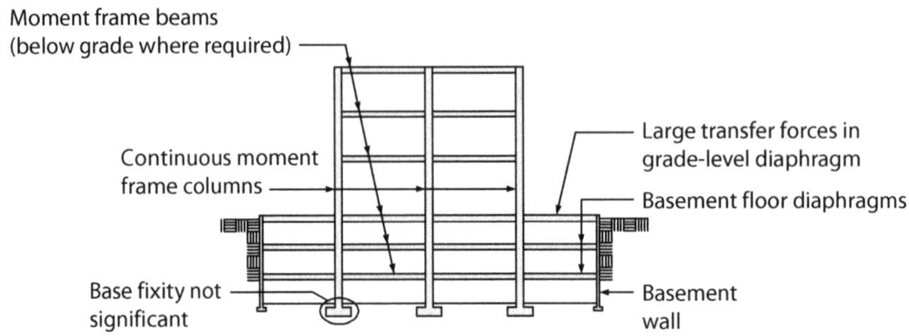

FIGURE 12.35 Moment frame extending through floors below grade.

and detailed to resist all the forces determined by the analysis. The foundation elements must also be capable of delivering the forces to the supporting soil.

Special moment frames sometimes extend through below-grade floors (Figure 12.35). Here we must consider the restraint and stiffness of the below-grade diaphragms and basement walls. In this condition, the model should represent the columns as continuous elements down to the footing. The type of rotational restraint at the column base will not have a significant effect on the behavior of the moment frame. Large forces will be transferred through the grade-level diaphragm to the basement walls, which are generally very stiff relative to the special moment frame.

12.6 Proportioning and Detailing Guidance

This section presents proportioning and detailing guidance for beams, columns, and beam-column joints of special moment frames. The text emphasizes the requirements of ACI 318 and accepted approaches to their implementation, supplemented by alternative recommendations on topics where the ACI 318 provisions may be deemed lacking. The guidance is presented in a sequence that parallels the typical design sequence, rather than a member by member sequence.

12.6.1 Beam Flexure and Longitudinal Reinforcement

Beam design follows a capacity-design approach that identifies where inelastic action is intended to occur and provides details to achieve intended inelastic response. Good practice is to design the beam so that plastic hinges occur at the faces of the beam-column joints (see Section 12.4.2). The beam plastic hinges should have a design moment strength not less than the factored design moment, that is, $\phi M_n \geq M_u$, where M_u is obtained from analysis of a numerical model of the building under the appropriate load combinations. Strength reduction factor ϕ is in accordance with Section 6.7. For beams reinforced in accordance with the provisions for special moment frame beams, as outlined below, the strength reduction factor is always $\phi = 0.9$.

During a design-level earthquake, yielding of the as-designed beams will determine the internal forces that develop throughout the structure. Because the design of other frame elements depends on the flexural strength of beams, the designer should take care to optimize each beam and minimize excess capacity.

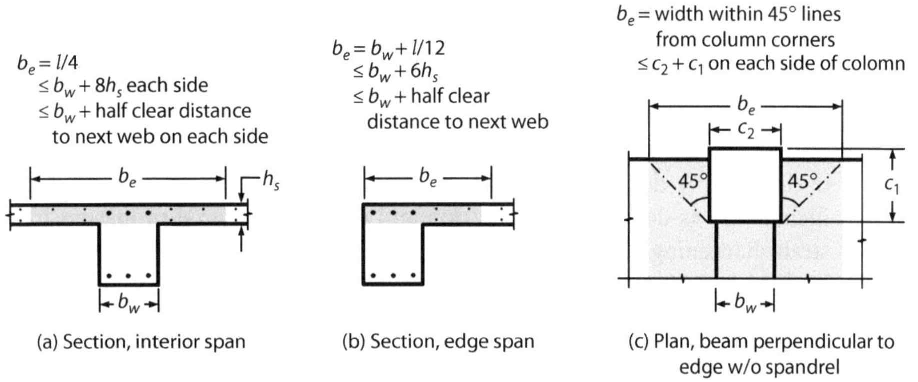

FIGURE 12.36 Definition of effective flange width. (*a*) ACI 318 effective flange width for interior beams with slab on both sides. (*b*) ACI 318 effective flange width for edge span beams with slab on one side. (*c*) ACI 352 effective flange width for beams framing perpendicular to a slab edge without spandrel. Where a substantial spandrel is present, the effective flange width of (*a*) or (*b*) can be used.

When a slab is cast monolithically with the beam, the slab acts as a flange, increasing the flexural stiffness and strength of the beam. As discussed in Section 6.10, developed slab reinforcement within the effective flange width (Figure 12.36) acts as beam flexural tension reinforcement and contributes to the beam flexural strength. ACI 318 is not explicit on how to account for this T-beam behavior when sizing the beam for required flexural strength, and different practices are used in different design offices.

One practice is to size the beam for the code required moment strength considering only the longitudinal reinforcement within the beam web and ignoring the slab contribution. Because the slab will participate with the beam in resisting flexure, the resulting beam will be flexurally stronger than necessary, leading consequently to higher demands throughout the building if the beam yields under earthquake loading.

An alternative practice, recommended here, is to size the beam for the design moment including developed longitudinal reinforcement within both the web and the effective flange width. The slab reinforcement may not be fully determined at the time the beams are designed, in which case the slab reinforcement ratio can be estimated based on experience with similar buildings, and checked after the design is completed.

A prevailing practice in some regions uses unbonded post-tensioned slabs cast monolithically with conventionally reinforced beams. Placing the unbonded strands outside the effective flange width does not mean those strands do not contribute to beam flexural strength. This is because, away from the slab edge, the post-tensioning produces a fairly uniform compressive stress field across the plate including the beam cross section. A reasonable approach is to calculate the average prestress acting on the combined slab-beam system, then apply this prestress to the T-beam cross section to determine the effective axial compression on the T-beam. This axial load, acting at the level of the slab, is used along with the beam longitudinal reinforcement to calculate the T-beam flexural strength. This recommendation applies only for interior connections that are far enough away from the slab edge so as to be fully stressed by the post-tensioning. It need not apply at an exterior connection close to the slab edge because slab post-tensioning will not effectively compress the beam at that location.

Once the beam is proportioned, the plastic moment strengths of the beam can be determined. The designer will use this moment strength to establish requirements for beam shear strength, beam-column joint strength, and column strength as part of the capacity-design process. Therefore, it should be calculated conservatively to include the as-designed cross section and expected material properties including strain-hardening. ACI 318 uses the probable moment strength M_{pr} for this purpose. Probable moment strength is calculated from conventional flexural theory considering the as-designed cross section, without strength reduction factor ϕ, and assuming reinforcement yield strength equal to at least $1.25f_y$. See Section 6.7.2 for an example of the calculation of M_{pr}. Although $1.25f_y$ is a low estimate of the stress that will develop in the longitudinal reinforcement of a beam plastic hinge, it is expected that overstrength in materials throughout the structure, as well as the use of strength reduction factors $\phi < 1$ for other parts of the structure, will provide adequate conservatism for design of capacity-protected members. Therefore, the factor 1.25 should be sufficient for general use. The designer can consider a larger factor on a case-by-case basis where especially critical design elements are encountered.

Flexural (longitudinal) reinforcement should be sized so that adequate development is available within beam-column joints (ACI 318 limits the bar diameter to $h_c/20$ at interior joints). ACI 318 limits the maximum longitudinal reinforcement ratio to 0.025, but 0.01 is more practical for constructability and to keep joint shear forces within reasonable limits. The usual lower limits for longitudinal reinforcement also apply. Furthermore, reinforcement quantities at the top and bottom of the beam should also be reasonably balanced to encourage complete reversal of curvature upon moment reversal. [If top reinforcement area significantly exceeds bottom reinforcement area, cracks that open when the beam is flexed in negative moment (top in tension) will not close when moment is reversed.]

Besides providing the required strength, the longitudinal reinforcement should provide continuity along the span. Bar cutoffs are permitted at sections of reduced design moment, but at least two bars should be effectively continuous top and bottom, and moment strength at any section should not drop below one-quarter of the maximum strength. Lap splices should not be used within joints, within $2h_b$ of the joint face, or at other sections where analysis indicates yielding may occur. Mechanical splices, if used, preferably are not located within $2h_b$ of the joint face. If mechanical splices are required within $2h_b$ of the joint face, they must be Type 2 mechanical splices capable of developing at least the specified tensile strength of the spliced bar (or actual strength if possible).

Figure 12.37 illustrates the moment strength and longitudinal reinforcement requirements of ACI 318.

12.6.2 Joint Shear and Anchorage

Once the flexural reinforcement in the beams has been determined, the next design step is to check the joint shear in the beam-column joints. Joint shear will often drive the size of the moment frame columns.

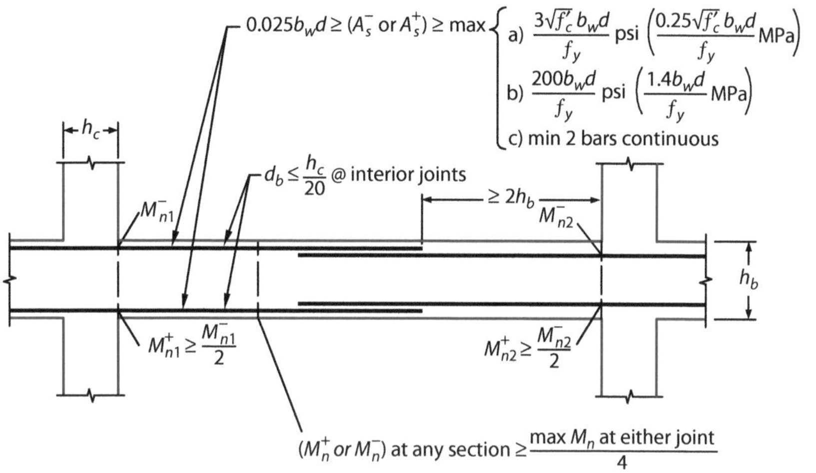

Figure 12.37 Beam flexural reinforcement requirements according to ACI 318.

ACI 318 contains *minimum requirements* for beam-column joints in buildings located in regions of highest seismicity; those requirements are adopted in most highly seismic jurisdictions of the United States. In contrast, ACI 352 (2002) provides *recommended practice* that generally equals or exceeds requirements of ACI 318. ACI 352 also includes connections for both highly seismic regions and less demanding regions. This text recommends following the recommendations of ACI 352, with minor exceptions as noted. The design procedure involves six steps, as follows:

1. Classify joint according to loading conditions and geometry.
2. Determine joint shear demands.
3. Size the connection for joint shear demands.
4. Develop beam and column longitudinal reinforcement.
5. Provide joint confinement.
6. Provide adequate strength and detailing in adjacent columns.

See Section 9.6 for complete discussion of the design approach for beam-column joints.

Detailing beam-column joints is an art requiring careful attention to several code requirements as well as construction requirements. Figure 12.38 shows example details for interior and exterior beam-column joints. Note that beam bars, possibly entering the joint from two different framing directions, must pass by each other and the column longitudinal bars. Joint hoop reinforcement, continuous through the height of the joint, adds to congestion. Large-scale drawings or physical mockups of beam-column joints should be prepared prior to completing the design so that adjustments can be made to improve constructability. Section 12.8 discusses this subject in more detail.

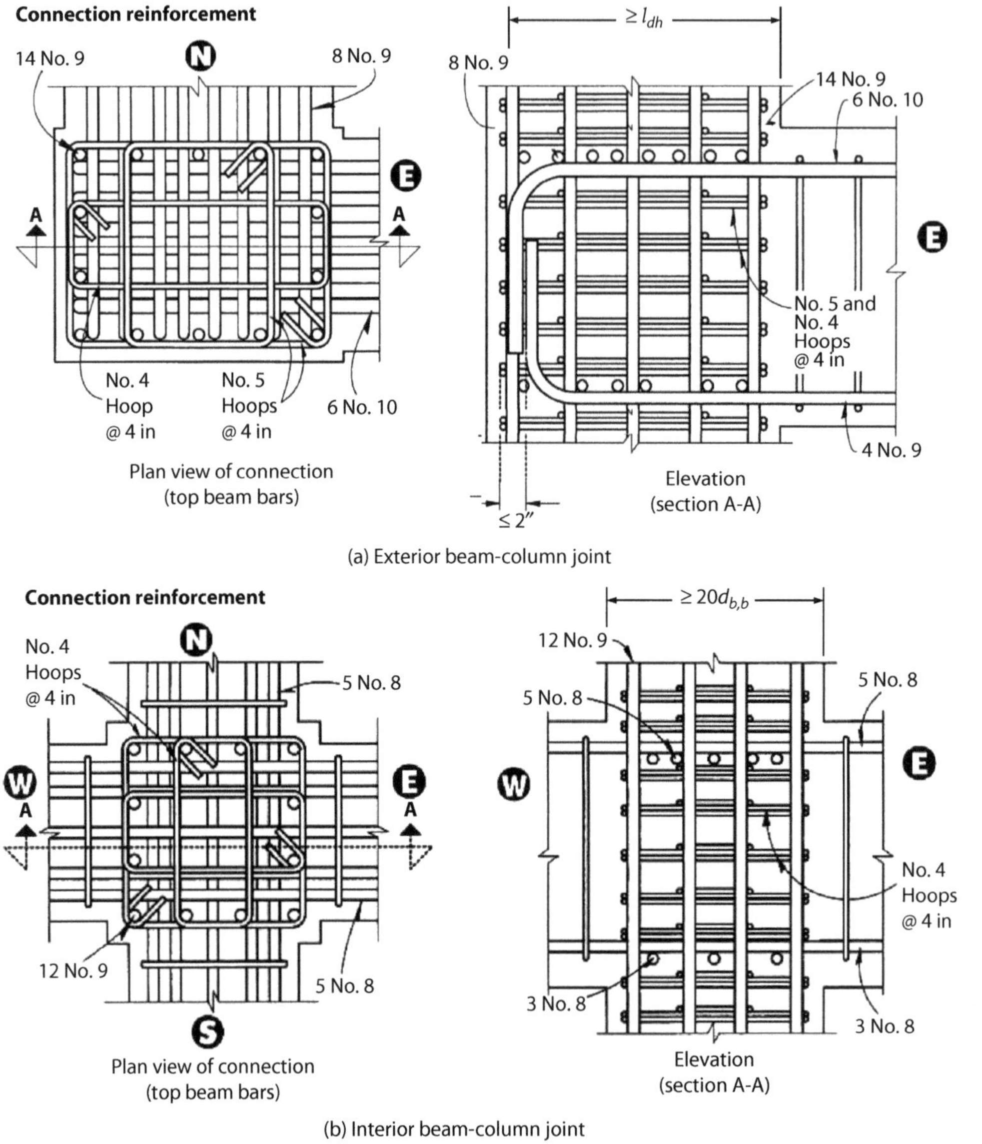

FIGURE 12.38 Example details for beam-column joints. (*After ACI 352, 2002, used with permission from American Concrete Institute.*)

All of the sketches shown in this chapter show the beam narrower than the column, but this is not always the case. Sometimes the beam and column are made the same width for constructability. This can create a problem in that the outermost beam bars will interfere with the outermost column bars. See Section 12.8.1 for additional discussion. In yet other cases, such as where story height and required headroom limits the beam depth, the beam may be wider than the column. In this case, beam longitudinal reinforcement outside the column width must be tied into

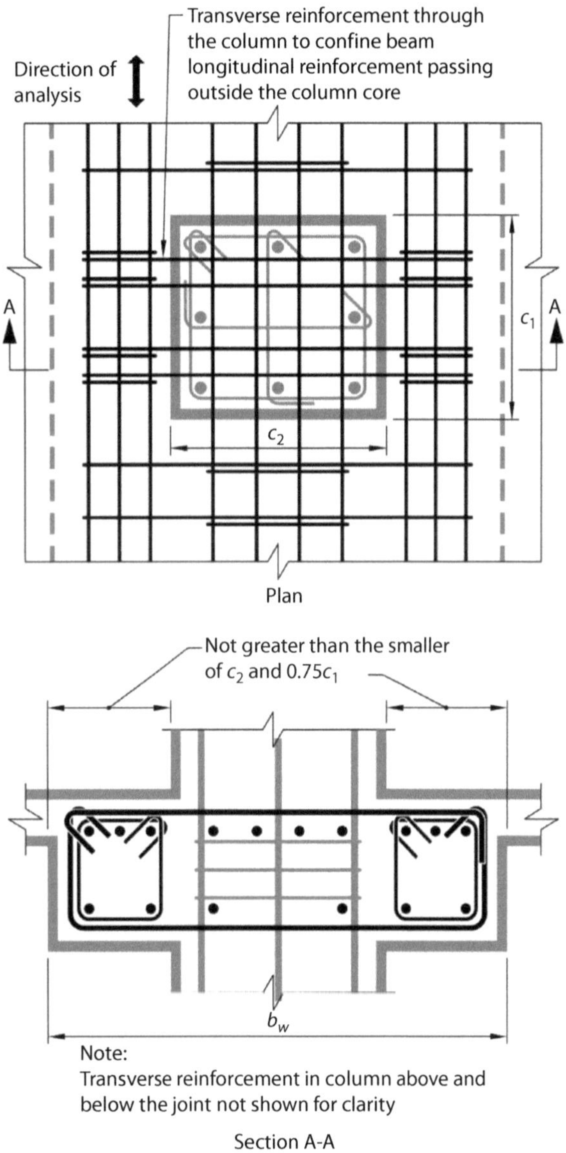

FIGURE 12.39 Wide beam details. (*After ACI 318, 2014, used with permission from American Concrete Institute.*)

the column with transverse reinforcement such that effective moment transfer occurs (Figure 12.39). Where a wide beam is used, the effective joint width should not be taken any greater than the overall width of the column.

12.6.3 Beam Shear and Transverse Reinforcement

The designer determines the beam design shear using the capacity-design approach as discussed in Section 12.4.3 and illustrated in Figure 12.19, considering frame sway both to the right and to the left. For a typical beam in a special moment frame, the resulting beam shears do not trend to zero near midspan as they would in a gravity-only beam. Instead, most beams of a special moment frame will have non-reversing shear demand along the length of a beam. If the shear does reverse along the span, the beam is one for which non-reversing beam plastic hinges will occur (see Section 12.4.2).

Typical practice for gravity-load design of beams is to take the design shear at d away from the column face. For special moment frames, the shear gradient typically is low such that the design shear at d is only marginally less than at the column face. Thus, for simplicity the design shear value usually is evaluated at the column face.

Beams in special moment frames must have either hoops or stirrups along their entire length. Hoops fully enclose the beam cross section and are provided to confine the concrete, restrain longitudinal bar buckling, improve bond between reinforcing bars and concrete, and resist shear. Stirrups, which generally are not closed, are used where only shear resistance is required.

To determine where hoops are required versus stirrups, we can divide the beam span into three different zones: the zones at each end of the beam where flexural yielding is expected to occur; the zone along lap-spliced bars, if any; and the remaining length(s) of the beam. See Figure 12.40.

The zone at each end, of length $2h_b$, must be well confined because this is where the beam is expected to undergo flexural yielding and this is the location with the highest shears. Therefore, codes require closely spaced closed hoops in this zone. Note that if flexural yielding is expected anywhere along the beam span other than the end of the beam, hoops must extend $2h_b$ on both sides of that yielding location. This latter condition is one associated with non-reversing beam plastic hinges (see Section 12.4.2), and is not recommended. Subsequent discussion assumes that this type of behavior is avoided by design.

The hoop reinforcement may be constructed of one or more closed hoops. Alternatively, U.S. practice permits it to be constructed of a typical beam stirrup with seismic hooks at each end closed off with a crosstie (or cap tie) (Figure 12.41). If the crosstie is confined by a slab on both sides, it can have 135° and 90° hooks at opposite ends that alternate along the beam length. If it is confined by a slab on one

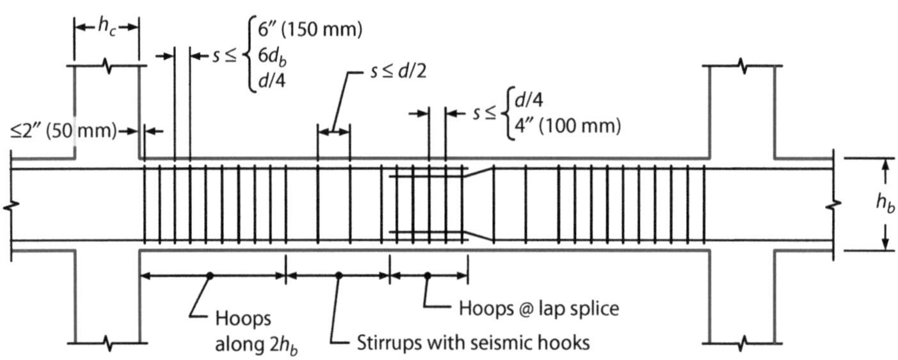

FIGURE 12.40 Hoop and stirrup locations and spacing requirements satisfying ACI 318.

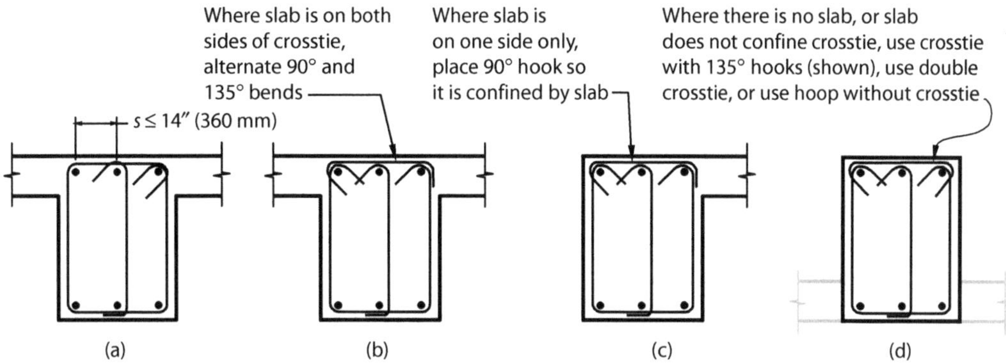

Notes:
1. Beam longitudinal bars are to be supported by hoops and crossties to resist buckling under deformation reversals. Hoop legs and crossties are arranged such that every corner and alternate longitudinal bar has lateral support provided by the corner of a hoop or crosstie with no bar farther than 6 in (150 mm) clear from a supported bar.
2. Horizontal spacing of crossties or legs of overlapping hoops not to exceed 14 in (360 mm).
3. Alternate crossties ( ) end for end unless otherwise noted.

Figure 12.41 Beam hoop reinforcement options: (*a*) hoop with vertical crosstie; (*b*) crosstie used to close a hoop where slab confines crosstie on both sides; (*c*) crosstie with 90° hook on same side as slab; (*d*) alternate detail where slab does not confine crosstie.

side, the 90° hook must be on the slab side where it will be confined, without alternating. If the crosstie is not confined by a slab, ACI 318 permits use of a crosstie with alternating 135° and 90° hooks, but it is preferable to use 135° hooks on both sides or 135°/90° crossties should be doubled up so that a 135° hook is at both ends.

Within potential plastic hinge regions, longitudinal bars will be subject to alternating tension and compression post-yield strain reversals. Hoops are required to support these bars against buckling. U.S. practice requires that at least alternate bars be supported by legs of hoops or crossties, and that no unsupported longitudinal bar be more than 6 in (150 mm) clear from a restrained bar. The maximum horizontal spacing between restrained longitudinal bars is 14 in (360 mm). Smaller spacing is recommended to improve confinement of the beam cross section.

The strength design requirement for shear is $\phi V_n \geq V_u$. In U.S. practice, $\phi = 0.75$ and nominal shear strength is $V_n = V_c + V_s$, with the following exception. Within $2h_b$ from the face of joints, or $2h_b$ on both sides of a section where flexural yielding is likely to occur, it is required to take $V_c = 0$ where both of the following conditions occur:

- The earthquake-induced shear force represents half or more of the required shear strength within the length $2h_b$.
- The factored axial compressive force, P_u, including earthquake effects is less than $A_g f_c'/20$.

By setting $V_c = 0$, additional transverse reinforcement is required with the intent of offsetting shear strength degradation associated with inelastic rotation cycles of the potential plastic hinge.

If beam longitudinal bars are lap-spliced, hoops must be provided to confine the length of the lap. According to ACI 318, corner and alternate longitudinal bars around the perimeter of the cross section at the lap are to be supported by legs of hoops or crossties, with no unsupported longitudinal bar more than 6 in (150 mm) from a supported longitudinal bar. Beam longitudinal bar lap splices are not to be used (a) within the joints; (b) within a distance of $2h_b$ from the face of the joint; and (c) where analysis indicates flexural yielding is likely due to inelastic lateral displacements of the frame. Generally, then, lap splices, if used, are placed near the midspan of the beam.

Summarizing, hoops are required along the beam end zones (where flexural yielding is expected) and along lap splices, with spacing limits as noted in Figure 12.40. Elsewhere, transverse reinforcement can be in the form of beam stirrups with seismic hooks at spacing not to exceed $d/2$.

Where hoops are provided at each end of the beam and along a splice, there may not be much length left where stirrups are acceptable. Because of this, and to prevent placement errors, it is practical to extend the hoop detail and spacing over the entire length of the beam. The design should conduct a quick quantity comparison to see the difference in the amount of detailed reinforcement. Both the weight of reinforcement and the number of pieces to be placed in the field affect the cost and should be considered when specifying the hoops and stirrups. If a design with hoops and stirrups of differing configurations and spacing is specified, the designer needs to conduct more rigorous observations to ensure that the ironworkers and special inspectors have a clear understanding of the placement requirements. These observations are most crucial early in the construction process when the first level of beams is constructed. Generally, after the first level the reinforcement pattern is appropriately replicated.

12.6.4 Column Design and Reinforcement

There are several strength checks associated with the columns of a special moment frame. As a first approximation, the column can be designed for the maximum factored gravity loads while limiting the area of longitudinal reinforcement to between 1% and 3% of the gross cross-sectional area. U.S. practice allows the longitudinal reinforcement to reach 6% of the gross section area, but this amount of reinforcement results in very congested splice locations. Mechanical couplers should be considered where the reinforcement ratio exceeds 3%.

Current U.S. practice permits tied column axial load as high as $0.80\phi P_o = 0.52P_o$, but good design practice would aim for lower axial loads. For axial load exceeding the balanced axial load, flexural yielding involves "yielding" of the compression zone, which can compromise ductility capacity. A reasonable design goal is to keep axial loads at or below the balanced point (or approximately $0.30 f_c' A_g$). Higher axial loads will require relatively larger amounts of confinement reinforcement if the column is to perform as intended, which can lead to constructability problems.

The seismic forces acting on a moment frame generally do not make large contributions to the axial load at interior columns. The designer should give special attention to the axial load in the exterior and corner columns because the gravity contribution may be small in comparison with the contribution from the seismic forces. See Section 12.4.3 for additional discussion on estimation of column axial loads.

In general, columns should be designed to satisfy the strong-column/weak-beam intent of the building code. In ACI 318, this requirement is stated as

$$\sum M_{nc} \geq \frac{6}{5} \sum M_{nb} \tag{12.11}$$

That is, at each beam-column connection the sum of the nominal moment strengths of the columns must be at least 1.2 times the sum of the nominal moment strengths of the beams framing into the joint (Figure 12.42*a*). The developed slab reinforcement within the effective flange width is to be included as beam flexural tension reinforcement when computing beam strength M_{nb}. Equation (12.11) must be verified independently for sway in both directions within a plane (e.g., East and West) and in each of the two principal framing directions (e.g., EW and NS). Column moment strength is dependent on the axial load. Therefore, if column axial loads vary for different load combinations or loading directions, the column moment strength used in Eq. (12.11) must be the smallest strength for the associated load combination (Figure 12.42*b*). The load combinations shown in Figure 12.42*b* are from ASCE 7. In most cases, the strong-column/weak-beam requirement controls the moment strength of the column.

ACI 318 does not require the column to satisfy the requirement $\phi M_{nc} \geq M_{uc}$, where M_{uc} is the column moment obtained from structural analysis under factored load combination including earthquake effect *E*. Such moments should be considered near the base, near the roof, and at other discontinuities.

The building code may contain provisions that exempt the column from satisfying Eq. (12.11) if the axial load under all load combinations is less than some minimum value. With the exception of roof connections, however, this

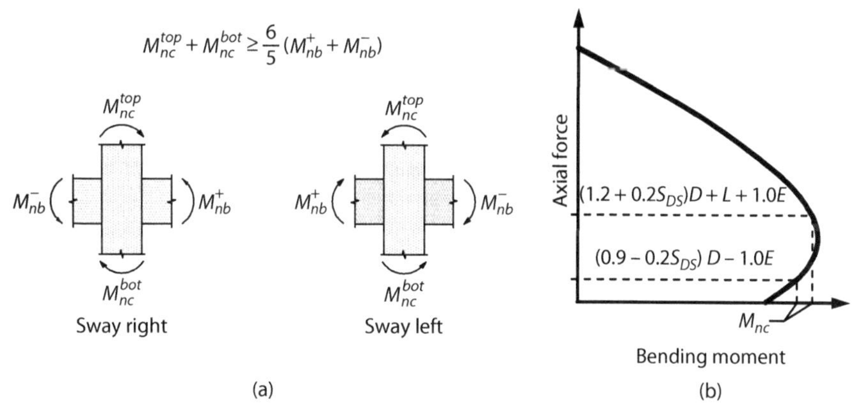

FIGURE 12.42 Strong-column/weak-beam design requirement. (*a*) Both sway right and sway left loading cases need to be considered. (*b*) Column axial force varies with load combinations and with earthquake loading direction, resulting in different column moment strengths. The sum of column strengths must exceed the sum of beam strengths for all load combinations.

exemption is not recommended because even lightly loaded columns, if flexurally weak, can form undesirable column sway mechanisms. At roof connections, it may be difficult to provide enough reinforcement in the column to satisfy Eq. (12.11). This is not a serious issue, as column yielding just below the roof is unlikely to lead to formation of an undesirable story mechanism. In this case, the detail should provide confinement along the length l_o below the roof-level beams regardless of the axial load level.

In still other cases it may not be practical to satisfy the strong-column/weak-beam provisions for a small number of columns. The strength and stiffness of such columns should not be included in the structural analysis model when determining the requirements for the seismic-force-resisting system. Columns not designated as part of the seismic-force-resisting system should be detailed accordingly (see Chapter 14).

Column transverse reinforcement is required to resist shear, confine lap splices, and confine zones where flexural yielding is anticipated. Therefore, the transverse reinforcement details may vary over height. Figures 12.43 and 12.44 show details required by ACI 318. Longitudinal bars should be well distributed around the perimeter of the cross section. Longitudinal bar lap splices, if any, must be located along the middle of the clear height and should not extend into the length l_o at the column ends. Such lap splices require closely spaced, closed hoops along the lap length. As an alternative, Type 2 mechanical splices can be used

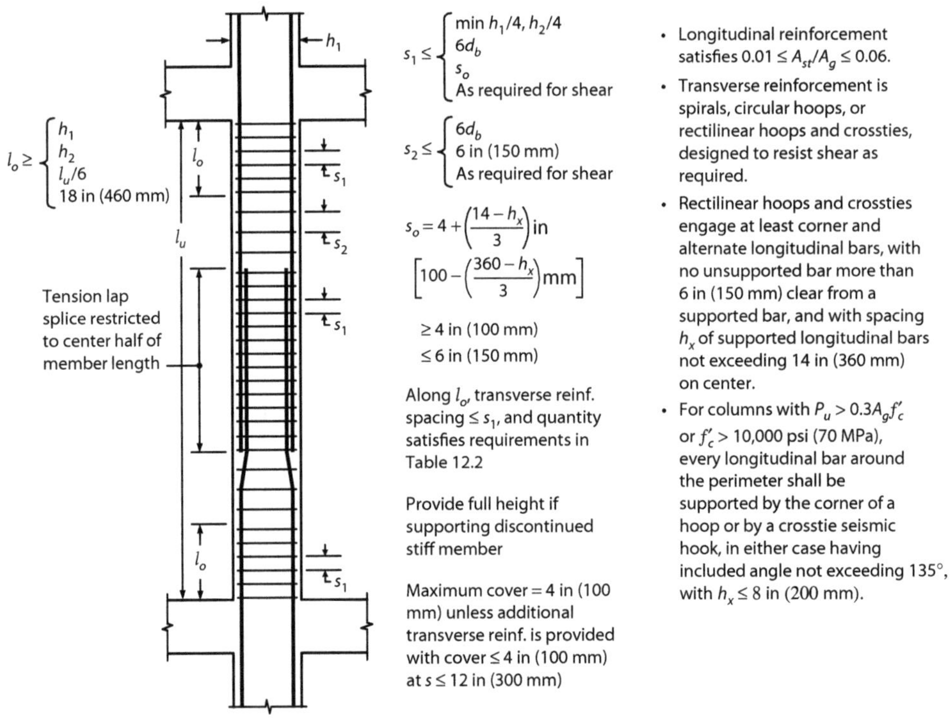

FIGURE 12.43 Column reinforcement requirements of ACI 318.

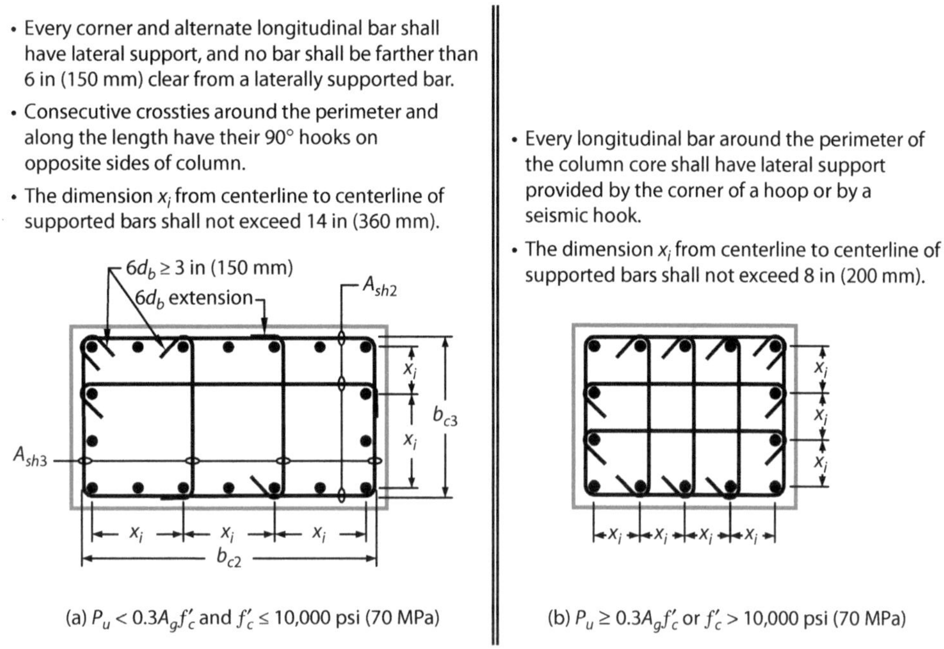

FIGURE 12.44 Column transverse reinforcement details.

anywhere along the column length, but preferably along the middle half of the column length. Type 1 mechanical splices may be permitted, but only if located at least twice the member depth from the face of the beams. Closely spaced hoops are also required along the length l_o measured from both ends to confine the concrete and restrain longitudinal bar buckling in case column flexural yielding occurs. Along the entire length, shear strength must be sufficient to resist the design shear forces, with maximum hoop spacing of $d/2$, where d is commonly taken equal to 0.8 times the column cross-sectional dimension h.

The designer should initially select column transverse reinforcement based on the confinement requirements. According to ACI 318-14, potential plastic hinge regions of special moment frame columns must have transverse reinforcement satisfying all the requirements listed in Table 12.2.

In Table 12.2, the concrete strength factor, k_f, and confinement effectiveness factor, k_n, are calculated according to Eqs. (12.12) and (12.13).

$$k_f = \frac{f'_c}{25{,}000} + 0.6 \geq 1.0, \text{ psi}$$

$$\left(k_f = \frac{f'_c}{172} + 0.6 \geq 1.0, \text{ MPa} \right) \tag{12.12}$$

$$k_n = \frac{n_l}{n_l - 2} \tag{12.13}$$

Transverse Reinforcement	Conditions	Applicable Expressions	
$\dfrac{A_{sh}}{sb_c}$ for rectilinear hoop	$P_u \le 0.3A_g f'_c$ and $f'_c \le 10{,}000$ psi (70 MPa)	Greater of (a) and (b)	$0.3\left(\dfrac{A_g}{A_{ch}} - 1\right)\dfrac{f'_c}{f_{yt}}$ (a) $0.09\dfrac{f'_c}{f_{yt}}$ (b)
	$P_u > 0.3A_g f'_c$ or $f'_c > 10{,}000$ psi (70 MPa)	Greatest of (a), (b), and (c)	$0.2k_f k_n \dfrac{P_u}{f_{yt}A_{ch}}$ (c)
ρ_s for spiral or circular hoop	$P_u \le 0.3A_g f'_c$ and $f'_c \le 10{,}000$ psi (70 MPa)	Greater of (d) and (e)	$0.45\left(\dfrac{A_g}{A_{ch}} - 1\right)\dfrac{f'_c}{f_{yt}}$ (d) $0.12\dfrac{f'_c}{f_{yt}}$ (e)
	$P_u > 0.3A_g f'_c$ or $f'_c > 10{,}000$ psi (70 MPa)	Greatest of (d), (e), and (f)	$0.35k_f \dfrac{P_u}{f_{yt}A_{ch}}$ (f)

TABLE 12.2 Transverse Reinforcement for Columns of Special Moment Frames

in which n_l = number of longitudinal bars or bar bundles around the perimeter of a column core that are laterally supported by the corner of hoops or by seismic hooks. According to ACI 318, the value of f_{yt} is limited to 100,000 psi (690 MPa). The form of some of the expressions here is similar to that of the Canadian Standard Association (CSA, 2004). See Elwood et al. (2009) for additional discussion of this and other codes.

The confinement equations for rectilinear hoops define the confinement in each principal direction and therefore must be checked in both principal directions of the column cross section. Thus, as illustrated in Figure 12.44, to determine total hoop leg area A_{sh2}, the dimension b_{c3} is substituted for b_c in Table 12.2, while to determine A_{sh3}, dimension b_{c2} is used. See Section 5.5.4 for additional discussion.

Once the confinement reinforcement is selected, the shear strength of the section needs to be checked. ACI 318 provisions for design shear force have some shortcomings, as discussed in Section 12.4.3. In lieu of the ACI 318 provisions, the procedure outlined in Section 12.4.3 is recommended. Summarizing, the design shear V_u is recommended to be taken equal to either of the following:

(a) V_u determined as $2M_{pr}/l_u$, where M_{pr} is probable moment strength of the column calculated without strength reduction factor, using steel yield stress equal to at least $1.25f_y$, and taking M_{pr} as the greatest moment over the range of anticipated column axial forces including earthquake effect. See Figure 12.23.

(b) V_u determined from Eq. (12.8).

The shear strength design requirement is expressed by $\phi (V_c + V_s) \geq V_u$, with $\phi = 0.75$. Consistent with beam design, in U.S. practice, V_c is set to zero over the length of l_o for any load combination for which both (a) and (b) occur:

(a) The column has factored axial force including earthquake effect $P_u \leq A_g f_c'/20$.

(b) Earthquake-induced shear force represents one-half or more of the total design shear.

As with beams, where column longitudinal bars are lapped there might not be much length left to take advantage of the more relaxed column tie spacing outside the l_o regions shown in Figure 12.43. For this reason it is common practice to call out uniform hoop spacing to prevent misplaced ties during construction. Where bars are not spliced at every floor, perhaps every other floor, more economy can be realized by specifying a larger spacing between the l_o regions. The benefit can be seen simply by counting the number of ties that can be saved as the spacing is relaxed.

If columns of a special moment frame extend below the base of the structure as shown in Figure 12.35, and those columns are required to transmit forces resulting from earthquake effects to the foundation, then those columns must satisfy the detailing and proportioning requirements for columns of special moment frames. In most conditions, the columns of a special moment frame will be carrying seismic forces over their entire height, so providing full-height ductile detailing is required.

Where a column frames into a strong foundation element or wall, it is highly likely that the column will yield at that connection under design earthquake loading. A conservative approach to detailing at such locations is recommended. A good approach is to increase the length of the confinement zone at the base to $1.5l_o$.

12.7 Additional Requirements

12.7.1 Special Inspection

Reinforced concrete special moment frames are complex structural elements whose performance depends on proper implementation of design requirements during construction. Therefore, wherever a special moment frame is used, regardless of the Seismic Design Category, continuous inspection of the placement of the reinforcement and concrete should be done by a qualified inspector. The inspector should be under the supervision of the licensed design professional responsible for the structural design or under the supervision of a licensed design professional with demonstrated capability for supervising inspection of construction of special moment frames. Continuous special inspection generally is interpreted to mean that the special inspector is on the site at all times observing the work requiring special inspection.

Generally, the special inspector is required to check work for conformance to the approved design drawings and specifications. Contract documents specify that the special inspector will furnish inspection reports to the building official, the licensed design professional, or other designated persons. Discrepancies are to be brought to the immediate attention of the contractor for correction, and then, if uncorrected, to the proper design authority and to the building official. A final signed report is to be submitted stating whether the work requiring special inspection was, to the best of the inspector's knowledge, in conformance with the approved plans and specifications and the applicable workmanship provisions of the Code.

12.7.2 Material Properties

Wherever a special moment frame is used, regardless of the Seismic Design Category, ACI 318 stipulates that materials conform to special requirements. These requirements are intended to result in a special moment frame capable of sustaining multiple inelastic deformation cycles without critical degradation. The following requirements apply.

Concrete

The specified compressive strength of concrete, f_c', is not to be less than 3000 psi (21 MPa). Additional requirements apply where lightweight concrete is used, but this text does not recommend lightweight concrete for special moment frames. Where high-strength concrete is used, the value of $\sqrt{f_c'}$ is restricted to an upper-bound value of 100 psi (0.7 MPa) for any beam or column shear strengths or anchorage/development strengths. The limit does not apply to beam-column joint shear strength or to development of bars at beam-column joints. Beam-column joint shear strengths calculated without the 100-psi (0.7-MPa) limit were conservative for laboratory tests having concrete compressive strengths up to 15,000 psi (100 MPa) (ACI 352). Based on local experience, some jurisdictions impose additional restrictions on the use of high-strength concrete.

Reinforcement

Inelastic flexural response is anticipated for special moment frames subjected to design-level earthquake shaking. Longitudinal reinforcement properties must be strictly controlled to promote ductile and predictable flexural behavior. To control the moment strength, the yield strength must at least meet the specified yield strength but not exceed the specified yield strength by a large margin. Strain-hardening is required so that inelastic action will be forced to spread along the member span.

In the United States, deformed reinforcement resisting earthquake-induced flexural and axial forces in frame members must conform to ASTM A706, Grade 60. According to this specification, the actual yield strength must not exceed the specified yield strength by more than 18,000 psi (124 MPa) and the ratio of the actual tensile strength to the actual yield strength must be at least 1.25. A706 also has excellent strain ductility capacity and chemical composition that makes it more suitable for welding. Alternatively, ASTM A615 Grades 40 and 60 (280 and 420) reinforcement is permitted if (a) the actual yield strength based on mill tests does not exceed f_y by more than 18,000 psi (124 MPa), (b) ratio of the actual tensile strength to the actual yield strength is at least 1.25, and (c) elongation meets requirements for A706 reinforcement. The optional use of A615 reinforcement sometimes is adopted because A615 reinforcement may be more widely available in the marketplace and may have lower unit cost.

Market forces and construction efficiencies sometimes promote the use of higher yield strength longitudinal reinforcement [e.g., Grade 75 (520)]. This reinforcement may perform suitably if the elongation and stress requirements match those of A706 reinforcement. Higher-strength reinforcement results in higher unit bond stresses and requires longer development and splice lengths. This reinforcement is not accepted on a routine basis, but requires special approval. See Kelly et al. (2013).

Higher-strength transverse reinforcement is permitted, and can reduce congestion problems especially for large members using higher-strength concretes. Where used, the value of f_{yt} used to compute the amount of confinement reinforcement should not exceed 100,000 psi (690 MPa). ACI 318 restricts the value of f_{yt} used in design of shear reinforcement to 60,000 psi (420 MPa) except 80,000 psi (550 MPa) is permitted for welded deformed wire reinforcement.

Mechanical Splices

Longitudinal reinforcement in special moment frames is expected to undergo multiple yielding cycles in prescribed locations during design-level earthquake shaking. If mechanical splices are used in these locations, they should be capable of developing nearly the tensile strength of the spliced bars. Outside yielding regions, mechanical splices, if used, can have reduced performance requirements.

In U.S. practice, mechanical splices are classified as either Type 1 or Type 2 mechanical splices, as follows: (a) Type 1 mechanical splices are capable of at least $1.25f_y$ in tension or compression, as required; (b) Type 2 mechanical splices are capable of developing at least the specified tensile strength of the spliced bar.

Where mechanical splices are used in beams or columns of special moment frames, only Type 2 mechanical splices are permitted within $2h$ from the column or beam face or from sections where yielding of the reinforcement is likely to occur as a result of inelastic lateral displacements. Either Type 1 or Type 2 mechanical splices are permitted in other locations.

Welding

Special moment frames are anticipated to yield when subjected to design-level earthquake ground motions, so special care is required where welding is done. Welded splices in reinforcement resisting earthquake-induced forces must develop at least $1.25f_y$ of the bar and should not be used within $2h$ from the column or beam face or from sections where yielding of the reinforcement is likely to occur as a result of inelastic lateral displacements.

Welding of stirrups, ties, inserts, or other similar elements to longitudinal reinforcement that is required by design is not permitted because cross-welding can lead to local embrittlement of the welded materials. Welded products should only be used where test data demonstrate adequate performance under loading conditions similar to conditions anticipated for the particular application. Some welded products used for transverse reinforcement are not capable of developing the reinforcement yield strength. Localized overstress in such reinforcement cannot be distributed to adjacent transverse reinforcement sets, and can lead to brittle failure. In general, such special products are not recommended here.

12.7.3 Additional System Design Requirements

Where special moment frames are used, certain other requirements must be followed. In some cases these additional requirements apply only in highly seismic regions (Seismic Design Categories D, E, and F). Structural diaphragms and foundations are required in any building as part of the seismic-force-resisting system (see Section 11.4 and Chapters 15 and 16).

As discussed in Section 12.2.1, a prevalent practice is to designate only part of the structural system to resist design seismic forces with the remainder not designated as part of the seismic-force-resisting system. Sometimes referred to as "gravity-only systems," those parts of the building not designated as part of the seismic-force-resisting system need to be capable of safely supporting gravity loads while they are simultaneously loaded and deformed by the building as it sways under the design earthquake ground motions. Failure to provide this capability has resulted in building collapses in recent past earthquakes. Chapter 14 presents requirements for gravity-only systems.

12.8 Detailing and Constructability Issues

A special moment frame relies on carefully detailed and properly placed reinforcement to ensure that it can maintain its strength through multiple deformation cycles beyond the yield deformation. Architectural requirements often push to get the beams and columns as small as possible, resulting in beams, columns, and joints that can become very congested. Early on in the design process, the designer should ensure that the required reinforcement not only fits within the geometric confines of the elements, but also can be properly placed in the field.

The text that follows is based on construction experiences, both good and bad, and draws from Wyllie and LaPlante (2003).

12.8.1 Longitudinal Bar Compatibility

When laying out the beam and column reinforcement, it is helpful to visualize the intersections among the various reinforcement lines. The column longitudinal bars are located around the perimeter of the column cross section, establishing vertical lines of column reinforcement around which the beam reinforcement must pass. Orthogonal beams framing into a joint have horizontal planes of longitudinal reinforcement that must not intersect one another; layers of bars from one beam must pass beneath or above layers of bars from the orthogonal beam. Figure 12.38 shows two three-dimensional sketches illustrating acceptable arrangements of reinforcement to avoid interferences. Figure 12.45(a) shows a well-coordinated joint with three beam bars passing through a column face with four vertical bars.

(a) (b)

Figure 12.45 Beam and column longitudinal reinforcement conflicts at beam-column joints. (*a*) Beam bars gently swept into the column core, with stirrup legs pulled in to contain the corner beam longitudinal reinforcement. (*Photograph used with permission from Magnusson Klemencic Associates.*) (*b*) Beam bars sharply swept into column core. Small corner bars added to hold stirrups. Bar congestion and the swept longitudinal bars caused an interference with the seismic hooks on stirrup legs, forcing the contractor to improperly bend the hooks so they are no longer anchored in the beam core. This should be avoided. (*Photograph used with permission from Nabih Youssef Associates.*)

The beam-column joint can be the critical design region. By keeping the column and beam dimensions large, beam and column longitudinal reinforcement ratios can be kept low and beam-column joint volumes kept large so that joint shear stresses are within limits. Large joints with low reinforcement ratios also help with placement of reinforcing bars and concrete.

Beams and columns always need longitudinal bars close to their faces and at corners to hold the stirrups or ties. Where the beam and column are the same width, these bars will conflict within the joint. This requires bending and offsetting one set of the bars, which will increase fabrication costs. Offsetting bars can also create placement difficulties and result in bar eccentricities that affect seismic performance. If the beam is at least 4 inches wider or narrower than the column (2 inches on each side), the bars can be detailed so that they are in different lines and thus do not need to be offset.

One option, pictured in Figure 12.45*a*, is to gently sweep the corner beam bars to the inside of the column corner bars. At the top of the beam, the vertical legs of the stirrup can be bent inward such that they can continue to engage the top corner beam bars. Corner bars will not fit tightly within the bends of the cap tie, but the hook extensions of the 135° hooks are normally long enough so they are still anchored into the core of the beam. One might consider using 135° hooks at both ends of the cap tie to improve the anchorage into the core of the beam. Similarly, the longitudinal beam bars at the bottom corners of the beam must be swept inward, and this will result on those bars no longer being held tightly within the corners of the stirrups.

Another option, pictured in Figure 12.45*b*, is to provide a smaller, discontinuous bar to support the stirrups at the edge of the beam. This requires additional reinforcement that is not considered in determining the beam moment strength and requires more pieces to be placed. The added reinforcement should be of small diameter so it does not create a large discontinuity in moment strength of the beam in the potential plastic hinge region. Attention should be paid to potential interferences between the longitudinal bars and the stirrup hooks. Additionally, the requirement to laterally support corner and alternate longitudinal bars should be provided for.

Making the beam wider or narrower than the column may create undesirable conditions along the exterior edge of a floor and may increase forming costs for both exterior and interior framing locations. Consideration needs to be given to the architectural condition along this exterior location. Even though different beam and column widths may work well for the structure, this may create a complicated enclosure detail that is more costly.

To support the beam hoops and stirrups, some of the top bars must be made continuous with lap slices or mechanical couplers near midspan. To meet the negative moment strength requirements, shorter bars passing through the column can be added to the continuous top bars.

Avoid multiple layers of longitudinal bars where possible, as this condition makes placement difficult, especially when two or more layers of top bars must be hooked down into the joint at an exterior column. If more than one layer of bars is required, it may be because the beam is too small; enlarge the beam if possible. This situation also occurs where lateral resistance is concentrated in a few moment frames, requiring large, heavily reinforced beams (Figure 12.46).

12.8.2 Beam and Column Confinement

Confinement of beams and columns is crucial to the ductile performance of a special moment frame. Usually confinement is provided by sets of hoops or hoops with crossties. Several examples are shown in the figures of this section.

For beams, it is permitted to make up hoops[2] from either continuously wound pieces of reinforcement closed by seismic hooks[3] or to use stirrups with seismic hooks closed by cap ties (Figure 12.41). For columns, it is required to use one or more hoops that extend continuously around the perimeter of the column (Figure 12.44). Crossties should not be used to close an open leg of a hoop in a

---

[2]A closed or continuously wound tie. A closed tie can be made up of several reinforcement elements each having seismic hooks at both ends. A continuously wound tie requires a seismic hook at both ends.

[3]A seismic hook is a hook on a stirrup or crosstie having a bend not less than 135°, except that circular hoops shall have a bend not less than 90°. Hooks have a $6d_b$ but not less than 3 in (75 mm) extension that engages the longitudinal reinforcement and projects into the interior of the stirrup or hoop. See Figure 12.44.

FIGURE 12.46 Beam-column joint having multiple layers of beam reinforcement hooked at back side of joint. Note the upturned beam (the slab is cast at the bottom face of the moment frame beam). For this detail, cap ties to complete the beam hoops should have 135° bends at both ends (see Figure 12.41).

column. Use of crossties with 90° bend at one end is not permitted for this purpose. Hoops made up of intersecting headed bars are not permitted. Hoops made up of cross-welded wire should not be used unless it can be shown that the weld is capable of developing the strength of the interconnecting pieces.

As shown in Figure 12.44, hoop sets for columns can be made up of hoops and crossties. For lower axial loads and concrete compressive strengths, crossties are permitted to have a seismic hook at one end and a 90° hook at the other end, provided the 90° hooks are alternated end for end along the longitudinal axis of the member and around its perimeter. For higher axial loads or higher concrete compressive strengths, seismic hooks are required on both ends of the crossties. The seismic hooks are preferred from a seismic performance perspective; alternating 135° and 90° hooks is a compromise that improves constructability. The concrete cover on beams and columns may spall off during response to ground shaking, exposing the stirrup and tie hooks. A 90° hook can more easily be bent outward from internal pressure. If this happens, the stirrup or tie effectiveness is reduced. In contrast, a 135° hook will remain anchored in the core of the member when the concrete cover spalls. In many situations there is no real cost premium for 135° hooks.

Another option besides crossties with hooks is to use headed reinforcement. It is important to ensure that the heads properly engage the longitudinal reinforcement that they hold in place. Special inspection of their final placement is

Figure 12.47 Column cage with hoops constructed from single reinforcing bar.

required. Headed bars cannot be used to form perimeter hoops. Yet another option is to use continuously bent hoops, that is, hoops constructed from a single piece of reinforcement (Figure 12.47). Whereas these hoops can result in reinforcement cages with excellent tolerances, the pre-bent shape limits field adjustments that may be required when interferences arise.

As shown in Figure 12.44, ACI 318 permits the horizontal spacing between legs of hoops and crossties to be as large as 14 in (360 mm) in some columns. Confinement can be improved by reducing this spacing. Better performance will be achieved if the restrained longitudinal bars are spaced around the perimeter no more than 6 or 8 in (150 or 200 mm) apart. Decreasing the horizontal spacing of crosstie legs permits an increase in vertical spacing while maintaining confinement effectiveness. The extra vertical spacing can reduce the total number of hoop sets and facilitate working between hoop sets. Because a typical hoop set comprises a three-layer stack of bars (crossties in one direction, then the hoop, then the crossties in the other direction), the actual clear spacing between hoop sets can be quite small.

The ties and stirrups should be kept to No. 4 (13) or No. 5 (16) bars. Number 6 (19) and larger bars have large diameter bends and are difficult to place.

Spirally reinforced columns are more ductile than columns with ties and are therefore better for extreme seismic loads. The spirals need to be stopped below the beam-column joint because it is very difficult, if not impossible, to integrate the spirals with the longitudinal beam reinforcement. Because transverse reinforcement is required to extend through the joint, the spirals can be replaced within the joint by circular hoop reinforcement.

12.8.3 Bar Splices

Lap splices of longitudinal reinforcement must be positioned outside intended yielding regions. Considering that column and beam ends, as well as lap splice lengths, all require closely spaced hoops, it commonly becomes simpler to specify closely spaced hoops along the entire beam or column length if lap splices are used. This is especially common for columns.

Large-diameter bars require long lap splices. In columns, these must be detailed so that they do not extend outside the middle half of the column length and do not extend into the length l_o at the end of the column. If longitudinal bars are offset to accommodate the lap splice, the offset should also be outside the length l_o (Figure 12.48*a*). Lap splices of the longitudinal reinforcement may create a congested area of the column as the number of vertical bars is doubled and the hoops must be tightly spaced. Splicing the vertical bars at every other floor as shown in Figure 12.48*b* will eliminate some of the congestion. Mechanical splices may also help reduce congestion.

Figure 12.48 Column reinforcement cages. (*a*) Lap splice and offset bars outside the specially detailed lengths l_o. (*Photograph used with permission from Englekirk Partners Consulting Structural Engineers.*) (*b*) Splicing longitudinal bars at every other floors to reduce splicing and congestion. (*Photograph used with permission from Magnusson Klemencic Associates.*)

12.8.4 Concrete Placement

Regardless of the effort to make sure the reinforcing bars fit together, reinforcement congestion is higher in the beams, columns, and joints than in other structural elements such as slabs. To help achieve proper consolidation of the concrete in these congested areas, the maximum aggregate size should be limited accordingly. Specifying 1/2-in (13-mm) maximum aggregate size is common for special moment frames. Sometimes small aggregate size will result in lower concrete strength, but other components of the concrete mixture can be adjusted to make up the lost strength.

Another key to well-consolidated concrete in congested areas is having a concrete mixture with a high slump. A slump in the range of 7 to 9 in (180 to 230 mm) may be necessary to get the concrete to flow in the congested areas.

Good consolidation can be difficult in highly congested areas because the reinforcement blocks insertion of the vibration equipment. Sometimes contactors will position internal vibration equipment prior to placing the reinforcement. Alternatively, external vibration may be considered if there is adequate access to all sides of the formwork.

Difficulties with vibration do not come into play if self-consolidating concrete is used. These concrete mixtures are extremely fluid and easily flow around congested reinforcement. There is a cost premium associated with the self-consolidating concrete itself. This premium diminishes with increasing strength. The formwork required to hold this type of concrete must also be much tighter than with a standard concrete mixture. Using self-consolidating concrete successfully is highly dependent on the experience and preference of the contractor. For this reason, only specify self-consolidating concrete in the structural documents when it has been previously discussed with the contractor.

References

Abdel-Fattah, B., and J.K. Wight (1987). "Study of Moving Beam Plastic Hinging Zones for Earthquake-Resistant Design of Reinforced Concrete Buildings," *Structural Journal*, Vol. 84, No. 1, pp. 31–39.

ACI 318-83 (1983). *Building Code Requirements for Reinforced Concrete (ACI 318–83) and Commentary*, American Concrete Institute, Farmington Hills, MI.

ACI 318 (2014). *Building Code Requirements for Structural Concrete (ACI 318–14) and Commentary*, American Concrete Institute, Farmington Hills, MI.

ACI 352 (2002). *Recommendations for Design of Beam-Column Connections in Monolithic Reinforced Concrete Structures (352R.02)*, American Concrete Institute, Farmington Hills, MI.

ASCE 7 (2010). *Minimum Design Loads for Buildings and Other Structures*, ASCE 7-10, American Society of Civil Engineers, Reston, VA.

ASCE 41 (2013). *Seismic Evaluation and Retrofit of Existing Buildings*, American Society of Civil Engineers, Reston, VA.

Blume, J.A., N.M. Newmark, and L.H. Corning (1961). *Design of Multistory Reinforced Concrete Buildings for Earthquake Motions*, Portland Cement Association, Chicago, IL, 318 pp.

CSA (2004). *Design of Concrete Structures*, CSA A23.3-04, Canadian Standards Association, 2004, Mississauga, Canada.

Elwood, K.J. (2014). Data in support of ACI 318-14 Code Change Proposal CH031. Personal communication.

Elwood, K.J., J. Maffei, and K.A. Riederer (2009). "Improving Column Confinement—Part 2: Proposed New Provisions for the ACI 318 Building Code," *Concrete International*, Vol. 31, No. 12, pp. 41–48.

Haselton, C.B., and G.D. Deierlein (2008). *Assessing Seismic Collapse Safety of Modern Reinforced Concrete Moment-Frame Buildings*, PEER 2007/08, University of California, Berkeley, CA, 274 pp.

Kelly, D.J., A. Lepage, D. Mar, J.I. Restrepo, J.C. Sanders, and A.W. Taylor (2013). *Use of High-Strength Reinforcement for Earthquake Resistant Concrete Structures*, NIST GCR 13-917-30, National Institute of Standards and Technology, Gaithersburg, MD, 230 pp.

Kelly, T. (1974). *Some Seismic Design Aspects of Multistorey Concrete Frames*, Master of Engineering Report, University of Canterbury, Christchurch, New Zealand, 163 pp.

Moehle, J.P., J.D. Hooper, and C.D. Lubke (2008). *Seismic Design of Reinforced Concrete Special Moment Frames: A Guide for Practicing Engineers*, NEHRP Seismic Design Technical Brief No. 1, NIST GCR 8-917-1, National Institute of Standards and Technology, Gaithersburg, MD.

NZS3101 (2006). *Concrete Design Standard, NZS3101:2006, Part 1 and Commentary on the Concrete Design Standard, NZS 3101:2006, Part 2*, Standards Association of New Zealand, Wellington, New Zealand.

Pantazopoulou, S.J., and X. Qi (1991). "Response of RC Frame under Lateral Loads," *Journal of Structural Engineering*, Vol. 117, No. 4, pp. 1167–1188.

Shahrooz, B.M., and J.P. Moehle (1987). *Experimental Study of Seismic Response of R.C. Setback Buildings*, EERC Report No. UCB/EERC-87/16, University of California, Berkeley, CA, 347 pp.

TBI (2010). *Guidelines for Performance-Based Seismic Design of Tall Buildings*, Report No. 2010/05, Pacific Earthquake Engineering Research Center, University of California, Berkeley, CA, 84 pp.

UBC (1973). *Uniform Building Code*, International Conference of Building Officials, Whittier, CA.

UBC (1983). *Uniform Building Code*, International Conference of Building Officials, Whittier, CA.

Visnjic, T. (2014). *Design Considerations for Earthquake-Resistant Reinforced Concrete Special Moment Frames*, Doctoral Dissertation, University of California, Berkeley, CA.

Visnjic, T., M. Panagiotou, and J.P. Moehle (2014). "Seismic Response of 20-Story Tall Reinforced Concrete Special Moment Resisting Frames Designed with Current Code Provisions," *Earthquake Spectra*, doi: http://dx.doi.org/10.1193/082112EQS267M.

Wyllie, L.A., and R.W. LaPlante (2003). *The Designer's Responsibility for Rebar Design*, The Structural Bulletin Series No. 1, Concrete Reinforcing Steel Institute, Schaumburg, IL, 16 pp.

CHAPTER 13

Special Structural Walls[1]

13.1 Preview

Earthquake-resistant concrete buildings are braced by either moment-resisting frames or structural walls. Chapter 12 addressed the design requirements for special moment frames. Chapter 13 addresses behavior and design requirements for structural walls intended for use in regions of highest seismicity. In U.S. design practice these are known as *special structural walls* or, equivalently, as *special shear walls*.

Special structural walls are proportioned and detailed to resist combinations of shear, moment, and axial force that arise as a building sways through multiple displacement cycles during strong earthquake ground shaking. Special proportioning and detailing requirements produce a wall capable of resisting strong earthquake shaking without critical loss of stiffness or strength. Special structural walls may include coupling beams. *Ordinary structural walls* also are defined in the building codes, but these are not permitted in regions of highest seismicity. *Intermediate precast structural walls* are permitted in regions of highest seismicity, but only in buildings up to 40 ft (12 m) tall, and typically for tilt-up construction. These latter systems are not covered in this chapter.

This chapter reviews the use of special structural walls, introduces the principles on which design is based, and discusses observations on structural wall behavior. It then describes in detail the analysis, design, and construction of special structural walls. Although the design requirements are drawn mainly from ACI 318, provisions of other building codes are introduced where applicable.

13.2 The Use of Special Structural Walls

13.2.1 Structural Walls in Buildings

Walls proportioned to resist combinations of shears, moments, and axial forces are referred to as *structural walls*. In U.S. design practice, a *special structural wall* is one satisfying requirements of ACI 318, Chapter 18 (*Earthquake-Resistant Buildings*),

[1]This chapter uses figures and text from Moehle et al. (2011). The significant writing contributions of Messrs. Ghodsi, Hooper, Fields, and Gedhada are gratefully acknowledged.

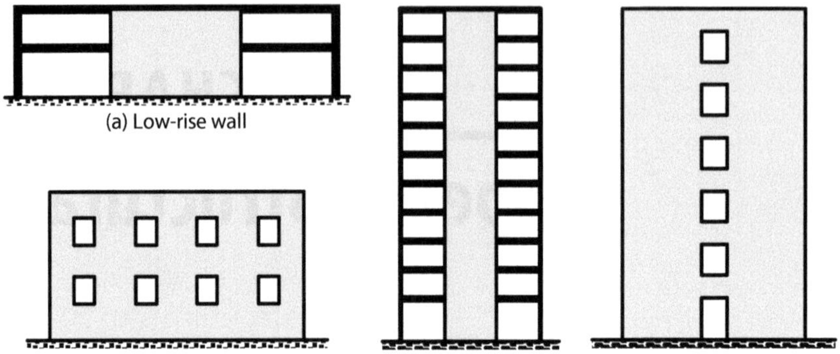

Figure 13.1 Sample structural wall elevations.

intended to result in strength and toughness required to resist earthquake effects in buildings assigned to Seismic Design Categories D through F.

Figure 13.1 illustrates some wall configurations. Solid walls are widely used to brace low-rise buildings. Sometimes walls are perforated with openings. In taller buildings, walls cantilever from a foundation to provide bracing over building height. In a coupled wall system, several window or door openings are aligned over height, with coupling beams or slabs connecting the separate walls.

13.2.2 When to Use Structural Walls

Selection of special structural walls as the primary seismic-force-resisting elements is influenced by considerations of seismic performance, functionality, constructability, and cost. For low- to mid-rise buildings, structural walls typically are more cost effective than other systems such as concrete special moment frames. Structural walls are used in concrete buildings with limited floor-to-floor heights or other architectural constraints that cannot accommodate frame beam depths. Structural walls provide considerable stiffness, improving performance or enabling longer spans while still satisfying code drift limits. Structural walls surrounding stairs and elevator cores serve a dual purpose of enclosing vertical shafts while providing efficient axial and lateral resistance.

For buildings in which special structural walls are the sole seismic-force-resisting system, ASCE 7 (2014) limits height to 160 ft (48.8 m) in Seismic Design Category D and E and 100 ft (30.5 m) in Seismic Design Category F. These heights can be increased to 240 ft (73.2 m) and 160 ft (48.8 m) if there is no extreme torsional irregularity (as defined in ASCE 7) and the shear in any line of walls does not exceed 60% of the total story shear. There is no height limit for a dual system combining walls with special moment frames capable of resisting at least 25% of prescribed seismic forces.

13.2.3 Wall Layout

Structural walls generally are stiff structural elements whose placement in a building can strongly affect building performance. Walls should be configured and located considering the range of loads the building will experience during its service life. The engineer and architect should work together to achieve a layout that meets structural, architectural, and programmatic requirements of the project. Aspects of wall layout and configuration that affect structural performance are summarized below.

Plan Layout

Walls should be well distributed within the building plan, with multiple walls resisting story shears in each principal direction. Preferably, long diaphragm spans are avoided because they cause large diaphragm moments, shears, and collector forces (Chapter 15). Furthermore, walls should be positioned such that their center of resistance is close to the center of mass, thereby avoiding induced torsion (Figure 13.2). Walls located near the perimeter may be preferred because they maximize torsional resistance.

Building codes contain provisions that quantify torsional irregularity, sometimes with penalties for large irregularities. Generally, these code requirements only consider linear-elastic response. Because inelastic response usually is expected, the designer might consider balancing torsional resistance at the strength level also. Thus, for loading in each principal direction, and for both linear-elastic response and for response at the strength level, wall shears should balance the total story shear without moment about the center of mass. Where identical walls are placed symmetrically, this objective is achieved automatically (Figure 13.2*b*). Additional design effort is required where walls of different cross sections are used (Figure 13.2*c*).

Tributary gravity loads help resist wall overturning moments, reducing reinforcement and foundation uplift demands. Thus, it may be advantageous to move walls inward from the perimeter (and away from adjacent columns) such that they support more gravity loads, as in Wall **e** in Figure 13.2, even though this reduces plan torsion resistance. Too much axial force, however, can result in undesirable compression-controlled flexural response with reduced drift capacity. A good plan layout balances these competing objectives.

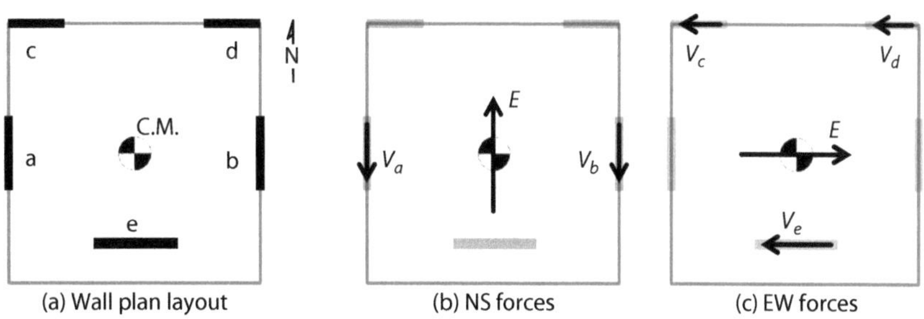

Figure 13.2 Example of wall plan layout. C.M. refers to center of mass.

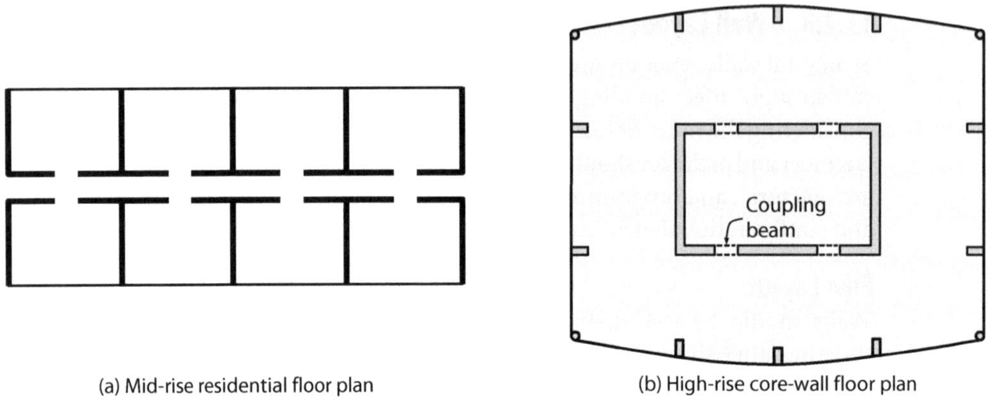

Figure 13.3 Typical floor plans in buildings with walls.

In buildings with post-tensioned slabs, stiff in-line walls resist slab elastic and creep shortening, sometimes with negative effect. Walls **c** and **d** (Figure 13.2) would resist slab shortening along line **cd**, such that post-tensioning force would tend to transfer from the slab and into the walls. Walls **a**, **b**, and **e** are well positioned to allow slab shortening.

Apart from the aforementioned considerations, placement of walls in buildings often is influenced by programmatic considerations and requirements for fire safety. In low-rise buildings, structural walls commonly are located as enclosures around stairs or elevators. In mid-rise residential buildings, walls define the corridors and residential units (Figure 13.3*a*). In high-rise core-wall buildings, a central core may enclose vertical shafts with occupied space surrounding the core (Figure 13.3*b*).

Vertical Discontinuities

Considerations of function and cost sometimes lead to wall openings and other wall discontinuities. These irregularities can lead to stress concentrations and localized lateral drift that may be difficult to quantify and accommodate in design, and in some cases may result in undesirable seismic response. Some irregularities should be avoided without further consideration, while others will require additional analysis and design effort.

In the past, demand for open space in the first story led to many older buildings in which walls from upper stories were discontinued in the first story, creating a weak first story (Figure 13.4*a*). These have performed poorly in past earthquakes (Figure 13.5). This configuration, classified by ASCE 7 as an Extreme Weak Story Irregularity, is no longer permitted in new buildings assigned to Seismic Design Categories D, E, and F.

Walls extending from the foundation and discontinued at some intermediate level (Figure 13.4*b*) are permitted by most codes, but the design may be penalized

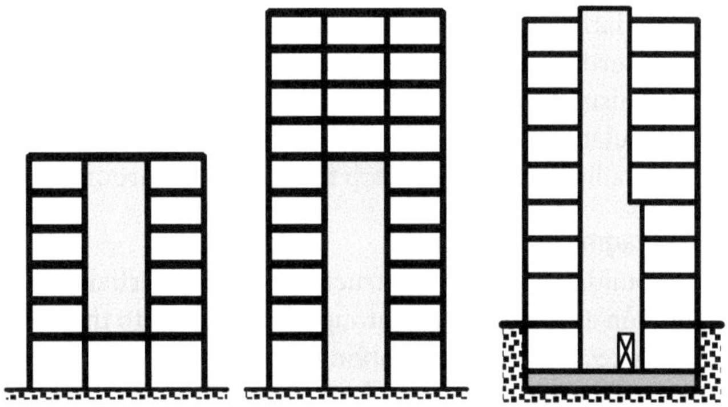

Figure 13.4 Wall vertical irregularities.

Figure 13.5 Permanent offset of weak first story due to discontinued wall, Olive View Hospital, 1971 San Fernando earthquake. (*Photograph by K. Steinbrugge, used with permission from the National Information Service for Earthquake Engineering, University of California, Berkeley.*)

by increased seismic design forces. It is preferred to have more gradual reduction in wall section (length, thickness, or both), as illustrated by Figure 13.4*c*. Such transitions in wall length or thickness may require details to enable the flow of forces near the transition. See Section 13.6.3.

Openings in walls disrupt the flow of forces and are best located in regular patterns that produce predictable force transfers. Figure 13.1*b* and *d* shows examples of regularly located wall openings. For such buildings, good design practice keeps vertical wall segments (piers) stronger than beams such that story failure mechanisms are avoided. Sometimes programmatic demands require openings in less regular pattern. Where these cannot be avoided, they require additional design and detailing effort to develop force transfers around openings.

Diaphragm Connectivity

In a building braced by structural walls, inertial forces generated by building vibration are transmitted through diaphragms to the walls, which in turn transmit the forces to the foundation. Good connections between diaphragms and structural walls are essential to the seismic force path. Chapter 15 discusses this subject in depth.

Programmatic requirements often locate diaphragm openings adjacent to structural walls, complicating the seismic force path. This can be especially acute at podium slabs where large wall forces may be transferred through the diaphragm to other stiff elements (Figure 13.6*a*). Good diaphragm transfer capacity is facilitated by solid diaphragms surrounding walls, rather than significantly perforated diaphragms (Figure 13.6*b*).

13.2.4 Wall Foundations

In low-rise buildings with long walls supporting sufficient gravity loads, spread footings may be adequate to resist design overturning moments. For higher overturning demands, pile foundations, possibly including tension tie-down capacity, can be used. More commonly, foundation elements are extended to pick up additional gravity loads that can help resist overturning. Figure 13.7*a* shows a grade beam acting as a foundation outrigger. Basement walls also can be proportioned to act as outrigger elements (Figure 13.7*b*). Alternatively, a wall extending into subterranean levels can use a horizontal force couple formed

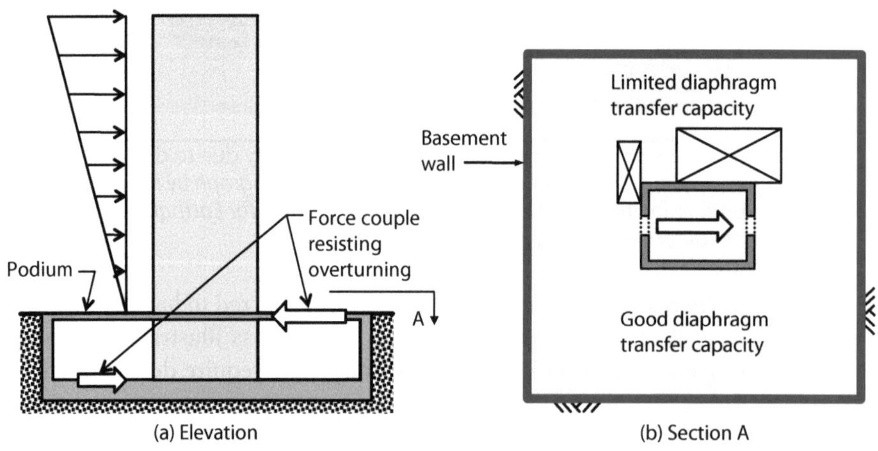

Figure 13.6 Force transfers between walls and diaphragms.

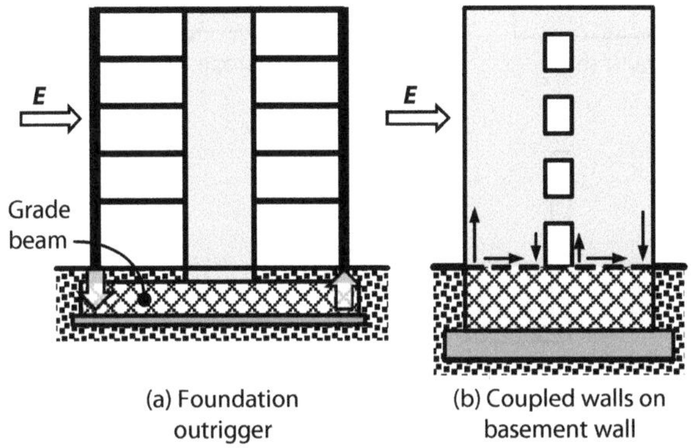

Figure 13.7 Various ways to spread overturning resistance.

between the grade-level diaphragm and diaphragms below to transfer the overturning moment to adjacent basement walls (Figure 13.7*c*). See additional discussion in Chapter 16.

If none of these solutions works, foundation rocking may need to be accepted. Most building codes do not recognize uplifting (rocking) walls as an accepted seismic-force-resisting system; in that case, either special approval is required or the wall cannot be counted on to provide seismic force resistance. Where used, uplifting walls can impose large deformation demands on adjacent framing members that should be accommodated through design.

13.2.5 Wall Configurations

Special structural walls can be configured in numerous ways (Figure 13.8). *Rectangular* cross sections are relatively easy to design and construct; however, very thin sections can have performance problems and should be avoided. "Bar bell" walls have boundary columns that contain longitudinal reinforcement for moment resistance, improve wall stability, and create an element to anchor beams framing into the wall. The boundary columns, however, may create an architectural impediment and increase forming costs. Intersecting wall segments can be combined to create flanged walls, including T, L, C, and I cross sections. *Core walls* enclose elevators, stairs, and other vertically extruded areas, with coupling beams connecting wall components over doorways. Considering lateral shear force in one direction, any wall segment aligned parallel to the lateral shear force is assumed to act as a web element resisting shear, axial force, and moment, while orthogonal wall segments are assumed to act as tension or compression flanges. The different segments also resist lateral forces acting in the orthogonal direction.

Walls with openings are considered to be composed of vertical and horizontal wall segments (Figure 13.9). A *vertical wall segment* is a wall segment bounded horizontally by two openings or by an opening and an edge. Similarly, a *horizontal wall segment* is bounded vertically by two openings or by an opening and an edge. Some walls, including some tilt-up walls, have narrow vertical wall segments that

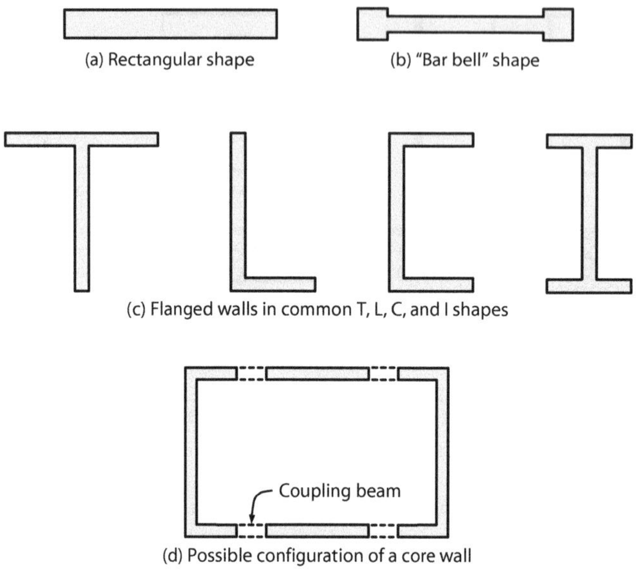

Figure 13.8 Various wall cross sections.

are essentially columns, but whose dimensions do not satisfy requirements of special moment frame columns. In consideration of these, ACI 318 defines a *wall pier* as a vertical wall segment having $l_w/b_w \leq 6.0$ and $h_w/l_w \geq 2.0$. The lower left vertical wall segment in Figure 13.9*b* may qualify as a wall pier. Special provisions apply to wall piers (Section 13.11).

The term *coupled wall* refers to a system in which cantilever walls are connected by coupling beams aligned vertically over wall height (Figure 13.10). The design goal is to develop a ductile yielding mechanism in the coupling beams over the height of the wall and flexural yielding at the base of the individual cantilever walls. In most configurations, most of the coupling beams will yield before the vertical wall segments yield. Depending on geometry and design forces, coupling beams can be configured as either conventionally reinforced or diagonally reinforced beams. Coupling beams sometimes are referred to as *link beams*. See Section 13.12.

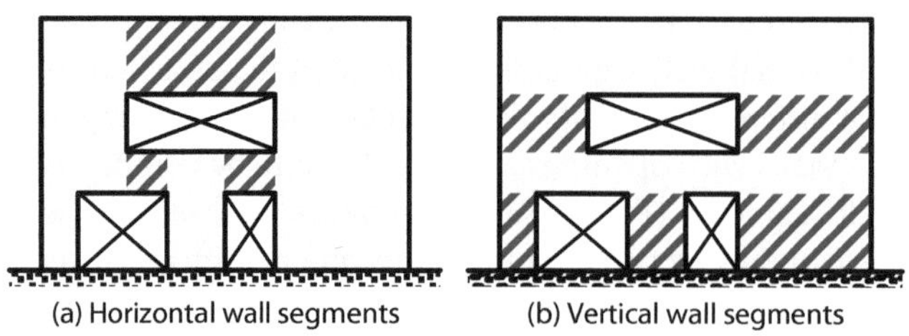

Figure 13.9 Vertical and horizontal wall segments (cross-hatched).

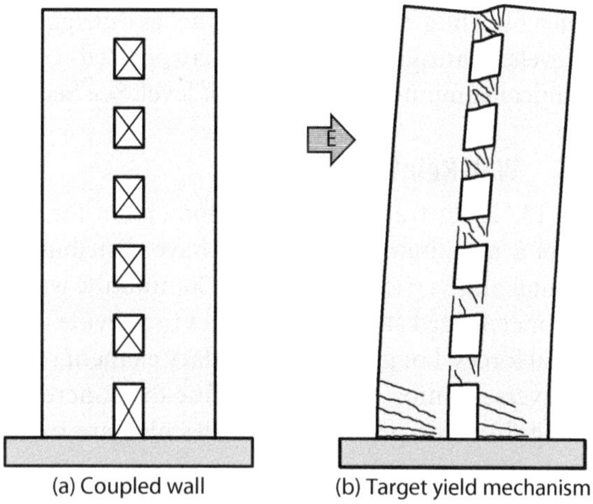

Figure 13.10 Coupled wall geometry and target yield mechanism.

Cap beams (deep beams) sometimes are used at the roof level above a stack of openings to increase coupling. In taller buildings, *outriggers* can be used to engage adjacent columns, thereby increasing building stiffness and reducing upper-story drifts. Outriggers should be designed as ductile "fuses" that protect the outrigger columns from being overloaded axially. Outrigger elements can be of reinforced concrete or structural steel. Figure 13.11 illustrates a 60-story core-wall building

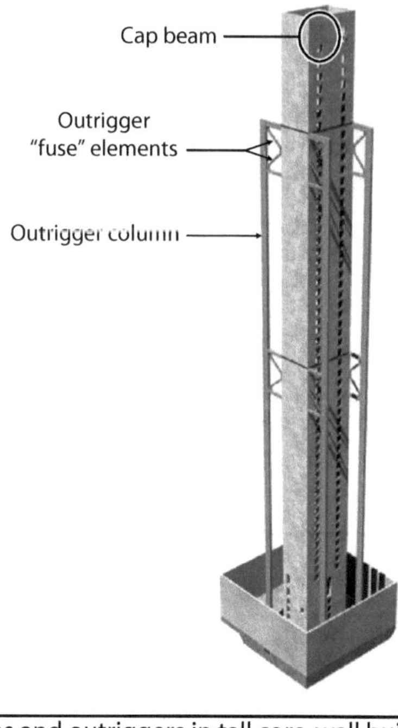

Figure 13.11 Cap beams and outriggers in tall core-wall building. (*Used with permission from Magnusson Klemencic Associates, Seattle.*)

in which buckling-restrained braces act as outriggers in the transverse direction at two levels. Outriggers can be incorporated conveniently in floors housing mechanical equipment or at the roof level. See Section 13.15.

13.2.6 Wall Reinforcement

Figure 13.12 illustrates typical reinforcement for a rectangular special structural wall. As a minimum, a wall must have distributed web reinforcement in both horizontal and vertical directions. Commonly, walls also have vertical reinforcement concentrated at wall boundaries to provide additional resistance to moment and axial forces. Longitudinal boundary element reinforcement usually is enclosed in transverse reinforcement to confine the concrete and restrain longitudinal bar buckling. In some practices, crossties also are used to restrain buckling of web vertical reinforcement.

In U.S. practice, the distributed web reinforcement ratios, ρ_l for vertical reinforcement and ρ_t for horizontal reinforcement, must be at least 0.0025, except ρ_l and ρ_t are permitted to be reduced if $V_u \le A_{cv}\lambda\sqrt{f'_c}$, psi ($A_{cv}\lambda\sqrt{f'_c}/12$, MPa). See Table 13.1. Reinforcement spacing each way is not to exceed 18 in (457 mm). At least two curtains (layers) of reinforcement are required if $V_u > 2A_{cv}\lambda\sqrt{f'_c}$, psi ($A_{cv}\lambda\sqrt{f'_c}/6$, MPa) or $h_w/l_w \ge 2.0$, in which h_w and l_w refer to height and length of the entire wall, respectively. Reinforcement ρ_t also is to be designed for wall shear forces. Finally, if $h_w/l_w \le 2$, ρ_l is not to be less than the provided ρ_t. ACI 318 has no requirements about which distributed reinforcement (vertical or horizontal) should be in the outer layer, although lap splices of vertical reinforcement will perform better if horizontal bars are placed outside the vertical bars as shown in Figure 13.12.

The portion along a wall edge or opening that is strengthened by longitudinal and transverse reinforcement is referred to as a *boundary element*. Where combined seismic and gravity loads result in high compressive demands on the edge, ACI 318 requires a *special boundary element*. These have closely spaced transverse reinforcement enclosing the vertical boundary bars to increase compressive strain capacity of core concrete and to restrain longitudinal bar buckling. See Section 13.8.1.

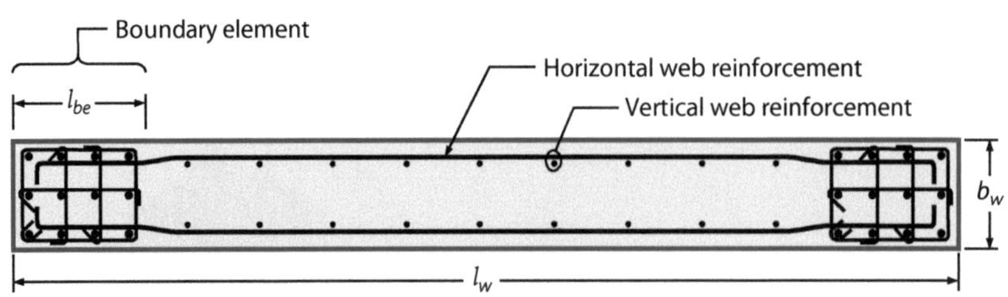

Figure 13.12 Typical reinforcement for rectangular wall.

Factored Shear, V_u	Wall Type	Type of Nonprestressed Reinforcement	Bar/wire Size	f_y, psi (MPa)	Minimum Longitudinal[†], ρ_l	Minimum Transverse, ρ_t
$\leq A_{cv}\lambda\sqrt{f'_c}$, psi ($A_{cv}\lambda\sqrt{f'_c}/12$, MPa)	Cast-in-place	Deformed bars	$\leq$ No. 5 (16)	$\geq$ 60,000 (420)	0.0012	0.0020
				< 60,000 (420)	0.0015	0.0025
			> No. 5 (16)	Any	0.0015	0.0025
	Precast[*]	Deformed bars or welded wire reinforcement	Any	Any	0.0010	0.0010
$> A_{cv}\lambda\sqrt{f'_c}$, psi ($A_{cv}\lambda\sqrt{f'_c}/12$, MPa)	Any	Any	Any	Any	0.0025	0.0025

[*]In one-way precast, prestressed walls not wider than 12 ft (3.7 m) and not mechanically connected to cause restraint in the transverse direction, the minimum reinforcement requirement in the direction normal to the flexural reinforcement need not be satisfied.

[†]Prestressed walls with an average effective compressive stress of at least 225 psi (1.55 MPa) need not meet the requirement for minimum longitudinal reinforcement, ρ_l.

TABLE 13.1 Minimum Reinforcement for Walls

13.2.7 Wall Proportioning

Walls should be proportioned to satisfy strength and drift limit requirements of the building code, unless an alternative approach is approved. According to ASCE 7, walls are designed for load combinations in which seismic forces E are determined using a force reduction factor R. The value of R depends on whether the wall is part of a Dual System ($R = 7$), a Building Frame System ($R = 6$), or a Bearing Wall System ($R = 5$). In a *Dual System*, special structural walls are combined with special moment frames (Chapter 12) that are designed to resist at least 25% of prescribed seismic forces. In a *Building Frame System*, special structural walls are designed to resist all seismic forces and an essentially complete space frame is designed to resist gravity loads. If there is not a complete space frame to support gravity loads, the system must be designed as a *Bearing Wall System*.

Different jurisdictions in the United States interpret the ASCE 7 provisions differently. San Francisco (DBI, 2009) declares the wall to be a bearing wall if it supports more than 5% of the entire building floor and roof loads in addition to self-weight. SEAW (2009) recommends designing a frame column into the wall boundary capable of supporting tributary gravity loads, such that $R = 6$ can be used regardless of the tributary loads to the wall. SEAOC (2008) recommends $R = 6$ without the need to add a frame column where confined boundary elements are provided. Check with the local jurisdiction to determine what is acceptable. Note that ACI 318 and ASCE 7 also use the term "bearing wall" for any individual wall that supports more than 200 lb/linear ft (2.92 kN/m) of vertical load in addition to self-weight. This definition of bearing wall should not be confused with the Bearing Wall System designation of ASCE 7.

Although cost considerations may suggest designing minimum-weight sections, such sections can be difficult to construct and might not perform well. Once the decision has been made to incorporate a wall in the building, formwork and reinforcement detailing will dominate costs. Selecting a thicker wall section is unlikely to have an appreciable effect on construction cost or functionality, but will reduce congestion and improve earthquake performance. Although ACI 318 has no prescriptive minimum thickness,[2] 8 in (200 mm) is a practical lower limit for special structural walls. Construction and performance generally are improved if wall thickness is at least 12 in (300 mm) where special boundary elements are used and at least 10 in (250 mm) elsewhere. Walls that incorporate coupling beams require minimum thickness around 14 in (350 mm) to accommodate reinforcement and required cover and bar spacing, although 16 in (400 mm) is a practical minimum where diagonally reinforced coupling beams are used. Flanges and enlarged boundary sections are helpful to stabilize boundaries and anchor reinforcement from adjacent members.

Detailed guidance on wall proportioning is provided in Section 13.7 and subsequent sections.

13.3 Principles for Design of Special Structural Walls

Buildings designed according to the seismic provisions of ACI 318 and ASCE 7 and similar building codes are intended to resist design-level earthquake motions through ductile inelastic response of selected members. For structural walls, the nature and extent of inelastic response will vary with wall strength and configuration. A good design anticipates the inelastic mechanism and provides proportions and details in the wall that will enable it to respond as intended. The following sections summarize the key principles for design of structural wall buildings.

In the discussion that follows, we will distinguish between slender and squat walls because of differences in resistance mechanisms and differences in behavior expectations. Slender walls ($h_w/l_w \geq 2$) tend to behave like flexural cantilevers. The preferred inelastic behavior mode of slender walls is ductile flexural yielding, without shear failure. In contrast, walls with very low aspect ratio ($h_w/l_w \leq 0.5$) tend to resist lateral forces through a diagonal strut mechanism in which concrete and distributed horizontal and vertical reinforcement resist shear. There is a transition in wall behavior for intermediate aspect ratios. Shear yielding of slender walls generally is considered unacceptable because it reduces inelastic deformation capacity below expected values. Shear yielding of very squat walls is often accepted because such walls tend to have high inherent strength and low ductility demand,

[2]ACI 318 requires walls to be at least 12 in (300 mm) thick if $h_w/l_w \geq 2$ and $c/l_w \geq 3/8$. See Section 13.8.

and because low-rise walls have low axial force and are unlikely to collapse axially as a result of shear distress.

13.3.1 Slender Walls

Select Intended Yield Mechanism

For slender walls, design should aim to achieve ductile flexural yielding at the wall base (Figure 13.13). For slender coupled walls, the target mechanism should include ductile yielding of coupling beams over wall height plus ductile flexural yielding at the wall base (Figure 13.10). Design should aim to avoid wall shear failure or failure of diaphragms and foundations.

Where a design intends for a wall to have a single critical section for moment and axial force, the distribution of flexural strength should be selected to promote significant yielding only at that critical section. One approach is to design the selected critical section to have flexural strength closely matching required strength, with some flexural overstrength provided at other locations (Figure 13.13). Building code provisions for bar development require longitudinal bars to extend well beyond the sections where they are required for strength, and this will tend to result in the desired strength distribution, but this should be checked as part of the design. Where this approach is used, the special details for ductile response can be concentrated around the selected critical section, with relaxed detailing elsewhere.

In some cases, alternative mechanisms have to be accepted. In tall buildings, in addition to the primary yielding mechanism, some wall flexural yielding in intermediate stories may be difficult to avoid; detail such locations so they are capable of moderate ductility capacity. In highly irregular walls, including walls with irregular openings, it can be difficult to precisely identify and control the

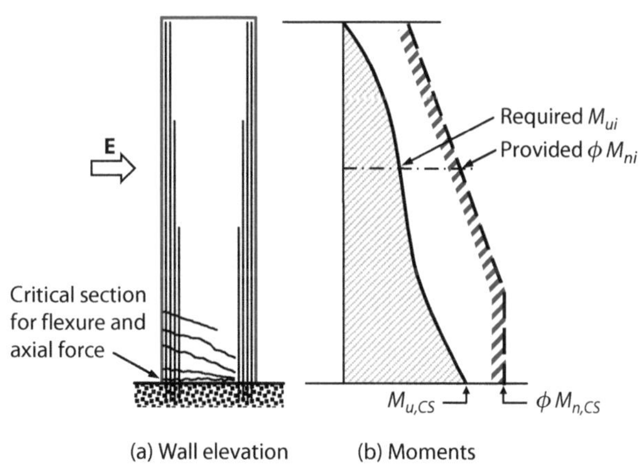

FIGURE 13.13 Provided versus required strength in a wall with a single critical section for moment and axial force. Note that the overstrength ratio $\phi M_n/M_u$ has the least value at the intended critical section, which will promote concentration of inelastic behavior near that critical section.

yielding mechanism. Some conservatism in the design of these systems can help achieve the desired performance.

Achieve Ductile Flexural Yielding

The wall around the intended critical section should be proportioned and detailed to be capable of multiple inelastic deformation cycles. Key factors for improving cyclic ductility are (a) keep global compressive and shear stresses low; (b) design confined, stable flexural compression zones; and (c) avoid splice failures.

A good wall design has axial force well below the balanced point, such that flexural tension reinforcement yields before the flexural compression zone reaches the compressive strain capacity. Using ACI 318 terminology, compression-controlled walls (concrete reaches strain 0.003 before extreme tension reinforcement yields) should be avoided. It is noteworthy that the 1997 Uniform Building Code (UBC 1997) limited wall axial force to $P_u \leq 0.35P_0$, which corresponds approximately to balanced axial force in a symmetric wall. ACI 318 does not have a similar limit, although the design requirements increase marginally for compression-controlled walls.

Although ACI 318 permits wall factored shear on individual wall segments as high as $V_u = 10\phi\sqrt{f_c'}A_{cv}$, psi ($0.83\phi\sqrt{f_c'}A_{cv}$, MPa), flexural ductility capacity for such walls is reduced compared with identical walls having lower shear. Good design practice aims for design shear, calculated considering flexural overstrength, not exceeding ~$4\phi\sqrt{f_c'}A_{cv}$ to $6\phi\sqrt{f_c'}A_{cv}$, psi ($0.33\phi\sqrt{f_c'}A_{cv}$ to $0.5\phi\sqrt{f_c'}A_{cv}$, MPa) so that ductility capacity is not overly compromised. Where flexural ductility demands are low, higher nominal shear stresses can be tolerated.

Inelastic flexural response might result in concrete compressive strains that exceed the unconfined crushing strain. If the compression zone lacks properly detailed transverse reinforcement, concrete crushing and vertical reinforcement buckling at a section can result in a locally weakened "notch" where deformations concentrate, leading to relatively brittle behavior (Figure 13.14). Transverse

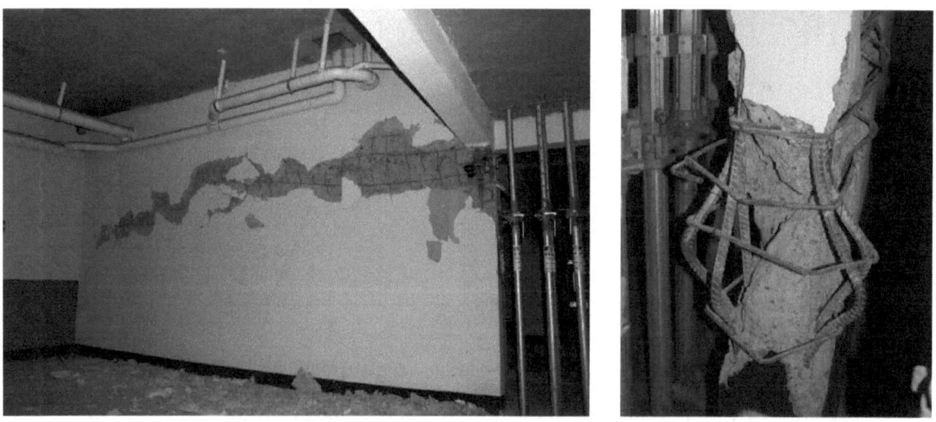

(a) Overall view of crushed wall

(b) Inadequately detailed transverse reinforcement at wall boundary

Figure 13.14 Concrete crushing and reinforcement buckling of inadequately confined wall, 2010 Chile earthquake.

reinforcement is necessary to confine the section, thereby enhancing concrete stress and strain capacity and restraining longitudinal bar buckling. Special boundary element transverse reinforcement should comprise closely spaced hoops with crossties engaging peripheral longitudinal bars (Figure 13.12). Additional ties might be provided to restrain buckling of vertical web reinforcement. In excessively thin walls, spalling of cover concrete can leave a relatively narrow core of confined concrete that may be unstable under compressive loading. Wall boundary thickness should be at least 12 in (300 mm) for sections requiring special boundary elements unless tests on representative sections demonstrate adequate performance for thinner sections. Concrete cover over confinement reinforcement should be minimized such that cover spalling, if it occurs, will not result in large reduction in section area. Good detailing practice also provides lateral support for every longitudinal bar in special boundary elements located within the intended hinge region.

Lap splices of vertical reinforcement can result in a locally strengthened section, such that yielding, if it occurs, may be shifted above or below the lap. Consequences of this shift should be considered. Lap splices subjected to multiple yielding cycles can "unzip" unless they are confined by closely spaced transverse reinforcement. For such splices, ACI 318 requires splice lengths at least 1.25 times lengths calculated for f_y in tension, with no requirement for confinement. Better practice either prohibits lap splices in the intended hinge zone or provides confining transverse reinforcement along the splice length.

Slender compression zones can be susceptible to overall buckling (Figure 13.15). The problem can be exacerbated if the section was yielded previously in tension due to loading in the opposite direction, leaving a more flexible pre-cracked section. Some building codes contain requirements intended to prevent lateral buckling

(a) 2010 Chile earthquake (b) 2011 Christchurch earthquake

Figure 13.15 Wall lateral buckling: (*a*) 2010 Chile earthquake (*photograph used with permission from F. Medina*); (*b*) 2011 Christchurch earthquake (*photograph used with permission from K. Elwood*).

(e.g., NZS3101, 2006; ACI 318, 2014). This text recommends $h_u/b \leq 10$ within the intended hinge region and $h_u/b \leq 16$ elsewhere, where h_u = unsupported wall height. See additional discussion in Section 13.4.3.

Avoid Shear Failure

Shear failure in a slender structural wall can lead to rapid strength loss at drifts below those anticipated in design. Shear failure also can compromise wall axial strength. This is especially so for walls resisting high shear forces [exceeding around $10\sqrt{f'_c}A_{cv}$, psi ($0.83\sqrt{f'_c}A_{cv}$, MPa)], because shear failure in such walls can occur by web crushing (Figure 13.16*a*). For relatively slender wall piers (Figure 13.16*b*), shear failure can create an inclined failure plane along which axial failure can occur. For these reasons, the engineer should design slender walls to avoid shear failure.

A capacity design approach should be used to identify the design shear associated with development of wall flexural strength. For structural walls, the problem is complicated in that the distribution of lateral inertial forces changes continuously as a building responds to an earthquake, such that lateral force and, hence, design shear, cannot be uniquely defined. Section 13.4.4 discusses this subject in greater detail.

Designing a wall to avoid shear failure requires consideration of several failure modes. Diagonal tension failure is evident in inclined cracks extending from flexural tension boundary through the wall web, and is controlled by horizontal and vertical web reinforcement. Diagonal compression failure is evident in web crushing near the flexural compression zone, and is controlled by limiting the maximum value of wall shear as a function of wall area and concrete compressive strength. Sliding shear failure is evident in horizontal cracks and sliding along construction joints, and is controlled by proper treatment of construction joints, including surface roughening and possibly intermittent shear keys, as well as placement of reinforcement across the potential sliding plane.

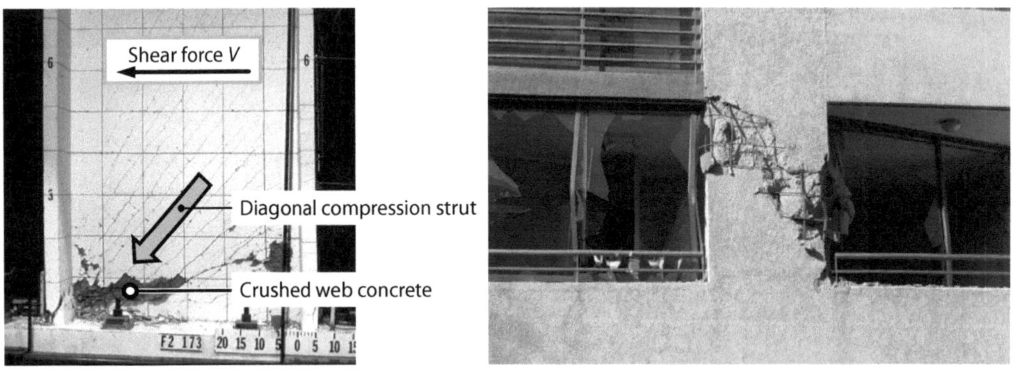

(a) Web crushing in flanged wall. After Oesterle et al., 1984.

(b) Shear damage and onset of axial failure in vertical wall segment, 2010 Chile earthquake.

Figure 13.16 Web crushing due to high shear force in laboratory test. (*a*) Web crushing in flanged wall. (*After Oesterle et al., 1984.*) (*b*) Shear damage and onset of axial failure in vertical wall segment, 2010 Chile earthquake.

13.3.2 Squat Walls

Behavior of squat (low-aspect-ratio) walls tends to be controlled by shear resistance mechanisms rather than by flexural yielding. Although displacement ductility capacity is reduced, this often is acceptable because buildings reinforced with squat walls tend to have high inherent strength with low inelastic demands. Two types of shear failure should be considered: shear yielding within the wall web and shear sliding at construction joints. Section 13.10 discusses these topics in greater detail.

13.3.3 Diaphragms and Foundations

The building code intent is that significant inelastic response will be limited to vertical framing elements of the seismic-force-resisting system (i.e., frames and walls) that are detailed for ductile response. Diaphragms and foundations, and their connections, are intended to remain essentially elastic. See Chapters 15 and 16 for additional discussion.

13.4 Observations on the Behavior of Special Structural Walls

13.4.1 Slender versus Squat Walls

Resistance mechanisms of structural walls can be idealized using the models shown in Figure 13.17. These were introduced in Chapter 7, and are repeated here briefly. See Section 7.12 for additional discussion of shear-resisting mechanisms of slender and squat walls.

The slender wall (Figure 13.17*a*) is idealized as a flexural cantilever that resists shear through a truss mechanism. Because the wall panel typically has long length and height, distributed reinforcement **a** is required to control web cracking. In U.S. practice, this reinforcement must provide reinforcement ratio of at least 0.0025 in both

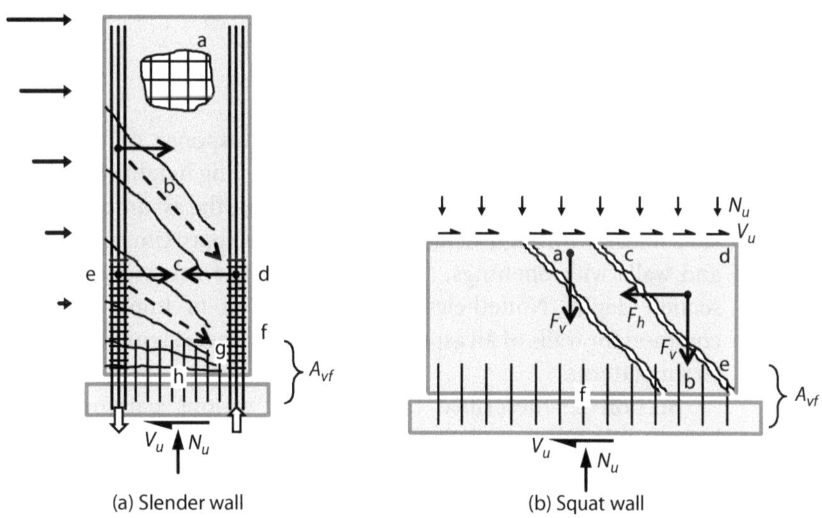

FIGURE 13.17 Resistance mechanisms in slender and squat walls.

vertical and horizontal directions at all sections, except where $V_u \le A_{cv}\lambda\sqrt{f'_c}$, psi ($A_{cv}\lambda\sqrt{f'_c}/12$, MPa). Shear force results in diagonal compression struts **b**. The horizontal component of these struts is resisted by horizontal shear reinforcement **c** that extends across the length of the section and is fully anchored at the boundaries (**d** and **e**). The vertical component of the diagonal compression struts is resisted by increments in flexural tension and compression forces of the boundary elements. Where shear stresses are high relative to the concrete strength, and where flexural yielding results in wide cracks in the flexural compression zone, the combination of diagonal compression and flexural compression can lead to diagonal compression failure at **g**, which might limit deformation capacity. Finally, shear must be transferred across construction joints such as **h** using a shear-friction mechanism. The preceding discussion assumes a wall with significant boundary elements. In a slender wall with light (or no) boundary element reinforcement, the mechanism is altered such that horizontal reinforcement transfers shear more gradually across the section.

The idealized resistance mechanism differs in a squat wall (Figure 13.17*b*). The mechanism is simplified by assuming no shear is resisted by concrete across crack surfaces. Thus, horizontal force equilibrium of segment **cde** requires distributed horizontal reinforcement providing force F_h. Moment equilibrium of segment **cde** about **e**, or segment **ab** about **b**, requires distributed vertical reinforcement providing force F_v. Thus, both vertical and horizontal reinforcement is required to resist shear in squat walls. In addition, shear must be transferred across construction joints such as **f** using a shear-friction mechanism. Reinforcement A_{vf} is most effective if distributed; thus, it may be preferred to distribute the flexural reinforcement uniformly without concentrated boundary elements.

It is usually feasible to design a slender wall to have a flexure-dominated yielding mechanism, without significant inelastic response in shear. In contrast, the distributed vertical reinforcement required for shear in a squat wall results in high moment strength, such that "shear yielding" is the more likely mechanism of inelastic response.

13.4.2 Flexural Response of Walls

Sections 6.9 and 6.10 discussed flexural response of walls. In walls that are relatively slender, usually defined as walls having height-to-length ratio exceeding 2.0, and that do not have significant openings, the assumption that flexural strains vary linearly through depth is a reasonable approximation. In more squat walls and walls with openings, axial strains may be markedly nonlinear through the section depth. Nonetheless, the assumption of linear variation of strains is common for walls of all aspect ratios except where significant openings distort the strain patterns.

Section 6.9 described the effects of various parameters on wall flexural behavior. It was noted that flexural strength of a symmetric wall is relatively insensitive to distribution of reinforcement, such that, from a flexural strength perspective, it is equally acceptable either (a) to provide minimum required distributed reinforcement with remaining longitudinal reinforcement concentrated at boundaries or (b) to distribute all the longitudinal reinforcement uniformly along the wall length. The latter approach may be preferable for squat walls, where displacement demands are relatively small and distributed reinforcement is required for shear resistance including sliding shear at construction joints. For slender walls, either approach can be used, but there are advantages to concentrating reinforcement at the boundaries, including the following:

- Concentrated boundary elements provide a convenient zone for securing transverse reinforcement for confinement, which is especially important at the wall boundaries where compressive strains are largest (Figure 13.18*a*).
- Curvature capacity is enhanced moderately with concentrated boundary reinforcement (Figures 6.42 and 13.18*c*), as the longitudinal reinforcement in the compression boundary resists more of the flexural compressive force, thereby relieving demands on concrete.
- Where reinforcement is uniformly distributed, the local reinforcement ratio at the wall edge is likely to be low. In extreme cases, the tensile strength of reinforcement at a cracked section may be less than the concrete tensile strength in adjacent sections, such that only one or few cracks form, leading to localized yielding (Figure 13.18*b*). By concentrating reinforcement at the boundary, the local reinforcement ratio at the wall edge is increased, promoting better distribution of flexural cracks, yielding over greater wall height, and increased displacement capacity. This observation is supported by analyses (Section 6.9.2), laboratory tests (Figure 6.41), and post-earthquake studies (Wood, 1989; SESOC, 2011).

ACI 318 requires boundary element longitudinal steel ratio of at least 0.0025 (the same as in the web). Eurocode 8 (2004) requires boundary element longitudinal steel ratio of at least 0.005.

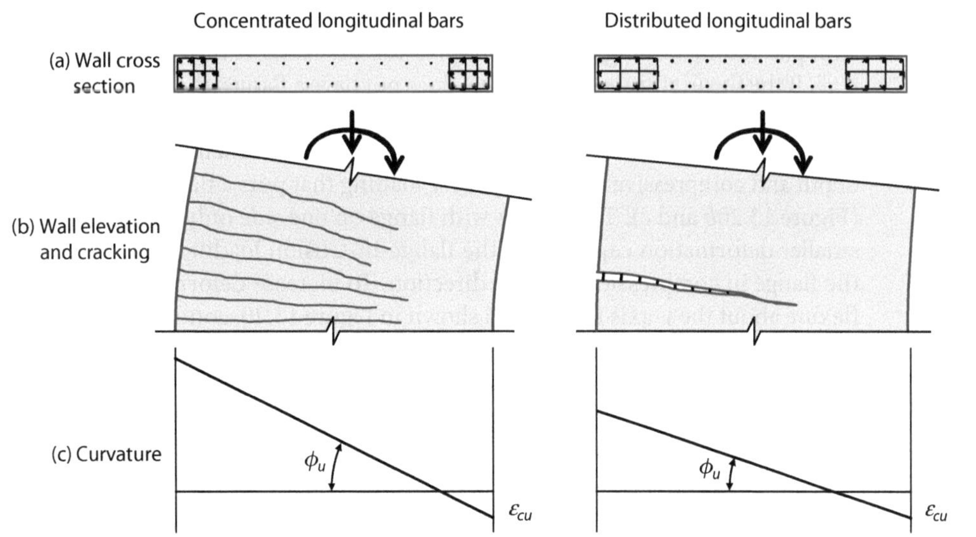

Figure 13.18 Potential effects of distribution of wall longitudinal reinforcement.

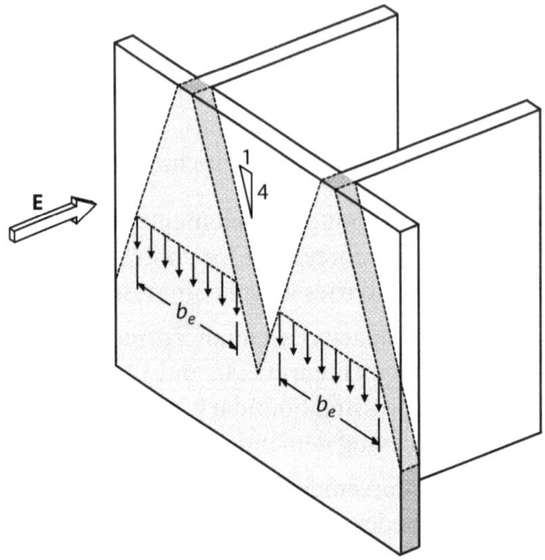

Figure 13.19 Effective flange width (illustrated for flange in tension).

Section 6.10.2 discussed effective flange width for flanged walls, noting that the flange contributes to flexural and axial response both as a compression element and as a tension element. Effective flange width varies with wall axial force, lateral deformation, and whether the flange is in tension or compression. Here we adopt the effective flange width of ACI 318 and Eurocode 8, that is, the effective flange width on each side of the web is equal to one-quarter of the height from the section under consideration to the top of the wall, but not greater than the actual width or half the distance to adjacent wall webs (Figure 13.19).

As discussed in Section 6.10.2, distributed reinforcement in a tension flange can add considerable flexural tension reinforcement, thereby increasing flexural strength and increasing flexural compression demands at the opposite end of the wall. If the flexural compression zone does not have a flange, the compression zone may be forced deep into the section, thereby increasing compressive strain demands for a given curvature demand (Figure 13.20*a*). Smaller compression zone depth and compressive strains occur for loading that puts a flange in compression (Figure 13.20*b* and *c*). Thus, walls with flange on one side only tend to have much smaller deformation capacity for the flange-in-tension loading direction than for the flange in compression loading direction. To increase deformation capacity for flexure about the y-axis for the wall shown in Figure 13.20, confinement is likely to be required at points **A** and **D**. Although not generally required by most building codes as a design check, flexure about a diagonal will produce large compressive strains in the corners where wall segments intersect, such that confinement should be provided at those locations as well (points **B** and **C** in Figure 13.20*d*).

Walls with axial force above the balanced point yield by compressive yielding of the concrete in the compression zone. Although provision of confinement reinforcement improves strain capacity of the concrete, premature compressive

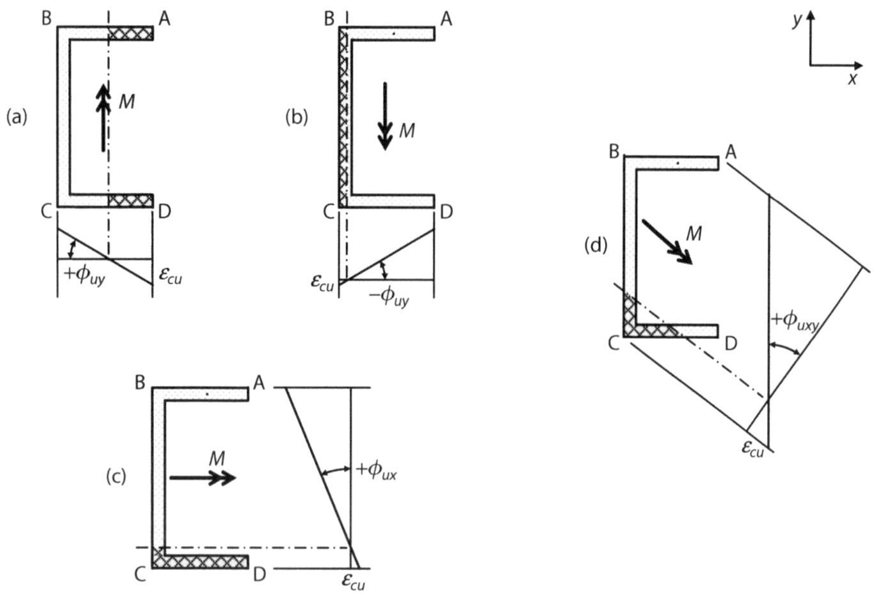

FIGURE 13.20 Flexural compression zones for confinement consideration.

yielding of the compression zone generally is considered undesirable because of reduced curvature capacity and required repairs in the event of wall yielding. Effect of axial force on curvature capacity of unconfined and confined walls is discussed in Section 6.9. Flexural behavior can be improved by keeping axial force below the balanced point. ACI 318 has no limit on axial stress [the 1997 Uniform Building Code had a limit of $P_u \leq 0.35P_0$, but this was not carried forward in the IBC (2012)]. Eurocode 8 limits the axial force to $P_u \leq 0.4P_0$. These limits may be acceptable for symmetric walls, but lower axial forces may be required in asymmetric walls (e.g., Figure 13.20) to achieve tension-controlled flexural behavior.

13.4.3 Stability of Flexural Compression Zone

Design practices prior to the 1990s favored rectangular walls with enlarged boundary elements (Figure 13.8*b*), contributing to stability of the flexural compression zone. More recently, prevailing practices in many countries favor rectangular sections without enlarged boundaries. The more slender flexural compression zones can be susceptible to inelastic lateral buckling as shown in Figure 13.15 (EERI, 2010, 2011).

Although global wall buckling occurs when the wall boundary is in compression, buckling may be influenced by residual tensile strain in the wall due to prior loading in the opposite direction (Corley et al., 1981; Paulay and Priestley, 1993; Chai and Elayer, 1999). To understand the underlying mechanism, consider a multi-story wall as shown in Figure 13.21. The foundation, floor diaphragms, and roof diaphragm provide lateral support at every floor level. Thus, the unsupported

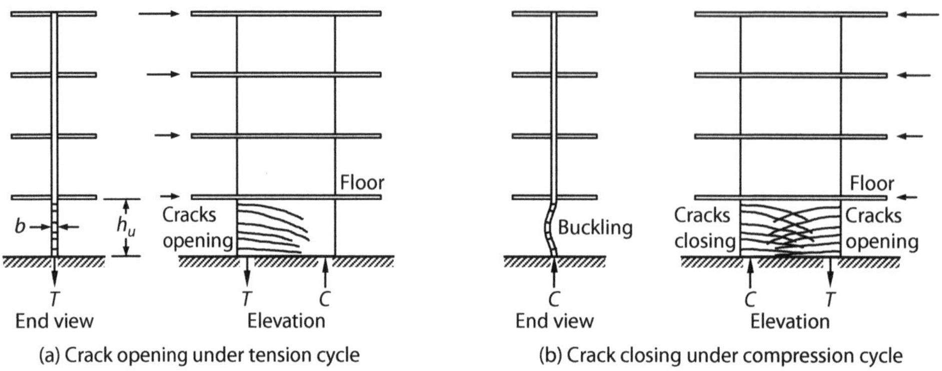

FIGURE 13.21 Lateral instability of wall boundary previously yielded in tension. (*After Chai and Elayer, 1999, used with permission from American Concrete Institute.*)

height of the wall boundary can be taken equal to the story clear height, h_u. An effective length kh_u can be defined based on the rotational restraints at the different floor levels. In the present analysis, which is concerned with very slender walls, k is taken equal to 0.5, representing an idealized condition of full fixity at the top and bottom of the wall unsupported length.

A typical wall boundary will be subjected to alternating tension and compression as the building responds to an earthquake (Figure 13.21). If the boundary yields in tension, cracked sections are produced, with crack width dependent on the amplitude of the reinforcement tensile strain ε_{sm} during the tension excursion. In a previously yielded wall, crack closure under deformation reversal may require yielding of the longitudinal reinforcement in compression. In a wall with two curtains of reinforcement, any slight asymmetry in the reinforcement will result in one curtain yielding before the other, leading to out-of-plane curvature and a tendency to buckle out of plane. In a wall with one curtain of reinforcement, out-of-plane curvature occurs even more readily. Whether the wall remains stable depends on the amplitude of the prior tensile strain ε_{sm} and the slenderness ratio h_u/b of the wall.

Buckling of prismatic reinforced concrete members under alternating axial tension-compression cycles was reviewed in Section 5.7.2. As derived in Section 5.7.2, the main variables affecting stability of an axially loaded prismatic member are (a) the slenderness ratio kh_u/b, (b) the maximum prior tensile strain ε_{sm} in longitudinal reinforcement, and (c) the number of curtains of longitudinal reinforcement. In general, slender wall sections should be reinforced with two curtains for longitudinal reinforcement to improve stability. For members with two curtains, the limiting slenderness ratio for stability of a prismatic member is

$$\frac{kh_u}{b_{cr}} = \frac{1}{0.7\sqrt{\varepsilon_{sm} - 0.005}} \tag{13.1}$$

We can use Eq. (13.1) to approximate stability conditions for a slender wall boundary. Assuming the wall boundary to be fixed at both ends, $k = 0.5$. Thus, Eq. (13.1) becomes

$$\frac{h_u}{b_{cr}} = \frac{1}{0.35\sqrt{\varepsilon_{sm} - 0.005}} \tag{13.2}$$

Results of Eq. (13.2) are plotted in Figure 13.22. Considering low-cycle fatigue, the maximum tensile strain normally accepted for longitudinal reinforcement is approximately $\varepsilon_{sm} = 0.05$. This suggests a maximum slenderness ratio of approximately 13. The ACI 318 limit for confined boundary elements is 16 (Figure 13.22).

The preceding derivations are based on an idealized wall boundary of limited length subjected to uniform compressive strain. Actual wall boundaries have strain gradient along the wall length, which would tend to support the wall boundary. This suggests that the preceding results are conservative for actual wall boundaries.

Moehle et al. (2011), writing about U.S. practice, recommended $h_u/b \leq 10$ within the intended hinge region and $h_u/b \leq 16$ elsewhere.

Eurocode 8 specifies minimum wall thickness of 8 in (200 mm) for confined parts of walls. Moreover, if the length of the confined part does not exceed the larger of $2b$ and $0.2l_w$, b should be at least $h_u/15$. Otherwise, b should be at least $h_u/10$.

According to NZ 3101, thickness of the wall boundary over the height of the plastic hinge but not less than the full height of the first story shall be at least:

$$b_m = \frac{\alpha_r k_m \beta (h_w/l_w + 2) l_w}{1700\sqrt{\xi_r}}$$

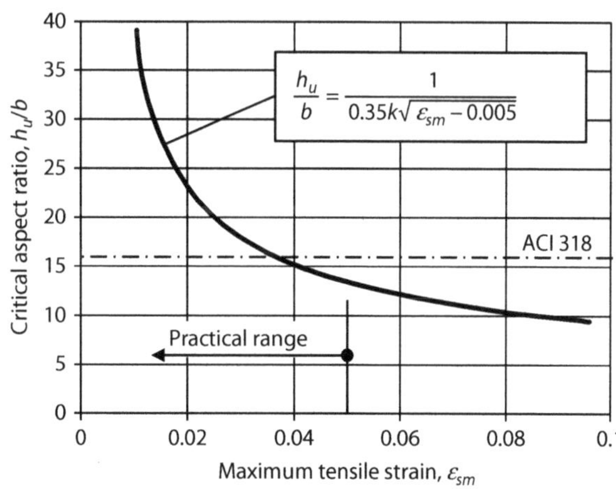

FIGURE 13.22 Critical slenderness ratio for wall boundaries.

in which $\alpha_r = 1$ for walls with two curtains of longitudinal reinforcement and 1.25 for walls with one curtain, $\beta = 7$ for ductile plastic regions, and $k_m = 1$, except for long walls it can be defined as

$$k_m = \frac{h_u}{(0.25 + 0.055h_w/l_w)l_w} \leq 1.0$$

$$\xi_r = 0.3 - \frac{\rho_l f_y}{2.5 f_c'} \geq 0.1$$

in which ρ_l refers to the local longitudinal reinforcement ratio in the wall boundary. These equations result in wall slenderness ratio h_u/b ranging from around 8 for slender, heavily reinforced walls to around 30 for squat, lightly reinforced walls.

13.4.4 Dynamic Response

A wall building responding dynamically to earthquake ground motions experiences transient patterns of lateral inertial forces. If the wall is designed to have a single critical section for moment and axial force, moments at that critical section are well constrained by the moment strength, but moments elsewhere, and shears in general, are less well constrained because of the dynamically changing lateral inertial forces. Capacity design remains a useful tool for establishing design forces, but it is insufficient for bounding the design forces because of uncertainty in the distribution of inertial forces.

To illustrate the aforementioned problem, consider the test results for a reduced-scale test structure reported by Eberhard and Sozen (1993). The test structure is an idealization of the framing system in one principal direction of a 10-story building (Figure 13.23*a*). The seismic-force-resisting system comprises a cantilever structural wall and two moment-resisting frames. The test structure is subjected to three shaking tests of increasing intensity, each of which drives the structure into the nonlinear range of response.

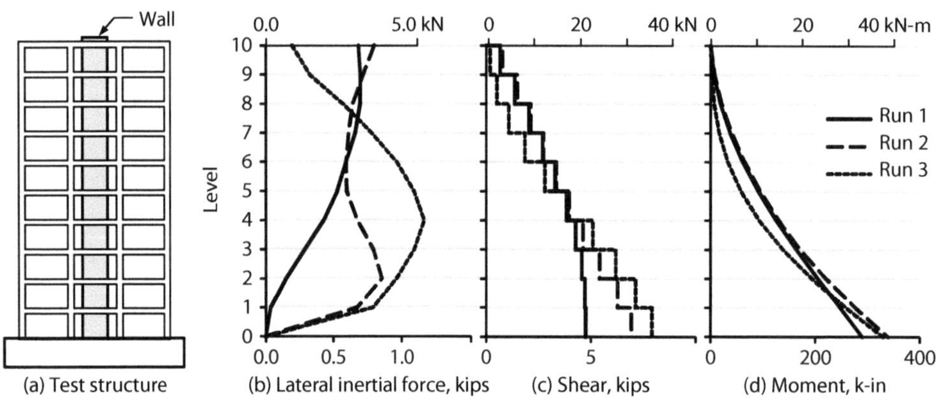

FIGURE 13.23 Distribution of lateral inertial forces, shears, and moments at times of maximum shear for frame-wall structure tested on shaking table. (*Data after Eberhard and Sozen, 1993.*)

Figure 13.23*b* shows the measured lateral inertial force distribution (product of acceleration and mass) at the time of maximum base shear in each test. For Run 1, the lateral force profile resembles the linear first-mode shape, with corresponding shears and moments (Figure 13.23*c* and *d*). Base moment strength is effectively reached during Run 1. Amplitude of the base motion is doubled for Run 2. Although lateral drifts (not shown) approximately double, base moment increases only moderately; this is because the wall is already yielding at the base, with only modest increase in resistance due to strain hardening as inelastic rotations increase. Note, however, that the lateral force profile at time of maximum base shear during Run 2 has roughly uniform shape, leading to significantly increased base shear. Similarly, for Run 3 the input motion is further amplified but again the base moment is almost unchanged. The lateral force profile, however, becomes increasingly weighted toward the base of the structure, producing an even higher base shear.

The base shear in the preceding example is determined by two factors. The first is moment strength at the base. The second is the distribution of lateral inertial forces. If lateral forces are approximately proportional to the first mode shape (Figure 13.24*a*), the centroid of lateral forces will be located at an effective height $h_{eff} \approx (2/3)h_w$, resulting in base shear $V_b = M_{pr,CS}/h_{eff} \approx 1.5M_{pr,CS}/h_w$. If lateral inertial forces become skewed toward the base (Figure 13.24*b*), with smaller h_{eff}, the base shear corresponding to development of $M_{pr,CS}$ increases. For the case of uniformly distributed lateral forces, $h_{eff} \approx (1/2)h_w$, resulting in base shear $V_b = M_{pr,CS}/h_{eff} \approx 2M_{pr,CS}/h_w$, representing a 33% increase over the first-mode shape.

The behavior illustrated in Figure 13.23 has been reported in numerous studies (e.g., Aoyama, 1986; Eibl and Kreintzel, 1988; Ghosh and Markevicius, 1990; Eberhard and Sozen, 1993; Kabeyasawa, 1993; Rodriguez et al., 2002; Panagiotou and Restrepo, 2011; Rejec et al., 2011). It can be understood conceptually by considering structural response as the sum of modal responses. As shaking

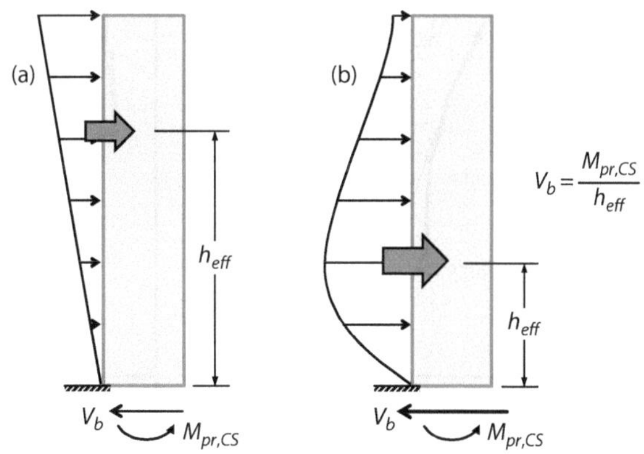

FIGURE 13.24 Relation among base moment, base shear, and distribution of lateral inertial forces.

intensity increases, deformations in each mode increase, and associated internal forces also increase unless limited by structural capacity. Thus, moment at the flexural critical section is limited, but other forces may continue to increase as shaking intensity increases. Eibl and Kreintzel (1988) suggested this behavior could be approximated as the sum of modal responses, with only first-mode force response limited by structural capacity. Applied in design, the first-mode force responses would be reduced by response modification coefficient R, whereas forces associated with other modes would not be reduced. This approach is adopted by Eurocode 8, as described later.

Figure 13.23 showed force profiles at the instant of maximum base shear. At other instances during shaking response, floor accelerations will combine to produce maximum shears and moments at other elevations. Figure 13.25 illustrates the envelope of wall moments and shears calculated for a 12-story wall for which flexural yielding is restricted to the base (Priestley and Amaris, 2003). In this example, $IR = 1$ defines the design-level shaking intensity. Note that wall moment at the base has effectively reached the moment strength for $IR = 0.5$ (i.e., half the design intensity), with only minor increase owing to hardening as intensity increases to $IR = 2.0$. Maximum wall moments at higher elevations, however, continue to increase as shaking intensity increases. Similarly, maximum wall shears at all levels increase progressively as shaking intensity increases.

Panagiotou and Restrepo (2009) report results of a study of 10-, 20-, and 40-story core-wall buildings subjected to impulsive earthquake ground motions. The walls were designed by three approaches: (1) The ACI 318 design provided

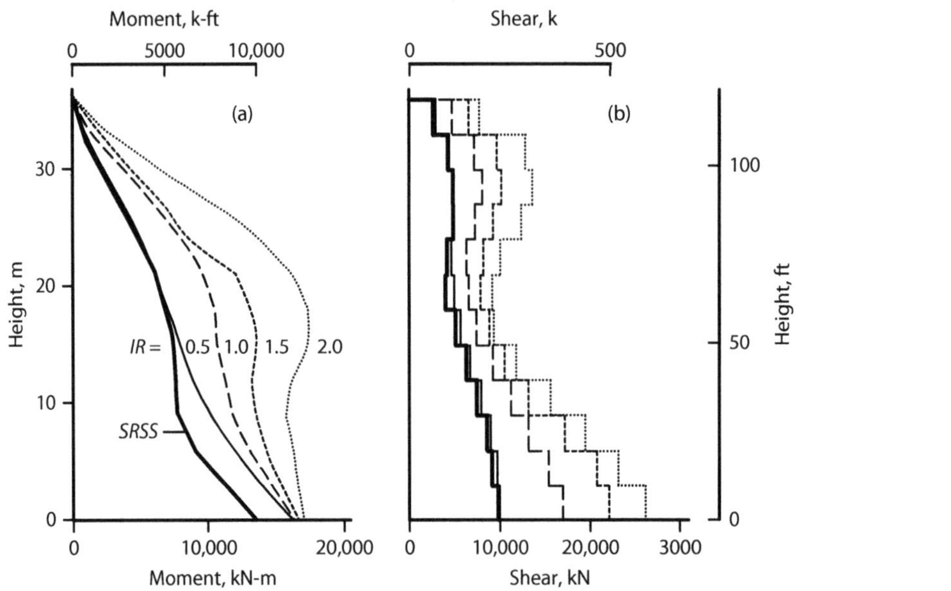

FIGURE 13.25 Calculated wall moments and shears over height of 12-story wall building under earthquake motions of varying intensity. Intensity Ratio (*IR*) is ratio between shaking intensity and design shaking intensity. (*After Priestley and Amaris, 2003.*)

strength over height that approximately followed results obtained using modal response spectrum analysis divided by response modification coefficient $R = 5$; (2) a single plastic hinge design had the same base moment strength as the ACI 318 design, but with infinite moment strength above the base to prevent yielding other than at the base; and (3) a double plastic hinge model allowed for plastic hinges at the base and at mid-height, with the remainder of the wall protected from flexural yielding. Figure 13.26*a* compares modal response spectrum moments, expected strengths based on the ACI 318 design, and moments calculated from nonlinear response history analysis, indicating the ACI 318 moment strength was reached or exceeded at several locations up the building height. Figure 13.26*b* compares modal response spectrum shears and shears calculated from nonlinear response history analysis for the three different design methods, indicating that actual shears can be much higher than modal response spectrum values. Figure 13.26*c* compares moments calculated from nonlinear response history analysis for the three different design methods, indicating that the ACI 318 and the double hinge design methods result in similar distributions, in both cases with moments lower than would be required to keep the wall linear above the base plastic hinge. Finally, Figure 13.26*d* compares curvature ductility demands calculated by nonlinear dynamic analysis for the three different design methods, indicating the ACI 318 and the double hinge design methods develop appreciable curvature above the base. Panagiotou and Restrepo (2009) recommend the double hinge design

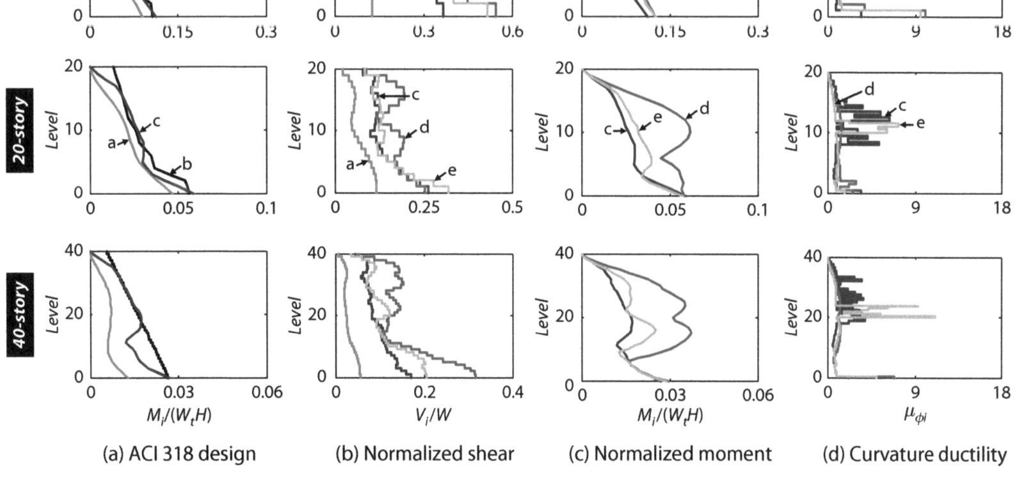

FIGURE 13.26 Calculated response of multi-story wall buildings designed by different techniques. (*After Panagiotou and Restrepo, 2009.*)

method because it effectively restricts yielding to predetermined levels, whereas the ACI 318 method results in wall flexural yielding at multiple levels.

The aforementioned studies suggest that expected moments and shears can exceed by a considerable margin the values obtained from modal response spectrum analysis divided by response modification coefficient R. At the time of this writing, there is considerable research and divergence of opinion on the best approach for design. The remainder of this section presents some recommendations for consideration.

Moment Design

Design for wall moment and axial force should consider wall flexural overstrength and wall dynamic response. The following design steps can be considered:

- Determine the wall design moments M_u over wall height based on analysis for design-level earthquake loading (Figure 13.27).
- Select the location for primary flexural yielding, usually at the wall base, and design that critical section to satisfy $\phi M_{n,CS} \geq M_{u,CS}$.
- Building codes require that multiple load combinations be considered for design, which invariably leads to some flexural overstrength at the critical section. Define a flexural overstrength factor as $\phi_o = M_{n,CS}/M_{u,CS}$, which is defined for the load combination that produces the highest value of $M_{n,CS}$.
- Amplify design moments over the wall height based on overstrength at the critical section. Resulting design moments are $M_u' = \phi_o M_u$.
- Further amplify the design moments by a dynamic amplification factor ω, leading to design moments $M_u'' = \omega \phi_o M_u$. Figure 13.27 illustrates this approach using a constant dynamic amplification factor ω, except the factored moment never need be taken greater than the nominal strength provided at the base. A value of $\omega = 1.25$ has been used in some designs.

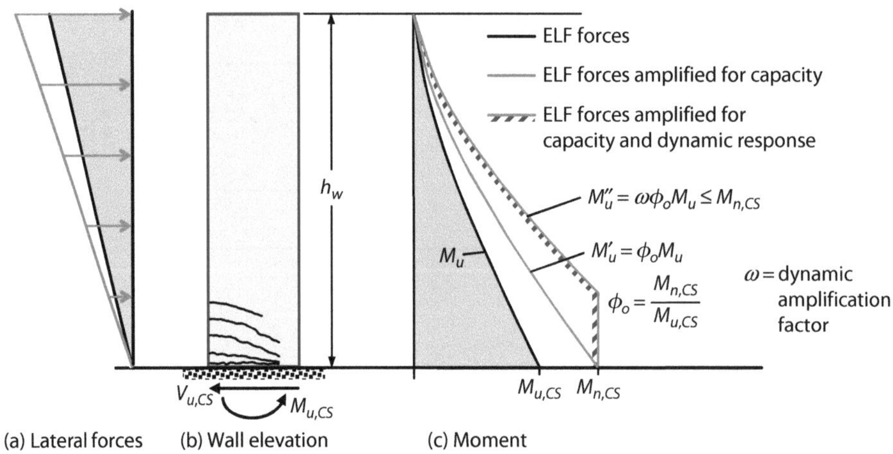

Figure 13.27 Wall design moment envelope illustrated for equivalent lateral force (ELF) design forces.

- Provide nominal moment strength (using same load combination including axial load as was used in defining ϕ_o) to satisfy $M_n \geq M_u''$ at every elevation outside the intended plastic hinge zone.

The recommended approach amplifies moment strength at all levels except the intended plastic hinge, such that primary yielding should occur where intended. Some yielding at levels above the intended plastic hinge also should be anticipated. In special cases, nonlinear dynamic analysis can be used to explore whether inelastic response in intermediate stories can be accommodated by the provided details.

Alternative approaches have been proposed. For cantilever walls, Eurocode 8 and NZS3101 require design moment diagram varying linearly from the expected moment at the base to zero at the top. Results in Figure 13.26 suggest this approach is insufficient to prevent wall yielding beyond the intended hinge zone. Priestley et al. (2007) recommend an alternative formulation of ω that varies with ductility demand, period, and elevation. This approach sometimes can result in design moments above the base exceeding those at the base, which can be problematic (Panagiotou and Restrepo, 2009). Panagiotou and Restrepo (2009) present an alternative that involves development of two plastic hinge zones, one at the base and another at mid-height.

Another alternative worth considering is to establish the target design strength as $M_u' = \phi_o M_u$, as shown in Figure 13.27. Carry out the moment strength design requiring $\phi M_n \geq M_u'$ at every section. Extend wall flexural reinforcement as required considering bar cutoff requirements (see Section 13.8.4), and then check that provided strength results in an adequate moment strength margin at every level above the intended critical section. Required bar extensions will usually result in a suitable strength margin.

Shear Design

Design for wall shear likewise should consider wall flexural overstrength and dynamic response. Here we review some procedures that have been proposed or that appear in current building codes.

In U.S. practice (ASCE 7), wall shears V_u are determined either from modal response spectrum analysis or equivalent lateral force (ELF) analysis, with seismic forces reduced by response modification coefficient R. If these shears are used for design without further modification, then ACI 318 requires $V_u \leq \phi V_n$, where $\phi = 0.6$. If the shears are amplified by overstrength factor $\phi_o = M_{n,CS}/M_{u,CS}$, then it is permitted to use $\phi = 0.75$. Considering that a yielding wall is likely to develop a moment closer to probable moment strength, overstrength factor $\phi_o = M_{pr,CS}/M_{u,CS}$ would be more appropriate. Typical values of the ratio $M_{pr,CS}/M_{u,CS}$ are 1.5 or greater, such that use of this amplification factor along with $\phi = 0.75$ generally will result in much more conservative design than the other option (unamplified shears with $\phi = 0.6$). The more conservative approach is recommended here. Note that ASCE 7 and ACI 318 do not consider amplification for dynamic effects, which is a shortcoming of the design approach that may result in unconservative designs.

Various approaches to account for dynamic amplification have been proposed. Aoyama (1986), Ghosh and Markevicius (1990), and Eberhard and Sozen (1993) proposed that wall base shear be estimated as the sum of mechanism shear for lateral forces proportional to first-mode shape plus a higher-mode shear equal to $D_m W A_{gmax}/g$, with values of D_m varying from 0.25 to 0.34. Substituting $\phi_o V_{u,CS}$ as an approximation to first-mode mechanism shear, the design base shear amplified for dynamic effects can be written as

$$V'_{u,CS} = \phi_o V_{u,CS} + D_m W A_{gmax}/g \tag{13.3}$$

in which $\phi_o = M_{pr,CS}/M_{u,CS}$, $V_{u,CS}$ = base shear from design lateral forces reduced by response modification coefficient R, W = seismic weight, A_{gmax} = peak ground acceleration (in consistent units of length/s^2), and g = gravity acceleration. From this, the dynamic amplification factor can be expressed as

$$\omega = V'_{u,CS}/\phi_o V_{u,CS} = 1 + \frac{D_m W A_{gmax}}{g\ \phi_o V_{u,CS}} \tag{13.4}$$

The dynamic amplification factor of Eq. (13.4) can be expressed more conveniently by replacing A_{gmax} by $S_a(T_C)/2.5$, in which $S_a(T_C)$ is design spectral acceleration in the constant acceleration region (Chapter 11), and approximating $V_{u,CS}$ as $\frac{W}{gR} S_a(T_1)$, in which $S_a(T_1)$ is design spectral acceleration at the first-mode period in the direction of the design base shear. Taking $D_m = 0.25$, Eq. (13.4) can be approximated as

$$\omega \approx 1 + 0.1 \frac{R}{\phi_o} \cdot \frac{S_a(T_C)}{S_a(T_1)} \tag{13.5}$$

Note that this expression applies only to base shear, not to shear at other elevations. For typical values $R = 5$, $\phi_o = 1.5$, and $S_a(T_C)/S_a(T_1) = 4$, Eq. (13.5) results in $\omega = 2.3$.

SEAOC (2008) recommends amplifying wall shears by overstrength factor $\phi_o = M_{pr,CS}/M_{u,CS}$ and by dynamic amplification factor ω as required by NZS3101, that is

$$\omega = 0.9 + n/10 \tag{13.6}$$

for buildings up to six stories, and

$$\omega = 1.3 + n/30 \le 1.8 \tag{13.7}$$

for buildings over six stories, in which n = number of stories. For structures designed by linear dynamic analysis, ω need not exceed the value given by

$$\omega = 1.2 + n/50 \tag{13.8}$$

For buildings with unusually tall stories, n should represent the equivalent number of stories.

Eurocode 8[3] adopts a formula proposed by Kreintzel (1990). According to the underlying principle, response in the first mode is limited by mechanism strength

[3]Notations and some definitions used here differ from those used in Eurocode 8. Hence, the expressions presented here do not match those given by Eurocode 8 and may produce slightly different results. The interested reader is referred to Eurocode 8 for the exact expressions used therein.

whereas response in higher modes is unaffected by yielding and instead is proportional to shaking intensity. Using the SRSS combination rule, design base shear can be expressed as

$$V'_{u,i} = \sqrt{\left(\frac{\phi_o V_{e,1}}{R}\right)^2 + \sum_{i=2}^{n}(V_{e,i})^2} \tag{13.9}$$

in which $\phi_o = M_{pr,CS}/M_{u,CS}$, $V_{e,1}$ = first-mode shear, R = response modification coefficient, and $V_{e,i}$ = shear due to mode i. The first term in parentheses can be considered design shear in the first translational mode parallel to the shear, modified by R, and increased to account for flexural overstrength at the base. The summation adds linear contributions from modes other than the first translational mode.

Assuming that only first and second modes contribute significantly to shear response, which is reasonable for low- to mid-rise buildings (Aoyama, 1986), Eq. (13.9) can be written as

$$V'_{u,i} = \frac{\phi_o V_{e,1}}{R}\sqrt{1+\left(\frac{V_{e,2}}{\phi_o V_{e,1}/R}\right)^2} = V_{pr,1}\sqrt{1+\left(\frac{R}{\phi_o}\cdot\frac{V_{e,2}}{V_{e,1}}\right)^2} \tag{13.10}$$

The term $V_{pr,1}$ can be considered the shear corresponding to development of probable moment strength under first-mode lateral loading profile and the square root term can be considered a multiplier to account for second-mode effect. For an elastic cantilever, ratio $V_{e,2}/V_{e,1}$ at the base of a wall can be expressed as $0.3S_a(T_2)/S_a(T_1)$, where $S_a(T_1)$ and $S_a(T_2)$ are design spectral accelerations at first- and second-mode translational periods, respectively, in the direction of the shear. For $R = 5$, $\phi_o = 1.5$, and $S_a(T_C)/S_a(T_1) = 4$, this results in dynamic amplification factor of 4.1, which is approaching the factor R used in design. Indeed, in some cases, the result of Eq. (13.10) can exceed the base shear corresponding to linear response, which is unreasonable. Consequently, it is necessary to place an upper limit on the result of Eq. (13.10).

The actual Eurocode 8 design equation differs moderately from Eq. (13.10). According to Eurocode 8, the design shear is

$$V'_{u,i} = \varepsilon V_{ui} \tag{13.11}$$

in which ε is a multiplier to account for combined dynamic and overstrength effects, defined for high ductility structures by

$$\varepsilon = q\cdot\sqrt{\left(\frac{\gamma_{Rd}}{q}\cdot\frac{M_{u,CS}}{M_{n,CS}}\right)^2 + 0.1\cdot\left(\frac{S_a(T_C)}{S_a(T_1)}\right)^2}\ \begin{cases}\le q\\ \ge 1.5\end{cases} \tag{13.12}$$

in which q = behavior factor (similar to R), γ_{Rd} = factor to account for overstrength due to steel strain-hardening, $S_a(T_C)$ is elastic acceleration response spectrum ordinate in the constant acceleration region, and $S_a(T_1)$ is elastic acceleration response spectrum ordinate at the building fundamental period in the direction of the design shear force.

Rejec et al. (2011) present an extensive analytical study of the Eurocode 8 procedure and find that it can be overly conservative in certain cases. Two improvements are recommended. First, the amplification factor only should apply to the first-mode design base shear (i.e., the linear elastic base shear in the first mode divided by q) rather than the design base shear (which is a combination of shears from multiple modes). Second, the upper limit of factor ε should be related to total shear force, not only shear force due to the first mode. With these improvements, the procedure was found to be reasonably accurate in estimating expected base shear over a range of building designs. Rejec et al. also note that Eurocode 8 uses Eq. (13.11) to amplify shear at all stories, whereas the coefficient 0.1 in Eq. (13.12) is valid only at the base.

Note that the Eurocode approach as well as the Reject et al. (2011) suggested modifications to it are intended to apply to wall buildings in which the lateral-force-resisting system is only cantilever walls. Studies show that these approaches might not produce useful results for buildings in which walls use outriggers or for which frame-wall interaction occurs.

13.4.5 Backstay Effects

In wall buildings with subterranean levels, the basement generally is built as a very stiff, six-sided box extending below grade (Figure 13.6). In addition to supporting building weight and retaining soils around the subterranean space, the box also may serve to resolve wall moments partially through a horizontal force couple formed between the podium diaphragm and subterranean diaphragms including the base mat. As discussed in Section 13.2.3 and Chapter 15, this can produce large forces in the diaphragm at the podium level. This action sometimes is referred to as the *backstay effect*. The term *flagpole effect* sometimes is used as well.

Figure 13.28 illustrates the backstay effect on wall shears and moments. In Figure 13.28*a*, the wall is isolated from the podium slab by a movement joint. Thus, wall shear is constant (ignoring additional inertial force in subterranean levels) and wall moment increases linearly to the foundation mat, where the mat is required to resist the entire wall shear and moment. In Figure 13.28*b*, the wall and podium slab are connected, developing a backstay force that may result in wall shear reversal with corresponding reduced wall moments. The connected

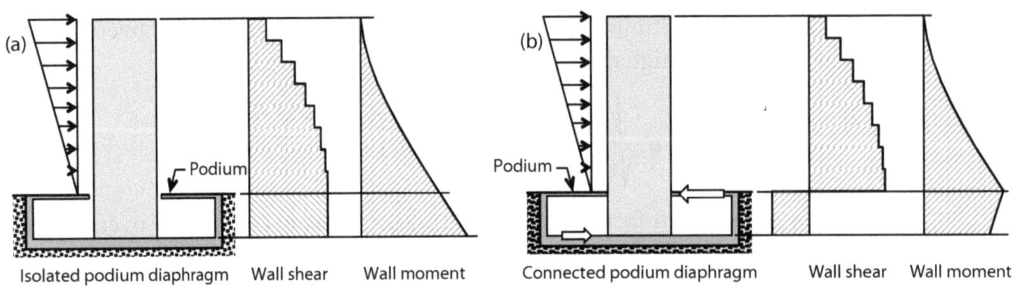

FIGURE 13.28 Backstay effect: (*a*) wall and podium diaphragm not connected; (*b*) wall and podium diaphragm connected.

condition (Figure 13.28*b*) generally is preferred because it provides a redundant force path (backstay force plus shear and moment at foundation mat) and because lateral displacements of the wall and podium diaphragm are compatible. The isolated podium (Figure 13.28*a*) generally is not preferred because the force path is not redundant (the foundation mat must resist the entire shear and moment) and because the wall and podium diaphragm are not constrained to have equal displacements. In this case, columns spanning the first story will experience amplified lateral displacements that must be considered in design.

The magnitude of the backstay effect depends in part on relative stiffness of the wall and the podium diaphragm—a relatively stiff diaphragm has larger backstay effect while a relatively flexible diaphragm has smaller backstay effect. Figure 13.29 illustrates effects of stiffness modeling on calculated results for a high-rise core-wall building. The building is analyzed using a nonlinear model for three different assumptions for shear modulus: (a) gross sections = G_c; (b) cracked sections = $0.5G_c$ for walls in the hinge zone, $0.75G_c$ for walls elsewhere, $0.5G_c$ for diaphragms; and (c) softened diaphragm = $0.5G_c$ for walls in the hinge zone, $0.75G_c$ for walls elsewhere, $0.1G_c$ for diaphragms. Figure 13.29 plots envelopes of wall shear for a single ground motion considering the different modeling assumptions. The effect on wall shears above the podium level is minimal. However, the effect on wall

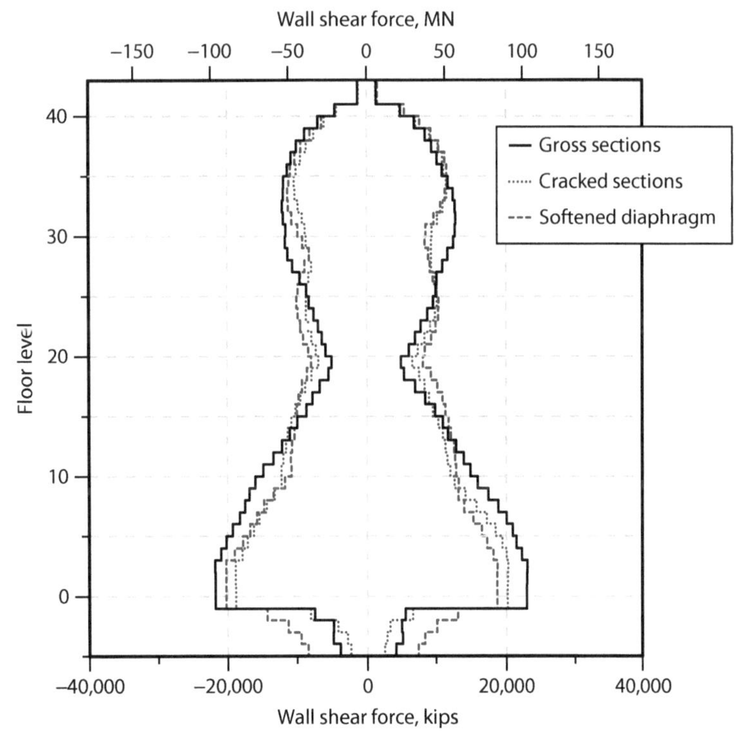

FIGURE 13.29 Effect of stiffness assumptions on calculated wall shears. (*After ATC 72, 2010, used with permission from J. Wallace.*)

shears below the podium level and on podium diaphragm forces (not shown) is significant. Recognizing the sensitivity of results to stiffness assumptions, some engineers conduct bounding analyses using different stiffness assumptions to obtain conservative estimates of shear forces.

13.4.6 Walls with Cap Beams and Outriggers

Cap beams are deep beams located at the top of a stack of openings in coupled walls, typically near the roof where mechanical equipment is located (Figure 13.11). Because of their high stiffness, cap beams can increase system coupling significantly, thereby reducing lateral drifts and increasing system torsional resistance. Cap beams generally attract large shear forces. If design shears exceed allowable values for coupling beams, beam depth can be decreased to reduce shear forces. Alternatively, the beam can be designed as a ductile fuse (e.g., a diagonally reinforced coupling beam), thereby limiting the design shear force. Section 13.12.1 provides additional discussion on coupling beam design.

Outriggers are structural elements that extend outward from a wall to engage other structural elements, thereby helping to resist overturning moments. Outriggers can be especially useful in foundations where they extend the effective base of the wall element, picking up additional gravity loads, and helping resist foundation overturning without developing tension (Figure 13.7*a*). In tall buildings responding in a predominantly flexural mode they can engage outbound columns to create a tension-compression couple, thereby alleviating wall moments and reducing lateral drifts (Figure 13.30). Although outriggers alleviate wall moments, they do little to resist horizontal shear. In tall core-wall buildings, the most effective elevation for outriggers is around two-thirds of the building height, but architectural considerations commonly do not permit this unless there is need for a mechanical floor at

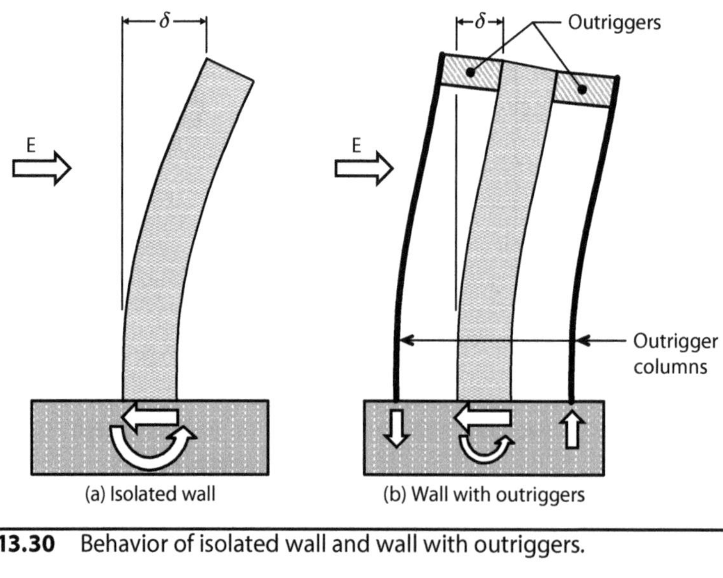

FIGURE 13.30 Behavior of isolated wall and wall with outriggers.

that elevation. Instead, it is more common to find outriggers at roof level where mechanical equipment is located. Figure 13.11 illustrates an example using multiple outriggers to stiffen the short direction of a core wall.

Outriggers can be constructed of reinforced concrete wall segments, steel braces, energy-dissipating elements, or combinations of these. Where they occur in occupied floors, multi-level outriggers with penetrations for circulation should be considered. Locations of penetrations should be selected to be compatible with the primary force path within the outrigger.

Design forces for outriggers should be determined using a combination of conventional structural analysis and capacity design. Outriggers spanning one or more floor heights are likely to engage floor diaphragms, which act as flanges to the outrigger. If this interaction is modeled in structural analysis software, realistic diaphragm stiffness should be used rather than rigid diaphragms, which overestimate outrigger stiffness. Extracting outrigger design forces can be complicated with some modeling approaches—usually it is preferable to obtain design forces from changes in outrigger column axial forces and apply those as point loads to sub-models of the outrigger cantilevering from the structural wall.

A capacity design approach should be used to achieve a ductile "fuse" in the outrigger while protecting outrigger columns from overload. Axial yielding of outrigger columns generally is not considered acceptable. Furthermore, moment transfer from an outrigger to a structural wall can result in very large panel zone shears in the wall, analogous to those that develop in a beam-column joint. The design should ensure that the structural wall is not overloaded locally by this transfer mechanism. See Section 13.15 for additional discussion.

Although they generally are provided to resist lateral forces, outriggers are very stiff structural elements that will participate in gravity load resistance. Outrigger construction should be sequenced to minimize dead load forces from stories above. Steel braces can be placed after completion of upper stories; likewise, concrete outriggers can have closure joints placed later in the construction schedule. In typical designs, columns have higher dead load stresses than walls, resulting in greater creep in columns. Unless there is a mechanism to adjust the long-term distribution of forces, dead loads from floors below may eventually hang from the outrigger due to creep effects.

13.4.7 Frame–Wall Interaction

Frames and walls respond differently to lateral forces, resulting in frame-wall interaction when they are combined in a building system. Frames deform in a predominantly "shear" mode, that is, inter-story drifts tend to be roughly proportional to the shear, with greater drifts in lower stories (Figure 13.31*a*). In contrast, cantilever walls deform in a predominantly "bending" mode, with greatest drifts in upper stories (Figure 13.31*b*). Combined in a building system, the floor and roof diaphragms impose displacement compatibility (Figure 13.31*c*). Consequently, the frame pulls back on the top of the wall, creating a force reversal near the top of the wall, and the frame is supported by the wall near the base, relieving the frame of most of its shear (Figure 13.31*d*).

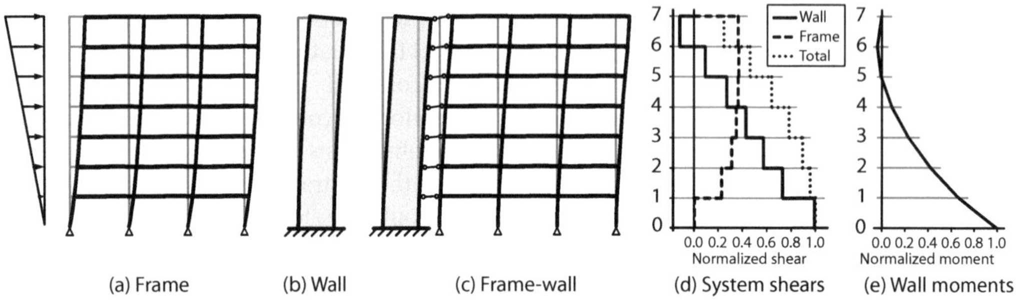

Figure 13.31 Frame–wall interaction.

Frame–wall interaction is sensitive to relative stiffness of the frame and wall, and will change during an earthquake due to cracking and subsequent yielding. Design actions should be determined using either sensitivity analyses or conservative interpretation of single analyses. For the example of Figure 13.31, building design should recognize that small calculated frame shears in the first story and small calculated wall moments in upper stories may change substantially with changes in stiffness. A conservative design will provide strength well in excess of the small values calculated.

13.5 Analysis Guidance

Chapter 11 described overall seismic design requirements for buildings, including layout of the structural system, required strength, and drift limits. Once a preliminary design has been completed, the structural engineer can develop a structural analysis model to determine seismic design actions for the structural system. The analysis model can be either a linear elastic model or it can model the inelastic load-deformation behavior of individual components. Linear elastic models, which are more commonly used, are the focus of discussion in this section.

13.5.1 Analysis Procedures

Many building codes provide three different linear analysis options for determining seismic forces in a special structural wall. These are the *equivalent lateral force procedure*, *modal response spectrum analysis*, and *seismic response history analysis*. Chapter 11 provided a brief overview of the first two of these procedures. Seismic response history analysis is seldom used with linear models.

The ELF (equivalent lateral force) procedure is the simplest of the linear analysis options and can be used effectively for basic low-rise structures. In U.S. codes, seismic base shear V is calculated using an approximate fundamental period, T_a, unless the building period is determined by structural analysis. Generally, analysis of moderate-to-tall structural wall buildings will show that the period is longer than the approximate period. In U.S. codes, the ELF procedure is not permitted for long-period structures (fundamental period T greater than 3.5 s) or structures with certain horizontal or vertical irregularities.

Modal response spectrum analysis often is preferred because it accounts for the elastic dynamic behavior of the structure. It is not unusual for the modal base shear to be less than the design base shear calculated with the ELF procedure. If the modal base shear is less than 85% of the ELF base shear, U.S. codes require that the modal base shear be scaled to 85% of the ELF base shear. Thus, the modal response spectrum procedure can result in cost savings relative to the ELF procedure.

For modal response spectrum analysis, a 3-D computer model typically is used, as it is an effective means of identifying effects of inherent torsion in the lateral system as well as directional interaction of flanged walls. For such analyses, code-prescribed accidental torsion forces typically are applied as static story torsions linearly combined with the dynamic results. It is common for the design load combination to include 100% of seismic response in one direction plus 30% in the orthogonal direction, and then to switch the load combination directions to 30%/100%. Multiple load combinations are required to bound orthogonal effects in both directions. To avoid excessive conservatism, the resulting structural wall demands typically are considered for each combination rather than being enveloped.

Interaction of all structural and nonstructural elements that affect the linear and nonlinear response of the structure to earthquake motions should be considered in the analysis. Important examples include interactions with masonry infills (partial or full height), architectural concrete walls, stairwells, cast-in-place stairways, and inclined parking ramps. It is not always necessary to include these elements in the global model; instead, the results of global analysis can be used to check whether interferences with nonstructural elements occur, and construction details can be modified as needed.

13.5.2 Stiffness Recommendations

When analyzing a structural wall, it is important to model appropriately the cracked stiffness of the wall and any coupling elements, as this stiffness determines the building periods, base shear, story drifts, and internal force distributions. ACI 318 prescribes three different options for approximating wall stiffness: (a) 50% of gross-section stiffness; (b) $I_e = 0.70I_g$ if uncracked or $0.35I_g$ if cracked, and $A_e = 1.0A_g$; or (c) more detailed analysis considering reduced stiffness under loading conditions. Actual stiffness of structural walls depends on reinforcement ratio, slip of reinforcement from foundations, foundation rotation, axial force, and other parameters. Flexural and axial stiffness values prescribed by ACI 318 are reasonable for many cases; effective shear modulus typically is as low as $G_c/10$ to $G_c/20$. ATC 72 (2010) provides additional guidance.

ACI 318 provides frame beam effective stiffness values, but these are not appropriate for typical coupling beams. Coupling beams are expected to sustain damage before significant yielding occurs in walls, leading to faster stiffness reduction. Coupling beam effective stiffness is further reduced because of concentrated end rotations associated with reinforcement slip from anchorage zones within the wall boundary. ATC 72 (2010) recommends effective stiffness $E_cI_e = 0.15E_cI_g$ with shear deformations calculated based on $G_c = 0.4E_c$ for $l_n/h \geq 2$ and $G_c = 0.1E_c$ for $l_n/h \leq 1.4$, with linear interpolation for intermediate aspect ratio.

The preceding recommendations intend to approximate secant stiffness to onset of yielding. Actual instantaneous stiffness varies with time as a structure oscillates at varying amplitude. A nonlinear analytical model can approximate these instantaneous stiffness changes with time, but at considerably greater expense in modeling and computation. For additional guidance, see Deierlein et al. (2010).

Floor diaphragms can be modeled adequately as rigid elements where effects of in-plane floor deformations are expected to be small. This is generally the case where diaphragm aspect ratio is small, where structural walls are evenly distributed across the diaphragm, and where there is not a significant stiffness discontinuity in the structural wall system or diaphragm. Where these conditions are not met, diaphragm in-plane flexibility should be modeled using realistic stiffness properties including expected cracking. This is especially important where the diaphragm transfers large shears, such as at setbacks and podium levels (Figure 13.6). For additional guidance, see Chapter 15.

13.5.3 Effective Flange Width

When a flanged wall undergoes drift, the flanges on both the tension side and the compression side participate in resisting axial force and moment. The structural analysis model should represent the stiffness (and strength for nonlinear models) associated with flange participation. Section 13.4.2 presents recommendations for wall effective flange width. Note that this effective width applies only to stiffness and moment strength calculations. For determination of tributary gravity loads, which resist uplift, use the full tributary flange width, not the effective flange width.

13.5.4 Foundation Modeling

Base restraint can have a significant effect on behavior of structural wall buildings. Building codes commonly permit the base of the structure to be modeled as fixed for the purpose of determining seismic loads. Alternatively, foundation flexibility can be modeled. Unlike a moment frame lateral system, which may be detailed to be fixed or pinned at its base, a structural wall always will be fixed to the supporting foundation element. For this reason, structural walls typically are modeled as having a fixed base, with no further foundation modeling. However, foundation elements supporting the structural wall need not be considered fixed with respect to the underlying soil. Foundation rocking and other soil–foundation–structure interaction effects can be considered, but are beyond the scope of this text. For buildings with multiple subterranean levels, a wall extending into the basement is more likely to be nearly fixed because it is "locked in" by the diaphragms, such that foundation rocking is less important.

13.5.5 Limit Analysis and Redistribution of Coupled Walls

Linear analysis, which typically is used to obtain design actions, does not recognize internal force redistribution that occurs as a building responds nonlinearly. Redistribution can be especially important for a coupled wall where interaction is strongly influenced by relative stiffness and strength of the parts of the coupled wall. Limit (or plastic) analysis can be useful to understand effects of nonlinear force redistribution. Consideration of redistribution also can result in greater efficiency in the coupled wall design compared with designs based solely on linear analysis results.

Consider a symmetric coupled wall system subjected to lateral forces (no vertical acceleration or gravity effects), as shown in Figure 13.32. Under the usual assumptions of linear analysis, identical walls 1 and 2 are assigned equal flexural, shear, and axial stiffness. If analyzed under these assumptions, the analysis results will indicate that walls 1 and 2 have identical shears and moments at every elevation, with wall 1 in tension and wall 2 in compression. It should be apparent, however, that wall 2 will be stiffer than wall 1 for this direction of lateral loading because of its higher axial compressive force. Consequently, the shears and moments in wall 2 should be higher than those in wall 1. In linear static analysis, we can adjust the stiffnesses of walls 1 and 2 to achieve more realistic results, but with linear dynamic analysis, where walls 1 and 2 are subjected to alternating axial compression and tension, this will not be feasible.

Suppose that the coupled walls and coupling beams are designed for actions obtained from linear analysis (Figure 13.32*a*). Now consider the coupled system in the fully plastic state (Figure 13.32*b*). All coupling beams develop their full plastic moment capacity M_{pr} with corresponding plastic shear V_{pr}. The plastic shears, in turn, act on the walls, summing over wall height to produce axial forces T_{pr1} and P_{pr2} at the wall base. Because V_{pr} exceeds design shear V_u for each beam (due to factors such as oversizing and material overstrength), axial tension force T_{pr1} will exceed the design value T_{u1} obtained from linear analysis. Considering *P-M* interaction, higher tension force will result in reduced moment strength for

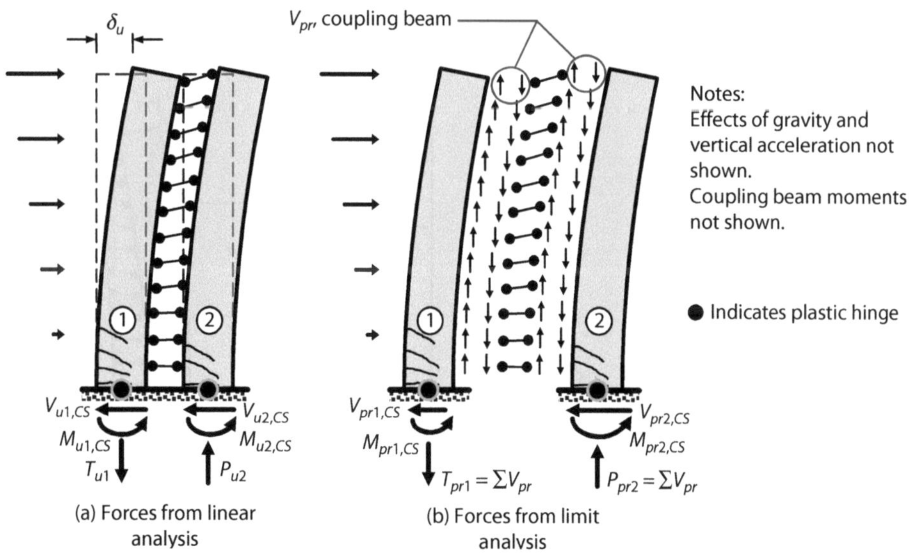

FIGURE 13.32 Coupled wall internal forces.

wall 1. Likewise, for wall 2, P_{pr2} will exceed P_{u2}, in this case leading to higher moment strength for wall 2. As a consequence of the higher stiffness and higher moment strength of the compression wall relative to the tension wall, shear forces will migrate from the tension wall to the compression wall. This effect is seldom considered in design practice in the United States.

Some redistribution of internal forces should be permissible in design of ductile structures such as coupled special structural walls. Eurocode 8 and NZS3101 allow redistribution of wall moments and shears in design. Although details of the procedures vary, they relate closely to those proposed in Paulay (1986), which are adapted for the recommendations made below.

In design of coupled special structural walls, it is acceptable to reduce the design moment in any wall by as much as 30% provided the total resistance of the building is not decreased. The reduction should apply to the wall pier that is in tension, with an equal increase in the wall pier that is in compression (Figure 13.33*a* and *b*, respectively). Wall shears also should be redistributed in proportion with the moment redistribution. Furthermore, it is acceptable to "average out" the coupling beam design shears (and moments), thereby resulting in greater economy by repetition of beam details over multiple stories. The maximum reduction in beam shear should not exceed 20% of the design shear obtained from linear analysis, and provided strength in adjacent stories should be increased such that total resistance provided by the beams is not decreased (Figure 13.33*c*)—that is, values of T_{u1} and P_{u2} should not decrease after redistribution is applied (Figure 13.33*d*).

The aforementioned reduction in shear and moment in the tension wall is desirable because it correctly reflects the trend of lateral force to redistribute from the tension wall to the compression wall. It also is desirable from the perspective of wall reinforcement and foundation design requirements, as the higher compressive force in the compression wall acts to provide additional wall and foundation moment resistance that could otherwise be difficult to provide in the tension wall.

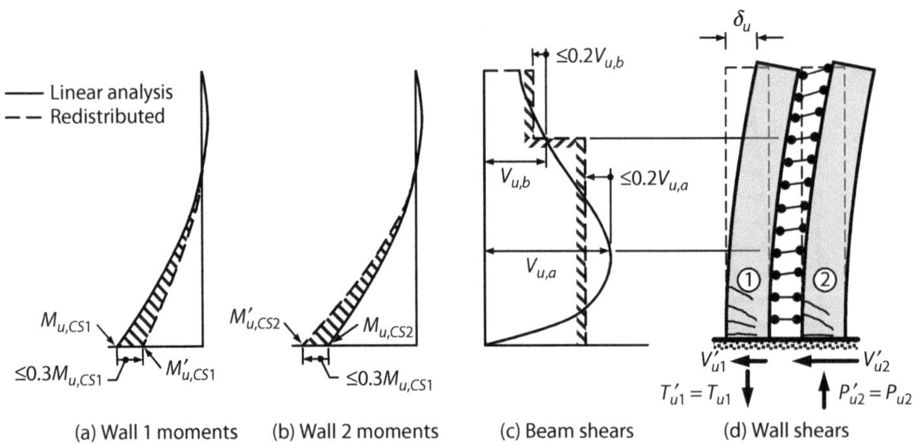

FIGURE 13.33 Redistribution in coupled walls. (*After Paulay, 1986, used with permission from Earthquake Engineering Research Institute.*)

13.6 Load and Resistance Factors for Wall Design

Building codes generally define design loads and load combinations applicable to design of structural systems, including special structural walls and coupling beams. In U.S. practice, load combinations require horizontal seismic effects to be evaluated in conjunction with vertical seismic effects, dead load, variable portions of live load, and other applied loads such as soil pressure, snow, and fluids. Section 11.5 presents loads and load combinations applicable to U.S. practice. The remainder of this section presents an abbreviated review of the load and resistance factors applicable to U.S. design practice for special structural walls and coupling beams, as represented in ASCE 7 and ACI 318.

In ASCE 7, the horizontal seismic effect is defined as $E_h = \rho Q_E$ and the vertical seismic effect is defined as $E_v = 0.2S_{DS}D$. In general, E_h and E_v must be applied in all combinations in both positive and negative directions. The two basic load combinations applicable to special structural wall design are

$$U = (1.2 + 0.2S_{DS})D + \rho Q_E + (0.5 \text{ or } 1.0)L + 0.2S \tag{13.13}$$

$$U = (0.9 - 0.2S_{DS})D + \rho Q_E + 1.6H \tag{13.14}$$

The load factor on L is permitted to equal 0.5 for all occupancies in which L is less than or equal to 100 psf (4.79 kPa), with the exception of garages or areas occupied as places of public assembly. Otherwise, the load factor on L is 1.0.

To define redundancy factor ρ, consider only vertical wall segments whose aspect ratio $h_w/l_w \geq 1$, where h_w = story height. If removal of one of these segments results in either a 33% reduction in story strength or an extreme torsional irregularity as defined in ASCE 7, $\rho = 1.3$. Otherwise, $\rho = 1.0$.

For combined moment and axial force in a wall, strength reduction factor ϕ is determined using the same procedure as is used for columns. For this purpose, ε_t is defined as net tensile strain in the extreme tension steel when the section reaches nominal strength ($\varepsilon_{cu} = 0.003$). If $\varepsilon_t \geq 0.005$, $\phi = 0.9$. If $\varepsilon_t \leq \varepsilon_y$ [taken as 0.002 for Grade 60 (420)], $\phi = 0.65$ for tied boundary elements or 0.75 for spiral reinforced boundary elements. The value of ϕ is interpolated for intermediate values of ε_t.

For wall shear including shear-friction, ACI 318 allows $\phi = 0.75$, except $\phi = 0.6$ if nominal strength V_n is less than the shear corresponding to development of wall nominal moment strength M_n. [This text recommends designing slender walls so the design shear strength (ϕV_n) is at least the shear corresponding to development of wall moment strength, preferably the probable moment strength. This typically is not practicable for squat walls; use of $\phi = 0.6$ usually is not a significant penalty for squat walls given their inherent strength.]

For conventionally reinforced coupling beams, $\phi = 0.75$ for shear and 0.9 for moment (except in unusual cases for which $\varepsilon_t < 0.005$). *For diagonally reinforced coupling beams*, $\phi = 0.85$ for shear (an explicit design check for moment strength is not required in diagonally reinforced coupling beams).

13.7 Preliminary Proportioning

Preliminary proportioning refers to initial structural wall sizing, before development of a detailed model of the building. In low- and mid-rise buildings, preliminary proportioning usually is controlled by required base shear. In taller buildings, preliminary proportioning may be controlled by drift limits.

13.7.1 Proportioning for Base Shear

Initial structural wall sizing typically considers building seismic base shear V_b versus wall design shear strength ϕV_n. Building seismic base shear V_b is determined from the applicable building code procedures (e.g., ASCE 7 in the United States) as discussed in Section 13.5.1. When considering preliminary shear demands for individual walls, several amplification factors should be considered:

1. In codes such as ASCE 7, redundancy factor ρ may amplify shear. See Section 13.6.
2. Torsion, both inherent and accidental, increases wall shear. Typical amplification factors, relative to the basic shear without torsion, are in the range 1.2 to 1.5.
3. Where shear is resisted by multiple vertical wall segments with different lengths, openings, and flanges, total shear will be distributed unevenly among the segments. The amplification factor for individual segments can vary widely.
4. Designing for multiple load combinations invariably will result in wall flexural overstrength. For slender walls where a flexural yielding mode is desired, wall shears should be amplified commensurately. A factor of approximately 1.4 is typical. See Section 13.3.1.
5. Dynamic effects can amplify wall shears in multi-story buildings. For slender walls where a flexural yielding mode is desired, dynamic amplification factor ω, defined in Section 13.4.4, can be applied. Factor ω tends to be larger for buildings with longer fundamental period.

The first three factors apply to most buildings, whereas the last two apply only to slender walls in multi-story buildings where flexural yielding is intended to be the controlling inelastic mechanism.

The estimated wall shear, including amplification factors noted above, should not exceed the limiting shear force of the building code. According to ACI 318, the average shear stress for all walls resisting a common shear force is not permitted to exceed $8\phi\sqrt{f'_c}$ psi ($0.67\phi\sqrt{f'_c}$ MPa), with no individual wall segment allowed to resist more than $10\phi\sqrt{f'_c}$ psi ($0.83\phi\sqrt{f'_c}$ MPa). If the amplification factor of the fourth item above is applied, $\phi = 0.75$; otherwise, $\phi = 0.6$ (see Section 13.6). As discussed in Section 7.12, improved wall behavior is achieved by targeting a lower design stress, in the range $4\phi\sqrt{f'_c}$ to $6\phi\sqrt{f'_c}$ psi ($0.33\phi\sqrt{f'_c}$ to $0.50\phi\sqrt{f'_c}$ MPa). A good first approximation of total required wall area in each direction is the amplified shear from the preceding paragraph divided by the target design shear stress.

13.7.2 Proportioning for Drift

Design of mid-rise and taller buildings may be controlled by drift limits. For such buildings, spectral displacement S_d can be obtained from the design response spectrum as $S_d = \left(\frac{T}{2\pi}\right)^2 S_a g = 9.8T^2 S_a$ in ($0.25T^2 S_a$ m). Because most such buildings fall in the period range where $S_a = S_{D1}/T$ (using the design response spectrum of Figure 11.8), this expression can be recast as $S_d = 9.8TS_{D1}$ in ($0.25TS_{D1}$ m). For a fixed-base, uniform cantilever, the fundamental period required to meet the story drift ratio limit IDR ($= \Delta_a/h_{sx}$) is approximately $T \le \frac{IDR}{21}\frac{h_n}{S_{D1}}$, where h_n and S_{D1} are in consistent units. The required total flexural stiffness of all walls is approximately $E_c I_e \ge 3.7 h_n W \left(\frac{S_{D1}}{IDR}\right)^2$. The basic assumptions of the expression are as follows: (a) fixed-base building, uniform over height, (b) flexural response in the first mode without torsion, and (c) inelastic drift can be estimated based on response of a linear oscillator having flexural stiffness $E_c I_e$. This expression can be used as a first approximation of required wall properties for drift control. Table 11.8 presents values for story drift limits.

13.8 Design of Slender Walls with Single Critical Section[4]

Walls or wall segments that are effectively continuous from base of structure to top of wall generally should be designed to have a single critical section for moment and axial force, as in Figure 13.13. Some discontinuity over wall height is acceptable provided the wall is proportioned so that the critical section occurs where intended. Coupled walls with stacks of openings as in Figures 13.10 and 13.11 also should be designed by this approach. Walls in tall buildings, where secondary plastic hinges may be unavoidable (see Section 13.4.4), also should use this method, although additional provisions to confine the wall around secondary hinge locations may be warranted. Where these conditions are not satisfied, the wall should be designed in accordance with Section 13.9.

13.8.1 Moment and Axial Force Design of Intended Plastic Hinge

Design for moment and axial force involves preliminary proportioning, boundary element transverse reinforcement layout, analysis for P-M strength, and iterations to optimize the layout considering coordination of boundary element vertical and horizontal reinforcement and section strength.

Preliminary Proportioning

For uncoupled rectangular wall sections, preliminary sizing of wall vertical reinforcement can be accomplished using the model of Figure 13.34, which assumes there is both distributed vertical reinforcement (stress resultant T_{s1}) and

[4]The term "critical section" generally refers to the cross section of a member that is proportioned to have the narrowest margin between provided and required strength, in this case, strength to resist combined moment and axial force. A wall designed for a single critical section is expected to sustain flexural yielding not only at the critical section, but also immediately above and below the critical section. This extended region should be detailed for the anticipated flexural demands.

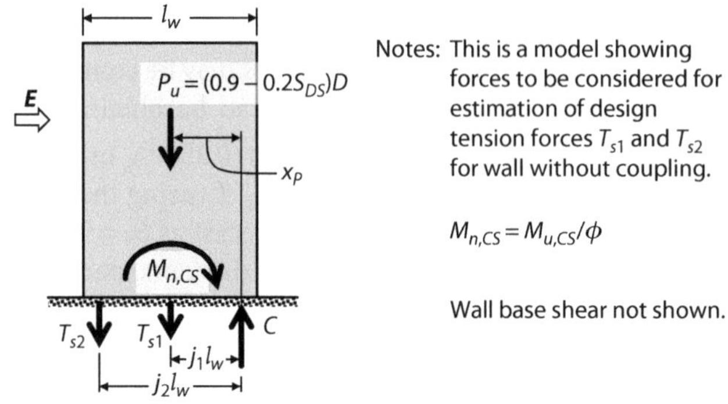

FIGURE 13.34 Model for initial selection of flexural tension reinforcement.

boundary vertical reinforcement (stress resultant T_{s2}). Summing moments about C results in

$$M_{n,CS} = P_u x_p + T_{s1} j_1 l_w + T_{s2} j_2 l_w \tag{13.15}$$

Magnitude and location of P_u are determined from tributary dead loads including self-weight for the load combination shown; knowing location of P_u, moment arm x_p can be approximated. In typical cases, the internal moment arms for distributed and concentrated reinforcement can be approximated as $j_1 l_w = 0.4 l_w$ and $j_2 l_w = 0.8 l_w$. One approach is to select minimum required distributed vertical reinforcement based on $\rho_l = 0.0025$, thereby approximately defining T_{s1}, and then use Eq. (13.15) to find boundary element tension force T_{s2} required to achieve target moment strength $M_{n,CS}$. (Note that squat walls sometimes require $\rho_l > 0.0025$; see Section 13.10.) Alternatively, if only distributed reinforcement is to be used, set $T_{s2} = 0$ and use Eq. (13.15) to solve for T_{s1}. Required distributed reinforcement is then $A_{st} = T_{s1}/f_y$, but not less than minimum required distributed reinforcement. For flanged sections, reinforcement within the effective flange width in tension contributes to T_{s2}.

The preceding discussion assumes wall moment strength is controlled by tensile yielding of vertical reinforcement, as recommended by this text. If moment strength is controlled by strength of the compression zone, a modified approach is required, and axial force P_u must be based on the load combination of Eq. (13.13).

In coupled walls, the preliminary proportioning approach is adjusted by including the additional axial force (either tension or compression, whichever controls) due to wall coupling. See Section 13.12 for additional discussion.

It is good design practice to provide hoop reinforcement to confine the most heavily strained portion of the flexural compression zone and to provide lateral support of vertical reinforcement (Figure 13.12). If boundary elements are required (described later in this section), ACI 318 requires them to extend horizontally from the extreme compression fiber a distance at least the greater of $c - 0.1 l_w$ and $c/2$, where c is the largest neutral axis depth calculated under combinations of P_u and M_u. Generally, P_u for this calculation is based on the load combination from Eq. (13.13). Figure 13.35 presents a chart for preliminary

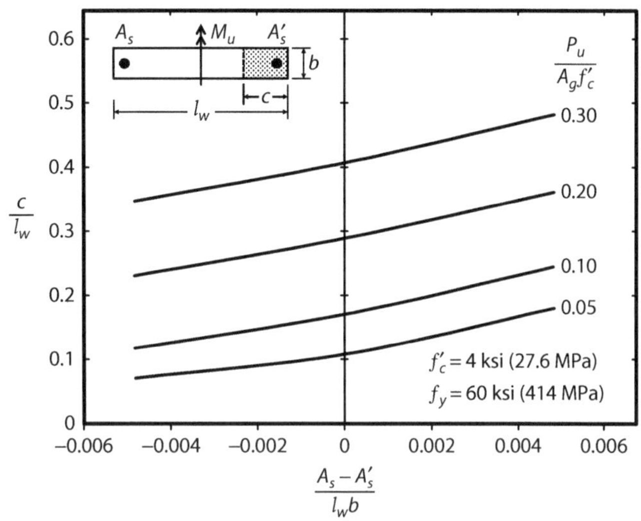

FIGURE 13.35 Approximate flexural compression depth. For flanged sections, adjust A_s, A_s', and b considering effective flange width.

estimation of neutral axis depth. If concentrated flexural tension reinforcement is provided in the boundary, it can be spread out within the confined region. If it is too congested, either proportions of the wall can be reconsidered, or the confined region can be extended farther into the flexural compression zone.

Having established preliminary proportions, the next step is to confirm *P-M* strength and neutral axis depth using section analysis.

P-M Strength Calculations

Strength calculations for structural walls resisting combined moment and axial force directly match the calculations for reinforced concrete columns. Specifically, the calculations assume linear strain distribution, idealized stress-strain relations for concrete and reinforcement, and material strain limits (e.g., ACI 318 limits concrete compressive strain to 0.003). All developed vertical reinforcement within effective flange widths, boundary elements, and the wall web must be included. *P-M* interaction software can facilitate the calculations.

For *P-M* calculations, the axial force P_u must be located correctly. Where axial force is based on tributary loads, with loads followed through the structure using hand calculations, the correct location of axial force usually is the centroid of loads tributary to the wall, including self-weight. Where a computer model is used to establish axial force, the correct location usually is the location reported from the computer model. Beware that some computer programs for *P-M* section analysis automatically place P_u at the geometric centroid of the section. If this location is incorrect, resulting moments must be corrected by adding or subtracting $M_u = P_u e$, where e is the eccentricity between correct and assigned location of P_u.

Because wall geometry and concrete strength typically are defined before detailed analysis, design for *P-M* resistance typically is a trial and error process using vertical reinforcement size and placement as the variables. Boundary element transverse reinforcement provides lateral support for vertical reinforcement, so design of both must be done in parallel. Boundary element detailing is considered next.

Boundary Elements

A *boundary element* is a portion along a structural wall edge or opening that is strengthened by longitudinal and transverse reinforcement. Where combined seismic and gravity loading results in high compressive demands on the edge, ACI 318 requires a *special boundary element*. Where compressive demands are lower, special boundary elements are not required, but boundary element transverse reinforcement still is required if the longitudinal reinforcement ratio at the wall boundary is greater than $400/f_y$, psi ($2.8/f_y$, MPa). For clarity, this text refers to these latter elements as *ordinary boundary elements* (a term not used in ACI 318). Figure 13.36 shows examples of special and ordinary boundary elements.

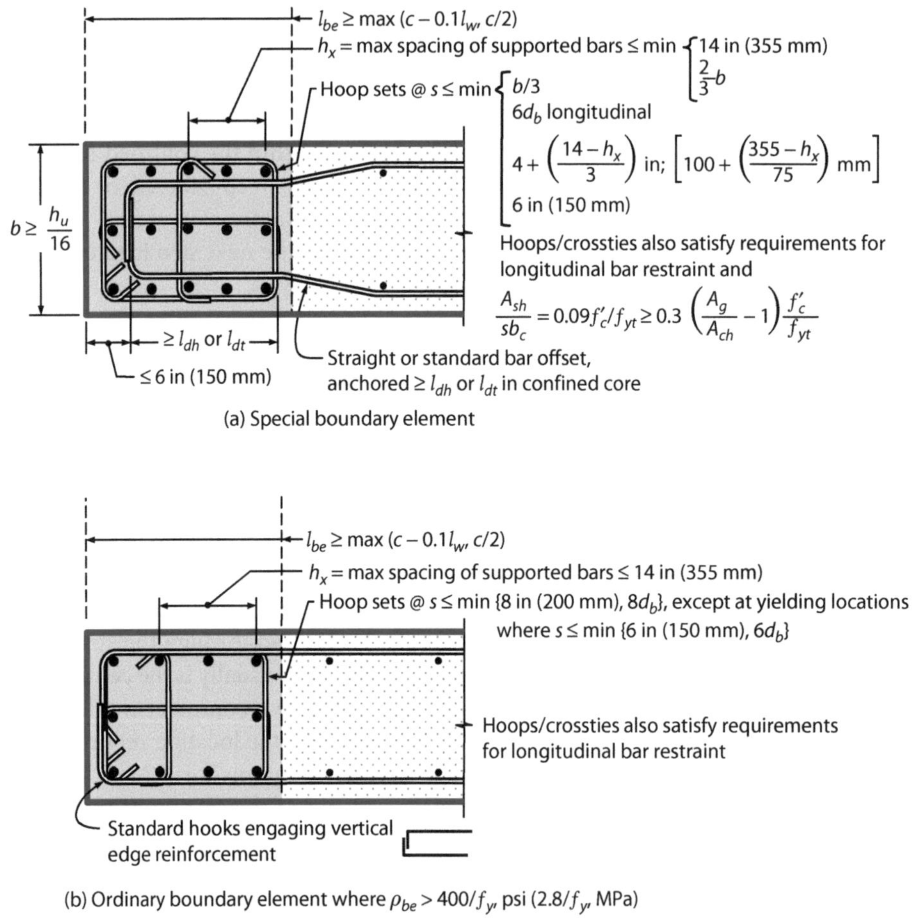

FIGURE 13.36 Special and ordinary boundary elements according to ACI 318.

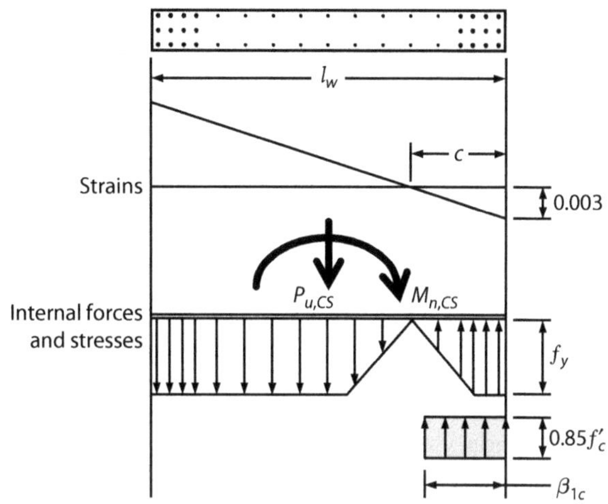

FIGURE 13.37 Calculation of neutral axis depth c.

ACI 318 uses a *displacement-based method* to determine whether special boundary elements are required. The seismic-force-resisting system is first sized and then analyzed to determine the top-level design displacement δ_u and corresponding maximum value of wall axial force P_u. The flexural compression depth c corresponding to nominal moment strength $M_{n,CS}$ under axial force P_u is then calculated using conventional P-M interaction analysis (Figure 13.37). If

$$c \geq \frac{l_w}{900(\delta_u / h_w)} \tag{13.16}$$

where h_w refers to total wall height from critical section to top of wall, then special boundary elements are required. Ratio δ_u/h_w in Eq. (13.16) is not to be taken less than 0.005.

Equation (13.16) is derived from the simplified model shown in Figure 13.38. It assumes that the target displacement is the value at MCE, or approximately

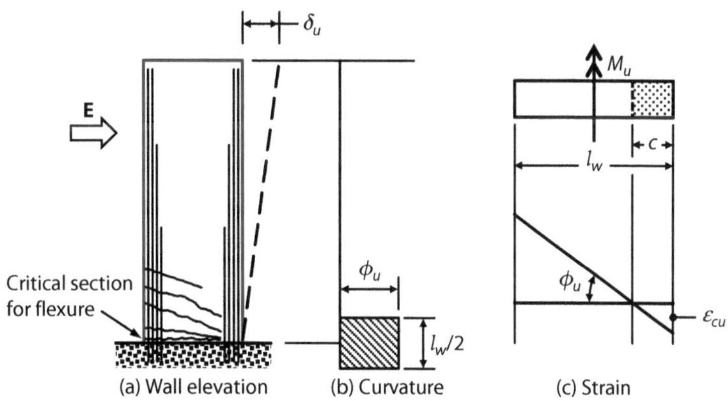

FIGURE 13.38 Model used to derive Eq. (13.16).

$1.5\delta_u$. Furthermore, displacement $1.5\delta_u$ is assumed to be due entirely to curvature ϕ_u centered on the wall critical section with plastic hinge length $= l_w/2$. Noting that $\phi_u = \varepsilon_{cu}/c$ and setting $\varepsilon_{cu} = 0.003$ result in $1.5\delta_u = 0.0015 l_w h_w / c$. Rearranging and rounding leads to the expression $c = \frac{l_w}{900(\delta_u/h_w)}$. If c exceeds this value, confinement is required.

In view of the uncertainties expressed in the previous paragraph, a preferred practice is to always provide special boundary elements within the intended plastic hinge zone of slender special walls designed to have a single critical section for moment and axial force, unless the design earthquake demands are especially low considering the seismic hazard or configuration of the structural system.

Tension-controlled flexural response is defined for $\varepsilon_t \geq 0.005$, which can also be expressed as $c \leq \frac{3}{8} l_w$, where c = depth of the flexural compression zone. Where special boundary elements are required and the calculated depth of the flexural compression zone is $c > \frac{3}{8} l_w$, inelastic response of the wall may produce significant inelastic compression demands on the flexural compression zone of the wall. Such walls should have a thickness that will promote ductile compressive response. ACI 318 requires wall thickness $b \geq 12$ in (300 mm) wherever special boundary elements are required in walls having $h_w / l_w \geq 2$ and $c > \frac{3}{8} l_w$.

Where special boundary elements are required by Eq. (13.16), ACI 318 requires that they extend vertically above and below the critical section a distance not less than the greater of l_w and $M_{u,CS}/4V_{u,CS}$. The limit l_w is based on the expectation that cover spalling in a well-confined section typically will not spread beyond a height approaching the section depth. The limit $M_{u,CS}/4V_{u,CS}$ defines the height above the critical section at which moment decreases to $0.75M_{u,CS}$, a value likely to be less than the spalling moment, assuming a straight-line moment diagram. Where the critical section occurs at or near the connection with a footing, foundation mat, pile cap, or other support, different requirements apply to the vertical extension of the special boundary element. See Figure 13.39 and subsequent discussion.

At the interface with a footing, foundation mat, pile cap, or other support, longitudinal reinforcement of structural walls must be fully developed in tension. Where yielding of longitudinal reinforcement is likely due to lateral drifts, ACI 318 requires development length to be calculated for $1.25f_y$; otherwise it is calculated for f_y. Where depth of foundation element precludes development of straight bars, standard hooks having l_{dh} calculated for $1.25f_y$ or f_y, as appropriate, are acceptable. The standard hook should extend full depth in most cases. See Figure 13.39.

Where a special boundary element terminates on a footing, foundation mat, or pile cap, ACI 318 requires that special boundary element transverse reinforcement extend at least 12 in (305 mm) into the foundation element. For any other support, or where a boundary element has an edge within one-half the footing depth from an edge of the footing (or mat or pile cap), the transverse reinforcement must extend into the support at least l_d, calculated for f_y in tension, of the largest longitudinal reinforcement. See Figure 13.39.

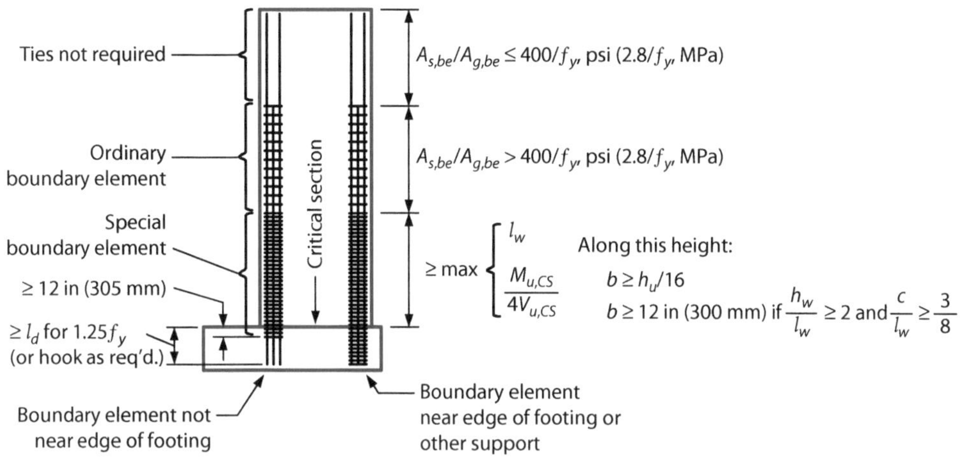

FIGURE 13.39 Boundary element extensions for walls designed by the *displacement-based method*, for critical section at foundation interface. For ordinary and special boundary element details, see Figure 13.36.

Where a special boundary element is required, ACI 318 requires it to extend horizontally from the wall edge a distance not less than the greater of $c - 0.1l_w$ and $c/2$ (Figure 13.36*a*). Flexural compression depth c is calculated at nominal moment strength $M_{n,CS}$ under maximum axial force P_u (Figure 13.37). In flanged sections, the special boundary element, if required, must include the effective flange width in compression and must extend at least 12 in (305 mm) into the web.

According to ACI 318, special boundary elements must have transverse confinement reinforcement satisfying the greater of the quantities given by Eqs. (13.17) and (13.18).

$$\frac{A_{sh}}{sb_c} \ge 0.09 f_c'/f_{yt} \tag{13.17}$$

$$\frac{A_{sh}}{sb_c} \ge 0.3\left(\frac{A_g}{A_{ch}} - 1\right) f_c'/f_{yt} \tag{13.18}$$

Based on the argument that spalling is unlikely along the inside face of the boundary element, it is permitted to calculate A_g as $A_g = b(l_{be} - \text{cover})$, that is, not counting the inside "cover" over the boundary element as part of A_g.

Because f_c' and f_{yt} typically are selected independently of boundary element requirements, the remaining variables in design of boundary element transverse reinforcement are confinement bar size and horizontal and vertical spacing of confinement hoop legs and crossties. The parameters for these variables are discussed in detail in "Vertical Reinforcement Layout."

The illustration of Figure 13.39 is for the case where special boundary elements are required at the critical section for moment and axial force. If special boundary elements are not required at the critical section, ACI 318 requires ordinary boundary elements if boundary element longitudinal reinforcement ratio $A_{s,be}/A_{g,be} > 400/f_y$, psi ($2.8/f_y$, MPa), where $A_{s,be}/A_{g,be}$ is the local ratio within the boundary element. Figure 13.36*b* shows requirements for ordinary boundary elements. If special boundary elements are not required at the critical section and $A_{s,be}/A_{g,be} \leq 400/f_y$, psi ($2.8/f_y$, MPa), ACI 318 permits the section to be detailed without ties enclosing the vertical reinforcement. This latter condition is seldom encountered in design of special structural walls, and the permitted detail is not recommended.

Vertical Reinforcement Layout

The process for laying out wall vertical reinforcement is iterative, considering requirements for *P-M* strength and boundary element transverse reinforcement. One approach is as follows:

1. Determine type of boundary element required (see preceding text). If none required, go to step 4.
2. For special or ordinary boundary elements, determine the required boundary element length l_{be} (Figure 13.36), with c estimated from Figure 13.35 or from *P-M* analysis.
3. Select trial boundary element transverse reinforcement size (No. 3, 4, or 5 (10, 13, or 16)) and vertical spacing. For special boundary elements, use Eqs. (13.17) and (13.18) to determine A_{sh}, from which the number of hoop and crosstie legs in each direction is determined. Check all vertical and horizontal spacing requirements of Figure 13.36, as applicable.
4. Select trial size and spacing of vertical reinforcement for the entire structural wall section. If a boundary element is required, spacing of verticals in the boundary element is dictated by hoop and crosstie arrangement from step 3, with corner and at least alternate verticals restrained by a hoop or crosstie. Verticals outside the boundary element are selected so as to provide the required ρ_l with spacing $s \leq 18$ in (457 mm).
5. Determine *P-M* strength. If provided strength is inadequate or over-conservative, refine bar sizes and repeat step 3 or 4. If acceptable, continue.
6. Use *P-M* analysis to check assumed boundary element extent in step 2. If inadequate or over-conservative, return to step 3 with new c. If acceptable, vertical reinforcement layout is complete.

Alternative iteration schemes also can lead to efficient designs. For example, some designers select boundary vertical reinforcement and spread it within required boundary length l_{be}, then lay out transverse reinforcement to support the verticals and confine the core, and iterate until all requirements are met.

In step 5 above, the basic design requirement of ACI 318 is the same as for columns, that is, all combinations of (M_u, P_u) must be less than the corresponding design values $(\phi M_n, \phi P_n)$. The value of ϕ is defined in Section 13.6. In addition, the maximum axial force cannot exceed the similar limit for columns. The usual approach is to use computer software to generate ϕP_n-ϕM_n interaction diagrams and then check that M_u, P_u pairs for all load combinations fall within the design limits. Section 13.6 discusses the relevant load combinations.

Modal Response Spectrum Force Combinations

As noted in Section 13.5.1, modal response spectrum analysis is a common method of determining wall design forces. This technique considers multiple vibration modes and combines the values of interest using either the SRSS or CQC method. Although results from each mode correctly indicate the sign of calculated quantities, SRSS and CQC results do not. For an uncoupled wall resisting lateral force in two orthogonal directions, there are four seismic load cases to be combined with non-seismic loads for the *P-M* check:

$$P_u + M_{ux} + M_{uy} \qquad P_u - M_{ux} + M_{uy}$$

$$P_u + M_{ux} - M_{uy} \qquad P_u - M_{ux} - M_{uy}$$

In contrast, the interactions in coupled walls result in significant induced axial forces. Consideration of all possible sign combinations results in eight possible seismic load cases:

$$+P_u + M_{ux} + M_{uy} \qquad +P_u - M_{ux} + M_{uy} \qquad -P_u + M_{ux} + M_{uy} \qquad -P_u - M_{ux} + M_{uy}$$

$$+P_u + M_{ux} - M_{uy} \qquad +P_u - M_{ux} - M_{uy} \qquad -P_u + M_{ux} - M_{uy} \qquad -P_u - M_{ux} - M_{uy}$$

Again, these seismic forces are combined with dead, live, soil, and snow loads per code-specified load combinations (Section 13.6) for final structural wall design.

Sometimes inspection of the eight possible sign combinations can identify combinations that are kinematically impossible and therefore require no further consideration. For example, consider the coupled planar structural wall shown in Figure 13.40. Lateral sway occurs with a single possible set of combined moments and axial forces. For the left-hand wall, axial tension occurs simultaneously with moment oriented so that maximum tension is induced on the left edge of that wall. The reverse combination occurs in the right-hand wall, where maximum compression is induced on the right edge of that wall. These axial and moment force sign pairings are determinant for these wall segments. Subtracting T_u from the left-hand wall or P_u from the right-hand wall would result in conditions that cannot occur; including these combinations would result in unnecessary wall flexural overstrength, which can cascade to increased design requirements elsewhere.

The sign–force combination of flanged and coupled structural walls is significantly more complex because of bi-directional interaction. Sign–force relationships revealed by an ELF analysis sometimes can help understand the range of sign–force possibilities from the modal response spectrum analysis results.

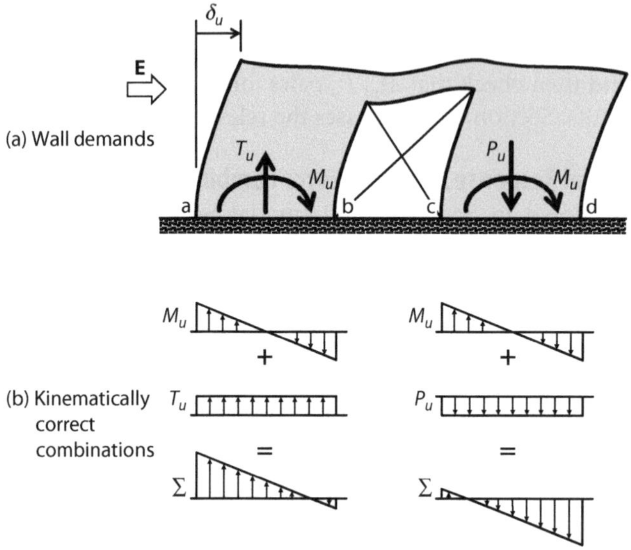

FIGURE 13.40 Elevation of laterally displaced coupled wall system.

13.8.2 Shear Design of the Intended Plastic Hinge

Design Shear

Overall proportions of structural walls in a building typically are determined by consideration of the seismic design forces or drift limits of the applicable building code (Section 13.7). Once vertical reinforcement is selected based on moment-axial strength requirements (Section 13.8.1), a capacity design approach should be used to establish design shear strength such that there is an acceptably low probability of shear failure. The approach should consider both probable moment strength of the wall and dynamic response of the structural system as discussed in Section 13.4.4.

MacGregor (1983) presents the basis for the load and resistance factor method currently used in U.S. codes. A safety margin is established to achieve an acceptable probability of ductile failure for the design service life. A larger safety margin then is set for brittle failure modes. For typical cases, the probability of brittle failure is approximately one quarter of the probability of ductile failure.

A similar approach was used in the *Guidelines for Performance-Based Seismic Design of Tall Buildings* (PEER, 2010). According to the Guidelines, for Maximum Considered Earthquake shaking, the target probability of reaching (ductile) flexural deformation capacity was set at 50%, while the target probability for reaching wall shear strength was set at 10%. Achieving this objective explicitly requires nonlinear analysis for large number of earthquake ground motions to establish response statistics. Where the statistics are available, the design objective can be achieved approximately by setting the design shear force equal to the mean plus one standard deviation value and the design shear strength equal to $\phi V_{n,e}$, in which $\phi = 0.75$ as specified in ACI 318 and $V_{n,e}$ = nominal shear strength according to ACI 318 but considering expected materials properties. For more typical cases where response statistics are not available, the Guidelines present an alternative approach.

Where design is according to the more traditional linear methods described in Section 13.5.1, a simplified approach is required. Design shear force should be defined considering flexural overstrength and dynamic effects as discussed in Section 13.4.4. Equations (13.6) through (13.8) are recommended to account for dynamic effect, although alternative approaches of Section 13.4.4 also can be acceptable. Design shear strength is defined using procedures of ACI 318, as described subsequently.

Design Shear Strength

Chapter 7 presents detailed information on shear strength of slender structural walls and introduces the ACI 318 design equations, which are used here. ACI 318 defines nominal shear strength as

$$V_n = A_{cv}(\alpha_c \lambda \sqrt{f_c'} + \rho_t f_y) \tag{13.19}$$

in which $A_{cv} = l_w b_w$; α_c is 3.0 in psi units (0.25 in MPa units) for $h_w/l_w \le 1.5$, is 2.0 psi (0.17 MPa) for $h_w/l_w \ge 2.0$, and varies linearly between these limits; and $\lambda = 0.75$ for all-lightweight concrete, 0.85 for sand-lightweight, and 1.0 for normalweight concrete. For design of an entire wall, ratio h_w/l_w refers to overall dimensions of the wall. For design of a vertical wall segment within a wall, the ratio refers to overall dimensions of the wall or the vertical wall segment, whichever ratio is greater. The intent is that a vertical wall segment never be assigned unit strength greater than that for the entire wall, although it can be assigned lower unit strength if its h_w/l_w is greater than that of the entire wall.

The basic design requirement is $\phi V_n \ge V_u$. Strength reduction factor ϕ is discussed in Section 13.6. This expression and the expression for V_n can be combined and solved for ρ_t, the required horizontal reinforcement ratio. Reinforcement composing ρ_t must be placed in two curtains if $V_u > 2A_{cv}\lambda\sqrt{f_c'}$, psi ($0.17A_{cv}\lambda\sqrt{f_c'}$, MPa), which almost always is the case. This text recommends always using two curtains within the hinge region of a slender wall because of greater stability than is obtained with one curtain (Section 13.4.3). The reinforcement must provide web reinforcement ratio not less than 0.0025 with maximum vertical spacing of 18 in (457 mm).

ACI 318 defines upper limits for shear strength of special structural walls. For all vertical wall segments resisting a common lateral force, combined V_n shall not be taken greater than $8\,A_{cv}\sqrt{f_c'}$, psi ($0.67A_{cv}\sqrt{f_c'}$, MPa), where A_{cv} is gross combined area of all vertical wall segments parallel to the applied shear. For any one of the individual vertical wall segments, V_n shall not be taken greater than $10A_{cv}\sqrt{f_c'}$, psi ($0.83A_{cv}\sqrt{f_c'}$, MPa), where A_{cv} is web area of the individual vertical wall segment. It is acceptable to interpret the common lateral force as either (a) entire story shear, in which case the combined area refers to all walls or vertical wall segments in the story, or (b) shear resisted by a single wall or a line of walls in a single plane, in which case the combined area refers to the area of walls or vertical wall segments in that plane. Lower shear stress is preferred, as this improves flexural deformation capacity of the wall.

Shear Reinforcement Detailing

If a special boundary element is required, ACI 318 requires horizontal shear reinforcement to extend to within 6 in of the wall edge and be anchored to develop f_y in tension within the confined core of the boundary element using standard hooks or heads (Figure 13.36*a*). One option is to extend the horizontal web reinforcement continuously to near the wall edge. Another option is to lap the horizontal web reinforcement with horizontal boundary element reinforcement such that the boundary element reinforcement serves as the wall shear reinforcement within the boundary element. This is only permitted if there is sufficient lap length and if horizontal boundary element reinforcement provides strength $A_{sh}f_{yt}/s$ parallel to the web reinforcement at least equal strength of the horizontal web reinforcement $A_v f_y/s$. In this case, it is permitted to terminate the horizontal web reinforcement without a standard hook or head. Note that according to this alternative, required reinforcement A_{sh} parallel to the web is the maximum of that required for confinement and shear; it is not necessary to sum the two requirements.

The option to lap web reinforcement with confinement reinforcement may be preferred in construction because it enables use of horizontal reinforcement that is either straight, or has offset bends, without hooks. Although this detail is widely used, its performance has not been verified by laboratory tests or field observations. Under lateral loading, the lap splice at one end of the wall will be in a flexural tension zone, which may weaken the lap. This text recommends avoiding lap splices of horizontal reinforcement in potential flexural tension zones and instead using the alternative detail with hooks or heads.

Another common practice is to lap splice the web horizontal reinforcement within the shear panel (roughly midway along the wall length). This detail should be avoided within the intended plastic hinge zone because integrity of the lap splice cannot be assured given expected inelastic response in flexure. Outside the intended hinge zone, this detail may be acceptable. The lap lengths must be determined considering the bars to be top-cast.

Where ordinary boundary elements are provided, the accepted detail for shear reinforcement is standard hooks that engage the edge reinforcement (Figure 13.36*b*). The preferred detail uses 135° hooks such that the tails extend into the confined core. ACI 318 allows use of U-stirrups that engage the edge reinforcement and lap with the horizontal reinforcement provided for shear. This detail is not recommended here because of concerns that the lap will be ineffective where wall flexure results in flexural tension transverse to the lap splice.

13.8.3 Shear-Friction Design of the Intended Plastic Hinge

In addition to web horizontal reinforcement to resist shear, shear-friction reinforcement is required wherever shear is transferred across an interface of two concrete volumes cast at different times. The intent is to prevent a sliding shear failure at such interfaces. This applies to the connection between walls and foundation and, for multi-story structural walls cast floor-by-floor, at the horizontal cold joint at each floor. This is especially important within or near yielding sections, as opening and closing of flexural cracks can degrade sliding resistance across crack surfaces.

Chapter 7 presents detailed information on transfer of shear across construction joints, and introduces the shear-friction concept. According to the concept, sliding resistance depends on interface roughness and clamping force across the interface. Where reinforcement is perpendicular to the sliding plane, nominal shear strength is

$$V_n = A_{vf} f_y \mu \tag{13.20}$$

$$V_n = (A_{vf} f_y + N_u)\mu \tag{13.20a}$$

$$V_n = (A_{vf} f_y - T_{u,net})\mu \tag{13.20b}$$

A_{vf} refers to distributed vertical reinforcement in the wall web; in a wall with boundary elements, A_{vf} can be calculated conservatively as if the distributed vertical web reinforcement continues uninterrupted into the boundary elements. Equation (13.20a) applies where permanent net compression force N_u acts perpendicular to the sliding plane, with N_u positive in compression. Equation (13.20b) applies where transient net tension force $T_{u,net}$ acts perpendicular to the sliding plane. Equation (13.20) gives conservative value of nominal shear strength. Chapter 7 provides alternative expressions that may give better estimates.

The basic design requirement is $\phi V_n \geq V_u$. Section 13.6 discusses strength reduction factor ϕ. Wall vertical reinforcement sized and located for *P-M* interaction resistance can serve double duty as shear-friction reinforcement. If that reinforcement proves insufficient to resist interface shear, additional distributed vertical dowels can be placed along the wall centerline, developed for f_y above and below the interface. Inclined bars can be more effective to resist sliding, although bars need to be inclined in both directions to resist alternating load directions.

According to ACI 318, V_n is not permitted to exceed the least of $0.2f_c'A_{cv}$, $(480 + 0.08f_c')A_{cv}$, and $1600A_{cv}$, psi [$0.2f_c'A_{cv}$, $(3.3 + 0.08f_c')A_{cv}$, and $11A_{cv}$, MPa].

In addition to reinforcement, ACI 318 requires that the interface be clean and free of laitance. If the surface is intentionally roughened to a full amplitude of ¼ in (6 mm), the friction coefficient can be taken as $\mu = 1.0\lambda$. Shear keys are an effective alternative where surface roughening to ¼ in (6 mm) amplitude cannot be achieved. Otherwise, friction is reduced and $\mu = 0.6\lambda$.

13.8.4 Requirements above the Intended Plastic Hinge

The region above the intended plastic hinge should be designed with overstrength that promotes flexural yielding at the intended plastic hinge. Details also should recognize that some flexural yielding may occur above the intended plastic hinge, especially for taller buildings, as described in Section 13.4.4.

Moment Strength, Splicing, and Bar Cutoffs

Design for moment and axial force above the intended plastic hinge follows the same procedures used in the flexural plastic hinge, except a capacity design approach should be used to establish required moment strength. Figure 13.27 illustrates the recommended approach, with details of the procedure described in the associated text. *P-M* interaction design follows the usual procedures, and typically involves considerations of *P-M* strength, development length, and bar cutoffs.

Vertical reinforcement can be terminated where it is no longer required to resist moment and axial force. For this purpose, ACI 318 refers to bar cutoff requirements for beams, as described in Chapter 8, but with $0.8l_w$ substituted for beam effective depth *d*. Thus, the basic requirements are that (1) terminated bars must be developed beyond points of maximum stress and (2) terminated bars must extend $0.8l_w$ beyond the point at which they are no longer required to resist moment and axial force. The $0.8l_w$ extension is in consideration of a potential tension shift effect where shear cracks occur. See Section 6.11.2.

Figure 13.41 illustrates application of the ACI 318 requirements to a wall, using a simplified moment envelope. Bars **a** provide design strength ϕM_n sufficient to resist M_u at the critical section for moment and axial force. If bars **b** are to be terminated, the requirements would be as follows: (i) bars **b** must be developed for $1.25f_y$ above the critical section requiring these bars (1.25 factor required by ACI 318 because they provide required strength at the intended plastic hinge) and (ii) bars **b** must extend at least $0.8l_w$ above the elevation where they are no longer required to resist moment and axial force (in this case, $0.8l_w$ above the point where continuing bars **c** provide design strength $\phi M_n = M_u$). This process can be continued up the wall height, as (iii) bars **d** must be developed for f_y above the critical section for bars **c**; (iv) bars **d** must extend at least $0.8l_w$ above the point where continuing bars **e** provide required strength; and (v) bars **e** must be developed for f_y above the critical section

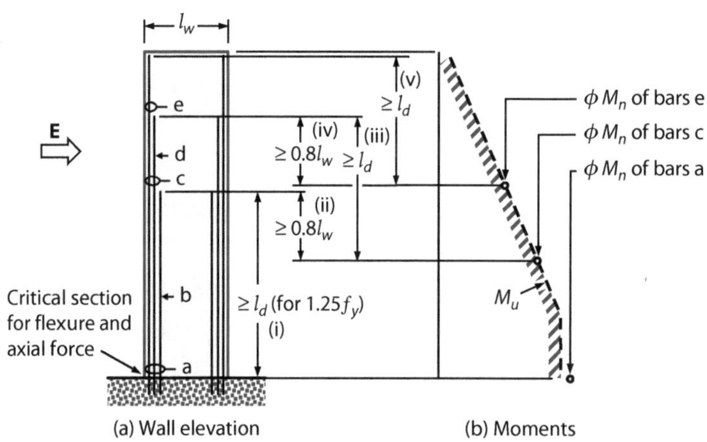

FIGURE 13.41 Bar cutoffs for vertical reinforcement for idealized M_u moment diagram.

for bars **e**. Except for the bars required to extend to the top of the wall, most of the bar cutoffs will be controlled by the requirement to extend bars $0.8l_w$ past the point where they are no longer required to resist moment and axial force, thereby simplifying design.

The aforementioned procedure seems unnecessarily onerous, especially considering that the wall moment diagram for a building responding to earthquake shaking is not accurately known, and dynamic response will adjust to any strength shortcomings. A practice used by many design offices is to extend bars l_d above the floor where the bars are no longer required to resist moment and axial force. This practice is not strictly in compliance with the aforementioned ACI 318 requirement, but it serves the intent to extend bars well past the point where they are no longer required for moment and axial force, and seems to be a reasonable approach for design.

ACI 318 contains development length requirements applicable where flexural reinforcement is terminated in tension zones. Although ACI 318 does not specifically exempt walls from these requirements, it is understood that they apply only to beams. Common engineering practice does not apply these requirements to design of special structural walls.

Shear Strength

Design for shear outside the intended flexural plastic-hinge zone should follow the same procedures that apply within the intended flexural plastic-hinge zone.

Confinement Reinforcement

As noted in Section 13.8.1 and Figure 13.39, ACI 318 requires special boundary elements to extend vertically above and below the critical section a distance not less than the greater of l_w and $M_{u,CS}/4V_{u,CS}$. Where the critical section occurs at or near the connection with a footing, foundation mat, pile cap, or other support, different requirements apply to the vertical extension of the special boundary element. See Section 13.8.1.

Outside the elevations where special boundary elements are required, ACI 318 requires ordinary boundary elements if the boundary element longitudinal reinforcement ratio $A_{s,be}/A_{g,be} > 400/f_y$, psi ($2.8/f_y$, MPa), where $A_{s,be}/A_{g,be}$ is the local ratio within the boundary element. Figure 13.36*b* shows requirements for ordinary boundary elements. Where $A_{s,be}/A_{g,be} \leq 400/f_y$, psi ($2.8/f_y$, MPa), ACI 318 permits the section to be detailed without ties enclosing the vertical reinforcement. See Figure 13.39.

As discussed in Section 13.4.4, very tall wall buildings might develop secondary flexural yielding near mid-height due to apparent higher-mode response. A challenge is that linear structural analysis, which is widely used, does not indicate directly whether such yielding might occur. Nonlinear dynamic analysis can provide insight into this issue. Some designers define an intermediate boundary element that satisfies all requirements for special boundary elements except the volume ratio is reduced by half; these intermediate boundary elements are extended into the potential secondary yielding zone. As a minimum, this text recommends that at least ordinary boundary elements extend through elevations that show high moment demands due to higher-mode response. Furthermore, to restrain longitudinal bar buckling, hoop spacing should not exceed $6d_b$ of the longitudinal bars at such locations.

13.9 Design of Walls without an Identified Critical Section

Walls that are not effectively continuous from base of structure to top of wall, including walls with irregular openings, cannot routinely be designed to have a single critical section for moment and axial force. Such walls are discouraged but cannot always be avoided. ACI 318 provides a stress-based approach for determining required confinement that is generally considered to be conservative. Additional considerations for irregular walls also apply—see Section 13.16.

Design for moment and axial force, shear, development, and detailing generally follows procedures outlined in Section 13.8, with one exception. Instead of using the *displacement-based method*, ACI 318 requirements for boundary elements are established using a *stress-based method*. First, the seismic-force-resisting system is sized and analyzed to determine axial forces and moments under critical load combinations. Using a gross-section model of the wall cross section, nominal stress at wall edges is calculated from

$$\sigma = P_u/A_g + M_{ux}/S_{gx} + M_{uy}/S_{gy} \tag{13.21}$$

Special boundary elements are required at an edge if nominal wall stress exceeds $0.2f'_c$. If a special boundary element is required, it must be continued vertically (upward and downward) until calculated compressive stress drops below $0.15f'_c$. See Figure 13.42. Although the *stress-based method* can be used for any wall, the

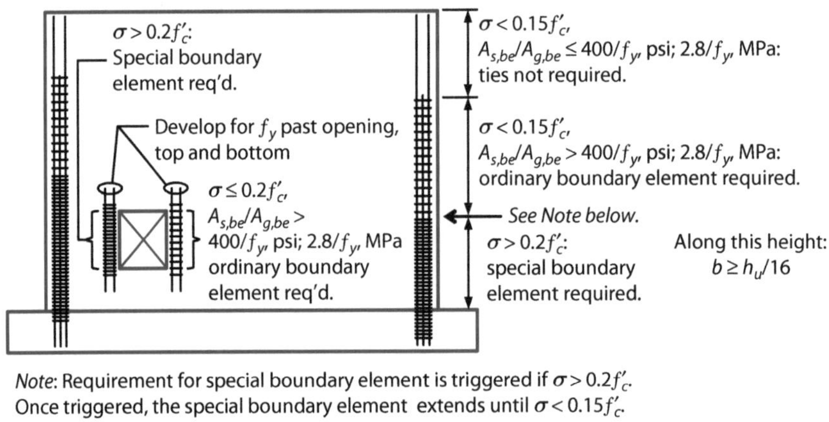

FIGURE 13.42 Boundary element requirements for walls designed by *stress-based method*. The sketch is for a squat wall, but the same provisions apply to slender walls. For extensions into foundations, see Figure 13.39 and associated text in Section 13.8.1. For ordinary and special boundary element details, see Figure 13.36.

preferred use is for irregular or discontinuous walls for which the *displacement-based method* does not apply.

Wall irregularities complicate determination of the yielding mechanism. Consequently, it can be more difficult to estimate flexural overstrength and corresponding overstrength factor for shear design. With some approximation and assumptions, reasonable estimates often can be made. Where the yield mechanism cannot be determined, it may become necessary to forego capacity design calculations and instead carry out the shear design using design shear forces directly from the structural analysis. In this case, use a conservative strength reduction factor, preferably less than the value $\phi = 0.6$ permitted by ACI 318.

13.10 Squat Walls

As noted in Section 13.4.1, low-aspect-ratio (or squat) walls tend to have high inherent flexural strength compared with shear strength, such that it can be difficult to achieve a flexural yielding mechanism for aspect ratio h_w/l_w less than around 1. Furthermore, squat walls tend to resist lateral forces through a diagonal strut mechanism that differs considerably from the flexural mechanism of a slender wall (Figure 13.17). For these reasons, the design approach and required details for squat walls differ from those of more slender walls. Design usually begins with shear design, followed by checking for shear-friction and then combined moment and axial force. The overall procedures are similar to those for slender walls, but with some notable differences.

Most squat walls are reinforced conventionally, that is, with vertical and horizontal reinforcement. An alternative approach uses diagonally oriented reinforcement. The two concepts are presented in the following sections.

13.10.1 Conventionally Reinforced Squat Walls

Design usually begins by considering shear strength requirements. Using ACI 318 procedures, nominal shear strength is defined by Eq. (13.19). In Eq. (13.19), α_c is 2.0 for $h_w/l_w \geq 2.0$ (psi units) [0.17 (MPa units)], is 3.0 (psi units) [0.25 (MPa units)] for $h_w/l_w \leq 1.5$, and varies linearly between these limits. The basic design requirement is $\phi V_n \geq V_u$. For many squat walls, especially those having $h_w/l_w < 1$, it will not be feasible to achieve shear strength greater than shear corresponding to development of moment strength, in which case the strength reduction factor is $\phi = 0.6$. Required horizontal reinforcement ratio ρ_t is determined from these expressions. As with slender walls, reinforcement composing ρ_t must be placed in two curtains if $V_u > 2A_{cv}\lambda\sqrt{f_c'}$, psi ($0.17A_{cv}\lambda\sqrt{f_c'}$, MPa). In addition, distributed horizontal reinforcement must provide web reinforcement ratio not less than 0.0025 with maximum vertical spacing of 18 in (457 mm). Finally, the upper limits of wall nominal shear strength [$8A_{cv}\sqrt{f_c'}$ and $10A_{cv}\sqrt{f_c'}$, psi ($0.67A_{cv}\sqrt{f_c'}$ and $0.83A_{cv}\sqrt{f_c'}$, MPa)] apply as noted in Section 13.8.2.

In a squat wall, distributed vertical reinforcement is as important as distributed horizontal reinforcement in resisting shear (Figure 13.17*b*). ACI 318 requires that reinforcement ratio ρ_l for distributed vertical reinforcement be at least equal to reinforcement ratio ρ_t for distributed horizontal reinforcement if $h_w/l_w \leq 2$.

Once the shear reinforcement design is completed, the next step is to check for shear-friction resistance at any construction joints where concrete is placed against hardened concrete. If additional reinforcement is required, either reinforcement ratio ρ_l can be increased or dowels can be added at the construction joint. See Section 13.8.3.

Next, the wall is checked for combined moment and axial force using *P-M* interaction analysis as outlined in Section 13.8.1. If vertical reinforcement is required in addition to the distributed reinforcement ρ_l provided for shear, then either add additional distributed reinforcement or add vertical reinforcement at the boundaries. These two approaches (distributed reinforcement or concentrated boundary reinforcement) are equally efficient in resisting moment, but distributed reinforcement is more effective in resisting sliding at construction joints.

Because very squat walls typically do not have a flexure-controlled yield mechanism, the requirements for boundary elements are usually checked using the stress-based approach [Eq. (13.21)]. In this case, requirements for boundary elements, if any, are illustrated in Figure 13.42. If a squat wall has a flexure-controlled mechanism, boundary element requirements are according to Section 13.8.1.

13.10.2 Diagonally Reinforced Squat Walls

Flexural cracking at the base of a squat wall creates a plane along which shear sliding can occur. Paulay et al. (1982) observed that vertical reinforcement crossing this interface loses its effectiveness to resist sliding if it yields due to flexure. Diagonally placed reinforcement can be more effective at resisting shear sliding (Figure 13.43). Diagonal reinforcement also contributes to wall moment resistance.

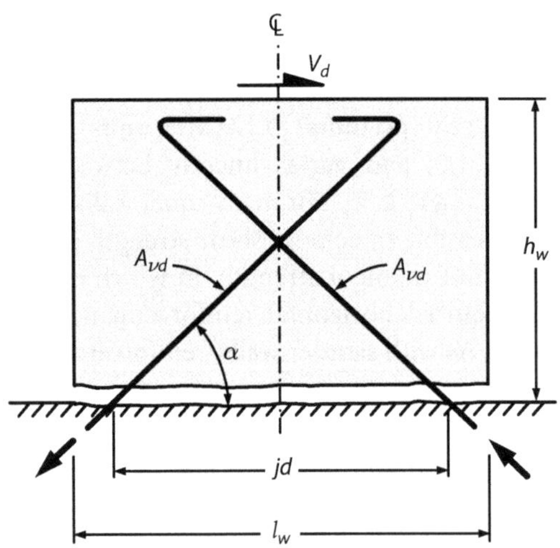

FIGURE 13.43 Squat wall with diagonal reinforcement. (*After Paulay et al., 1982, used with permission from American Concrete Institute.*)

From first principles, we can write the contributions of diagonal reinforcement to shear and moment resistance as

$$V_{nd} = 2A_{vd} f_y \cos\alpha \tag{13.22}$$

$$M_{nd} = A_{vd} f_y jd \sin\alpha \tag{13.23}$$

Where ductile flexural yielding is intended, the added moment strength due to the diagonal bars may be undesirable. In such cases, the diagonal bars can be configured to intersect at the base (i.e., $jd = 0$). Initially, prior to onset of shear sliding, both diagonals would act as flexural tension reinforcement, resulting in moment strength enhancement $M_{nd} \approx A_{vd} f_y l_w \sin\alpha$. If shear sliding were induced, the force in one of the diagonals would become compressive, in which case the moment resistance theoretically would drop to zero while shear resistance was according to Eq. (13.22). Behavior of squat walls with this reinforcement detail has not been demonstrated through laboratory testing.

Paulay et al. (1982) also recommend that the diagonal bars extend above the wall base at least $h_w/2$ or $l_w/2$, whichever is less, to avoid shear failure occurring above the diagonals.

13.11 Wall Piers

In U.S. practice, a *wall pier* is defined as a relatively narrow vertical wall segment that is essentially a column, but whose dimensions do not satisfy requirements of special moment frame columns. According to ACI 318, a vertical wall segment is to be considered a wall pier if $l_w/b_w \le 6.0$ and $h_w/l_w \ge 2.0$, where b_w, l_w, and h_w refer to dimensions of the vertical wall segment. Design of wall piers follows the usual requirements for vertical wall segments, but additional requirements apply as noted below.

ACI 318 requires wall piers to satisfy splice, confinement reinforcement, and shear strength requirements applicable to columns of special moment frames. Alternatively, wall piers with $(l_w/b_w) > 2.5$ can be designed as follows:

- Design shear force V_u is either the shear corresponding to development of M_{pr} at both ends or Ω_0 times the shear determined by analysis of the structure for design load combinations including earthquake effects. Design strength ϕV_n is calculated according to the usual provisions for walls. Although not required by ACI 318, it would be prudent to reduce shear strength if the section has net tension, similar to requirements for columns of special moment frames.
- Transverse reinforcement should be in the form of hoops except where only one curtain of distributed shear reinforcement is provided [permitted only if $V_u \le 2A_{cv}\lambda\sqrt{f'_c}$, psi $(0.17A_{cv}\lambda\sqrt{f'_c}$, MPa)], in which case it is permitted to use single-leg shear reinforcement with 180° bends at each end engaging boundary vertical reinforcement. Maximum spacing of transverse

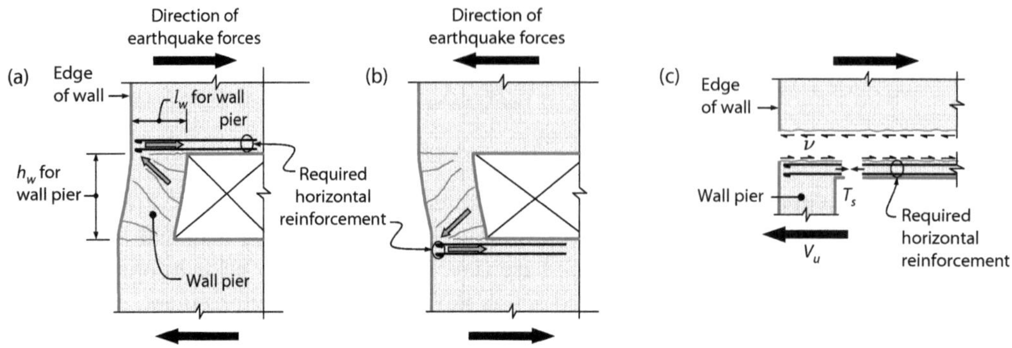

Figure 13.44 Reinforcement required above and below a wall pier at the edge of a wall.

reinforcement is 6 in (150 mm). Transverse reinforcement must extend at least 12 in (300 mm) above and below the clear height of the wall pier.

- The requirements for special boundary elements are checked according to Section 13.9.

Wall piers at the edge of a wall require horizontal reinforcement extending into adjacent wall segments above and below the wall pier, proportioned to transfer the design shear force from the wall pier into adjacent wall segments (Figure 13.44). Figure 13.44*a* and *b* shows the required reinforcement for loading in two opposite directions. Figure 13.44*c* shows free-body diagrams that can be used to determine the area and length for the required horizontal reinforcement for the loading shown in Figure 13.44*a*. First, determine the shear in a wall pier, V_u, based on analysis of the building. This force will be transmitted into the longer wall segment above the wall pier, which presumably is designed for a lower unit shear force per unit length v. The total length of the added reinforcement is based on the requirement to transfer the force V_u at the lower unit shear force v. Hence, the length of the added reinforcement is equal to $l = V_u/v$. Knowing the length of the added bars and the unit shear force v, the tension force T_s can be calculated, leading to the required reinforcement area. For additional discussion on walls with openings, see Section 13.16.1.

13.12 Coupled Wall Systems

Design of coupled special structural walls introduces design complexities beyond those encountered for uncoupled walls. Coupling beams often have relatively low aspect ratio and high deformation demands, requiring special details to achieve ductile performance. Coupling between walls results in axial force variations complicating their design. Coupling beam-wall connections require additional attention to avoid conflicts in reinforcing bar placement.

A first consideration in design of coupled wall systems is structural analysis to determine design moments, shears, and axial forces for the coupling beams and walls. See Section 13.5.5 for discussion of coupled wall analysis including force redistribution.

13.12.1 Coupling Beams

Design requirements for coupling beams vary depending on their aspect ratio l_n/h and seismic demands. ACI 318 classifies coupling beams into three categories. As a practical matter, a fourth category for very deep beams is added here. Figure 13.45 illustrates the design space for these categories.

a. *Coupling beams with* $l_n/h \geq 4$ should satisfy proportioning and detailing requirements for beams of special moment frames, except certain dimensional limits are exempted. Such beams are considered too shallow for efficient use of diagonally placed reinforcement as allowed for deeper beams. Instead, moment reinforcement is placed horizontally at top and bottom of the beam.

b. *Coupling beams with* $l_n/h < 2$ *and* $V_u > 4\lambda\sqrt{f_c'}A_{cw}$, psi ($0.33\lambda\sqrt{f_c'}A_{cw}$, MPa) should be reinforced with two intersecting groups of diagonally placed bars symmetrical about midspan, unless it can be shown that loss of stiffness and strength of the coupling beams will not impair vertical-load-carrying ability of the structure, egress from the structure, or integrity of nonstructural components and their connections to the structure. Implicit in the exception is the requirement for the engineer to demonstrate that the seismic-force-resisting system satisfies code strength and drift requirements in the absence of the excepted coupling beams.

c. *Other coupling beams* not falling within the limits of the preceding two bullets can be reinforced as either conventionally reinforced special moment frame beams or diagonally reinforced beams. In Figure 13.45, beams falling to the right of the dashed line likely can be designed efficiently as special

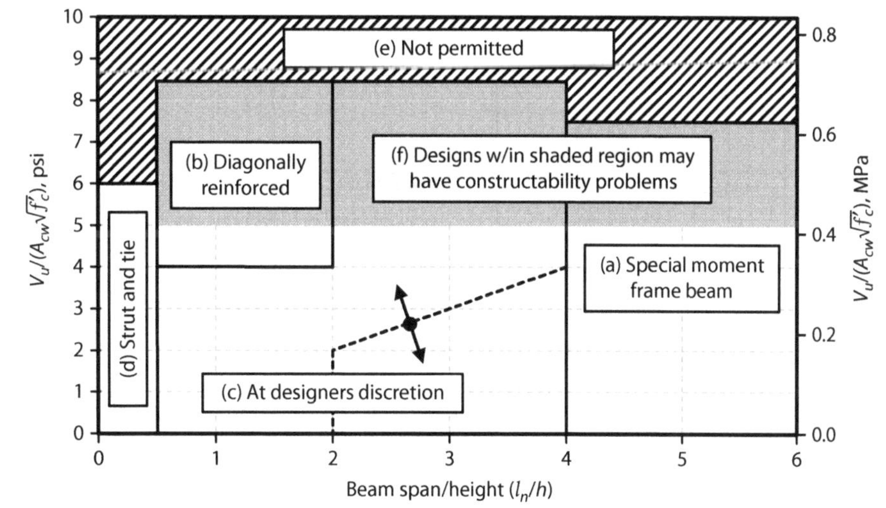

FIGURE 13.45 Coupling beam design space.

moment frame beams, whereas those to the left probably are better designed with diagonal reinforcement.

d. Very low aspect ratio beams may be better designed using a strut-and-tie model.

The darkly shaded area of Figure 13.45 defines the upper limit on beam design shear stress according to ACI 318. The lightly shaded area indicates designs that are permitted by ACI 318 but that may have constructability problems because of reinforcement congestion.

Coupling Beam Designed as Special Moment Frame (Conventionally Reinforced) Beam

Coupling beams designed as special moment frame beams have longitudinal reinforcement placed horizontally at top and bottom of the beam and hoop reinforcement that confines the end regions. Figure 13.46 illustrates typical details. Because l_n/h is relatively small, longitudinal bars cannot be lapped and it may be easier to use closed hoops over the entire beam span rather than only $2h$ at each end. Skin reinforcement, if any, typically is terminated after short extension into the wall [~6 in (150 mm)]; alternatively, it can be developed into the wall, in which case it contributes to beam moment strength.

For beams reinforced with top and bottom longitudinal reinforcement, moment and shear strengths are calculated according to conventional procedures. For moment, the design requirement is $\phi M_n \geq M_u$, where M_u is determined from building analysis under design load combinations, and $\phi = 0.9$. For shear, the requirement is $\phi V_n \geq V_e$, where V_e is determined from equilibrium of the beam assuming it develops M_{pr} at both ends with distributed load w_u acting along the span (Figure 13.47). M_{pr} is probable moment strength, calculated using conventional assumptions except longitudinal reinforcement yield strength is assumed equal to $1.25f_y$. Within $2h$ from member ends, shear strength is based on $V_c = 0$, that is, $V_n = V_s = A_v f_{yt} d/s$, with an upper bound of $V_n = 10\sqrt{f'_c}A_{cw}$, psi ($0.83\sqrt{f'_c}A_{cw}$, MPa). Strength reduction factor for shear is $\phi = 0.75$.

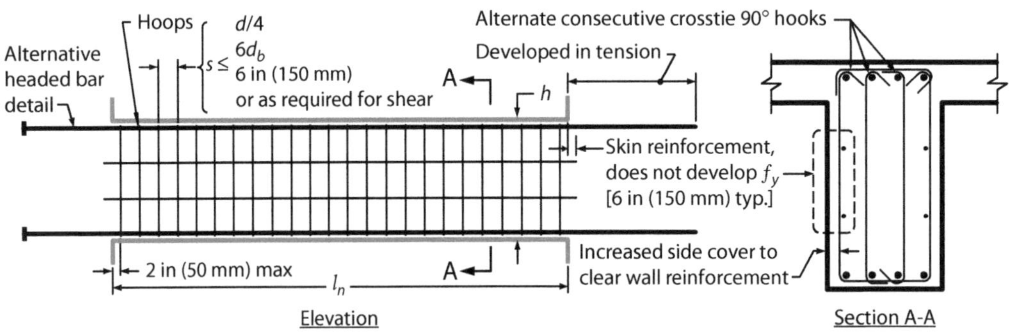

Figure 13.46 Details for conventionally reinforced coupling beams.

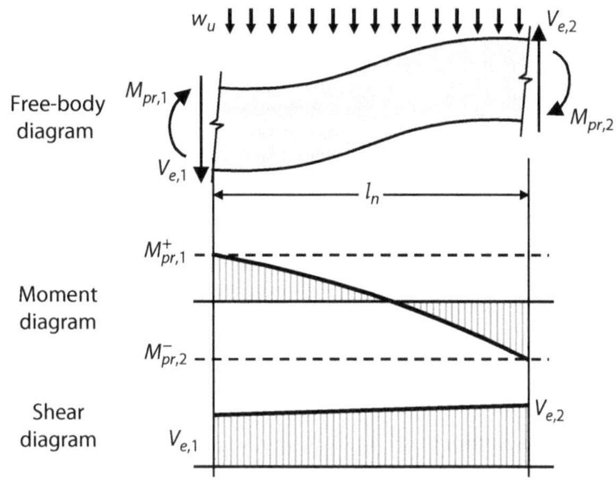

FIGURE 13.47 Design shear for conventionally reinforced coupling beam. Reversed loading case also must be considered.

Coupling Beam Designed as Diagonally Reinforced Beam

Figure 13.48 shows typical details for a coupling beam reinforced with two intersecting groups of diagonally placed bars symmetrical about midspan. Each group of diagonal bars consists of a minimum of four bars provided in two or more layers. The diagonal bars are required to extend straight into the wall a distance at least 1.25 times the development length for f_y in tension. A challenge is avoiding interference between the diagonal bars and boundary element transverse and longitudinal reinforcement. If an adjacent wall opening or edge (e.g., at top of wall) requires the diagonal bar extension to be bent, additional reinforcement is required to resist the unbalanced force resulting from the change in reinforcement direction, similar to the requirement for offset bars in columns. This detail should be avoided where practicable. Minimum wall thickness to accommodate both wall and coupling beam reinforcement is around 14 in (350 mm), although 16 to 18 in (400 to 450 mm) is more practical.

Two reinforcement options are used in U.S. practice. The first option is to confine individual diagonals using hoops and crossties such that corner and alternate diagonal bars are restrained in a hoop or crosstie corner (Figure 13.48*a*). Confinement reinforcement along the entire diagonal length must be made of hoops with or without crossties, configured to support the diagonal bars, with spacing around the perimeter not exceeding 14 in (356 mm). The confinement reinforcement must satisfy Eqs. (13.17) and (13.18), assuming each diagonal as an isolated column with minimum cover over the diagonal cage. Maximum permitted hoop spacing along the diagonal is $6d_b$ of the diagonal bars. Confinement reinforcement can be difficult to place along the free lengths of the diagonals and even more difficult where the diagonals intersect at midspan or enter the wall boundaries. In addition, distributed longitudinal and transverse reinforcement is required around the beam perimeter with total area in each direction not less than $0.002b_ws$ and spacing not exceeding 12 in (300 mm). Longitudinal bars should be terminated no more than 6 in (150 mm) into the wall boundary so that they do not contribute appreciably to beam moment strength.

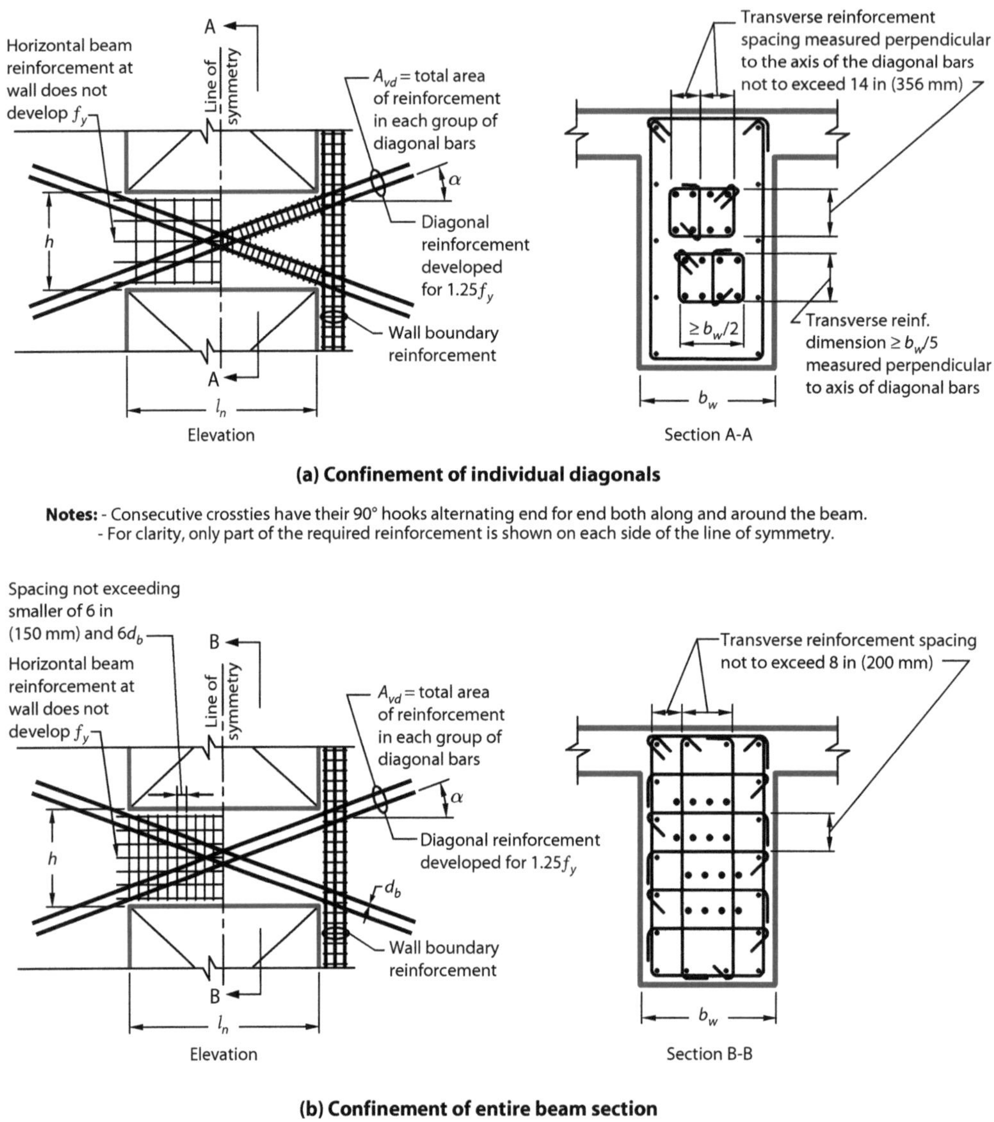

Figure 13.48 Alternative details for diagonally reinforced coupling beams. (*After ACI 318, 2014, used with permission from American Concrete Institute.*)

The second option is intended to ease construction difficulties commonly encountered with the first option. By this option, hoops and crossties confine the entire beam cross section (Figure 13.48*b*). Confinement reinforcement along the entire beam length must satisfy the volumetric ratio requirements required by Eqs. (13.17) and (13.18), with maximum spacing along the beam span not exceeding 6 in (150 mm) and $6d_b$ of the diagonal bars, and with spacing of crossties or legs of hoops around the beam cross section not exceeding 8 in (200 mm). Although total amount of confinement reinforcement may be greater with this second option, increased material costs often are more than offset by reduced labor costs.

A diagonally reinforced coupling beam can be idealized as a truss with tension and compression diagonals along the axes of the diagonally placed reinforcement (Figure 13.49). Vertical equilibrium of the truss defines shear strength V_n as

$$V_n = 2A_{vd} f_y \sin\alpha \le 10\sqrt{f_c'}A_{cw}, \text{ psi } (0.83\sqrt{f_c'}A_{cw}, \text{ MPa}) \tag{13.24}$$

The inequality at the right side of Eq. (13.24) is not from equilibrium but instead expresses the upper bound permitted by ACI 318, similar to the limit on wall shear (Section 13.8.2). Equation (13.24) requires determination of reinforcement angle α. At least two layers of reinforcement are required in each diagonal bundle, so more than minimum cover is required to the centroid of the bundle. A good starting point is to assume centroidal depth at the critical section is $jd = h - 8$ in (200 mm), from which α can be estimated.

The basic strength design requirement for a diagonally reinforced coupling beam considers only shear; moment resistance is automatically provided by the idealized truss (Figure 13.49). The design requirement is $\phi V_n \ge V_u$, where V_u is determined from building analysis under design load combinations, and $\phi = 0.85$.

Main reinforcement must be fully developed in adjacent wall segments. Coupling beams can be expected to undergo large rotations, leading to large reinforcement tensile strains. Therefore, bars should be developed for $1.25f_y$ in tension. Where bars are developed in special boundary elements of walls, ACI 318 development lengths should be adequate. Where boundaries are not specially confined, development lengths should be increased by a factor of 1.3. If horizontal or diagonal bars have more than 12 in (300 mm) of fresh concrete cast below, which is the usual condition, they should be considered top-cast bars with additional length factor of 1.3.

Headed reinforcement sometimes is used to shorten development lengths and facilitate construction. It should be noted that slip of reinforcement from adjacent wall segments is an important component of overall deformation capacity of a coupling beam. Consequently, short headed bar anchorage might reduce deformation capacity of a coupling beam. Extending the headed bar beyond minimum development length l_{dt} will improve coupling beam deformation capacity.

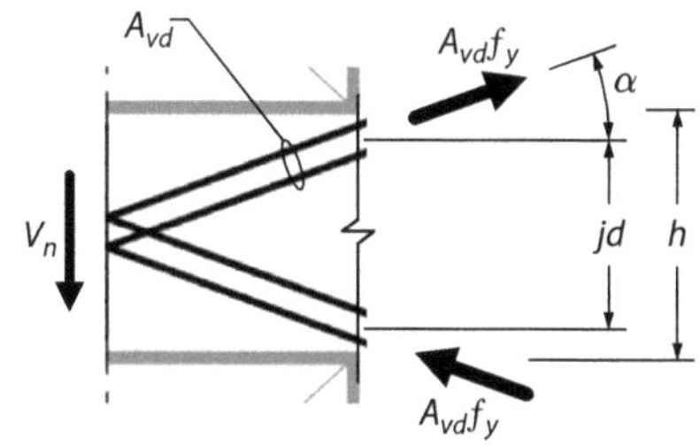

FIGURE 13.49 Free-body diagram of half-span of diagonally reinforced coupling beam. Gravity loads not shown.

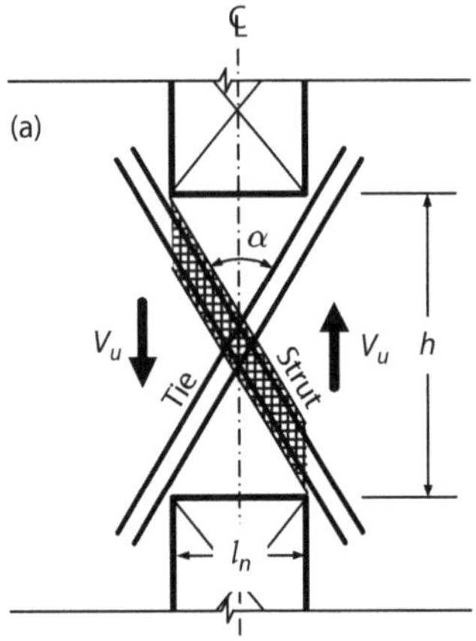

Figure 13.50 Very low aspect ratio beams with diagonal reinforcement.

The preceding text presents details prescribed by ACI 318. Several alternative detailing approaches have been proposed for use, including hybrid beams combining elements of conventionally reinforced and diagonally reinforced beams, alternative arrangements of diagonally oriented reinforcement, and steel coupling beams. It is advisable to check with the local jurisdiction to determine acceptability of alternative designs.

Very Low Aspect Ratio Beams

Very low aspect ratio ($l_n/h \leq 0.5$) coupling beams are likely to attract high shear forces because of their high stiffness. To achieve ductile behavior, one option is to design the beams with diagonal reinforcement, as in Figure 13.50*a*. Strut-and-tie modeling concepts suggest that the angle α between diagonal compression strut and tension tie can be taken as small as 25°, which would correspond to $l_n/h \approx 1/4$. Performance of diagonally reinforced beams with this aspect ratio has not been verified by tests. Another alternative shown in Figure 13.50*b* maintains a more typical angle between the diagonals, but also has not been verified by tests.

Some designers use strut-and-tie models directly, as in Figure 13.51. Note that, for the model shown, horizontal ties are in tension for loading in both directions. Yielding would result in progressive elongation of the ties and deterioration of the shear-resisting mechanism. For this reason, the ties should be designed to remain in the linearly elastic response range.

13.12.2 Coupled Walls

Under lateral loading, coupling between walls causes variations in wall axial force in addition to moment and shear (Figure 13.32). Assuming the two coupled walls resist equal moment (without moment redistribution), the tension wall will require more flexural tension reinforcement than the compression wall. Considering lateral loading in the opposite direction, the tension and compression walls switch. Thus, individual walls designed for alternating lateral forces are likely to have asymmetric

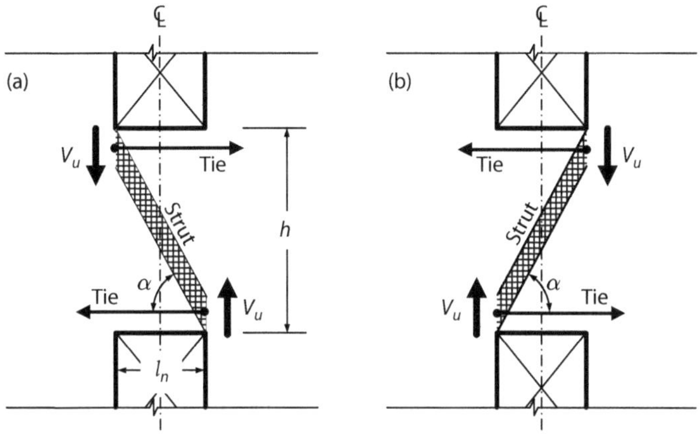

FIGURE 13.51 Very low aspect ratio beam design using strut-and-tie model.

FIGURE 13.52 Characteristic coupled wall cross sections.

boundary elements such as shown in Figure 13.52. The amount of asymmetry can be reduced if moment redistribution is considered, as outlined in Section 13.5.5.

Figure 13.53 illustrates the *P-M* capacity check for a pair of coupled walls symmetric about the system centerline. The solid curve corresponds to *P-M* nominal strength, with the right side applicable to the compression wall and the

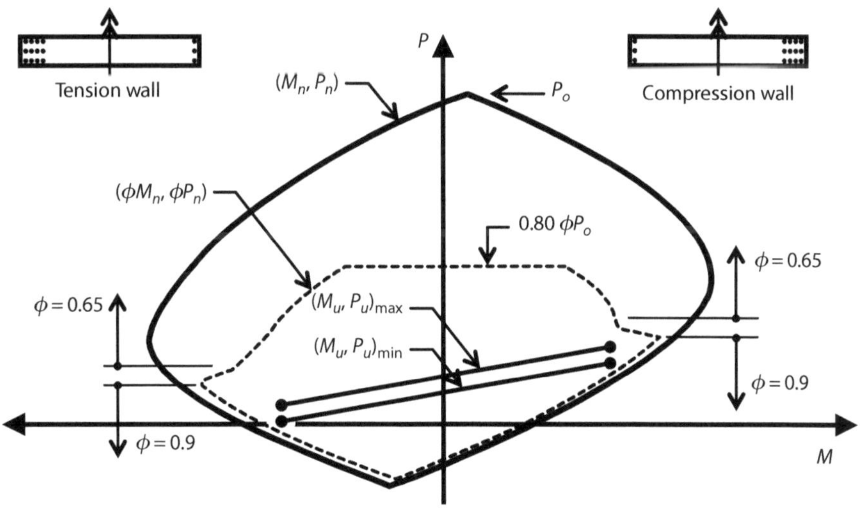

FIGURE 13.53 *P-M* capacity check for coupled walls.

left side applicable to the tension wall. The dashed curve is design strength (nominal strength reduced by strength reduction factor ϕ). The range of *P-M* demands under design load combinations including earthquake load is shown by the two inclined lines. This is an example of a well-designed wall with axial force well below the balanced point.

13.13 Wall Panel Zones

The term *panel zone* refers to a region of a structural wall (or other element) subjected to relatively uniform shear, with or without normal stresses. Figure 13.54 illustrates some examples of panel zones. Because a panel zone is a wall region in which forces from adjacent wall segments are resolved, the nominal shear stress in a panel zone may be much greater than the nominal shear stress in the adjacent wall segments. The effect is similar to that observed for beam-column joints (Chapter 9).

Shear resistance in a panel zone can be idealized using a conventional shear model or using a strut-and-tie model. In Section 7.12.4, an empirical equation was proposed for use where a conventional shear model is used. That expression provides a fairly accurate estimate of panel zone shear strength for walls subjected to monotonic loading. Data for panel zones subjected to stress reversals are lacking. It seems likely that the upper bound on the shear strength would be reduced under such loading conditions. Therefore, this book recommends reducing the upper limit on the shear strength to 0.8 times the limit in Chapter 7. Thus, the recommended shear strength is

$$
\begin{aligned}
v_n &= 3\sqrt{f_c'} + \rho_{smin} f_y \le 0.2 f_c', \text{ psi} \\
&= 0.25\sqrt{f_c'} + \rho_{smin} f_y \le 0.2 f_c', \text{ MPa}
\end{aligned}
\tag{13.25}
$$

in which ρ_{smin} = the lesser of the distributed reinforcement ratios in the vertical and horizontal directions.

For design, ACI 318 limits the shear stress on a panel zone to $v_u \le \phi 10\sqrt{f_c'}$, psi ($\phi 0.83\sqrt{f_c'}$, MPa), where $\phi = 0.75$ for capacity-protected panels and 0.60 for panels

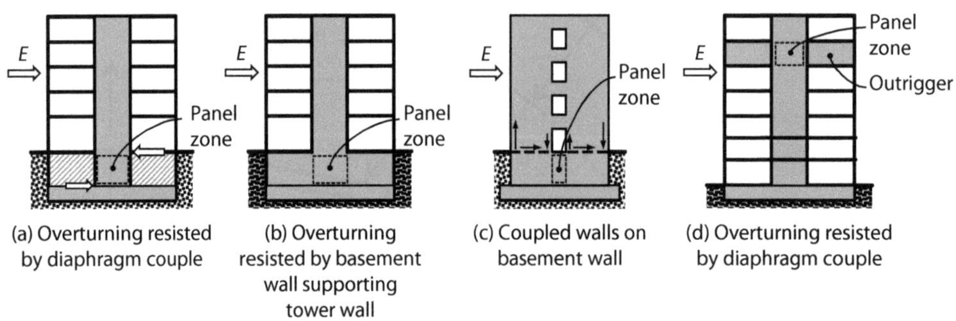

FIGURE 13.54 Panel zones in structural walls.

that are not capacity-protected. Recalling that this shear stress limit is based on diagonal crushing within the flexural hinging zone, it seems reasonable to permit a greater limit within panel zones resisting fairly uniform shear with limited inelasticity. This book suggests that it may be permissible to increase the shear stress limit where the following conditions are satisfied: (a) panel zone shear demand v_u is determined using a capacity design approach; (b) strength is in accordance with the requirement $v_u \le \phi v_n$, where v_n is calculated using Eq. (13.25) and $\phi = 0.75$; and (c) the panel zone is confined by transverse reinforcement as required for special boundary elements.

Sections 13.14, 13.15, and 13.16.1 present some examples of wall panels in buildings.

13.14 Wall Transfer at Podium and Subterranean Levels

Wall force transfer at podium or subterranean levels leads to back-stay forces discussed in Section 13.4.5. The segment of the wall immediately below the transfer level can be idealized as a panel zone. Such panel zones can be subjected to relatively large shear forces. Figure 13.55 illustrates the panel zone forces for a rectangular cross-section wall that extends into a basement, with moment equilibrium provided mainly by the horizontal force couple acting between the grade-level slab and the foundation. This panel zone can be analyzed using the concepts of Section 13.13. Note that one of the roles of the wall boundary element reinforcement is to transfer the required panel zone shears uniformly along the vertical boundaries of the panel zone. Thus, it is important to extend the wall boundary element reinforcement along the full depth of the panel zone and down to the foundation.

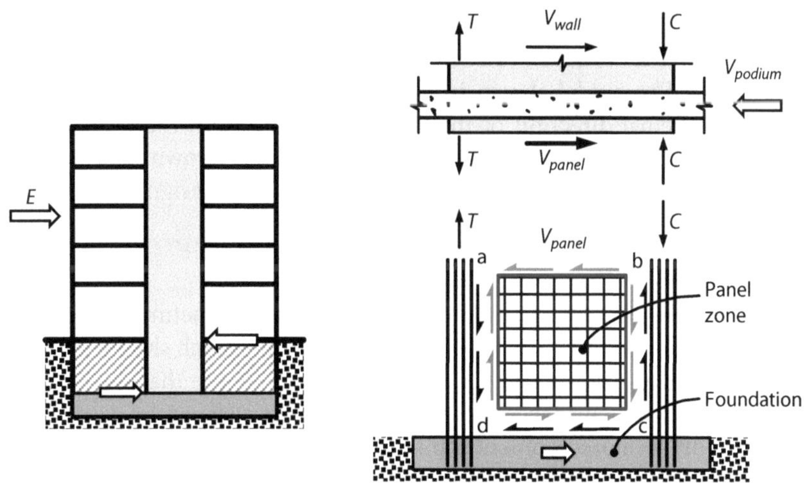

(a) Overturning resisted by diaphragm couple

(b) Blow-up elevation view showing wall shear, podium force, chords, and shear panel

FIGURE 13.55 Force transfer at podium level for rectangular wall.

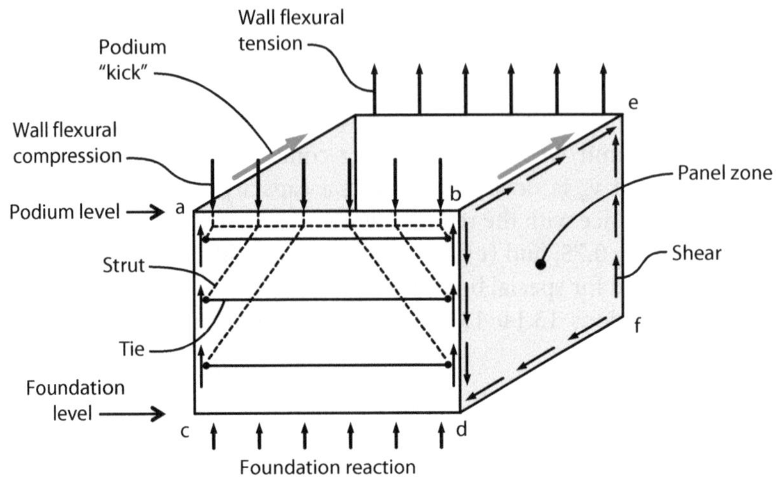

Figure 13.56 Force transfer at podium level for core wall. Forces not transmitted directly to the foundation are transmitted to the side walls (web) of the core, as idealized by a strut-and-tie model.

Force transfer of core walls at podium or subterranean levels is more complicated. In the idealization shown in Figure 13.56, flexural compression and tension enter the subterranean portion of the core through the flanges. Part of the flange force is resisted directly by the foundation, consistent with the relative stiffnesses of the structural and foundation systems. The remainder of the flange force is transferred through shear to the side walls (webs) of the core. A strut-and-tie idealization can be used to show the requirements for struts within the flange, horizontal ties within the flange, and shear flow at the intersection between the flange and the web. The idealized resistance mechanism induces shear within the panel zone, which can be designed using the concepts described in Section 13.13.

The model shown in Figure 13.56 considers lateral loading along only one principal direction of the core wall. Loading in the transverse direction would induce forces to be superimposed on those shown. Although such actions are apparent, they are not routinely recognized in design.

13.15 Outriggers

As noted in Section 13.4.6, outriggers are sometimes used in tall buildings to reduce building drifts. A capacity design approach should be used (a) to achieve a ductile "fuse" in the outrigger and (b) to ensure that the outrigger column axial force capacity exceeds the maximum force that can be delivered by the outrigger. Furthermore, moment transfer from an outrigger to a structural wall can result in very large wall panel zone shears that must be considered in design.

Consider the outrigger element illustrated in Figure 13.57. The outrigger extends multiple stories with staggered penetrations to facilitate circulation. Strut-and-tie models (one for each direction of lateral loading) are convenient for

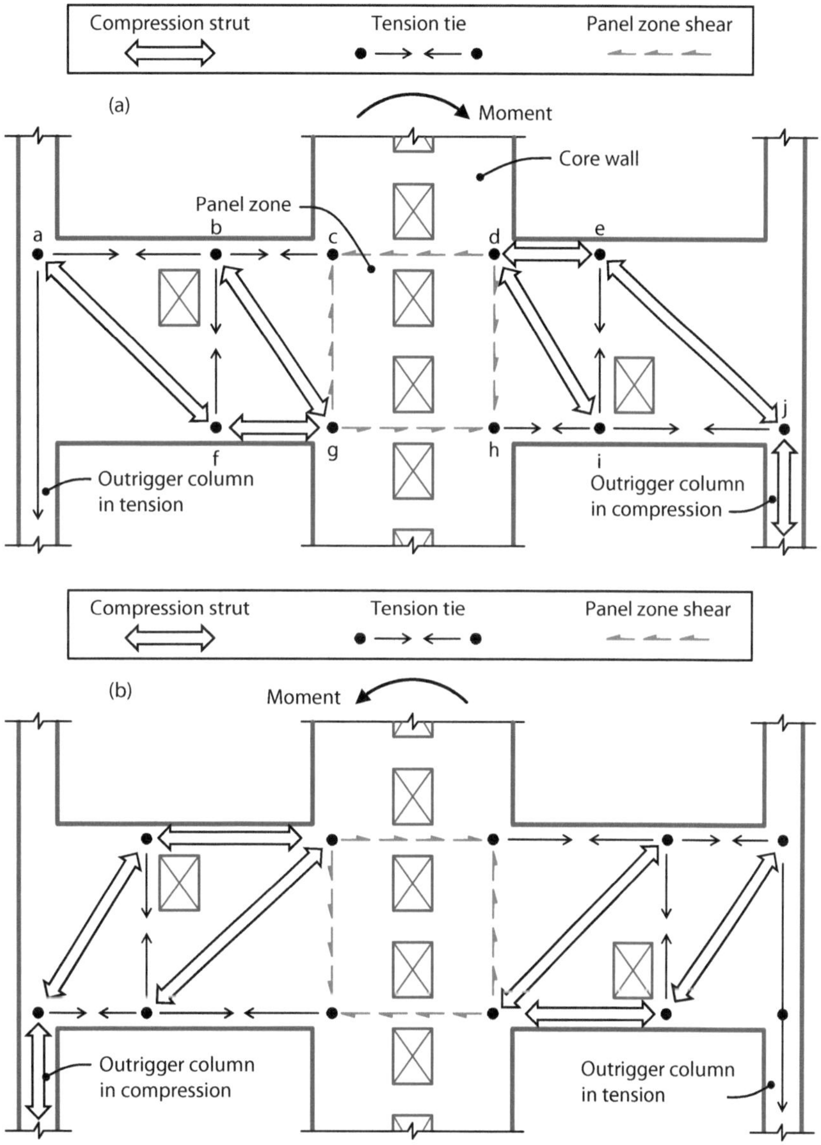

FIGURE 13.57 Outrigger strut-and-tie model.

analysis given the irregular opening pattern. Preferably, the model is defined such that reinforcement that yields in tension for loading in one direction also yields in compression for loading in the opposite direction. Otherwise, yield strains may ratchet up with progressive cycling, resulting in large beam elongation and degradation. For the truss model shown in Figure 13.57, acceptable behavior may result if the fuse is maintained within zones **bcfg** and **dehi**, as top and bottom chords that yield in tension for loading in one direction can be designed to yield

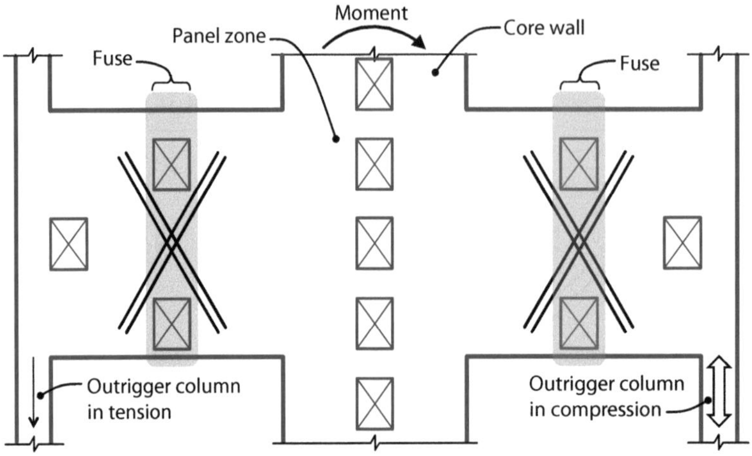

Figure 13.58 Outrigger with diagonally reinforced fuse.

in compression for loading in the opposite direction. Tensile yielding of reinforcement along lines **ab**, **bf**, **ei**, or **ij** would be problematic, as there is no compression strut along those lines to reverse the tensile strains for loading in the opposite direction.

The strut-and-tie model also identifies forces applied to wall panel zone **cdgh**. Note that panel zone shears generally are opposite to the shears on the core wall due to lateral loading. The maximum shear stress on the panel zone, based on algebraic sum of core-wall shear and panel zone shear, must not exceed the limiting shear stress for the wall. See Section 13.13 for discussion on shear stress limits of panel zones.

Diagonal reinforcement also may be effective in developing a ductile fuse (Figure 13.58).

13.16 Geometric Discontinuities

13.16.1 Walls with Openings

Coupled Walls

Coupled walls in which the openings extend over full height and terminate at a foundation mat generally are not considered to have a geometric discontinuity, although the segment of foundation mat between vertical wall segments must be designed for shears and moments transmitted from the system. Sometimes a stack of vertical openings is interrupted by solid wall segments at the roof level, at mechanical stories, or at basement walls, and these solid segments can be subjected to large demands that must be considered in design. Figure 13.59 illustrates the case where coupled walls terminate on a basement wall that spreads forces into foundation elements. The segment of basement wall between the two coupled walls is a discontinuity region or panel zone subjected to high stresses under lateral loading. Severe damage to this region has been observed following past earthquakes (e.g., Figure 13.59*a*).

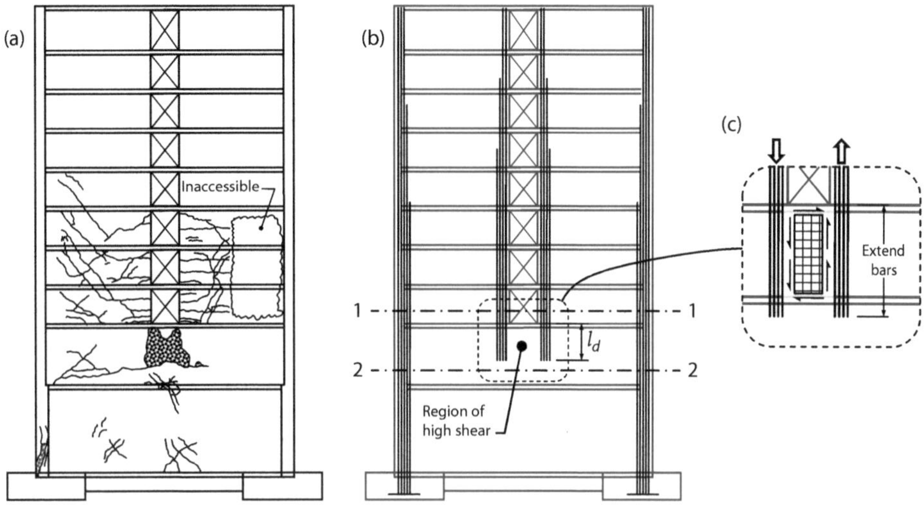

Figure 13.59 (*a*) Damage in wall from 1989 Loma Prieta, California earthquake; (*b*) inappropriate termination of coupled wall boundary reinforcement; and (*c*) improved detail.

Story-by-story analysis, as is commonly done in design, might miss the discontinuity. Section cut 1-1 (Figure 13.59*b*) would identify the need for boundary reinforcement at both sides of the openings, but section cut 2-2 just below the openings might suggest that the solid wall requires boundary reinforcement only at the exterior edges. Given this interpretation of the design actions, a common mistake is to terminate the interior boundary bars l_d below the stack of openings, without recognizing the shear in the panel zone. An improved approach analyzes the force transfer mechanism in the panel zone and extends boundary bar development length such that panel shear stresses are limited to an acceptable value (Figure 13.59*c*).

Resolution of internal forces in the panel zone of Figure 13.59*c* merits a closer look. As shown in Figure 13.60, if tension and compression forces T_1 and C_1 are at

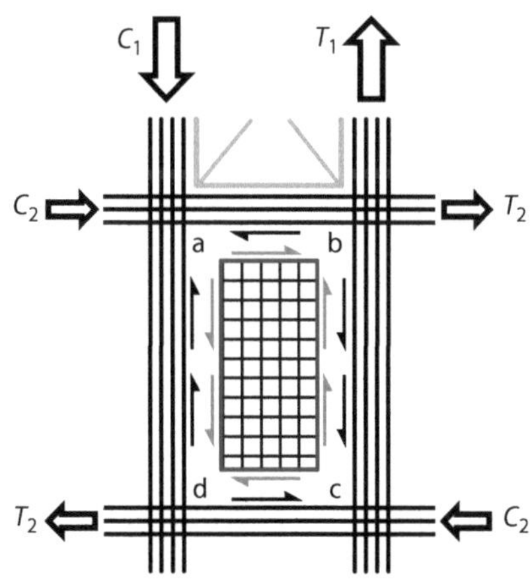

Figure 13.60 Panel zone shear and tension/compression chords.

least partially resolved within panel zone **abcd**, then some panel zone distributed shear reinforcement likely is required. Furthermore, shear stresses acting along horizontal surfaces **ab** and **cd** require placement of chord reinforcement above and below the panel zone to resist forces C_2 and T_2. It may be convenient to define panel zone dimensions such that these tension and compression chord forces are resisted by reinforcement in floor slabs located just below the stack of openings and one level lower. To limit shear stresses to acceptable values, however, may require distributing the forces deeper within the basement wall.

Tension and compression forces T_2 and C_2 identified in Figure 13.60 need to be resolved in adjacent wall panels. One useful solution is obtained by assuming the coupled walls are weightless, lightly coupled, and of equal dimension l_w (Figure 13.61). Shear forces coming from the walls above also are ignored. Assuming $T_2 = C_2$, horizontal force equilibrium of the chord along **ab** requires $T_2 = C_2 = v_1 l_h b_w / 2$. Horizontal force equilibrium of chords along **ef** and **ij** likewise requires $T_2 = C_2 = v_2 l_w b_w$. Equating these relations and rearranging results in $v_2 = \frac{l_h}{2l_w} v_1$. Thus, $v_1 \gg v_2$ for typical geometries. Turning attention to the vertical chords at the boundaries of the openings, vertical force equilibrium of the tension chord requires $T_1 = (v_1 + v_2) h_s b_w = \left(1 + \frac{l_h}{2l_w}\right) v_1 h_s b_w$. Solving for v_1, the shear stress in panel zone **abcd** is

$$v_1 = \frac{1}{1 + l_h / 2l_w} \left(\frac{T_1}{h_s b_w} \right) \tag{13.26}$$

Considering typical geometries $l_h = 6$ ft (1.8 m) and $l_w = 20$ ft (6.1 m), the multiplier on average shear stress in Eq. (13.26) is 0.9. Thus, most of the shear is resolved in

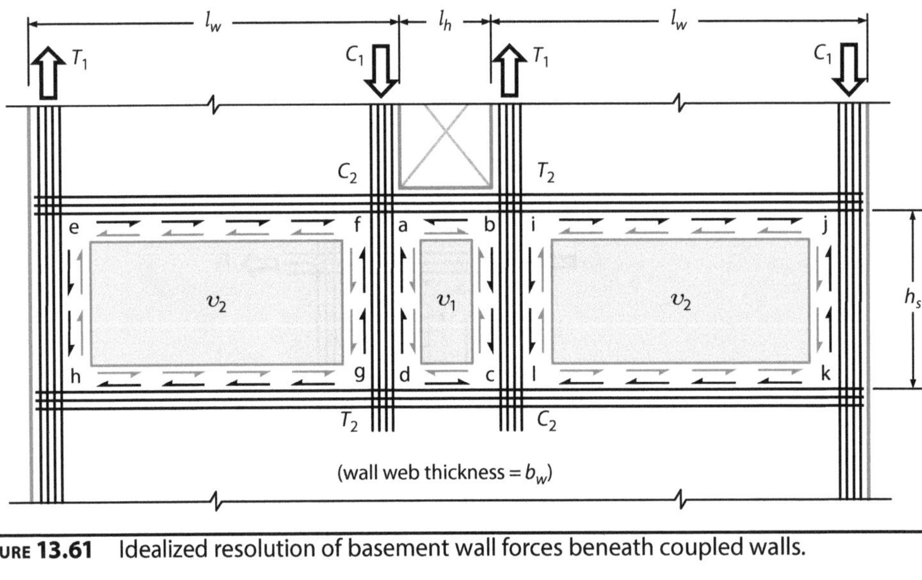

FIGURE 13.61 Idealized resolution of basement wall forces beneath coupled walls.

the panel zone beneath the stack of openings, with much lower shear stress in the wider panels to either side.

The preceding solution satisfies equilibrium within the stated limitations, but it does not recognize required deformation compatibility among adjacent wall segments. Importantly, shear distortion of panel **abcd** likely would result in shear transfer to the panel immediately below panel **abcd**. Thus, a better design solution would extend some of the boundary bars below panel **abcd** to engage additional depth of the solid wall. Naeim et al. (1990) present results of linear finite element analyses for a range of geometries, providing further insight.

Based on the preceding results and judgment, the following design approach is recommended. Two design zones are established below the stack of openings (Figure 13.62): Zone 1 extends below the stack the greatest of $1.5l_h$, l_d of the boundary bars, and h_s, but not less than the length required to transfer shear stress τ_1 within acceptable stress limits; Zone 2 extends the rest of the way to the foundation. All boundary bars should extend through Zone 1 and at least half the boundary bars should extend all the way to the foundation. Proportion and reinforce Zone 1 for panel zone shear equal to at least $1.25A_sf_y/2$: Factor 1.25 accounts for overstrength and divisor 2 recognizes that half the boundary steel is extended into Zone 2. Design Zone 2 for the remainder of the panel zone shear. Design chord reinforcement above and below Zone 1 consistent with the shear stresses assumed to be resisted within Zones 1 and 2. Chord bars can be terminated progressively along the length.

Building codes are not explicit about shear stress limits for wall panel zones below stacks of openings. Naeim et al. (1990) recommend a shear stress limit of $15\phi\sqrt{f'_c}$, psi ($1.25\phi\sqrt{f'_c}$, MPa), provided special confinement is used wherever shear stress exceeds $10\phi\sqrt{f'_c}$, psi ($0.83\phi\sqrt{f'_c}$, MPa). See Section 13.13 for the limits recommended by this book.

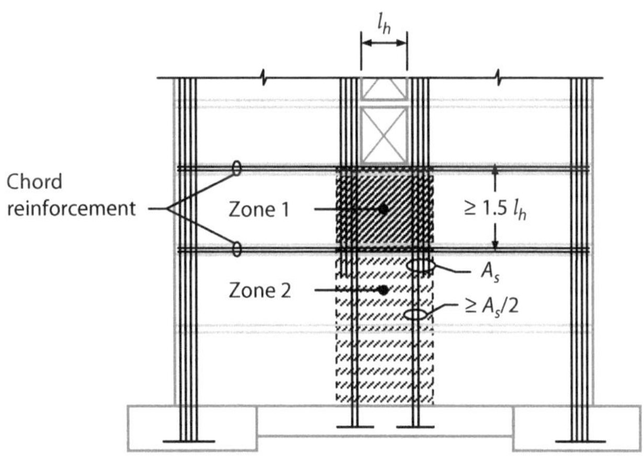

FIGURE 13.62 Reinforcement detailing beneath stacks of openings in coupled walls.

Isolated Openings

Where isolated openings occur in multi-story walls, design should consider transfer of axial, moment, and shear forces around the opening. Usually, design can be accomplished using conventional *P-M* and shear analysis adjusted to recognize presence of the opening. Strut-and-tie modeling can be useful to understand flow of forces around the opening, as well as required reinforcement details. Taylor et al. (1998) present an example verified in laboratory tests.

P-M interaction analysis must explicitly account for the change in vertical force paths. A common example occurs at a wall opening with solid panels above and below (Figure 13.63). For design of the solid panel immediately above and below the opening, *P-M* interaction analysis should exclude a portion of wall stacked with the opening. However, because the wall is solid elsewhere, it can be assumed that plane sections remain plane. The effect of the opening likely is negligible beyond approximately l_h above and below the opening, where l_h is the width of the opening (Figure 13.63).

Wall compression or tension zones with large openings should be designed to transmit the wall normal stresses around the opening. Figure 13.64 illustrates the use of a strut-and-tie model to visualize the flow of forces around an opening in a compression flange of a core wall.

Design also needs to consider transfer of shear forces around the opening. One procedure is to determine how much shear is to be carried by each of the vertical wall segments, design horizontal reinforcement to drag horizontal shears between the vertical wall segments and the solid segments above and below, and then design each of the vertical wall segments for the assigned shear. Figure 13.65 illustrates the procedure. In this example, it is assumed that the narrow vertical wall segment to the right of the opening carries no shear, such that all the shear along **bc** or **ef** must be dragged to the solid panel (**ab** or **de**) of the wall. Additional

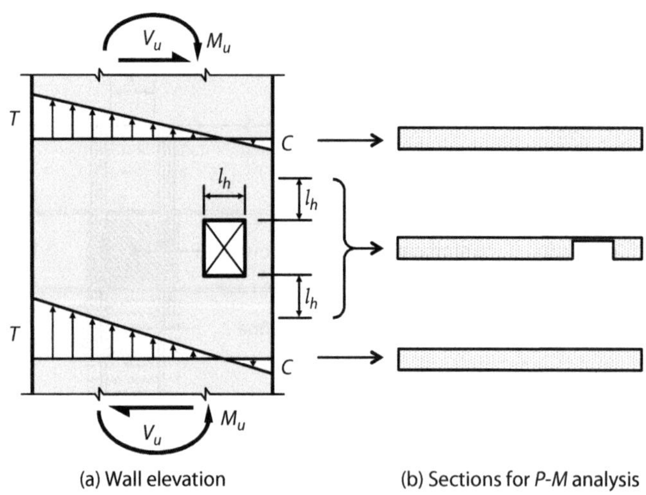

Figure 13.63 *P-M* analysis at irregular opening.

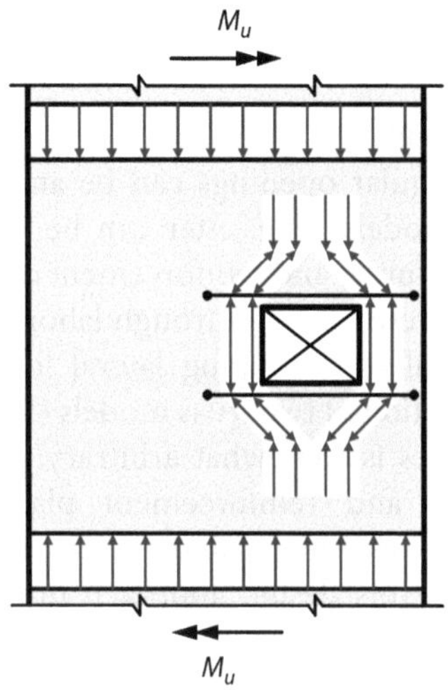

FIGURE 13.64 Strut-and-tie model to represent flow of forces around a compressed flange.

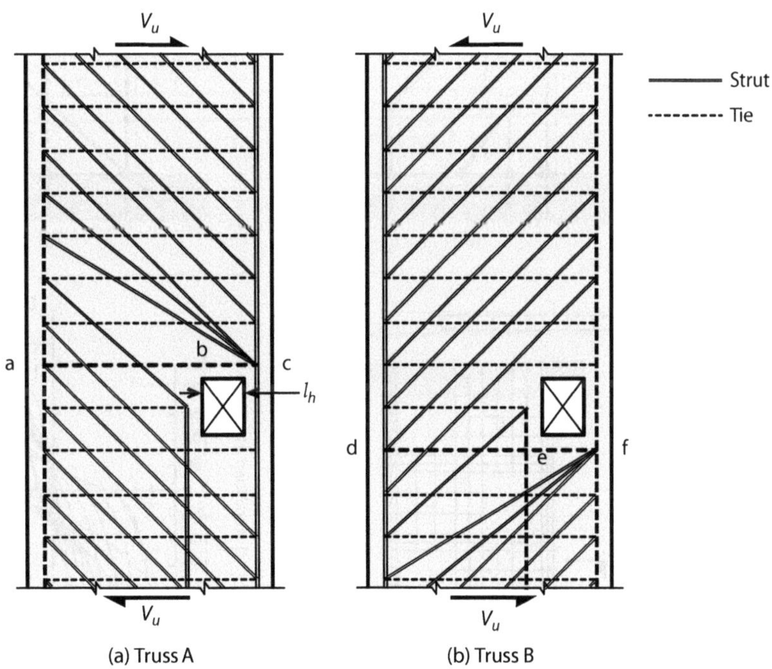

FIGURE 13.65 Truss model for shear transfer around an opening.

horizontal reinforcement is required above the opening (**ac**) and below the opening (**df**), depending on the direction of horizontal shear.

Irregular Openings

Walls with irregular openings can be analyzed using finite element models or strut-and-tie models. The latter can be particularly helpful for understanding flow of internal forces and reinforcement requirements. Yañez et al. (1992) present examples that were studied through laboratory tests. Figure 13.66 illustrates one of their test walls. Alternating lateral forces *E* were applied at a single level, requiring definition of two truss models (Figure 13.66*a* and *b*). The exact location of struts and ties is somewhat arbitrary, but should be guided by linear-elastic stress patterns and reinforcement placement preferences (Figure 13.66*c*). Laboratory tests on this wall and walls with other opening patterns demonstrated effectiveness of this design approach in achieving moderate ductility capacity (Figure 13.66*d*).

Figure 13.67 illustrates strut-and-tie modeling for a multi-story wall with distributed lateral forces applied at each level, as would occur where lateral forces are transferred from floor and roof diaphragms. (The lateral force at the top level is

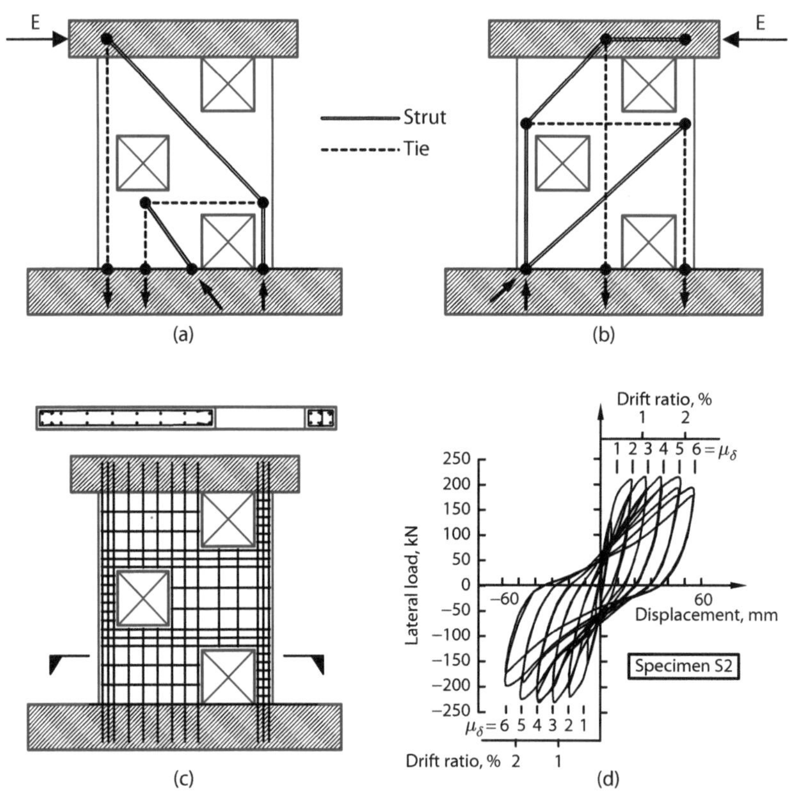

FIGURE 13.66 Squat wall with distributed openings. (*After Yañez et al., 1992.*)

shown as 3, which, for example, could be 3 klf or 3 kN/m.) A simple strut-and-tie model is developed and solved as shown in Figure 13.67*b*.

The strut-and-tie model forces, when combined with other loads as required by the governing code, can be used directly to design the wall. According to the strut-and-tie model of ACI 318, the strength reduction factor for struts, ties, nodal zones, and bearing areas is $\phi = 0.75$. That factor is adopted here, except $\phi = 0.60$ for wall shear and $\phi = 0.65$ for wall piers in axial compression. Thus, design of tension reinforcement along tie **im** in Figure 13.67*c* would require reinforcement area $A_s = T_u/\phi f_y$, with $\phi =$ 0.75. Design of the wall pier along **lp** for axial compression would be according to $P_u \le \phi P_{n,max} = 0.80\phi[0.85 f_c'(A_g - A_{st}) + f_y A_{st}]$, with ϕ = 0.65. Some codes (e.g., ASCE 7) may consider wall pier **lp** as an element supporting a discontinuous wall, in which case design forces are amplified by Ω_0 unless a capacity design approach is used to bound the axial force. Additional detailing requirements may apply (see Section 13.11). Also, although wall pier **lp** has no shear according to the strut-and-tie model, a prudent design would assign nominal shear to the wall pier equal to

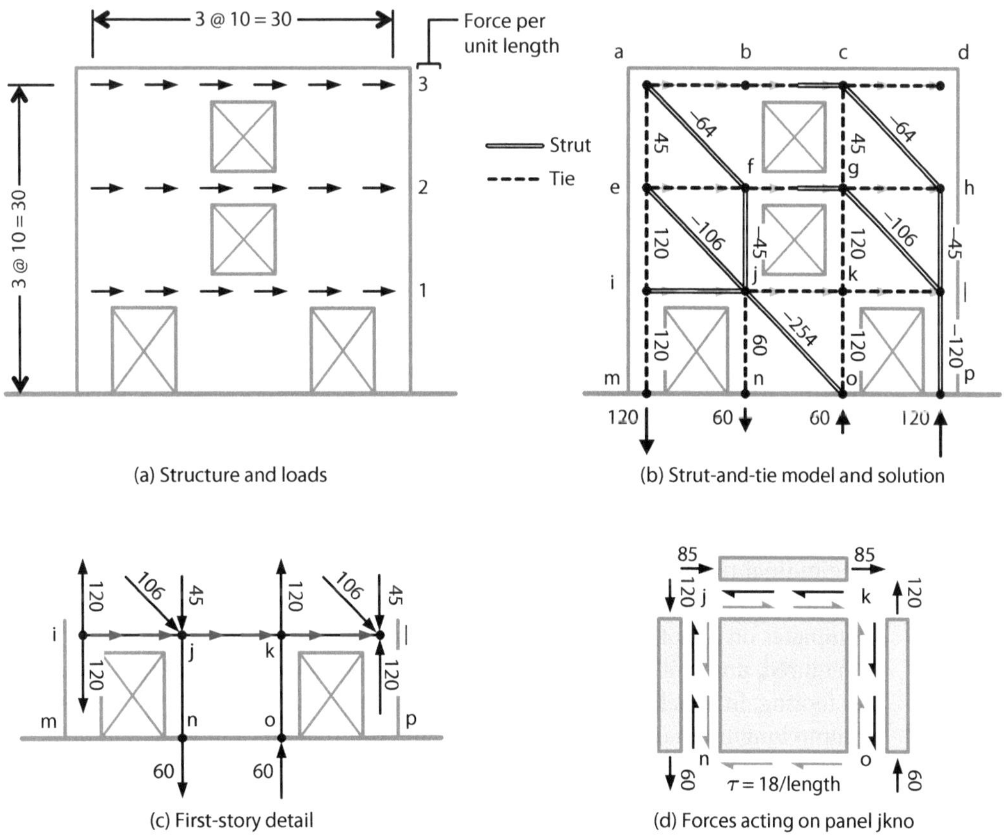

FIGURE 13.67 Squat wall with irregular openings under distributed lateral forces. All units are consistent, lateral forces are in force per unit length, strut forces are negative for compression, and tie forces are positive for tension.

$V_{ui} = \frac{A_{cwi}}{\sum_i A_{cwi}} V_u$, in which V_{ui} is shear in wall pier i, A_{cwi} is shear area of wall pier i, the summation is over all walls at that elevation, and V_u is total shear at that level.

Design of panel **jkno** requires detailed consideration of forces acting around that panel (Figure 13.67*d*). Panel shear force is equal to the absolute value of the horizontal component of diagonal strut force −254, which is 180. Assuming it acts over length of 10, the shear per unit length is 18. This unit shear acts around all four sides of the panel, and is resolved in boundary elements (or chords) located along the free edges. Each of these chords needs to be designed for the tensile or compressive forces shown. Panel **jkno** can be designed as a squat wall segment as described in Section 13.10, a panel zone as described in Section 13.13, or using a strut-and-tie model (Chapter 7).

Additional Details at Openings and Top of Walls

Vertical and horizontal reinforcement that is interrupted by openings or by the top of the wall should be terminated with standard hooks. Alternatively, the bars can be capped with U-bars having same size and spacing as the terminated bars. Where hooks on large bars do not fit within the wall thickness, smaller bars at reduced spacing can be used.

Cracks tend to form at corners of wall openings. In addition to minimum reinforcement, it is good practice (and some codes require) to place bars around the opening or diagonally across each corner. Around window, door, and similar sized openings, ACI 318 requires at least two No. 5 (No. 16) bars in walls having two layers of reinforcement in both directions and one No. 5 bar in walls having a single layer of reinforcement in both directions. Such bars are to be developed for f_y in tension at the corners of the openings.

13.16.2 Columns Supporting Discontinuous Walls

A column or wall pier supporting a discontinuous structural wall (e.g., member **lp** in Figure 13.67) can be subjected to compressive overload due to axial force and moment transfer from the discontinuous wall. For columns, ACI 318 requires full-height column confinement for all stories beneath the discontinuous wall if axial force related to earthquake effect exceeds $A_g f'_c/10$ (or $A_g f'_c/4$ where design forces have been magnified to account for overstrength of the vertical elements of the seismic-force-resisting system). The confinement reinforcement must extend upward into the discontinuous wall at least the development length of the longitudinal reinforcement. If the column terminates on a wall, the confinement reinforcement must extend the same distance downward into the wall below. If it terminates on a footing or mat, 12 in (300 mm) extension into the footing or mat is required, unless it terminates within one half the footing depth from an edge of the footing, in which case it must extend at least l_d (calculated for f_y) of the largest column longitudinal reinforcement. Similar requirements apply for wall piers.

13.16.3 Thickness Transitions

Wall reinforcement must be detailed to accommodate changes in wall thickness over height. Where the thickness change is less than the lesser of 6 in (150 mm) and $l_{sp}/5$, the dowel bar and the terminated wall reinforcement can be considered to form a non-contact lap splice (Figure 13.68*a*). Transverse reinforcement is recommended

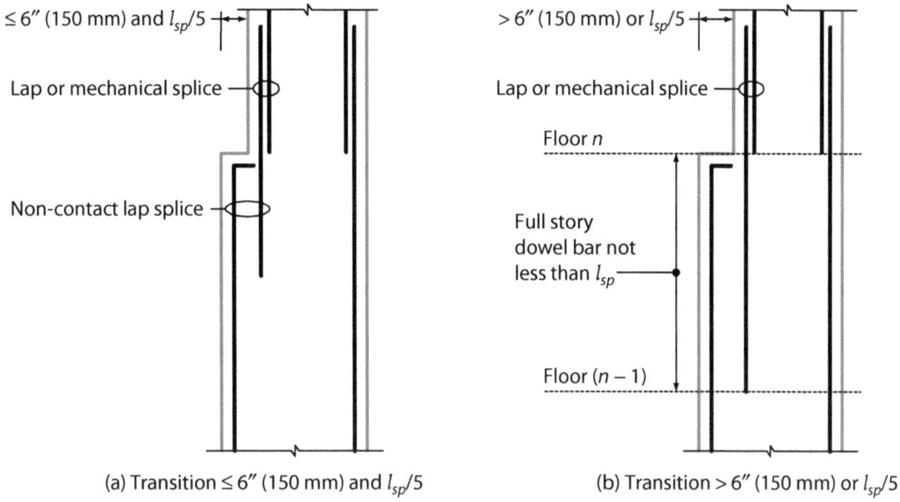

FIGURE 13.68 Vertical reinforcement detail at wall thickness transitions. Wall shown in section.

to confine the splice at such locations. Where the transition is greater than the lesser of 6 in (150 mm) and $l_{sp}/5$, the lap between the dowel bar and the terminated wall reinforcement cannot be considered a non-contact lap splice. An acceptable detail is to extend the dowel bar the full depth of the story below the thickness transition, but not less than l_{sp} (Figure 13.68*b*). By so doing, the terminated wall reinforcement is considered to provide required section strength at floor level $n - 1$, while the dowel reinforcement provides required section strength at floor level n, with a lap occurring between the bars over the thickness transition.

13.16.4 Foundation Steps

Where walls are flush with the edge of a foundation or foundation step, confinement reinforcement should be extended along the edge at least l_d, calculated for f_y in

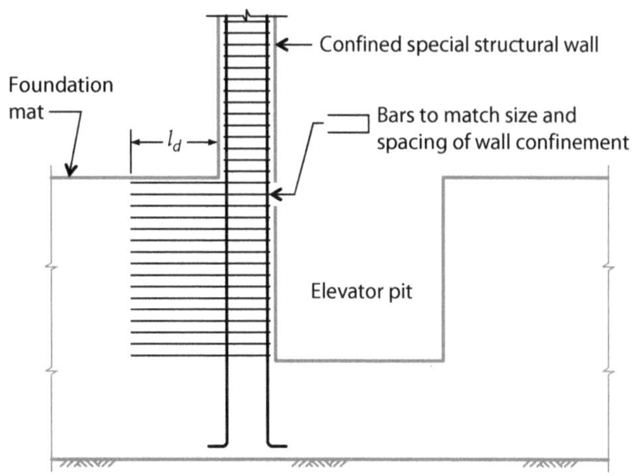

FIGURE 13.69 Detail where wall terminates at an elevator pit or other foundation step.

tension, of the largest longitudinal reinforcement. A common occurrence is at an elevator pit, where good practice continues the longitudinal and transverse reinforcement at least along the full depth of the pit (Figure 13.69).

13.17 Additional Requirements

13.17.1 Special Inspection

As with special moment frames, proper construction of special structural walls, using good quality materials, is essential to ensuring that a building, once constructed, complies with the requirements of the code and the approved design. In U.S. construction practice, special inspection is required to foster proper construction. See the special inspection requirements in Section 12.7.1.

13.17.2 Material Properties

Concrete and Shotcrete

Wherever a special structural wall is used, regardless of the Seismic Design Category, ACI 318 requires that materials conform to special requirements. Minimum specified compressive strength is $f_c' = 3000$ psi (20.7 MPa) for structural concrete. Additional requirements apply for lightweight concrete. Where high-strength concrete is used, the value of $\sqrt{f_c'}$ should be restricted to an upper-bound value of 100 psi (0.69 MPa) for any shear strengths or anchorage/development strengths. Some jurisdictions impose additional restrictions on the use of high-strength concrete.

The code may contain restrictions on use of shotcrete for structural walls because of concerns about encasement of reinforcing bars. Preconstruction tests may be required to demonstrate adequacy.

Reinforcement

Deformed reinforcement resisting earthquake-induced moment and axial forces in special structural walls and coupling beams should conform to minimum requirements for strength and ductility. In the United States, reinforcement must conform to ASTM A706 Grade 60 (420). Alternatively, ASTM A615 Grades 40 and 60 (280 and 420) are permitted if A706 stress and strain requirements are met. The optional use of A615 reinforcement sometimes is adopted because it may be more widely available and less expensive. Higher-strength longitudinal reinforcement, including ASTM A706 Grade 80 (550), is prohibited by ACI 318 because of insufficient test data to demonstrate its use and concerns about higher bond stresses and increased buckling tendency. Alternative materials (including high-strength reinforcement) are permitted if results of tests and analytical studies are presented in support of its use.

Higher-strength reinforcement up to 100,000 psi (690 MPa) nominal yield strength is permitted for design of transverse reinforcement in the United States. This reinforcement can reduce congestion problems especially for members using higher strength concrete. Where used, the value of f_{yt} used to compute the amount of confinement reinforcement is not to exceed 100,000 psi (690 MPa), and the value of f_{yt} used in design of shear reinforcement is not to exceed 60,000 psi (420 MPa) except 80,000 psi (550 MPa) is permitted for welded deformed wire reinforcement. The intent of the code requirement is to limit width of shear cracks.

Mechanical Splices

Longitudinal reinforcement in special structural walls is expected to undergo multiple yielding cycles in prescribed locations during design-level earthquake shaking. If mechanical splices are used in these locations, they should be capable of developing nearly the tensile strength of the spliced bars. Outside yielding regions, mechanical splices, if used, can have reduced performance requirements.

ACI 318 classifies mechanical splices as either Type 1 or Type 2 as follows: (a) Type 1 mechanical splices are to be capable of $1.25f_y$ in tension or compression, as required; (b) Type 2 mechanical splices are to develop the specified tensile strength of the spliced bar. Where mechanical splices are used in special structural walls, only Type 2 mechanical splices are permitted within a distance equal to twice the member depth from sections where reinforcement yielding is likely to occur as a result of inelastic lateral displacements. Either Type 1 or Type 2 mechanical splices are permitted in other locations.

Welding

Welded splices in reinforcement resisting earthquake-induced forces should develop at least $1.25f_y$ of the bar and should not to be used within a distance equal to twice the member depth from sections where reinforcement yielding is likely to occur as a result of inelastic lateral displacements. Stirrups, ties, inserts, or other similar elements should not be welded to longitudinal reinforcement that is required by design because cross-welding can lead to local embrittlement of the welded materials. Welded products should be used only where test data demonstrate adequate performance under loading conditions similar to conditions anticipated for the particular application.

13.17.3 Additional System Design Requirements

Where special structural walls are used, U.S. codes may impose other requirements for the structural system. In some cases these additional requirements apply only in highly seismic regions (Seismic Design Categories D, E, and F). Structural diaphragms and foundations are required as part of the seismic-force-resisting system in any building (see Section 11.4 and Chapters 15 and 16). Where used as part of a dual system, a special moment frame must satisfy requirements of Chapter 12. Gravity framing requirements are discussed in Chapter 14. Anchoring to concrete must be according to ACI 318 Appendix D. Note that those provisions do not apply to design of anchors in portions of structural walls that are intended to yield during design-level shaking.

13.18 Detailing and Constructability Issues

13.18.1 Reinforcement Cage Fabrication

Although a structural wall is considered a singular element, reinforcement modules within the wall may be pre-tied and hoisted into place as separate pieces (Figure 13.70). The pre-tied modules are spliced to create a fully interlocked

Figure 13.70 Pre-tied modules (some modules encircled). (*Used with permission from Magnusson Klemencic Associates.*)

reinforcement cage prior to closing the forms and casting the wall. The splices should comprise horizontal bars extending along the wall length rather than short bars that extend between the individual cages with minimum splice lengths.

13.18.2 Boundary Element Confinement

As discussed in Section 13.8.1, the extent of a boundary element is linked integrally to the size and spacing of the vertical reinforcement within. Furthermore, a vertical bar is required in the corner of each hoop or crosstie bend. For this reason, it may be convenient first to determine the desired confinement layout prior to selecting vertical reinforcement. Fortunately, the confinement quantity and layout are defined by a closed form equation that is independent of design forces.

Confinement variables typically at the designer's discretion are confinement bar size, and horizontal and vertical spacing of confinement hoop legs and crossties. Large diameter confining bars are desirable to reduce congestion, but bars larger than No. 5 (16) are impractical because of required space for bar bends and hook tails. For higher strength steel, there also can be a limit to what bar size is bendable with locally available equipment.

Confinement reinforcement typically is made up of hoops and crossties. Crossties should not be used to close an open leg of a hoop in a wall boundary. Use of crossties with 90° bend at one end is not permitted for this purpose. Hoops made up of intersecting headed bars are not permitted. However, headed bars can be used instead of crossties if care is taken to ensure that the heads are large enough and properly placed to engage the longitudinal bars. Hoops made up of cross-welded wire should not be used unless it can be shown that the weld is capable of developing the strength of the interconnecting pieces.

Horizontal spacing of confinement legs, and hence the spacing of vertical reinforcement within the boundary element, typically will be much tighter [4 to 8 in (100 to 200 mm)] than desired for the remainder of the wall. It is common to select vertical bar spacing within a boundary element that is a divisor of the vertical bar spacing in the unconfined portion of the wall. For example, if 12 in (300 mm) spacing of vertical reinforcement is considered practical for the unconfined wall, the spacing of vertical bars within the boundary element should be 6 in or 4 in (150 mm or 100 mm). This is beneficial because, as vertical boundary bars drop off at higher elevations, the remaining bars align with and can be spliced to the 12 in (300 mm) grid.

Boundary element reinforcement very much resembles a ductile column within the structural wall. A representative boundary element at the end of a planar wall is shown in Figure 13.71. Note that each crosstie has a 90° and a 135° hook, and these must be alternated end for end along both the length and the height.

Some flanged walls require confinement throughout the flange, in which case confinement should extend at least 12 in (300 mm) into the wall web (Figure 13.72). For very long confined boundary regions, one approach is to provide closely spaced confinement reinforcement in both directions at wall ends, with only closely spaced thru-wall crossties along the middle extent of the wall. This is allowable if the more widely spaced horizontal shear reinforcement in the web provides required confinement stress across the thickness.

FIGURE 13.71 Boundary confinement for planar wall. (*Left photograph used with permission from J. Wallace, right photograph used with permission from Magnusson Klemencic Associates.*)

FIGURE 13.72 Boundary confinement for wall flange. (*Used with permission from Magnusson Klemencic Associates.*)

Structural wall longitudinal reinforcement should extend into supporting elements and be fully developed for f_y in tension, except it should be developed for $1.25f_y$ at locations where yielding of longitudinal reinforcement is likely to occur as a result of lateral displacements. Where boundary elements are provided, equivalent horizontal confinement should be extended into the support. For structural walls on shallow foundations, this confinement should be extended 12 in (300 mm) into the footing or mat. For structural walls supported by all other elements, or where the edge of the boundary element is within one-half footing depth from an edge of the footing, the confinement should extend into the support a distance equal to the development length of the largest vertical bar in the boundary. The critical subset of this category is boundary elements landing flush with the edge of a foundation or significant foundation step. This commonly occurs for structural walls that enclose elevator cores. The elevator pit dimensions commonly require a significant depression on one side of the structural wall. For this condition, it is recommended that the base of the depression be considered as the base of the structural wall. Vertical bars are therefore developed below the depression, and confinement is continued through the depth of the depression (Figure 13.69).

13.18.3 Bar Compatibility

The interface between wall ends and coupling beams can be difficult to detail and construct. Main coupling beam reinforcement must extend into the wall end a

Figure 13.73 Mockup to study anchorage of diagonal reinforcement in heavily reinforced boundary element.

distance sufficient to fully develop the bar. Bar compatibility becomes especially challenging where diagonal bars must extend into a heavily confined section. Full-scale preconstruction mockups can help identify solutions for particularly challenging designs (Figure 13.73).

Development of coupling beam longitudinal reinforcement requires that it be placed inside the wall vertical reinforcement. For conventionally reinforced beams, this results in side cover to beam longitudinal reinforcement around 3.0 in (75 mm) (Figure 13.46, Section A-A). Stirrups and cap-ties must be detailed for this increased cover so that the corner longitudinal bars are firmly placed in stirrup bends. This decreased available width must also be considered when verifying clear horizontal spacing between longitudinal bars, a necessary measure to facilitate concrete placement and consolidation.

13.18.4 Bar Splices

Reinforcement in structural walls should be developed or spliced for f_y in tension. At locations where yielding of longitudinal reinforcement is likely to occur as a result of lateral displacements, development and lap splice lengths of longitudinal reinforcement should be 1.25 times values calculated for f_y in tension. Lap splices, mechanical splices, and welded splices are permitted, with lap splices being the most common. Although not required by some building codes, lap splices should be confined by transverse reinforcement at locations where yielding of reinforcement is anticipated. As for all elements of concrete construction, reinforcing bars larger than No. 11 (36) should not be lap spliced in structural walls. Mechanical and welded splices should be according to Section 13.17.2.

The first splice of vertical reinforcement typically occurs immediately above the foundation, where wall longitudinal reinforcement laps with dowel bars. These dowels provide the critical mechanism of transferring tension forces from the structural wall to the foundation. All vertical reinforcement must be extended into the foundation a depth sufficient to be fully developed for tension. For constructability purposes, it is recommended that dowels with 90° hooks extend to the bottom of the foundation where they can be tied firmly to foundation bottom reinforcement.

For structural walls with two curtains of reinforcement, it is preferred for the vertical reinforcement to be inside the horizontal reinforcement. This arrangement improves splice strength and buckling restraint for the verticals.

Horizontal reinforcement is always treated as "top-cast" reinforcement, requiring $\psi_t = 1.3$ for all development and lap splice length calculations. Splice locations might not be finalized until the contractor has determined the breakdown of pre-tied segments and the overall erection sequence including formwork operability. For structural walls with pre-tied segments, horizontal reinforcement has the additional function of tying the pieces together in the final arrangement (Figure 13.70).

13.18.5 Miscellaneous Detailing Issues

As significant obstructing elements, structural walls must be coordinated closely with mechanical, electrical, and plumbing (MEP) designs to enable routing and distribution of these systems. Although it is preferable to spatially separate structural walls from the other trades, it is often necessary to provide blockouts and sleeves to allow for minor penetration of the structural walls. It is recommended to identify early those areas that are not available for penetrations, typically boundary elements, coupling beams, and the development zone of coupling beams in wall ends. Where penetrations occur, it is important to provide trim reinforcement around all edges. The exact layout and size of trim reinforcement should be selected to provide a complete load path for all local forces and to inhibit cracking of the walls from the sides of the penetrations.

Transfer of diaphragm forces between slabs and structural walls ideally occurs in a distributed manner along the length of the wall and collector elements. Collector reinforcement typically comprises large quantities of longitudinal reinforcement. Collector forces must be fully resolved into the wall, requiring embedment in excess of a typical development length when the wall horizontal reinforcement is insufficient to provide a complete splice. See additional discussion in Chapter 15.

When steel elements are framed to structural walls, the connection detail typically takes the form of an embedded steel plate with deformed bars or headed studs welded to that plate and developed into the backing structural wall. This is a frequent occurrence for structural walls enclosing and forming an elevator core. Steel members will be required to separate multi-bank elevators, and to support elevator and counterweight rails. These members must be attached to the structural walls in very precise locations. To allow for tolerance in placement of embedded steel connection plates, it is recommended to oversize the plates to allow for misplacement up to 3 in (75 mm) without compromising integrity of the connection.

13.18.6 Concrete Placement

Similar to column construction, placement of structural wall concrete in high-aspect-ratio forms (height/width) inevitably includes issues of concrete drop height, blind vibration, practical lift heights, and selection of a mixture with appropriate flowability. These issues need to be clearly discussed and coordinated with the contractor to ensure that the final product is fully consolidated, monolithic, and isotropic.

The intersection of slabs and structural walls is a region in which the placement sequence and resulting concrete strength needs to be considered, especially where wall concrete strength significantly exceeds slab concrete strength. For multi-story construction, structural walls are typically cast to the underside of the slab above. The slab is cast over the top of the wall, and the wall construction resumes above. This results in a plane of slab concrete placed through the structural wall. ACI 318 allows column concrete compressive strength to be 1.4 times that of the floor system, but that provision is intended to apply only for axial force transmission in columns. Some jurisdictions deem this applicable for structural walls. Preferably, application of this provision will be limited to moment and axial force transmission where the wall is confined by slabs on all sides. Applying this for shear goes beyond the code intent and is not recommended.

The standard remedy is to place higher strength concrete in the slab over the top of the structural wall, extending 2 ft (600 mm) beyond the face of the wall. This method, typically called puddling, must be carefully scheduled with the slab pour to ensure that the high strength concrete is well integrated with the remainder of the slab.

The slip-form method of constructing structural walls eliminates this weakened plane at the structural wall-to-slab intersection. In this and other similar wall forming techniques, the structural wall is cast continuously through the depth of the slab, construction joints not with standing. Although this method avoids the potential for insufficient concrete strength in the structural wall within the slab thickness, the slab-to-wall connection must be detailed to accommodate all load transfer mechanisms. This critical location must transfer vertical shear from gravity forces in the slab and in-plane horizontal shear from diaphragm

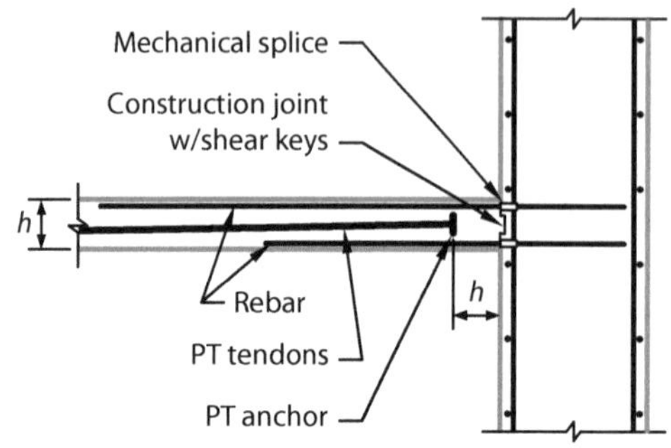

Figure 13.74 Structural wall continuous at slab connection.

forces, and it must maintain integrity during drift-induced rotation of the slab-to-wall connection. For post-tensioned slabs, tests have demonstrated that an acceptable detail is to provide nonprestressed dowels extending from the wall into the slab, with the dead end of tendons held away from the slab–wall interface a distance equal to the slab depth (Klemencic et al., 2006). Figure 13.74 shows an example of this connection. If the dowels are connected to the wall with mechanical splices, these should be Type 2 mechanical splices.

References

ACI 318 (2014). *Building Code Requirements for Structural Concrete (ACI 318-14) and Commentary*, American Concrete Institute, Farmington Hills, MI.

Aoyama, H. (1986). "Earthquake Resistant Design of Reinforced Concrete Frame Buildings with Flexural Walls," *Proceedings, 2nd US-Japan Workshop on Improvement of Building Seismic Design and Construction Practice*, San Francisco, CA, pp. 101–129.

ASCE 7 (2010). *Minimum Design Loads for Buildings and Other Structures*, American Society of Civil Engineers, Reston, VA, 658 pp.

ATC 72 (2010). *Modeling and Acceptance Criteria for Seismic Design and Analysis of Tall Buildings*, Applied Technology Council. Also available as PEER Report No. 2010/111 at http://peer.berkeley.edu/tbi/wp-content/uploads/2010/09/PEER-ATC-72-1_report.pdf, 222 pp.

Chai, Y.H., and D.T. Elayer (1999). "Lateral Stability of Reinforced Concrete Columns under Axial Reversed Cyclic Tension and Compression," *ACI Structural Journal*, Vol. 96, No. 5, pp. 780–789.

Corley, W.G., A.E. Fiorato, and R.G. Oesterle (1981). "Structural Walls," *Special Publication 72*, American Concrete Institute, Farmington Hills, MI, pp. 77–131.

DBI (2009). "Structural Bulletin SB 09-09," Department of Building Inspection, City and County of San Francisco.

Deierlein, G.G., A.M. Reinhorn, and M.R. Willford (2010). *Nonlinear Structural Analysis for Seismic Design*, NEHRP Seismic Design Technical Brief No. 4, National Institute of Standards and Technology, Gaithersburg, MD, 32 pp.

Eberhard, M.O., and M.A. Sozen (1993). "Behavior-based Method to Determine Design Shear in Earthquake-Resistant Walls," *Journal of Structural Engineering*, Vol. 119, No. 2, pp. 619–640.

EERI (2010). "The Mw 8.8 Chile Earthquake of February 27, 2010," EERI Special Earthquake Report—June 2010, Earthquake Engineering Research Institute, Oakland, CA, 20 pp.

EERI (2011). "The M 6.3 Christchurch, New Zealand, Earthquake of February 22, 2011," EERI Special Earthquake Report—May 2011, Earthquake Engineering Research Institute, Oakland, CA, 16 pp.

Eibl, J., and E. Kreintzel (1988). "Seismic Shear Forces in RC Cantilever Shear Walls," *Proceedings, 9th World Conference on Earthquake Engineering*, Paper 9-1-1, Tokyo-Kyoto, Japan.

Eurocode 8 (2004). *Eurocode 8: Design of Structures for Earthquake Resistance, Part 1, General Rules, Seismic Actions and Rules for Buildings*, Comité Européen de Normalisation, European Standard EN 1998-1:2004, Brussels, Belgium.

Ghosh, S.K., and V. Markevicius (1990). "Design of Earthquake Resistant Shearwalls to Prevent Shear Failure," *Proceedings, 4th US National Conference on Earthquake Engineering*, Palm Springs, Earthquake Engineering Research Institute, Oakland, CA, Vol. 2, pp. 905–913.

IBC (2012). *International Building Code*, International Code Council.

Kabeyasawa, T. (1993). "Ultimate-State Design of Wall-Frame Structures," in *Earthquake Resistance in Reinforced Concrete Structures, A Volume Honoring Hiroyuki Aoyama*, Department of Architecture, University of Tokyo, Japan, pp. 431–440.

Kreintzel, E. (1990). "Seismic design shear forces in RC cantilever shear wall structures," *European Earthquake Engineering*, Vol. 3, pp. 7–16.

Klemencic, R., J.A. Fry, G. Hurtado, and J.P. Moehle (2006). "Performance of Post-tensioned Slab-Core Wall Connection," *PTI Journal*, Vol. 4, No. 2, pp. 7–23.

MacGregor, J.G. (1983). "Load and Resistance Factors for Concrete Design," *ACI Journal*, Vol. 80, No. 4, pp. 279–287.

Moehle, J.P., T. Ghodsi, J.D. Hooper, D.C. Fields, and R. Gedhada (2011). *Seismic Design of Cast-in-Place Concrete Special Structural Walls and Coupling Beams: A Guide for Practicing Engineers*, NEHRP Seismic Design Technical Brief No. 6, National Institute of Standards and Technology, Gaithersburg, MD, 37 pp.

Naeim, F., B. Schindler, J.A. Martin, and S. Lynch (1990). "Hidden Zones of High Stress in Seismic Response of Structural Walls," *Proceedings, 1990 Structural Engineers Association of California Convention*, pp. 402–422.

NZS3101 (2006). *Concrete Design Standard, NZS3101:2006, Part 1 and Commentary on the Concrete Design Standard, NZS 3101:2006, Part 2*, Standards Association of New Zealand, Wellington, New Zealand.

Oesterle, R.G., J.D. Aristizabal-Ochoa, K.N. Shiu, and W.G. Corley (1984). "Web Crushing of Reinforced Concrete Structural Walls," *ACI Journal*, Vol. 81, No. 3, pp. 231–241.

Panagiotou, M., and J.I. Restrepo (2009). "Dual-Plastic Hinge Design Concept for Reducing Higher-Mode Effects on High-Rise Cantilever Wall Buildings," *Earthquake Engineering and Structural Dynamics*, Vol. 38, pp. 1359–1380.

Panagiotou, M., and J.I. Restrepo (2011). "Displacement-Based Method of Analysis for Regular Reinforced-Concrete Wall Buildings: Application to a Full-Scale 7-Story Building Slice Tested at UC–San Diego," *Journal of Structural Engineering*, Vol. 137, No. 6, pp. 677–690.

Paulay, T. (1986). "The Design of Ductile Reinforced Concrete Structural Walls for Earthquake Resistance," *Earthquake Spectra*, Vol. 2, No. 4, pp. 783–823.

Paulay, T., and M.J.N. Priestley (1993). "Stability of Ductile Structural Walls," *ACI Structural Journal*, Vol. 90, No. 4, pp. 385–392.

Paulay, T., M.J.N. Priestley, and A.J. Synge (1982). "Ductility in Earthquake Resisting Squat Shearwalls," *ACI Journal*, Vol. 79, No. 4, pp. 257–269.

PEER (2010). *Guidelines for Performance-Based Seismic Design of Tall Buildings*, Report No. PEER-2010/05, Pacific Earthquake Engineering Research Center, University of California, Berkeley, CA, 84 pp.

Priestley, M.J.N., and A. Amaris (2003). "Dynamic Amplification of Seismic Moments and Shear Forces in Cantilever Walls," *Proceedings, Fib Symposium, Concrete Structures in Seismic Regions*, Athens, Greece.

Priestley, M.J.N., G.M. Calvi, and M.J. Kowalsky (2007). *Displacement Based Seismic Design of Structures*, IUSS Press, Pavia, Italy, 721 pp.

Rejec, K., T. Isaković, and M. Fischinger (2011). "Seismic Shear Force Magnification in RC Cantilever Structural Walls Designed According to Eurocode 8," *Bulletin of Earthquake Engineering*, DOI 10.1007/s10518-011-9294-y, 20 pp.

Rodriguez, M.E., J.I. Restrepo, and A.J. Carr (2002). "Earthquake-Induced Floor Horizontal Accelerations in Buildings," *Earthquake Engineering and Structural Dynamics*, Vol. 31, pp. 693–718.

SEAOC (2008). "Reinforced Concrete Structures," Article 9.01.010, *SEAOC Blue Book—Seismic Design Recommendations*, Seismology Committee, Structural Engineers Association of California, Sacramento, CA.

SEAW (2009). *Special Reinforced Concrete Shear Walls as Building Frame Systems for Mid- and High-Rise Buildings*, White Paper 1-2009, Earthquake Engineering Committee, Structural Engineers Association of Washington, Olympia, WA.

SESOC (2011). "Practice Note—Design of Conventional Structural Systems Following the Canterbury Earthquakes," Version No. 4—21 December, 2011, Structural Engineering Society of New Zealand, 40 pp.

Taylor, C.P., P.A. Cote, and J.W. Wallace (1998). "Design of Slender Reinforced Concrete Walls with Openings," *ACI Structural Journal*, Vol. 95, No. 4, pp. 420–433.

UBC (1997). *Uniform Building Code*, International Conference of Building Officials, Whittier, CA.

Wood, S.L. (1989). "Minimum Tensile Reinforcement Requirements in Walls," *ACI Structural Journal*, Vol. 86, No. 5, pp. 582–591.

Yañez, F., R. Park, and T. Paulay (1992). "Seismic Behavior of Walls with Irregular Openings," *Proceedings, 10th World Conference on Earthquake Engineering*, Madrid, Vol. 6, pp. 3303–3308.

CHAPTER 14
Gravity Framing

14.1 Preview

Gravity framing is a term used to describe structural framing that is proportioned to have strength and stiffness as required for gravity loads, but is not considered as part of the lateral-force-resisting system. Although it is not designed as part of the seismic-force-resisting system, gravity framing must be capable of supporting the gravity loads while the building sways under earthquake motions, and therefore its design requires consideration of seismic effects. For a building designed using performance-based procedures, a common practice is to include the gravity framing in the nonlinear model of the building such that the response of the entire system can be most accurately represented. For these reasons, it is important to understand how gravity framing is designed and how it participates in the seismic response of a building, even if it is not considered part of the seismic-force-resisting system.

This chapter describes the historic development and current use of gravity framing in building construction, reviews the principles for design, presents analysis guidance, and reviews the key design and construction requirements. The presentation includes beam-column, slab-column, and slab-wall gravity frames. Precast concrete gravity framing is used in some structures, but it is not considered in this chapter.

14.2 The Use of Gravity Framing

14.2.1 Historic Development

Seismic design requirements first entered the main body of U.S. building codes in 1933, the year of the 1933 Long Beach earthquake. The earliest seismic codes did not contain provisions delineating requirements for the seismic-force-resisting system and for gravity framing. Nonetheless, it was not uncommon for an engineer to assign all of the seismic forces to limited parts of the structure, with the remainder of the structure designed for gravity loads only. This practice was formally recognized in 1967 when the Uniform Building Code (UBC, 1967) permitted rigid elements that are assumed not to be part of the lateral-force-resisting system, on the condition that their effect on the action of the system must be considered and provided for in the design. When requirements for seismic-force-resisting frames were introduced into the building codes in the 1970s, it was common to design all frame lines to be part of the lateral-force-resisting system. There are, however, examples from the 1970s in which portions of a building were designated as seismic-force-resisting frames with the remaining frames designed only for gravity loads. This practice became increasingly common in the United States in the 1980s and was commonplace by the mid-1990s. In 1983, ACI 318 introduced requirements for frame members that were not proportioned to resist forces induced by earthquake motions. Procedures for checking the gravity framing were not well understood, however, and many engineers seemingly ignored the requirements.

Some frames designed as gravity-only frames did not perform well in the 1994 Northridge earthquake, leading to more stringent requirements for proportioning and detailing of those frames in ACI 318-95. Commonly, the detailing requirements for the gravity-only beam-column frames approach those for special moment frames, tending to some economy by including those components in the seismic-force-resisting system if they can be made to satisfy all of the applicable requirements. Provisions for slab-column gravity frames were introduced in ACI 318-05.

Whereas the practice of designating portions of the building as "gravity-only" framing is now commonplace in the United States, this practice is not widely adopted in some other parts of the world. Instead, the prevailing viewpoint is that the entire system participates in seismic resistance and, therefore, it should all be designed as part of the seismic-force-resisting system.

14.2.2 Example Applications

Gravity framing is used in buildings of all shapes and sizes, with all types of seismic-force-resisting systems. This section discusses three illustrative examples.

Figure 14.1*a* depicts a low-rise parking structure braced by structural walls located around the building perimeter. Typical gravity framing for this application comprises one-way slabs supported on beams that span the longer distance between columns. Either cast-in-place or precast construction can be used. Parking structure framing is complicated by the inclined ramps, which create a split diaphragm. The resulting diaphragm has relatively large span-to-depth ratio. Diaphragm in-plane deformations may increase displacement demands on the gravity framing relative to the smaller in-plane lateral drifts of the structural walls. Furthermore, the inclined ramps result in varying clear heights between columns shared by the ramps and the parking decks (framing line B in Figure 14.1*a*). Under lateral drift, these "captive columns" may sustain comparatively higher shear forces, making them susceptible to shear and axial failures. To solve this potential problem, the ramps and the parking decks can be supported by two structurally separate (or split) columns at each support location (framing line C in Figure 14.1*a*).

Figure 14.1*b* depicts a building braced by special moment frames around the building perimeter. In some buildings, the special moment frames in

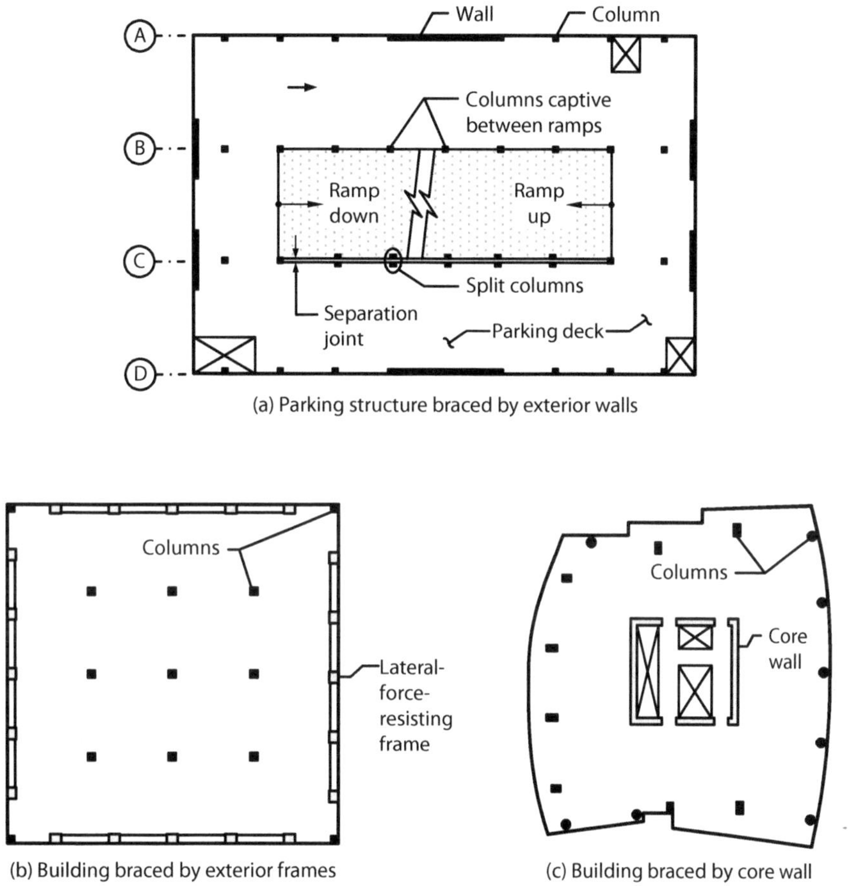

FIGURE 14.1 Plan views of three building framing systems using gravity framing: (*a*) parking structure (generic); (*b*) 21-story perimeter frame (Los Angeles); (*c*) 44-story tall core wall (Seattle).

orthogonal directions share a corner column. Such framing, however, can create design challenges for the corner columns. As an alternative, the frames can stop one bay short of the corner column, as shown, thereby reducing biaxial bending and overturning actions on the corner columns. Special moment frames can also be located inboard from the perimeter. Gravity framing commonly comprises cast-in-place flat-plate framing, reinforced either by bonded, nonprestressed deformed reinforcement or by unbonded post-tensioned strands. This framing system is used in mid- to high-rise buildings up to 40 stories tall.

Figure 14.1*c* depicts a building braced by core walls. Gravity framing commonly comprises unbonded, post-tensioned flat plates. This framing system is used in high-rise buildings to 60 or more stories.

(a) Column failure (b) Aerial photograph of collapsed building

Figure 14.2 Damage and collapse of the Kaiser parking garage in the 1994 Northridge earthquake. (*Left photograph by R. Reitherman, used with permission from the National Information Service for Earthquake Engineering, University of California, Berkeley. Right photograph used with permission from Earthquake Engineering Research Institute.*)

14.2.3 Performance of Gravity Framing in Past Earthquakes

Gravity framing in many buildings has performed well in past earthquakes. The present discussion, however, focuses on observed failures, as these illustrate design deficiencies for which particular attention may be required. Such failures were especially noted following the 1994 Northridge, California earthquake.[1] Four Northridge earthquake examples are presented here. An additional example from the 2010 Chile earthquake illustrates damage to slab-wall framing.

The Kaiser West Los Angeles parking garage was a five-story, cast-in-place, post-tensioned concrete structure (Corley et al., 1996). The lateral system comprised shear walls at the east and west ends and perimeter moment frames along the north and south perimeter. The failure is thought to have initiated in the interior gravity columns, which had widely spaced hoops. As these columns lost vertical load-carrying capacity, the floors collapsed pulling the exterior walls of the structure inward. Figure 14.2 illustrates an exterior gravity column failure and the overall building collapse.

The California State University, Northridge south parking garage was a three-story structure (Corley et al., 1996). The lateral system comprised perimeter special moment frames. Interior columns had widely spaced transverse reinforcement, and sustained axial failures under combined lateral and vertical loads. Similar to

[1]Gravity failures have not been prominent in other past earthquakes, although they have been observed. Collapse of the Four Seasons Apartments in the 1964 Alaska earthquake (Benuska and Clough, 1973) may have been due to failure of post-tensioned lift-slab connections. Collapse of some waffle-slab systems in the 1985 Mexico earthquake may have been due to punching shear failures of waffle-slab frames (Meli and Rodriguez, 1988).

the Kaiser parking garage, axial failure of the interior columns led to floor collapses that pulled the exterior frames of the structure inward. Figure 11.25 illustrates interior column and overall structure failures.

Bullock's department store was a three-story building (Youssef et al., 1996). The structural system consisted of a two-way concrete joist (waffle) floor system with column drop panels and cast-in-place concrete columns. It had been retrofitted with discontinuous shear walls. The failure included punching of the slab-column connections, leading to extensive collapse of the floor system (Figure 14.3). There was no evidence of slab reinforcement through the columns at the punched locations.

The Four Seasons office building had a four-story, cast-in-place structure (Sabol, 1996). The lateral system comprised perimeter moment frames. The interior framing was an unbonded post-tensioned flat plate with drop capitals supported on columns. Several interior slab-column connections experienced punching failures, but the building did not collapse. Figure 14.4 illustrates the framing system and observed damage.

Several shear wall buildings experienced damage to floor slabs during the 2010 Chile earthquake. Figure 14.5 shows a typical example in which the floor slab acted as a coupling element between walls that were separated by door openings or hallways. Typical damage included flexural cracking and, in some cases, slab punching around the wall boundary elements.

FIGURE 14.3 Damage and partial collapse of Bullock's department store in 1994 Northridge earthquake. (*Used with permission from the National Information Service for Earthquake Engineering, University of California, Berkeley.*)

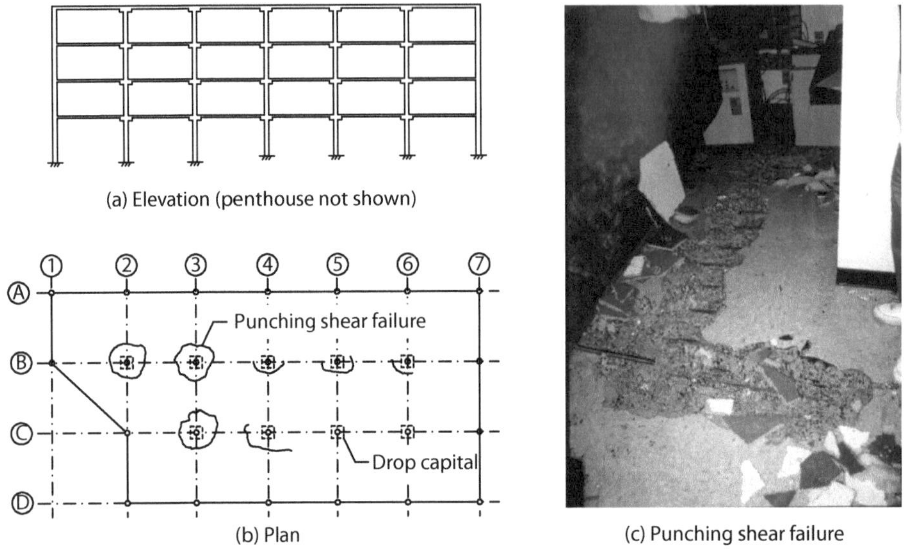

FIGURE 14.4 Damage to unbonded post-tensioned slab-column frame in 1994 Northridge earthquake. (*After Sabol, 1996.*)

FIGURE 14.5 Punching shear failure of a slab acting as a coupling element between two structural walls, 2010 Chile earthquake. View upward toward ceiling showing wall boundary and damaged slab.

14.3 Principles for Design of Gravity Framing

The main concern for gravity framing is damage that leads to the loss of the gravity-load-carrying capacity of the framing. Thus, the following principles should be foremost in gravity framing design:

1. Control deformation demands on gravity framing.
2. Confine column sections where yielding is expected.
3. Avoid shear and axial failures.

The following subsections provide additional details.

14.3.1 Control Deformation Demands

Gravity systems are rarely a problem in buildings with good drift control. For example, low-rise buildings with ample structural walls and stiff diaphragms generally have small lateral drifts and thereby protect gravity framing from damage. Buildings braced by frames and buildings with flexible diaphragms may have relatively larger lateral drifts that can result in damage to the gravity framing. High-rise buildings with structural walls also are susceptible to drift-induced damage, especially in upper stories where rigid-body rotations of the walls can result in relatively large story drift. Interferences with nonstructural components such as stairs or infill walls or with structural components such as inclined ramps should be accounted for in determining the drift demand and capacity.

14.3.2 Confine Column Sections Where Yielding Is Expected

Where the design deformation demands are high and the columns are weaker than the beams, inelastic flexural response of the columns should be anticipated. Such columns should be confined by closely spaced, well-configured transverse reinforcement that enables the development of ductile plastic hinges. Column confinement is especially important for columns having axial force near or above the balanced point. Where column longitudinal bars are lap-spliced immediately above the floor, transverse reinforcement should also be provided to improve lap splice performance.

14.3.3 Avoid Shear and Axial Failures

Shear failure of columns can result in relatively brittle behavior, possibly including axial failure, especially where transverse reinforcement is light and axial loads are high. Good practice is to provide transverse reinforcement that is sufficient to resist the shear corresponding to development of probable moment strengths in the columns or in the members framing into the columns.

Shear failure of beams generally is less critical than shear failure of columns. Nonetheless, the provided shear strength should not be less than the shear that is induced when the gravity framing is subjected to expected lateral displacements.

Where slab-column framing is used, slab shear reinforcement should be provided unless the expected lateral drifts and vertical connection shears are low. The slab-column connection should be detailed with slab bottom reinforcement passing through the column cage. In the event that shear failure occurs, this reinforcement can suspend the floor slab from the column, thereby avoiding progressive collapse that cascades from one connection to others.

14.4 Analysis Guidance

14.4.1 Analysis Procedure

If a building is conceptualized as having gravity framing elements that are distinct from the lateral-force-resisting elements, then the usual approach for design is to develop an analysis model in which lateral resistance is provided only by the seismic-force-resisting system, with no lateral resistance from the gravity framing elements. Reactive mass of the entire building, however, including the mass associated with the gravity framing, must be represented in the model. Gravity loads tributary to the lateral-force-resisting system are assigned to that system. Additionally, where *P*-delta effects are significant, the *P*-delta effects associated with the gravity framing should be included. Usually, this can be done by introducing a "leaning column" that has no lateral resistance but that supports the gravity loads that are tributary to the gravity-framing system (Figure 14.6). As the structural model sways, the leaning column will develop *P*-delta moments that will increase the lateral forces applied to the lateral-force-resisting system.

The preceding approach applies to buildings designed by prescriptive provisions of some building codes, including ASCE 7 (2010). As described in Chapter 11 of this text, the basic procedures of ASCE 7 use a linear-elastic model that represents the effective stiffness of the seismic-force-resisting system deformed to near the yield point. Effects of nonlinear response are represented in part by using a series of factors, including R, C_d, I_e, and Ω_0. Using those procedures, the design displacements are the displacements from an equivalent elastic analysis (using un-reduced forces) multiplied by $C_d I_e / R$. As discussed in Chapter 11, this book recommends that $C_d I_e / R$ not be taken less than 1.0. See additional discussion in Chapter 11.

Where performance-based designs are adopted, the analytical model usually considers nonlinear force-displacement relations of components and materials. In some approaches, only the lateral-force-resisting system is represented in the analytical model, with a leaning column to represent *P*-delta effects of the gravity-framing system. In other approaches (e.g., TBI, 2010), both the lateral system and the gravity-framing system are included in the analytical model. This latter approach generally produces a better analytical representation of the overall building behavior. It also allows for force and deformation demands on the gravity system to be obtained directly from the overall structural analysis model.

If the gravity framing is not included in the analytical model of the building framing system, then an additional analysis step is required to identify the effects of the lateral displacements on the gravity framing. For this purpose, an analytical

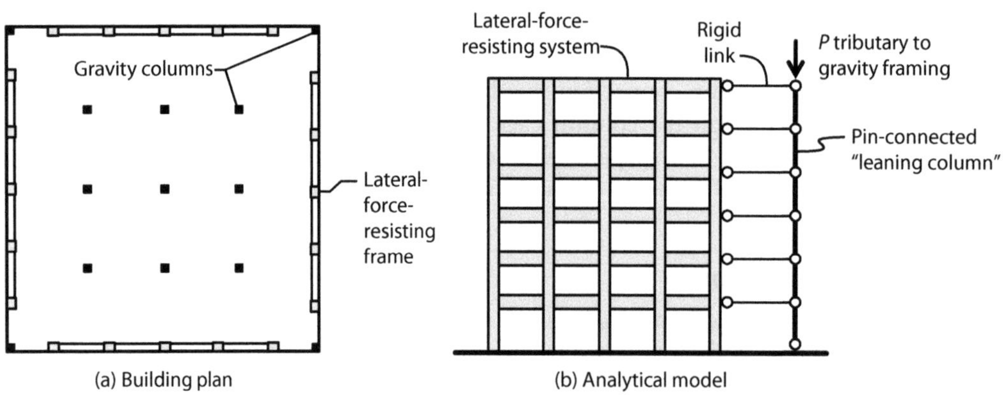

FIGURE 14.6 Planar analytical model for building braced by perimeter frames.

model of the gravity framing is developed and subjected to lateral displacements obtained from the seismic analysis of the lateral system. ASCE 7 and ACI 318 require that the structural integrity of the gravity system be checked only at displacements corresponding to the Design Earthquake (DE) shaking level. Preferably, however, the gravity framing should be checked at lateral displacements corresponding to the Maximum Considered Earthquake (MCE) shaking level, or approximately 1.5 times DE displacements.

Regardless of the approach taken, the analytical model should conservatively bound the expected displacement amplitudes that will be imposed on the gravity framing, and should include, as appropriate, the effects of concrete cracking, foundation flexibility, and deformation of floor and roof diaphragms.

14.4.2 Stiffness Recommendations

The analysis model for the seismic-force-resisting system should represent the effective stiffness considering the design loads. Where a linear model is used, the effective stiffness should approximate the cracked-section stiffness of the framing members. Where a nonlinear model is used, effects of concrete cracking and reinforcement yielding can be modeled directly in the nonlinear analytical model. See Chapters 12 and 13 for recommendations.

The analysis model of the gravity framing similarly should consider the effects of cracking on member stiffness. Where beam-column framing is used, the model should include flexibility of the beams, columns, and beam-column joints. See Chapter 12 for recommended procedures.

Slab-column framing can be modeled using the effective slab-width model of Section 10.6. In typical designs, however, it is not necessary to calculate the earthquake-induced forces in slab-column connections. Instead, the integrity of slab-column gravity framing can be evaluated using a simple check of drift and gravity shear in accordance with the procedure in Section 14.5.5.

Stiffness models for slab-wall framing depend on the framing geometry at the slab-wall connection. Where a slab frames into a core wall in a taller building, the stiffness of the wall sections for out-of-plane bending typically is much larger than the stiffness of the slab. The slab can be modeled as a beam having width equal to the wall width plus an effective width on either side of the wall. The author is not aware of studies for this effective width. Based on findings for slab-column connections, it is recommended that the effective width on either side of the wall be estimated as $l_1/6$. Where slabs act as coupling elements between rectangular walls, the effective width can be based on the model in Section 10.10.

14.5 Design Guidance

14.5.1 Design Actions

Components of the gravity-framing system should be designed to support gravity loads and any other loads as the framing is subjected to expected lateral displacements. According to ACI 318, members not designated as part of the seismic-force-resisting system are to be evaluated for gravity load combinations of $(1.2D + 1.0L + 0.2S)$ or $0.9D$, whichever controls, acting simultaneously with the design displacement, δ_u. The load factor on the live load, L, is permitted to be reduced to 0.5 except for garages, areas occupied as places of public assembly, and all areas where L is greater than 100 psf (4.8 kPa).

Note that ACI 318 does not require consideration of vertical seismic loading. The common view is that most designs are not highly susceptible to effects of vertical accelerations, although such effects should not be ignored where they are important. Additionally, some structural systems may develop significant axial forces due to earthquake-induced displacements. For example, in tall buildings the gravity columns can accumulate significant axial forces due to overturning effects, as depicted in Figure 14.7 (Yang et al., 2010). Such effects should be considered for tall buildings.

The design displacement, δ_u, is usually interpreted in building codes as the displacement corresponding to the DE. The gravity system, however, should be capable of supporting applied loads under maximum expected displacements. This book recommends checking gravity framing under displacements due to MCE effects, which under ASCE 7 provisions are approximately 1.5 times the DE displacements.

14.5.2 Columns

According to ACI 318, the requirements for cast-in-place columns (and beams) depend on the magnitude of moments and shears induced in those members when subjected to the design displacement δ_u. Where the induced moments and shears do not exceed the design moment and shear strengths (ϕM_n and ϕV_n), the requirements given in Figure 14.8*a* apply. Where the induced moments or shears exceed the corresponding design strengths, the requirements given in Figure 14.8*b* apply. If the

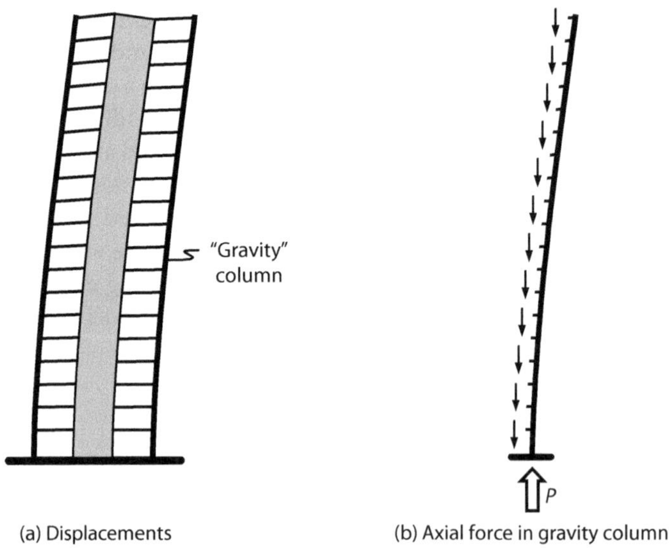

FIGURE 14.7 Axial force in gravity column due to overturning actions in tall building.

- Longitudinal reinforcement satisfies $0.01 \le A_{st}/A_g \le 0.06$.
- Transverse reinforcement is spirals, circular hoops, or rectilinear hoops and crossties, designed to resist shear corresponding to M_{pr}.
- Rectilinear hoops and crossties engage at least corner and alternate longitudinal bars, with no unsupported bar more than 6 in (150 mm) clear from a supported bar, and with spacing of supported bars h_x within cross section not exceeding 14 in (350 mm) on center. Where $P_u \ge 0.3A_g f'_c$ or $f'_c \ge$ 10,000 psi (70 MPa), every longitudinal bar around the perimeter of the column core shall have lateral support provided by the corner of a hoop or by a seismic hook, with h_x not exceeding 8 in (200 mm).

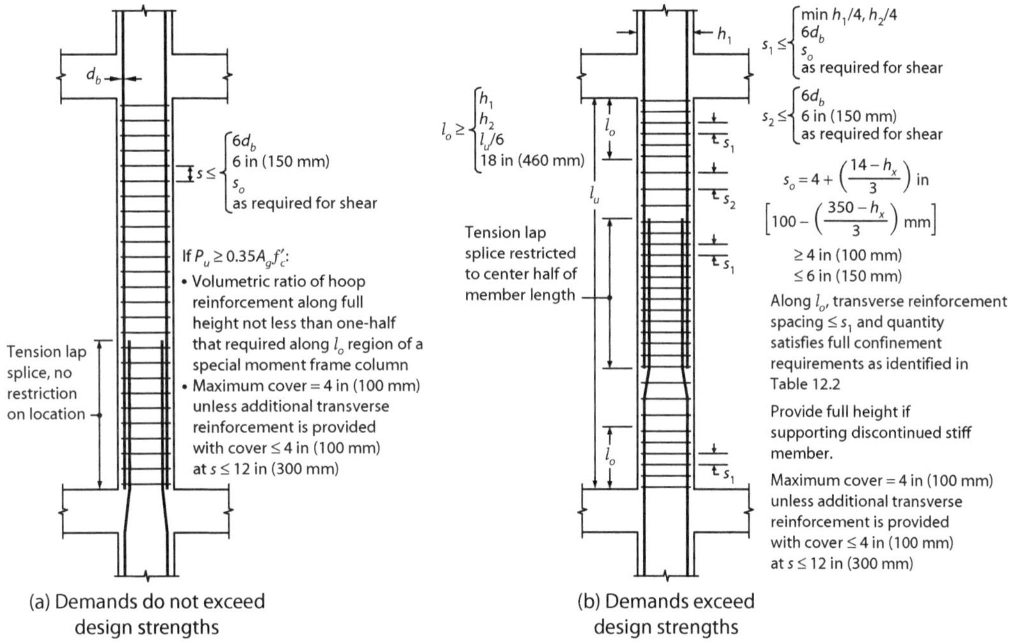

FIGURE 14.8 Requirements for columns not part of the seismic-force-resisting system. *Note:* If effects of displacements on demands are not calculated, then (*b*) applies.

effects of δ_u are not checked explicitly, then the design is to be based on the assumption that induced moments and shears exceed the design moment and shear strengths, such that the requirements given in Figure 14.8*b* apply.

Regardless of the magnitude of induced moments and shears, ACI 318 requires that columns be designed to have shear strength satisfying $\phi V_n \ge V_e$. Nominal shear strength V_n is calculated in accordance with procedures for columns of special moment frames, that is, $V_n = V_s + V_c$, with $V_c = 0$ along length l_o if both (a) the earthquake-induced shear force represents one-half or more of the maximum required shear strength, and (b) the factored axial compressive force including earthquake effects is less than $A_g f'_c/20$. Shear force V_e is determined using the maximum probable moment strengths M_{pr} at the ends of the column associated with the range of axial forces acting on the column. Alternatively, the column shear need not exceed the shear associated with development of probable moment strengths M_{pr} in the beams or slabs framing into the joints. The requirement to design for M_{pr} is due to the potentially high consequences of column shear failure and the low confidence in code-based design displacements. Where improved estimates of lateral displacements provide a high degree of confidence that column shear failure will not occur, it may be reasonable to relax the requirements.

Note that the permitted location of lap splices depends on the anticipated column demands. Where demands are relatively low, lap splices are designed as tension splices and can be located anywhere along the member length. Where demands are higher, lap splices are designed as tension splices and must be located within the middle half of the column length. The intent is to avoid a lap splice in the potential plastic hinge region. ACI 318 is unclear about the definition of column length, and a common interpretation is that column length refers to floor-to-floor height rather than clear height. Some engineers interpret the ACI 318 provision as requiring that the splice be centered at the mid-height, but be permitted to extend outside the middle half of the column length if necessary to accommodate the required lap splice length. Given that the end of the column is confined by closely spaced transverse reinforcement, and that the moment strength of the column is not being relied on for seismic resistance, this practice seems reasonable and should be permitted for gravity columns. The same allowance should not apply to columns designated as part of the seismic-force-resisting system.

Where demand forces exceed design strengths, some additional requirements apply. Materials (including concrete, reinforcement, welded splices, and mechanical splices) should conform to minimum requirements for special moment frames. Columns supporting reactions from discontinued stiff members, such as walls, should have full confinement (as described in Figure 14.8*b*) over their full height at all levels beneath the discontinuity if the factored axial compressive force in the column, related to earthquake effect, exceeds $A_g f_c'/10$ (or $A_g f_c'/4$ where design forces have been amplified to account for overstrength of the vertical elements of the seismic-force-resisting system).

14.5.3 Beams

According to ACI 318, the requirements for cast-in-place beams depend on the magnitude of moments and shears induced in those members when subjected to the design displacement, δ_u. Where the induced moments and shears do not exceed the design moment and shear strengths (ϕM_n and ϕV_n), the requirements given in Figure 14.9*a* apply. Where the induced moments or shears exceed the corresponding design strengths, the requirements given in Figure 14.9*b* apply. If effects of δ_u are not explicitly checked, then the design is to be based on the assumption that induced moments and shears exceed the design moment and shear strengths, such that the requirements given in Figure 14.9*b* apply.

Where demands exceed design strengths, ACI 318 requires that beams be designed to have shear strength satisfying $\phi V_n \geq V_e$. Nominal shear strength V_n is calculated in accordance with procedures for beams of special moment frames, that is, $V_n = V_s + V_c$, with $V_c = 0$ along length l_o if both (a) the earthquake-induced shear force represents one-half or more of the maximum required shear strength, and (b) the factored axial compressive force including earthquake effects is less than $A_g f_c'/20$. Shear force V_e is calculated using the probable moment strengths M_{pr} at

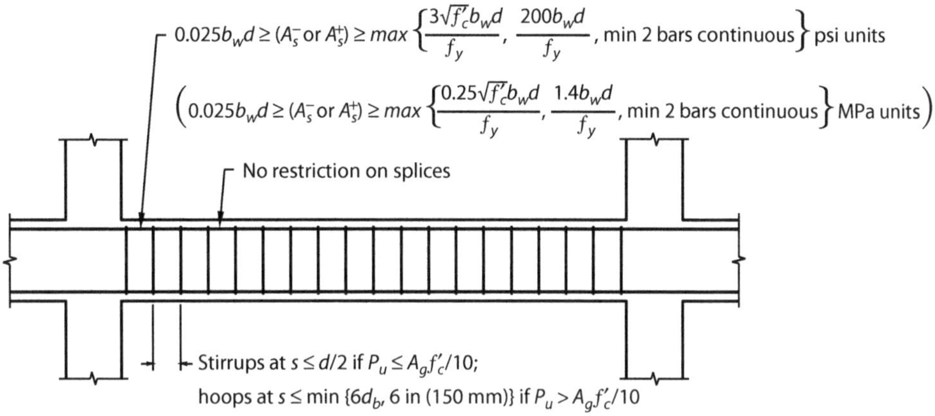

(a) Demands do not exceed design strengths

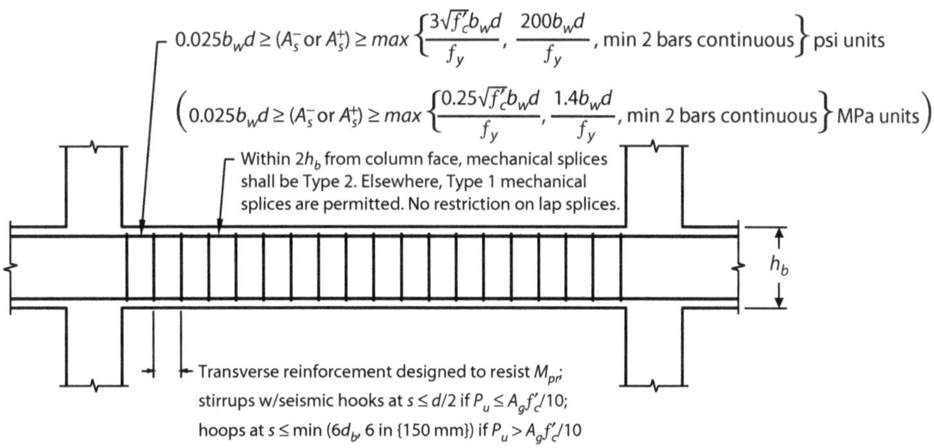

(b) Demands exceed design strengths

FIGURE 14.9 Requirements for beams not part of the seismic-force-resisting system. *Note:* If effects of displacements on demands are not calculated, then (*b*) applies.

the ends of the beam. In addition, materials (including concrete, reinforcement, welded splices, and mechanical splices) should conform to minimum requirements for special moment frames.

14.5.4 Beam-Column Joints

For beam-column joints of the gravity-framing system, ACI 318 does not have specific requirements for checking joint shear demands under load combinations including earthquake effects. Instead, the emphasis is on providing joint transverse reinforcement to control cracking, confine the joint, and support column longitudinal reinforcement. The specific requirements depend on whether the demands exceed capacities of the adjacent framing members, as described next:

Where Earthquake-Induced Demands in Beams and Columns Do Not Exceed Design Strengths

In this case, the design of the beam-column joints need only satisfy requirements for non-seismic conditions. According to ACI 318, joint transverse reinforcement is not required for interior connections confined by beams on four faces. Other connections are to have transverse reinforcement in each principal direction satisfying

$$A_v \geq 0.75\sqrt{f'_c}\frac{b_w s}{f_{yt}} \text{ but not less than } 50\frac{b_w s}{f_{yt}}, \text{ psi}$$

$$A_v \geq 0.062\sqrt{f'_c}\frac{b_w s}{f_{yt}} \text{ but not less than } 0.35\frac{b_w s}{f_{yt}}, \text{ MPa} \tag{14.1}$$

According to ACI 352 (2002), exterior joints should have at least two layers of transverse reinforcement located between the top and bottom levels of beam longitudinal reinforcement of the deepest member framing into the joint. The center-to-center tie spacing or spiral pitch should not exceed 12 in (300 mm). To facilitate placement of transverse reinforcement in Type 1 joints, cap or split ties may be used, provided the lap length is sufficient to develop the tie yield strength. Where required, configuration of ties or spirals in the joint should satisfy the usual requirements for tied or spiral columns. The transverse reinforcement can be eliminated in interior joints for which (a) beams frame into all four sides of the joint and (b) each beam width is at least 3/4 of the column width and leaves no more than 4 in (100 mm) of the column width uncovered on either side of the beam. It can also be eliminated in the direction parallel to the beams in the case of a joint with beams framing into two opposite sides of the joint and for which each beam width is at least three quarters of the column width and leaves no more than 4 in (100 mm) of the column width on either side of the beam. ACI 352 also recommends use of vertical joint reinforcement for "roof" connections in which (a) the joint has a free horizontal face at the discontinuous end of a column and (b) discontinuous beam reinforcement is the nearest longitudinal reinforcement to the free horizontal face. Such reinforcement, however, seems unnecessary in gravity framing.

ACI 318 does not have specific requirements for checking joint shear demands. Presumably, the joints would be designed to be adequate under factored load combinations excluding earthquake effects. ACI 352 design recommendations for Type 1 joints would be suitable for this purpose. See Section 9.6 for additional discussion.

Where Earthquake-Induced Demands in Columns Exceed Design Strengths

In this case, ACI 318 requires that joints be reinforced with transverse reinforcement equivalent to that required along the length l_o of special moment frame columns. Where beams frame into all four sides of the joint and where each beam width is at least three-fourths of the column width, the amount of transverse reinforcement can be reduced by half, and the spacing can be increased to 6 in (150 mm) within the overall depth h of the shallowest framing beam. These requirements are the same as the transverse reinforcement requirements for joints of special moment frames where the beam width does not exceed the column width. Committee 352 has somewhat relaxed demands, but these are seldom considered where ACI 318 provisions are required.

14.5.5 Slab-Column Framing

The primary concerns for slab-column framing are punching shear failure and progressive collapse due to punching at multiple connections. To address these concerns, ACI 318 has provisions for shear reinforcement and for structural integrity reinforcement, as described next.

Shear Reinforcement

Chapter 10 described how the drift capacity of a slab-column connection without shear reinforcement is sensitive to the magnitude of the vertical shear being supported by the connection. ACI 318 approximates the relation between the drift ratio capacity and the gravity shear using the bilinear relation shown in Figure 14.10. If a connection has combined shear and story drift ratio falling below the bilinear relation given in Figure 14.10, shear reinforcement is not required. Otherwise, shear reinforcement is required. For this purpose, the story drift ratio is taken as the larger of the story drift ratios of the adjacent stories above and below the slab-column connection. Gravity shear stress v_{ug} is the factored nominal shear stress on the slab critical section for two-way action due to gravity loads without moment about the slab critical section (i.e., $v_{ug} = V_{ug}/b_o d$). The quantity v_c is the shear stress capacity of the critical section attributable to concrete (i.e., $v_c = V_c/b_o d$, where V_c is defined in Chapter 10), and ϕ = 0.75. This requirement should be checked at critical sections adjacent to columns, capitals, drop panels, and drop capitals. Note that ACI 352.1 (2011) recommends reducing the term V_c by factor $C_v = 0.75$ for Type 2 connections (those required to sustain gravity loads under earthquake-induced inelastic deformations).

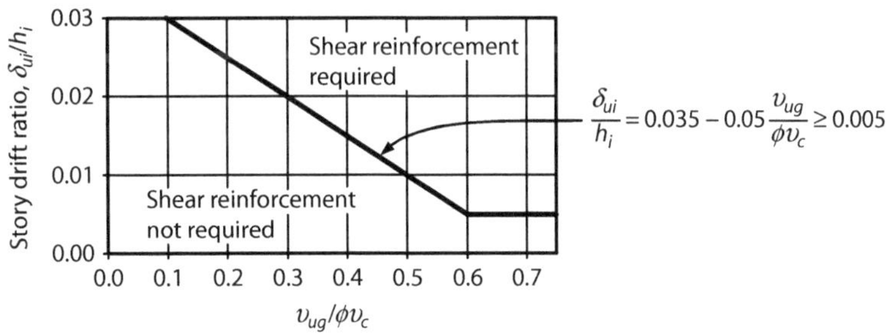

FIGURE 14.10 Criterion to determine whether slab shear reinforcement is required. (*After ACI 318, 2014, used with permission from American Concrete Institute.*)

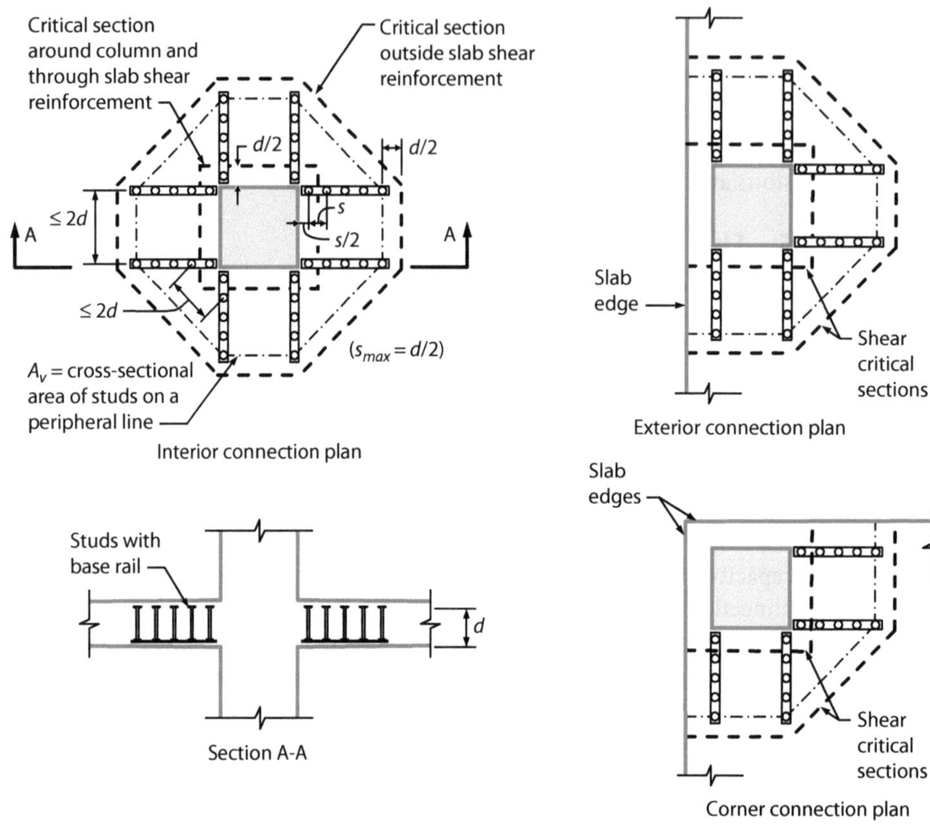

FIGURE 14.11 Stud rail details for slab-column connections requiring shear reinforcement. (*After ACI 318, 2014, and ACI 352.1, 2011, used with permission from American Concrete Institute.*)

Where shear reinforcement is required, stud rails are the most common form of reinforcement. According to ACI 318, the stud rails are required to provide $v_s \geq 3.5\sqrt{f_c'}$, psi ($v_s \geq 0.29\sqrt{f_c'}$, MPa) at the critical section b_o around the column, capital, drop panel, or shear capital, and must extend at least $4h$ from the face of the support or change in thickness in both principal directions. Recommended details are shown in Figure 14.11. The spacing between successive lines of shear reinforcement that surround the column should not exceed $s = d/2$ measured in a direction perpendicular to the column face.[2] ACI 318 requires that the first peripheral line of headed studs be placed no farther than $d/2$ from the column face. ACI 352.1, however, recommends that the first line of headed studs be placed no farther than $s/2$ from the column face, where s is the typical spacing away from the column. Spacing between adjacent shear

[2]ACI 318 permits the maximum spacing of $s = 0.75d$ in some cases, but that spacing is not recommended here.

reinforcement elements, measured on the perimeter of the first two peripheral lines of shear reinforcement, should not exceed $2d$.

Detailing for Structural Integrity

The overall structural integrity of a building can be greatly enhanced by relatively minor changes in detailing of the reinforcement. For slab-column connections, this is generally achieved by providing continuous bottom reinforcement passing through the column core. For unbonded post-tensioned construction, an alternative is to layer the tendons so as to provide a secondary load path. Additional details can also improve overall performance, as described in the following paragraphs.

Figure 14.12 illustrates some typical details in *conventionally reinforced* concrete slab-column connections. For interior connections, at least two of the main bottom bars in each direction should be continuous through the column core to provide structural integrity. At exterior connections, those bars that are perpendicular to the slab edge should be anchored to develop f_y in tension at the exterior support. The requirement to use two bars in each direction is based on equilibrium requirements for slabs of usual proportions supporting residential or office loading. Some engineers routinely use the alternative recommendations of Section 10.9 to determine the required continuous bottom reinforcement. Those alternative recommendations should be considered at least for slabs with unusually long spans or high loads.

ACI 352.1 recommends additional details. For interior connections, at least two of the main top slab bars should be continuous through or anchored within the column cage in each direction (Figure 14.12). At exterior columns, either an edge beam should be provided or else an effective edge beam should be formed within the slab depth to resist torsion that arises from moments acting about an axis parallel to the edge. To form the effective beam within the slab depth, top slab reinforcement perpendicular to the slab edge should be hooked and lapped with the bottom bars, thereby forming an effectively continuous loop of reinforcement. According to ACI 352.1, spacing of this reinforcement should not exceed $0.75h$. Tests reported in Hwang and Moehle (2000), however, used spacing as wide as $2h$ without adverse effect. To complete the effective beam, at least four reinforcing

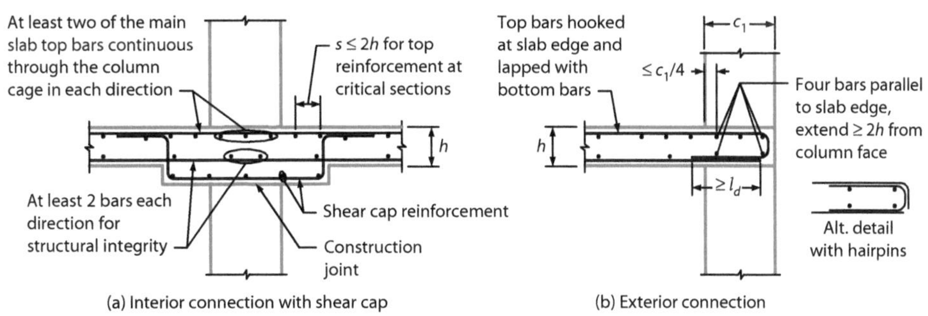

(a) Interior connection with shear cap

(b) Exterior connection

FIGURE 14.12 Typical details for reinforced concrete slab-column connections.

bars should be provided parallel to the slab edge (Figure 14.12*b*). For the purpose of torsional resistance, conceptually these four bars should be located within the slab bars that are perpendicular to the slab edge. In typical slab detailing, however, all bars in a given direction are placed either in the outer or inner layer of reinforcement, such that the four bars can be within the perpendicular bars along two opposite edges but not along the other edges. Tests reported in Hwang and Moehle (2000) showed that either detail performs adequately.

Punching shear tests of slab-column connections with shear capitals similar to those shown in Figure 14.12*a* indicate that the corners of unreinforced capitals can detach and drop from the connection (Larsen, 1989). Nominal reinforcement is recommended to avoid this failure mechanism.

For *prestressed connections*, ACI 318 permits structural integrity requirements to be met by either one of two options:

1. A minimum of two of the post-tensioning tendons in each direction are required to pass through or be anchored within the column core at a connection. Outside column and shear cap faces, these two structural integrity tendons are to pass under any orthogonal tendons in adjacent spans. Where the two structural integrity tendons are anchored within the column core, the anchorage is to be located beyond the column centroid and away from the anchored span. This arrangement in effect produces a "net" to suspend all the tendons (and the slab) from the column in the event of punching shear failure.
2. Alternatively, the slab is required to have continuous bottom reinforcement passing through the column core and anchored at exterior supports. The area of bottom reinforcement in each direction must not be less than the greater of $\frac{4.5\sqrt{f'_c}}{f_y}bd$, psi ($\frac{3\sqrt{f'_c}}{8f_y}bd$, MPa) and $300bd/f_y$, psi ($2.1bd/f_y$, MPa), where b is the width of the column face through which the reinforcement passes. These bars should extend at least the bar development length beyond the column or shear cap face.

Whether option 1 or 2 is selected varies among design offices, with some regional preferences. Some design offices routinely elect to use both options.

Figure 14.13 illustrates a detail that uses both options. To accomplish this detail requires careful attention to the sequence of tendon placement. First the two tendons in the distributed direction are placed, followed by all of the banded tendons, with the remaining distributed tendons placed on top of the banded tendons. Where this arrangement is used, required top bonded reinforcement in the distributed direction is placed under the banded tendons, and reinforcement in the banded direction is placed at the elevation of the banded tendons. Note that this arrangement places all of the banded reinforcement at the maximum depth. The two distributed tendons that pass through the column are at reduced depth, but most of the remaining distributed tendons can also be placed at full depth. In addition to the tendons and top reinforcement, the detail shown also includes continuous bottom reinforcement for structural integrity.

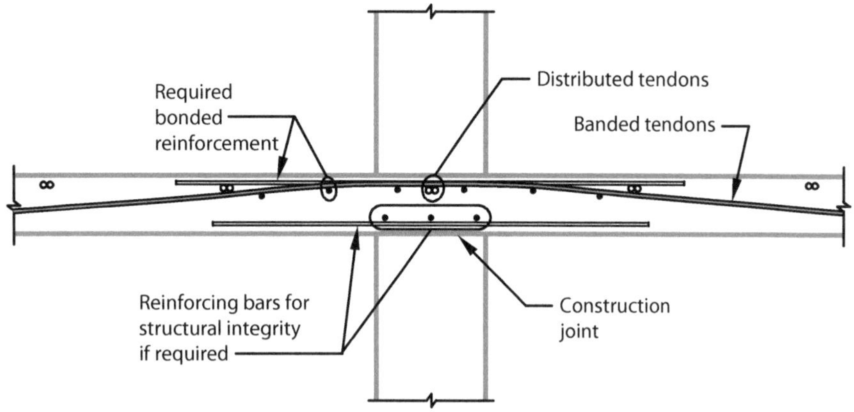

FIGURE 14.13 Typical arrangement of post-tensioned and bonded deformed reinforcement for post-tensioned concrete slab-column connection.

A simpler placement scheme, commonly adopted in some regions, places all the banded reinforcement first, with the distributed reinforcement above the banded reinforcement. Although the effective depth of the banded reinforcement is reduced in this scheme, the effect of the reduced depth can be considered in design. Where this detail is adopted, structural integrity must be provided by continuous bottom reinforcement as required by option 2 above.

Figure 14.14 illustrates typical tendon and reinforcement placement at interior and exterior connections.

14.5.6 Slab-Wall Framing

Figure 14.15 illustrates typical details for unbonded post-tensioned construction framing into a wall out of plane, including the case where the slab frames into a core wall. Figure 14.15*a* illustrates a detail for the case where the wall is cast to the

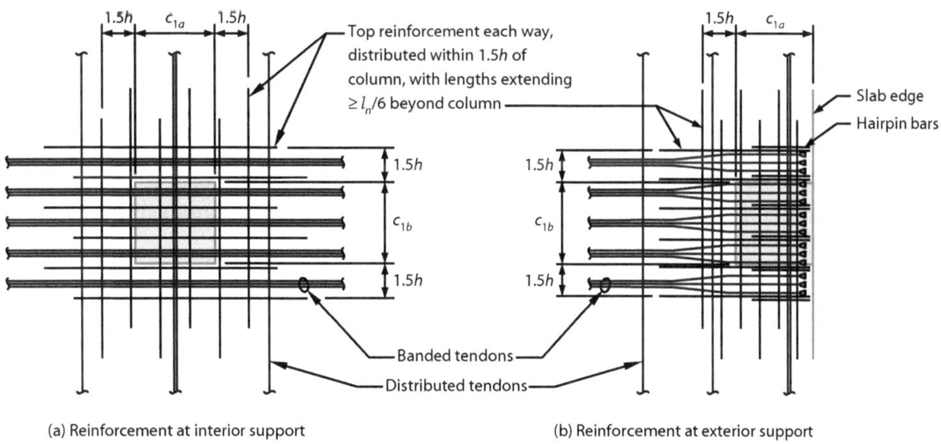

FIGURE 14.14 Typical details of post-tensioned concrete slab-column connections.

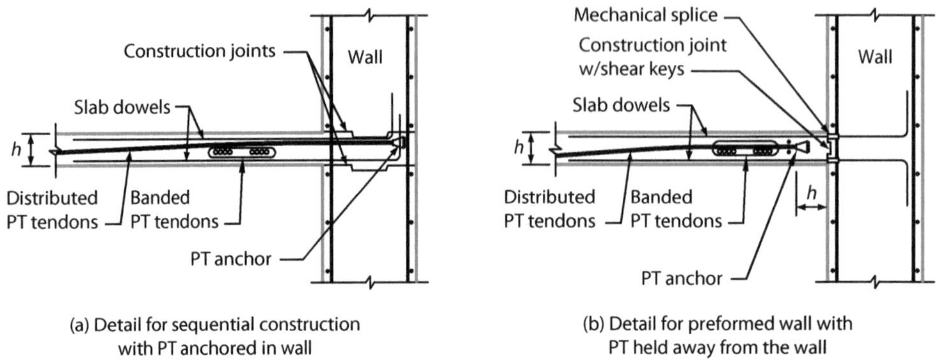

FIGURE 14.15 Typical details for slab-wall connections for the slab framing perpendicular to the wall web or flange.

bottom of the slab, with horizontal construction joints between the slab and wall segments. In this case, it is preferable to anchor the tendon within the wall thickness, unless the slab and tendon continue into the next bay. Where the slip-form construction method is used, a vertical joint is formed between the slab and wall (Figure 14.15*b*). In this case, it may be possible to anchor the tendon at the back face of the wall. An alternative detail, which may be preferred based on construction considerations, is to anchor the tendon within the slab, with continuity provided by slab dowels mechanically anchored to the wall. In this case, it is preferable to hold the tendon anchors about one slab thickness from the face of the wall to provide a length for inelastic rotations in the event of large earthquake drift demands. Klemencic et al. (2006) present test results demonstrating adequate performance of this latter detail.

14.5.7 Wall Piers

Wall piers not designated as part of the seismic-force-resisting system should be detailed to enable them to support gravity loads under earthquake-induced deformations and forces. According to ACI 318, wall piers are column-like vertical segments of structural walls. Wall piers having $l_w/b_w \le 2.5$ are considered to behave essentially as columns, and should be detailed in accordance with requirements for columns (Section 14.5.2). It can be difficult to design the wall segment for shears corresponding to the development of M_{pr}. Therefore, it is permitted to design for shear equal to Ω_0 times the shear obtained from analysis of the building under expected displacements.

According to ACI 318, for wall piers having $l_w/b_w \ge 2.5$, alternative provisions can be applied. Design shear is determined by the procedure of the preceding paragraph. Design shear strength ϕV_n and distributed shear reinforcement are in accordance with requirements for walls. Transverse reinforcement is to be in the form of hoops except it is permitted to use single-leg horizontal reinforcement parallel to l_w where only one curtain of distributed shear reinforcement is provided. Single-leg horizontal reinforcement should have 180° bends at each end that engage wall pier boundary longitudinal reinforcement. Vertical spacing of transverse reinforcement is not to exceed 6 in (150 mm). Transverse reinforcement is to extend at least 12 in (305 mm) above and below the clear height of the wall pier. Special boundary elements are to be provided if stress calculated based on $\sigma = P_u/A_g + M_{ux}/S_{gx} + M_{uy}/S_{gy}$ exceeds $f_c'/5$.

14.6 Additional Requirements

14.6.1 Special Inspection

As with special moment frames and special structural walls, proper construction of gravity framing, using good quality materials, is essential to ensuring that a building, once constructed, complies with the requirements of the code and the approved design. In U.S. construction practice, special inspection is required to foster proper construction. See the special inspection requirements in Section 12.8.1.

14.6.2 Material Properties

Concrete and reinforcement should be in conformance with the general requirements of the building code. Where earthquake-induced demands exceed design capacities, or where demands are not calculated, materials, welded splices, and mechanical splices should be in conformance with the requirements for special structural systems. According to ACI 318, specified compressive strength of concrete must be at least 3000 psi (21 MPa) and specified compressive strength of lightweight concrete must not exceed 5000 psi (35 MPa). Longitudinal reinforcement must be ASTM A706 Grade 60 (420) or ASTM A615 Grades 40 or 60 (280 or 420) with equivalent mechanical properties. Type 1 mechanical splices and welded splices are not permitted within a distance equal to twice the member depth from a column or beam face or from critical sections where yielding of the reinforcement is likely to occur as a result of inelastic lateral deformations. Type 2 mechanical splices are permitted at any location. Welding of stirrups, ties, inserts, or other similar elements to longitudinal reinforcement that is required by design is not permitted.

14.7 Detailing and Constructability Issues

Similar to a special moment frame, gravity-only framing relies on carefully detailed and properly placed reinforcement to ensure that it can maintain its vertical-load-carrying capacity through multiple deformation cycles beyond the yield deformation. Architectural requirements often push to get the columns as small as possible, resulting in congested columns supporting axial loads above the balanced point. Performance of the column relies on properly placed transverse reinforcement to confine the core concrete, provide lateral support to longitudinal reinforcement, and resist shear.

Beam-column gravity frames require many of the same detailing and construction considerations as special moment frames. The reader is referred to

Section 12.8 for discussion about longitudinal bar compatibility, beam and column confinement, bar splices, and concrete placement. For columns whose earthquake-induced demands do not exceed design strengths, column lap splices can be placed just above the floor level. For columns with earthquake demands exceeding design strengths, ACI 318 requires lap splices to be located in the middle one-half of the column length. It is common practice, however, to permit long laps to extend outside the middle half. Mid-height lap splices require additional crane time during construction, to support the cage until it is securely tied in place. To reduce crane time, and to prevent downward slippage of the cage during construction, some detailers extend four of the longitudinal bars the full story height such that the cage can rest on the floor slab below (Figure 14.16). This practice seems reasonable for gravity columns, as they are well confined (Figure 14.8) and are not required to sustain lateral strength under deformation reversals. The detail using extended longitudinal bars would not be appropriate for columns that are part of the seismic-force-resisting system.

Figure 14.17 shows an example of an exterior slab-column connection in unbonded post-tensioned flat-plate construction with tendons banded in one direction. Detailing of the connection region requires consideration of anchorage hardware located within the column core, as well as bonded reinforcement to resist bursting stresses at anchorages. Figure 14.18 illustrates an interior connection with stud rails.

Figure 14.16 Gravity column with longitudinal reinforcement extended to support the reinforcement cage during construction. (*Used with permission from KPFF Consulting Engineers.*)

FIGURE 14.17 Unbonded post-tensioned flat-plate construction.

FIGURE 14.18 Interior connection of unbonded post-tensioned flat-plate construction with stud rails.

Figure 14.19 Connection of slab and core wall. Note the anchorage of unbonded tendons outside the wall, with continuity provided by shear keys and deformed reinforcement connected to the wall through form savers.

In many high-rise towers, the core walls are constructed ahead of the columns and floor slabs to expedite construction. When this technique is used, a vertical cold joint is introduced at the slab–wall interface. One option is to embed sleeves in the core wall and anchor the tendons on the opposite side of the wall. A common alternative practice is to locate the anchors for the unbonded post-tensioning cables within the slab immediately adjacent to the wall. The slab-wall connection is made through reinforcing dowels placed near the top and bottom of the slab, with Type 2 mechanical couplers at the slab–wall interface. The connection should be supplemented with intermittent shear keys to improve shear transfer between the slab/diaphragm and the wall. Figure 14.19 illustrates this latter connection detail. See Klemencic et al. (2006) for additional discussion.

References

ACI 318 (2014). *Building Code Requirements for Structural Concrete (ACI 318-14) and Commentary*, American Concrete Institute, Farmington Hills, MI.

ACI 352 (2002). *Recommendations for Design of Beam-Column Connections in Monolithic Reinforced Concrete Structures (352R.02)*, American Concrete Institute, Farmington Hills, MI, 37 pp.

ACI 352.1 (2011). *Guide for Design of Slab-Column Connections in Monolithic Concrete Structures*, American Concrete Institute, Farmington Hills, MI, 28 pp.

ASCE 7 (2010). *Minimum Design Loads for Buildings and Other Structures*, American Society of Civil Engineers, Reston, VA, 608 pp.

Benuska, L., and R.W. Clough (1973). "Dynamic Analyses of Building Failures," in *The Great Alaska Earthquake of 1964—Engineering*, National Academy of Sciences, Washington, DC, pp. 283–307.

Corley, W.G., L. Cluff, S. Hilmy, W. Holmes, and J. Wight (1996). "Concrete Parking Structures," in *Northridge Earthquake Reconnaissance Report, Vol. 2, Earthquake Spectra*, Earthquake Engineering Research Institute, Oakland, CA, Supplement to Volume 11, pp. 75–98.

Hwang, S.-J., and J.P. Moehle (2010). "Vertical and Lateral Load Tests of a Nine-Panel Flat-Plate Frame," *ACI Structural Journal*, Vol. 97, No. 1, pp. 193–210.

Klemencic, R., J.A. Fry, G. Hurtado, and J.P. Moehle (2006). "Performance of Post-Tensioned Slab-Core Wall Connections," *PTI Journal*, Vol. 4, No. 6, pp. 7–23.

Larsen, T. (1989). *Experimental Study of the Shear Strength of Post-Tensioned Flat-Plates with Column Shear Capitals*, CE 299 Report, Department of Civil Engineering, University of California, Berkeley, CA. Available at http://nisee.berkeley.edu/documents/elib/www/documents/201005/CE299/larsen-shear-capitals.pdf

Meli, R., and M. Rodriguez (1988). "Seismic Behavior of Waffle-Flat Plate Buildings," *Concrete International*, Vol. 10, No. 7, pp. 33–41.

Sabol, T.A. (1996). "Flat Slab Failure in Ductile Concrete Frame Building," in *1994 Northridge Earthquake Buildings Case Studies Project*, California Seismic Safety Commission, pp. 167–187.

TBI (2010). *Guidelines for Performance-Based Seismic Design of Tall Buildings*, Report No. 2010/05, Pacific Earthquake Engineering Research Center, University of California, Berkeley, CA, 84 pp.

UBC (1967). *Uniform Building Code*, International Council of Building Officials.

Yang, T.Y., G. Hurtado, and J.P. Moehle (2010). "Seismic Behavior and Modeling of Flat-Plate Gravity Framing in Tall Buildings," *Proceedings, 9th US National Conference on Earthquake Engineering*, Toronto.

Youssef, N., J. Vinkler, and B. Prakash (1996). "Bullock's Department Store, Northridge Fashion Center," in *1994 Northridge Earthquake Buildings Case Studies Project*, California Seismic Safety Commission, pp. 63–82.

CHAPTER 15

Diaphragms and Collectors[1]

15.1 Preview

Building structures generally comprise a three-dimensional framework of structural elements configured to support gravity and lateral loads. Although the complete three-dimensional system acts integrally to resist loads, we commonly conceive of the seismic-force-resisting system as being composed of vertical elements, horizontal elements, and the foundation (Figure 15.1). The vertical elements extend between the foundation and elevated levels, providing a continuous load path to transmit gravity and seismic forces from the upper levels to the foundation. The horizontal elements typically consist of diaphragms including collectors. Diaphragms transmit inertial forces from the floor system to the vertical elements of the seismic-force-resisting system. They also tie the vertical elements together, and thereby stabilize and transmit forces among these elements as may be required during earthquake shaking. Diaphragms are thus an essential part of the seismic-force-resisting system, and require design attention by the structural engineer to ensure the structural system performs adequately during an earthquake.

In the United States, diaphragms are required to be designed as part of the seismic-force-resisting system of every new building assigned to Seismic Design Categories (SDC) B, C, D, E, or F of the International Building Code (IBC, 2012). Although they can consist of truss elements or horizontal diagonal bracing, in most cases diaphragms are constructed as essentially solid, planar elements. Concrete diaphragms can be conventionally reinforced or prestressed, and can be of cast-in-place concrete, a topping slab on a metal deck or on precast concrete, or interconnected precast concrete elements without topping. The scope of this chapter is restricted to cast-in-place concrete diaphragms, either conventionally reinforced or prestressed. However, many of the concepts presented here apply equally to other diaphragm types.

[1]This chapter uses figures and text from Moehle, Hooper, and Kelly (2010). Permission by NIST for use of this material, as well as the significant writing contributions of Messrs Hooper and Kelly, is gratefully acknowledged.

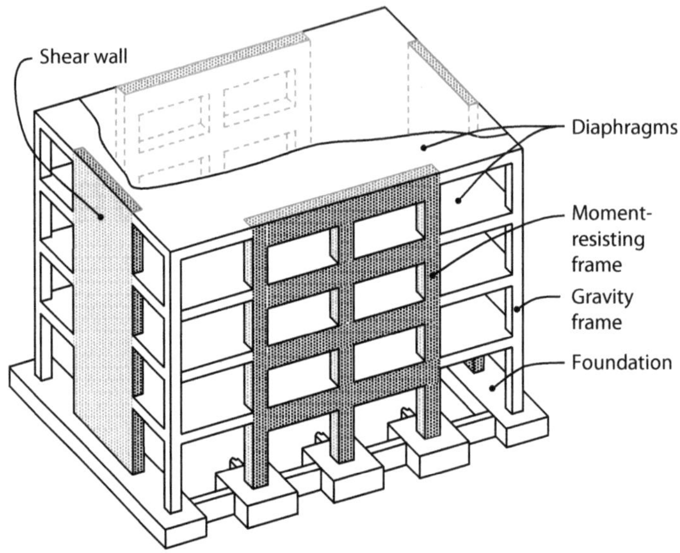

Figure 15.1 Sketch of a basic building system comprising horizontal spanning elements (diaphragms), vertical spanning elements (walls and frames), and foundation.

In the United States, design requirements for concrete diaphragms are contained in the IBC, which establishes general regulations for buildings; ASCE 7 (2010), which focuses on determination of design forces; and ACI 318 (2014), which focuses on proportioning and detailing requirements. Although the philosophy and general approach to diaphragm design have been relatively unchanged in recent past years, the specific details of the building codes have changed with successive editions. The designer should refer to the legally adopted code to determine the specific requirements enforced for a particular project.

This chapter defines diaphragms and their parts, explains diaphragm behavior and design principles, and presents guidance and requirements for diaphragm analysis, proportioning, detailing, and construction.

15.2 The Roles of Diaphragms

As partially illustrated by Figure 15.2, diaphragms resist forces from several types of actions:

- *Diaphragm in-plane forces:* Lateral forces from load combinations including wind, earthquake, and horizontal fluid or soil pressure generate in-plane shear, axial, and bending actions in diaphragms as they span between, and transfer forces to, vertical elements of the lateral-force-resisting system. For wind loading, lateral force is generated by wind pressure acting on building cladding, and is transferred by diaphragms to the vertical elements. For earthquake loading, inertial forces are generated within the diaphragm and tributary portions of walls, columns, and other elements, and then transferred

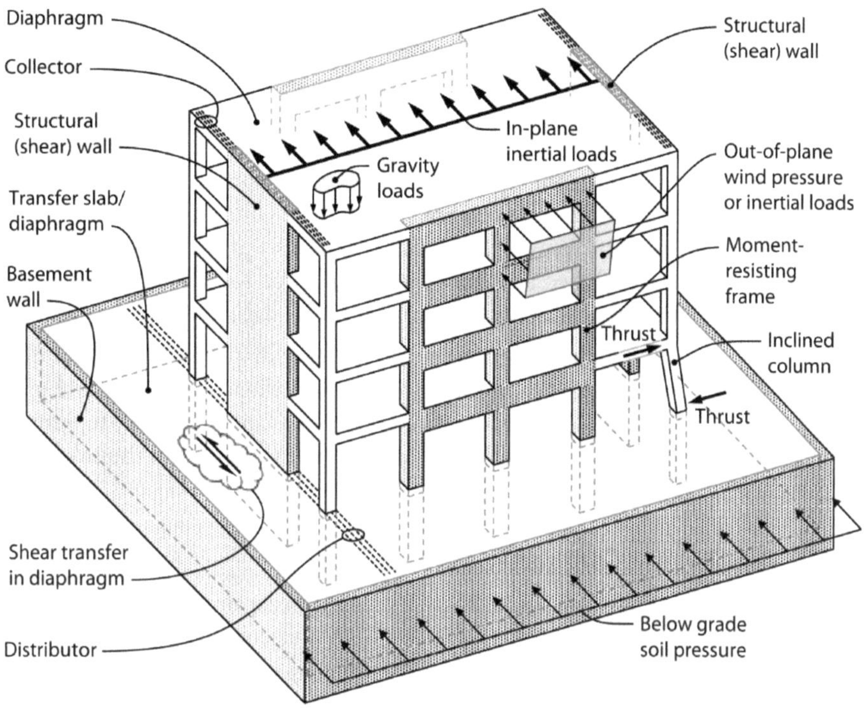

Figure 15.2 Roles of diaphragms.

by diaphragms to the vertical elements. For buildings with subterranean levels, lateral forces are generated by soil pressure bearing against the basement walls; in a typical system, the basement walls span vertically between floors also serving as diaphragms, which in turn distribute the lateral soil forces to other force-resisting elements.

- *Diaphragm transfer forces:* Vertical elements of the lateral-force-resisting system may have different properties over their height, or their planes of resistance may change from one story to another, creating force transfers between vertical elements. A common location where planes of resistance change is at the grade level of a building with an enlarged subterranean plan; at this location forces may transfer from the narrower tower into the basement walls through a "podium" diaphragm.
- *Connection forces:* Wind pressure acting on exposed building surfaces generates out-of-plane forces on those surfaces. Similarly, earthquake shaking can produce inertial forces in vertical framing and nonstructural elements such as cladding. These forces are transferred from the elements where the forces are developed to the diaphragm through connections.
- *Column bracing forces:* Architectural configurations sometimes require inclined columns, which can result in large horizontal thrusts acting within

the plane of the diaphragms due to gravity and overturning actions. The thrusts can act in different directions depending on orientation of the column and whether it is in compression or tension. Where these thrusts are not balanced locally by other elements, the forces must be transferred into the diaphragm so they can be transmitted to other suitable elements of the lateral-force-resisting system. Such forces are common and may be significant with eccentrically loaded precast concrete columns that are not monolithic with adjacent framing. The diaphragm also provides lateral support to columns not designed as part of the lateral-force-resisting system by connecting them to other elements that provide lateral stability for the structure.

- *Diaphragm out-of-plane forces:* Most diaphragms are part of floor and roof framing and therefore support gravity loads. The general building code may also require consideration of out-of-plane forces due to wind uplift pressure on a roof slab and vertical acceleration due to earthquake effects.

15.3 Diaphragm Components

Different parts of a diaphragm include the diaphragm slab, chords, collectors (also known as drag struts or distributors), and connections to the vertical elements. These different parts can be identified by considering the load path in a simple diaphragm.

Figure 15.3 illustrates a simplified model of how a diaphragm resists in-plane loads. (See Section 15.5.2 for additional diaphragm models.) In this illustration, a solid rectangular diaphragm spans between two end walls and resists uniformly distributed in-plane lateral loading. We can idealize the diaphragm as a simply supported beam spanning between two supports, with reactions and shear and moment diagrams as shown (Figure 15.3*b*). Bending moment M_u can be resisted by a tension (T_u) and compression (C_u) couple (Figure 15.3*c*). The components at the diaphragm boundary acting in tension and compression are known as the *tension chord* and the *compression chord*, respectively.

If the diaphragm moment is resisted by tension and compression chords at the boundaries of the diaphragm as shown in Figure 15.3*a*, then equilibrium requires that the diaphragm shear be distributed uniformly along the depth of the diaphragm as shown in Figure 15.3*c*. If the shear is uniformly distributed, tension and compression elements called *collectors* are required to "collect" this shear and transmit it to the walls. A collector can transmit all of its forces into the ends of the walls as shown on the right side of Figure 15.4*a*, or the collector can be spread into the adjacent slab as shown on the left side of Figure 15.4*a*. Section 15.5.3 discusses the effective width of a collector spread into a slab.

Figure 15.4*b* illustrates how the collector tension and compression forces are determined. Starting at a free end, the tension or compression force increases linearly as shear is transferred uniformly into the collector. If it is assumed that the collector force is transferred to the wall uniformly along its length, then the collector force variation is as shown along **bc**.

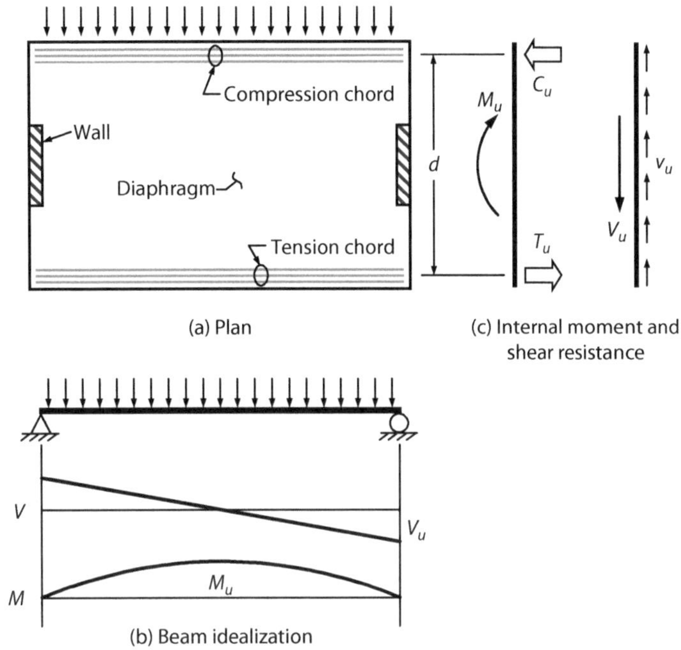

Figure 15.3 Tension and compression chords.

Diaphragms also transfer load among vertical elements of the seismic-force-resisting system. A common example is where a wall intersects a podium slab in a building with subterranean levels. In this case, shear is transferred from the wall into the diaphragm and from there to other elements such as basement walls. The element transferring the force from the wall to the diaphragm is a collector, but sometimes is referred to as a distributor. See Figure 15.5. As used in this text, a *collector* is an element that takes distributed load from the diaphragm and delivers it to a vertical element, whereas a *distributor* takes force from a vertical element and distributes it into the diaphragm.

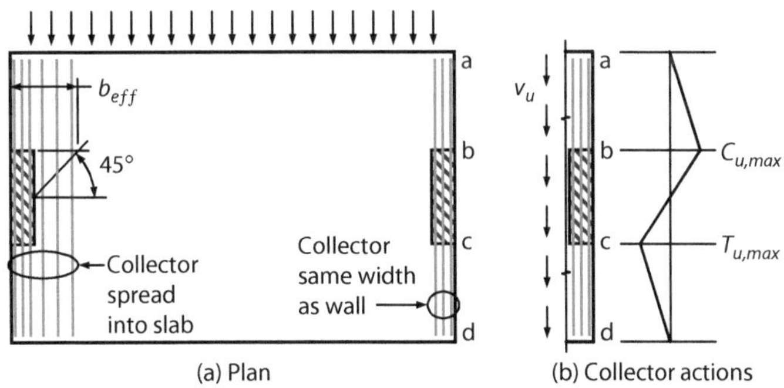

Figure 15.4 Collectors.

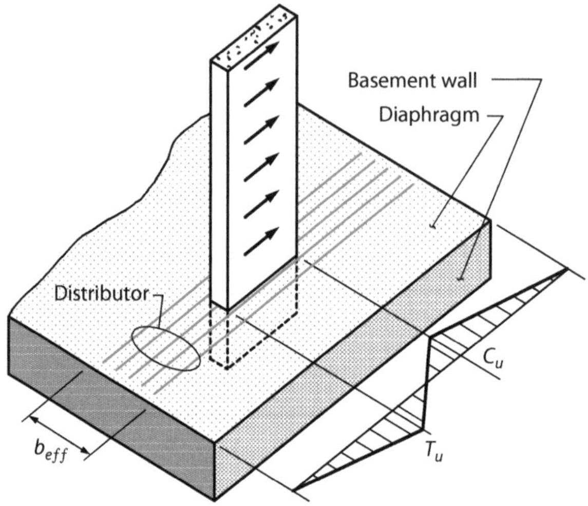

FIGURE 15.5 Distributor at podium transfer level.

15.4 Diaphragm Behavior and Design Principles

15.4.1 Dynamic Response of Buildings and Diaphragms

From fundamental studies of structural dynamics (e.g., Chopra, 2012), we know that the dynamic response acceleration of an oscillator subjected to earthquake ground motion varies with time. For multi-story buildings, the time-varying combinations of multiple vibration modes produce different acceleration histories at each floor. This is illustrated in Figure 15.6, which plots measured responses for a nine-story test structure tested on a shaking table. Instances of peak acceleration and peak base-shear response are identified by open circles. Note that the peak acceleration at each floor is approximately 0.5*g*, but that the peaks occur at different times. Each floor should be designed to resist the inertial force corresponding to the peak response acceleration for that floor. It would be overly conservative, however, to design the vertical elements of the seismic-force-resisting system for the sum of all of the individual peaks, because the peak floor responses occur at different times.

In consideration of the preceding observation, building codes generally contain two different sets of design lateral forces, one for the overall structural system and another for the individual diaphragms. Figure 15.7 illustrates the design lateral forces from ASCE 7 as follows:

1. One set of inertial design forces, F_x, is applied to the design of the vertical elements of the seismic-force-resisting system.
2. A second set of inertial design forces, F_{px}, is applied to the design of the diaphragms.

In general, the diaphragm inertial design forces F_{px} are equal to or greater than the overall system inertial design forces F_x. Methods for determining the design forces

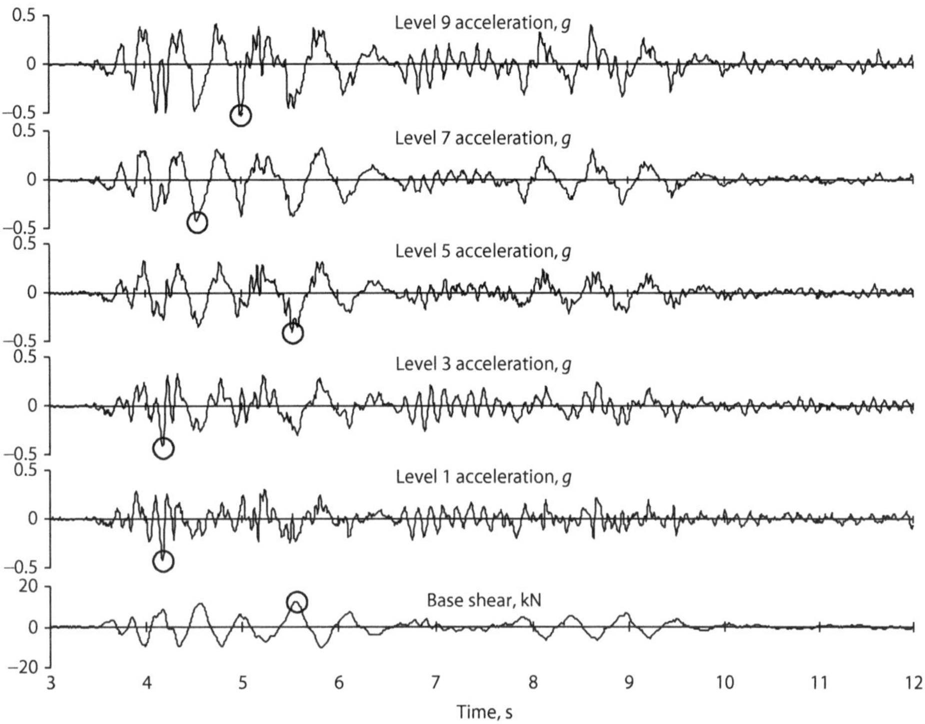

FIGURE 15.6 Measured acceleration and base-shear histories for a nine-story test structure. (*After Moehle and Sozen, 1980.*)

vary depending on whether static or dynamic methods of analysis are used. The interested reader is referred to ASCE 7.

In addition to resisting inertial forces (the product of the mass tributary to the diaphragm and the floor acceleration), diaphragms must also be able to transfer forces between different vertical elements of the seismic-force-resisting system as required by displacement compatibility. For example, frames and walls acting independently have different displacement profiles under lateral loads. If interconnected

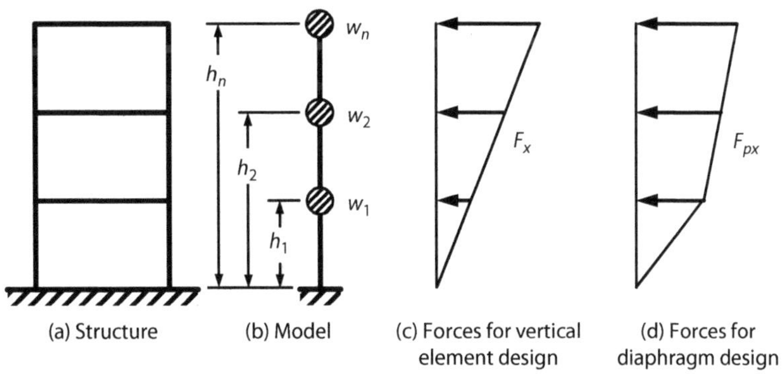

FIGURE 15.7 Design forces for vertical elements and diaphragms.

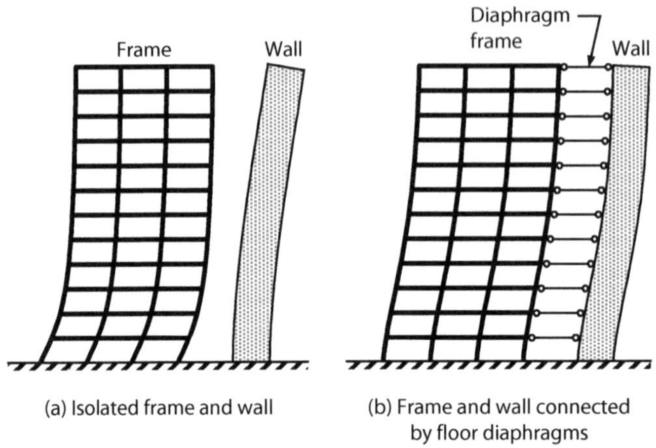

FIGURE 15.8 Diaphragms develop transfer forces by imposing displacement compatibility between different vertical elements of the seismic-force-resisting system.

by a diaphragm, the diaphragm develops internal forces as it imposes displacement compatibility (Figure 15.8). Almost all buildings have force transfers of this type that should be investigated and considered in design. Considering only diaphragm actions due to F_{px} is, in general, not sufficient.

Large diaphragm transfer forces should be anticipated at offsets or discontinuities of the vertical elements of the seismic-force-resisting system. Figure 15.9 shows a building with vertical discontinuities at (a) a setback in the building profile and (b) a podium level at grade. If the diaphragms are modeled as rigid elements in a computer analysis of the building, unrealistically large transfer

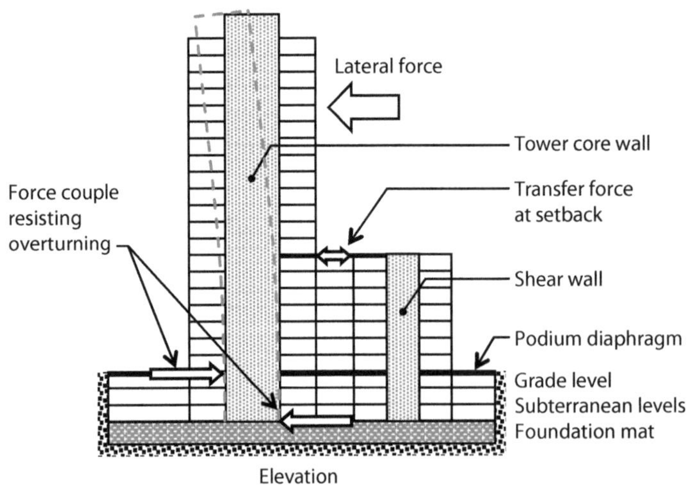

FIGURE 15.9 Diaphragm transfer forces at irregularities in the vertical elements of the seismic-force-resisting system.

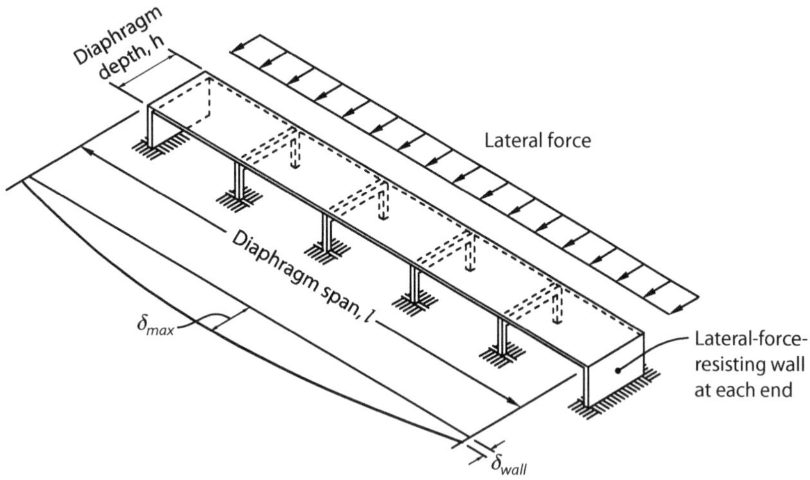

Figure 15.10 Example of diaphragm that might not be considered rigid in its plane.

forces might be calculated at the levels of the discontinuities. At such locations, and perhaps for one or several floors below the discontinuity, more realistic diaphragm forces can be obtained by modeling diaphragm flexibility. Openings in diaphragms should be arranged so that they do not disrupt distributors carrying large transfer forces at these levels.

Where slender diaphragms span between widely spaced lateral-force-resisting elements, the effects of diaphragm deformations on the framing members should be considered. Both the design forces and the design displacements can be affected by diaphragm flexibility (Figure 15.10).

A typical configuration in parking structures uses the diaphragm as parking surface and ramp, with the diaphragm split longitudinally. Other functionality considerations in parking structures typically result in long distances between vertical elements of the seismic-force-resisting system. Consequently, diaphragm segments tend to be relatively long and narrow. Lateral deformations in these flexible diaphragms contribute to dynamic response and can result in diaphragm displacements significantly exceeding displacements of the vertical elements (Fleischman et al., 2002). Design of gravity columns needs to accommodate the increased displacements. In addition, the inclined ramps can act as unintended diagonal braces that interrupt the intended framing action of the vertical elements and result in considerable axial load in the diaphragm. Expansion joints can relieve this action if provided at every level. See SEAOC (2009).

15.4.2 Intended and Observed Behavior

One of the principles of earthquake-resistant design is to maintain a relatively stiff and damage-free diaphragm that is capable of tying together the vertical elements of the seismic-force-resisting system. Thus, diaphragms are designed for essentially linear behavior; that is, minor nonlinearity may be acceptable but significant inelastic response, if it occurs at all, will be restricted to the vertical elements. To achieve this goal, seismic design of a diaphragm should clearly identify the load paths to the vertical elements, and should aim to provide diaphragm strength along that load path at least equal to the maximum force that can be developed by the vertical elements.

Design approaches for cast-in-place diaphragms have been relatively effective in limiting diaphragm damage, with few cases of observed damage following earthquakes. Some connections between diaphragms and shear walls fractured during the 1994 Northridge earthquake (Corley et al., 1996), leading to code changes for collector design. Other types of concrete diaphragms, especially precast diaphragms with or without topping slabs, require greater attention to proportions and details to achieve the goal of essentially elastic behavior.

15.5 Analysis Guidance

15.5.1 Design Lateral Forces

In U.S. practice, seismic design of diaphragms is required for all buildings in SDC B through F. ASCE 7 contains the main provisions. The design must consider lateral seismic forces F_x, diaphragm design forces F_{px}, and any transfer forces associated with response under the design seismic loading.

The lateral seismic forces F_x are determined in the analysis of the vertical elements of the seismic-force-resisting system (Figure 15.7*c*). These forces typically are determined from either an equivalent lateral force or modal response spectrum analysis, and represent the overall building design lateral force distribution, the sum of which results in the total design base shear V_b.

As discussed in Section 15.4.1, the lateral seismic forces F_x do not necessarily reflect the estimated maximum force induced at a particular diaphragm level. Thus, building codes also require the diaphragm to be designed for the diaphragm design force F_{px} (Figure 15.7*d*). Associated design requirements typically are evaluated by applying F_{px} to one floor at a time rather than all floors simultaneously, using either simplified models (Section 15.5.2) or the overall building model. Sabelli et al. (2009) discuss approaches to diaphragm analysis considering the overall building model.

Diaphragms must also be designed to resist the transfer forces that develop due to framing interaction among different vertical elements of the seismic-force-resisting system, as well as any other forces such as those induced by hydrostatic pressures and sloping columns. Where diaphragms are modeled as semi-rigid finite elements, section cuts through the diaphragm can be used to determine the forces transferring between the vertical element and diaphragm. Where diaphragms are modeled as rigid, this approach cannot be used. Instead, section cuts through the vertical elements above and below the diaphragm can be made; the force transferred between the diaphragm and wall is the difference in these forces (Figure 15.11). This procedure works directly for equivalent lateral force and seismic response history procedures. When modal response spectrum analysis

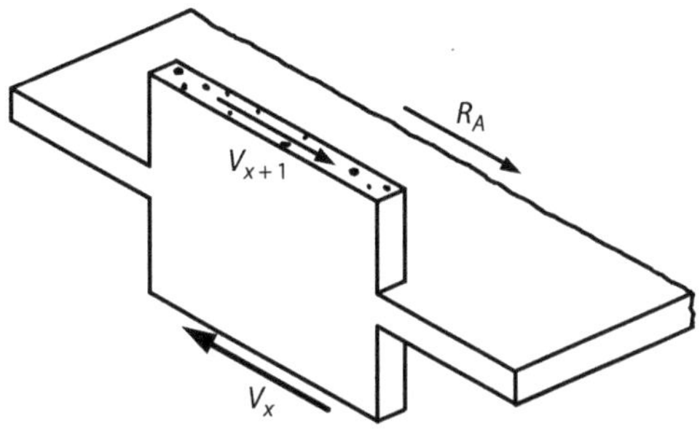

FIGURE 15.11 Section cuts through vertical elements to determine forces transferred between diaphragm and vertical element.

is used, the transfer force for each vibration mode must be determined, and then the design value can be calculated by combining the individual modal values using the SRSS or CQC methods.

Failure of connections between diaphragms and walls in the 1994 Northridge earthquake (Corley et al., 1996) triggered code changes to increase the design forces for collectors, with the intent that they be designed to be a strong link in the load path. According to ASCE 7, for structures assigned to SDC C through F, collectors including splices and connections to resisting elements are required to resist load combinations that include an overstrength factor Ω_0 (see Section 11.5). Codes may also impose minimum and maximum design forces.

When design actions are determined using linear response spectrum or response history analysis, the resulting accelerations can be used directly to determine diaphragm inertial design forces F_{px}. According to U.S. practice, the accelerations should be scaled by I_e/R. (If forces are measured directly from the finite element analysis, it is not always clear how to scale the results, as the ability to separate transfer forces from inertial forces can be compromised.) Numerical and laboratory studies (Rodriguez et al., 2007) show that floor accelerations and associated diaphragm actions are underestimated if linear-elastic response quantities are scaled by I_e/R. Better correlation is obtained by using modal spectral response combinations in which only the first-mode responses are scaled by I_e/R.

When design actions are determined using nonlinear response history analysis, diaphragm accelerations and the resulting forces can be determined directly from the analysis. It may be acceptable to use these forces directly for the design of collectors and their connections as an alternative to using Ω_0-amplified forces, provided the analysis and design approach is established so as to achieve the intent of the code that the collector not be the weak link in the load path. The model should represent the expected material strengths, rather than the specified strengths, and dispersion in the calculated results should be considered in determining the design value, U. The design strength of the collector itself should be based on the conventional approach, that is, $U \leq \phi S_n$. By appropriate selection of the design demands and strengths, an acceptably low probability of failure can be achieved.

Capacity-based design is another way to determine diaphragm design forces. This approach uses the maximum force that can be delivered to a diaphragm by the framing system as the design force, and the reliable resistance as the design strength. Engineers should consider expected material properties, multiple failure mechanisms, multiple load patterns, and appropriate strength calculation procedures so that the resulting demands and capacities safely cover the range of combinations that reasonably can be expected.

It is common practice to model reinforced concrete diaphragms as rigid in an analysis model if the span-to-depth ratio is less than or equal to 3 and if there are no significant horizontal irregularities. In all other cases, the flexibility of the diaphragm should be considered. Stiffness reduction associated with diaphragm cracking is commonly approximated by applying a stiffness modifier to the diaphragm in-plane gross-section stiffness properties. Stiffness modifiers for reinforced concrete diaphragms commonly fall in the range of 0.15 to 0.50 when analyzing the building for code-level earthquake demands. In some cases, it may be prudent to "bound" the solution by analyzing the structure using both the lower and upper range of diaphragm stiffnesses, and selecting the design values as the largest forces from the two analyses.

The recommendations of the preceding paragraph are considered to represent good practice. Regardless, ACI 318 permits the diaphragm to be modeled using any set of reasonable and consistent assumptions. Thus, the rigid diaphragm model is commonly used for cast-in-place diaphragms regardless of aspect ratio.

15.5.2 Diaphragm Modeling and Analysis Approaches

Internal forces in a diaphragm can be calculated using approaches that range from simple hand analysis to complex computer analysis. The analysis needs only to be as complex as necessary to represent how lateral forces flow through the building including the diaphragms. For regular buildings in which lateral resistance is provided by similar vertical elements distributed throughout the floor plan, simple models are often adequate for determining the diaphragm forces. For buildings with irregularities or with dissimilar vertical elements, significant force transfers may occur among the vertical elements at various levels, requiring more complex models to determine the diaphragm design forces.

For smaller buildings with regular geometries, two lines of vertical elements in a given direction, and continuous vertical elements from foundation to roof, a simple model such as the *equivalent beam model* can be suitable for determining design forces. If three or more lines of vertical elements are present in a given direction and there are no major discontinuities in the vertical elements, the *equivalent beam-on-springs model* may be appropriate for determining diaphragm demands. For buildings with torsion or significant force transfers between vertical elements, it may be necessary to analyze a complete model of the building to adequately identify the forces acting between the diaphragm and vertical elements. These forces can be used with the *corrected equivalent beam model* to assess diaphragm design actions. Finally, any building with diaphragm discontinuities may require more complex models such as a *finite element model* or *strut-and-tie model*.

Equivalent Beam Model

This model treats the diaphragm as a horizontal beam spanning between idealized rigid supports. The rigid supports represent vertical elements such as shear walls. Figure 15.3 is a representation of the equivalent beam model. For the case shown, the beam is simply supported, as the walls are at the far ends of the diaphragm. If the walls were inboard of the diaphragm edges, the beam representing the diaphragm cantilevers beyond the supports. Shear and moment diagrams are established for the equivalent beam by the usual methods.

Equivalent-Beam-on-Springs Model

The equivalent-beam-on-springs model envisions the diaphragm as a beam supported by flexible supports (Figure 15.12). It is most suitable in buildings for which force transfers are minimal, such as single-story buildings or multi-story buildings with similar vertical elements. For this model, the diaphragm may be treated either as a rigid beam or as a beam with flexural and shear stiffness properties. The stiffness of each spring is a representation of the stiffness of the supporting vertical element. A simple computer model or hand calculations may be used to determine the shear and moment diagrams. For a multi-story building with dissimilar vertical elements, such that significant force transfers occur through the diaphragm, the equivalent-beam-on-springs model may not be suitable. In such cases, a simple

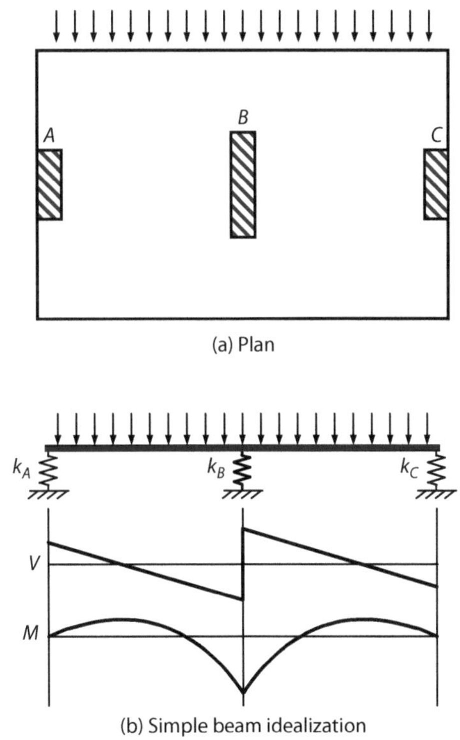

Figure 15.12 Equivalent-beam-on-springs model.

computer model with beam elements to represent the diaphragms and vertical elements, or a complete model of the building, may be suitable alternatives.

Corrected Equivalent Beam Model

The corrected equivalent beam model can be used to approximate diaphragm actions where there is significant interaction among vertical elements of the seismic-force-resisting system. Such effects may occur where vertical elements of different stiffness interact or where vertical irregularities or building torsion occur. When a computer analysis is done, the forces transferred to the diaphragm at a vertical element can be obtained by appropriate section cuts. For smaller buildings without irregularities, the reactions may be determined using the direct inertial force, F_x (or F_{px}), with due consideration of the effects of plan torsion. Referring to Figure 15.13, the diaphragm forces to the vertical elements are computed as follows:

$$R_i = F_x \frac{k_{ix}}{\sum k_{ix}} + F_x e_x \frac{e_i k_i}{J_r} \tag{15.1}$$

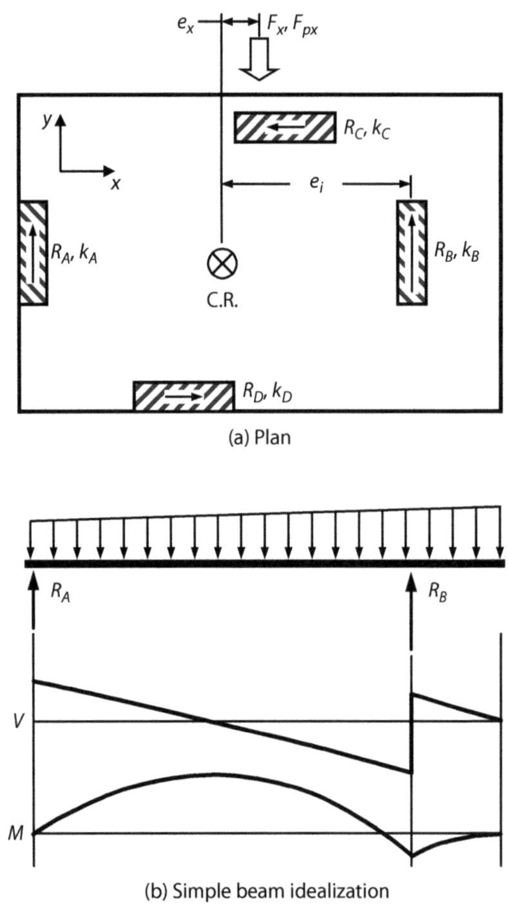

FIGURE 15.13 Corrected equivalent beam model.

where R_i is the force acting between the diaphragm and vertical element i, F_x is the story force, k_{ix} is the stiffness of vertical element i in the x direction, e_x is the perpendicular distance between the center of rigidity and the center of mass, e_i is the perpendicular distance between the center of rigidity and the stiffness k_i of vertical element i, and J_r is the polar moment of inertia computed as follows:

$$J_r = \Sigma e_i^2 k_i \tag{15.2}$$

To approximate the actions within the diaphragm, the forces R_i acting between the diaphragm and the vertical elements in the direction under consideration are summed (in Figure 15.13, this would be $R_A + R_B = F_x$) and their centroid is determined. For a rectangular diaphragm of uniform mass, a trapezoidal distributed force having the same total force and centroid is then applied to the diaphragm. The resulting shears and moments (Figure 15.13*b*) are acceptable for diaphragm design.

Finite Element Model

Finite element modeling of a diaphragm can be useful for assessing the force transfer among vertical elements, force transfer around large openings or other irregularities, and behavior of ramps in parking garages. Figure 15.14 shows an example of an irregularly shaped diaphragm that warrants use of finite element modeling. To adequately model the diaphragm flexibility, finite element meshing typically needs to be 1/10 to 1/5 of the bay length or wall length. If section cuts are made through the diaphragm model to determine the shear distribution within the diaphragm, the finite element mesh at and near the section cut should be moderately fine. Stiffness may be reduced to account for cracking effects (see Section 15.5.1).

Strut-and-Tie Model

Strut-and-tie models can be used to idealize the flow of force through a diaphragm. Such models have not been used extensively for the overall design of a diaphragm, although they can be useful for this purpose. Instead, strut-and-tie models are

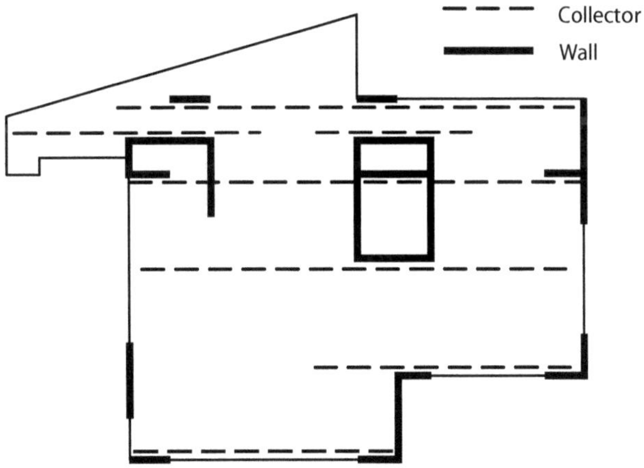

Figure 15.14 Irregularly shaped diaphragm. (*After Mohr and Harris, 2009.*)

more often used to identify force paths and reinforcement layouts around discontinuities. Where used, distributed reinforcement to control cracking should be at least 0.0025 times the gross slab area in each direction.

15.5.3 Idealized Load Paths within the Diaphragm

Flexural Behavior

Diaphragms typically are designed using classical beam theory assuming plane sections remain plane even though the proportions may be more like those of a deep beam. Traditionally, moment demands are resisted by tension and compression chords located close to opposite outer edges of the diaphragm. The chord forces, C_u in compression and T_u in tension, are computed as

$$C_u = T_u = M_u/d \tag{15.3}$$

Using this approach, the in-plane shear stress is uniform across the depth of the diaphragm, with value V_u/td.

Some codes permit the use of distributed reinforcement to resist the diaphragm moment. If this is done, moment strength is calculated using the traditional approach in which strain varies linearly through the depth, with stresses appropriately corresponding to strains. With distributed reinforcement, development of moment strength may require large tensile strains and potentially unacceptable cracking near the tension edge. A good rule of thumb is that the required flexural tension reinforcement should be located within the outermost quarter of the diaphragm depth, or there should be boundary reinforcement plus distributed reinforcement. See Figure 15.15.

If distributed reinforcement away from a diaphragm edge is used to resist moment, the unit shear stress is not constant through the diaphragm depth but

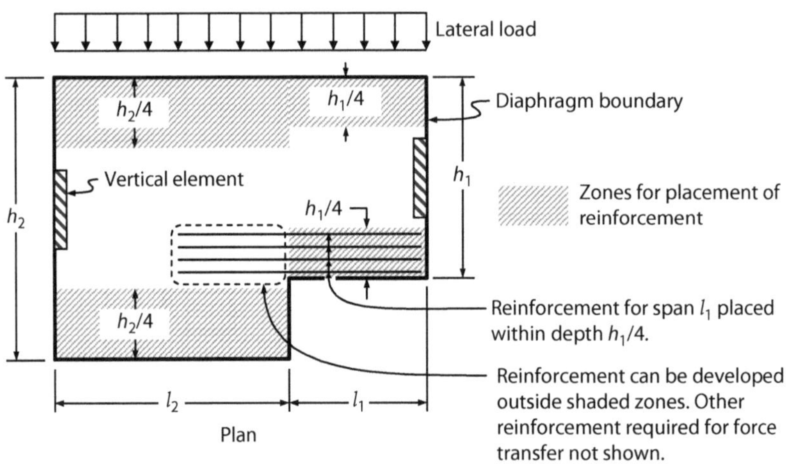

Figure 15.15 Recommended locations of nonprestressed reinforcement resisting tension due to moment and axial force.

instead varies gradually through the depth with a peak value exceeding V_u/td. For example, with uniformly distributed flexural tension reinforcement, the peak diaphragm shear stress approaches $2V_u/td$. The diaphragm should be designed for higher shear stresses where they occur. If chord reinforcement is distributed within the outermost quarter of the diaphragm depth, as recommended here, it is acceptable to assume uniform shear stress of V_u/td.

Collectors

Collectors are tension and compression elements that gather (collect) shear forces from diaphragms and deliver them to the vertical elements. Collectors also deliver forces from vertical elements into the diaphragm as shown in Figure 15.5. This latter type of collector, referred to as a distributor, is required where forces are redistributed among vertical elements such as at a podium level (Figure 15.2). Collectors can be in the form of beams or a zone of reinforcement within a slab (Figure 15.4*a*). Wide sections of slabs used as collectors are referred to as distributed collectors.

In some cases, tension and compression collectors can be fit within the width of the vertical element, in which case all of the tension and compression force is transferred into the vertical element at the boundary. In this case, only the diaphragm shear along the length of the vertical element needs to be transferred through shear-friction. In other cases, the collector will be wider than the width of the vertical element, such that only part (or none) of the collector force is transferred directly into the boundary, with the rest being transferred through shear-friction to the side of the vertical element. In the latter case, shown in Figure 15.16, the collector is defined to include the compression portion (points **a** to **b**), the tension portion (points **c** to **d**), and the shear transfer portion along the length of the wall (points **b** to **c**).

There are no building code requirements governing the effective width b_{eff} of a collector (Figure 15.16). SEAOC (2005) suggests b_{eff} of the collector should not

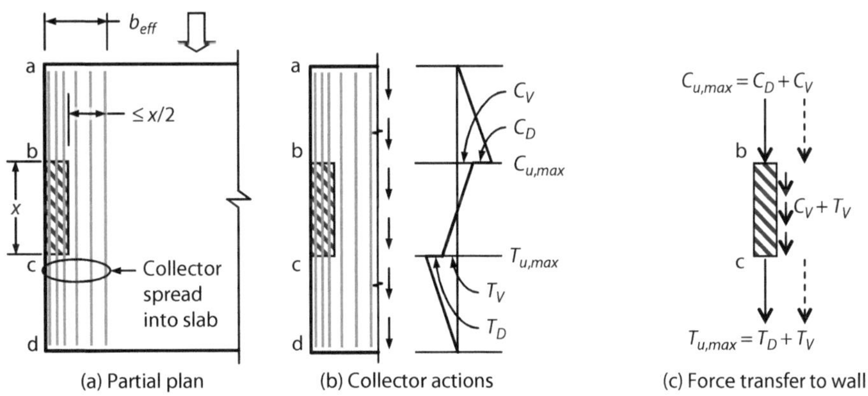

Figure 15.16 Force transfer where a collector is wider than the vertical element to which it transmits diaphragm shears. In (*b*), the collector actions are shown to have a step function at the wall boundary for the purpose of illustrating the force transfer directly into the wall, even though the force transfer into the wall must occur gradually.

exceed the vertical element width plus a width on both sides of the vertical element equal to half the contact length between the diaphragm and the vertical element (Figure 15.16*a*).

Collectors may be designed extending across the full depth of a diaphragm or may extend for only a portion of the depth (Sabelli et al., 2009). *Full-depth collectors*, as shown in Figures 15.4*a* and 15.16, are required where the full depth of the diaphragm is used to resist diaphragm moment because in this case diaphragm shear exists through the full depth.

Partial-depth collectors (Figure 15.17) rely on slab reinforcement provided for gravity loads to collect diaphragm shear, such that shear stresses are more concentrated near the vertical element rather than being distributed through the diaphragm depth. This leads to a shorter collector length and smaller collector forces. In some cases, all of the force might theoretically be transferred directly to the wall without a collector, but to control cracking near the ends of the wall it is recommended to extend the collector into the diaphragm at least a bay width or 25 ft (8 m), whichever is greater.

Design of partial-depth collectors involves establishing a load path within the diaphragm that optimizes use of existing slab reinforcement and added collector reinforcement. The load path is similar to that of a dapped beam. In both cases, reinforcement is required at the end of the full-depth section that collects the shear and transfers it to the reduced depth section at the end of the member. In a diaphragm, this reinforcement comprises distributed slab reinforcement referred to as a *secondary distributed collector*. The combination of the collector and secondary distributed collector sometimes is referred to as a *distributed collector*.

For the partial-depth collector shown in Figure 15.17, the following design considerations apply. Near the diaphragm midspan, the full depth of the diaphragm

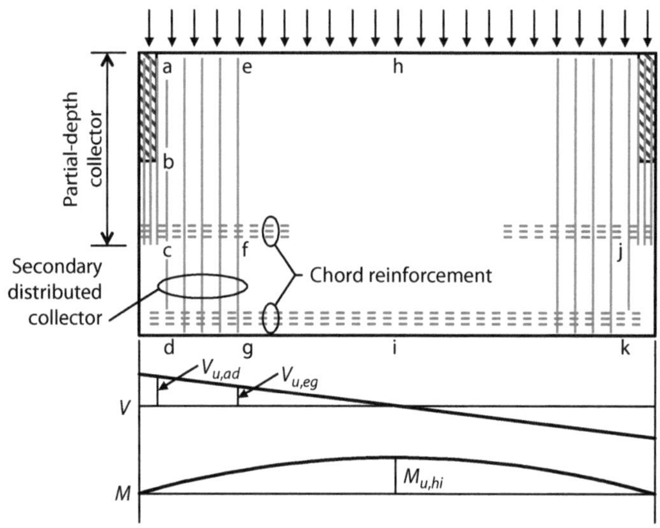

FIGURE 15.17 Partial-depth collector.

is used to resist diaphragm moment and shear. Thus, at midspan the required chord tension force is $T_u = M_{u,hi}/d$. Along **eg**, the diaphragm must be sufficient to transfer uniform shear stress $v = V_{u,eg}/d$. The secondary collector reinforcement must be sufficient to transmit in tension the shear picked up along **fg** plus additional diaphragm inertial forces in region **cdfg**. Along **ac**, the diaphragm reinforcement must be sufficient to transfer $V_{u,ad}$ as uniformly distributed shear. Partial-depth collector reinforcement along **bc** must be sufficient to carry the shear picked up along **bc**. And, finally, chord reinforcement along **cf** must be capable of resisting the diaphragm moments along that length assuming the effective depth of the diaphragm is reduced to the length **ef**. (Note that if the secondary distributed collector was not included in the design, the effective depth of the diaphragm for moment calculations would be reduced to the length **ef** for the entire diaphragm span, requiring larger area of chord reinforcement, and distributed steel would still be required in the region **cdjk** to transmit inertial loads developed in that region to the shallower effective beam of depth **ef**. Treating the diaphragm as a shallower beam of depth **ef** also could result in wide cracks forming at the extreme tension edge of the diaphragm as it is flexed under lateral load.)

Force Transfer to Vertical Elements

Diaphragm shear is transferred to vertical elements by tension and compression acting on the boundaries of the vertical element and by shear-friction acting along the length of the vertical element. Where collector reinforcement enters a vertical element, the collector reinforcement must extend into the vertical element a length sufficient to transfer the force to the parts of the vertical element. For example, in Figure 15.18, each column resists one-third of the total collector force. Therefore, one-third of the collector reinforcement can be cut l_d past the first column, another one-third can be cut l_d past the second column, and the remaining one-third must be developed at the last column. Collectors that extend the entire length of a vertical element, but not less than l_d, ensure that the force is transferred from the collector to the element without further consideration.

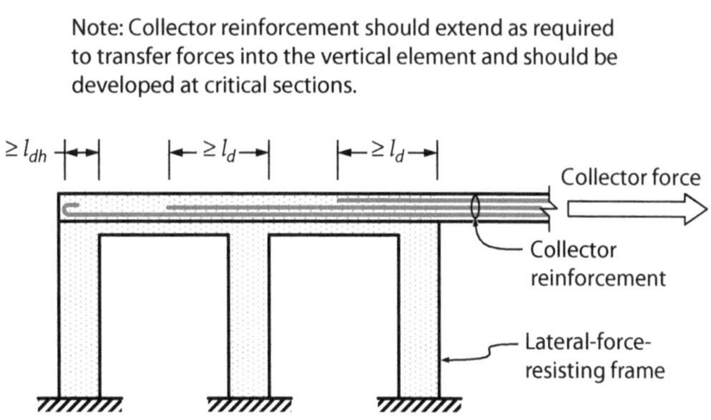

FIGURE 15.18 Transfer of collector force directly to shear wall.

If the collector is wider than the vertical element such as shown in Figure 15.16, portions of the collector force, C_D and T_D, may be transferred directly to the vertical element boundaries and the remaining portions, C_V and T_V, are transferred to the vertical element by shear-friction along the length of the vertical element. This portion of the collector force is represented in Figure 15.16*c* by the distributed force along the length of the wall. The design shear-friction force is the combination of the collector force along the vertical element, which is amplified by the overstrength factor, Ω_0, and the direct diaphragm shear force, v_u, along the length of the vertical element. Slab reinforcement perpendicular to the vertical element is typically added to serve as this shear-friction reinforcement.

Large Openings

For a diaphragm with small openings (on the order of one or two diaphragm thicknesses for typical diaphragms), common practice is to place reinforcement on either side of the opening having cross-sectional area equal to the area of reinforcement disrupted by the opening, with no other special analysis. For a larger opening, the diaphragm must be designed to transfer the forces around the opening.

Design of a diaphragm with a large opening is analogous to the design of a beam with an opening. Consider the diaphragm in Figure 15.19. One approach is to assume the reinforcement labeled **L** redistributes uniform shear left of the opening to the portions of the diaphragm above and below the opening in proportion with their relative stiffness. The reinforcement labeled **R** distributes the shear from above and below the opening to a uniform shear to the right of the opening. The reinforcement labeled **T** and **B** resists the local moment within the section above and below the opening. This moment is sometimes approximated as $V_T l/2$ (or $V_B l/2$), which is correct if the inflection point is at the center of the length of the opening. It is prudent to assume that the inflection point may vary, which will increase the moment. If a finite element analysis is being used, section cuts can identify local forces and a hand analysis approach can be used to check the results.

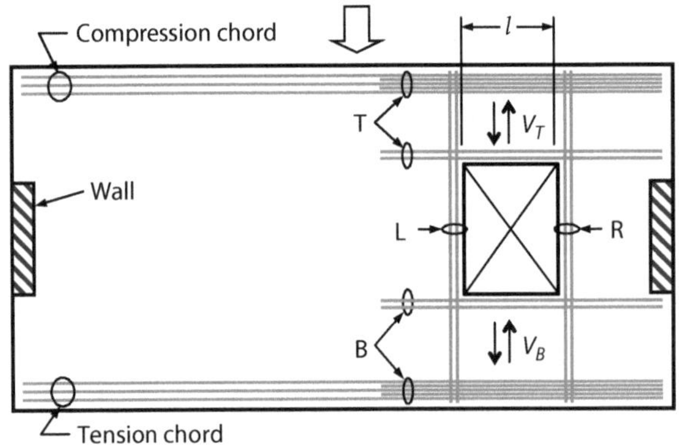

Figure 15.19 Diaphragms with large openings. (*Drawing by D. Kelly, from Moehle et al., 2010.*)

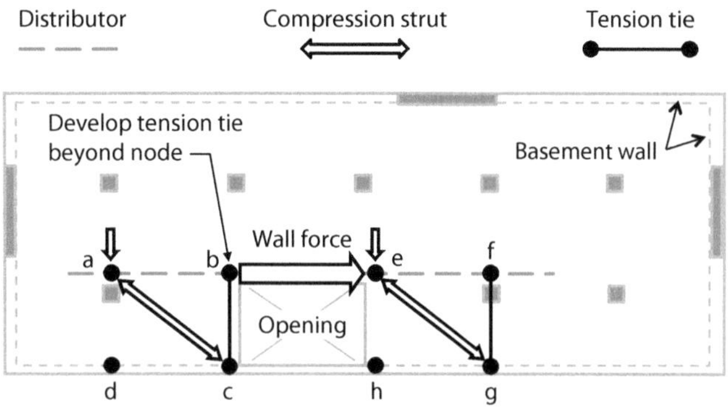

FIGURE 15.20 Strut-and-tie model of the force transfer between a basement wall and a diaphragm with an opening.

Where collector/distributor forces have to be transferred around large openings, strut-and-tie models can be used to inform reinforcement layouts. Consider the example of Figure 15.20, in which force from a wall is delivered to a diaphragm adjacent to an opening and must be transferred to nearby basement walls. Distributor reinforcement is provided to transfer the force to the diaphragm. The force transfer to the basement walls can be visualized as occurring through compression struts acting at an angle between about 30° and 60° relative to the wall force. Considering **abcd** as a free-body diagram, moment equilibrium about **d** requires tension tie **bc**, which must be developed into the adjacent diaphragm segment. Moment equilibrium about **c** could not be provided by a tension tie from **a** to **d** because the tension tie would have to be anchored to the basement wall out of plane, which is inappropriate; instead, moment equilibrium about **c** is provided by a compressive force from the adjacent diaphragm segment at **a**. (Force reversal, not shown, would reverse the diagonal compression struts and require tension ties at **ad** and **eh**.) Section 15.6.6 provides additional discussion on how to detail the required reinforcement.

15.5.4 Displacement Compatibility for Flexible Diaphragms

Flexible diaphragms will experience mid-span displacements due to inertial loading in addition to the drift experienced by the vertical elements of the seismic-force-resisting system (Figure 15.10). Components not designated as part of the seismic-force-resisting system, such as gravity beams and columns, out-of-plane wall bending, slab-column and slab-wall connections, and cladding attachment, should be evaluated for displacement compatibility based on the additional displacement of the diaphragm. In some cases, it may be appropriate to include critical elements of the gravity system in the building lateral model to explicitly evaluate forces developed due to displacement compatibility.

15.5.5 Ramps

Ramps and sloping diaphragms can create unique challenges when evaluating design forces as they may connect between different stories of a structure. In some cases, story shear can migrate out of the vertical elements of the seismic-force-resisting system through the ramp in the form of shear or axial forces. This additional force should be considered in the design of sloping diaphragms. Engineering practice varies with respect to how to treat these conditions in an analysis model. Idealizing a sloping diaphragm as a flat, continuous element might not correctly identify such forces, and might lead to overstating the stiffness of the diaphragm at a particular location. The potential implications of the modeling assumptions for ramps should be considered when determining whether to explicitly include sloping diaphragms in an analysis model. For additional guidance, see SEAOC (2009).

15.5.6 Diaphragm Slabs-on-Ground

Slabs-on-ground sometimes are used as diaphragms to redistribute horizontal earthquake force from vertical elements. This is done where the foundation directly below a vertical element does not have adequate soil friction and passive bearing resistance for the horizontal force at the base of the vertical element. The slab-on-ground diaphragm redistributes some of the horizontal load to locations where additional resistance to sliding is obtained. Friction below the slab-on-ground diaphragm and below other foundation elements, as well as passive bearing on these other foundation elements, provides the added resistance to sliding. See Chapter 16 for discussion of foundation design.

Slabs-on-ground that serve as diaphragms are considered structural slabs. Although these slabs typically do not need to be reinforced for the moment caused by loads on the surface of the slab, they must be reinforced for the in-plane shear and moment. These slabs must also meet the minimum reinforcement requirements for a structural slab.

15.6 Design Guidance

15.6.1 Load and Resistance Factors

Diaphragms are designed for load combinations that include effects of horizontal seismic forces, vertical seismic forces, dead load, variable portions of live load, and other loads such as soil pressures, snow, and fluids. See Chapters 1 and 11 for load combinations.

In general, diaphragms and collectors are permitted to be designed for seismic forces applied independently in each of the two orthogonal directions. Some codes (e.g., ASCE 7), however, require consideration of the interaction of orthogonal effects for structures having non-parallel systems or certain types of plan irregularities. Where an equivalent lateral force or modal response spectrum analysis is used, the common approach is to combine 100% of the effects in one primary direction with 30% of the effects in the other direction. If a response-history analysis is performed, orthogonal pairs of ground motion histories are applied simultaneously. Although not generally required by the building code, common practice is to consider orthogonal combinations for all diaphragm and collector designs. This approach is adopted here.

Once the design forces are determined, reinforced concrete diaphragms and their connections must be designed to resist all shears, moments, and axial forces, including effects of openings and other discontinuities. In U.S. practice, diaphragms in buildings assigned to SDC D through F must satisfy the provisions of ACI 318 Chapter 18 (*Earthquake-Resistant Structures*). For buildings assigned to SDC B or C, the general "non-seismic" requirements of ACI 318 apply, including the requirements of Chapter 12 (*Diaphragms*).

To reduce the likelihood of inelastic shear behavior in diaphragms, the strength reduction factor ϕ for diaphragm shear should not exceed ϕ used for shear design of the vertical elements of the seismic-force-resisting system. ACI 318 requires this for structures assigned to SDC D, E, or F, or special moment frames or special structural walls in any SDC. For example, if all the vertical elements of the seismic-force-resisting system are shear walls that use a value of $\phi = 0.75$ for shear, the value of ϕ for diaphragm shear design also is 0.75; if the shear walls use a value of $\phi = 0.6$ for shear, as is required if capacity design concepts are not applied in the wall design, then the value of ϕ for diaphragm shear design is also 0.6.

15.6.2 Chord Longitudinal and Confinement Reinforcement

Section 15.5.3 described the calculation of chord forces when simplified beam models are used to approximate diaphragm internal forces. Where nonprestressed reinforcement is concentrated near the edge of the diaphragm, the reinforcement area of the tension chord is calculated as $A_s = \frac{1}{\phi}\frac{T_u}{f_y}$, where $\phi = 0.9$. Typically the chord reinforcement is placed within the middle third of the slab or beam thickness, so as to minimize interference with slab or beam longitudinal reinforcement and reduce contributions to slab and beam moment strength.

Where chord reinforcement is positioned within a special moment frame beam, the chord and the beam typically are oriented to resist orthogonal effects, such that the same reinforcing bars can resist moment for loading in one direction and chord tension for loading in the orthogonal direction. Where orthogonal effects are combined using the 100%/30% combination rule, with the beam in the X direction, the longitudinal reinforcement is the larger of that required to resist actions due to the combinations (a) $1.0X + 0.3Y$ and (b) $0.3X + 1.0Y$. This approach complicates the design process, but results in a more efficient design with lower system overstrength than would occur if the reinforcement areas were determined independently and simply added. Note also that, if the beam also serves as a collector, the beam has to be proportioned taking the tension and/or compression into account.

In U.S. practice, bonded tendons may be used as reinforcement to resist collector forces, diaphragm shear, or diaphragm flexural tension provided they are proportioned such that the stress due to design earthquake forces does not exceed 60,000 psi (414 MPa). Unbonded, unstressed tendons are not permitted to resist collector, shear, or flexural tension because of concerns about wide cracks and excessive diaphragm flexibility resulting from earthquake loads. Pre-compression from unbonded tendons is permitted to resist diaphragm forces if a seismic load path is provided.

The use of pre-compression warrants additional discussion. In the typical case of a prestressed floor slab, the prestressing is required, at a minimum, to resist the factored load combination $1.2D + 1.6L$, where D is dead load and L is live load. L may have been reduced as permitted by the general building code. For wind or earthquake design, however, the gravity load to be resisted by prestressing is reduced because the governing load combination is $1.2D + f_1L + (W \text{ or } E)$, where f_1 is either 1.0 or 0.5 depending on the nature of live load L. Thus, only a portion of the effective prestress is required to resist the reduced gravity loads. The remainder of the effective prestress can be used to resist in-plane diaphragm moments. Additional moment, if any, is resisted by added reinforcement. Expressed algebraically, the percentage of prestressing available as pre-compression is

$$\left[1-\left(1.2+\frac{f_1}{D/L_{red}}\right)\Bigg/\left(1.2+\frac{1.6}{D/L_{red}}\right)\right]\times 100 \tag{15.4}$$

For the case of $D = 120$ psf (5.75 kPa), $L = 40$ psf (1.92 KPa) reducible by 40%, and minimum permitted prestress of 125 psi (0.86 MPa), the pre-compression stress f_{pc} available to resist earthquake effects is 14.5 psi (0.10 MPa). If prestressing is higher than 125 psi (0.86 MPa), as is common, higher pre-compression stress will be available. The diaphragm in-plane moment that can be resisted by the pre-compression is calculated using the diaphragm gross-section properties, i.e., $M = f_{pc}S_m$. Additional moment, if any, must be resisted by bonded reinforcement.

Experience suggests that compression chords do not, in general, require confinement reinforcement, with one exception. High axial compressive stresses can develop in struts that transfer diaphragm forces due to shear or moment around openings or other discontinuities (e.g., Figure 15.21). According to ACI 318, if the

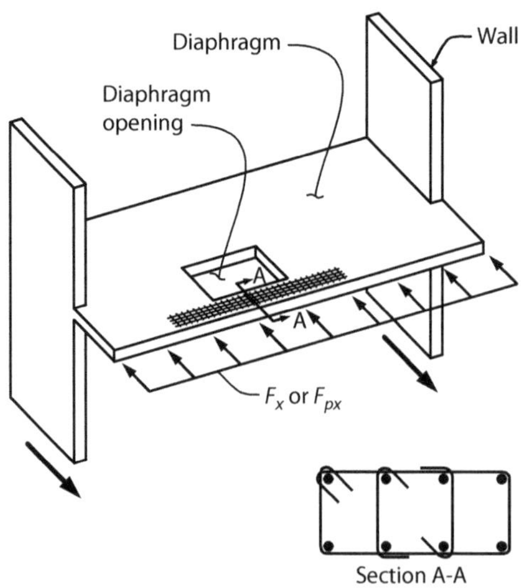

Figure 15.21 Confinement reinforcement in axial struts around openings. (*Used with permission from American Concrete Institute.*)

calculated compressive stress on a strut exceeds $0.2f_c'$, confinement reinforcement satisfying requirements for special boundary elements of special structural walls is required. The required confinement reinforcement, including hoops with required seismic hook dimensions, can be difficult to fit within typical slab depths, and may require increased depth. Where required, confinement reinforcement should be continued into the slab beyond the strut at least the tension development length of the longitudinal reinforcement but not less than 12 in (300 mm).

15.6.3 Diaphragm Shear Strength

Every section of a diaphragm must have a design shear strength per unit length not less than the factored shear force per unit length. For diaphragms having chord reinforcement located near the extreme flexural tension edge of the diaphragm, the factored shear V_u is uniformly distributed through the diaphragm depth. The design shear strength in this case is given by ϕV_n, where $\phi = 0.6$ or 0.75 as discussed in Section 15.6.1, and V_n is given by

$$V_n = A_{cv}(2\lambda\sqrt{f_c'} + \rho_t f_y), \text{ psi} \left[A_{cv}\left(\frac{\lambda\sqrt{f_c'}}{6} + \rho_t f_y\right), \text{MPa} \right] \tag{15.5}$$

in which uniformly distributed slab reinforcement of steel ratio ρ_t is positioned perpendicular to the diaphragm moment reinforcement (i.e., parallel to the shear force) (Figure 15.22). Cross-sectional dimensions must be selected to satisfy Eq. (15.6).

$$V_u \le 8\phi A_{cv}\sqrt{f_c'}, \text{ psi } [0.66\phi A_{cv}\sqrt{f_c'}, \text{MPa}] \tag{15.6}$$

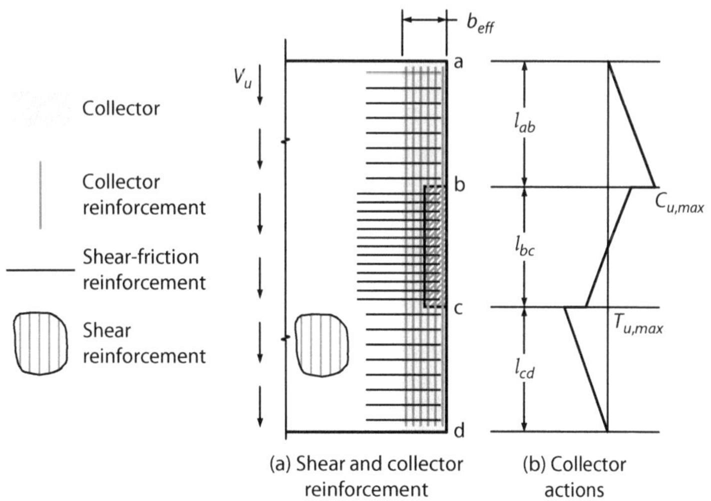

FIGURE 15.22 Shear reinforcement, collector reinforcement, and shear-friction reinforcement.

Where the factored shear stress varies throughout the section and is given on a unit basis (e.g., kips/in), the design unit shear strength, calculated by substituting the unit area for A_{cv} in Eq. (15.5), must exceed the factored unit shear at every section.

Shear reinforcement can be placed anywhere within the slab thickness within required cover limits. Some designers specify a continuous mat of bottom reinforcement that satisfies both the diaphragm shear and slab moment requirements, although other options are applicable. It is acceptable to use the same reinforcement to resist both diaphragm out-of-plane moments as well as in-plane shears rather than summing the two reinforcement requirements.

The minimum reinforcement ratios for structural diaphragms correspond to the required amount of temperature and shrinkage reinforcement. For nonprestressed diaphragms, reinforcement spacing each way must not exceed 18 in (457 mm). This maximum spacing requirement does not apply to prestressed or post-tensioned systems, except for one-way prestressed slabs the spacing limit applies to the reinforcement in the nonprestressed direction.

15.6.4 Force Transfer (Including Collector Forces) to Vertical Elements

Force transfer between diaphragms and vertical elements generally occurs through a combination of collectors (including their connections) and shear-friction. Design forces in collectors and their connections are determined both by the design seismic forces and by the selected load path. For reference in this discussion, consider the diaphragm and collector illustrated in Figure 15.22, with collector-to-wall detail shown in Figure 15.23.

Shear force transfers from the diaphragm to the collector through shear-friction along the length of the collector. The strength requirement is $\phi V_n \geq V_u$, where ϕ is the strength reduction factor as defined in Section 15.6.3, and $V_n = A_{vf} f_y \mu$. For monolithic construction, $\mu = 1.4$.

The collector itself must be designed for compression and tension forces that develop along its length, including the overstrength factor Ω_0 where applicable. For the system shown in Figure 15.22, the maximum design compression force in collector segment **ab** is $C_{u,max} = \Omega_0 v_u t l_{ab}$ and the maximum design tension force in

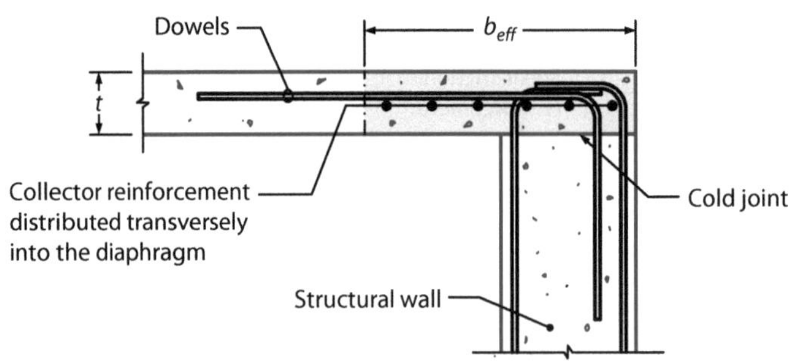

FIGURE 15.23 Detail of collector-to-wall connection.

collector segment **cd** is $T_{u,max} = \Omega_0 v_u t l_{cd}$. For compression forces, the collector is designed so that $\phi P_{n,max} = 0.80\phi[0.85 f_c'(A_g - A_{st}) + f_y A_{st}] \geq C_{u,max}$, where $\phi = 0.65$. The expression for $P_{n,max}$ is the maximum axial force permitted for a compression member in ACI 318. For tension forces, the collector is designed so that $\phi f_y A_{st} \geq T_{u,max}$, where $\phi = 0.90$.

Required cross-sectional dimensions of the collector can be determined by either tension or compression limits described in the preceding paragraph. In many cases, however, alternative detailing considerations will determine the dimensions, as follows. In structures assigned to SDC D, E, or F, ACI 318 requires that collector elements have transverse confinement reinforcement if the compressive stress based on the gross cross section exceeds $0.2f_c'$ for standard load combinations and $0.5f_c'$ for load combinations with overstrength factor Ω_0. The latter case generally controls. The transverse confinement reinforcement is required until the compressive stresses are below $0.15f_c'$ or $0.4f_c'$ for standard load combinations and load combinations with overstrength factors, respectively. Collector cross-sectional dimensions are commonly sized to avoid triggering these requirements for transverse confinement reinforcement.

For tension loading, ACI 318 requires the collector to be sized so that longitudinal reinforcement at splices and anchorage zones satisfies either (1) or (2):

1. Center-to-center spacing is at least three longitudinal bar diameters, but not less than 1.5 in (38 mm), and concrete clear cover is at least two and one-half longitudinal bar diameters, but not less than 2 in (50 mm).
2. Transverse reinforcement provides A_v not less than the greater of $0.75\sqrt{f_c'}\frac{b_w s}{f_{yt}}$, psi $(0.062\sqrt{f_c'}\frac{b_w s}{f_{yt}}$, MPa) and $(50 b_w s)/f_{yt}$, psi $((0.35 b_w s)/f_{yt}$, MPa).

These requirements are intended to reduce the possibility of bar buckling and to provide adequate bar development conditions in the vicinity of splices and anchorage zones.

For the detail shown in Figure 15.23, the entire collector force, amplified by overstrength factor Ω_0, plus any diaphragm shear along the length of the wall, must be transferred to the wall through shear-friction across the cold joint. For the example of Figure 15.22, the total force is $C_{u,max} + T_{u,max} + v_u t l_{bc}$. (Note that the terms $C_{u,max}$ and $T_{u,max}$ include the factor Ω_0 but the term $v_u t l_{bc}$ does not include factor Ω_0.) Dowels crossing the cold joint can provide the required shear-friction reinforcement. If the interface is clean and free of laitance, and is intentionally roughened to an amplitude of ¼ in (6 mm) as required by ACI 318, the friction coefficient is $\mu = 1.0\lambda$.

If the wall is slip-formed, with the diaphragm cast later, some of the collector force enters the wall boundary directly through compression and tension. The portion of the collector force outside the width of the wall plus the shear developed along the length of the wall are transferred into the side face of the wall through shear-friction. Provision of shear keys is recommended to enable use of a friction coefficient $\mu = 1.0\lambda$.

When designing shear-friction reinforcement, out-of-plane (due to gravity loads) and in-plane (due to seismic loads) effects must be combined using the appropriate load combination. Connections between diaphragms and vertical elements of the seismic-force-resisting system must also be capable of resisting forces associated with out-of-plane loading of the vertical elements. Where equivalent lateral force or modal response spectrum analysis is used, orthogonal earthquake effects are combined using the 100%/30% rule.

15.6.5 Reinforcement Development

ACI 318 requires that all reinforcement used to resist collector forces, diaphragm shear, or flexural tension be developed or spliced for f_y in tension. Reductions in development or splice length for calculated stresses less than f_y are not permitted. Type 2 splices are required where mechanical splices are used to transfer forces between the diaphragm and the vertical elements of the seismic-force-resisting system.

15.6.6 Special Cases

Openings Adjacent to Vertical Elements

Architectural requirements sometimes dictate openings adjacent to major walls that act as part of the seismic-force-resisting system. This can create force transfer challenges, especially at podium levels where large forces may need to be transferred from a vertical element in one plane to another vertical element in another plane. The preferred approach is to work with the architect to plan locations of openings so that they do not interfere with major force transfers. Where openings in critical locations cannot be avoided, workable solutions can sometimes be designed.

Figure 15.20 illustrates the force transfer from a shear wall to a basement wall where the two walls were separated by a large opening. Figure 15.24 illustrates how reinforcement might be detailed in the diaphragm segment to the left of the opening. For the distributor in tension, uniform shear flow can be assumed between the distributor and diaphragm segment along **ab**, resulting in shear stresses acting on the diaphragm segment as shown in Figure 15.24*b*. Moment equilibrium of the segment requires equal shear stresses along faces **ab**, **bc**, **ad**, and **cd**. Shear reinforcement, uniformly distributed in both directions, is required to resist this applied shear. Shear-friction reinforcement is required along **cd** to transfer the shear to the basement wall. Edge **bc** requires a tension tie to collect the shear stresses along that edge; that tension tie needs to be developed into the adjacent diaphragm. A compression reaction near point **a** completes the equilibrium requirements. Similar but opposite action occurs for the collector in compression (Figure 15.24*c*).

Openings commonly occur adjacent to structural walls, as shown in Figure 15.25. Such openings may disrupt the flow of forces from the diaphragm into the wall. In the example shown, the design of the diaphragm adjacent to the wall should consider diaphragm in-plane shear, shear-friction transfer at the wall-diaphragm interface,

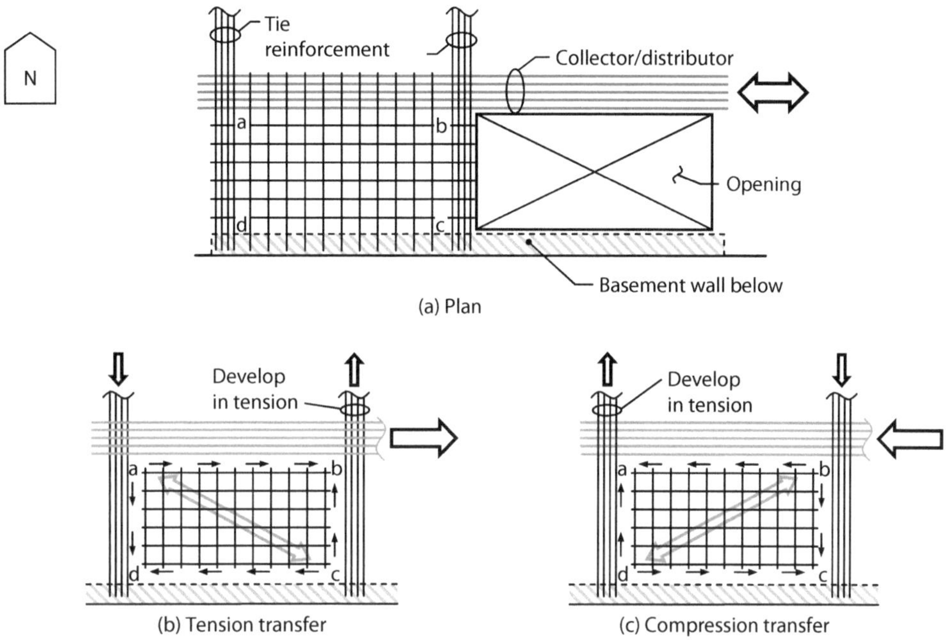

FIGURE 15.24 Reinforcement to transfer collector/distributor force around an opening.

and diaphragm in-plane moment. Transfer of all of the collector force into the narrow width of the wall might result in a highly congested region within the wall. To reduce congestion, a common approach is to distribute the collector reinforcement into a greater portion of the wall height. In effect, the collector that was initially distributed horizontally in the diaphragm has been "turned" so that it is now vertically distributed in the wall. Strut-and-tie models can help in following the flow of forces from the diaphragm into the wall, and can help ensure that reinforcement required for the force transfer is not overlooked. Once the slab portion of the collector has been "turned into the wall," the collector must extend into the wall at least a development length but not less than the length required to transfer the force to the wall horizontal reinforcement. Extending the collector along the full wall length is often the preferred practice.

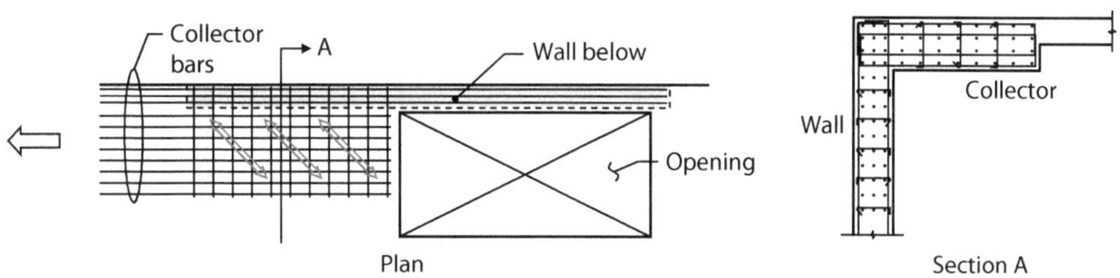

FIGURE 15.25 Turning a collector into a wall near an opening.

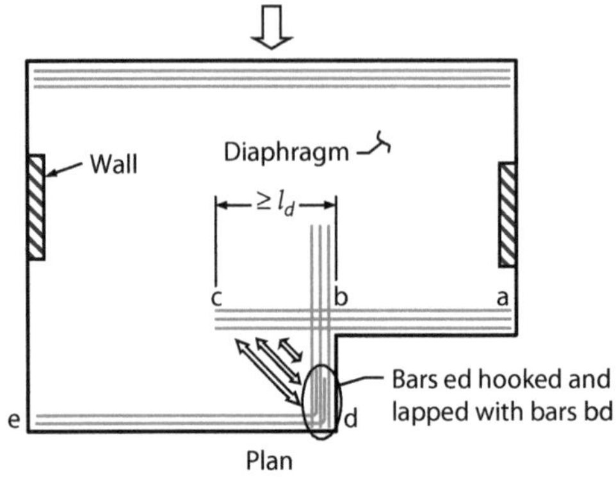

FIGURE 15.26 Reinforcement for diaphragm with a re-entrant corner.

Re-entrant Corners

At re-entrant corners of diaphragms such as shown in Figure 15.26, either tension chord **ac** can extend across the full width of the diaphragm or the chord can follow the profile of the diaphragm around the re-entrant corner. In the latter case, the tension chord reinforcement **ac** needs to be developed into the diaphragm. The developed length **bc** transfers force to the diaphragm, creating shear in panel **bcd**. Chord reinforcement **bd** collects this shear (similar to the problem illustrated in Figure 15.24) and develops it into the diaphragm. Finally, tension chord **ed** can be designed based on the corresponding moment and effective depth. Because there is considerable moment at **d**, chord reinforcement **de** should be hooked at **d** and lapped with adjacent reinforcement **bd**. Specific reinforcement details (area and placement) will depend on the specific geometry of the diaphragm.

Steps and Depressions

Where steps or depressions occur, reinforcement must be provided to transfer the design forces through the offset. Figure 15.27 illustrates collector reinforcement

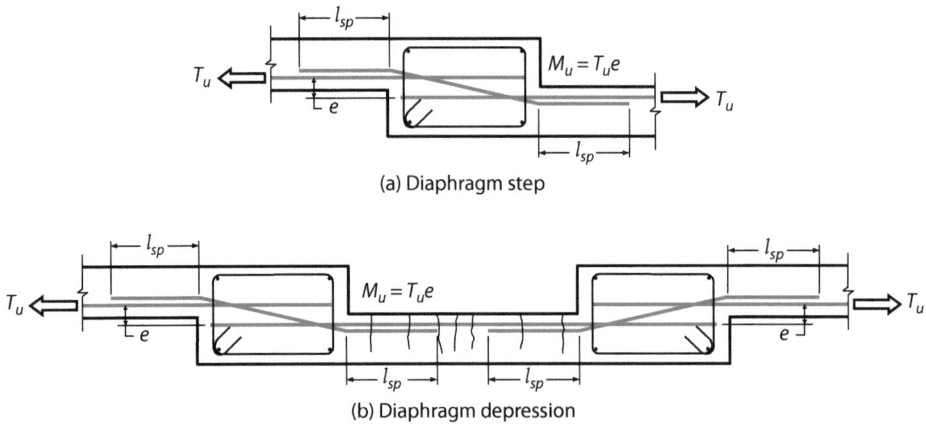

FIGURE 15.27 Steps and depressions. (*a*) Diaphragm step; (*b*) Diaphragm depression, showing flexural cracking of the depression that could be induced by the eccentric loading.

passing through a step and a depression. To the extent practicable, collector reinforcement should be placed so as to minimize the eccentricity on opposite sides of the step when in tension; if the collector also transmits compression, the eccentricity will be determined by the gross dimensions of the offset. If collector reinforcement cannot be placed straight, either it can be bent or bent bars can be spliced with the main bars. The vertical force created by offset bars should be resisted by hoop legs; see analogous provisions for offset column bars. In addition, any eccentricity in the collector bars creates a moment $T_u e$ that must be resolved within the structure. If there is a wall at this location oriented perpendicular to the collector, the wall may be able to resist the moment by out-of-plane bending. Alternatively, the overlapped section of the step can be reinforced as a beam to transmit moment through torsion to adjacent columns, though this can be problematic because of challenging reinforcement details and because of the large twist that may be associated with torsion. If the diaphragm is transmitting shear across the step, hoop reinforcement can resist the applied shear through shear-friction at the interface.

Extension of Collectors/Distributors into Diaphragms

Large transfer forces at podium slabs can require long distributors whose elongations may result in displacements that are incompatible with design assumptions. The problem can be exacerbated in podium slabs with multiple openings or with short transverse distances to basement walls. Consider the idealized podium slab plan of Figure 15.28. The wall along line **B** should transfer its force through the podium slab to the east of the openings and into basement wall along line **A**. Distributor **ab** is provided for this purpose. The rate of transfer will be determined by the shear strength of the podium slab to the north of distributor **ab**, and this will determine the required distributor length. If distributor bar cutoffs are staggered as the distributor force decreases, the strain in the distributor steel is likely to be near the yield strain along the entire length, which in a long distributor can result in significant elongation. For a 100-ft (30-m) long distributor, the elongation along **ab** is as much as 0.002 × 1200 in = 2.4 in (60 mm).

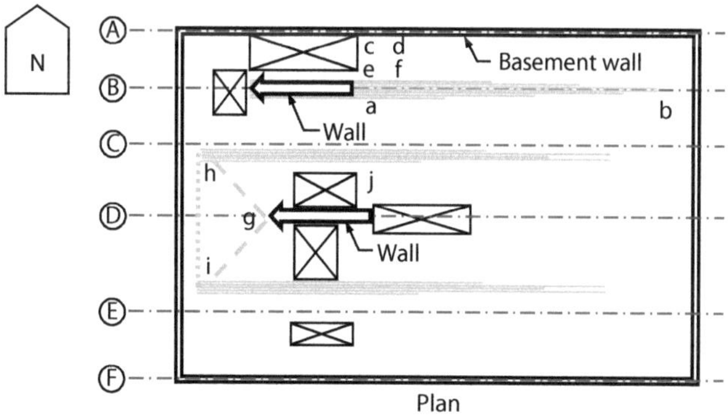

Figure 15.28 Challenging design conditions created by long load paths.

Assuming fixity at point **b**, the movement at point **a** is thus 2.4 in (60 mm). This creates two problems. First, shear strains in slab segment **cdef** may cause slab damage. Second, the wall forces may find a stiffer load path through the shear wall down to the diaphragm level below, producing diaphragm shears that have not been considered in design. A similar situation occurs for the wall along line **D**. In this case, a long load path, first in compression to points **h** and **i**, then in tension through the long distributors, may result in diaphragm deformation incompatibilities at locations such as point **j**.

The problems of the podium slab in Figure 15.28 are exacerbated by multiple openings constraining load paths out of the walls and into the podium. Such problems should be identified early in the design process so that architectural adjustments can be pursued.

15.7 Additional Requirements

15.7.1 Material Properties

In U.S. practice, the minimum permitted specified compressive strength, f_c', is 2500 psi (17 MPa) for structural concrete including diaphragms. Where diaphragms are cast monolithically with portions of special moment frames or shear walls for a structure assigned to SDC D, E, or F, the minimum f_c' is 3000 psi (21 MPa). This is typically not an issue as f_c' of 4000 to 6000 psi (28 to 41 MPa) is commonly specified for floor systems.

For some structures, f_c' of moment frames or shear walls is higher than that of the diaphragm system. U.S. practice allows column concrete f_c' to be 1.4 times that of the floor system, but this is intended to apply only for axial force transmission and therefore should not be applied to walls or columns of the seismic-force-resisting system. Walls and moment frames commonly are located along the edges of the building slab or along openings, where the concrete is not confined by adjacent concrete on all sides. Additionally, these elements have high shear stress that must be transferred through the floor, requiring the higher strength.

For shear walls, the higher wall strength can be maintained using a jump core or flying form system for the wall construction to precede the floor construction. Where concrete for the portion of the wall or moment frame through the thickness of the floor system is placed with concrete for the floor system, the higher strength concrete should be puddled at these elements and extended 2 ft (0.6 m) into the slab as allowed for columns.

Lightweight concrete, which can reduce floor mass, is not commonly used because of lack of availability in many markets and concerns about reduced performance capability.

According to U.S. practice, the values of f_y and f_{yt} used in design of reinforcement resisting shear are not permitted to exceed 60,000 psi (420 MPa), except up to 80,000 psi (550 MPa) is permitted for welded deformed wire reinforcement. The intent of the stress limits is to control the width of shear cracks.

Reinforcement for chords and collectors is limited by the general requirements for bonded reinforcement of the building code, with two exceptions:

1. Where chord or collector reinforcement is placed within beams including the effective flange of special moment frames, and therefore acts as beam flexural reinforcement, the chord or collector reinforcement must satisfy restrictions for special moment frame reinforcement.
2. Where bonded tendons are used, U.S. practice limits the stress from design earthquake forces to 60,000 psi (420 MPa). Although stress in other chord and collector reinforcement is not limited, consideration should be given to deformation compatibility between tension chords, collectors, and the floor slab. High tensile stress and strains in collectors and chords can result in excessive cracking that will migrate into the slab.

15.7.2 Inspection Requirements

Reinforced concrete diaphragms and their chords and collectors are a part of the seismic-force-resisting system. Proper construction of diaphragms and their elements is paramount to ensuring that the structure will perform as intended during a major earthquake.

In an effort to ensure proper construction, special inspections are required for most concrete buildings in U.S. practice. The IBC requires that the design professional for a building prepare a statement of special inspections identifying the required inspections for construction of the building. The statement is to include inspection requirements for seismic-force-resisting systems in structures assigned to SDC C, D, E, or F. The only exception to this requirement is for reinforced concrete buildings that are less than 25 ft (8 m) in height above the grade plane and that are located on a site with design spectral response acceleration at short periods, S_{DS}, less than or equal to 0.5*g*. Diaphragms and their elements provide resistance to prescribed seismic forces; therefore, diaphragms are part of the seismic-force-resisting system and should be identified on the statement of special inspections.

Concrete for diaphragms also requires special inspections, including:

- Verifying use of required design mix.
- Sampling fresh concrete for strength test specimens, performing slump and air content tests, and determining concrete temperature at time of placement.
- Concrete placement.
- Maintenance of specified curing temperature and techniques.
- Grouting of bonded prestressing tendons that are part of the seismic-force-resisting system.

Structural observations by the registered design professional are required for all structures assigned to SDC D, E, or F whose height is greater than 75 ft (23 m). Shorter structures of high occupancy categories or SDC E also require structural observations. Observing diaphragm components is recommended as part of this practice.

15.7.3 Bracing Columns to Diaphragms

Diaphragms brace columns where they connect. Where the required force is not specified in the code, a force of 2% to 4% of the column axial load is generally considered sufficient. For low and midrise cast-in-place concrete buildings this check is rarely made as the inherent strength of the diaphragm-to-column connection easily provides this strength. For tall buildings with large heavily loaded columns, this check should be made. For these columns, the diaphragm check should include bearing stresses at the face of the column, adequacy of diaphragm reinforcement anchored into the column at edge conditions, and adequate diaphragm buckling strength to resist the bracing force. These requirements and recommendations also apply to precast buildings with cast-in-place diaphragms.

Inclined columns require a more rigorous check of forces at the diaphragm-to-column interface. At the top and bottom of the inclined portion of a column, there is a horizontal component of force that the diaphragm must resist and deliver to vertical elements of the seismic-force-resisting system (Figure 15.2). The magnitude of this horizontal component depends on the inclination of the column. Where architecturally feasible, the column inclination from vertical should not exceed about 15° (i.e., 1 horizontal to 4 vertical). Larger inclination angles are not generally proscribed by the building code, but large diaphragm thrusts and challenges in adequately reinforcing the column, diaphragm, and diaphragm-column connection should be anticipated. Where the column axial loads are low and the inclination approaches vertical, the slab may be capable of resisting the horizontal component. The slab may also be adequate at intermediate levels where the inclined column passes without a change in direction. At these intermediate levels only the incremental vertical force added to the column at that level creates a horizontal thrust for which the connection must be designed. For highly loaded inclined columns and columns with inclines exceeding 15° from vertical, it may be necessary to thicken the slab or provide a beam to transfer the thrust from the inclined column.

15.7.4 Interaction of Diaphragm Reinforcement with Vertical Elements

Chord and collector reinforcement of diaphragms is often located in beams that are part of special moment frames or within slabs adjacent to those beams. This reinforcement will likely not be stressed to its yield strength from chord or collector forces during the earthquake at the same time the moment frame beam is fully yielding. However, deformation compatibility will typically dictate that the chord or collector reinforcement will yield along with the beam (the reinforcement will strain as the beam flexes). Therefore, this chord or collector reinforcement will add moment strength to the beam that must be considered when checking strong-column weak-beam requirements, the design shear force for the beam, and beam-column joint demands. (A strict interpretation of the ACI 318 provisions is that this reinforcement need not be included in beam and beam-column joint shear calculations if it is located in the beam effective flange width rather than within the beam web; however, the preferred approach is to include it in all cases where it is located within the effective beam width.) Similar considerations should be made when designing coupling beams for shear walls.

Collector or chord compressive forces can also increase the moment strength of beams as the axial load is likely below the balanced point. In determining the compressive force of a chord to add to a beam, only 30% of the chord force is likely required as the chord force is usually caused by earthquake forces orthogonal to forces loading the moment frame. Collector forces that act on beams are likely caused by the same earthquake force that loads the moment frame. Therefore, 100% of the collector force is likely required to be considered when designing the beam. Similar considerations apply to chord and collector tension forces. These axial forces should be considered when evaluating the strong-column weak-beam requirements and when determining the beam design shear force and design shear strength. If the compressive force causes axial stress in the beam greater than $0.1f_c'$, the beam should be provided with confinement reinforcement similar to that required of columns.

Similar considerations apply to the design of intermediate moment frames.

Collectors and chords are designed to respond linearly under axial tension and compression, but where these elements enter the boundaries of shear walls they may be subjected to significant flexure as the walls rock back and forth during an earthquake. Where possible, the reinforcement of these elements should be located to minimize flexural yielding. This may be achieved by using shallower members or by placing the main collector or chord reinforcing bars near the mid-depth. For structures assigned to SDC D, E, or F, transverse confining reinforcement is recommended at these locations to improve compressive capacity of the concrete and buckling resistance of the reinforcement.

15.8 Detailing and Constructability Issues

15.8.1 Diaphragm Reinforcement

Many concrete slabs are designed to have a continuous bottom mat of uniformly distributed reinforcement. For this reason, transverse reinforcement for diaphragm shear resistance is commonly incorporated into the bottom mat. In heavily reinforced diaphragms that are thick slabs, a continuous top and bottom mat of reinforcement is often provided. Designers should specify the required lap splice and development length of the reinforcement in the construction documents, as diaphragm reinforcement splice and development requirements may exceed what is otherwise required.

In post-tensioned slabs, the location of diaphragm reinforcement needs to be coordinated with the locations of PT strands and anchorages. Designating layers within the slab depth for diaphragm reinforcement and PT strands is an effective method of minimizing conflict. Slab design needs to take into account the actual location of the layers of reinforcement if this approach is used.

Welded wire fabric is not generally used for diaphragm reinforcement in cast-in-place slabs because the reinforcement provided for gravity support uses standard reinforcing bars. The use of welded wire fabric for diaphragm reinforcement is typically limited to topping slabs over precast concrete systems or steel metal decks.

15.8.2 Collector and Chord Detailing

Collector and chord reinforcement is often located in the mid-depth of the slab. In structures assigned to SDC D, E, and F, U.S. practice requires center-to-center spacing at least $3d_b$, but not less than 1.5 in (38 mm), and concrete clear cover at least $2.5d_b$, but not less than 2 in (50 mm). Otherwise, transverse reinforcement is required.

Connections of collector reinforcement to vertical elements of the seismic-force-resisting system are often congested regions. In many cases, numerous large diameter bars are required to be developed into confined boundary zones of shear walls as shown in Figure 15.29*a*. Designers should study these connections in detail to ensure adequate space exists. In many cases, increased slab thickness or beams are required to accommodate reinforcement detailing at the connections. Figure 15.29*b* shows where a beam was created to accommodate the collector reinforcement. Designers should also consider the slab depth provided where large collectors intersect. Multiple layers of large diameter reinforcing bars can result in excessive congestion. Similarly, designers should be aware of locations where collectors intersect concrete beam longitudinal reinforcement.

Long collectors (Figure 15.30) can accumulate strains over their length resulting in displacements that may be incompatible with the modeling assumptions. The accumulated displacements also might exceed displacement capacities of adjacent components. Designers can consider providing additional collector reinforcement to reduce the strain and associated collector elongation. Providing confinement reinforcement around the collector can increase the ductility of the concrete locally, but it will not address potential problems associated with incompatible deformations. Redesigning the force transfer system should also be considered.

Where collector (or chord) reinforcement is required at a location coincident with a beam, the chord reinforcement can be placed within the beam. Beam transverse reinforcement, if properly detailed, can also serve as collector (or

(a) Collector connection to shear wall boundary zone

(b) Beam for large collector

Figure 15.29 Collector detailing. (*Used with permission from Magnusson Klemencic Associates.*)

Figure 15.30 A long collector with confinement reinforcement. (*Used with permission from Simpson Gumpertz & Heger.*)

chord) confinement. If chord reinforcement does not fit entirely within the beam width, then the effective diaphragm depth should be based on the actual distribution and location of the chord reinforcement.

15.8.3 Confinement

Transverse (confinement) reinforcement may be required in collectors or elements transferring axial forces around openings or other discontinuities. Required seismic hook dimensions can make reinforcement detailing difficult in typical slab depths. Designers can increase the width of the collector within the slab or the thickness of the slab until the compressive stresses are low enough that confinement is not required. Alternatively, beams with sufficient dimensions can be added to facilitate the required confinement detailing. Where confinement is not required by the code, the designer still may consider adding some transverse reinforcement to improve connection toughness at critical locations. Figures 15.30 and 15.31 show examples of added transverse reinforcement that is not in the form of closed hoops, yet will result in improved collector behavior.

Figure 15.31 Confinement of a collector. (*Used with permission from Simpson Gumpertz & Heger.*)

15.8.4 Shear Transfer

Transfer between diaphragms and vertical elements of the seismic-force-resisting system can be accomplished in a number of ways. Shear-friction reinforcement can be provided for transfer of forces along the length of the vertical element. The designer should consider the location of the construction joint when selecting the shear-friction coefficient μ for design. Where the vertical elements of the seismic-force-resisting system are cast in advance of the slabs, or vice versa, shear keys should be considered in order to achieve a value of $\mu = 1.0$. Shear-friction reinforcement must be developed at each critical failure plane.

Compressive collector forces can be transferred via direct bearing at wall ends. An appropriate effective slab bearing area should be considered, and compressive stresses in the slab should be evaluated to determine if confinement is required. Tensile forces can be transferred through collector reinforcement that is developed in both the diaphragm and the vertical element. The embedment length of tension collectors should consider the assumed shear stress distribution and capacity within each element. See Figure 15.18.

15.8.5 Mechanical Splices

Large diameter diaphragm and collector reinforcing bars are commonly spliced using mechanical couplers. Because lap splices of No. 14 (No. 43) bars or larger are prohibited by U.S. codes, mechanical couplers are required. For smaller diameter bars, mechanical splices may be used to reduce reinforcement congestion. Mechanical splices used to transfer forces between the diaphragm and the vertical elements of the seismic-force-resisting system are required to be Type 2. Mechanical couplers considered for a project should be approved for use (in the United States, they should have current ICC approval). Detailing and placement must provide the required cover over the coupler body, which typically has a diameter larger than the diameter of the bars being coupled.

Contractors may elect to cast vertical elements of the seismic-force-resisting system in advance of the slabs. Mechanical splice and anchorage devices facilitate this construction method. Form savers are commonly used for anchorage of shear-friction and collector reinforcement. Designers should pay attention to the size of the form savers, as they are typically large relative to the bar being anchored. Figure 15.32 shows shear-friction reinforcement connections to a shear wall with form savers.

15.8.6 Conduits and Embedded Services

The placement of electrical conduit, plumbing sleeves, and other services within a reinforced concrete slab is common practice. Figure 15.33 shows an extreme example of embedded conduit within a reinforced concrete diaphragm. Designers should consider these and similar items that can reduce the capacity and stiffness of the diaphragm. Typical details showing restrictions on the placement of conduits and embedded services and associated supplemental reinforcement are helpful, but often are insufficient to cover many of the conditions that occur in a project. Contract documents should require the contractor to provide detailed

FIGURE 15.32 Form savers for dowel anchorage. (*Used with permission from Magnusson Klemencic Associates.*)

layout drawings of the conduits and other embedded services well in advance of concrete placement. This allows the designer time to review the impact to the diaphragm and to provide supplemental instructions, including additional reinforcement if required. Pre-construction meetings can provide an opportunity for the designer to review placement and reinforcement requirements in detail prior to construction. The location of nonstructural items embedded in the slab should also be considered with respect to collector layout, as conflicts can be problematic.

15.8.7 Location of Construction Joints

Construction joints create weakened planes within a diaphragm. They can also impact development and splices of reinforcement. Shear-friction reinforcement is commonly provided across construction joints to maintain continuity of the diaphragm in shear. The impacts to the continuity and development of chord and collector reinforcement at construction joints should also be understood. As with conduits, typical details, limitations, and instructions should be clearly detailed in

FIGURE 15.33 Embedded conduit. (*Used with permission from Magnusson Klemencic Associates.*)

the contract documents. Contract documents should also require that contractors provide detailed construction joint layout drawings well in advance of concrete placement.

References

ACI 318 (2014). *Building Code Requirements for Structural Concrete (ACI 318-14) and Commentary*, American Concrete Institute, Farmington Hills, MI.

ASCE 7 (2010). *Minimum Design Loads for Buildings and Other Structures*, American Society of Civil Engineers, Reston, VA, 658 pp.

Chopra, A.K. (2012). *Dynamics of Structures: Theory and Applications to Earthquake Engineering*, 4th ed., Prentice Hall, Englewood Cliffs, New Jersey, 992 pp.

Corley, W.G., L. Cluff, S. Hilmy, W. Holmes, and J. Wight (1996). "Concrete Parking Structures," in *Northridge Earthquake Reconnaissance Report, Vol. 2, Earthquake Spectra*, Earthquake Engineering Research Institute, Oakland, CA, Supplement to Volume 11, pp. 75–98.

Fleischman, R.B., K.T. Farrow, and K. Eastman (2002). "Seismic Performance of Perimeter Lateral System Structures with Highly Flexible Diaphragms," *Earthquake Spectra*, Vol. 18, No. 2, pp. 251–286.

IBC (2012). *International Building Code*, International Code Council.

Moehle, J.P., J.D. Hooper, and D.J. Kelly (2010). *Seismic Design of Cast-in-Place Concrete Diaphragms, Chords, and Collectors: A Guide for Practicing Engineers*, NEHRP Seismic Design Technical Brief No. 3, National Institute of Standards and Technology, Gaithersburg, MD.

Moehle, J.P., and M.A. Sozen (1980). *Experiments to Study Earthquake Response of R/C Structures with Stiffness Interruptions*, Civil Engineering Studies SRS-482, University of Illinois at Urbana-Champaign.

Mohr, B.A., and S.K. Harris (2009). "A New Method for Collector Design in Stiff Diaphragms," *Proceedings, 2009 SEAOC Convention*, San Diego, CA, pp. 339–343.

Rodriguez, M.E., J.I. Restrepo, and J.J. Blandón (2007). "Seismic Design Forces for Rigid Floor Diaphragms in Precast Concrete Building Structures," *Journal of Structural Engineering*, Vol. 133, No. 11, pp. 1604–1615.

Sabelli, R., W. Pottebaum, and B. Dean (2009). "Diaphragms for Seismic Loading," *Structural Engineer*, Part 1, January, pp. 24–29; Part 2, February, pp. 22–23.

SEAOC (2005). *Using a Concrete Slab as a Collector*, Seismology and Structural Standards Committee, Structural Engineers Association of California, Sacramento, CA, 15 pp.

SEAOC (2009). "Concrete Parking Structures," *The SEAOC Blue Book: Seismic Design Recommendations*, Structural Engineers Association of California, Sacramento, CA, www.seaoc.org/blubook, 9 pp.

CHAPTER 16

Foundations

16.1 Preview

Foundations must be designed to transmit forces from the superstructure into the supporting soil within acceptable deformations while providing adequate safety against bearing or uplift failure. Design should consider both the structural elements of the foundation (including spread footings, mat foundations, piles, ties, and other foundation elements) and the soil that supports those elements. Building codes may permit seismic design of buildings based on a rigid foundation assumption, or may allow for (or require) modeling of soil–foundation–structure interaction, which can modify input ground motions, effective damping, and effective stiffness of the soil-foundation-structure system. Although foundation design is mainly the responsibility of the structural engineer, it relies on information provided by the geotechnical engineer, and benefits from good dialogue between the two disciplines.

This chapter describes typical foundation elements used in earthquake-resistant buildings; introduces concepts of soil–structure interaction; summarizes geotechnical information required for seismic design of foundations; discusses foundation performance objectives; and presents analysis, design, and detailing guidance for various foundation elements and systems. Although foundation elements can be made of materials other than structural concrete, the discussion here is limited to structural concrete elements, including cast-in-place and precast, and including conventionally reinforced and prestressed elements. Emphasis will be on structural engineering aspects of foundation design. For geotechnical aspects, the reader is referred to texts on geotechnical engineering.

This chapter references the design requirements of U.S. building codes and guidelines, among others. In the United States, the three main reference codes at the time of this writing are ACI 318 (2014), ASCE 7 (2010), and the IBC (2012). Although most concrete provisions are typically contained in ACI 318, for foundation design both ASCE 7 and the IBC introduce several structural concrete provisions. Furthermore, the IBC does not adopt Chapter 14 of ASCE 7, which is the chapter containing most but not all of the structural concrete foundation provisions. Consequently, most of the ASCE 7 provisions do not apply, and the IBC provisions take precedence over ACI 318 provisions where conflicts exist. Where U.S. code requirements are presented in this chapter, they refer to those that govern in most jurisdictions of the United States.

16.2 Foundation Elements in Earthquake-Resisting Buildings

Foundations are generally categorized as being either shallow foundations or deep foundations. Shallow foundations are often used where the underlying soils near the surface are sufficient to resist foundation forces without excessive settlement, bearing failure, or uplift. Deep foundations are used where the soils near the surface are not adequate for applied loads, which may occur where shallow soils are poor or where taller structures introduce large foundation forces, where certain geotechnical hazards such as liquefaction and lateral spreading potential exist, or where adjacency and surcharge issues favor deeper force transfers. In general, shallow foundations are more economical to construct than deep foundations. For both shallow and deep foundations, it may be necessary to tie the foundation elements together near the surface so that individual foundation elements do not become separated due to soil movement. Connections between individual elements can also be designed to enable the elements to act together as an integral unit. Grade beams, structural slabs-on-ground, or both can be used for this purpose.

The choice of foundation type is usually recommended by the geotechnical engineer based on considerations of the structural loading and the geotechnical characteristics of the site. Primary considerations include:

- *Settlement and differential settlement:* This is usually a consideration in determining allowable soil bearing pressures for long-term loading, and in determining whether isolated foundation elements or a continuous mat foundation are preferable as shallow foundation elements, or whether deep foundation elements are preferred.
- *Bearing capacity failure:* This is usually a consideration for overturning resistance under seismic loading, although a large margin of safety against bearing failure under sustained loads is also provided.
- *Frost damage:* The foundation must have adequate depth to protect against freezing.
- *Adverse soil conditions:* These include expansive soils and soil materials that are reactive with normal concretes.

The following paragraphs introduce types of shallow and deep foundations, with a brief discussion of their resistance mechanisms.

16.2.1 Shallow Foundations

Figure 16.1 illustrates various spread footings. *Isolated spread footings* can be used to support concentric axial force without shear and moment, as nominally occurs for gravity columns. Isolated footings can also support combined gravity and lateral load effects. *Combined footings* are commonly used where the extension of a footing beyond a column is limited, as often occurs at property lines. By combining two (or more) columns on a single footing, the footing can be

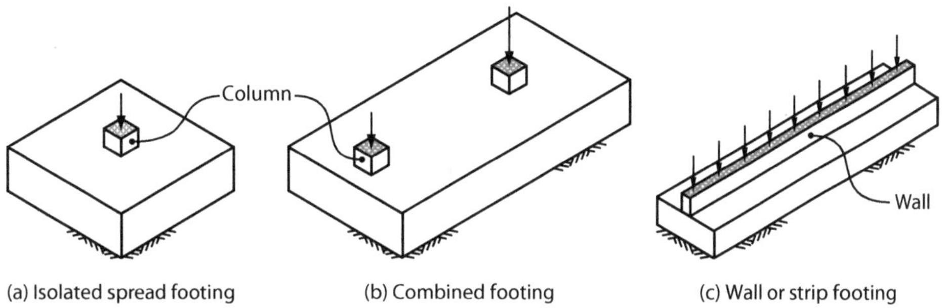

Figure 16.1 Spread footings.

proportioned such that the centroid of its area coincides with the centroid of the sustained loads, thereby producing uniform bearing under the footing. Combined footings are also used where large overturning moments under lateral loads exceed the capacity of isolated spread footings. *Wall (or strip) footings* are used to support structural walls. Usually, the wall and wall footing are centered so as to produce uniform bearing under sustained loads. For exterior basement walls, however, an eccentric wall footing extending only toward the interior of the excavation is common.

A *mat foundation* is a continuous, thick, reinforced concrete element that supports multiple columns or walls (Figure 16.2). Mat foundations may be suitable where soil and load conditions result in unacceptably large differential settlements across the foundation plan, but where conditions are not so poor as to require a deep foundation. Where a basement is below the groundwater table, a mat foundation together with basement walls can act as a "bathtub" to keep the basement dry and to resist hydrostatic uplift forces. Mat foundations can also become economical as an alternative to individual footings where the sum of areas

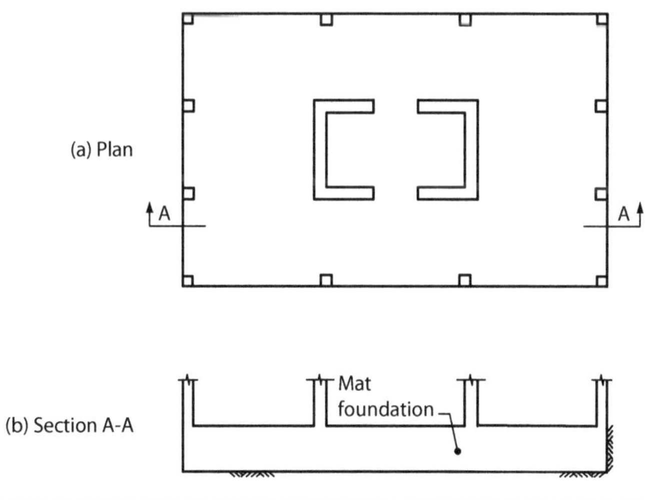

Figure 16.2 Mat foundation.

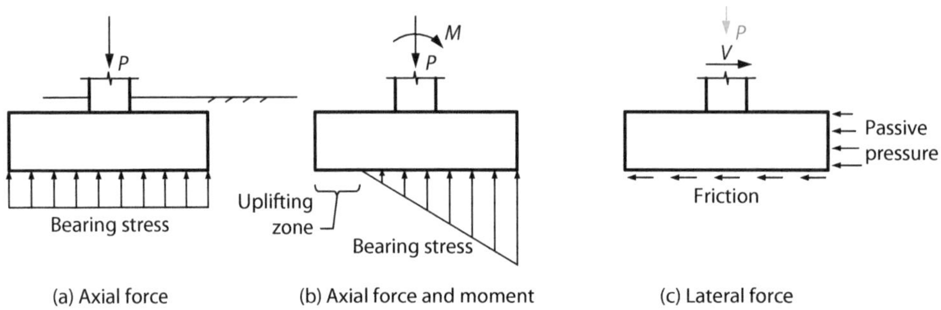

Figure 16.3 Resistance mechanisms for spread footings.

of individual footings becomes a large fraction of the total building footprint, or where multiple grade beam ties are required between individual foundation elements. Mat foundations may also be beneficial where large overturning moments from lateral forces require a large bearing area or capability to resist uplift.

Figure 16.3 illustrates resistance mechanisms applicable to all types of shallow foundations. Concentric axial force results in nominally uniform bearing stresses on the soil. Axial force and moment result in eccentric bearing stress. Figure 16.3*b* illustrates the condition where stresses due to moment exceed the bearing stresses due to axial load; in such conditions, uplift occurs because the soil–foundation interface cannot resist tension. Horizontal forces may be resisted by a combination of friction and passive pressure (Figure 16.3*c*). Where available friction and passive pressure resistance is inadequate, the footing can be tied to adjacent foundation elements to increase resistance.

16.2.2 Deep Foundations

One of the most common types of deep foundations is the *pile foundation*. Piles can be precast or cast-in-place concrete, or can be made of other materials. Precast piles must be driven into place, while cast-in-place piles are cast in pre-excavated holes. Various types of piles include:

- *End-bearing pile:* This type of pile derives most of its capacity from a bearing stratum on which the tip bears. End-bearing piles are used where soil media closer to the surface are soft or prone to settlement or liquefaction, and where a suitable bearing stratum is at reasonable depth. Sometimes end-bearing piles are designed ignoring skin friction along the length of the pile. However, where the softer soil may settle, the pile must be designed for down-drag from skin friction.
- *Friction pile:* This type of pile derives its resistance primarily from friction or adhesion along the length of the pile. They are commonly used where a bearing stratum is too deep to be usable. A pile that resists tension does so by friction and would be considered a friction pile.

- *Combined end-bearing and friction pile:* This type of pile derives its resistance from a combination of end bearing and friction.
- *Batter pile:* This is a pile that is driven at an angle with respect to vertical, generally for the purpose of resisting horizontal forces.

Micropiles (or minipiles) are deep foundation elements constructed using steel casing, often reinforced with high-strength threaded bars. Diameters typically are in the range of 3 to 12 in (70 to 300 mm). Typically, the casing is advanced to a design depth using a drilling technique, and then reinforcing steel is inserted and grouted within the casing. The reinforcement can be bonded at depth and then prestressed to improve uplift stiffness and usable resistance.

Piers are large, reinforced concrete members, similar to piles. They are also sometimes referred to as *drilled shafts*, *bored piles*, *barrettes*, or *drilled caissons*, depending on their details and method of construction. A large pier is sometimes called a *caisson*.

A single pile can support load from a single element. More commonly, however, multiple piles are tied together by a pile cap to support one or more superstructure elements. In tall buildings with inadequate soils near the surface, a foundation mat can be supported by large numbers of piles.

Figure 16.4 illustrates resistance mechanisms for pile foundations. Figure 16.4*a* shows piles in which axial load is resisted through the combined action of end bearing and friction. Figure 16.4*b* illustrates forces that develop under combined moment and lateral force. Moment resistance is provided primarily by a tension-compression

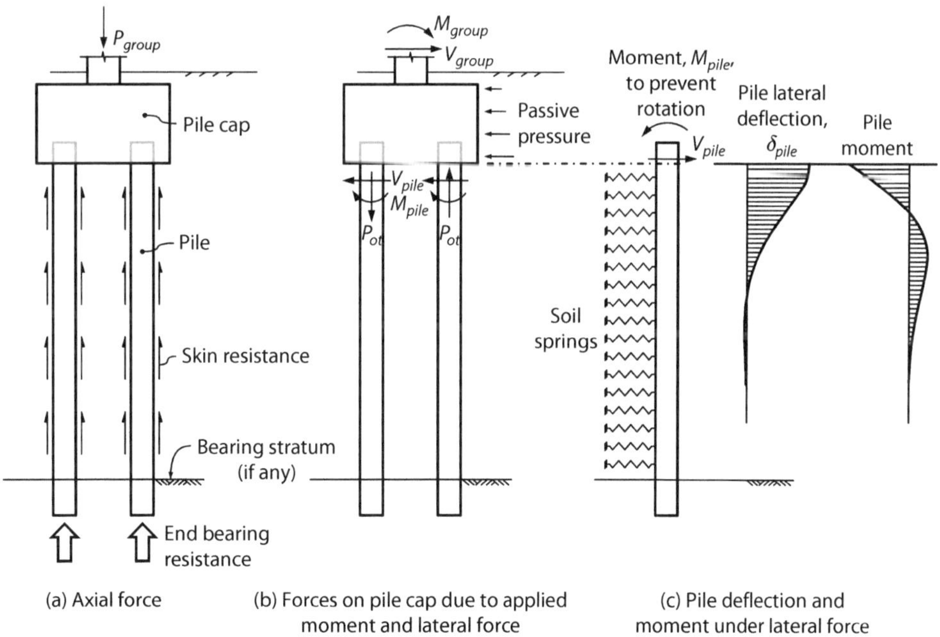

FIGURE 16.4 Resistance mechanisms for pile foundations.

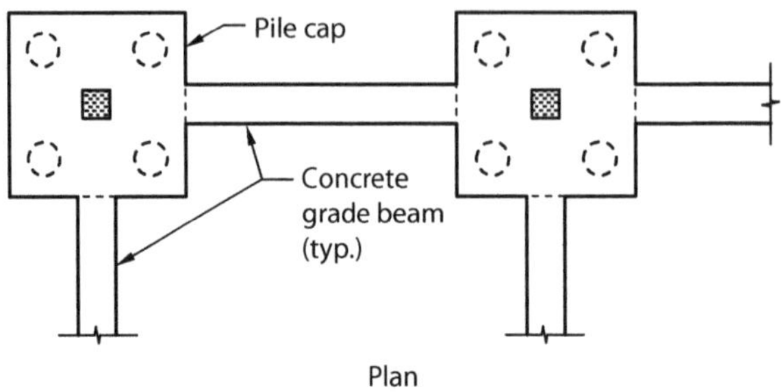

FIGURE 16.5 Grade beam connecting pile caps.

couple ($T = C = P_{ot}$). Lateral force is resisted by a combination of passive pressure against the pile cap plus lateral resistance from the piles. (Soil can settle beneath a pile cap. Consequently, frictional resistance and soil bearing under pile caps is generally ignored.) A geotechnical engineer typically will conduct an analysis (often referred to as a *p-y* analysis, where p is the lateral force on a pile per unit length, and y is the horizontal soil displacement). The outcomes of a *p-y* analysis include the relation between horizontal force V_{pile} and horizontal displacement δ_{pile} of the pile, and the pile shear and moment profiles. Figure 16.4*c* illustrates a model and results for a *p-y* analysis where the pile is fixed at its connection with the pile cap. Maximum resistance commonly is limited by acceptable lateral displacement at the top of the pile.

Other foundation options exist, including combinations of shallow and deep foundation elements, as in a piled raft foundation. These are not considered here.

16.2.3 Grade Beams and Structural Slabs-on-Ground

In regions of high seismicity, a good practice and requirement of many building codes is to use *ties* to interconnect individual pile caps, drilled piers, or caissons. In addition, individual spread footings founded on soft clay or soils vulnerable to failure or collapse under seismic loading (ASCE 7 Site Class E or F) should be interconnected by ties. The ties can be provided by individual *grade beams* (Figure 16.5), reinforced concrete beams within slabs-on-ground, or reinforced concrete slabs-on-ground. Competent rock, hard cohesive soils, or very dense granular soils may be suitable to tie foundation elements together without the need for specially constructed ties, but this approach is rarely used.

The previous paragraph describes a grade beam or slab-on-ground as a tension or compression tie that interconnects individual foundation elements. Grade beams can also be reinforced to transmit moment and shear between adjacent foundation elements, thereby increasing the overturning resistance of the combined system.

16.3 Soil–Structure Interaction

The seismic response of a building is affected by interactions among three linked systems: the structure, the foundation, and the soil underlying and surrounding the foundation. Analysis for *soil–structure interaction* (SSI), also referred to as soil–foundation–structure interaction (SFSI), evaluates the collective response of these systems to earthquake ground motion.

SSI effects are categorized as inertial interaction effects, kinematic interaction effects, and soil-foundation flexibility effects:

- *Inertial interaction:* A vibrating structure develops base shears and moments, which, when transferred from the foundation to the surrounding soil, result in translation and rotation of the soil–foundation interface. This leads to two principal effects. First, the vibration period of the structure-foundation-soil system is elongated relative to the fixed-base period. Second, deformations of the soil lead to increased system damping due to both hysteretic response of the soil and radiation damping (i.e., waves generated at the foundation–soil interface radiate away from the foundation, carrying energy away from the system). These effects can be represented directly using a finite-element model that includes the soil and the structure in a combined analysis. Alternatively, inertial effects can be approximated using available solutions for idealized soil-structure configurations and properties. ASCE 7 and FEMA (2009) contain prescriptive provisions with equations for approximating effects of inertial interaction on period and damping.
- *Kinematic interaction:* Free-field ground motions vary both temporally and spatially, and generally decrease in intensity with depth from the free surface. Across a rigid foundation, however, only a single motion is possible. Thus, a rigid foundation enforces an averaging of the incoming motions, with the result that peaks of the free-field motion are reduced. This effect is commonly referred to as *base-slab averaging*. Essentially the same effect occurs without a rigid foundation if the elevated floors are stiff concrete slabs. Base-slab averaging primarily affects the high-frequency components of the free-field motion. If the foundation is located well below the grade level, as occurs in buildings with subterranean levels, then the effective motion at depth may be lower than the motion that occurs at the surface. This latter kinematic interaction effect is referred to as the *embedment effect*. Kinematic effects are not included in ASCE 7, but are adopted in ASCE 41 (2013) and FEMA (2005).
- *Soil-foundation flexibility effects:* Forces acting between the soil and foundation elements result in deformations of the foundation. Such effects can be especially important in the design of large mat foundations, or grade beams connecting individual footings or pile caps.

These effects reduce input ground motion, increase damping, and increase the structural vibration period. Because these effects tend to reduce design response quantities (except possibly displacements), it is usually conservative to ignore SSI in design and instead treat a building as if it has a fixed base. Increasingly, however, SSI is being considered in the assessment of existing buildings and performance-based design of new buildings. For more detailed discussion of these effects and their implementation, see ASCE 41 (2013), FEMA (2009), Klemencic et al. (2012), and Stewart et al. (2013).

16.4 Geotechnical Investigation Report

Collaboration between the structural engineer and geotechnical engineer can lead to better geotechnical and foundation information, which in turn can lead to improvements in the overall analysis and design of the structure. Typical types of required information are summarized below.

The geotechnical engineer will generally require the following information:

- Project location.
- Description of the building, including height, footprint size, basement depth, foundation systems under consideration, structural occupancy or "risk" category, and approximate fundamental period.
- Applicable building codes, standards, and design guidelines.
- Foundation loads, including both allowable stress design combinations and factored load combinations.
- Indication of whether geotechnical foundation design will be based entirely on allowable stresses, or whether a load and resistance factor design (LRFD) approach will be used for some aspects of design.
- Restrictions on differential and total settlement; commonly the geotechnical engineer will make recommendations on these parameters, but project-specific limitations should be conveyed to the geotechnical engineer where required.
- Seismic performance objectives, including hazard levels that are considered.
- Approach to structural modeling, whether site-specific response spectra will be required, whether ground motions for response history analysis will be required, and preferred methods of scaling seed ground motions as applicable.
- Information on whether soil–structure interaction will be included in the design, and the degree of sophistication anticipated.

The structural engineer will generally require the following information:

- Geological description of the site.
- General soil conditions, including soil boring logs and field and laboratory test results.
- Measured (or estimated) and design groundwater elevation, and recommendations for dewatering, if necessary.
- Maximum frost depth, if any.
- Site seismicity information, including information necessary for calculating code design values (such as site class), and assumptions and methods for developing response spectra.
- Ground motion records, if required for response history analysis, indicating methods used to select seed motions; scaling methods; methods of

considering fault-normal/fault-parallel motions; and measures of ground motion suitability, including comparison with original motions, response spectra, and durations.

- Identification of potential geological and seismic hazards, including slope stability, liquefaction and lateral spreading, differential settlement, and surface faulting. If such hazards are present, include recommendations for hazard mitigation.
- If soil–structure interaction is going to be considered in design, recommendations on the effects of base slab averaging, embedment, and soil damping and dashpots. Additional SSI information may be required for specific foundation types, as noted subsequently.
- Foundation recommendations, including recommending the type of foundation system, anticipated differential and total settlements, drainage requirements, advisability of using on-site materials, compaction criteria, hydrostatic uplift forces, required moisture barriers, and determination of whether soil design parameters can be increased for short-term loading such as wind or earthquake loads.
- For shallow foundations, the following information is usually required:
    - *Allowable soil bearing pressures* under dead, dead plus live, and dead plus live plus wind/seismic cases, and the assumed factor of safety. Identify whether bearing pressures are a function of loaded area.
    - *Ultimate soil bearing capacities* for designs in which an LRFD approach will be applied to geotechnical foundation design under load combinations including earthquake. Identify whether these are a function of loaded area.
    - Coefficient of friction for sliding resistance, passive resisting pressure, and whether these can be combined. Provide allowable stress values. Also provide ultimate values where an LRFD approach will be applied to the geotechnical foundation design.
    - Where more exact displacement analysis is required, and for mats in general, vertical and horizontal modulus of subgrade reaction, spring constants, and dashpots, for both static and transient loadings. Lower and upper bound values as may be required for bounding analyses. Effects of mat size on provided quantities. Guidance on iteration procedures to account for consolidation where needed.
    - Anticipated factors of safety for design values.
- For deep foundations, the following information is usually required:
    - Size and type of deep foundation elements recommended.
    - Allowable end bearing and skin friction values at allowable levels and at ultimate levels.

- Where applicable, depths to bearing stratum, and recommendations for penetration into the stratum.
- Allowable uplift loads.
- Minimum element spacing and minimum time between the placement of elements.
- Allowable lateral loads and corresponding displacements for both fixed and pinned pile head conditions, plus moment and shear diagrams over the height of piles consistent with the displacement and end fixity conditions. Report both expected values and upper/lower-bound estimates.
- Information on pile group effects for both axial and lateral loading.
- Special construction and inspection requirements.
- Load test requirements (if any), and also integrity testing requirements (if any).

- For excavations, the following information is usually required:
    - Maximum slope of embankments during construction, and soil treatment necessary to stabilize slopes during construction.
    - Underpinning and shoring recommendations.
    - Tie-back recommendations.
- For retaining walls, including basement walls, the following information is usually required:
    - Design values for cantilever and restrained retaining walls, including soil pressures (active, at-rest, and passive), seismic and non-seismic surcharge data, and recommended approach for designing for seismic soil pressures.
    - Recommendations on use of basement walls to resist lateral loads, including passive pressure on walls that are perpendicular to lateral forces and friction along walls that are parallel to lateral forces.
- Recommendations for slabs-on-ground.
- Existing building conditions, if any.

For additional discussion on communication between the structural engineer and geotechnical engineer, see, for example, Stewart et al. (2013) and Klemencic et al. (2012).

16.5 Foundation Performance Objectives and Design Values

Foundation performance under sustained gravity loads should consider settlement (both total and differential), soil bearing capacity, and structural capacity of the foundation elements. Foundation failure under sustained gravity loads is usually associated with excessive total or differential settlement of the foundation elements, rather than soil bearing or foundation element structural failure. Thus, design practice for geotechnical foundation design (i.e., the stresses on the soils) has been based on allowable stresses under service load combinations, with the soil allowable stresses limited to values that will control settlements to acceptable values while providing a large factor of safety against bearing failure.

Wind and seismic loads induce transient effects on the soils, such that the primary concern is bearing failure rather than long-term settlement. For this reason, allowable stresses specified for gravity loads usually are allowed to be increased by one-third for service-level load combinations including wind or seismic effects. Larger increases for dynamic effects may be considered where recommended by the geotechnical engineer.

Although U.S. building codes currently emphasize allowable stress design for geotechnical foundation design, design practice has been moving toward an LRFD approach for seismic design. The NEHRP Recommended Seismic Provisions (FEMA, 2009) includes commentary and a resource paper outlining an acceptable approach, and the NEHRP Recommended Seismic Provisions: Design Examples (FEMA, 2012) presents foundation design examples. That approach is adopted in this chapter, with some minor revisions.

Elastic foundation behavior usually is intended under sustained loads, with uplift of shallow foundations usually avoided. For wind loads, elastic response still is the usual intent, but some uplift is permitted by some design offices. For large transient loads, such as those that occur under design seismic loads, some uplift and nonlinear response of soils may be acceptable, provided overturning resistance requirements are satisfied. Subsequent sections on specific foundation elements address these topics in greater detail.

Performance expectations under seismic loads are not prescribed by most codes. Practices vary depending on the experiences and philosophy of the individual design office, as well as the expectations of the client. For example, on the West Coast of the United States, typical practice for residential and office occupancies (Risk Category II of ASCE 7) is to design the foundation elements to resist forces determined from the code analysis with forces reduced by response modification coefficient R as permitted for design of the superstructure, without any consideration of superstructure overstrength. A foundation that is designed by this practice can be expected to experience some inelastic response when subjected to design-level earthquake shaking. This practice is accepted by many design offices (and apparently the profession in general) because of the paucity of field experience demonstrating inadequate building performance due to failure of the foundation system under earthquake shaking. The exceptions would be large differential movements where foundation elements are inadequately tied together, or gross foundation failure due to soil liquefaction.

Other design offices generally follow the procedure of the preceding paragraph, but apply some additional conservatism to design for brittle failure modes, including design for shear and axial compression. This practice allows for some inelastic response of foundations, with the intent that inelastic response be restricted to ductile response modes. In this text, this approach is recommended as a minimum.

Some design offices, however, consider inelastic response of foundation elements unacceptable because of limited ductility capacity in foundation elements and because of the challenges associated with inspection and repair of foundations following an earthquake. These offices generally follow a *strong foundation* design philosophy in which the foundation elements are designed to resist the demands that the superstructure is capable of delivering to the foundation. The strong foundation design approach not only produces a foundation that is less likely to be damaged, but also results in more predictable and controlled behavior of the entire structural system, as nonlinear response is restricted to superstructure elements whose behavior is better understood and whose components can be detailed for anticipated actions. Therefore, this approach may be especially appropriate for foundations supporting buildings with higher performance objectives, including hospitals and other facilities that must be operational shortly following an earthquake. Some sophisticated clients or facility-specific standards may require the use of the strong foundation design approach.

To limit nonlinear response in foundation elements, different techniques are sometimes used. Techniques include:

- Designing to avoid failure modes of low ductility, such as shear failure. One method of accomplishing this is to design for required moment strength, then increase design shear forces by flexural overstrength factor ϕ_o, with ϕ_o taken equal to 1.25 or greater.
- Detailing minimum reinforcement beyond code minimum requirements to promote ductile rather than brittle (under-reinforced) behavior.
- Using reduced R factors for the foundation relative to the R factors used for the superstructure. For example, a common practice in some offices is to use $R = 6$ for foundation elements where design of the superstructure is based on $R = 8$. Note that this is not quite equivalent to applying a global overstrength factor of $8/6 = 1.33$, as the factor R applies only the earthquake-related forces.
- Using code overstrength factor Ω_0. See Chapter 11 for Ω_0 factors from ASCE 7.
- Designing the foundation to be capable of developing the strength of the superstructure elements framing into the foundation.
- Obtaining foundation design forces from results of nonlinear response history analyses of the superstructure.

In any design approach, design of foundation elements for shear is based on the design shear strength ϕV_n. According to ACI 318, for structures that are designed to resist earthquake effects, E, using (a) intermediate precast structural walls in Seismic Design Category D, E, or F, (b) special moment frames, or (c) special structural walls, ϕ is required to be 0.60 if the nominal shear strength of the member is less than the shear corresponding to the development of the nominal flexural strength of the member. Otherwise, $\phi = 0.75$. It is unclear that this provision was intended to apply to foundation design. Common practice is to use $\phi = 0.75$ for shear design of foundations, without consideration of whether a capacity design approach was used to determine the shear forces.

The following sections present design considerations for different foundation elements, including discussion of elastic and inelastic response characteristics, proportioning, and reinforcement detailing. These considerations can be helpful in selecting a performance objective for foundation system design.

16.6 Spread Footings

16.6.1 Behavior and Analysis Considerations

Design of spread footings should consider intended behavior. Figure 16.6 illustrates various options.

Figure 16.6*a* illustrates the hinged column concept, in which a special detail is provided to reduce column moment strength at the connection with the footing, thereby reducing moment transfer. For example, the concrete section can be reduced at the interface by providing an elastomeric material that reduces contact around a central shear key. The shear key is detailed with small diameter dowel bars providing continuity and shear transfer capacity through shear-friction. This detail is not commonly implemented in buildings, but finds widespread use in earthquake-resistant bridge construction.

Figure 16.6*b* illustrates a yielding column with a strong footing. The column details should enable the development of required rotations. Appropriate details may include closely spaced transverse reinforcement and continuous column longitudinal reinforcement. For columns that are part of the seismic-force-resisting system, either lap splices should be removed from the potential yielding zone or Type 2 mechanical splices should be provided, and hooks on longitudinal bars should be oriented toward the center of the column. For columns not intended as part of the seismic-force-resisting system, lap splices of longitudinal bars may (or may not) be acceptable, and hooks on longitudinal bars can be oriented away from the center of the column.

Figure 16.6*c* illustrates a rocking footing in which rotation requirements are accommodated through footing uplift and rocking. This behavior requires

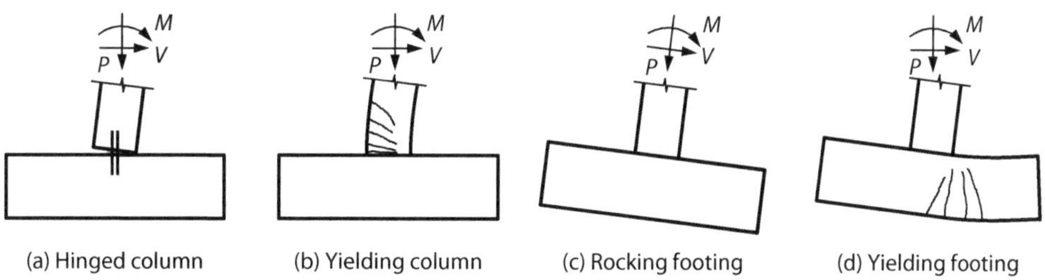

Figure 16.6 Schematic behavior options for spread footings.

checking the soil bearing capacity to determine if the footing will remain stable. See Section 16.6.2 for additional discussion.

Figure 16.6*d* illustrates a yielding footing. Unless overlying soil is capable of holding the footing down and forcing yielding under negative bending moment (top in tension), yielding of the footing will occur in only one direction. Consequently, the footing will develop a "bowl" shape after several yielding cycles, reducing its ability to resist rocking motions. Based on this behavior aspect, some engineers opt not to use yielding footings. Many other engineers prefer this yielding mechanism to yielding of the column. Where this option is selected, a capacity design approach should be employed to promote flexural yielding rather than more brittle shear failure. Following the approach commonly used for special moment frames, a reasonable approach is to assume the footing will develop a probable moment strength approximately 1.25 times the nominal moment strength. An overstrength factor can thereby be developed, which in turn can be applied for shear design.

Spread footings usually are modeled as rigid elements supported on soil. Soil stresses are usually assumed to vary linearly across the footing depth, except nonlinear soil pressure variation is sometimes considered for load combinations including earthquake effects (see Section 16.6.2). Where nonlinear soil pressure or uplift is permitted, the footing analysis problem is nonlinear, such that principles of linear superposition no longer apply. Consequently, the footing must be analyzed for each load combination to determine acceptability. Analysis for individual load cases, followed by superposition of the effects of those load cases, is incorrect under such conditions.

16.6.2 Geotechnical Considerations

Preliminary plan dimensions of footings usually are determined by consideration of allowable bearing pressure under service-level dead plus live loads. To minimize differential settlements, target bearing pressures should be similar for all footings. To minimize rotation of individual footings, sustained dead plus live loads should be concentric with the centroid of the footing, thereby producing uniform bearing stresses under individual footings as in Figure 16.7*b*. It may not always be possible to achieve a uniform stress condition, in which case it is permissible to have nonuniform stresses under individual footings, as in Figure 16.7*c*. Uplift under sustained loads (Figure 16.7*d*) is typically considered unacceptable.

To check whether bearing pressures under sustained loads are within allowable bearing pressure limits, the eccentricity $e = M/P$ is calculated, where e = eccentricity of load relative to the geometric centroid of the footing plan area, and M and P are moment and axial force under service load combinations. For $e = 0$, bearing stresses are uniform (Figure 16.7*b*). For $0 < e \leq h/6$, bearing stresses are linearly distributed with no uplift (Figure 16.7*c*). For $h/6 < e \leq h/2$, uplift occurs (Figure 16.7*d*). The footing is unstable for $e > h/2$. Figure 16.7*b*–*d* presents formulas for calculating bearing pressures. The preceding cases are for eccentricity that is measured parallel to dimension h of a rectangular footing. Similar expressions can be developed for biaxial eccentricity.

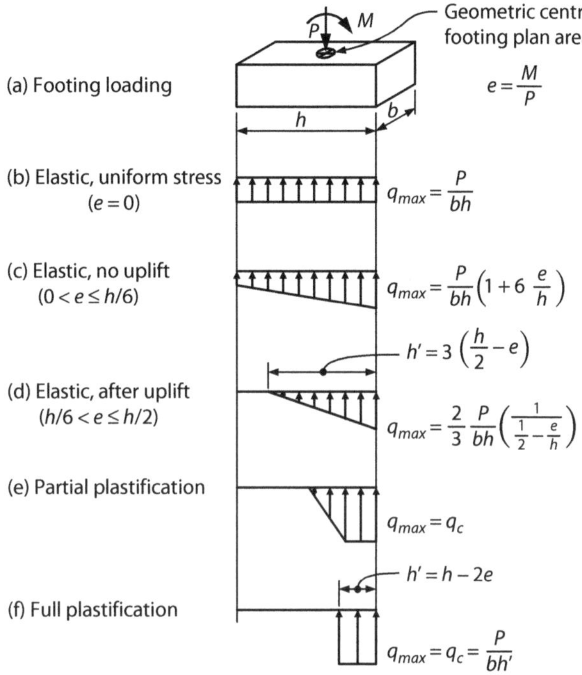

FIGURE 16.7 Soil pressure distributions (based on the usual simplifying assumption that a rigid footing develops linearly varying bearing pressures under elastic loading conditions).

Under transient loads for which linear elastic response is intended, such as combined gravity and wind/seismic loads, some uplift of individual footings may be permissible, provided the bearing stresses are within allowable values (Figure 16.7*d*). For service-level load combinations including wind or seismic loads, it may be permissible to increase allowable stresses based on consideration of loading rate, as noted in Section 16.5.

As noted previously, many codes emphasize allowable stress design, but some design offices have begun to use LRFD for load combinations including seismic effects. Where this approach is used, the loading should be based on factored load combinations intended for strength design, and the resistance should be based on ultimate foundation load capacities reduced by appropriate strength reduction factors, ϕ. According to the NEHRP Recommended Provisions, Part 3, Resource Paper 4 (FEMA, 2009), if ultimate foundation load capacities are determined based on geotechnical site investigations including laboratory or in situ tests, ϕ factors equal to 0.8 for cohesive soils and 0.7 for cohesionless soils should be used for vertical, lateral, and rocking resistance for all foundation types. If ultimate foundation load capacities are determined based on full-scale field testing of prototype foundations, ϕ factors equal to 1.0 for cohesive soils and 0.9 for cohesionless soils are recommended.

Under the LRFD approach, analysis for bearing stresses depends on whether elastic soil response is intended, or whether inelastic soil response is considered

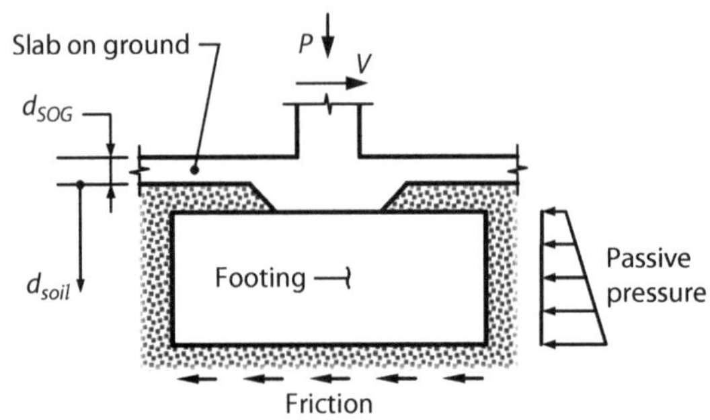

FIGURE 16.8 Friction and passive pressure acting on footing.

acceptable. Where elastic soil response is intended, the maximum stress should be calculated in accordance with Figure 16.7*c* or *d*, as appropriate. Usually, however, it is acceptable to consider inelastic soil response, in which case the maximum resistance can be calculated considering full plasticity of the soil (Figure 16.7*f*). The intermediate response identified as partial plastification (Figure 16.7*e*) is permitted but is not generally checked, as there is no advantage over simply checking the resistance based on full plastification.

Horizontal shear is transferred from a footing to the surrounding soil through friction at the base and passive bearing pressure along the leading face (Figure 16.8). The geotechnical report should indicate whether friction and passive pressure can be combined (i.e., assumed to act at their maximum values simultaneously) or whether they can only be considered separately or in combination at reduced values. Where the allowable stress method is used, allowable friction and passive pressures should be specified. Where seismic design will use the LRFD method, ultimate friction and passive resistance should be specified.

The frictional resistance is proportional to the normal force associated with column axial load, overlying soil, and footing weight. The minimum value of normal force associated with each design load combination should be used to determine frictional resistance. Friction coefficients around 0.4 to 0.65 are typical. If service-level friction coefficients are specified for use with sustained loads, then an increase in friction coefficient may be permitted for load combinations including wind and seismic loads based on consideration of load rate effects. If ultimate friction values are specified, then the design friction force is equal to $\phi\mu P$, in which ϕ = the strength reduction factor as defined previously, μ = the friction coefficient, and P = the minimum bearing force on the bottom of the footing consistent with the design load combination.

Passive pressure may be expressed using a passive pressure coefficient K_p, in which case the passive pressure at depth d_{soil} is $K_p\gamma d_{soil}$, where γ = unit density of soil. Alternatively, passive pressure may be expressed directly as weight per unit volume, in which case the passive pressure at any depth is the product of the d_{soil} and the weight per unit volume. For rock, passive pressure may be expressed as a constant at the upper surface, increasing linearly with depth. If service-level passive pressure is specified for use with sustained loads, then an increase in passive pressure may be permitted for load combinations including wind and seismic loads based on consideration of load rate effects. If ultimate passive pressure is specified, then the design passive pressure is reduced by the strength reduction factor, ϕ, as defined previously.

Where isolated footings are inadequate for bearing pressures or lateral forces, combined footings and grade beams can be used to spread the loads out into the foundation. These subjects are discussed in greater detail in Sections 16.6.4, 16.6.5, and 16.9. The NEHRP Recommended Seismic Provisions: Design Examples (FEMA, 2012) presents complete design examples for spread footings, including combined footings.

16.6.3 Footing Design and Reinforcement Details

Whereas geotechnical foundation design may be done by either the allowable stress or strength design methods, structural design of spread footings is always done by the strength design method. Footing design follows the usual procedures, which are covered in most textbooks on reinforced concrete design. This text provides only a brief review of the key points.

Footing depth and details are generally selected to simplify construction. As is common for gravity load design, footing depth is usually established so that shear reinforcement is not required. Both one-way shear and two-way shear requirements are checked for footings supporting columns or boundary elements. Wall (or strip) footings are subjected to bending in one direction, so only one-way shear is checked.

For one-way shear, the critical section is located a distance d away from the column face, where d = the average of the effective depths of the footing bottom reinforcement in the two orthogonal directions (Figure 16.9*a*). For steel columns with base plates, the critical section is measured a distance d away from lines halfway between the face of the column and the edge of the base plate. In U.S. practice, design shear strength is ϕV_n, in which $\phi = 0.75$, $V_n = V_c + V_s$, and V_c is defined as

$$V_c = 2\lambda\sqrt{f'_c}b_w d,\ \text{psi}\ (0.17\lambda\sqrt{f'_c}b_w d,\ \text{MPa}) \tag{16.1}$$

Shear reinforcement is not required if $V_u \leq \phi V_c$.

For two-way shear, the critical section is located a distance $d/2$ away from a concrete column face, or $d/2$ away from lines halfway between the face of a steel column and the edge of its base plate (Figure 16.9*b*). In U.S. practice, design shear strength is $\phi V_n = \phi(V_c + V_s)$, in which $\phi = 0.75$. Where shear reinforcement is not used, V_c is defined by the least of

$$V_c = \left(2 + \frac{4}{\beta}\right)\lambda\sqrt{f'_c}b_o d,\ \text{psi}\ \left[0.17\left(1 + \frac{2}{\beta}\right)\lambda\sqrt{f'_c}b_o d,\ \text{MPa}\right] \tag{16.2}$$

$$V_c = \left(\frac{\alpha_s d}{b_o} + 2\right)\lambda\sqrt{f'_c}b_o d,\ \text{psi}\ \left[0.083\left(\frac{\alpha_s d}{b_o} + 2\right)\lambda\sqrt{f'_c}b_o d,\ \text{MPa}\right] \tag{16.3}$$

$$V_c = 4\lambda\sqrt{f'_c}b_o d,\ \text{psi}\ [0.33\lambda\sqrt{f'_c}b_o d,\ \text{MPa}] \tag{16.4}$$

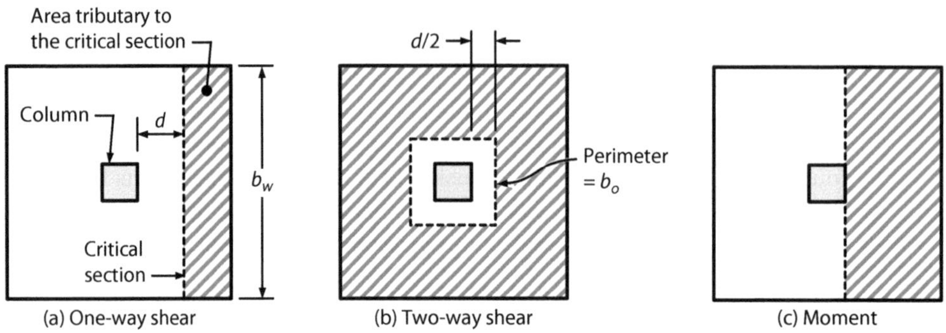

FIGURE 16.9 Critical sections for footing design for the case of concrete columns.

in which β = ratio of long side to short side of the column (or equivalent ratio of sides halfway between a steel column and its steel base plate); α_s = 40, 30, or 20 for interior, edge, or corner column locations relative to the edges of the footing; and b_o = perimeter of the critical section around the column. Where shear reinforcement is used, V_c is not to be taken greater than $V_c = 2\lambda\sqrt{f_c'}b_o d$, psi [$0.17\lambda\sqrt{f_c'}b_o d$, MPa] and V_n is not to exceed $V_n = 6\sqrt{f_c'}b_o d$, psi [$0.5\sqrt{f_c'}b_o d$, MPa].

Design of an isolated spread footing for moment strength considers the footing as a one-way flexural member first in one principal direction and then in the orthogonal direction. The critical section for concrete columns is at the face of the column, while for steel columns the critical section is halfway between the face of the column and the edge of the base plate (Figure 16.9c). In U.S. practice, for footings of uniform thickness, A_{smin} in each principal direction should result in a ratio of reinforcement area to gross concrete area of at least

(a) 0.0018 where Grade 60 (420) reinforcement is used, or

$$\text{(b)}\ \frac{0.0018\times 60{,}000}{f_y},\ \text{psi}\left[\frac{0.0018\times 420}{f_y},\ \text{MPa}\right]\ \text{where higher strength reinforcement is used} \tag{16.5}$$

At critical sections where flexural yielding is considered to be likely, additional reinforcement may be preferred to ensure that flexural strength is at least equal to the cracking strength. Some specify $M_n \geq 1.2M_{cr}$ for this case. A more conservative approach, which is not generally applied, is to satisfy the minimum steel area as required for beams, that is

$$A_{s,min} = \frac{3\sqrt{f_c'}}{f_y}b_w d \geq \frac{200}{f_y}b_w d,\ \text{psi}\left(\frac{0.25\sqrt{f_c'}}{f_y}b_w d \geq \frac{1.4}{f_y}b_w d,\ \text{MPa}\right) \tag{16.6}$$

Spacing should not be greater than 450 mm (18 in).

In wall (or strip) footings and two-way square footings, reinforcement generally is distributed uniformly across the entire width of the footing. In two-way rectangular footings, reinforcement in the long direction generally is distributed uniformly across the width of the footing. In the short direction, a portion of the total reinforcement, $\gamma_s A_s$, is distributed uniformly across a strip (centered on the centerline of the column or boundary element) having width equal to the length of the short side of the footing, with the remainder of the required reinforcement, $(1 - \gamma_s)A_s$, distributed uniformly outside the central strip; $\gamma_s = 2/(\beta+1)$, where β is the ratio of the length of the long side of the footing to the length of the short side of the footing. An alternative is to provide the more dense reinforcement uniformly through the footing to avoid multiple spacings. Reinforcement is to be developed for tension or compression, whichever applies, with critical sections as described previously for moment reinforcement.

Longitudinal reinforcement of columns and structural walls that resist forces induced by earthquake effects is required to extend into the footing, mat, or pile cap, and be developed for tension at the interface (Figure 16.10). The usual requirement is to provide development length for f_y, except the development length should be for $1.25f_y$ where a wall yielding region occurs at the interface. For a wall boundary element, hooked bars, if required, are permitted to be hooked away from the center of the boundary element. For columns designed assuming fixed-end conditions at the foundation, hooked bars, if required, should have the 90° hooks near the bottom of the foundation with the tail of the hook oriented toward the center of the column. This requirement is intended to promote development of a diagonal compression strut across the joint, as required for beam-column joint shear resistance (Chapter 9). Caltrans (2003) uses an alternative approach for footing design for bridge structures.

ACI 318 requires that transverse reinforcement of a wall boundary element extend into a footing, mat foundation, or pile cap at least 12 in (305 mm) (Figure 16.10). Where a column or boundary element edge occurs within $h/2$ of the edge of a footing, mat foundation, or pile cap, transverse reinforcement occurring just above the interface is to be extended into the footing along the required development length of the longitudinal bars. This latter requirement is intended to reinforce the edge of the footing and prevent splitting failure from application of concentrated forces at the free edge.

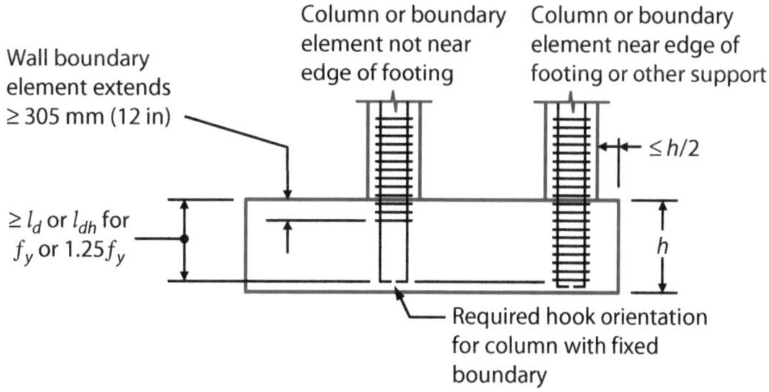

Figure 16.10 Required extensions of column and boundary element reinforcement into footings, mat foundations, and pile caps.

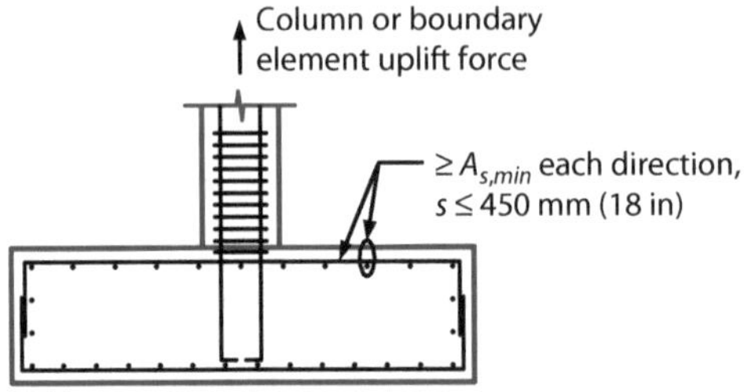

FIGURE 16.11 Top reinforcement required for uplift in footings, foundation mats, and pile caps.

Where earthquake effects create uplift forces in boundary elements of special structural walls or columns, ACI 318 requires longitudinal reinforcement in the top of the footing, mat foundation, or pile cap to resist actions resulting from the design factored load combinations (Figure 16.11). Uplift resistance can be provided by overlying soil, footing weight, shear from grade beams framing into the footing, shear in combined footings, tension piles, or combinations of these. The provided reinforcement should not be less than the minimum required by Eq. (16.6), although sometimes less is specified considering the low consequences of cracking. Notwithstanding these ACI requirements, a good practice is to provide at least nominal top reinforcement for all footings supporting buildings subjected to earthquake effects, whether uplift forces are present or not.

Some engineers routinely hook all footing longitudinal reinforcement, overlap the tails of the hooks along the outside faces, and provide skin reinforcement for thick footings, as shown in Figure 16.12. This creates a closed cage to contain the core concrete and improve integrity of the footing. This is not, however, standard practice.

16.6.4 Combined Footings

Combined footings are footings supporting two or more columns or boundary elements. Combined footings are common where site or property line restrictions require one column near the edge of a footing, as shown in Figure 16.12*a*. If an

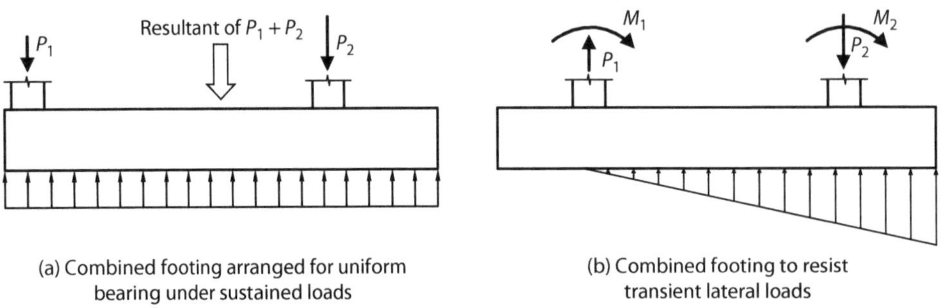

FIGURE 16.12 Design conditions for combined footings.

isolated footing was used, the soil pressures would be highly eccentric, resulting in long-term footing rotation. By combining two column loads on a combined footing, the centroid of the sustained loads and of the combined footing area can be designed to coincide, thereby producing uniform bearing pressure.

Combined footings are also common where overturning moments from lateral loads exceed bearing capacities of reasonably sized isolated footings. A combined footing increases overturning capacity, and may reduce footing dimensions to reasonable values. Where this solution is insufficient, mat foundations, deep foundations, or various foundation outrigger systems can be considered. Subsequent sections describe these other systems.

16.6.5 Foundation Ties

ASCE 7 and the IBC require that individual spread footings that are founded on soil classified as Site Class E or F be interconnected by ties. The same requirement applies to individual pile caps, drilled piers, or caissons for Site Class C and above. Ties are required to have a design strength in tension or compression at least equal to $0.1S_{DS}P_u$, where S_{DS} = 5%-damped design, spectral response acceleration parameter at short periods, and P_u = the larger pile cap or column factored dead plus factored live load in accordance with the load combination including E. According to the IBC, the provided tie strength need not exceed one-quarter of the smaller footing design gravity load. As an alternative to individual ties, the tie force can be provided by reinforced concrete beams within slabs-on-ground or by the slab-on-ground itself. ASCE 7 permits the tie requirement to be waived where competent rock, hard cohesive soils, very dense granular soils, or other approved means provide confinement to the individual footings, pile caps, drilled piers, or caissons. Some engineers disallow the use of competent soil in lieu of tie beams in buildings, but would consider that approach for some structures with wide column spacing, such as parking structures.

ACI 318 provides detailing requirements for grade beams and slabs-on-ground applicable to buildings in the highest seismic risk categories (D, E, and F). Grade beams designed to act as horizontal ties between pile caps or footings are required to have longitudinal reinforcement that is continuous through footings. At discontinuous locations, such as perimeter footings, the longitudinal reinforcement is to be developed in the footing, preferably providing straight or hooked bar development beyond the centerline of the column or boundary element. The smallest cross-sectional dimension should be at least the smaller of (a) the clear spacing between connected columns divided by 20 and (b) 18 in (460 mm). Closed ties are required at a spacing not to exceed the lesser of one-half the smallest cross-sectional dimension and 12 in (300 mm). Figure 16.13*a* shows required details, in this example for a grade beam at the same upper elevation as the footing, but below the slab-on-ground. The connection detail in Section A-A is described later in this section.

According to ACI 318, grade beams designed as beams to resist moments due to framing action between adjacent footings or pile caps should have details consistent with the details used in special moment frame beams in the superstructure.

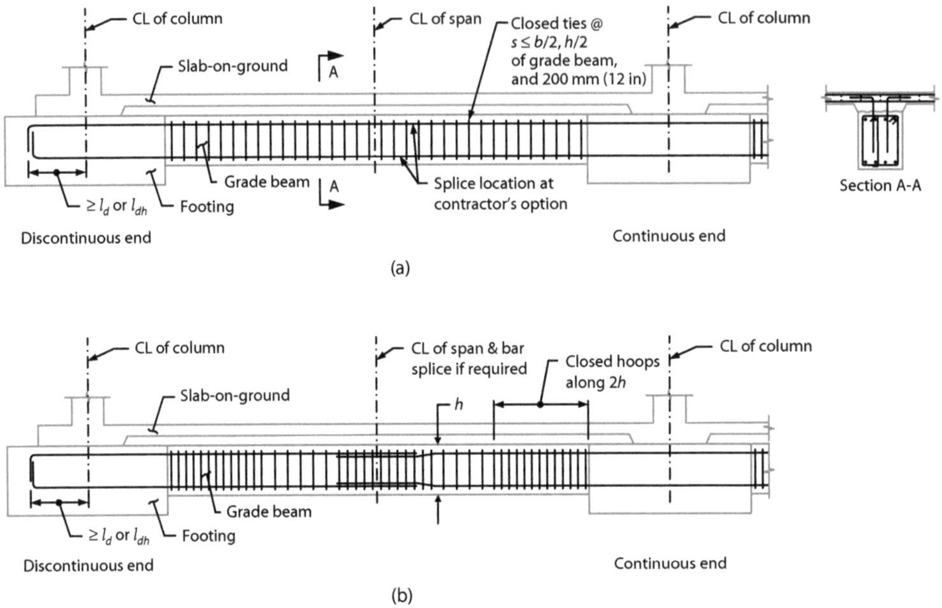

FIGURE 16.13 Dimensional and detailing requirements for grade beams: (*a*) grade beam designed as a tension or compression tie; (*b*) grade beam designed to resist moment from column or wall.

Figure 16.13*b* illustrates the required details. The IBC includes an exception to this detailing requirement for grade beams designed to resist the maximum forces that can be imparted from the superstructure to the grade beam. Maximum forces can be from nonlinear analysis, limit analysis, or from code-based linear analysis with forces amplified by overstrength factor Ω_0.

Sometimes it is advantageous to integrate the grade beam with the slab-on-ground. Figure 16.14 illustrates one concept for this detail. In this example, the grade beams are wider than the frame column, so the column hoops are discontinued at the top of the grade beam and an enlarged column cage is formed below that level and extending into the footing. An alternative detail consists of thickening the slab down to the top of the footing and providing vertical hairpins between the top of the footing and the tie beam. The details should be consistent with the intended framing action of the column-foundation system.

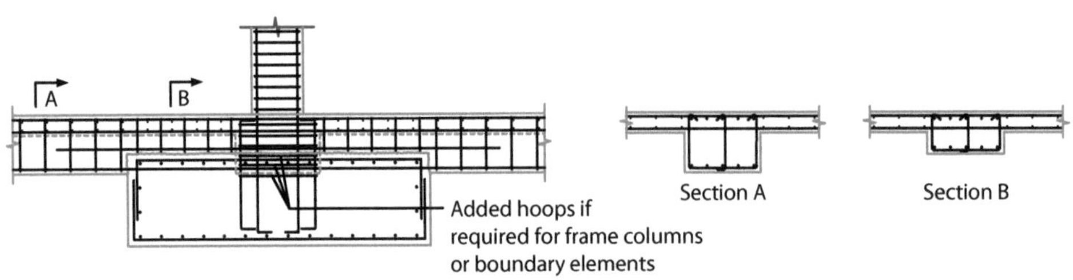

FIGURE 16.14 Possible detailing for grade beam designed as tension or compression tie, with top elevation above footing elevation. Reinforcement for orthogonal grade beam details not shown.

Slabs-on-ground that resist seismic forces from walls or columns that are part of the seismic-force-resisting system should be designed as structural diaphragms (Chapter 15). Shear transfer from a slab-on-ground can be accomplished through grade beams. Figure 16.13*a*, Section A-A shows a connection between a structural slab-on-ground using dowels to transfer horizontal shear. Figure 16.14 illustrates grade beams integral with the slab-on-ground. Where a slab-on-ground acts as a structural diaphragm, the contract documents should clearly indicate that the slab-on-ground is a structural diaphragm and part of the seismic-force-resisting system.

16.7 Mat Foundations

The following discussion includes brief consideration of intended performance and behavior, structural analysis, geotechnical considerations, and structural proportioning and detailing for mat foundations. For additional discussion, see Klemencic et al. (2012).

16.7.1 Behavior and Analysis Considerations

Mat foundations are, in a sense, large combined footings that support multiple vertical elements of the structural system. For smaller mat foundations, behavior and analysis can be similar to that for isolated or combined footings. This includes modeling the mat as a rigid footing element. As mat foundation plan dimensions increase, out-of-plane deformations of the foundation become increasingly important and should be considered in modeling and analysis.

Possible behavior modes for mat foundations are similar to those for spread footings (Figure 16.6). Pinned connections between vertical elements and the mat foundation, although feasible, are somewhat unusual in practice. A rocking mat foundation with linear superstructure is also unusual, although some uplift and foundation rocking is acceptable under transient loads. Main behavior options are yielding superstructure, yielding mat, or a combination of both.

A common approach is to design the mat foundation and the superstructure for the same design load combinations, including using the same response modification coefficient R for earthquake-related forces. This implies that some inelastic response may occur in the foundation. To promote reduced degree of inelastic response in the foundation, the structural engineer sometimes will increase some or all of the foundation design forces. This is a common approach for tall buildings designed by performance-based design procedures (TBI, 2010). Section 16.5 describes some techniques used to limit inelastic response in foundation elements.

Soil–structure interaction effects can be pronounced in mat foundations. Mat foundations typically extend across large footprints, leading to kinematic effects that reduce (especially) higher frequency input motions. In buildings with subterranean levels, embedment effects also reduce the free-field motions. Therefore, it is common to adjust the input motions for buildings with mat foundations based on considerations of soil–structure interaction (Section 16.3).

As mentioned previously, mat foundation deformations can have an important effect on the soil bearing pressures and settlements, both for sustained gravity loads and for transient wind and earthquake loads. To represent these effects, large mat foundations and the supporting soils commonly are modeled using finite-element analysis. For strength design of the mat itself, it is typical to consider gross-section properties, with a thick plate element formulation to include effects of shear deformations. For settlement evaluation, a sensitivity study is sometimes done using reduced stiffness to account for mat cracking. Stiff superstructure elements such as shear walls and basement walls affect overall stiffness of the mat. Therefore, the analysis model should include elements to represent the in-plane stiffness of these walls. Where elevator pits create depressions in the mat, this effect should also be modeled.

Finite-element analysis generally results in nonuniform bearing pressure on the soil under the mat (see Section 16.7.2 for additional discussion). This nonuniform pressure is to be used for design of the mat itself. Because bearing pressures and potential uplift may be a nonlinear function of the applied loads, superposition of individual load cases is not permitted. Instead, each load combination must be analyzed separately, and the design then must consider the envelope of the results.

Although a complete model of the superstructure, the foundation, and the soils can be developed and analyzed, this is not always done. A common approach is to analyze a simplified model of the soil-foundation-structure system, and then apply the resulting foundation forces to a more detailed finite-element model of the mat foundation and supporting soil. Figure 16.15*b* and *c* illustrates simplified models that are sometimes used. Soil spring and dashpot properties are recommended by the geotechnical engineer. See Section 16.7.2 for additional discussion on geotechnical aspects.

Regardless of the sophistication of the analytical model, there remain many unknowns and approximations in both soil and structural properties. It is advisable to check the sensitivity of the results (bearing pressures, settlements, moments, and shears) to the assumptions of the analytical model. Sometimes bounding analyses

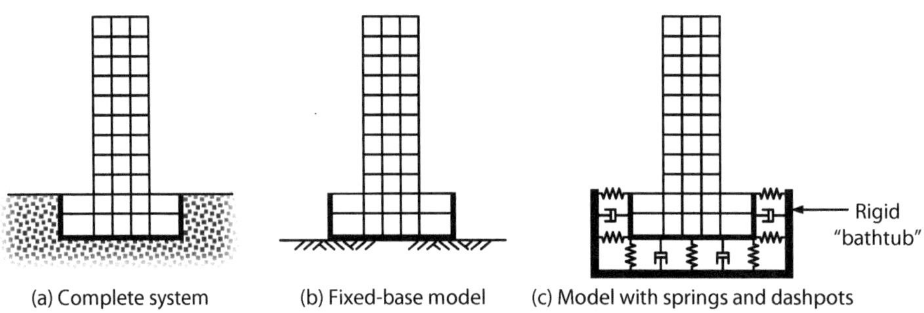

Figure 16.15 Schematic illustration of building with subterranean levels: (*a*) complete system, (*b*) fixed-base model, and (*c*) model with soil stiffness and damping represented by springs and dashpots. (*After TBI, 2010, used with permission from University of California, Berkeley.*)

are done, in which best estimate properties, and then "upper" and "lower" bound properties, are assumed, with the design based on the envelope of results obtained.

16.7.2 Geotechnical Considerations

Preliminary sizing of a mat foundation often is based on analysis using the rigid footing assumption and allowable bearing pressure. See methods presented in Section 16.6.2. Suitability of the preliminary size can be checked using the rigid footing assumption for smaller mat foundations, or using finite-element analysis considering mat and soil flexibility.

For finite-element analysis, the subgrade modulus is needed. Dishing of the mat foundation can result in differential deformations of the soil and variable bearing pressures. Under some conditions, this may require variable subgrade moduli. Iterations may be required to ensure a no-tension solution. Preferably, the software is capable of automating these iterations. The iterations are needed for each load combination, as the solutions involve nonlinear stress distributions that cannot be superimposed.

When the mat foundation analysis is completed, resulting bearing pressure distributions and deformation plots should be reviewed by the geotechnical engineer to verify that calculated bearing pressures fall within an acceptable range and are compatible with the deformations. Some local peak overstress is commonly accepted, provided the average over a defined region is within acceptable values. The geotechnical engineer may also use the results to calculate settlements, and may revise subgrade moduli based on results. Iteration with the geotechnical engineer continues until acceptable and compatible bearing pressures and displacements are achieved.

The design also needs to provide adequate lateral force resistance. Sliding resistance can be provided by friction beneath the mat and by passive resistance of surrounding soil acting against the mat and the basement walls. Sliding and passive resistance are mobilized at different displacements. Thus, the geotechnical engineer needs to specify whether both friction and passive resistance can be mobilized at the same time.

16.7.3 Mat Foundation Design and Reinforcement Details

Design for shear and moment resistance is similar to design of footings. Where finite-element analysis is used, the shears and moments need to be integrated over strips of defined width to produce strip design shears and moments. Some local overstress often can be made acceptable by averaging across strips, although caution is needed to avoid overlooking local zones of high stress.

As with footings, the traditional approach is to select mat foundation thickness so as to avoid the use of shear reinforcement. Although current building codes are silent on this issue, this approach may be inappropriate for seismic designs where thick concrete elements will be subjected to multiple stress reversals. Some research (Reineck et al., 2003) suggests that thick concrete elements without shear reinforcement have reduced unit shear strength (possibly due to wide inclined cracks and reduced aggregate interlock). By providing shear reinforcement, crack width is better controlled, such that the unit shear strength is not reduced with increasing thickness. Based on these considerations, many designers routinely provide mat foundation shear reinforcement unless shear stresses are well below limits prescribed by current codes. This text recommends this latter approach.

In U.S. practice, one-way shear is checked in accordance with ACI 318. There is no specific provision on effective width to be used. Some designers use the full width of mat. If, however, the seismic resistance is concentrated in one or few vertical elements, then the effective width may be less than the full width. Klemencic et al. (2012) suggest that the effective width should not exceed the width of the structural member producing the shear plus one mat thickness on either side of that member. For one-way shear, ACI 318 defines the design shear strength as $2\phi\sqrt{f_c'}bd$, psi ($0.17\phi\sqrt{f_c'}bd$, MPa) for sections without shear reinforcement. Based on consideration of reduced unit shear strength for deep sections, this text recommends that the design shear strength be $1\phi\sqrt{f_c'}bd$, psi ($0.083\phi\sqrt{f_c'}bd$, MPa) for sections without shear reinforcement. Where minimum shear reinforcement is provided, design shear strength can be based on $\phi(V_c + V_s)$, with $\phi V_c = 2\phi\sqrt{f_c'}bd$, psi ($0.17\phi\sqrt{f_c'}bd$, MPa). Minimum shear reinforcement is defined by ACI 318 as

$$A_{v,min} = 0.75\sqrt{f_c'}\frac{b_w s}{f_{yt}} \geq 50\frac{b_w s}{f_{yt}}, \text{ psi}$$

$$\left[0.062\sqrt{f_c'}\frac{b_w s}{f_{yt}} \geq 0.35\frac{b_w s}{f_{yt}}, \text{ MPa}\right] \tag{16.7}$$

Spacing of shear reinforcement should not exceed $d/2$ in both directions.

Two-way shear strength should be checked around components such as walls, braced frames, and moment-resisting columns. Design for two-way shear can be in accordance with ACI 318, but this text recommends reducing unit shear strength to one-half of the values expressed in Eqs. (16.2) through (16.4) for thick sections without shear reinforcement. Where at least minimum shear reinforcement is provided in accordance with Eq. (16.7), design two-way shear stress capacity can be based directly on Eqs. (16.2) through (16.4), except V_c is not to be taken greater than $V_c = 2\lambda\sqrt{f_c'}b_o d$, psi ($0.17\lambda\sqrt{f_c'}b_o d$, MPa) and V_n is not to be taken greater than $V_n = 6\sqrt{f_c'}b_o d$, psi ($0.5\sqrt{f_c'}b_o d$, MPa).

For moment design, design strips should be made wide enough to smooth out local moment peaks and facilitate the placement of reinforcement. It is common to select strip widths equal to bay widths. Moment strength design can be in accordance with the usual procedures of ACI 318. Some mat foundations have multiple layers of reinforcement. Effective depths should consider the multiple layers.

Design of moment reinforcement should satisfy all requirements for bar development, curtailment, spacing, and cover. The minimum code spacing of 1 in (25 mm) or $1d_b$, whichever is less, is impractical for construction of a thick mat. Proper consolidation requires that a vibrator be dropped through the full depth of the mat. Zones no less than 6 in (150 mm) square are required for this purpose

(Klemencic et al., 2012). Top layers require tall chair supports or "standees." It may be possible to use shear reinforcement for this purpose, provided the shear reinforcement extends through the full depth as is required by ACI 318 and other codes (see below).

Minimum shrinkage and temperature reinforcement is required in both directions. For Grade 60 (420) reinforcement, the minimum steel ratio is 0.0018 each way, with spacing not to exceed 18 in (460 mm). This reinforcement can be placed either top or bottom, or both. Many structural engineers split this steel between the top and bottom faces. If uplift occurs, then minimum reinforcement is required as shown in Figure 16.11.

Laps of mat moment reinforcement are common given the size of typical mat foundations. A good practice is to specify the lap splice lengths (accounting for top bar effects of the top layers of reinforcement), but to leave the location of splices as a contractor's option.

Shear reinforcement should extend as close as possible to tension and compression faces of the mat and should hook around longitudinal reinforcement (ACI 318) (Figure 16.16). A typical configuration is a single-leg stirrup having 135° hook on one end and 90° hook on the opposite end. Shear reinforcement typically is placed after the moment steel is in place. Fishing in bars with 90° hooks can be difficult. The use of bars with heads on the bottom end and 135° or 90° hooks on the top end can improve constructability at increased material cost.

Dowels will be required to connect walls and columns to the mat foundation. Connections can be by either lap splices or mechanical splices, depending on performance, construction, and cost considerations. Dowels should extend into the mat at least the required development or hook length. Some engineers are concerned that a short development or hook length into a deep mat foundation could result in a cone pullout failure unless mat reinforcement adjacent to the dowels can carry the forces deeper into the mat. Some engineers recommend carrying the dowel bars all the way to the bottom of the mat. In addition to addressing the pullout failure concern, this option may improve constructability, as hooks on these bars can rest atop the bottom layers of reinforcement rather than being suspended partway into the mat (Figure 16.16).

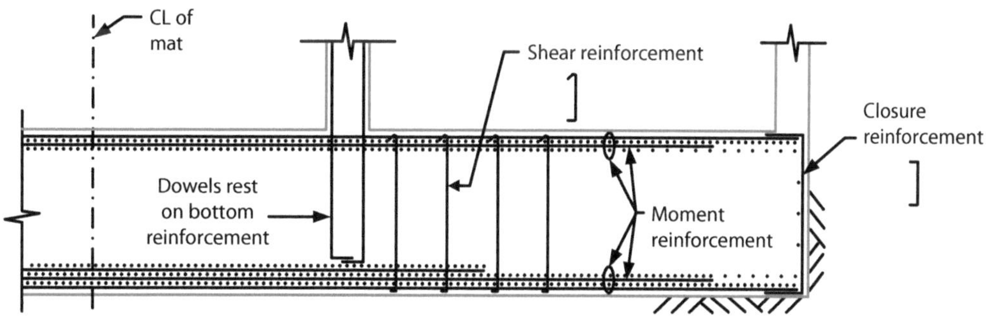

Figure 16.16 Illustration of selected mat foundation reinforcement, including moment reinforcement, shear reinforcement, closure reinforcement, and extension of column or boundary element dowels to bottom of mat.

Although not required by the building code, a good practice is to provide vertical closure reinforcement and skin reinforcement along any free edges of the mat (Figure 16.16).

As noted in the discussion of footings, ACI 318 requires wall boundary element transverse reinforcement to extend into a footing, mat foundation, or pile cap at least 12 in (305 mm) (Figure 16.10). In thick mats with multiple layers of top reinforcement, this is impractical and does not seem necessary from a performance perspective. Where a column or boundary element edge occurs within $h/2$ of the edge of a mat foundation, transverse reinforcement occurring just above the interface is to be extended into the mat along the required development length of the longitudinal bars. This latter requirement is intended to reinforce the edge of the mat and prevent splitting failure from concentrated forces applied near the free edge. Common locations where this applies are at the perimeter of a mat foundation and around elevator pits.

16.8 Deep Foundations

16.8.1 Behavior and Analysis Considerations

Deep foundation elements should be designed and constructed to withstand deformations and forces from earthquake ground motions and from structural response. Deformations should include both free-field soil strains (possibly including soil liquefaction effects and possibly modified for soil–structure interaction) and deformations due to loads imposed from the superstructure to the foundation. The pile shaft and connection with the pile cap should be capable of resisting the shears and moments associated with these deformations. Bearing actions to be considered include pile tip resistance, shaft frictional resistance, pile axial strength, and the connection between the shaft and the pile cap.

A pile subjected to lateral forces from the superstructure derives its lateral resistance from passive bearing pressure against the leading face of the pile cap as well as passive pressure against and friction alongside the pile shaft. As a result of the lateral forces acting transverse to the pile shaft, a pile develops shears and moments along its length. Figure 16.17*a* illustrates a characteristic moment profile that might develop for the case of a pile that is fixed to the pile cap, with pile cap rotation restrained. Points of potential flexural yielding are adjacent to maximum moment sections. The most vulnerable location is at the connection with the pile cap, where both shears and moments reach their highest values. Some release of the moment at the pile cap can be achieved by debonding the pile longitudinal reinforcement over a short distance within the pile cap. Inelastic flexural response can also occur around the maximum moment section at depth, although this region is less vulnerable because moment demands typically are lower and because inelastic response of the pile can spread both above and below the critical section, thereby enhancing the pile plastic rotation capacity. According to Bobet et al. (2001), the most commonly observed damage is in the pile shaft just below the pile cap.

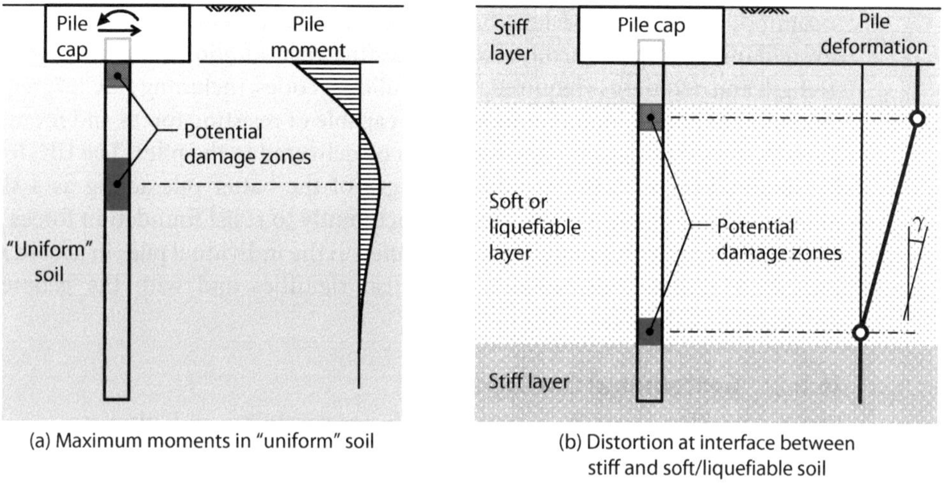

FIGURE 16.17 Potential damage zones due to lateral deformations.

Piles are commonly used where soft or liquefiable layers occur near the surface, with firmer bearing material at depth. Under site response to earthquake loading, the soft layers tend to sustain greater shear distortion than the stiff layers. This can lead to pile damage at the interfaces between the stiff and soft layers (Figure 16.17*b*). Where a layer liquefies but does not flow laterally, the pile needs to be capable of spanning through the "fluid" layer without buckling. Compaction of liquefied layers can also lead to considerable down-drag (friction along the pile shaft due to settling soils) that should be considered in design. Where a layer liquefies and spreads laterally, very large distortion can occur across the liquefied zone, leading to pile damage at the interfaces with the non-liquefied layers. This type of damage has been observed in several past earthquakes (Bobet et al., 2001).

Batter (or raked) piles are piles installed at an angle relative to vertical (Figure 16.18). Because the axial stiffness of a pile is much greater than its lateral stiffness, a batter pile provides much greater lateral resistance than an equivalent vertical pile. As such, batter piles can attract very large forces that must be

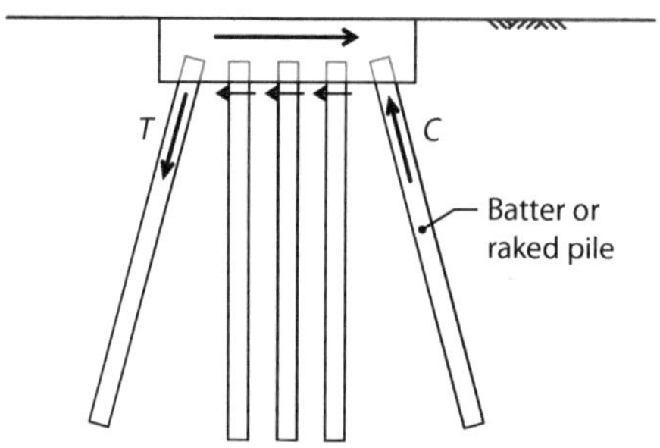

FIGURE 16.18 Batter piles. Induced moment not shown.

accommodated in the design. Batter piles in wharf structures have sustained severe damage in past earthquakes, suggesting that additional attention to their strength and detailing is required. Some building codes, including ASCE 7, require that batter piles and their connections be capable of resisting forces and moments considering the maximum forces that can be delivered to the piles. The IBC limits the required force to the nominal strength of the batter pile acting as a short column. Where vertical and batter piles act jointly to resist foundation forces as a group, the design forces should be distributed to the individual piles in accordance with their relative horizontal and vertical rigidities and with the geometric distribution of the piles in the group.

16.8.2 Geotechnical Considerations

Design of piles needs to consider both serviceability and ultimate capacity. Serviceability checks are done for unfactored loads. Foundation capacity checks have also traditionally been done under service loads, but LRFD approaches, recognizing expected strengths of the foundation element, are increasingly used for seismic design. This latter approach is especially important for existing buildings, where direct consideration of foundation performance at strength levels can lead to considerable cost savings. Some design offices are using the LRFD approach for seismic design of new buildings. ASCE 41 (2013) contains strength analysis procedures applicable to existing buildings in the United States. The NEHRP Recommended Provisions, Part 3, Resource Paper 4 (FEMA, 2009) describes a tentative LRFD approach for new buildings, and the NEHRP Recommended Seismic Provisions: Design Examples (FEMA, 2012) presents complete design examples for piles. Section 16.6.2 presents strength reduction factors for use with the LRFD approach.

Figure 16.19 illustrates forces acting on a pile cap supported by four piles. Vertical load P_{group} acting on the cap is resisted by axial force P_{pile} in the individual piles. Lateral force V_{group} acting on the cap is resisted by a combination of passive pressure $V_{passive}$ acting on the leading face of the pile cap and shear V_{pile} in the individual piles. Moment M_{group} acting at the top of the pile cap is resisted by a combination of the moments M_{pile} in the individual piles, the force couple due to axial force P_{ot} in the individual piles, and the shears V_{pile} in the individual piles and $V_{passive}$, each acting through its respective moment arm. The largest contribution

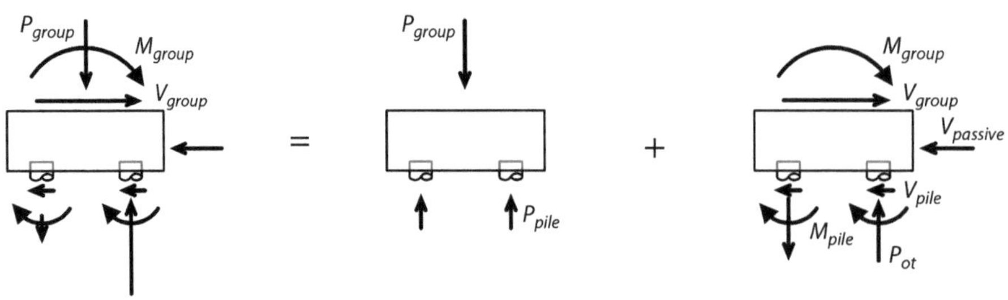

Figure 16.19 Free-body diagrams showing forces acting on a four-pile pile cap. (*After FEMA, 2012.*)

to moment resistance typically is the force couple resulting from axial force in the individual piles. The structural engineer determines each of these quantities based on information provided by the geotechnical engineer and ensures that each is within the allowable or ultimate limits.

Compressive axial load capacity of a single pile depends on skin friction, end bearing, or both. Piles within a group interact, resulting in a reduction of individual pile capacity. U.S. codes require group effects to be considered where the center-to-center spacing of piles in a group is less than three times the minimum cross-sectional dimension of the individual piles (IBC, 2012). Tensile axial load capacity (or uplift capacity) of a single pile depends on skin friction along the body of the pile plus the weight of the pile. Tensile capacity of a group of piles might also be limited by the weight of the piles, weight of soil within the pile group, and skin friction around the pile group (Caltrans, 2003). The geotechnical engineer should provide either allowable values or expected capacities, including group effects where relevant.

Moment resistance of a pile group depends in part on moments acting on individual piles, but the primary source of moment resistance is the force couple provided by compressive and tensile axial forces P_{ot} in the individual piles (Figure 16.19). In the allowable stress method, the pile cap generally is considered rigid with pile forces varying linearly across the length of the pile cap. In the LRFD method, individual piles can be assumed to have reached their tensile or compressive load capacities, without the need for a strain-compatibility analysis. For very large pile caps, or for mats supported by piles, flexibility of the cap (or mat) should be considered, and a strain-compatibility analysis should be used to determine overturning resistance.

Lateral resistance is provided by passive pressure $V_{passive}$ acting on the front face of the pile cap and by individual pile shears V_{pile}. The shear force (and moment) acting on individual piles is determined by a nonlinear analysis of the combined pile-soil system considering the properties of the pile and soil as well as the end fixity conditions of the pile. U.S. codes require group effects to be considered where the center-to-center spacing of piles is less than eight times the minimum cross-sectional dimension of the pile (IBC, 2012). The results of the analysis include the relation between pile shear and lateral displacement at the loaded end as well as the distribution of displacements, shears, and moments along the pile length. Figure 16.4*c* illustrates qualitative results from this type of analysis. For design purposes, lateral resistance commonly is limited by permissible lateral displacements at the top of the pile. Maximum permissible lateral displacements typically range from 0.5 to 2.0 in (10 to 50 mm), depending on performance objectives for the building. Given the permissible lateral displacement, the geotechnical engineer can estimate the corresponding lateral shear at the top of a pile or pile group, the corresponding pile moments, and the passive pressure acting on the pile cap. This shear is the effective design shear resistance of the pile group. Commonly, the permissible lateral displacement will be insufficient to develop the full passive pressure capacity on the face of the pile cap. The analysis, therefore, should also provide information on the amount of passive pressure that can be considered together with the pile shears for design purposes. This subject is discussed in ASCE 41 (2013) and FEMA (2012).

16.8.3 Pile Design and Reinforcement Details

Pile design is done following the strength design method. The pile is to be provided axial, moment, and shear strength in accordance with the calculated design forces. Additionally, reinforcement detailing is selected to enable the development of inelastic flexural response at locations where yielding is likely under design loads or deformations, such as the locations illustrated in Figure 16.17. As noted in Section 16.1, pile foundation design requirements applicable to the United States can be found in ACI 318, ASCE 7, and the IBC. The different code provisions have conflicting requirements, complicating interpretation of the provisions. At the time of this writing, the provisions of the IBC, which may adopt some of the ACI 318 provisions, generally govern, whereas the ASCE 7 provisions do not apply. The following discussion first considers requirements for uncased concrete piles, followed by provisions that are applicable to other specific pile types. Only requirements for structures assigned to Seismic Design Categories D, E, and F are considered.

Uncased Concrete Piles

Figure 16.20 illustrates the general arrangement of reinforcement for uncased, reinforced concrete piles. Special transverse reinforcement should be provided

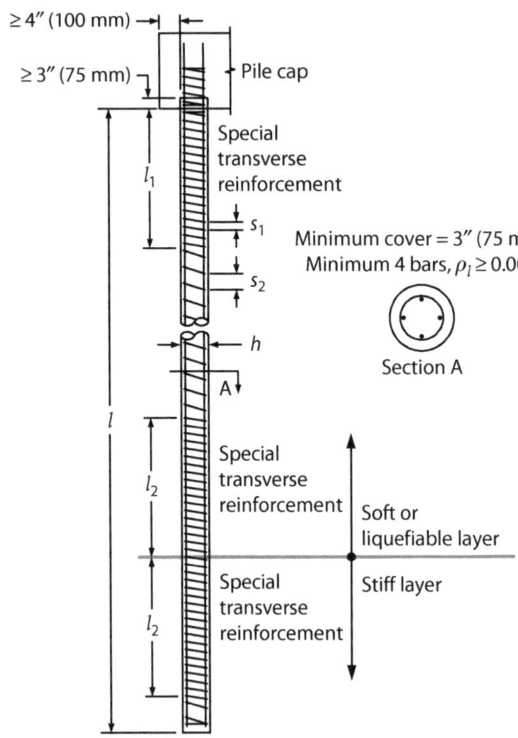

Figure 16.20 Requirements for uncased, reinforced concrete piles.

along several pile diameters immediately beneath the pile cap where inelastic action is most likely to occur. Special transverse reinforcement should also extend above and below any interfaces between hard/stiff strata and soft/liquefiable strata.

According to ACI 318, the special transverse reinforcement is equivalent to that required for columns of special moment frames. Specifically, for spiral-reinforced sections, the special transverse reinforcement must satisfy Eqs. (16.8) and (16.9).

$$\rho_s \geq 0.45\left(\frac{A_g}{A_{ch}} - 1\right)\frac{f_c'}{f_{yt}} \tag{16.8}$$

[1]

$$\rho_s \geq 0.12\frac{f_c'}{f_{yt}} \tag{16.9}$$

For rectangular hoop-reinforced sections, the special transverse reinforcement must satisfy Eqs. (16.10) and (16.11).

$$\min\left(\frac{A_{sh2}f_{yt}}{b_{c3}s}, \frac{A_{sh3}f_{yt}}{b_{c2}s}\right) \geq 0.3\left(\frac{A_g}{A_{ch}} - 1\right)f_c' \tag{16.10}$$

$$\min\left(\frac{A_{sh2}f_{yt}}{b_{c3}s}, \frac{A_{sh3}f_{yt}}{b_{c2}s}\right) \geq 0.09f_c' \tag{16.11}$$

Longitudinal spacing s_1 of hoop sets is the same as that specified for the end region of columns of special moment frames. Note that ACI 318 does not require special transverse reinforcement for piles in foundations supporting one- and two-story stud bearing wall construction.

IBC (2012) does not require application of Eq. (16.8) for any construction. Also, where transverse reinforcement is required, transverse reinforcing ties shall be a minimum of No. 3 (10) bars for up to 20-in (500-mm) diameter piles and No. 4 (13) bars for piles of larger diameter.

The length l_1 of the confinement reinforcement beneath a pile cap varies with soil conditions, and is different in the different building codes. According to ACI 318, l_1 must be at least equal to the greater of five times the member cross-sectional dimension and 6 ft (1.8 m). According to the IBC, l_1 must be at least three times the least element dimension. (In highly rectangular elements, such as can occur with barrettes, this text recommends using not less than twice the longer element dimension.) For piles in Site Class E or F soils, the IBC requires the special confinement to extend at least seven times the least element dimension below the pile cap.

Longitudinal reinforcement should be provided where required by analysis. According to ACI 318, piles, piers, and caissons resisting tensile forces are required to have continuous longitudinal reinforcement over the length that is subjected to

---

[1]Note that the IBC does not require the application of this equation for any construction, and it is generally not used in the United States.

design tensile forces. The longitudinal reinforcement is to be detailed to transfer tensile forces within the pile cap to supported structural members. According to the IBC, minimum longitudinal reinforcement comprises four longitudinal bars with minimum longitudinal reinforcement ratio of 0.005. The minimum reinforcement must extend along a length below the top of the element not less than the greatest of the following:

1. One-half of the element length;
2. A distance of 10 ft (3048 mm);
3. Three times the least element dimension; and
4. The distance from the top of the element to the point where the design cracking moment exceeds the required moment strength.

Some design offices routinely provide minimum longitudinal reinforcement along the full pile length.

Pile longitudinal reinforcement must pass through at least the bottom mat of pile cap reinforcement. Interferences can be reduced by always using even numbers of longitudinal bars in piles.

According to ASCE 7 and the IBC, splices of deep foundation elements should be designed to develop at least the lesser of (a) the nominal strength of the element and (b) the moments, shears, and axial forces associated with development of superstructure overstrength. Superstructure overstrength can be determined from nonlinear response history analysis, limit analysis, or code-based linear analysis using load combinations with the overstrength factor Ω_0.

ACI 318 does not require special transverse reinforcement at interfaces between soft and hard strata, although the commentary suggests providing it. For Site Class E and F soils, the IBC requires longitudinal and special transverse reinforcement to extend a distance l_2 = seven times the minimum member dimension from interfaces of strata that are hard or stiff and strata that are liquefiable or are composed of soft- to medium-stiff clay.

In addition to the zones of special confinement identified in Figure 16.20, ACI 318 also requires special confinement of piles in soil that is not capable of providing lateral support, or in air and water, along the entire unsupported length plus the greater of $5h$ of the pile and 6 ft (1.8 m) into competent soil at both ends.

According to the IBC, along reinforced lengths that are not located within lengths l_1 and l_2, spacing s_2 of transverse reinforcement is permitted to be increased to the least of 12 longitudinal bar diameters, $h/2$, and 12 in (300 mm).

Precast Concrete Piles

Precast concrete piles are generally driven into the soil. The driving process potentially results in high tensile stresses due to wave reflection. Thus, nonprestressed precast piles are generally not used. Building codes, nonetheless, contain provisions for their design.

Because precast piles are driven into the soil, there is some uncertainty regarding the final elevation of the pile tips. ACI 318 requires that the length of provided transverse reinforcement be sufficient to account for potential variations in the elevation of the pile tips. The IBC requires minimum longitudinal reinforcement ratio of 0.01 throughout the length. Transverse reinforcement must be closed ties or spirals with a minimum 3/8-in (9.5-mm) diameter. Spacing of transverse reinforcement is not to exceed the lesser of eight times the diameter of the smallest longitudinal bar and 6 in (150 mm) within a distance of $3h$ of the bottom of the pile cap. Spacing of transverse reinforcement is not to exceed 6 in (150 mm) throughout the remainder of the pile.

Precast Prestressed Piles

The principles for the design of precast prestressed piles are similar to those for cast-in-place, nonprestressed piles. The IBC contains a series of prescriptive requirements for precast prestressed piles that differ in detail from the requirements for cast-in-place nonprestressed piles. The interested reader is referred to the IBC for additional details.

Micropiles

According to the IBC, a micropile is required to have a permanent steel casing extending from the top of the micropile down to the point of zero curvature. For structures assigned to Seismic Design Category D, E, or F, the micropile is to be considered as an alternative system, for which supporting documentation and test data must be submitted to the building official for review and approval.

16.8.4 Pile Cap Design and Reinforcement Details

In general, the detailing requirements for spread footings and mat foundations apply to pile caps. See Sections 16.6.3 and 16.7.3.

Design of pile anchorage into a pile cap should consider the combined effects of axial forces due to uplift and bending moments due to fixity to the pile cap. According to ASCE 7, piles resisting uplift forces or providing rotational restraint should be anchored into the pile cap as follows:

1. In the case of uplift, the anchorage should be capable of developing the least of (a) the nominal tensile strength of the longitudinal reinforcement in a concrete pile (or the nominal tensile strength of a steel pile), (b) 1.3 times the pile pullout resistance, and (c) the axial tension force resulting from the seismic load effects including overstrength factor Ω_0. The pile pullout resistance is taken as the ultimate frictional or adhesive force that can be developed between the soil and the pile plus the pile weight.
2. In the case of rotational restraint, the anchorage should be designed to resist the axial and shear forces and moments resulting from the seismic load effects including overstrength factor Ω_0 or should be capable of developing the full axial, bending, and shear nominal strength of the pile.

Where tension forces induced by earthquake effects are transferred between a pile cap or mat foundation and a precast pile by reinforcing bars grouted or post-installed in the top of the pile, ACI 318 requires that the grouting system be demonstrated by test to develop at least $1.25f_y$ of the bar.

Connections between batter piles and pile caps have sustained extensive damage in past earthquakes. Therefore, ACI 318 requires that pile caps incorporating batter piles be designed to resist the full compressive strength of the batter piles acting as short columns. ASCE 7 requires that batter piles and their connections be designed considering the maximum forces that can be delivered to or resisted by the piles. The IBC limits the required force to the nominal strength of the batter pile acting as a short column.

Where piles, piers, or caissons resist tensile forces, longitudinal reinforcement should be detailed to transfer tensile forces into the pile cap and from there into the supported structural members. Where the vertical lateral-force-resisting elements are columns, the pile cap moment strengths should exceed column moment strength.

The design approach for a pile cap depends on its aspect ratio. According to ACI 318, where the distance between the axis of any pile and the axis of a column is more than twice the distance between the top of the pile cap and the top of the pile, the pile cap can be designed by conventional methods of checking sectional shear and moment (with special rules on computation of shear in the vicinity of piles). For thicker pile caps, alternative requirements apply. For such cases, it is common to use the strut-and-tie method for design.

Figure 16.21 illustrates strut-and-tie models for a four-pile pile cap under different loads. Figure 16.21*a* shows that the strut-and-tie model must be three-dimensional

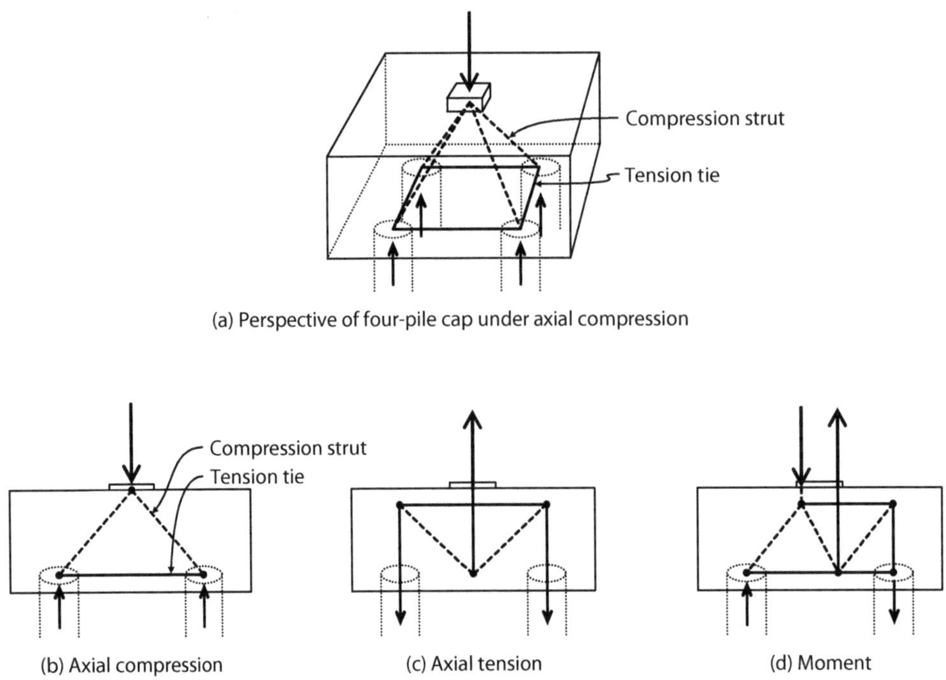

FIGURE 16.21 Strut-and-tie models for pile caps: (*a*) 3-D perspective view; (*b to d*) 2-D views.

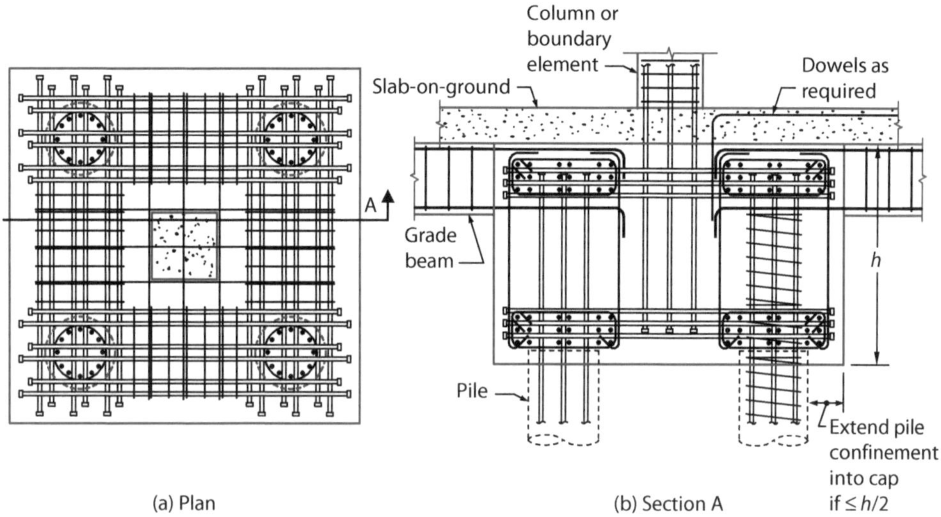

Figure 16.22 Illustration of pile cap reinforcement to resist axial compression and uplift forces.

for this pile arrangement. Figure 16.21*b* shows a side view of the same strut-and-tie model for pure axial load. Note the tension ties at the bottom of the pile cap have constant tensile force up to the center of the piles; the bars must be fully developed using straight development lengths, standard hooks, or headed reinforcement. Figure 16.21*c* illustrates a strut-and-tie model for uplift. Note the requirement for a tension tie at the top of the pile cap for uplift loading. Finally, Figure 16.21*d* illustrates a strut-and-tie model for a pile cap resisting pure moment. For each loading, the axial force in the piles can be solved using equilibrium. The strut and tie forces required to equilibrate those reactions are then readily derived. Adebar and Zhou (1996), Klein (2002), and Widianto and Bayrak (2010) present examples of the use of strut-and-tie models for pile cap design.

Figure 16.22 illustrates a possible reinforcement arrangement for a pile cap designed to resist both downward compressive force and uplift. Headed reinforcement is used to anchor the tension ties within the relatively small pile cap plan dimensions.

16.8.5 Foundation Ties

In U.S. practice, foundation ties are required between individual pile caps, drilled piers, or caissons for all structures assigned to Seismic Design Categories D, E, or F. See Section 16.6.5 for additional details.

16.9 Combined Footings and Outriggers to Increase Overturning Resistance

Combined footings and foundation outriggers are commonly used to increase overturning resistance beyond that provided by isolated spread footings or pile caps. The concept of combined footings was introduced in Section 16.6.4.

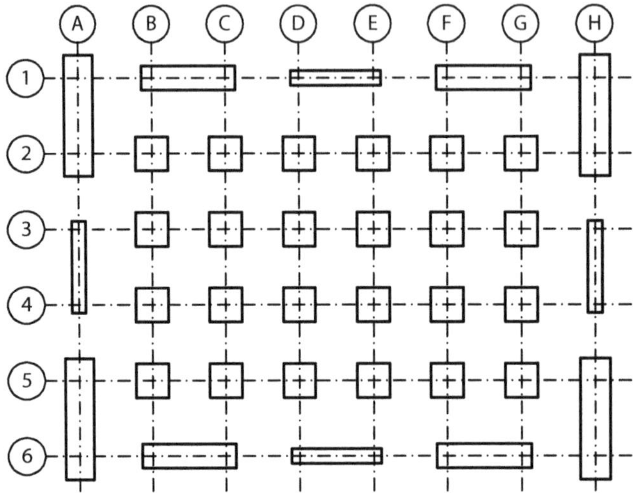

Figure 16.23 Foundation plan for perimeter moment-resisting frame system with interior gravity framing. (*After FEMA, 2012.*)

Figure 16.23 illustrates how combined footings might be arranged around the perimeter of a building using perimeter moment-resisting frames. In this case, each combined footing supports two columns. In addition to the added resistance provided by the increased footing area, uplift resistance is considerably increased because the gravity loads that are supported by each of the columns help hold the footing down.

Grade beams can serve a similar purpose by connecting individual footings together. Figure 16.24 illustrates some framing options that can be considered. Figure 16.24*a* illustrates a condition in which the combined beam and supporting footing are designed to be stronger than the supported column, resulting in a strong footing and yielding column. Figure 16.24*b* illustrates the option where the combined beam and supporting footing are designed to be weaker than the supported column. The expected performance can be difficult to ensure with this

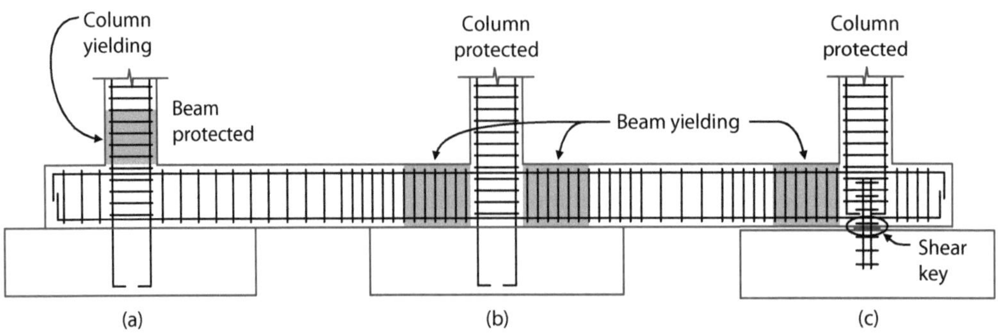

Figure 16.24 Alternative yielding mechanisms for columns interconnected by grade beams: (*a*) strong beam, weak column; (*b*) strong column, weak beam; (*c*) strong column, weak beam, with shear key to reduce moment transfer to footing.

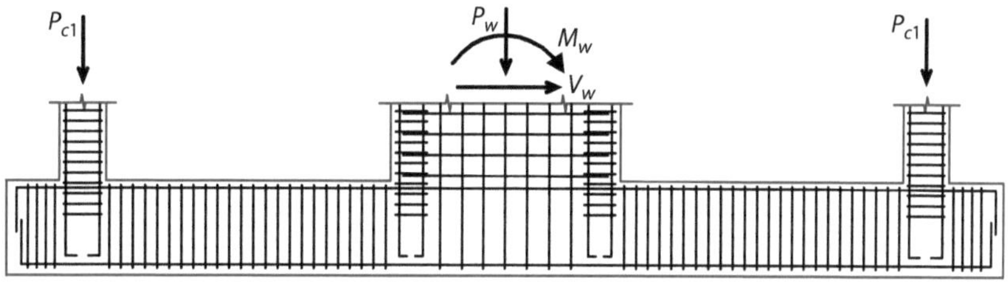

FIGURE 16.25 Grade beam acting as a combined footing and foundation outrigger.

configuration, as there is considerable uncertainty about the moment resistance of the footing and some uncertainty about where beam hinging will occur. Increased certainty of performance can be achieved using a pinned connection between the footing and the beam-column system above (Figure 16.24*c*). This detail requires additional attention during construction, and is seldom used.

Foundation outriggers become particularly important in buildings with structural walls or braced frame systems. Figure 16.25*c* illustrates the use of a large, wide grade beam to act both as a combined footing and as an outrigger system that increases the wall overturning resistance by picking up adjacent columns.

16.10 Buildings with Subterranean Levels

Subterranean levels in buildings provide a ready load path for spreading lateral forces from vertical elements with concentrated force resistance outward to the overall footprint of the building. This subject has been discussed in Chapters 13 and 15.

Figure 16.26 illustrates the case of a structural wall supported directly by a basement wall. This solution assumes horizontal force resistance is provided by friction under the basement, but an alternative solution would include passive bearing pressure against the leading basement wall and possibly friction along the basement walls that are parallel to the lateral force. Where a traditional linear elastic solution to the problem is pursued, and where uplift or other soil nonlinearity does not occur under the combined loads, it can be convenient to separate the vertical loads (Figure 16.26*b*) and the lateral loads (Figure 16.26*c*).

To visualize the force flow under moment transfer, consider the internal forces of Figure 16.27*a*. Moment resistance of the shear panel **efgh** is achieved through vertical shear on faces **eg** and **fh**, tension and compression couples on faces **eg** and **fh**, horizontal shear along **ef** and **gh**, and a small amount of bearing pressure beneath the panel along **gh**. Magnitudes of tension and compression couples are determined largely by the required moment resistance of basement wall panels **abcd** and **ijkl**. A usual design goal is to spread the horizontal force resistance uniformly along the base of the wall (plus possibly some passive bearing pressure and side wall friction); this determines the amount by which the tension force T_L exceeds compression force C_L, and the amount by which C_R exceeds T_R.

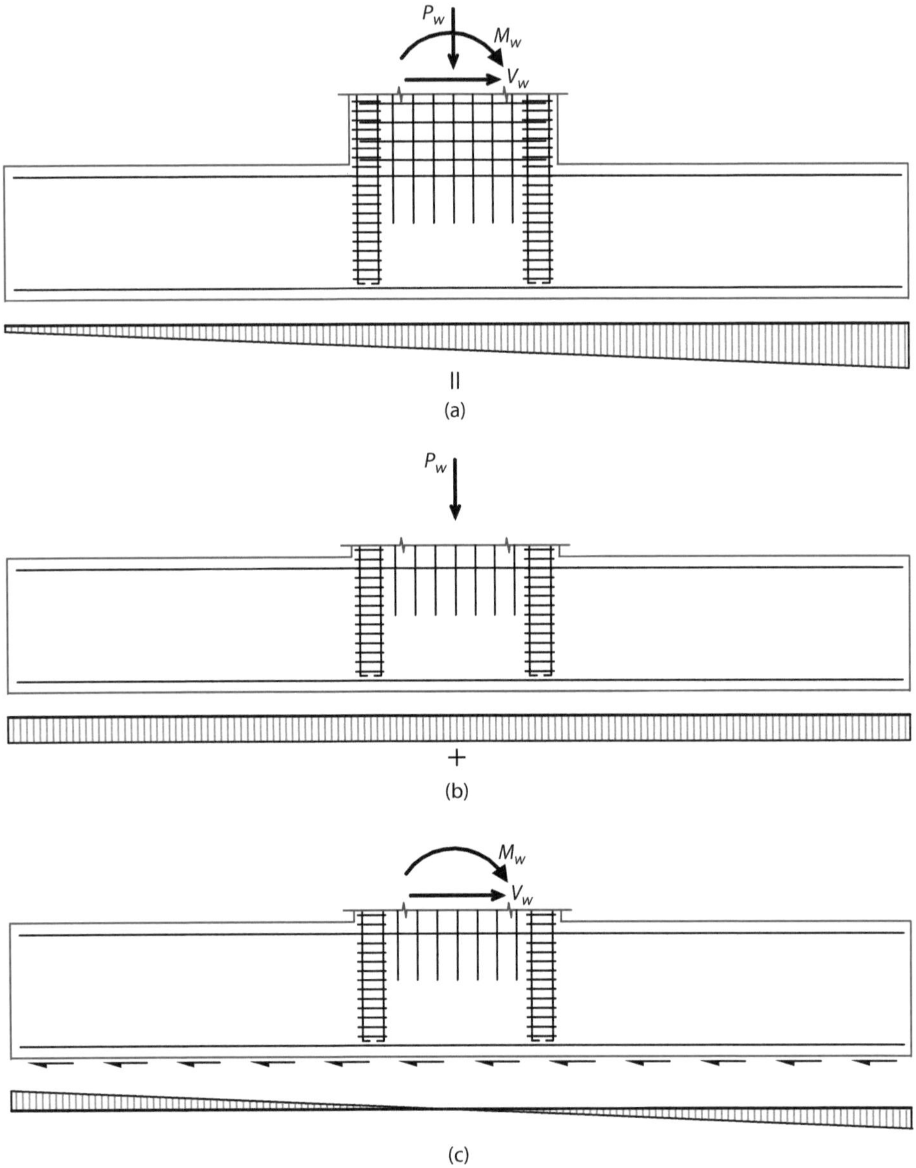

FIGURE 16.26 Wall supported by basement wall. Superposition is valid only where soil bearing pressure is linear without uplift under the combined loading.

For the case of a shear wall centered on the basement wall, with no passive pressure against the basement wall, no friction along the side walls, and no uplift, the vertical shears V_L and V_R are

$$V_L = V_R = 1.5\left(\frac{M_w + V_w h_f}{l_f}\right)\left[1 - \left(\frac{l_w}{l_f}\right)^2\right] \tag{16.12}$$

For typical values of l_w/l_f, this can be approximated as $V_L = V_R = 1.5\left(\frac{M_w + V_w h_f}{l_f}\right)$, which is consistent with the traditional solution of parabolic shear flow for a rectangular section.

The tension-compression couples made up of C_L, T_L, C_R, and T_R depend on the overturning moment, wall shear, foundation wall depth, wall lengths, and the position of the shear wall along the length of the basement wall. Each case has to be analyzed based on the individual conditions. The calculations generally are not difficult, but they can be tedious.

Figure 16.27*b* suggests an alternative solution using a strut-and-tie model. This approach can be used as an approximation for the case of linearly varying soil bearing pressure, for cases of nonlinear bearing (plastic soil behavior or uplift), or for pile-supported basement walls. In the latter case, a simple solution is obtained by assuming that the piles have reached their ultimate capacities, and to work the problem backward from there to determine maximum resistance.

The solutions of Figure 16.27 imply that all the overturning resistance is provided by a planar basement wall in line with the shear wall. In a typical basement, the perimeter walls extend continuously around the foundation, creating a box. The entire box can be used for overturning resistance, although some limits on the "flange" width should be considered in very wide foundations.

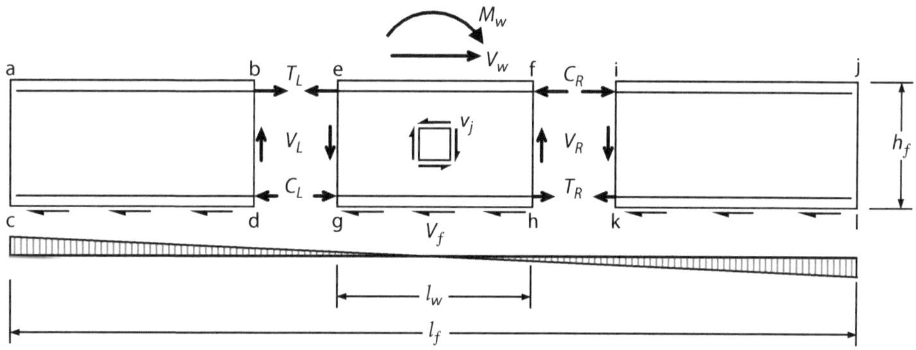

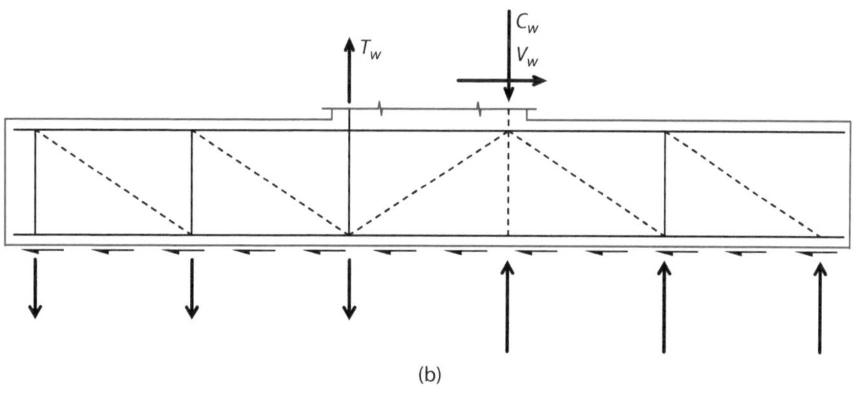

FIGURE 16.27 Equilibrium considerations for wall supported by basement wall.

Proportioning and detailing of panel zone **efgh** require additional consideration. Wall boundary element reinforcement generally is required to extend down through the height of the basement wall so that tension and compression forces of the boundary element can be transferred into the wall. The panel zone typically develops high joint shear stress v_j (Figure 16.27*a*), similar to that occurring in a beam-column joint. The magnitude of the shear stress can most readily be determined by calculating the vertical joint shear stress, which is identically equal to the horizontal joint shear stress. Building codes do not explicitly consider design of this region. Strict interpretation of ACI 318 would suggest that the panel zone shear should not exceed $10\phi\sqrt{f'_c}l_w b_w$, psi ($0.83\phi\sqrt{f'_c}l_w b_w$, MPa). Confinement of this zone, including closely spaced distributed horizontal reinforcement, distributed vertical reinforcement, and crossties through the wall thickness may be advisable where shear stresses are high. See additional discussion on panel zones in Sections 13.13 and 13.14.

Where walls are not supported directly on basement walls, the lower level floor slabs can act as diaphragms to transfer horizontal forces out of the shear wall and into the basement walls (Figure 16.28). Note that a portion of the overturning resistance can be taken out directly through moment in the base mat supporting the wall, with additional resistance provided by a tension-compression couple in the floor slabs. Analysis to determine the relative magnitude of the various forces usually is done considering cracked-section stiffness of the diaphragms, gross-section stiffness (or nearly so) of the basement walls, and appropriate stiffness of the base mat and subgrade. Because of the large uncertainties in stiffnesses and strengths, bounding analyses are commonly done to bound the solution. Distributors (collectors) generally are designed to distribute the diaphragm shear stresses uniformly through the length of the diaphragm, although shorter collectors sometimes are used where the diaphragm shear strength is adequate considering the shorter transfer length. See Chapter 15 for additional discussion of diaphragms.

U.S. building codes require collectors and distributors to be designed for Ω_0-amplified forces. This factor clearly applies to the collectors of Figure 16.28.

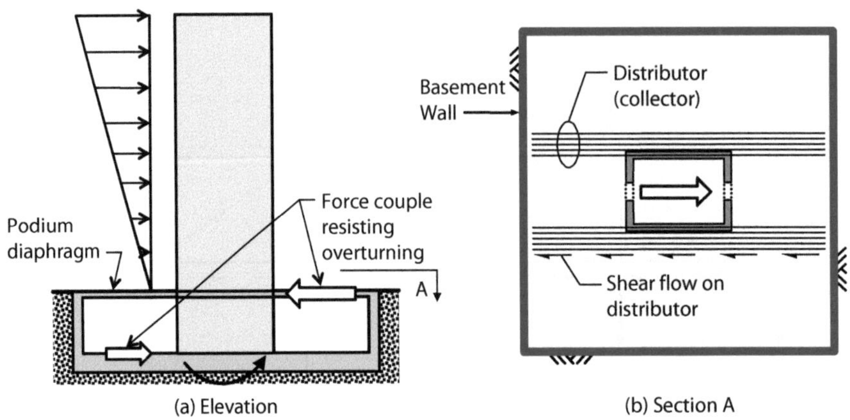

Figure 16.28 Force transfer in buildings with subterranean levels.

Although the building code is open to interpretation regarding the design of reinforcement for a basement wall directly supporting a shear wall, it is the opinion of the author that the Ω_0 factor applies for the design of the distributors whose forces are represented by T_L, C_L, T_R, and C_R in Figure 16.27. The alternative of designing for reduced forces corresponding to the system capacity always applies.

16.11 Basement Walls

16.11.1 Behavior and Design

Basement walls are designed for out-of-plane pressure due to soil bearing along the wall height. Because a basement wall is restrained against translation at the foundation, the grade level, and intermediate subterranean floor slabs, it is considered from a geotechnical perspective to be a "non-yielding" wall. As such, design for load combinations including only gravity loads considers at rest soil pressures, which increase with depth (Figure 16.29). The structural model of the wall usually considers the wall to be pinned at the foundation level, continuous at intermediate levels, and pinned at grade level (unless the grade level slab is much thicker than the wall).

Design for seismic loads is somewhat controversial at the time of this writing. Kramer (1996) presents traditional models for seismic earth pressures. Commonly used models are the Mononobe–Okabe model (for retaining walls without restraint at the top) and the Wood model (for basement walls). Recent tests (Sitar et al., 2012) suggest that the Mononobe–Okabe model is more appropriate to use for basement walls as well, although some downward adjustment of forces is commonly recommended. Lew (2012) reports that there is no evidence of engineered retaining walls or basement walls being damaged in earthquakes due to excessive seismic earth pressures, but stops short of recommending major changes from current

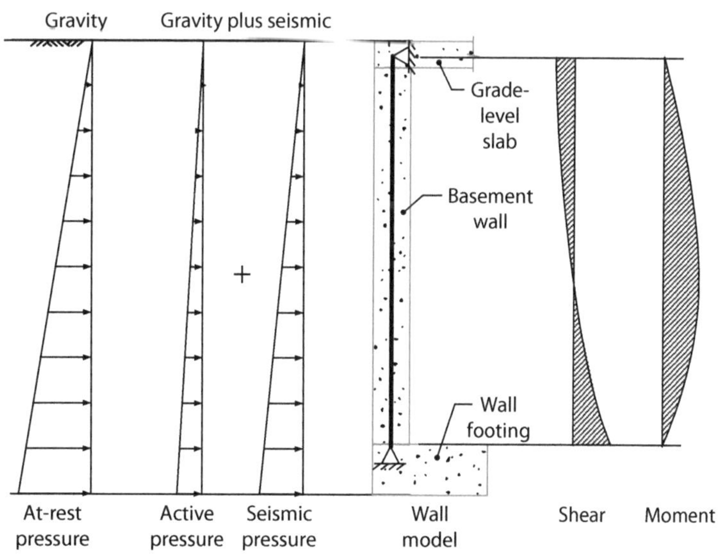

FIGURE 16.29 Out-of-plane loading for one-story basement wall.

practices. Lew (2012) recommends using a modified Mononobe–Okabe model for seismic earth pressures to be combined with the active earth pressure from gravity loads (combining with at-rest pressure is inappropriate for this load combination).

This text recommends that the structural engineer and geotechnical engineer discuss the appropriate approach to determining design earth pressures.

Out-of-plane earth pressures are resisted primarily by the wall footing and the slabs at each level. Therefore, a conservative approach is to consider the basement wall as a one-way element spanning vertically between levels. Shrinkage and temperature reinforcement provided in the horizontal direction usually will be sufficient for the two-way moments that develop near the ends of the walls. Note that the floor levels act as diaphragms to transfer the bearing pressure to orthogonal walls, creating in-plane shear in those walls. Basement walls also must be designed for in-plane shear due to other lateral forces transferred from the superstructure to the basement level. High in-plane shear may require horizontal reinforcement exceeding the requirements for shrinkage and temperature effects. In general, the interaction between in-plane and out-of-plane shear is not considered because of the different failure plans associated with each action.

16.11.2 Reinforcement Detailing

Figure 16.30 illustrates representative reinforcement of a one-story basement wall. Reinforcement is provided for out-of-plane moment and in-plane shear. Reinforcement provided in each in-plane direction should be at least as

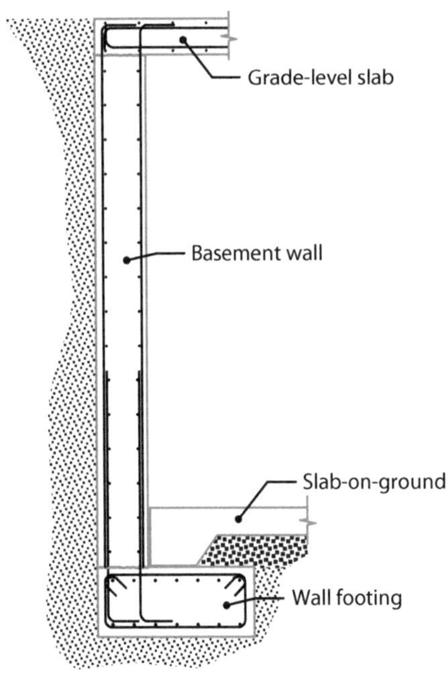

Figure 16.30 Representative details for basement wall.

required for shrinkage and temperature effects. Wall thickness is as required for out-of-plane shear.

It is not uncommon for the footing to be inboard of the foundation, which facilitates construction and provides a counteracting moment to help resist the out-of-plane moment produced by the soil pressure acting inward on the wall.

References

ACI 318 (2014). *Building Code Requirements for Structural Concrete (ACI 318-14) and Commentary*, American Concrete Institute, Farmington Hills, MI.

Adebar, P., and L.Z. Zhou (1996). "Design of Deep Pile Caps by Strut-and-Tie Models," *ACI Structural Journal*, Vol. 93, No. 4, pp. 437–448.

ASCE 7 (2010). *Minimum Design Loads for Buildings and Other Structures*, American Society of Civil Engineers, Reston, VA, 608 pp.

ASCE 41 (2013). *Seismic Evaluation and Retrofit of Existing Buildings*, American Society of Civil Engineers, Reston, VA.

Bobet, A., R. Salgado, and D. Loukidis (2001). *Seismic Design of Deep Foundations*, Publication FHWA/IN/JTRP-2000/22, Joint Transportation Research Program, Indiana Department of Transportation and Purdue University, West Lafayette, Indiana, 111 pp. doi: 10.5703/1288284313311.

Caltrans (2003). *Bridge Design Specifications, Section 4—Foundations*, California Department of Transportation, Sacramento, CA, 70 pp.

FEMA (2005). *Improvement of Nonlinear Static Seismic Analysis Procedures*, FEMA 440, prepared by the Applied Technology Council for the Federal Emergency Management Agency, Washington, DC.

FEMA (2009). *NEHRP Recommended Seismic Provisions for New Buildings and Other Structures*, FEMA P-750/2009 Edition, prepared by the Building Seismic Safety Council of the National Institute of Building Sciences for the Federal Emergency Management Agency, Washington, DC.

FEMA (2012). *NEHRP Recommended Seismic Provisions: Design Examples*, FEMA P-751/September 2012 Edition, prepared by the Building Seismic Safety Council of the National Institute of Building Sciences for the Federal Emergency Management Agency, Washington, DC.

IBC (2012). *International Building Code*, International Code Council.

Klein, G.J. (2002). "Example 9—Pile Cap," *Examples for the Design of Structural Concrete with Strut-and-Tie Models*, Special Publication No. 208, American Concrete Institute, Farmington Hills, MI, pp. 213–223.

Klemencic, R., I.S. McFarlane, N.M. Hawkins, and S. Nikolaou (2012). *Seismic Design of Reinforced Concrete Mat Foundations: A Guide for Practicing Engineers*, NEHRP Seismic Design Technical Brief No. 7, NIST GCR 12-917-22, National Institute of Standards and Technology, Gaithersburg, MD, 29 pp.

Kramer, S.L. (1996). *Geotechnical Earthquake Engineering*, Prentice-Hall, Inc., Upper Saddle River, New Jersey, 653 pp.

Lew, M. (2012). "Recent Findings on Seismic Earth Pressures," *The Structural Design of Tall and Special Buildings*," Wiley Online Library, doi: 10.1002/tal.1052.

Reineck, K.H., D.A. Kuchma, K.S. Kim, and S. Marx (2003). "Shear Database for Reinforced Concrete Members without Shear Reinforcement," *ACI Structural Journal*, Vol. 100, No. 2, pp. 240–249.

Sitar, N., R. Geraili Mikola, and G. Candia (2012). "Seismically Induced Lateral Earth Pressures on Retaining Structures and Basement Walls," Geotechnical Engineering State of the Art and Practice Keynote Lectures from GeoCongress 2012, *Geotechnical Special Publication No.* 226, American Society of Civil Engineers, Reston, VA.

Stewart, J.P., C.B. Crouse, T. Hutchinson, B. Lizundia, F. Naeim, and F. Ostadan (2013). *Soil-Structure Interaction for Building Structures*, NIST GCR 12-917-21, National Institute of Standards and Technology, Gaithersburg, MD, 268 pp.

TBI (2010). *Guidelines for Performance-Based Seismic Design of Tall Buildings*, Report No. 2010/05, Pacific Earthquake Engineering Research Center, University of California, Berkeley, CA, 84 pp.

Widianto and O. Bayrak (2010). "*Example 11: Deep Pile Cap with Tension Piles*," *Special Publication No.* 273, American Concrete Institute, Farmington Hills, MI, pp. 11.1–11.22.

Notation

a	=	shear span (distance from maximum moment to inflection point), Chapters 6, 7, and 8
A_b	=	area of an individual bar or wire, Chapters 4 and 8
A_{bl}	=	area of longitudinal reinforcing bar, Chapters 5 and 8
A_{brg}	=	net bearing area of the head of stud, anchor bolt, or headed deformed bar, Chapter 8
A_{bt}	=	area of transverse reinforcing bar, Chapters 5 and 8
A_c	=	area of concrete in compression or tension member, Chapter 5
A_c	=	area of concrete section resisting shear transfer, Chapter 7
A_{ch}	=	cross-sectional area of a structural member measured to the outside edges of transverse reinforcement, Chapters 4, 5, 9, 12, 13, 14, and 16
A_{cover}	=	cross-sectional area of cover concrete, Chapter 5
A_{cs}	=	cross-sectional area at one end of a strut in a strut-and-tie model, taken perpendicular to the axis of the strut, Chapter 7
A_{cv}	=	gross area of concrete section bounded by web thickness and length of section in the direction of shear force considered, Chapters 7, 13, and 15
A_{cw}	=	area of concrete section of an individual pier, horizontal wall segment, or coupling beam resisting shear, Chapters 7 and 13
A_e	=	effective shear area of cracked section (truss model), Chapter 7
A_g	=	gross area of concrete and steel section, Chapters 5, 6, 7, 9, 12, 13, 14, 15, and 16
$A_{g,be}$	=	area of boundary element, Chapter 13
A_{gmax}	=	peak ground acceleration, Chapter 13
A_i	=	area of element i of fiber model, Chapter 6
A_j	=	effective cross-sectional area within a joint in a plane parallel to plane of reinforcement generating shear in the joint, Chapter 9
A_{jh}	=	cross-sectional area of each hoop set in the direction parallel to the joint shear, Chapter 9
A_{nz}	=	area of a face of a nodal zone or a section through a nodal zone, Chapter 7
A_s	=	area of nonprestressed longitudinal reinforcement on one face of beam or wall, Chapters 2, 6, 7, 8, 12, 13, 14, 15, and 16
A'_s		area of nonprestressed longitudinal reinforcement on face of beam or wall opposite A_s, Chapters 6, 12, and 13
$A_{s1}, A_{s2}, \ldots, A_{sn}$	=	area of nonprestressed longitudinal reinforcement in layers 1, 2, …, n, Chapter 5
$A_{s,be}$	=	area of boundary element longitudinal reinforcement, Chapter 13

A_{sh}, A_{sh2}, A_{sh3} = total cross-sectional area of transverse reinforcement (including crossties) within spacing s and perpendicular to dimension b_c. A_{sh2} and A_{sh3} refer to areas perpendicular to b_{c3} and b_{c2}, respectively, Chapters 4, 5, 6, 12, 13, 14, and 16

A_{si} = total area of surface reinforcement at spacing s_i in the ith layer crossing a strut, with reinforcement at an angle α_i to the axis of the strut, Chapter 7

A_{sm} = area of slab bottom longitudinal reinforcement along each principal direction that is continuous through or anchored in the column core, Chapter 10

$A_{s,\min}$ = minimum area of flexural reinforcement in beams and footings, Chapters 6 and 16

A_{sp} = area of spiral or circular hoop reinforcement, Chapters 4 and 7

A_{st} = total area of nonprestressed longitudinal reinforcement, Chapters 4, 5, 6, 7, 13, 14, and 15

A_{sx}, A_{sy}, A_{sz} = distributed reinforcement oriented in x, y, and z directions, respectively, Chapter 7

A_{tp} = area of prestressed steel in a tie, Chapter 7

A_{tr} = total cross-sectional area of all transverse reinforcement within spacing s that crosses the potential plane of splitting through the reinforcement being developed or spliced, Chapter 8

A_{ts} = area of nonprestressed reinforcement in a tie, Chapter 7

A_v = effective shear area of cross section, Chapter 6

A_v = area of shear reinforcement within spacing s, Chapters 7, 10, 13, and 14

A_{vd} = total area of reinforcement in each group of diagonal bars in a diagonally reinforced coupling beam or diagonally reinforced wall, Chapters 7 and 13

A_{vf} = area of shear-friction reinforcement, Chapters 7, 13, and 15

$A_{v,min}$ = minimum area of shear reinforcement within spacing s, Chapters 7 and 16

b = width of compression face of member, Chapters 5, 6, 12, 13, and 16

b = overall width of rectangular footing in direction perpendicular to lateral load or eccentricity e, Chapter 16

b_1 = dimension of the critical section b_o measured in the direction of the span for which moments are determined, Chapter 10

b_2 = dimension of the critical section b_o measured in the direction perpendicular to b_1, Chapter 10

b_c = column cross-sectional dimension perpendicular to dimension h_c, Chapter 9

b_c, b_{c2}, b_{c3} = cross-sectional dimension of member core measured to the outside edges of the transverse reinforcement composing area A_{sh}. b_{c2} and b_{c3} refer to directions 2 and 3, Chapters 4, 5, 6, 9, 12, 13, and 16

b_{ch} = cross-sectional dimension of member core measured to the outside edges of the transverse reinforcement composing area A_{sh}, Chapter 9

b_{cr} = critical thickness of prismatic member or wall boundary, Chapters 5 and 13

b_e = effective flange width, Chapters 6, 12, and 13

b_e = two-way slab effective width, Chapter 10

b_{eff} = effective width of collector, Chapter 15

b_{flange} = total width of flange in flexural member, Chapter 6

BI = beam reinforcement index, Chapter 9

b_o = perimeter of critical section for two-way shear in slabs, Chapters 10, 14, and 16

b_s = thickness of strut, measured out-of-plane in a two-dimensional structure, Chapter 7

b_{slab} = width of slab in which longitudinal reinforcement is placed for moment transfer between slab and column, Chapter 10

b_w = web width, or diameter of circular section, Chapters 6, 7, 9, 12, 13, 14, 15, and 16

c = distance from extreme compression fiber to neutral axial, Chapters 1, 6, and 13

c = cross-sectional dimension of square or circular column, Chapter 10

c = distance from centroid of slab critical section to location where shear stress v_u is calculated, Chapter 10

C = flexural compression force balancing flexural tension force T or T_s, Chapters 7, 8, 9, 10, and 12

C = ratio of in-place concrete strength to standard cylinder compressive strength, Chapters 4, 5, and 6

C = compression force on column, pile, or wall boundary, Chapters 13 and 16

C = effective clear cover for developed bar, smallest of c_s, c_b, and $c_c/2$, Chapter 8

c_1 = dimension of rectangular or equivalent rectangular column measured in the direction of the span for which beam moments are being determined, Chapters 10, 12, and 14

c_2 = dimension of rectangular or equivalent rectangular column measured in the direction perpendicular to c_1, Chapter 12

C_1 = ratio of maximum relative displacement of a yielding oscillator to the maximum relative displacement of a linear oscillator having the same initial period, Chapter 11

C_1, C_2 = chord forces acting around panel zone, Chapter 7

c_{AB} = distance from face AB of slab critical section to centroid of slab critical section, Chapter 10

c_b = smaller of (a) the distance from center of a bar or wire to nearest concrete surface, and (b) one-half the center-to-center spacing of bars or wires being developed, Chapter 8

C_c = compressive stress resultant of (unconfined) concrete, Chapters 5, 6, 9, and 13

c_{c1}	=	clear cover over developed reinforcing bar in the 1 direction, Chapter 8
c_{c2}	=	clear cover over developed reinforcing bar in the 2 direction, Chapter 8
c_{c3}	=	clear spacing between developed reinforcing bars, Chapter 8
C_{cc}	=	compressive stress resultant of confined concrete, Chapters 5 and 6
C_d	=	deflection amplification factor, Chapters 11 and 14
C_D	=	factored collector compressive force transferred directly to edge of vertical element, Chapter 15
C_e	=	elastic seismic coefficient, Chapter 11
C_L, C_R	=	compressive forces in distributors, Chapter 16
C_{pr}, C'_{pr}	=	flexural compression forces balancing flexural tension forces T_{pr} and T'_{pr}, Chapter 12
C_s	=	seismic response coefficient, Chapters 11 and 12
C_s, C_s', C_s''	=	compressive stress resultants in longitudinal reinforcement areas A_s, A_s', A_s'', Chapters 5, 6, 8, 9, and 13
C_{s1}, C_{s2}, …, C_{sn}	=	compressive stress resultants of longitudinal reinforcement in layers 1, 2, …, n, Chapter 5
C_{sh1}, C_{sh2}, …, C_{shn}	=	compressive forces in steel layers 1, 2, …, n due to unrestrained shrinkage strain, Chapter 5
c_{sp}	=	distance from extreme unspalled compression fiber to neutral axial, Chapter 6
C_u	=	factored compressive force at section, Chapters 13 and 15
$C_{u,max}$	=	maximum value of C_u, Chapter 15
C_v	=	factored collector compressive force transferred through shear-friction to vertical element, Chapter 15
C_v	=	coefficient to account for slab shear strength reduction as function of lateral loading, Chapters 10 and 14
C_{vx}	=	vertical distribution factor for design seismic forces, Chapter 15
C_w	=	flexural compression force at wall boundary, Chapter 16
d	=	distance from extreme compression fiber to centroid of longitudinal tension reinforcement, Chapters 5, 6, 7, 8, 9, 10, 11, 12, 13, 14, 15, and 16
D	=	dead loads, or related internal moments and forces, Chapters 1, 11, 12, 13, 14, and 15
D	=	damage in fatigue loading, Chapter 2
D	=	diameter of confined core, measured to outside of spiral or hoops, Chapter 4
D	=	inside diameter of bend in standard hook, Chapter 8
d'	=	distance from extreme compression fiber to centroid of longitudinal compression reinforcement, Chapters 6 and 10
D'	=	cross-sectional dimension of confined concrete core measured between the centerline of the peripheral hoop or spiral, Chapter 7
d''	=	depth of concrete core of column measured center-to-center of peripheral rectangular hoop, circular hoop, or spiral, Chapter 5

d_b = nominal diameter of bar, Chapters 2, 5, 6, 7, 8, 9, 10, 12, 13, 14, 15, and 16
$d_{b,b}$ = nominal diameter of beam longitudinal bar passing through joint, Chapter 9
$d_{b,c}$ = nominal diameter of column longitudinal bar passing through joint, Chapter 9
d_{bl} = nominal diameter of longitudinal bar, Chapters 5 and 8
d_{bt} = nominal diameter of transverse bar, Chapter 8
D_i = fraction of total damage in half cycle i in fatigue loading, Chapter 2
D_m = coefficient for estimating contribution of higher modes to wall base shear, Chapter 13
d_{SOG} = total depth of slab-on-ground, Chapter 16
d_{soil} = depth of soil, Chapter 16
d_t = distance from extreme compression fiber to centroid of extreme layer of longitudinal tension reinforcement, Chapter 1
e = eccentricity, Chapters 6, 13, 15, and 16
e = eccentricity between beam and column centerlines, Chapter 9
E = load effects of earthquake, or related internal moments and forces, Chapters 1, 7, 11, 12, 13, 15, and 16
E = modulus of elasticity, Chapters 1, 5, and 6
E_c = modulus of elasticity of concrete, Chapters 1, 3, 4, 5, 6, 7, 11, and 13
E_h = force effect obtained from application of horizontal seismic forces to the structural model, Chapters 11 and 13
e_i = eccentricity of vertical element i relative to center of rigidity, Chapter 15
E_m = effect of horizontal seismic forces including structural overstrength, Chapter 11
E_{mh} = effect of horizontal seismic forces including structural overstrength, Chapter 11
E_r = double (or reduced) modulus, Chapter 5
E_s = modulus of elasticity of steel, Chapters 1, 2, 4, 5, 6, 7, and 13
E_{sh} = strain-hardening modulus, Chapter 2
E_t = instantaneous tangent modulus, Chapter 5
E_v = force effect associated with vertical ground motion acceleration, Chapters 11, 13, and 15
e_x = eccentricity of diaphragm design lateral force relative to center of rigidity, Chapter 15
f = vibration frequency, s^{-1}, Chapter 11
f = stress, Chapter 3
f = form factor accounting for nonuniform shear stress across section depth, Chapter 7
F = load due to fluids with well-defined pressures and maximum heights, Chapter 1
f_1, f_2, f_3 = stresses in the 1, 2, and 3 directions, Chapters 3, 4, 5, and 6

F_1, F_2 = radial forces per unit length developed by lap-spliced reinforcement, Chapter 8
f_{1e}, f_{2e} = effective confinement stresses in the 1 and 2 directions, Chapters 4 and 6
F_a = short-period site coefficient (at 0.2 s period), Chapter 11
f_c = concrete stress, Chapters 4, 5, and 6
f_c' = specified compressive strength of concrete, Chapters 1, 3, 4, 5, 6, 7, 8, 9, 10, 12, 13, 14, 15, and 16
f_{c1}, f_{c2}, f_{c3} = principal concrete stresses in the 1, 2, and 3 directions, Chapter 3
f_{c1max} = maximum principal compressive stress capacity in the 1 direction, Chapter 3
f_{cc} = confined concrete compressive stress, Chapter 5
f_{cc}' = confined concrete compressive strength, Chapters 4, 5, and 6
$f_{cc,\ in\ situ}'$ = in situ compressive strength of confined concrete, Chapter 6
f_{cd} = axial compressive stress in diagonal compression strut, Chapter 7
f_{ce} = effective compressive strength of concrete in a strut or a nodal zone, Chapter 7
$f_{c,\ in\ situ}'$ = in situ compressive strength of unconfined concrete, Chapter 6
f_{cr} = critical longitudinal reinforcement stress at buckling, Chapter 5
f_{cr}' = required average compressive strength to ensure that no more than the permissible proportion of tests will fall below the specified compressive strength, used as a basis for selection of concrete proportions, Chapter 3
F_d = force in diagonal compression strut, Chapter 7
f_{emin} = lesser of the effective confinement stresses, Chapters 4 and 6
F_h, F_v = horizontal and vertical forces in reinforcement required to equilibrate sections of squat wall, Chapter 13
F_j = design inertial force at level j, Chapter 11
F_n = nominal strength of strut, tie, nodal zone, or bearing area in strut-and-tie model, Chapter 7
F_{nn} = nominal strength at face of a nodal zone, Chapter 7
F_{ns} = nominal strength of a strut, Chapter 7
F_{nt} = nominal strength of a tie, Chapter 7
f_{pc} = precompression stress available to resist earthquake effects, Chapter 15
f_{pc} = compressive stress in concrete (after allowance for all prestress loses) at centroid of cross section resisting externally applied loads, Chapter 10
F_{px} = diaphragm design force, Chapter 15
$F_{px,max}$ = upper limit to F_{px}, Chapter 15
$F_{px,min}$ = lower limit to F_{px}, Chapter 15
f_{py} = specified yield strength of prestressing steel, Chapter 7
f_r = modulus of rupture of concrete, Chapters 3 and 6
f_s = steel stress, Chapters 2, 4, 5, 6, 8, and 9
f_s' = stress in steel area A_s', Chapter 6
f_{se} = effective stress in prestressed steel after allowance for all prestress losses, Chapter 7

f_{sm}	=	maximum tensile stress in steel, Chapters 5, 8, and 13
$f_{s,max}$	=	maximum tensile stress in steel, Chapters 6 and 8
f_{sp}	=	splitting tensile strength of concrete, Chapter 3
f_{st}	=	stress developed in transverse reinforcement confining a lap splice, Chapter 8
f_{su}	=	ultimate stress capacity (tensile strength) of steel, Chapter 2
$f_{su,actual}$	=	actual (or measured) ultimate stress capacity (tensile strength) of steel, Chapter 2
f_t	=	tensile strength of concrete, Chapters 5, 7, and 9
F_t	=	tension force resisted by tension tie, Chapter 7
F_u	=	required strength of strut, tie, nodal zone, or bearing area in strut-and-tie model, Chapter 7
F_v	=	long-period site coefficient (at 1.0 s period), Chapter 11
f_x, f_z	=	normal stresses acting on boundary of panel zone, Chapter 7
F_x	=	portion of the seismic base shear, V_b, induced at level x, Chapter 15
f_y	=	specified yield strength of reinforcement, Chapters 1, 2, 4, 5, 6, 7, 8, 9, 10, 11, 12, 13, 14, 15, and 16
F_y	=	oscillator yield force, Chapter 11
$f_{y,actual}$	=	actual (or measured) yield strength of reinforcement, Chapters 2 and 8
$f_{y,lower}$, $f_{y,upper}$	=	lower and upper yield stress of reinforcement, Chapter 2
f_{yt}	=	specified yield strength f_y of transverse reinforcement, Chapters 2, 4, 5, 6, 7, 8, 9, 10, 12, 13, 14, 15, and 16
g	=	acceleration due to gravity, Chapters 7, 11, 12, and 13
G_c	=	concrete shear modulus, Chapters 7 and 13
G_{eff}	=	effective shear modulus of concrete, Chapter 6
h	=	overall thickness or height of member, Chapters 5, 6, 7, 8, 10, 11, 12, 13, 14, 15, and 16
h	=	overall length of rectangular footing in direction of lateral load or eccentricity e, Chapter 16
H	=	load due to lateral earth pressure, ground water pressure, or pressure of bulk materials, or related internal moments and forces, Chapters 1, 11, and 13
h^*	=	effective modal height, Chapter 11
h'	=	length of bearing in footing with uplift, Chapter 16
h_b	=	overall thickness of beam, Chapters 1, 9, 12, and 14
h_c	=	overall dimension of rectangular or equivalent rectangular column measured in the direction of the span for which moments in columns or beams, or shear in joints, are being determined, $= c_1$, Chapters 1, 9, and 12
h_{eff}	=	effective height for lateral forces, Chapter 13
h_f	=	height of basement wall, Chapter 16
h_i	=	story height of story level i, Chapters 1 and 14
h_j	=	the height above the base to level j, Chapter 11

h_j	=	overall dimension of joint in direction of forces/displacements generating joint shear, $= h_c$, Chapters 9 and 12
h_n	=	height above the base to the top of the building, Chapters 1 and 15
h_s	=	thickness of slab or flange, Chapters 6 and 12
h_{sx}	=	story height below level x, Chapters 11 and 12
h_w	=	height of entire wall from base to top, or clear height of wall segment or wall pier considered, Chapters 7 and 13
h_x	=	maximum center-to-center spacing of longitudinal bars laterally supported by crossties or hoop legs on all faces of the column, Chapters 12, 13, and 14
h_x	=	height above the base to level x, Chapter 15
I	=	moment of inertia of section about centroidal axis, Chapters 6 and 11
I_{cr}	=	moment of inertia of cracked section about centroidal axis, Chapter 6
IDR	=	drift ratio limit, $= \Delta_a/h_{sx}$, Chapter 13
I_e	=	importance factor, Chapters 1, 11, 12, and 15
I_e	=	effective moment of inertia for calculation of deflection, Chapters 6, 7, 12, 13, and 14
I_g	=	moment of inertia of gross concrete section about centroidal axis, neglecting reinforcement, Chapters 1, 6, 7, 11, 12, 13, and 14
IR	=	ratio of shaking intensity to design-level intensity, Chapter 13
j	=	coefficient defining the internal moment arm in a flexural member, Chapters 6, 7, 8, 9, and 13
J_c	=	property of slab critical section analogous to polar moment of inertia, Chapter 10
JI	=	joint transverse reinforcement index, Chapter 9
J_r	=	moment of inertia of stiffness of wall vertical elements about plan centroid of wall stiffness, Chapter 15
k	=	ratio of depth of flexural compression zone to effective depth, Chapters 6 and 11
k	=	coefficient representing the effect of minimum confining stress f_3 on increase in stress capacity f_{c1max}, Chapter 3
k	=	effective length factor, Chapters 5 and 13
k	=	distribution exponent for design seismic forces, Chapter 15
k	=	shear strength degradation coefficient, Chapter 7
k	=	ratio of cracked section (truss model) to gross section shear stiffness, Chapter 7
k or k_i	=	stiffness coefficient or stiffness coefficient of element i, Chapter 15
K	=	ratio of confined concrete compressive strength to strength of unconfined concrete, Chapter 6
K	=	stiffness coefficient, Chapter 11
k_{dyn}	=	dynamic increase factor for concrete under high strain rate loading, Chapter 4
k_e	=	confinement effectiveness coefficient, Chapters 4 and 6

k_f = concrete strength factor used in confinement requirement, Chapter 12
k_M = factor used to modify lap splice length based on moment gradient, Chapter 8
k_n = confinement effectiveness factor used in confinement requirement, Chapter 12
k_p = coefficient in confinement equation, $= P/P_o$, Chapter 12
K_p = passive pressure coefficient, Chapter 16
k_{tr} = factor used to modify lap splice length based on size of transverse reinforcement, Chapter 8
K_{tr} = transverse reinforcement index, Chapter 8
K_{vcr} = shear stiffness of truss model, Chapter 7
K_{vg} = shear stiffness gross section, Chapter 7
l = span length of beam, one-way slab, or longitudinal bar; clear projection of cantilever; span of diaphragm or diaphragm segment; length of compression member, Chapters 1, 5, 6, 11, 12, and 15
l = distance between cracks in tension member, Chapter 5
l = span of square slab panel, Chapter 10
l = length of pile, Chapter 16
L = live loads, or related internal moments and forces, Chapters 1, 10, 11, 12, 13, 14, and 15
l_1 = length of span in direction that moments are being determined, measured center-to-center of supports, Chapters 10 and 14
l_1 = length just below pile cap along which special transverse reinforcement is required, Chapter 16
l_2 = length of span in direction perpendicular to l_1, measured center-to-center of supports, Chapter 10
l_2 = length above and below interface between soft and hard soil strata along which special transverse reinforcement is required, Chapter 16
l_a = anchorage length over which bond stress occurs, Chapters 6 and 8
l_b = length of beam, measured center-to-center of the joints in a frame, or from inflection point in one span to inflection point in adjacent span, Chapters 9 and 12
l_{be} = length of boundary element, Chapter 13
l_c = length of compression member in a frame, measured center-to-center of the joints in the frame, Chapters 9 and 12
l_d = development length in tension, Chapters 7, 8, 13, 14, 15, and 16
l_{dc} = development length in compression of deformed bars and deformed wire, Chapter 8
l_{dc} = length of tensile stress field in a corner beam-column joint, Chapter 9
l_{dh} = development length in tension of deformed bar or deformed wire with a standard hook, measured from critical section to outside end of hook (straight embedment length between critical section and start of hook [point of tangency] plus outside radius of bend), Chapters 7, 8, 9, 12, 13, and 16

l_{dt} = development length in tension of headed deformed bar, measured from the critical section to the bearing face of the head, Chapters 7, 8, 9, and 13
l_{eff} = effective length of lap splice, Chapter 8
l_f = length of basement wall, Chapter 16
l_h = horizontal dimension of opening, Chapters 10 and 13
l_{min} = minimum length between cracks required to form a new crack, Chapter 5
l_n = length of clear span measured face-to-face of supports, Chapters 1, 7, 9, 12, 13, and 14
l_o = length, measured from joint face along axis of column, over which special transverse reinforcement must be provided, Chapters 12 and 14
l_o = length between inflection points in axially loaded compression member, Chapter 5
l_p = plastic hinge length, Chapters 6, 11, and 12
L_r = roof live load, or related internal moments and forces, Chapter 1
L_{red} = the effect of live load reduced based on tributary area, Chapter 15
l_{sc} = required lap splice length for deformed bars and deformed wire in compression, Chapter 8
l_{sp} = lap splice length in tension, Chapters 8, 13, and 15
l_{st} = required lap splice length for deformed bars and deformed wire in tension, Chapter 8
l_u = unsupported length of compression member, Chapters 5, 8, 12, and 14
l_w = length of entire wall or length of segment of wall considered in direction of shear force, Chapters 6, 7, 13, 14, and 16
m = empirical coefficient in Coffin–Manson relation, Chapter 2
m = empirical coefficient in equation for effective moment of inertia, Chapter 6
m = mechanical reinforcement ratio, Chapters 5 and 13
M = mass, Chapter 11
M = moment, Chapters 1, 5, 6, 7, 8, 9, 11, 12, 13, 15, and 16
M = empirical coefficient in Coffin–Manson relation, Chapter 2
M = moment magnitude, Chapter 11
$M_{0.004}$ = flexural moment corresponding to reaching compressive strain 0.004 on cover concrete, Chapter 6
M^* = effective modal mass, Chapter 11
M_a = maximum moment in member due to applied loads, either service loads if analysis is for service load effects or design loads if analysis is for design load effects, Chapters 6 and 11
M_b = flexural moment in a beam, Chapters 9 and 12
$M_{b,MRSA}$ = overturning demand of the structural system based on modal response spectrum analysis, measured at the base, Chapter 12

$M_{b,\Omega}$ = overturning strength of the structural system measured at the base, Chapter 12

M_c = flexural moment in a column, Chapters 9 and 12

M_{cr} = cracking moment, Chapters 6 and 11

$M_{first\ yield}$ = flexural moment at first yield of the longitudinal reinforcement or the flexural moment corresponding to reaching concrete compressive strain 0.002, whichever occurs first, Chapter 6

M_{group} = overturning moment applied to a pile group, Chapter 16

M_i, M_i' = moment strengths calculated for cover intact and spalled, respectively, Chapter 6

M_n = nominal moment strength at section, Chapters 1, 6, 7, 9, 12, 13, 14, and 16

$M_{n,CS}$ = nominal moment strength at critical section for moment and axial force, Chapter 13

M_{pile} = moment on individual pile, Chapter 16

M_{pr} = probable flexural strength of members, with or without axial load, determined using the properties of the member at the section where strength is calculated assuming tensile stress capacity in the longitudinal bars of at least $1.25f_y$ and a strength reduction factor, ϕ, of 1.0, Chapters 1, 6, 7, 9, 11, 12, 13, and 14

$M_{pr,CS}$ = probable moment strength at wall critical section for flexure and axial force, Chapter 13

M_r = radial slab moment, Chapter 10

M_{tr} = portion of slab moment balanced by support, Chapter 10

M_u = factored moment at section, or moment at section based on applicable load combination, Chapters 1, 6, 12, 13, and 15

M_u' = factored moment at section amplified for overstrength, Chapter 13

M_u'' = factored moment at section amplified for overstrength and dynamic response, Chapter 13

$M_{u,CS}$ = factored moment at critical section for flexure and axial force, Chapter 13

$M_{u,CS}'$ = factored moment at critical section for flexure and axial force after adjustment based on moment redistribution, Chapter 13

M_{ux}, M_{uy} = factored moments about x and y axes, Chapters 13 and 14

M_w = moment on wall, Chapter 16

m_x = moment due to application of unit virtual force, Chapter 6

M_x = moment about y axis, Chapter 12

M_y = moment about x axis, Chapter 12

M_y = yield moment, Chapters 6 and 11

M_θ = circumferential slab moment, Chapter 10

N = number of stories, Chapters 1, 11, and 12

n = modular ratio, E_s/E_c, Chapters 4, 5, 6, and 7

n = number of items, Chapters 5, 8, and 9

n	=	designation for the level that is uppermost in the main portion of the building, Chapters 1, 13, and 15
N	=	axial force normal to cross section occurring simultaneously with shear, taken positive for compression and negative for tension, Chapter 7
N	=	radially oriented clamping force provided by arc of circular hoop or spiral reinforcement per hoop spacing or spiral pitch s, Chapter 8
$\overline{N}$	=	average field standard penetration resistance for the top 100 ft (30 m), Chapter 11
N_f	=	number of full cycles in fatigue loading, Chapter 2
N_{fi}	=	number of full cycles at fracture in fatigue loading for strain amplitude ε_{ai}, Chapter 2
n_l	=	number of longitudinal bars in column cross section supported by crossties or legs of hoops, Chapters 4 and 12
n_u	=	unit normal stress N_u/A_g for wall, Chapter 7
N_u	=	factored axial force normal to cross section occurring simultaneously with V_u or T_u; to be taken as positive for compression and negative for tension, Chapters 7 and 13
p	=	lateral force on a pile per unit length, Chapter 16
P	=	axial compressive force, Chapters 1, 5, 6, 7, 8, 9, 12, 13, 14, and 16
P	=	concentrated force applied to member, Chapters 6 and 7
P_{c1}, P_{c2}	=	axial forces on columns 1 and 2, Chapter 16
P_e	=	column axial force due to earthquake effect, Chapter 12
$P_{e,code}$	=	column axial force obtained from modal response spectrum analysis for code design response spectrum reduced by R/I_e, Chapter 12
P_g	=	design axial force due to gravity loads only, Chapter 12
PGA	=	peak ground acceleration, Chapter 11
P_{group}	=	axial load applied to a pile group, Chapter 16
PI	=	plasticity index, Chapter 11
p_j	=	joint average axial stress, Chapter 9
P_n	=	nominal axial strength of cross section, Chapters 6 and 13
$P_{n,max}$	=	maximum nominal axial force permitted considering accidental eccentricity, Chapters 13 and 15
P_o	=	nominal axial strength at zero eccentricity, Chapters 5, 6, 9, 12, and 13
P_o'	=	axial load resistance of compression member at arbitrary point after spalling of cover concrete, Chapter 5
P_{oo}	=	axial load strength of compression member after spalling of cover concrete and full development of confined core compressive strength, Chapter 5
P_{ot}	=	axial force developed in pile due to overturning moment, Chapter 16
P_{pile}	=	axial force in a pile, Chapter 16
P_{pr}	=	axial compression force in coupled wall pier corresponding to development of M_{pr} in coupling beams, Chapter 13

P_u = factored axial force at section, or axial force at section based on applicable load combination, Chapters 1, 6, 12, 13, 14, and 16

$P_{u,CS}$ = factored axial force at critical section for flexure and axial force, Chapter 13

P_w = axial force on wall, Chapter 16

q = behavior factor in Eurocode 8, Chapter 13

q = load per unit area, Chapter 10

q_c = bearing pressure under footing, Chapter 16

Q_E = effect of horizontal seismic (earthquake-induced) forces, Chapter 13

q_{max} = maximum value of soil bearing pressure, Chapter 16

q_u = factored load per unit area, Chapter 10

r = variable used in confined concrete stress–strain relation, Chapter 4

r = inside radius of bend in standard hook, Chapter 8

R = response modification coefficient, Chapters 11, 12, 13, 15, and 16

R = reaction, Chapter 8

R = radius of circular hoop or spiral reinforcement, Chapter 8

R = response modification coefficient from ASCE 7, Chapters 11, 14, and 15

R_0 = ratio of expected building strength to design building strength, Chapter 11

R_d = response modification considering inelastic response only, Chapter 11

$R_{D/L}$ = ratio of uniformly distributed dead to live load, Chapter 15

R_{eff} = effective response modification coefficient (elastic base shear divided by expected base shear strength), Chapter 11

R_i = reaction force in diaphragm at vertical element i, Chapter 15

s = center-to-center spacing of items, such as longitudinal reinforcement or transverse reinforcement, Chapters 4, 5, 6, 7, 8, 9, 10, 12, 13, 14, 15, and 16

S – snow load, or related internal moments and forces, Chapters 1 and 11

s' = effective buckling length for longitudinal bar buckling over n hoop spacings or spiral wraps, Chapter 5

s_1 = center-to-center spacing of column transverse reinforcement within length l_o and along lap splices, Chapters 12 and 14

s_1 = spacing of transverse reinforcement along lengths of pile l_1 and l_2, Chapter 16

S_1 = mapped MCE, 5-percent-damped, spectral response acceleration parameter at a period of 1 s, Chapter 11

s_2 = center-to-center spacing of column transverse reinforcement along lengths outside the lengths l_o and the lengths of lap splices, Chapters 12 and 14

s_2 = center-to-center spacing of legs of hoops or crossties measured horizontally along the length of a wall, Chapter 4

s_2 = spacing of transverse reinforcement between lengths l_1 and l_2 of a pile, Chapter 16

s_a	=	slip of reinforcement from anchorage, Chapters 6 and 8
S_a	=	pseudo-acceleration, Chapters 1, 11, 13, and 15
SA	=	maximum absolute (total) acceleration of a linear oscillator, Chapter 11
$s_{critical}$	=	hoop spacing or spiral pitch corresponding to developing the critical buckling stress f_{cr}, Chapter 5
S_d	=	maximum displacement of a linear oscillator relative to the ground, Chapters 11 and 13
S_{D1}	=	design, 5-percent-damped, spectral response acceleration parameter at a period of 1 s, Chapters 11 and 13
S_{DS}	=	design, 5-percent-damped, spectral response acceleration parameter at short periods, Chapters 11, 12, 13, 15, and 16
S_{gx}, S_{gy}	=	section modulus of gross section, Chapters 13 and 14
s_i	=	center-to-center spacing of reinforcement in the ith layer adjacent to the surface of the member, Chapter 7
S_m	=	elastic section modulus, Chapter 15
S_{M1}	=	the MCE, 5-percent-damped, spectral response acceleration at a period of 1 s adjusted for site class effects, Chapter 11
S_{MS}	=	the MCE, 5-percent-damped, spectral response acceleration at short periods adjusted for site class effects, Chapter 11
S_n	=	nominal strength of member or cross section, Chapter 1
s_o	=	spacing variable for column transverse reinforcement dependent on h_x, Chapters 12 and 14
S_S	=	mapped MCE, 5-percent-damped, spectral response acceleration parameter at short periods, Chapter 11
$\bar{S}_u$	=	average undrained shear strength in top 100 ft (30 m), Chapter 11
S_v	=	pseudo-velocity, Chapter 11
SV	=	maximum velocity of a linear oscillator relative to the ground, Chapter 11
s_x, s_y, s_z	=	spacing of distributed reinforcement measured in the x, y, and z directions, Chapter 7
t	=	time, Chapter 12
t	=	thickness of steel shell or diaphragm slab, Chapters 4 and 15
T	=	the fundamental period of the building (also referred to as T_1), s, Chapters 1, 11, 12, 13, and 15
T	=	axial tensile force, Chapters 5, 13, and 16
T	=	flexural tension force in beam, slab, column, or diaphragm, Chapters 6, 9, 10, and 15
T_0	=	period in seconds corresponding to initiation of the constant acceleration portion of the design response spectrum, Chapter 11
T_1	=	fundamental vibration period, Chapter 13
T_1, T_2	=	chord forces acting around panel zone, Chapter 7

T_a	=	approximate fundamental period of the building as determined by building code, Chapters 12 and 13
T_D	=	factored collector tension force transferred directly to edge of vertical element, Chapter 15
T_L	=	long-period transition period used to define design response spectrum, Chapter 11
T_L, T_R	=	tensile forces in reinforcement of distributors, Chapter 16
T_n	=	nominal tensile strength, Chapter 5
T_{pr}	=	probable tensile force in beam longitudinal reinforcement A_s when beam develops associated probable moment, Chapter 12
T_{pr}	=	axial tensile force in coupled wall pier corresponding to development of M_{pr} in coupling beams, Chapter 13
T'_{pr}	=	probable tensile force in beam longitudinal reinforcement A'_s when beam develops the associated probable moment, Chapter 12
T_s, T_{s1}, T_{s2}, …, T_{sn}	=	tensile force in longitudinal reinforcement, or in longitudinal reinforcement layers 1, 2, …, n, Chapters 5, 6, 7, 8, 9, and 12
T_S	=	period in seconds at transition between constant acceleration and constant velocity portions of the design response spectrum, Chapter 11
T_{s1}	=	tensile force resultant associated with distributed vertical reinforcement in a wall, Chapter 13
T_{s2}	=	tensile force resultant associated with boundary element vertical reinforcement in a wall, Chapter 13
T_u	=	factored tensile force at section, Chapters 13 and 15
$T_{u,max}$	=	maximum value of T_u, Chapter 15
$T_{u,net}$	=	transient net tensile force, Chapter 13
T_v	=	factored collector tensile force transferred through shear friction to vertical element, Chapter 15
T_w	=	flexural tension force at wall boundary, Chapter 16
T_y	=	tensile force at yield stress f_y, Chapter 6
u	=	bond stress between reinforcement and concrete, Chapters 5, 6, 7, and 8
U	=	result of load combination, Chapters 1 and 13
$\overline{u}$	=	average (uniform) bond stress, Chapters 6, 8, and 9
v	=	shear stress, Chapters 6, 7, and 13
V	=	shear force, Chapters 1, 6, 7, 8, 9, 11, 15, and 16
V_a	=	shear resisted by aggregate interlock, Chapter 7
V_b	=	base shear, Chapters 11, 12, 13, and 15
V_b	=	shear in beam, Chapter 9
v_c	=	nominal shear stress resisted by concrete, Chapters 7, 9, and 14
V_c	=	nominal shear strength provided by concrete, Chapters 7, 10, 12, 13, 14, and 16
V_{code}	=	column shear calculated for factored load combinations including code-specified seismic force effects, Chapter 12

V_{col}, V_{column} = shear in column, Chapters 1, 9, and 12
v_{cr} = nominal shear stress at onset of concrete cracking, Chapter 9
V_{cr} = shear at onset of shear cracking in squat wall, Chapter 7
V_d = dowel force acting between longitudinal reinforcement and concrete, Chapters 6 and 7
V_e = design shear force corresponding to the development of the probable moment strength of the member, $= V_{pr}$, Chapters 12, 13, and 14
$V_{e,i}$ = shear due to mode i, Chapter 13
V_f = shear corresponding with development of flexural strength, Chapter 7
V_f = horizontal shear resisted by shear friction between footing and soil, Chapter 16
V_g = shear force on the slab critical section for two-way action due to gravity loads, Chapter 10
V_{group} = lateral shear force applied to a pile group, Chapter 16
v_j, v_{jh}, v_{jv} = joint average shear stress, subscripts h and v indicating horizontal and vertical, Chapters 9 and 16
V_j, V_{jh} = horizontal beam-column joint shear force, Chapters 9 and 12
V_{jv} = vertical beam-column joint shear force, Chapter 9
V_L, V_R = vertical shear forces in basement wall, Chapter 16
V_{MRSA} = shear force from linear modal response spectrum analysis, Chapter 12
v_n = nominal shear strength per unit length, Chapter 13
v_n = nominal shear stress per unit area, Chapters 7 and 13
V_n = nominal shear strength, Chapters 1, 7, 9, 10, 12, 13, 14, 15, and 16
$V_{n,e}$ = nominal shear strength calculated considering expected materials properties, Chapter 13
$V_{n,max}$ = upper limit on V_n for structural walls, Chapter 7
V_o = shear strength provided by concrete for slab critical section subjected to uniform shear stress, Chapter 10
V_p = contribution of axial force to shear resistance, Chapter 7
V_p = vertical component of effective prestress force at section, Chapter 10
$V_{passive}$ = shear resistance due to passive bearing pressure, Chapter 16
V_{pile} = shear force on individual pile, Chapter 16
V_{pr} = shear in member based on development of M_{pr} at critical sections for flexure and axial force, Chapters 12 and 13
$V_{pr,1}$ = shear corresponding to development of probable moment strength under first-mode lateral loading profile, Chapter 13
$V_{pr,CS}$ = shear corresponding to development of probable moment strength, Chapter 13
v_s = nominal shear stress resisted by steel reinforcement, $= V_s/b_w d$ or $V_s/b_o d$, Chapters 7 and 14
V_s = nominal shear strength provided by shear reinforcement, Chapters 6, 7, 10, 12, 13, 14, and 16

$\overline{v}_s$ = average soil shear wave velocity at small shear strains in upper 100 ft (30 m), Chapter 11

v_u = factored shear stress, Chapters 10 and 15

V_u = factored shear force at section, Chapters 1, 7, 10, 12, 13, 15, and 16

V_u = maximum shear force measured in test, Chapter 7

V_{uc} = shear force for defining secant stiffness past yielding in a squat wall, Chapter 7

$V_{u,CS}$ = design shear acting at the critical section for moment and axial force, Chapter 13

$V'_{u,CS}$ = wall base shear modified for dynamic effects, Chapter 13

v_{ug} = factored nominal shear stress on the slab critical section for two-way action due to gravity loads, Chapter 14

V_{ug} = factored nominal shear force on the slab critical section for two-way action due to gravity loads, Chapter 14

$V'_{u,i}$ = design shear, Chapter 13

V_w = shear force on wall, Chapter 16

V_w = shear force in structural wall, Chapter 16

V_x = shear at level x, Chapter 15

V_y = yield force for an oscillator or structure, Chapter 11

w = moisture content, percent, Chapter 11

w = load per unit length of beam or one-way slab, Chapter 11

w = relative slip between reinforcement and concrete, Chapter 5

W = effective seismic weight of the building, Chapters 1, 11, 12, and 13

W = wind load, or related internal moments and forces, Chapter 1

w/c = water to cementitious materials ratio, Chapter 3

w_c = unit weight of normalweight concrete or equilibrium density of lightweight concrete, Chapter 3

w_{cr} = crack width, Chapter 5

w_j = portion of W that is located at or assigned to level j, Chapter 11

w_{px} = weight tributary to the diaphragm at level x, Chapter 15

w_s = effective strut width in strut-and-tie model, Chapter 7

w_t = effective tie width in strut-and-tie model, Chapter 7

w_u = factored load per unit length of beam or slab, Chapters 1, 9, and 12

w_x = portion of effective seismic weight of the building that is located at, or assigned to, level x, Chapter 15

x = smaller of the dimensions from face of beam to edge of column, measured perpendicular to beam span, Chapters 9 and 12

x = variable used in confined concrete stress–strain relation, Chapter 4

x = contact length between diaphragm and vertical element, Chapter 15

X = principal horizontal axis of building, Chapter 15

x_p = horizontal distance between compression force resultant and axial force P_u, Chapter 13

y = horizontal displacement of soil under lateral force p from pile, Chapter 16

Y = principal horizontal axis of building, Chapter 15

z = perpendicular distance from neutral axis, Chapter 6

α = ratio at a beam-column joint of the sum of column nominal flexural strengths to beam nominal flexural strengths, Chapter 12

α = factor to represent longitudinal reinforcement overstrength, typically 1.25, Chapters 8, 9, and 12

α = angle between transverse reinforcement leg and the normal to the section cut, Chapter 4

α = angle defining the orientation of reinforcement, Chapters 7 and 13

α = central angle between adjacent pairs of lap-spliced bars in a circular cross section, Chapter 8

α = effective width coefficient for stiffness of slab-column connections, Chapter 10

α_1 = coefficient defining the stress intensity for the equivalent rectangular stress block, Chapter 6

α_b = factor to account for different bond conditions in beam-column joints as function of ratio of beam tension reinforcement area to beam compression reinforcement area, Chapter 9

α_c = coefficient defining the relative contribution of concrete strength to nominal wall shear strength, Chapters 7 and 13

α_c = displacement modification factor to account for shear deformations, Chapter 7

α_d = factor to account for different bond conditions in beam-column joints as function of ductility demand, Chapter 9

α_f = factor to account for different bond conditions in one-way and two-way beam-column joints, Chapter 9

α_i = angle between the axis of a strut and the bars in the ith layer of reinforcement crossing that strut, Chapter 7

α_L = load factor on live load L in some load combinations, Chapter 1

α_s = constant used to calculate V_c in slabs and footings, Chapters 10 and 16

β = beam-column joint strength factor related to load history, Chapter 9

β = ratio of long to short dimensions of column cross section or other reaction area, and ratio of long to short sides of footing, Chapters 10 and 16

β_1 = factor relating depth of equivalent rectangular compressive stress block to neutral axis depth, Chapters 5, 6, and 13

β_{cr} = factor to account for effects of slab cracking on lateral stiffness of slab-column connection, Chapter 10

β_n = factor to account for the effect of the anchorage of ties on the effective compressive strength of a nodal zone, Chapter 7

β_p = factor used to calculate V_c in prestressed slabs, Chapter 10

β_s	=	factor to account for effect of cracking and confining reinforcement on effective compressive strength of concrete in a strut, Chapter 7
γ	=	shear strength factor reflecting geometry of beam-column connection, Chapters 9 and 12
γ	=	distance from concrete compression resultant C_c to wall centerline, Chapter 5
γ	=	unit density of soil, Chapter 16
Γ	=	modal participation factor, Chapter 11
γ_1	=	parameter for effect of in-plane geometry on strength of beam-column connections, Chapter 9
γ_2	=	parameter for effect of out-of-plane geometry on strength of beam-column connections, Chapter 9
γ_e	=	parameter for effect of beam eccentricity on beam-column joint shear strength, Chapter 9
γ_f	=	factor used to determine the moment transferred by slab flexure at slab-column connections, Chapter 10
γ_P	=	factor to recognize that not all beams develop M_{pr} over building height, Chapter 12
γ_{Rd}	=	factor to account for overstrength due to steel strain-hardening, Chapter 13
γ_s	=	factor determining the distribution of moment reinforcement in rectangular footing, Chapter 16
γ_v	=	factor used to determine the moment transferred by eccentricity of shear at slab-column connections, Chapter 10
δ	=	displacement, Chapters 1, 5, 6, 7, and 13
Δ	=	roof displacement, Chapters 11 and 12
Δ	=	displacement during a test, Chapters 2, 6, and 7
Δ_a	=	allowable story drift, Chapter 11
δ_c	=	out-of-plane displacement corresponding to concrete crushing in buckling compression member, Chapter 5
Δ_c	=	shortening of full-length diagonal compression strut in truss model, Chapter 7
Δ_e	=	peak displacement for a linear oscillator or structure, Chapter 11
δ_f	=	displacement due to flexure, Chapter 6
Δf_c	=	change in concrete stress along length Δx, Chapter 5
Δf_p	=	increase in stress in prestressed steel due to factored loads, Chapter 7
Δf_s	=	change in steel stress along prescribed length, Chapters 5 and 9
Δ_j	=	story drift for story j, Chapter 11
δ_{pile}	=	lateral displacement of pile at loaded end, Chapter 16
δ_s	=	displacement due to slip of reinforcement from anchorage, Chapter 6
Δ_s	=	elongation of transverse reinforcement in truss model, Chapter 7
Δ_t	=	target story drift, Chapter 11

ΔT_c, ΔT_{s1}, ΔT_{s2}, …, ΔT_{sn}	=	change in concrete and steel stress resultants required to equilibrate stress resultants due to unrestrained shrinkage strain, Chapter 5
δ_u	=	design displacement, Chapters 11, 13, and 14
δ_v	=	displacement due to shear, Chapter 6
$\Delta V/V$	=	volumetric strain, Chapter 3
δ_x	=	design deflection of level x, Chapter 11
Δx	=	short length of member, Chapters 5, 7, and 8
δ_{xe}	=	deflection of level x as determined by elastic analysis using code design forces reduced by R/I_e, Chapter 11
δ_y	=	yield displacement for an oscillator, member, or structure, Chapters 6, 7, and 11
Δ_y	=	yield displacement for an oscillator, member, or structure, Chapter 7
δ_y'	=	effective yield displacement for an oscillator, member, or structure, Chapters 7 and 11
$\Delta\varepsilon_{ai}$	=	strain increment during fatigued loading, Chapter 2
$\Delta\varepsilon_s'$	=	strain loop width, Chapter 2
ε	=	strain, Chapters 3, 5, 6, 7, and 12
ε	=	wall shear multiplier to account for combined dynamic and overstrength effects, Chapter 13
$\dot{\varepsilon}$	=	strain rate, s^{-1}, Chapters 3 and 4
ε_0	=	strain at peak concrete stress, Chapters 3, 4, 5, and 6
ε_1, ε_2, ε_3	=	strains in the 1, 2, and 3 directions, Chapters 3 and 4
ε_a	=	strain amplitude in fatigue loading, Chapter 2
ε_{ai}	=	strain amplitude during cycle i in fatigue loading, Chapter 2
ε_c	=	concrete strain, Chapters 1, 4, and 6
ε_{cc}	=	strain at peak of the confined concrete stress–strain relation, Chapters 4 and 6
ε_{cmax}	=	maximum strain in compressed concrete of flexural compression zone, Chapter 6
ε_{cr}	=	critical longitudinal reinforcement strain at buckling, Chapter 5
ε_{cu}	=	maximum strain of confined concrete, Chapters 4, 5, 6, and 13
ε_l	=	longitudinal strain, Chapter 2
ε_{max}	=	maximum strain in strain range in fatigue loading, Chapter 2
ε_{min}	=	minimum strain in strain range in fatigue loading, Chapter 2
ε_p	=	strain range from strain at stress reversal (from tension to compression) to strain at buckling, Chapter 5
ε_{res}	=	residual strain in steel, Chapter 5
ε_s	=	steel strain (of steel area A_s if applicable), Chapters 2, 5, 6, and 13
ε_s'	=	strain of steel area A_s', Chapter 6
ε_{sh}	=	strain at onset of strain hardening, Chapter 2
ε_{sh}	=	unrestrained shrinkage strain, Chapter 5

ε_{sm} = maximum tensile strain in a member for it to remain stable upon loading in compression, Chapters 5 and 13

ε_{sp} = spalling strain, Chapter 6

ε_{su} = strain at steel fracture or, alternatively, strain at stress f_{su}, Chapters 2 and 4

ε_t = net tensile strain in extreme layer of longitudinal tension steel at nominal strength, excluding strains due to effective prestress, creep, shrinkage, and temperature, Chapters 1, 6, and 13

ε_{ty} = value of net tensile strain in the extreme layer of longitudinal tension reinforcement used to define a compression-controlled section, Chapter 1

ε_y = yield strain, Chapters 2, 5, 11, 12, and 13

ς = softening coefficient, Chapter 3

η = coefficient defining fraction of total shear resisted by transverse reinforcement, Chapter 6

η = factor to reduce the post-spalling strength contribution of the longitudinal reinforcement if buckling occurs, Chapter 5

η = beam-column joint strength factor related to connection geometry, Chapter 9

θ = rotation, Chapters 1, 6, and 11

θ = angle between diagonal compression member and longitudinal axis in a truss model, Chapter 7

θ_a = rotation corresponding to maximum applied moment, Chapter 11

θ_{cr} = rotation corresponding to cracking, Chapter 11

θ_p = plastic rotation, Chapter 11

θ_y = rotation corresponding to yield moment, Chapter 11

κ = coefficient used to define location of tension reinforcement, Chapter 5

κ = coefficient used to define joint shear strength, Chapter 9

λ = modification factor reflecting the reduced mechanical properties of lightweight concrete, all relative to normalweight concrete of the same compressive strength, Chapters 6, 7, 8, 9, 10, 13, 15, and 16

λ_c = coefficient to adjust in-place concrete strength as function of standard cylinder compressive strength, Chapters 4 and 5

μ = coefficient of friction, Chapters 7, 13, 15, and 16

μ_δ = displacement ductility, Chapter 11

μ_ϕ = curvature ductility, Chapters 6, 11, and 12

ν = Poisson's ratio, Chapters 2, 3, and 4

ξ = ratio of out-of-plane displacement δ to thickness b, Chapter 5

ρ = redundancy factor based on the extent of structural redundancy present in a building, Chapters 13 and 15

ρ = steel ratio, A_s/bd, Chapters 6 and 9

ρ' = compression steel ratio, A_s'/bd, Chapter 6

ρ_b = balanced reinforcement ratio, Chapter 6

ρ_{be} = ratio of area of boundary element longitudinal reinforcement to gross area of boundary element, Chapter 13

ρ_f = longitudinal reinforcement ratio in flange, Chapter 6

ρ_j = volume of joint transverse reinforcement in direction parallel to joint shear divided by joint volume, Chapter 9

ρ_l = ratio of area of distributed longitudinal reinforcement to gross area perpendicular to that reinforcement, Chapters 5, 7, 13, and 16

ρ_l' = ratio of area of longitudinal reinforcement to gross area of confined core perpendicular to that reinforcement, Chapter 5

ρ_s, ρ_{s1}, ρ_{s2}, ρ_{s3} = ratio of volume of spiral, circular hoop, or rectilinear transverse reinforcement including crossties to total volume of core confined by the reinforcement (measured out-to-out of reinforcement). ρ_{s1}, ρ_{s2}, and ρ_{s3} refer to volume ratio of rectilinear transverse reinforcement including crossties in the 1, 2, and 3 directions, Chapters 4, 5, 6, 9, 12, 14, and 16

ρ_{smin} = lesser of ρ_{sx} and ρ_{sz}, Chapters 7 and 13

ρ_{sx}, ρ_{sz} = steel ratio of distributed reinforcement parallel to the x and z directions, Chapters 7 and 13

ρ_t = ratio of area distributed transverse reinforcement to gross concrete area perpendicular to that reinforcement, Chapters 7 and 13

ρ_{vd} = steel ratio for one group of diagonal bars, $= A_{vd}/b_w h$, Chapter 7

ρ_w = ratio of A_s to $b_w d$, Chapter 7

ρ_w = lesser of ρ_t and ρ_l in squat wall, Chapter 7

σ = standard deviation of compressive strengths, Chapter 3

σ = axial stress, Chapters 6, 13, and 14

σ_1 = maximum principal tensile stress, Chapter 9

σ_x, σ_y = normal stresses in transverse and axial directions, taken positive in tension, Chapter 7

τ_{xy} = shear stress, Chapter 7

ϕ = strength reduction factor, Chapters 1, 6, 7, 9, 10, 11, 12, 13, 14, 15, and 16

ϕ = curvature, Chapters 6 and 13

ϕ_{cr} = cracking curvature, Chapter 6

$\phi_{first\ yield}$ = flexural curvature corresponding to $M_{first\ yield}$, Chapter 6

ϕ_j = modal shape value at level j, Chapter 11

ϕ_{max} = maximum curvature, Chapters 5 and 13

ϕ_o = overstrength factor, Chapters 12, 13, and 16

ϕ_u = ultimate curvature, Chapters 6, 11, 12, and 13

ϕ_{ux}, ϕ_{uy}, ϕ_{uxy} = curvature about axis x, axis y, or an angle to axes x-y, Chapter 13

ϕ_y = yield curvature, Chapters 6, 11, and 12

ψ_c = factor used to modify hook development length based on concrete cover, Chapter 8

ψ_e	=	factor used to modify development length based on reinforcement coating, Chapters 8 and 9
ψ_r	=	factor used to modify development length based on confinement reinforcement, Chapter 8
ψ_s	=	factor used to modify development length based on reinforcement size, Chapter 8
ψ_t	=	factor used to modify development length based on reinforcement location, Chapters 8 and 13
ω	=	circular frequency, rad/s, Chapter 11
ω	=	dynamic amplification factor, Chapters 12 and 13
Ω_0	=	amplification factor to account for overstrength of the seismic-force-resisting system, Chapters 11, 12, 14, 15, and 16

Index

C

E

F

G

H

I

K

L

M

N

O

P

Printed by Libri Plureos GmbH in Hamburg,
Germany

9 780071 839440